Lecture Notes in Computer Science 9950

More information about this series at http://www.springer.com/series/7407

Akira Hirose · Seiichi Ozawa
Kenji Doya · Kazushi Ikeda
Minho Lee · Derong Liu (Eds.)

Neural Information Processing

23rd International Conference, ICONIP 2016
Kyoto, Japan, October 16–21, 2016
Proceedings, Part IV

Editors
Akira Hirose
The University of Tokyo
Tokyo
Japan

Kazushi Ikeda
Nara Institute of Science and Technology
Ikoma
Japan

Seiichi Ozawa
Kobe University
Kobe
Japan

Minho Lee
Kyungpook National University
Daegu
Korea (Republic of)

Kenji Doya
Okinawa Institute of Science and Technology Graduate University
Onna
Japan

Derong Liu
Chinese Academy of Sciences
Beijing
China

ISSN 0302-9743 ISSN 1611-3349 (electronic)
Lecture Notes in Computer Science
ISBN 978-3-319-46680-4 ISBN 978-3-319-46681-1 (eBook)
DOI 10.1007/978-3-319-46681-1

Library of Congress Control Number: 2016953319

LNCS Sublibrary: SL1 – Theoretical Computer Science and General Issues

Printed on acid-free paper

This Springer imprint is published by Springer Nature
The registered company is Springer International Publishing AG
The registered company address is: Gewerbestrasse 11, 6330 Cham, Switzerland

Preface

This volume is part of the four-volume proceedings of the 23rd International Conference on Neural Information Processing (ICONIP 2016) held in Kyoto, Japan, during October 16–21, 2016, which was organized by the Asia-Pacific Neural Network Society (APNNS, http://www.apnns.org/) and the Japanese Neural Network Society (JNNS, http://www.jnns.org/). ICONIP 2016 Kyoto was the first annual conference of APNNS, which started in January 2016 as a new society succeeding the Asia-Pacific Neural Network Assembly (APNNA). APNNS aims at the local and global promotion of neural network research and education with an emphasis on diversity in members and cultures, transparency in its operation, and continuity in event organization. The ICONIP 2016 Organizing Committee consists of JNNS board members and international researchers, who plan and run the conference.

Currently, neural networks are attracting the attention of many people, not only from scientific and technological communities but also the general public in relation to the so-called Big Data, TrueNorth (IBM), Deep Learning, AlphaGo (Google DeepMind), as well as major projects such as the SyNAPSE Project (USA, 2008), the Human Brain Project (EU, 2012), and the AIP Project (Japan, 2016). The APNNS's predecessor, APNNA, promoted fields that were active but also others that were leveling off. APNNS has taken over this function, and further enhances the aim of holding technical and scientific events for interaction where even those who have extended the continuing fields and moved into new/neighboring areas rejoin and participate in lively discussions to generate and cultivate novel ideas in neural networks and related fields.

The ICONIP 2016 Kyoto Organizing Committee received 431 submissions from 38 countries and regions worldwide. Among them, 296 (68.7 %) were accepted for presentation. The first authors of papers that were presented came from Japan (100), China (78), Australia (22), India (13), Korea (12), France (7), Hong Kong (7), Taiwan (7), Malaysia (6), United Kingdom (6), Germany (5), New Zealand (5) and other countries/ regions worldwide.

Besides the papers published in these four volumes of the Proceedings, the conference technical program includes

- Four plenary talks by Kunihiko Fukushima, Mitsuo Kawato, Irwin King, and Sebastian Seung
- Four tutorials by Aapo Hyvarinen, Nikola Kazabov, Stephen Scott, and Okito Yamashita,
- One Student Best Paper Award evaluation session
- Five special sessions, namely, bio-inspired/energy-efficient information processing, whole-brain architecture, data-driven approach for extracting latent features from multidimensional data, topological and graph-based clustering methods, and deep and reinforcement learning
- Two workshops: Data Mining and Cybersecurity Workshop 2016 and Workshop on Novel Approaches of Systems Neuroscience to Sports and Rehabilitation

The event also included exhibitions and a technical tour.

Kyoto is located in the central part of Honshu, the main island of Japan. Kyoto formerly flourished as the imperial capital of Japan for 1,000 years after 794 A.D., and is presently known as "The City of Ten Thousand Shrines." There are 17 sites (13 temples, three shrines, and one castle) in Kyoto that form part of the UNESCO World Heritage Listing, named the "Historic Monuments of Ancient Kyoto (Kyoto, Uji and Otsu Cities)." In addition, there are three popular, major festivals (Matsuri) in Kyoto, one of which, "Jidai Matsuri" (The Festival of Ages), was held on October 22, just after ICONIP 2016.

We, the general chair, co-chair, and Program Committee co-chairs, would like to express our sincere gratitude to everyone involved in making the conference a success. We wish to acknowledge the support of all the sponsors and supporters of ICONIP 2016, namely, APNNS, JNNS, KDDI, NICT, Ogasawara Foundation, SCAT, as well as Kyoto Prefecture, Kyoto Convention and Visitors Bureau, and Springer. We also thank the keynote, plenary, and invited speakers, the exhibitors, the student paper award evaluation committee members, the special session and workshop organizers, as well as all the Organizing Committee members, the reviewers, the conference participants, and the contributing authors.

October 2016

Akira Hirose
Seiichi Ozawa
Kenji Doya
Kazushi Ikeda
Minho Lee
Derong Liu

Organization

General Organizing Board

JNNS Board Members

Honorary Chairs

Shun-ichi Amari	RIKEN
Kunihiko Fukushima	Fuzzy Logic Systems Institute

Organizing Committee

General Chair

Akira Hirose	The University of Tokyo, Japan

General Co-chair

Seiichi Ozawa	Kobe University, Japan

Program Committee Chairs

Kenji Doya	OIST, Japan
Kazushi Ikeda	NAIST, Japan
Minho Lee	Kyungpook National University, Korea
Derong Liu	Chinese Academy of Science, China

Local Arrangements Chairs

Hiroaki Nakanishi	Kyoto University, Japan
Ikuko Nishikawa	Ritsumeikan University, Japan

Members

Toshio Aoyagi	Kyoto University, Japan
Naoki Honda	Kyoto University, Japan
Kazushi Ikeda	NAIST, Japan

Shin Ishii	Kyoto University, Japan
Katsunori Kitano	Ritsumeikan University, Japan
Hiroaki Mizuhara	Kyoto University, Japan
Yoshio Sakurai	Doshisha University, Japan
Yasuhiro Tsubo	Ritsumeikan University, Japan

Financial Chair

Seiichi Ozawa — Kobe University, Japan

Member

Toshiaki Omori — Kobe University, Japan

Special Session Chair

Kazushi Ikeda — NAIST, Japan

Workshop/Tutorial Chair

Hiroaki Gomi — NTT Communication Science Laboratories, Japan

Publication Chair

Koichiro Yamauchi — Chubu University, Japan

Members

Yutaka Hirata	Chubu University, Japan
Kay Inagaki	Chubu University, Japan
Akito Ishihara	Chukyo University, Japan

Exhibition Chair

Tomohiro Shibata — Kyushu Institute of Technology, Japan

Members

Hiroshi Kage	Mitsubishi Electric Corporation, Japan
Daiju Nakano	IBM Research - Tokyo, Japan
Takashi Shinozaki	NICT, Japan

Publicity Chair

Yutaka Sakai — Tamagawa University, Japan

Industry Relations

Ken-ichi Tanaka	Mitsubishi Electric Corporation, Japan
Toshiyuki Yamane	IBM Research - Tokyo, Japan

Sponsorship Chair

Ko Sakai	University of Tsukuba, Japan

Member

Susumu Kuroyanagi	Nagoya Institute of Technology, Japan

General Secretaries

Hiroaki Mizuhara	Kyoto University, Japan
Gouhei Tanaka	The University of Tokyo, Japan

International Advisory Committee

Igor Aizenberg	Texas A&M University-Texarkana, USA
Sabri Arik	Istanbul University, Turkey
P. Balasubramaniam	Gandhigram Rural Institute, India
Eduardo Bayro-Corrochano	CINVESTAV, Mexico
Jinde Cao	Southeast University, China
Jonathan Chan	King Mongkut's University of Technology, Thailand
Sung-Bae Cho	Yonsei University, Korea
Wlodzislaw Duch	Nicolaus Copernicus University, Poland
Tom Gedeon	Australian National University, Australia
Tingwen Huang	Texas A&M University at Qatar, Qatar
Nik Kasabov	Auckland University of Technology, New Zealand
Rhee Man Kil	Sungkyunkwan University (SKKU), Korea
Irwin King	Chinese University of Hong Kong, SAR China
James Kwok	Hong Kong University of Science and Technology, SAR China
Weng Kin Lai	Tunku Abdul Rahman University College, Malaysia
James Lam	The University of Hong Kong, SAR Hong Kong
Kittichai Lavangnananda	King Mongkut's University of Technology, Thailand
Min-Ho Lee	Kyungpoor National University, Korea
Soo-Young Lee	Korea Advanced Institute of Science and Technology, Korea
Andrew Chi-Sing Leung	City University of Hong Kong, SAR China
Chee Peng Lim	University Sains Malaysia, Malaysia
Chin-Teng Lin	National Chiao Tung University, Taiwan
Derong Liu	The Institute of Automation of the Chinese Academy of Sciences (CASIA), China

Chu Kiong Loo	University of Malaya, Malaysia
Bao-Liang Lu	Shanghai Jiao Tong University, China
Aamir Saeed Malik	Petronas University of Technology, Malaysia
Danilo P. Mandic	Imperial College London, UK
Nikhil R. Pal	Indian Statistical Institute, India
Hyeyoung Park	Kyungpook National University, Korea
Ju. H. Park	Yeungnam University, Republic of Korea
John Sum	National Chung Hsing University, Taiwan
DeLiang Wang	Ohio State University, USA
Jun Wang	Chinese University of Hong Kong, SAR Hong Kong
Lipo Wang	Nanyang Technological University, Singapore
Zidong Wang	Brunel University, UK
Kevin Wong	Murdoch University, Australia
Xin Yao	University of Birmingham, UK
Li-Qing Zhang	Shanghai Jiao Tong University, China

Advisory Committee Members

Masumi Ishikawa	Kyushu Institute of Technology
Noboru Ohnishi	Nagoya University
Shiro Usui	Toyohashi University of Technology
Takeshi Yamakawa	Fuzzy Logic Systems Institute

Technical Program Committee

Abdulrahman Altahhan
Sabri Arik
Sang-Woo Ban
Tao Ban
Matei Basarab
Younes Bennani
Ivo Bukovsky
Bin Cao
Jonathan Chan
Rohitash Chandra
Chung-Cheng Chen
Gang Chen
Jun Cheng
Long Cheng
Zunshui Cheng
Sung-Bae Cho
Justin Dauwels
Mingcong Deng
Kenji Doya
Issam Falih
Tetsuo Furukawa
Kuntal Ghosh
Anupriya Gogna
Hiroaki Gomi
Shanqing Guo
Masafumi Hagiwara
Isao Hayashi
Shan He
Akira Hirose
Jin Hu
Jinglu Hu
Kaizhu Huang
Jun Igarashi
Kazushi Ikeda
Ryoichi Isawa
Shin Ishii
Teijiro Isokawa
Wisnu Jatmiko
Sungmoon Jeong
Youki Kadobayashi

Keisuke Kameyama
Joarder Kamruzzaman
Rhee-Man Kil
DaeEun Kim
Jun Kitazono
Yasuharu Koike
Markus Koskela
Takio Kurita
Shuichi Kurogi
Susumu Kuroyanagi
Minho Lee
Nung Kion Lee
Benkai Li
Bin Li
Chao Li
Chengdong Li
Tieshan Li
Yangming Li
Yueheng Li
Mingming Liang
Qiao Lin
Derong Liu
Jiangjiang Liu
Weifeng Liu
Weiqiang Liu
Bao-Liang Lu
Shiqian Luo
Hongwen Ma
Angshul Majumdar
Eric Matson
Nobuyuki Matsui
Masanobu Miyashita
Takashi Morie
Jun Morimoto
Chaoxu Mu
Hiroyuki Nakahara
Kiyohisa Natsume
Michael Kwok-Po Ng
Vinh Nguyen
Jun Nishii
Ikuko Nishikawa
Haruhiko Nishimura
Tohru Nitta
Homma Noriyasu
Anto Satriyo Nugroho
Noboru Ohnishi
Takashi Omori
Toshiaki Omori
Sid-Ali Ouadfeul
Seiichi Ozawa
Paul Pang
Hyung-Min Park
Kitsuchart Pasupa
Geong Sen Poh
Santitham Prom-on
Dianwei Qian
Jagath C. Rajapakse
Mallipeddi Rammohan
Alexander Rast
Yutaka Sakaguchi
Ko Sakai
Yutaka Sakai
Naoyuki Sato
Shigeo Sato
Shunji Satoh
Chunping Shi
Guang Shi
Katsunari Shibata
Hayaru Shouno
Jeremie Sublime
Davor Svetinovic
Takeshi Takahashi
Gouhei Tanaka
Kenichi Tanaka
Toshihisa Tanaka
Jun Tani
Katsumi Tateno
Takashi Tateno
Dat Tran
Jan Treur
Eiji Uchibe
Eiji Uchino
Yoji Uno
Kalyana C. Veluvolu
Michel Verleysen
Ding Wang
Jian Wang
Ning Wang
Ziyang Wang
Yoshikazu Washizawa
Kazuho Watanabe
Bunthit Watanapa

Juyang Weng
Bin Xu
Tetsuya Yagi
Nobuhiko Yamaguchi
Hiroshi Yamakawa
Toshiyuki Yamane
Koichiro Yamauchi
Tadashi Yamazaki
Pengfei Yan
Qinmin Yang
Xiong Yang
Zhanyu Yang
Junichiro Yoshimoto
Zhigang Zeng
Dehua Zhang
Li Zhang
Nian Zhang
Ruibin Zhang
Bo Zhao
Jinghui Zhong
Ding-Xuan Zhou
Lei Zhu

Contents – Part IV

Applications

Theory and Algorithms

Applications

Classifying Human Activities with Temporal Extension of Random Forest

Shih Yin Ooi(✉), Shing Chiang Tan, and Wooi Ping Cheah

Faculty of Information Science and Technology, Multimedia University, Jalan Ayer Keroh Lama, 75450 Melaka, Malaysia
{syooi, sctan, wpcheah}@mmu.edu.my

Abstract. Sensor-Based Human Activity Recognition (HAR) is a study of recognizing the human's activities by using the data captured from wearable sensors. Avail the temporal information from the sensors, a modified version of random forest is proposed to preserve the temporal information, and harness them in classifying the human activities. The proposed algorithm is tested on 7 public HAR datasets. Promising results are reported, with an average classification accuracy of ~98 %.

Keywords: Human activity · Classification · Random forest · Temporal sequences · Machine learning

1 Introduction

Human Activity Recognition (HAR) is a study of recognizing the human's activities by considering the locomotion as well as the environmental conditions. HAR is gaining enormous attention due to the emerging of smart homes, mobile devices, wearable sensors, smart fabrics, and assistive robotics [1]; across the markets of United States, Europe, Japan and Korea. In the literature, HAR is widely studied from two approaches: (1) vision-based, and (2) sensor-based [2]. Vision-based HAR has been well explored especially in the field of computer vision. The merit of this paradigm is its ability to supply intuitive and rich information. However, it also requires loads of memory resources to store the high dimensional images, and video sequences [3]. Alternatively, one may opt for sensor-based HAR, which is primarily relying on sensor devices to monitor the human activity. The captured sensor data are usually a set of sequential data, mainly consisting of locomotors state changes. The merit of this paradigm is its simplicity in real-world application. It makes the human monitoring possible by just using the small and transportable units, such as wearable sensors, smart phones, smart fabrics, etc. The main interest of this paper is to devise a classification method in predicting the human activities, with the approach of sensor-based HAR.

2 Related Works

As discussed in the aforementioned section, the captured sensor data are mainly sequential data, which may carrying along the vast amount of temporal information. Thus, many works have been devoted to extract these temporal information and further

A. Hirose et al. (Eds.): ICONIP 2016, Part IV, LNCS 9950, pp. 3–10, 2016.
DOI: 10.1007/978-3-319-46681-1_1

utilizing them in prediction task through the data mining and machine learning approach. Jatobá et al. [4] conducted an empirical evaluation for six machine learning techniques. The results shown that ANFIS and CART were outperformed the rest, tested on an online activity monitoring system.

One of the benchmark techniques to extract temporal information is Hidden Markov Models (HMM). Mannini et al. [5] proposed an indoor-outdoor pedestrian navigation system with a wearable sensor. The main classifier used was HMM, reported with the average classification accuracy of $\sim$95 % on an acceleration waveforms dataset made available by [6]. San-Segundo et al. [7] proposed a HARS by utilizing HMM as the main learning model. The proposed method was outperformed [8] when performing on a public dataset – UCI Human Activity Recognition Using Smartphones [9]. However, instead of evaluating HMM capability as a classifier, the main contribution of [7] is to examine the capability of HMM in segmenting the record physical activities. The work reported with Segmentation Error Rate of 2.1 %.

Another favorite learner is support vector machines (SVM) [10]. Anguita et al. [8, 9] is one of the active groups which employed SVM as the main learner in recognizing the human activities, based on the sensor input from smartphones. To not over abusing the smartphone battery consumption, they modified the conventional SVM with fixed-point arithmetic. The work was reported with precision of 89 %. Variants of neural network techniques are widely exploited too [11, 12]. Ronao et al. [11] proposed a deep convolutional neural network (convnet) for a sensor-based HARS. Tested on an in-house dataset, the accuracy of 95.75 % was reported. Wu et al. [12] proposed a mixed-kernel based weighted extreme learning machine (MK-WELM) to tackle the issue of imbalanced data especially for abnormal actions. Tested on variants setting for several UCI datasets, MK-WELM has shown its superiority in handling imbalanced data (the best accuracy $\sim$98 %), by combatting both the benchmarked ELM and weighted ELM algorithms.

Considering the avail of temporal information from sensor data, there were several works handling the case with time series approach [13, 14]. Recently, Liu et al. [15, 16] developed a time series pattern dictionary (shapelets) to mine various sensor data from multiple sensors. Tested on an in-house dataset, the proposed method reported an accuracy of 96.54 %. This findings substantiated the usefulness of temporal information from the sensor data. Thus, we see the potential to device a temporal-based classifier in the context of sensor-based HAR.

2.1 Motivation and Justification

Random forest (RF) [17] is one of the most successful implementation on bagging approach. It is proven to be robust in handling different classification tasks, and there are variants of extension made on it. The integration of temporal learning methodology in the architecture of RF has been made possible in our founding paper [18]. Thus, in this paper, we would like to test its suitability in classifying the human activities, which are generally in temporal fashion.

3 Temporal Extension of Random Forest

The extension mechanism of RF is depicted in Fig. 1. The extension is conducted from two perspectives: (1) *Temporal Sampling Mechanism*: to rearrange the sensor data by considering its sequence of happening (will be further discussed in Sect. 3.2), and (2) *Temporal Randomization*: to modify the randomization procedure in RF to preserve temporal sequence of captured sensor data (will be further discussed in Sect. 3.3).

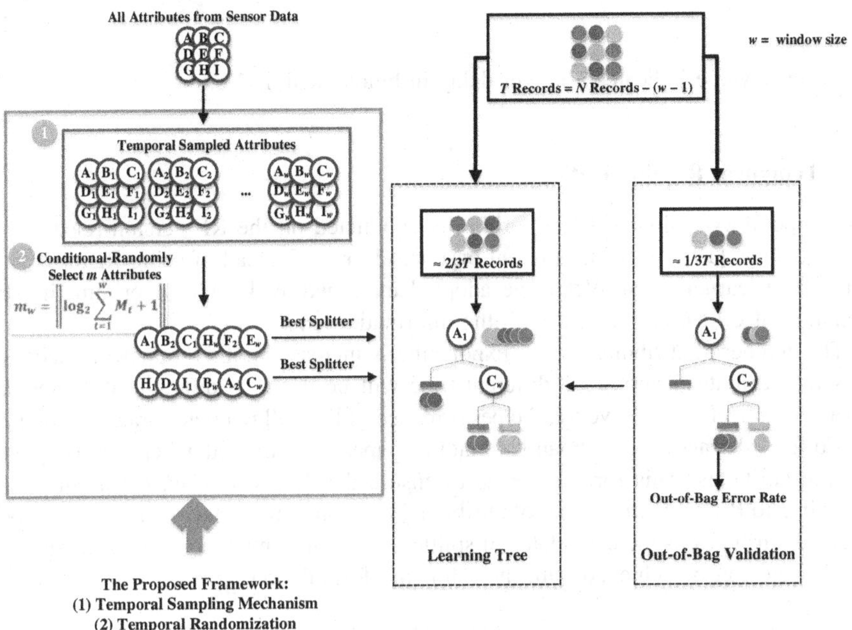

Fig. 1. A single tree formation in the temporal extension of RF

3.1 Preliminary

In the founding paper of RF [17], Breiman defined it as:

Definition 1. A RF classifier is a collection of multiple tree-based classifiers, defined as $\{h(x, \Theta_k), k = 1, \ldots, L\}$ where $\{\Theta_k\}$ is a set of independently distributed random vectors, and each tree $h(x, \Theta_k)$ casts a unit vote for the most popular class at input x.

3.2 Temporal Sampling Mechanism

We believed that most of the performed human activity will be affected by his/her previous activities, i.e. if a person was requested to climb stairs for 5 times, the motion will be vary throughout the time, such as, the speed of climbing will slower because of tiredness. Thus, to enforce the classifier will consider the previous-timestamped sensor attributes when classifying a performed activity, a set of consecutive records can be merged based on the desired time window (w) observations. A set of sensor attributes is represented as $A = \{a_1, a_2, a_3, \ldots a_M\}$. To make the observation of a set of activities performed within a time window of w, the merged set of sensor attributes can be defined as $\tilde{A}$:

$$\tilde{A}_j = A_1||A_2||A_3||\cdots||A_{w=j}, \text{ where}$$
$$\{(a_1, \ldots a_M)^{w=1} \in A_1; (a_1, \ldots a_M)^{w=2} \in A_2; (a_1, \ldots a_M)^{w=3} \in A_3; \cdots; (a_1, \ldots a_M)^{w=j} \in A_j\}; \quad (1)$$

the larger w value indicates a longer delay in human activity' effects.

3.3 Temporal Randomization

The temporal extension of RF algorithm is conducted on the RF version released in Weka package [19], but with different base tree learner. Instead of using Random tree as the base learner as of Weka, we adopted unpruned Weka J48 algorithm (implementation of C4.5), normalized with the information gain ratio.

The number of attributes will be exponentially increased in this framework because the same attribute occurred on different time will be deem as different attributes, as explained in Sect. 3.2. However, the performance of RF will not deteriorated as long as the Eq. (2) (defined by Breiman's founding paper) is enforced when choosing the splitting attributes. This formulation is to ensure the diversity of each learning tree. According to Breiman, m number of attributes to be randomly selected from M number (total) of attributes as the candidates of splitter, value of m must be very much smaller than M, and can be achieved with the following formulation:

$$m = ||\log_2(M) + 1|| \quad (2)$$

As indicated in our founding paper [18], the immediate application of Eq. (2) is inappropriate when a temporal observation is a necessity. For an instance, if all attributes are chosen from the similar window, i.e.: a_2, b_2, and c_2, then the desired temporal observation is totally omitted in this context. Thus, the Eq. (2) implementation must restrained with a condition, which can be expressed as below:

$$m_w = \left\|\log_2 \sum_{t=1}^{w} M_t + 1\right\|, \textbf{ with the condition of } \left\|m_w - \frac{m_w}{w}\right\| \notin A_w \quad (3)$$

The condition enforced that at least 50 % of the selected attribute splitters must not come from the same time window.

4 Experimental Evaluation

4.1 Data Sets and Experimental Setup

Experimental tests have been conducted on 7 public HAR databases from UCI Machine Learning Repository (http://archive.ics.uci.edu/ml/datasets), summarized as in Table 1. The experiments are evaluated on a 10-fold cross validation.

Table 1. Summary of experimental datasets

	Dataset	#Instances	#Attributes	#Classes
D1	Activity recognition from single chest-mounted accelerometer	402843 *(for 3 users)*	5	7
D2	ADL recognition with wrist worn accelerometer dataset	79438	4	14
D3	Daily and sports activities dataset	142500	46	19
D4	EMG physical action dataset	128886	9	20
D5	User identification from walking activity	149332	5	22
D6	Smartphone-based recognition of human activities and postural transitions data set	10929	561	12
D7	Indoor user movement prediction from RSS data set	13197	4	2

4.2 Experimental Results

Due to the ways of dataset collection are different for these 7 datasets, 5 observation windows (from $w = 1$ to $w = 5$) are recursively tested on each dataset to observe their performances, until the error rates stop growing. The respective *Classification Accuracies* are projected in Fig. 2. As explained in Sect. 3.2, $w = 1$ indicates there is no temporal or delayed effect, and the delayed impacts can be observed when $w > 1$.

As depicted in Fig. 2, all HAR datasets have achieved higher classification accuracies when the window size is increased. The classification accuracies for almost all datasets are obviously improved when the observation window is >1. However, it is impossible to determine a fixed optimal number of time windows for these 7 datasets, because the delay impacts caused by different individuals are different and very subjective to their psychomotor ability as well as the way of collecting the dataset, i.e. the used devices, the interval time of observation, performed activities, etc.

This observation convinced our hypothesis on the delayed and temporal effects in HAR datasets. More empirical results are detailed in Table 2.

4.3 Performance Comparison

To substantiate the importance of temporal value in human activity recognition, the results are experimentally compared with Weka RF. Obviously, the temporal extension of RF is surpassing the classification ability of RF in these 7 datasets as shown in Table 3.

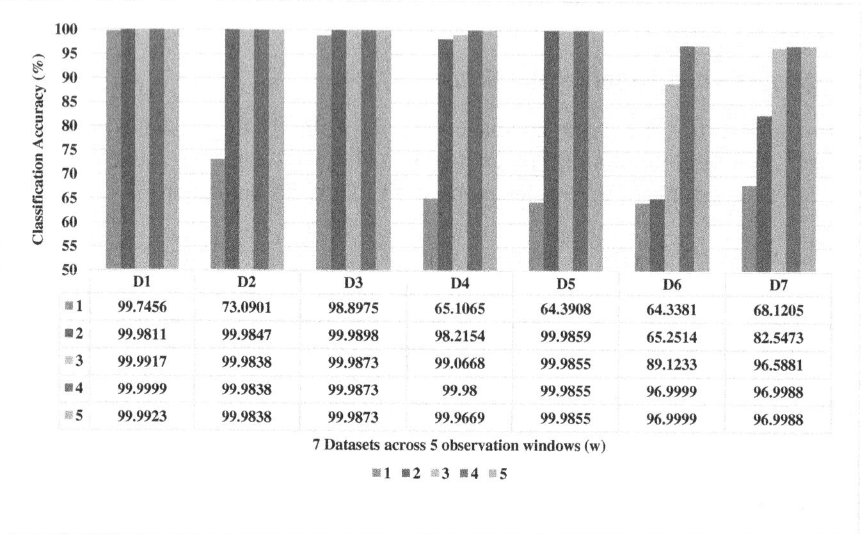

Fig. 2. Classification accuracies for 5 windows in 7 HAR datasets with 10-fold cross validation

Table 2. Weighted average of True Positive Rate (TPR), False Positive Rate (FPR), Precision, Recall, and Area under ROC Curve (AUC) for 7 HAR datasets on respective best-performed windows (*w*) size

Dataset	Weighted average					
	Best window (w)	*TPR*	*FPR*	*Precision*	*Recall*	*AUC*
D1	4	1	0	1	1	1
D2	2	1	0	1	1	1
D3	2	1	0	1	1	1
D4	4	1	0	1	1	1
D5	2	1	0	1	1	1
D6	4	0.890	0.012	0.960	0.890	0.960
D7	4	0.822	0.008	0.940	0.822	0.940

Furthermore, to compare the performance of proposed paradigm with two relevant works, Anguita et al. [8], and San-Segundo et al. [7], we adopted the similar dataset which was used in their works (the used dataset is ***D6***). To ensure the fairness in comparison, the original dataset setting is adopted, which are 70 % training data and 30 % testing data (this presetting data is available directly from UCI). The result is tabulated in Table 4.

From Table 4, our proposed algorithm managed to plot on 96 % of precision, better than [8], and slightly poor than [7]. However, the recall rate of our proposed algorithm is similar to [8], but reasonably poor than [7]. This is mainly due to [7] had performed a very delicate consideration during the feature extraction steps, so that the recognition

Table 3. Performance comparison of our proposed paradigm *(with best performed w)* and RF based on classification accuracy (%), Kappa statistic and out-of-bag (OOB) error, with number of trees = 10

Dataset	Temporal extension of RF			RF		
	Classification accuracy (%)	*Kappa statistic*	*OOB error*	*Classification accuracy (%)*	*Kappa statistic*	*OOB error*
D1	**99.9999***	0.9999	0.0065	99.7457	0.9967	0.0097
D2	**99.9847***	0.9999	0.0077	73.0897	0.6583	0.2788
D3	**99.9898***	0.9999	0.0086	98.8975	0.9884	0.0242
D4	**99.9800***	0.9998	0.0084	65.1067	0.6123	0.4419
D5	**99.9859***	0.9998	0.0052	64.3908	0.6134	0.3947
D6	**96.9999***	0.9842	0.0113	64.3645	0.8255	0.5421
D7	**96.9988***	0.9851	0.0043	69.5897	0.7222	0.4671

Table 4. Performance comparison of our proposed paradigm with Anguita et al.'s and San-Segundo et al.'s works on UCI dataset of Smartphone-Based Recognition of Human Activities

Weighted average	Anguita et al. [8]	San-Segundo et al. [7]	Our proposed model
Precision	0.89	0.98	0.96
Recall	0.89	0.98	0.89

segmentation process will not be affected by a specific user's intra-variances in this dataset, where our work is solely focus on classifier. Thus, its sensitivity might slightly hampered in this case.

5 Conclusion

In this paper, a temporal extension of RF is adopted to classify human activity. The temporal version of RF model is an integration framework of two schemes: (1) Temporal Sampling: to re-arrange all sensor attributes, so that a temporal observation from the machine learning perspective is made possible; (2) Temporal Randomization: to enforce the RF randomization procedure to be conducted in a specific temporal manner. The promising results of all datasets substantiated the usefulness of temporal RF in HARS.

References

1. Rashidi, P., Mihailidis, A.: A survey on ambient-assisted living tools for older adults. IEEE J. Biomed. Health Inf. **17**(3), 579–590 (2013)
2. Chen, L., Hoey, J., Nugent, C.D., Cook, D.J., Yu, Z.: Sensor-based activity recognition. IEEE Trans. Syst. Man Cybern. Part C Appl. Rev. **42**(6), 790–808 (2012)

3. Guan, Q., Li, C., Guo, X., Wang, G.: Compressive classification of human motion using pyroelectric infrared sensors. Pattern Recogn. Lett. **49**, 231–237 (2014)
4. Jatobá, L.C., Großmann, U., Kunze, C., Ottenbacher, J., Stork, W.: Pattern recognition methods for classification of physical activity. In: 30th Annual International Conference of the IEEE Engineering in Medicine and Biology Society, pp. 5250–5253 (2008)
5. Mannini, A., Sabatini, A.M.: Machine learning methods for classifying human physical activity from on-body accelerometers. Sensors **10**(2), 1154–1175 (2010)
6. Bao, L., Intille, S.S.: Activity recognition from user-annotated acceleration data. In: Ferscha, A., Mattern, F. (eds.) PERVASIVE 2004. LNCS, vol. 3001, pp. 1–17. Springer, Heidelberg (2004)
7. San-Segundo, R., Lorenzo-Trueba, J., Martínez-González, B., Pardo, J.M.: Segmenting human activities based on HMMs using smartphone inertial sensors. Pervasive Mob. Comput. **30**, 84–96 (2016)
8. Anguita, D., Ghio, A., Oneto, L., Parra, X., Reyes-Ortiz, J.L.: Energy efficient smartphone-based activity recognition using fixed-point arithmetic. Special session in ambient assisted living: home care. J. Univ. Comput. Sci. **19**(9), 1295–1314 (2013)
9. Anguita, D., Ghio, A., Oneto, L., Parra, X., Reyes-Ortiz, J.L.: A public domain dataset for human activity recognition using smartphones. In: European Symposium on Artificial Neural Networks, Computational Intelligence and Machine Learning, pp. 24–26 (2013)
10. Reyes-Ortiz, J.-L., Oneto, L., Samà, A., Parra, X., Anguita, D.: Transition-aware human activity recognition using smartphones. Neurocomputing **171**, 754–767 (2015)
11. Ronao, C.A., Cho, S.-B.: Human activity recognition with smartphone sensors using deep learning neural networks. Expert Syst. Appl. **59**, 235–244 (2016)
12. Wu, D., Wang, Z., Chen, Y., Zhao, H.: Neurocomputing mixed-kernel based weighted extreme learning machine for inertial sensor based human activity recognition with imbalanced dataset. Neurocomputing **190**, 35–49 (2016)
13. Chamroukhi, F., Mohammed, S., Trabelsi, D., Oukhellou, L., Amirat, Y.: Joint segmentation of multivariate time series with hidden process regression for human activity recognition. Neurocomputing **120**, 633–644 (2013)
14. Sanchez-Valdes, D., Alvarez-Alvarez, A., Trivino, G.: Dynamic linguistic descriptions of time series applied to self-track the physical activity. Fuzzy Sets Syst. **1**, 1–20 (2015)
15. Liu, L., Peng, Y., Liu, M., Huang, Z.: Sensor-based human activity recognition system with a multilayered model using time series shapelets. Knowl. Based Syst. **90**, 138–152 (2015)
16. Liu, L., Peng, Y., Wang, S., Liu, M., Huang, Z.: Complex activity recognition using time series pattern dictionary learned from ubiquitous sensors. Inf. Sci. **340–341**, 1–17 (2016)
17. Breiman, L.: Random forests. Mach. Learn. **45**(1), 5–32 (2001)
18. Ooi, S.Y., Tan, S.C., Cheah, W.P.: Temporal Sampling Forest (TSF): an ensemble temporal learner. Soft. Comput. (2016). doi:10.1007/s00500-016-2242-7
19. Witten, I.H., Frank, E., Hall, M.A.: Data Mining: Practical Machine Learning Tools and Techniques, 3rd edn. Morgan Kaufmann Publishers Inc., Burlington (2011)

Echo State Network Ensemble for Human Motion Data Temporal Phasing: A Case Study on Tennis Forehands

Boris Bačić(✉)

School of Engineering, Computer and Mathematical Sciences, Auckland University of Technology, Auckland, New Zealand
boris.bacic@aut.ac.nz

Abstract. Temporal phasing analysis is integral to ubiquitous/"smart" coaching devices and sport science. This study presents a novel approach to autonomous temporal phasing of human motion from captured tennis activity (3D data, 66 time-series). Compared to the optimised Echo State Network (ESN) model achieving 85 % classification accuracy, the ESN ensemble system demonstrates improved classification of 95 % and 100 % accurate phasing state transitions for previously unseen motions without requiring ball impact information. The ESN ensemble model is robust to low-sampling rates (50 Hz) and unbalanced data sets containing incomplete data time-series. The demonstrated achievements are applicable to exergames, augmented coaching and rehabilitation systems advancements by enabling automated qualitative analysis of motion data and generating feedback to aid motor skill and technique improvements.

Keywords: Computational Intelligence (CI) · Sport and rehabilitation · Biomechanics · Augmented Coaching Systems (ACS) · Data analytics · Human Motion Modelling and Analysis (HMMA)

1 Introduction

Over the recent years, there has been an increase in the popularity of exergames, and ubiquitous and wearable devices that promote an active lifestyle and health monitoring functions. Components such as Inertial Measurement Units (IMUs) and video sensors are found in mobile technology but also in augmented coaching devices that may be attached to the human body or sport equipment (e.g. www.xsens.com, www.shimmersensing.com, www.babolatplay.com, and www.zepp.com). In general, to develop augmented coaching system and technology (ACST) for a specific sport, the underlying technology should be able to capture, process and quantify specific movement parameters that happened over time in an on-line or off-line fashion. Whether the purpose is to analyse swing parameters or to track leg movement, it is common for developed solutions [1] to rely on proprietary sport-specific algorithms for *temporal phasing* analysis that could extract motion patterns around the impact vibration or sound (e.g. tempo and velocity at impact). However, there is little evidence about future developments of ACST that could provide intuitive feedback to aid sport

A. Hirose et al. (Eds.): ICONIP 2016, Part IV, LNCS 9950, pp. 11–18, 2016.
DOI: 10.1007/978-3-319-46681-1_2

technique similar to a coach [2–5]. This paper supports the author's vision that future ACST will combine multi-disciplinary approaches from sport science and qualitative analysis of human motion with data science and computational intelligence to provide feedback aimed at improving motor skill acquisition and technique associated with a specific motion pattern. This paper presents a novel modelling solution to temporal phasing analysis and indexing utilising captured real-life 3D motion data time-series that can learn from data and analyse previously unseen tennis forehands. The presented generic artefacts of achieved autonomous temporal phasing analysis are integral to various models of Qualitative Movement Diagnostics (QMD) in sport science, containing preparation, action zone and follow-through motion sequence associated with ball impact or projectile release, with kicking, throwing or striking action [6].

1.1 Related Work and Prior Studies

The Echo State Network (ESN), Liquid State Machine (LSM) and Spiking Neural Network (SNN) models are examples of the "third generation" of biologically inspired neural networks [7] for temporal and spatial problem areas that are also considered as alternatives to traditional algorithmic, neuro-fuzzy, heuristic- and feature extraction-based approaches [8, 9]. Jaeger's ESN model [8] is based on recurrent neural networks which does not require signal-to-spike train conversions for its operation. For an ESN's training task, it is common to perform adjustment the output layer, while leaving the input layer and reservoir unchanged. The presented work is linked to and expands the earlier studies in the context of Human Motion Modelling and Analysis (HMMA) and tennis [5, 10, 11].

Prior to modern development of the third generation of neural networks, the finite-state automata theory and finite-state machines (FSM) were introduced in the early days of computing science to address temporal and spatial problem areas with a distinct historical example known as the Turing machine. In this study, it is contended that a combination of modern and 'old' approaches can advance computer science, showing that the ESN ensemble model orchestration can be achieved by implementing sequential control logic to advance classifier performance and reduce the need for training data. The sequential logic of FSM eliminates the need for supervised learning from data as the domain expert knowledge required for orchestration can be expressed in the state transitions control mechanism. The expected benefits from such system configurations include potential improvements in classification tasks and robustness to unbalanced data sets; avoiding false positive event classifications; and preventing output state transitions that are not possible in real-life.

1.2 Temporal Phasing Analysis: Sport Science and Tennis Backgrounds

Temporal phasing analysis is integral part of coaching practice, rehabilitation and sport science with the early observational model [12] introduced in 1984. Temporal phasing combined with computational intelligence is related to the areas of: surveillance, exergames and ACST design [5]. In today's tennis, swings are executed with a variety of diverse paces, rhythms, stances and individual preferences (e.g. backswing and swing preparation timings). In sport science, there is a known phenomenon of *experts' disagreements* on observed motion [6]. While the descriptive rules are more sport-specific and decision boundaries may vary between experts, phasing analysis is common to sport science and integral to a number of models of qualitative analysis of human motion. Typical examples of temporal phasing analysis in tennis are: (a) the use of replays during media coverage, during (b) coaching sessions for analysis and feedback; and (c) in real-time scenarios, where a coach can also rely on his/her temporal phasing expertise while observing a player's motion and communicating short keywords as *attention cuing*. In this paper it is contended that temporal phasing will be implemented in the near-future generations of ACST and exergames which will provide feedback to aid motor skill learning and technique acquisition. For the experimental design in this paper, it is important to notice that there is no ground truth or agreed measure for determining the exact start of the forehand swing event (as opposed to vibration or sound around the impact with the ball). Personalisation and the diversity of forehand swings may be considered as one of the challenges in coaching practice. The expert's decision in this paper was that the preparation phase starts at the racquet transitions from backward to forward movement.

2 Experimental Setup: Data Collection and Visualisation of Temporal Phasing for Supervised Machine Learning

The data set was acquired in a laboratory setting using a nine fixed-location camera motion capture system (SMART-e 900 eMotion/BTS) sampling at 50 Hz, and a set of 22 retro-reflective markers attached to anatomical landmarks on player's body. The relatively small data set (21 forehand samples) was considered of sufficient size for the purpose of this study without warranting the need for additional synthetic data. Temporal phasing of all data frames into four categories was achieved visually using 3D animated stick figure (Fig. 1). Temporal phasing output classes (0, 1, 2, 3) are labelled as ('non-event', 'preparation', 'action', 'follow-through').

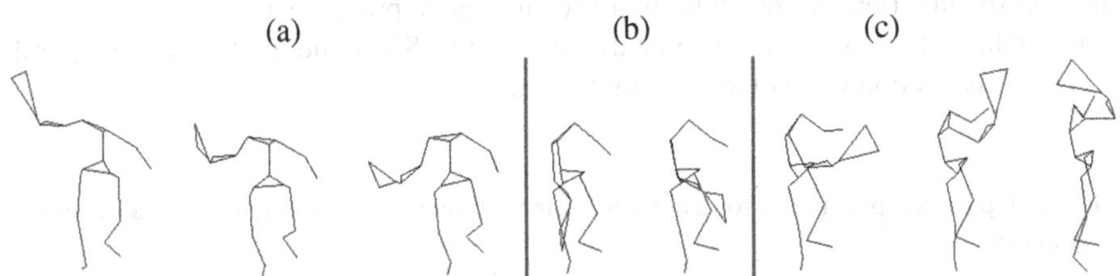

Fig. 1. Temporal phasing analysis of a forehand swing event. Distinct temporal phases for output classes (1, 2, 3) cover: (a) *preparation*, (b) *action* zone around estimated intended impact, and (c) *follow-through* post-impact recovery. The sequence of 3D stick figures is not shown at equal time distance.

3 Data Analysis and Modelling

The motion data set includes a variety of 'good' and 'bad' tennis swings, executed with diverse swing speed and from diverse stances. All tennis swings were performed by a certified tennis coach (the author of the paper) and reviewed by three other coaches. The motion data set does not include ball impact information (e.g. impact vibration pattern). The output class distribution (Fig. 2) is considered as unbalanced.

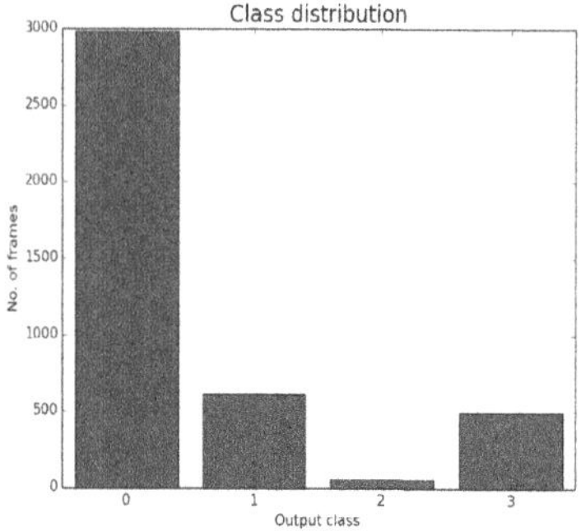

Event duration analysis: [no. of frames]

Action zone event = minority output class 2

Minimum: 2

Maximum: 3

Forehand event = output class 1+2+3

Minimum: 42

Maximum: 71

Median: 53

Standard deviation: 7.61

Range (max – min): 29

Fig. 2. Uneven class distribution where minority class typically lasts for two (or three frames for a slower swing) and majority class duration lasts for over 800 samples.

Preparing motion data for the optimised ESN model (Table 1) includes the pre-processing steps: (a) removal of the static marker time series (used for capture volume referencing and orientation purposes); (b) replacing missing values (NaN) with adjacent frames' average values; and (c) linear normalisation within the interval [−1, 1].

Given the properties of experimental data (temporal and spatial, unbalanced output class distribution, short duration of minority class, diversity of captured foreheads), the modelling task may be perceived as too complex. As a design decision in this study, it was decided to perform one grid search (Table 1) for parameter optimisation for all four data streams (leaving potential room for further improvements).

Preliminary findings (Fig. 3) show the optimised ESN model performing temporal phasing classification of 3D tennis motion data.

Table 1. Optimised parameters for leaky reservoir ESN model, utilising tanh(·) for all neurons activation function

Parameter description	Optimal value	Increment	Range
Number of neurons	800	x +=100	[400, …, 1000]
Leak rate	0.9	x +=0.1	[0.5, …, 1]
Spectral radius	0.9	x +=0.1	[0.5, …, 1]
Input scaling	0.0001	$x=10^{-y}$, y+=1	[0.00001, …, 0.1]
Ridge parameter	0.0001	$x=10^{-y}$, y+=1	[0.00001, …, 0.1]

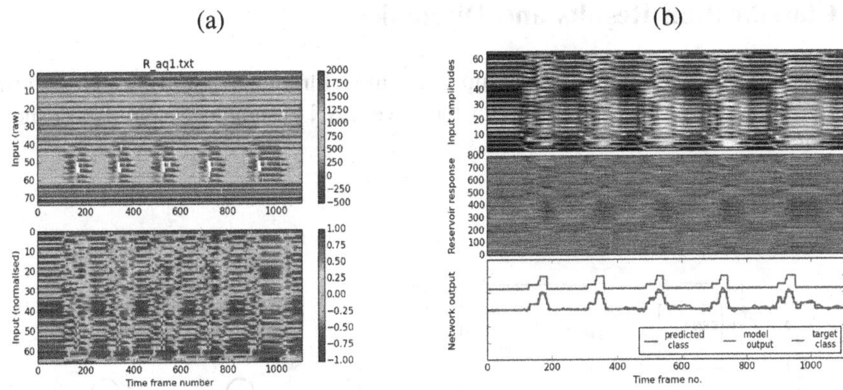

Fig. 3. Steps involved in processing and ESN classification of motion data time-series: (a) removal of static reference markers, input data pre-processing with (NaN) replacements and normalising to [−1,1]; and (b) a single ESN model for temporal phasing of forehand events.

Table 2. Error importance for temporal phasing from observed ESN model performance

Error importance	Description
Low	Start of preparation event
Low	End of follow-through
Moderate	Start of action zone
Moderate	End of action zone
High	False event detection
High	Impossible state transition.

The attribution of observed issues (Fig. 3) linked to a range of classification errors that warranted an ESN ensemble approach is summarised in Tables 2 and 3.

Based on visualisation of ESN 2 performance (similar to Fig. 3b), the developed ESN ensemble system (Fig. 4a and b) required only minor adjustments of the clipping threshold (CT) parameter in order to improve ESN 1 (CT = 0.55) and ESN 2 (CT = 0.4) readout signals conversion into their discrete outputs (Fig. 4a, c and d).

Table 3. Problem summary and error analysis that warranted an ensemble approach

Key issue	Error description
• Unbalanced output class distribution • Relatively low data sampling rate (50 Hz) • Incomplete marker trajectories • Varying duration of diverse forehands • Predicted output class state transitions are in disagreement with domain expertise • ESN dynamic behaviour producing 'echoed' and false state transitions • Matching ESN readouts in the minority class	• Occasional appearance of missing fast moving marker trajectories as a broken time-series in captured motion data • Event recognition associated with output class 2 • The system produces states transitions among the output classes that appear implausible to domain experts (**Fig. 4**)

4 Classification Results and Discussion

All ESN models used the first 400 samples for model training (Fig. 4). Table 4 shows improved classification performance for the novel ESN ensemble system over a single ESN model.

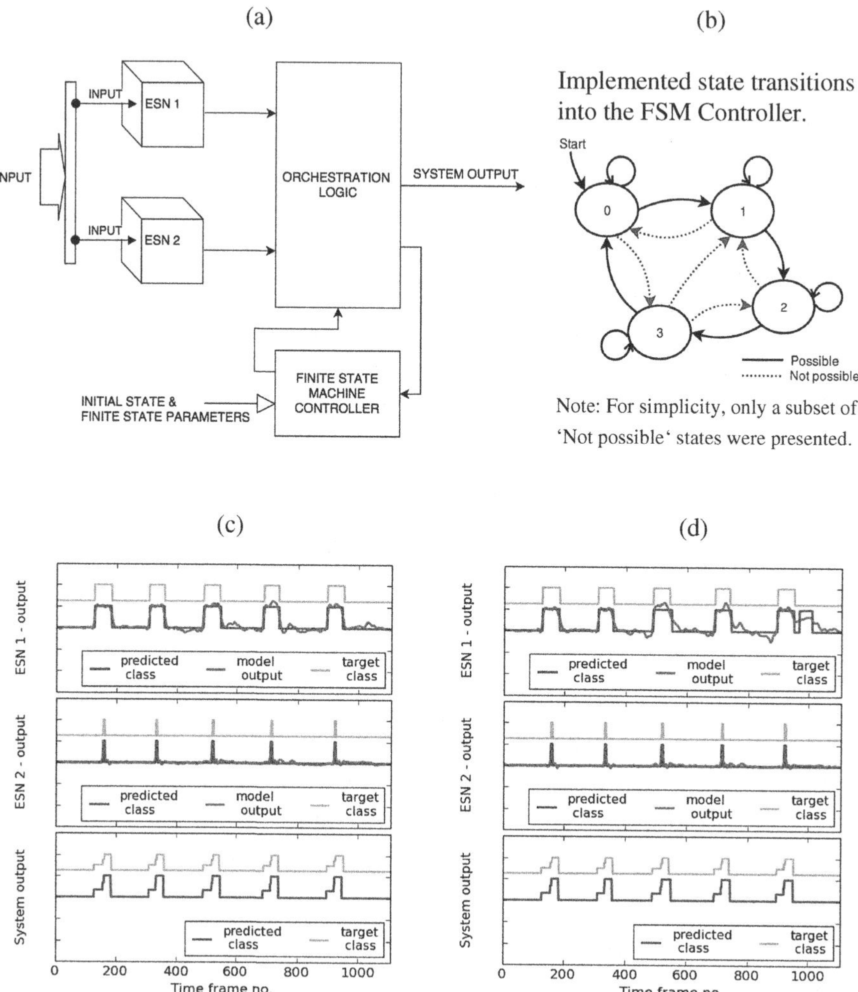

Fig. 4. Produced ESN ensemble system for autonomous temporal phasing of motion data (a); implemented state transition logic (b) into FSM controller required for orchestration of constituent ESN modules. Interim results of the system output visualisation show (c) typical system outputs and (d) robustness to occasional random classification errors (ESN1 - output, last false event) by FSM orchestration controller. ESN 2 event was extended (in training data) by 2 frames while the system output was reduced by two frames for improved accuracy at 50 Hz.

Table 4. Classification results for a single ESN and an ESN ensemble system over 12 repetitions

Experimental model	Data stream median accuracy [%] 1	2	3	4	Average class performance [%]	Average accuracy (Range)	Processing time (Range)
Single ESN	79.9	91.1	84.3	87.3	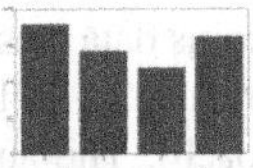	85.30% (14.14%)	1.46 s (0.22 s)
ESN ensemble	96.7	94.1	95.0	95.7		95.32% (3.81%)	3.20 s (0.7 s)

The achieved accuracy was above expectations, given the inclusion of all markers' trajectories, fast interpolation and normalisation, and a relatively small ESN model training portion (approx. 40 %). Regarding the validity of the exact classification of start and stop of event there are possible uncertainties in regards to expert's decision boundaries. However, it must be noted that the expectation is that, even with missing distal marker trajectories, tennis professionals should be able to identify the frame(s) with the most likely intended impact with the ball utilising the interactive functions of the developed animated 3D stick figure player [5]. The processing times were measured on a PC running Linux Ubuntu 14.04, Python 2.7 interpreter using single CPU processing with the hardware specification: CPU 4.2 GHz, RAM 16 GB/2.4 GHz, and a solid-state disk. The data set was considered similar to Microsoft's Kinect 2 in terms of occasional loss of marker time series information and relatively low sampling frequency.

5 Conclusions, Recommendations and Future Work

The presented ESN ensemble approach for autonomous temporal phasing of forehand swings required a relatively small portion of training data (approximately 40 %) and domain expertise to achieve high classification accuracy (95 %) on a relatively small data set. The ESN ensemble model has also achieved 100 % on forehand detection accuracy and temporal phasing state transitions, eliminating the occasional random occurrence of false output states from the ESN constituent models. System configuration combining ESN ensemble handling and a finite-state machine (FSM) control mechanism for orchestration purposes showed improvements compared to a single optimised ESN model, in particular: (a) in avoiding false positive event classifications; and (b) output state transitions that are not possible in real life but could otherwise be incorrectly produced from temporal and spatial data. The produced artefacts (models, architecture, and FSM controller for ESN ensemble orchestration) contribute to computer science by combining old (FSM) with modern approaches (third generation of neural networks), enabling multi-disciplinary scientific advancements for similar temporal and spatial problem areas such as computational intelligence, data analytics and sport science (e.g. qualitative movement diagnostics). From the experimental results, it is evident that FSM orchestration is faster than the ESN approach and requires no training data. Future work will be extended to diverse human motion data contexts.

Acknowledgements. The author wishes to express his appreciation to developers of Oger Toolbox (http://organic.elis.ugent.be/organic/engine) and Spyder IDE (https://pythonhosted.org/spyder/) utilised in this study. The tennis data was obtained in the Peharec polyclinic for physical therapy and rehabilitation, Pula (Croatia) in collaboration with Petar Bačić (biomechanics lab specialist and professional tennis coach). The author also wishes to express his sincere appreciation to Dr. Stefan Schliebs and Dr. Russel Pears for their valuable comments and insights.

References

1. Lightman, K.: Silicon gets sporty. IEEE Spectr. **55**, 48–53 (2016)
2. Bačić, B.: Bridging the gap between biomechanics and artificial intelligence. In: XXIV International Symposium on Biomechanics in Sports - ISBS 2006, Salzburg, Austria, pp. 371–374 (2006)
3. Bacic, B.: Evolving connectionist systems for adaptive sports coaching. Neural Inf. Process. – Lett. Rev. **12**, 53–62 (2008)
4. Bačić, B.: Connectionist methods for data analysis and modelling of human motion in sporting activities. Ph.D. thesis, School of Computer and Mathematical Sciences, AUT University, New Zealand, Auckland (2013)
5. Bačić, B.: Prototyping and user interface design for augmented coaching systems with MATLAB and Delphi: implementation of personal tennis coaching system. In: MATLAB Conference 2015, Auckland (2015)
6. Knudson, D.V.: Qualitative Diagnosis of Human Movement: Improving Performance in Sport and Exercise. Human Kinetics, Champain (2013)
7. Maass, W.: Networks of spiking neurons: the third generation of neural network models. Neural Netw. **10**, 1659–1671 (1997)
8. Jaeger, H., Lukoševičius, M., Popovici, D., Siewert, U.: Optimization and applications of echo state networks with leaky-integrator neurons. Neural Netw. **20**, 335–352 (2007)
9. Paugam-Moisy, H.: Spiking neuron networks a survey. In: IDIAP Research Institute (2006)
10. Bacic, B.: Echo state network for 3D motion pattern indexing: a case study on tennis forehands. In: Bräunl, T., et al. (eds.) PSIVT 2015. LNCS, vol. 9431, pp. 295–306. Springer, Heidelberg (2016). doi:10.1007/978-3-319-29451-3_24
11. Bacic, B.: Extracting player's stance information from 3D motion data: a case study in tennis groundstrokes. In: Mori, S., et al. (eds.) PSIVT 2015 Workshops. LNCS, vol. 9555, pp. 307–318. Springer, Heidelberg (2016). doi:10.1007/978-3-319-30285-0_25
12. Gangstead, S.K., Beveridge, S.K.: The implementation and evaluation of a methodical approach to qualitative sports skill analysis instruction. J. Teach. Phys. Educ. **3**, 60–70 (1984)

Unregistered Bosniak Classification with Multi-phase Convolutional Neural Networks

Myunggi Lee[1], Hyeogjin Lee[1], Jiyong Oh[1], Hak Jong Lee[2], Seung Hyup Kim[2], and Nojun Kwak[1](✉)

[1] Graduate School of Convergence Science and Technology, Seoul National University, Seoul, Korea
{myunggi89,hjinlee,yong97,nojunk}@snu.ac.kr
[2] Department of Radiology, College of Medicine, Seoul National University, Seoul, Korea
{hakjlee,kimshrad}@snu.ac.kr

Abstract. Deep learning has been a growing trend in various fields of natural image classification as it performs state-of-the-art result on several challenging tasks. Despite its success, deep learning applied to medical image analysis has not been wholly explored. In this paper, we study on convolutional neural network (CNN) architectures applied to a Bosniak classification problem to classify Computed Tomography images into five Bosniak classes. We use a new medical image dataset called as the Bosniak classification dataset which will be fully introduced in this paper. For this data set, we employ a multi-phase CNN approach to predict classification accuracy. We also discuss the representation power of CNN compared to previously developed features (Garbor features) in medical image. In our experiment, we use data combination method to enlarge the data set to avoid overfitting problem in multi-phase medical imaging system. Using multi-phase CNN and data combination method we proposed, we have achieved 48.9 % accuracy on our test set, which improves the hand-crafted features by 11.9 %.

Keywords: Medical image · Bosniak classification · Deep convolutional neural network · Unregistered medical image

1 Introduction

Deep Learning has made a significant breakthrough in natural image classification systems and has become a powerful tool in various fields of artificial intelligence. In particular, the use of deep Convolutional Neural Network (CNN) in computer vision problems has been an essential. In the recent progress in deep learning, many deep convolutional neural network architectures have been proposed, such as AlexNet [1], VGGNet [2] and GoogLeNet [3] which generate

This work was supported by Brain Fusion Program of Soul National University (800-20140265).

A. Hirose et al. (Eds.): ICONIP 2016, Part IV, LNCS 9950, pp. 19–27, 2016.
DOI: 10.1007/978-3-319-46681-1_3

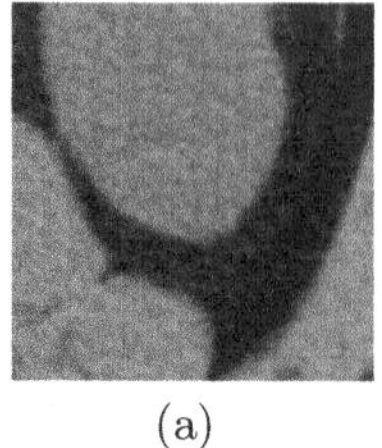
(a)

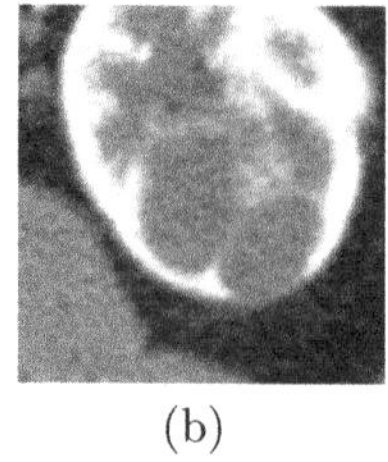
(b)

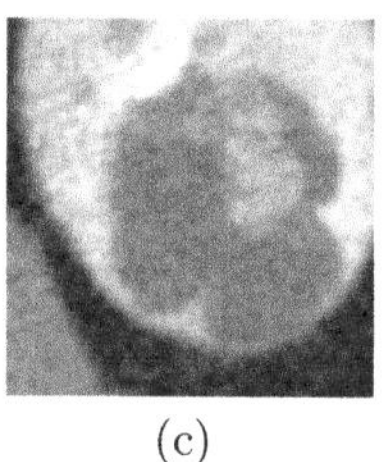
(c)

Fig. 1. Example of three phases of CT images. (a) Before contrast injection, (b) 50 s after contrast injection and (c) 5 min after contrast injection of ROI cropped from CT images.

state-of-the-art results on a variety of challenging tasks. One of the advantages brought by deep learning model is the high-level features produced by the top layers of the model compared to the previous hand-crafted features.

Though deep learning has made great advances in natural image classification problems, its potential in the field of medical image has not been completely explored because it is difficult to obtain large enough data in certain medical image problems and the DICOM[1] files are hard to annotate manually. For the training of CNN, it usually requires hundreds of thousands of data to achieve reasonable results avoiding overfitting. For this reason, the amount of publicly available medical images is increasing rapidly. Nevertheless, it is difficult to gather a large quantity of such medical images because of patient privacy and security policies. Moreover, regardless of policies, the hardness of annotating such a unique data format leads to an exhaustive work. In spite of the issues above, we have seen many approaches using deep learning applied to medical image problems [4,5].

In this paper, we propose a multi-phase CNN trained in a supervised fashion with labeled data to automatically diagnose the Bosniak categories. The Bosniak classification system of renal cystic masses divides renal cystic masses into five categories based on image characteristics on contrast-enhanced Computed Tomography (CT). We manually annotate regions of interest (ROI) for CT images of three phases which will be explained in Sect. 2. Using our data augmentation schemes, each sample of unregistered ROI triplets[2] is fed to the proposed multi-phase CNN structure. The proposed multi-phase CNN has achieved 48.9 % classification accuracy which enhances both a single CNN and a hand-crafted feature like Gabor Features plus SVM by more than 10 % respectively.

[1] Digital image and Communications in Medicine (DICOM) is a standard for handling, storing, printing, and transmitting information in medical image.

[2] The term 'unregistered' is used to indicate that the three images from different phases have different shapes and sizes of lesions.

2 Bosniak Classification Problem

The Bosniak classification is a practical and accurate method to evaluate renal cystic lesions which divides renal cystic masses into five categories according to image characteristics on contrast-enhanced CT. It is helpful for doctors in predicting a risk of malignancy and suggesting either follow up or treatment.

CT image criteria makes it possible to analyse renal cysts' contour and contents, presence of septations or calcifications, and enhancement after contrast agent injection. The categories are determined by characteristics of kidney region CT images from the three phases which are 'before contrast injection', '50 s after contrast injection' and '5 min after contrast injection' as shown in Fig. 1. Combining all the detailed characteristics of each phase, we are able to classify which Bosniak category the patient belongs to. One of the challenges of Bosniak classification with renal cystic lesions is that diverse characteristics of such lesions per Bosniak class as well as overall analysis of three phases are needed for accurate classification [6].

There has been some approaches to automatically classify Bosniak class in machine learning fashion [7,8]. Most previous works in medical image focus on hand-crafted features like Gabor features, which are unable to comprehensively represent low-variant medical images for classification [9]. Moreover, as far as we know, it has not been reported the use of CNNs in Bosniak classification problem as well as the analysis of unregistered input images with the multi-phase CNN approach.

3 Algorithm

3.1 Data Acquisition

The database we use in our experiment is CT images of patients from Seoul National University BunDang Hospital (SNUH) with an Institutional Review Board (IRB) approval. The dataset maintains overall 371 patients' CT images. Each patient has three DICOM files which is taken from 'before contrast-enhanced injection', '50 s after injection' and '5 min after injection'.

Table 1. Data distribution of Bosniak classification database

Category	# Patient	Pre	50 sec	5 min	Total
Bosniak I	112	798	799	807	2404
Bosniak II	145	1102	1163	1139	3404
Bosniak IIF	26	302	296	295	893
Bosniak III	45	274	298	307	879
Bosniak IV	43	390	394	394	1178
Total	371	2866	2950	2942	8758

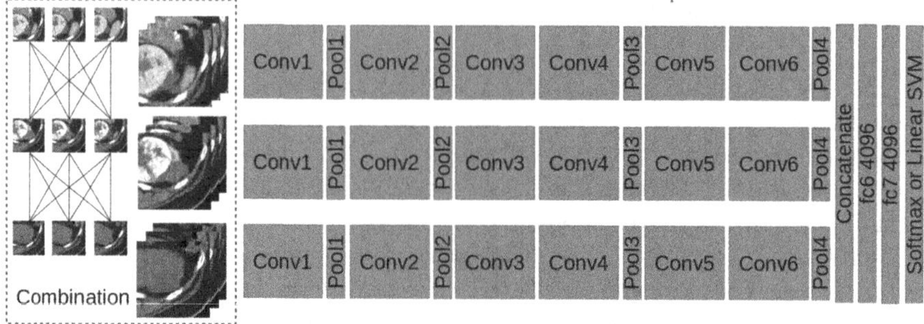

Fig. 2. Overall architecture and data combination used in our experiment. Our proposed architecture has three independent CNNs (not sharing weights), each CNN has 9 layers. The first 6 layers has convolution operation following by pooling operation. Due to the padded convolutions for the third and fifth layers, the output feature map sizes are equal to the input feature map size. All the feature maps extracted from layer 6 (pool4) are concatenated followed by a fully-connected layer. The concatenated feature map is encoded into fully-connected layer with 28,800 dimension.

To classify Bosniak class, we manually annotate several ROIs using our annotation software. In slice-by-slice images from DICOM files, we cropped ROIs that have characteristics of complex renal cysts as shown Fig. 1. The number of cropped images per patient varies from 2 to 28. Overall 8,758 image patches were extracted from DICOM files. However, as it can be seen from Table 1, the number of image patches per class is highly unbalanced and not enough for training a CNN. Also, these three-phase image patches are unregistered, which makes it more difficult to classify accurately.

3.2 Data Augmentation

In order to train our proposed approach, we enrich our data set by applying data augmentation techniques to each image patches. We also applied a new input strategy, namely data combination for each patient. The image cropped from DICOM files has different sizes that varies from 40 by 40 to 160 by 160. First we prepare two sets of image sizes by resizing each image to 128 by 128 and 256 by 256. Then, we generate 10 times more image patches by random translations and horizontal reflections. The resulting input image sizes to the CNN are 112×112 and 227×227, respectively.

Another approach to enrich our dataset is the data combination method specifically designed for our application. For each patient we have more than 2 images per phase from which the image patches are cropped in slice-by-slice manner. With multi-phase CNN architecture taken into consideration, we proceed the inter-phase data combination to enlarge our data set, that is, choosing one image from each phase for a combination. For example, suppose a patient has two pre-injection, three 50 s, four 5 min after injection patches, then we get

24 ($2 \times 3 \times 4$) different input combinations fed to three-stream CNN architecture we propose. Figure 2 illustrates overall architecture and data combination method used in our experiment.

3.3 Multi-phase Convolutional Neural Network

In this section, we describe the algorithm used in our experiment and methods we used to evaluate the Bosniak Classification system. Our goal is to predict Bosniak class according to CT images of three phases. We should consider overall analysis of the characteristics of ROIs of three phases which are cropped from CT images.

A single CNN model consists of multiple convolution and sub-sampling (pooling) layers. Each of the layers is performed by using different convolutional operation. The result of convolution defines the features of input data which can be interpreted as corners, curves, lines and so on. After each convolution layer, a pooling layer follows to reduce the dimension of each feature map. After repeating these two operations, the feature map from the last pooling layer is followed by several fully-connected MLPs (multi-layer perceptron) whose output is a vector fed to a classifier. Given our own dataset size and the characteristics of low-variance in medical image, we choose to implement the modified AlexNet [1] for our specific application. In our application, we applied consecutive convolution layers in two places, (1) layer 3 and 4 and (2) layer 5 and 6 in order to smooth the noise in the feature map. To avoid overfitting, we apply 70 % drop-out to fully-connected layer [10]. The architecture of the single network for input image size of 128×128 is depicted in Table 2.

We propose a multi-phase CNN which is a slight variant of the multi-scale CNN architecture used in [11]. Originally, the multi-scale convolutional net contains multiple copies of a single network (not sharing the weights) that are applied to a Laplacian pyramid version of the images. Being slightly different from this, our method applies ROIs of three phases as the input data. More precisely, the three images of the different phases are fed into three independent single networks whose architectures are as described in Table 2. The last pooling layer (layer 5) produces the feature maps of each single network. Then the three feature maps of each phase are concatenated. After the concatenated feature

Table 2. Single neural network architecture (for input image size of 128×128)

Layer	1	2	3	4	5	6	7	8	9
Stage	Conv + max	Conv + max	Conv	Conv + max	Conv	Conv + max	Full	Full	Full
# Channel	32	64	96	96	128	128	4096	4096	5
Filter size	5 × 5	3 × 3	3 × 3	3 × 3	3 × 3	3 × 3	-	-	-
Conv.Stride	2	2	1	1	1	1	-	-	-
Pooling size	2	2	-	2	-	2	-	-	-
Pooling stride	2	1	-	2	-	2	-	-	-
Padding size	0	0	1	0	1	0	-	-	-
Spatial input	112 × 112	54 × 54	26 × 26	26 × 26	12 × 12	12 × 12	5 × 5		

maps are encoded into fully-connected layer (layer 7), the extracted features from the last fully-connected layer are trained using softmax with stochastic gradient descent method.

4 Experiment

In this part, we will fully introduce the dataset used in our application and set-ups for our experiment. Using the algorithm described in Sect. 3, we have a comprehensive analysis on the Bosniak classification problem using different strategies of the input data fed to our proposed CNN architecture. We will also show how we deal with unregistered low-variance CT images and discuss how to avoid the overfitting problem caused by relatively small dataset. Furthermore, we compare the proposed method with the traditional Gabor feature extraction plus SVM classification. In this paper, the classification accuracy is measured by counting the number of correctly classified unregistered ROI triplets.

4.1 Single Convolutional Networks

Before implementing our multi-phase CNN, we experiment on single-phase-only data with different single CNN architectures and different input conditions.

First of all, we compare our model with AlexNet and NIN [12] fine-tuned with pre-training on ImageNet dataset. In many cases, fine-tuning on ImageNet pre-trained model performs slightly better than training from scratch. However, we have figured out that the trained features with ImageNet pre-trained model are not suitable in the case of medical image problem as it can be seen in Table 3. Using our method depicted in Table 2, the single phase architecture performed around 5 % better than the conventional network architectures. It is interesting to see that single network with 5-min-only input data has slightly better performance than the other phases. It can be seen that 5-min data has more significant features compared to the others.

Table 3. Accuracy of single network with each phase input

Method	Pre	50 sec	5 min
Alexnet-256 (from scratch)	27.09	30.26	28.50
Alexnet-256(fine tuning)	22.34	26.72	25.83
NIN-256(fine tuning)	27.04	30.45	29.15
Ours-256	32.15	34.98	35.63
Ours-128	32.26	31.02	37.21

4.2 Multi-phase Convolutional Networks

In the multi-phase CNN experiment, we conducted three different experiments using two different sizes (128×128 and 256×256) of training data sets. We also compare our method against the features extracted from Gabor filter trained with linear and nonlinear kernel SVM classifiers. Gabor filters have rich applications in image processing [13,14], especially in feature extraction for the texture analysis. In our experiment, Gabor filters with different frequencies and different orientation directions have been used to extract complex features from cropped CT images. We extract 1,000 dimensional Gabor features per phase and concatenate features from the three phases together. In this way, we have feature vectors with 3,000 dimension for each input data. Training with SVM with linear and kernel method, we obtain 28.1 % and 37.0 % classification accuracies, respectively. Compared to Gabor features trained with SVM, our proposed method shows 48.9 % accuracy, which is 11.9 % better (Table 4).

Table 4. Accuracy of multiple network with each phase input. The 256 and 128 means the size of input data. Comb means using our data augmentation method of data combination.

Method	Accuracy
Gabor & linear SVM	28.1
Gabor & kernel SVM	37.0
Ours-256 & comb.	47.6
Ours-128 & comb.	48.9

4.3 Implementation Details

The training is completed on Titan-X GPU with caffe. The weights in the networks are initialized randomly with Gaussian distribution. They are then updated by stochastic gradient descent, accompanied by a momentum term of 0.9 and an L_2 weight decay of 1×10^{-5}. The learning rate is 1×10^{-4} and is decreased by a factor of 0.5 in each 10,000 steps. Drop-out with a rate of 0.7 is employed on the fully connected layers (6th and 7th) in the classifier.

5 Conclusion

In this paper, we implemented multi-phase convolutional neural networks to solve the Bosniak classification problem which classifies unregistered CT images into five different Bosniak classes. The performance of the proposed multi-phase CNN shows an improvement over the single-phase CNNs, demonstrating that the high-level features of the multi-phase CNN provide a robust representation of the three-phase input images since the three input images are all useful for the classification.

We also proposed data combination method specially designed for our application. The method efficiently enlarges our training data set and avoids overfitting problem in our application. In this way, the registration of different phase images is not needed. This is particularly important in CT image analysis, where registration is a big challenge due to non-rigid deformations. Although the performance of the multi-phase CNN is better than the single CNNs, it is still not satisfactory and there is more room to be improved. We can infer the reason for the relatively low performance as below: (1) The lack of sufficient training data in Bosniak IIF, III and IV leads to an overfitting in our architecture. (2) The class-imbalance causes a bias on the classification accuracy. (3) Because different people annotated the images, the quality of ROI patches are irregular. Despite the issues mentioned above, we want to emphasize that our results can be used as a baseline in the Bosniak classification problem in the future works.

References

1. Krizhevsky, A., Sutskever, I., Hinton, G.E.: Imagenet classification with deep convolutional neural networks. In: Advances in Neural Information Processing Systems, pp. 1097–1105 (2012)
2. Simonyan, K., Zisserman, A.: Very deep convolutional networks for large-scale image recognition, arXiv preprint arXiv:1409.1556 (2014)
3. Szegedy, C., Liu, W., Jia, Y., Sermanet, P., Reed, S., Anguelov, D., Erhan, D., Vanhoucke, V., Rabinovich, A.: Going deeper with convolutions. In: Proceedings of the IEEE Conference on Computer Vision and Pattern Recognition, pp. 1–9 (2015)
4. Xu, Y., Mo, T., Feng, Q., Zhong, P., Lai, M., Chang, E.I., et al.: Deep learning of feature representation with multiple instance learning for medical image analysis. In: IEEE International Conference on Acoustics, Speech and Signal Processing (ICASSP), pp. 1626–1630. IEEE (2014)
5. Cireşan, D.C., Giusti, A., Gambardella, L.M., Schmidhuber, J.: Mitosis detection in breast cancer histology images with deep neural networks. In: Mori, K., Sakuma, I., Sato, Y., Barillot, C., Navab, N. (eds.) MICCAI 2013, Part II. LNCS, vol. 8150, pp. 411–418. Springer, Heidelberg (2013)
6. Miranda, C.M.N.R.D., Maranhão, C.P.D.M., Santos, C.J.J.D., Padilha, I.G., Farias, L.D.P.G.D., Rocha, M.S.D.: Bosniak classification of renal cystic lesions according to multidetector computed tomography findings. Radiol. Bras. **47**(2), 115–121 (2014)
7. Lee, Y., Kim, N., Cho, K.-S., Kang, S.-H., Kim, D.Y., Jung, Y.Y., Kim, J.K.: Bayesian classifier for predicting malignant renal cysts on MDCT: early clinical experience. Am. J. Roentgenol. **193**(2), W106–W111 (2009)
8. Curry, N.S., Cochran, S.T., Bissada, N.K.: Cystic renal masses: accurate Bosniak classification requires adequate renal CT. Am. J. Roentgenol. **175**(2), 339–342 (2000)
9. Soares, J.V., Leandro, J.J., Cesar Jr., R.M., Jelinek, H.F., Cree, M.J.: Retinal vessel segmentation using the 2-D gabor wavelet and supervised classification. IEEE Trans. Med. Imaging **25**(9), 1214–1222 (2006)
10. Srivastava, N., Hinton, G., Krizhevsky, A., Sutskever, I., Salakhutdinov, R.: Dropout: a simple way to prevent neural networks from overfitting. J. Mach. Learn. Res. **15**, 1929–1958 (2014)

11. Farabet, C., Couprie, C., Najman, L., LeCun, Y.: Learning hierarchical features for scene labeling. IEEE Trans. Pattern Anal. Mach. Intell. **35**(8), 1915–1929 (2013)
12. Lin, M., Chen, Q., Yan, S.: Network in network, arXiv preprint arXiv:1312.4400 (2013)
13. Kyrki, V., Kamarainen, J.-K., Kälviäinen, H.: Simple gabor feature space for invariant object recognition. Pattern Recognit. Lett. **25**(3), 311–318 (2004)
14. Yang, M., Zhang, L.: Gabor feature based sparse representation for face recognition with gabor occlusion dictionary. In: Daniilidis, K., Maragos, P., Paragios, N. (eds.) ECCV 2010, Part VI. LNCS, vol. 6316, pp. 448–461. Springer, Heidelberg (2010)

Direct Estimation of Wrist Joint Angular Velocities from Surface EMGs by Using an SDNN Function Approximator

Kazumasa Horie[1(✉)], Atsuo Suemitsu[2], Tomohiro Tanno[1], and Masahiko Morita[1]

[1] University of Tsukuba, 1-1-1 Tennodai, Tsukuba, Ibaraki, Japan
{horie,tanno,mor}@bcl.esys.tsukuba.ac.jp
[2] Sapporo University of Health Sciences, 2-1-15 Nakanuma Nishi 4, Higashi-ku, Sapporo, Hokkaido, Japan
sue@sapporo-hokeniryou-u.ac.jp

Abstract. The present paper proposes a method for estimating joint angular velocities from multi-channel surface electromyogram (sEMG) signals. This method uses a selective desensitization neural network (SDNN) as a function approximator that learns the relation between integrated sEMG signals and instantaneous joint angular velocities. A comparison experiment with a Kalman filter model shows that this method can estimate wrist angular velocities in real time with high accuracy, especially during rapid motion.

Keywords: Surface electromyogram · Angular velocity estimation · Selective desensitization neural network

1 Introduction

Surface electromyograms (surface EMGs) are electrical activities recorded using skin surface electrodes. Surface EMGs are produced by skeletal muscles and contain information about a motion and its purpose. Recently, methods have been proposed for recognizing human motions from surface EMGs in order to develop an EMG-based human-machine interface. We can classify these methods into two types: classification of motion types and estimation of the joint angles.

Methods that can recognize complex hand motions in real time have been proposed (e.g., [FO1]). However, these methods can recognize only the summary of the motion purpose and classify only six to eight types of motion with high accuracy. We cannot develop a user-friendly interface that reflects the user's intentions adequately with only classification methods.

Joint-angle estimation methods (e.g., [KK1]) can recognize more detailed purposes of motion than classification methods. However, these methods cannot estimate joint angles during rapid motion. The non-linear relationship between the isometric muscle strength and the joint angle during rapid motion makes it

A. Hirose et al. (Eds.): ICONIP 2016, Part IV, LNCS 9950, pp. 28–35, 2016.
DOI: 10.1007/978-3-319-46681-1_4

difficult for these methods to model the motion. Thus, these methods cannot be applied to a practical interface.

In addition, the estimation method of joint angles and velocities using a Kalman filter (KF) [AK1] has been proposed. This method, however, was applied only to slow motion (about 30 deg/s). No method has been proposed for estimating the angular velocities of fast-moving joints such as the wrist (wrist angular velocities may reach 1,900 deg/s at a maximum). Furthermore, the KF may not have ability to express the dynamics of wrist rotation around the roll axis because of the non-linearity of the relation between the surface EMGs and the angular velocity.

In this study, we model the relationship between multichannel surface EMGs and wrist joint angular velocities using a function approximator and estimate the latter from the former. This method will enable an interface to use motion-purpose information which could not previously be obtained.

We choose a selective desensitization neural network (SDNN) as the function approximator. Recent studies have shown that the SDNN has excellent learning and generalization abilities. For example, Nonaka et al. applied an SDNN to the approximation of a two-variable function and showed that the SDNN can learn a non-linear and discontinuous function with a small training sample [NM1]. These features will enable the proposed method to learn the complicated relation between surface EMGs and the instantaneous joint angular velocities.

The following sections describe the SDNN and our proposed method. We also verified the efficacy of the proposed method through the estimation of wrist angular velocities.

2 Selective Desensitization Neural Network

SDNNs are known to provide good approximations of a wide range of functions using a small set of training data samples [NM1]. In this section, we explain how a function $y = f(\mathbf{x})$ is approximated based on an SDNN given a continuous-valued input vector $\mathbf{x} = (x_1, \ldots, x_m)(m \geqq 2)$.

The input layer of the SDNN consists of m neural groups. Each group has n units that represent an input signal x_i, i.e., the input signal is represented in a distributed manner by the activity patterns of the units. If the range of possible values for input signals is divided into q ranges, each interval corresponds to different activity patterns which are configured using equal numbers, $+1$ and -1. The pattern changes gradually as the variable value changes progressively, resulting in a high correlation between the patterns when the values are close and a low correlation when the values are far apart. The correlation between patterns of the values that are far apart is 0.

The middle layer comprises $m(m-1)$ $(= {}_m\mathrm{P}_2)$ neural groups ($G^{1,1}$, $G^{1,2}, \ldots, G^{m,m-1}$). The units in $G^{\mu,\nu}$ are connected with both of the units in G^{μ} and G^{ν}, which are located in the input layer. This realizes a procedure called desensitization, which neutralizes the output of the units, regardless of their input and inner potential. For example, if the ith units of G^{μ} are desensitized by the jth unit of G^{ν}, the output of the ith unit of $G^{\mu,\nu}$ is given by

$$g_i^{\mu,\nu} = \frac{g_j^{\nu}+1}{2} \cdot g_i^{\mu}. \tag{1}$$

where g_i^{μ} is the output of the ith unit of G^{μ} and g_j^{ν} is the output of the jth unit of G^{ν}. This creates a pattern in which half of the units are 0 and the rest are the same as $G^m u$ (either $+1$ or -1).

The output layer of the SDNN comprises l units. Each of the units is connected to all of the units in the middle layer. The output of the ith unit in the output layer is calculated by

$$\begin{aligned} y^i &= H\left(\textstyle\sum_{\mu,\nu(\mu\neq\nu)}\sum_{j=1}^{n} w_{i,j}^{\mu,\nu} \cdot g_j^{\mu,\nu} + h_i\right) \\ H(u) &= \begin{cases} 1 \ (u>0) \\ 0 \ (otherwise) \end{cases} \end{aligned} \tag{2}$$

where $w_{i,j}^{\mu,\nu}$ is the synaptic weight from the jth unit of the $G^{\mu,\nu}$ in the middle layer and h_i is a threshold. The final output y of the SDNN is determined based on the number of units with an output of 1 in the output layer. Learning is achieved by error-correction training (the p-delta learning rule [AW1]) in this network.

3 Proposed Method

Our proposed method has three components: surface EMG acquisition, signal preprocessing, and function approximation (Fig. 1).

3.1 Surface EMG Acquisition

The surface EMGs are measured at 10 points on the forearms (Fig. 2). A relatively large number of sensors is required, but it is not necessary to position the sensors accurately on specific muscles. The surface EMGs are measured and sampled at 1 kHz using Personal-EMG (Oisaka Electronic Device Ltd.) equipment and a 12-bit A/D converter.

3.2 Signal Preprocessing

Surface EMGs are not suitable for use as input signals because of their fluctuations. Thus, it is necessary to extract features by signal preprocessing. In the proposed method, the integral of the surface EMG (IEMG) and its mean over the previous 300 ms (average IEMG: AIEMG) are obtained as features.

IEMGs can be obtained easily with little time lag using a low-pass filter. Thus, the IEMG is well-suited for a human-machine interface that requires a rapid response. Unfortunately, this signal is not directly related to joint angular velocities; it is related to the force generated by muscles. It is difficult to achieve

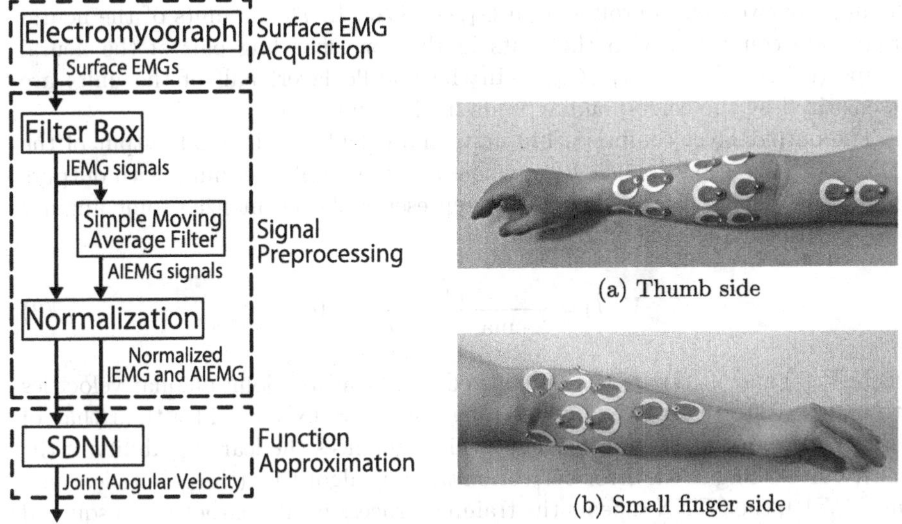

Fig. 1. Structure of the proposed method. **Fig. 2.** Arrangement of EMG sensors.

accurate estimation with only IEMGs. AIEMG is the mean of the IEMG over the previous 300 ms. We consider that the AIEMG corresponds to muscle contraction speed. AIEMG is well-suited for estimation of joint angular velocities. However, there is a large time lag between the AIEMGs and the actual joint angular velocity itself. Therefore, the combined use of both of these features can be expected to enhance the accuracy and reduce the adverse effects of the time lag between the AIEMGs and the actual angular velocities.

In the proposed method, a filter box (Oisaka Electronic Device Ltd.) and a simple-moving-average (SMA) filter are used to convert the surface EMGs into features. Both features are normalized against their maximum values during each time step before being used as inputs for the SDNN.

3.3 Function Approximation

As a component of function approximation, the SDNN models the relation between the features and the joint angular velocities. The SDNN is constructed as shown in Fig. 3 and used as a function approximator. Note that the SDNN is prepared for each axis of rotation, but all SDNNs have the same structure and parameters.

The input layer comprises 10 neural groups $(G_1, \cdots, G_{10})$ and 10 other neural groups $(G_{11}, \cdots, G_{20})$, which represent the IEMGs and the AIEMGs, respectively. The activity patterns of the units are determined from the input signals using the parameters $n = 96, q = 96$.

The middle layer has two parts P_1 and P_2, which comprise 90 and 20 neural groups, respectively. The units of the neural groups in P_1 are connected with the units of two neural groups from $G_1, \cdots, G_{10}$. In P_2 the units of the neural groups are connected with the units in the corresponding pair of the neural groups $((G_1, G_{11}), \cdots, (G_{10}, G_{20}))$. In the middle layer, half of the units are desensitized by the corresponding units in the input layer.

The output layer comprises 140 units and calculates the final output of the SDNN by $0.01k - 0.2$ ($[-0.2, 1.2]$), where k is the number of units with outputs of 1 in the output layer. The output represents the normalized joint angular velocity, which is calculated by

$$V_n(t) = \frac{V_m(t)}{2 \cdot \max_{t'} |V_m(t')|} + 0.5 \tag{3}$$

where $V_n(t)$ and $V_m(t)$ are the normalized and measured joint angular velocities at time t, respectively. The output range of the SDNN is wider than that of the normalized joint angular velocity, which enhances the learning ability of the SDNN. According to the error-correction training algorithm (the p-delta learning rule [AW1]), the SDNN repeats the training process until the root mean squared error (RMSE) is sufficiently low or a specific number of iterations have been completed.

4 Experiment

To evaluate the proposed method, we performed a comparison experiment with a Kalman filter (KF) model [AK1]. The KF model did not use the AIEMGs as input because a preliminary experiment showed that they decrease the estimation accuracy. To implement the KF model, Matlab and the System Identification Toolbox were used.

4.1 Method

A three-axis gyroscope (MP-G3-2000B, MicroStone Co. Ltd.) was used to measure the wrist angular velocities. The gyroscope was mounted on the back of the right hand (Fig. 4) and measured the angular velocities around the pitch and roll axes. The velocities were normalized against the maximum values during each time step.

To obtain measurements of wrist angular velocities and surface EMGs, eight male subjects (24 ± 2 years old) were asked to execute flexion-extension and supination-pronation repeatedly at different speeds for 10 s. The subjects repeated this task nine times for each motion, and finally, we obtained 18 (2 motions $\times$ 9 repetitions) samples in total. Next, we selected six samples (three samples for each motion) to train the corresponding SDNN and performed three-fold cross-validation.

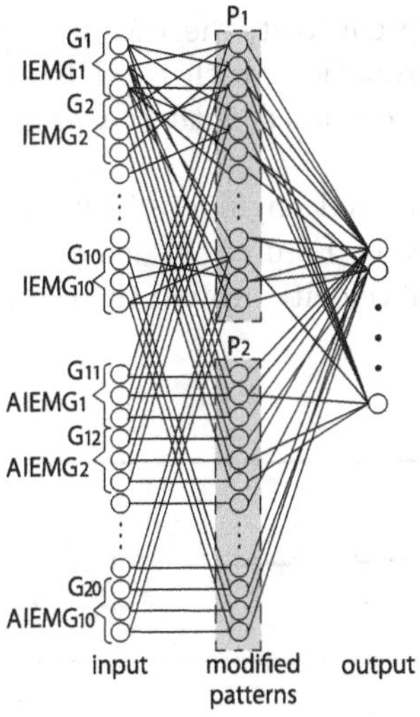

Fig. 3. Structure of the SDNN used in the proposed method.

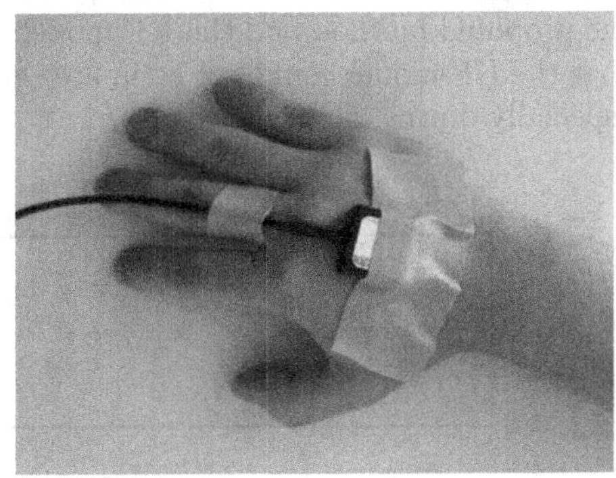

Fig. 4. Installed gyroscope.

4.2 Results

The experimental results are summarized in Table 1. The RMSE values around the pitch axis (corresponding to flexion and extension) with the proposed method and the KF model were 79.7 and 97.7 deg/s, respectively. Each system could estimate the approximate angular velocities around the pitch axis. There were

Table 1. Root mean squared error among the estimated angular speeds (deg/s).

(a) Around the pitch axis									
	Subject								
	A	B	C	D	E	F	G	H	Ave.
SDNN	63.9	60.0	84.8	89.5	76.3	60.4	113.3	89.6	79.7
KF	91.7	43.1	137.1	142.5	108.7	49.9	127.3	81.4	97.7
SVR(*1)	75.4	73.6	88.8	103.1	94.3	65.5	124.6	91.7	89.6
(b) Around the roll axis									
	Subject								
	A	B	C	D	E	F	G	H	Ave.
SDNN	115.3	105.7	122.4	140.1	124.3	133.6	134.8	100.5	121.9
KF	243.3	150.9	265.8	247.9	220.5	272.0	285.0	176.7	232.8
SVR(*1)	143.1	121.2	127.9	178.4	135.2	152.1	167.3	104.9	141.3

(*1) RMSE obtained by using a SVR as a function approximator (Reference).

no significant differences between the methods. In contrast, the RMSE around the roll axis (corresponding to supination and pronation) with the KF model (232.8 deg/s) were two times higher than those obtained with the proposed method ($p < 0.01$: Wilcoxon signed-rank test).

Figures 5 and 6 show the example of the estimated angular velocities with the proposed method and the KF model, respectively. Figure 6(b) demonstrates that the KF model could not estimate the angular velocity around the roll axis, especially during rapid motion.

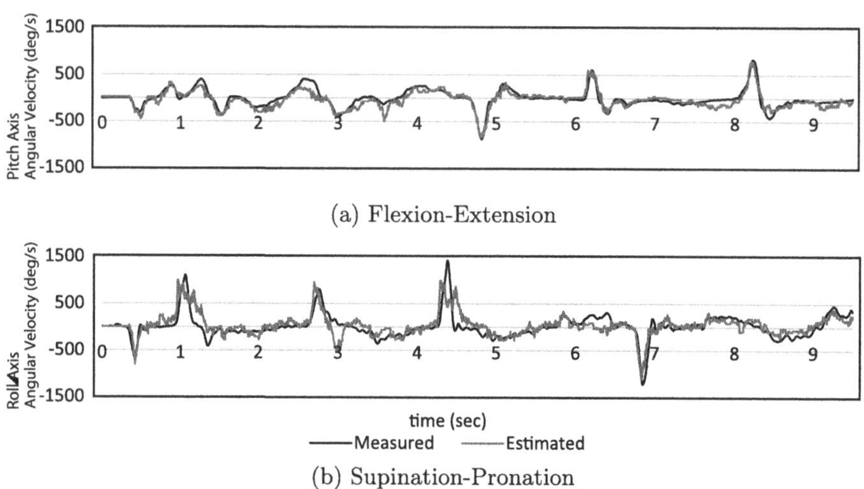

(a) Flexion-Extension

(b) Supination-Pronation

Fig. 5. Example of joint angular velocity estimation with the proposed method.

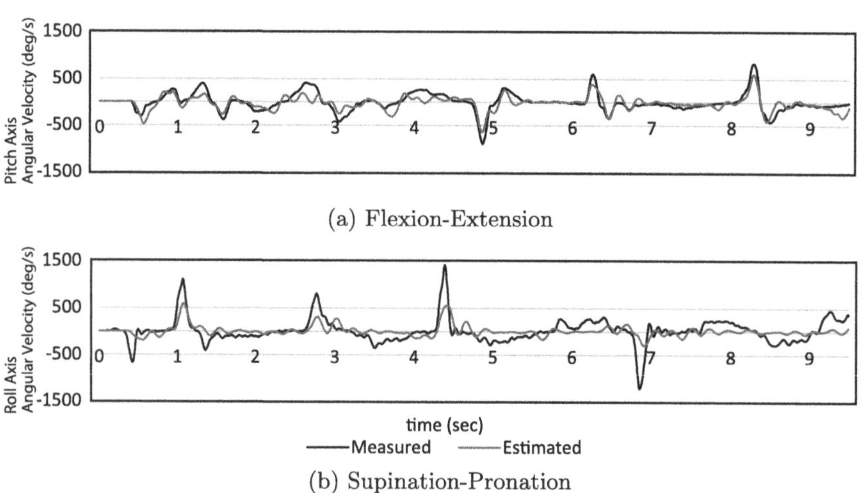

(a) Flexion-Extension

(b) Supination-Pronation

Fig. 6. Example of joint angular velocity estimation with the KF.

The estimation accuracy of the KF model depends on the linearity of the relation between the IEMG signals and the angular velocities. The relation between the signals and the joint angular velocities around the roll axis may be very complicated because pronation and supination require the coordination of multiple muscles (flexor carpi radialis, pronator teres, etc.), making it difficult for the KF model to solve this problem.

In contrast, the SDNN has adequate ability to learn the non-linearity relationship. Thus, the proposed methods can handle rapid motion or rotation around the roll axis. The difference in learning ability between the KF and the SDNN caused the variation in estimation accuracy.

5 Conclusion

In this study, we proposed a method for directly estimating the wrist angular velocities around the pitch and roll axes from multichannel surface EMGs using SDNNs. The experimental results show that the proposed method can estimate the approximate angular velocities of the wrist. The proposed method will be useful for recognizing the purpose of the rapid motion.

In future research, we aim to improve the estimation accuracy and to implement an EMG-based device using the proposed method.

Acknowledgment. This work was supported partly by JSPS KAKENHI grant numbers 22300079 and 24700593 and by Tateishi Science and Technology Foundation grant number 2157011.

References

[FO1] Fukuda, O., Tsuji, T., Kaneko, M., Otsuka, A.: A human-assisting manipulator teleoperated by EMG signals and arm motions. IEEE Trans. Robot. Autom. **19**(2), 210–222 (2003)

[KK1] Koike, Y., Kawato, M.: Estimation of arm posture in 3D-space from surface EMG signals using a neural network model. IEICE Trans. Inf. Syst. **E77–D**(4), 368–375 (1994)

[AK1] Artemiadis, P.K., Kyriakopoulous, K.J.: EMG-based control of a robot arm using low-dimensional embeddings. IEEE Trans. Robot. **26**(2), 393–398 (2010)

[NM1] Nonaka, K., Tanaka, F., Morita, M.: The capability of selective desensitization neural networks at two-variable function approximation. IEICE Tech. Rep. Neurocomputing **111**(96), 113–118 (2011)

[AW1] Auer, P., Burgsteiner, H., Maass, W.: A learning rule for very simple universal approximators consisting of a single layer of perceptrons. Neural Netw. **21**(5), 786–795 (2008)

Data Analysis of Correlation Between Project Popularity and Code Change Frequency

Dabeeruddin Syed, Jadran Sessa, Andreas Henschel, and Davor Svetinovic(✉)

Department of Electrical Engineering and Computer Science, Masdar Institute of Science and Technology, P.O. Box 54224, Abu Dhabi, UAE
{dsyed,jsessa,ahenschel,dsvetinovic}@masdar.ac.ae

Abstract. Github is a source code management platform with social networking features that help increase the popularity of a project. The features of the GitHub like watch, star, fork and pull requests help make a project popular among the developers, in addition to enabling them to work on the code together. In this work, we study the relation between the project popularity and the continual code changes made to a GitHub project. The correlation is found by using the metrics such as the number of watchers, pull requests, and the number of commits. We correlate the time series of code change frequency with the time series of project popularity. As a result, we have found that projects with at least 1500 watchers each month have a strong positive correlation between the project popularity and frequency of code changes. We have also found that the number of pull requests is 73.2 % more important to the popularity of a project than the number of watchers.

Keywords: Data analytics · Mining software repositories · Open-source development

1 Introduction

The socialized pull-request model [1] inspires software evolution and maintenance into crowd-based development. A conventional contribution process in GitHub involves following steps. First, a contributor finds an attractive project by following some well-known developers and watching their projects [2]. Second, in order to contribute, a developer either creates his own repository or forks an existing base repository. GitHub maintains both repositories. The idea behind this is to allow a software developer to continue working without making any changes to the base repository [3]. This approach avoids any of the merge conflicts with the work of the other developers and provides required flexibility. When their work is finished, the contributor sends the patches from the forked repository to its source by a pull-request [4]. This action usually opens a discussion among the developers before the changes are accepted.

The most notable aspect of GitHub that contributes to the popularity of a project is the social networking which allows projects to grow by enabling

A. Hirose et al. (Eds.): ICONIP 2016, Part IV, LNCS 9950, pp. 36–43, 2016.
DOI: 10.1007/978-3-319-46681-1_5

many developers to effectively work together on a project [5]. The interaction between the developers, in turn, helps for the frequent code changes to the projects. The other features that help to popularize an open source project are that the GitHub members can follow other members and also subscribe to various project repositories as watchers. These features make the GitHub to stand out among the many other conventional repository-hosting online software-development platforms [6].

This work focuses on exploring the following three hypothesis. The correlation between the popularity of a project and frequency of code change for highly popular projects may be strong, moderate or weak and may be positive or negative, and there is no evidence to prove that it stands in a particular category of correlation.

Highly and consistently popular projects (with at least 1500 watchers every month and projects in the top 12 programming languages) have strong and positive correlation between the project popularity and the frequency of code change. The popularity of a project will bring about the awareness of the project among the development community and that in turn, leads to larger number of developers working together resulting in more frequent changes in the source code.

The number of pull requests is at least 50 % more important to the popularity than the number of watchers. We calculate the number of pull requests and the number of watchers for highly and consistently popular open source projects on GitHub, compare their values, and use them as metrics for the project popularity.

2 Related Work

Borges et al. [7] performed studied the correlation of open source software popularity with the metrics such as stars, forks and usage of the software. They proposed a framework for tracking popularity of GitHub systems, and stressed out the importance of stars as a measure of a system's popularity. Weber et al. [8] tried to find the differences between popular and unpopular GitHub projects written in Python. They found out that popular projects have more documentation, and that in-code features are of more significance than author metadata features. Lee et al. [9] analyzed the effect of the top developers on project popularity. They found that the top developers are reputed to bring the expertise in the software development and as such have many followers.

Dabbish et al. [5] studied the transparent nature of GitHub through its features like watching and following. Sheoran et al. [10] examined the behavior of watchers and their involvement in the projects they watched. They found out that having watchers is of great importance to the growth of the project because substantial amount of contributors started as project watchers. Yang et al. [11] discussed the developer commit patterns in GitHub in their research paper and mined the relationship between the commit patterns and the code evolution. Bissyande et al. [12] conducted a study about the popularity, interoperability and impact of different programming languages in 100,000 open source projects.

McDonald and Goggins [13] found that the contributor growth, community involvement and visible activity are the key to the success of a project, code quality and the code changes in the open source projects.

3 Research Method

We first define the popularity and code change metrics in order to find out the correlation between project popularity and code change frequency.

Similar to the work done by Aggarwal et al. [14], we interpret the indicators of project popularity to be number of watchers. We use pull requests in addition, as Dabbish et al. [5] conclude that pull requests are significant indicator of activity and project popularity as well. Pull requests are more important to popularity than the watchers as pulls show the developers to be actively involved in the project, rather than just be passive watchers.

$$ProjectPopularity = (\#Watchers) + (\#PullRequests)^2 \tag{1}$$

In order to define the code change frequency metric, we need to take two parameters in consideration. The first one is the number of lines of code that have been changed. The second parameter is the frequency of source code change, which serves as an indicator of the activity for contributors. Therefore,

$$Change = Nlog(\sum_{i=1}^{N} C_i + 1) \tag{2}$$

where N represents the number of commits per project, while C_i denotes total number of lines changed in source code.

We considered randomly selected open source projects from GitHub and filtered out top 1 % of the highly and consistently popular projects by use of the top programming languages, which have at least 1500 watchers increment per month with an assumption that the projects with top programming languages are popular. We acquired data about the number of watchers, commits and pull requests for these projects, resulting in about 1.8 million entries. The data collected was over the period of the past two years. We collected data with 102,000 pull requests and 1,660,000 watchers and the datasets of every month were individually modeled.

For evaluating the hypotheses, we have considered only those projects which are at least one year old. Initially, most of the software engineering projects do not have much change in code or that in popularity. Hence, in order to examine the first hypothesis, we had to discover the correlation between time-series of project popularity and frequency of code change and for that, we have used the Pearson's and Spearman's correlation coefficients along with the Welch's t-test. We did it by finding the correlation with delay between those two time-series, i.e., by defining cross correlation as follows [14]:

$$\rho_{cross} = \arg\max \rho(CodeChange_{t+\tau}, Popularity_{\tau})$$

where τ represents time delay for the popularity to take effect on the frequency of source code change, and ρ_{cross} is a maximum correlation value for $\tau \in [1, 11]$.

4 Results and Analysis

Figure 1 shows the distribution of the programming languages across the projects and the total number of watchers of the projects grouped together according to the programming language in which they are written.

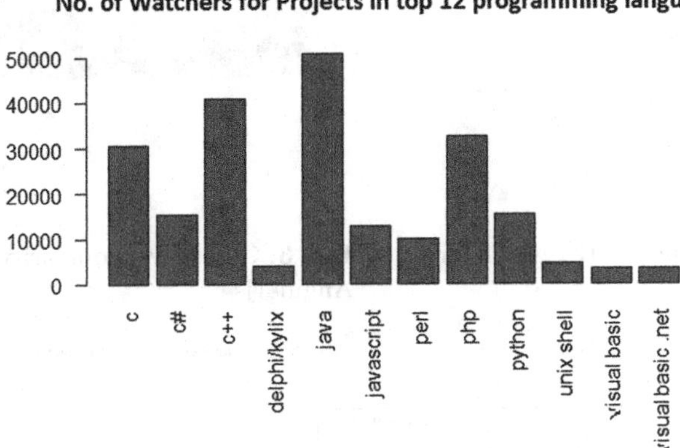

Fig. 1. Watchers by programming languages (top systems, from top 12 - programming languages)

Among the considered projects with top programming languages, we have further filtered them into top 1 % and top 10 % projects. The top 1 % of the projects being called very popular projects with at least increase in the number of watchers by 1500 every month and top 10 % being called popular projects.

We restrict our analysis to the top 1 % of the very popular projects which are given from the Table 1. These are the projects which are at least one year old and have consistent increase in the popularity.

We use the correlation analysis as mentioned in the methodology section on the resultant dataset of 96 projects which represent top 1 % of the highly and consistently popular projects. The Fig. 4 represents the time series of the values of popularity and frequency of code change in terms of required metrics for a project named Angularjs (Figs. 2, 3 and 5).

We have inspected the correlation coefficient values between the metrics of popularity and frequency of code change by Pearson's and Spearman's methods. Sometimes it is possible to get negative values for them indicating that the projects had many pull requests pending for years whereas the high correlation values for projects would indicate a good record of merging the pull requests and issuing commits on a regular basis.

All projects we have analyzed show positive values for the correlation coefficient calculated by Pearson's method and Spearman's method, and none of the values are negative as we have considered only the very popular projects. Out of 91 projects, 2 projects have correlation coefficient values between 0 and 0.3

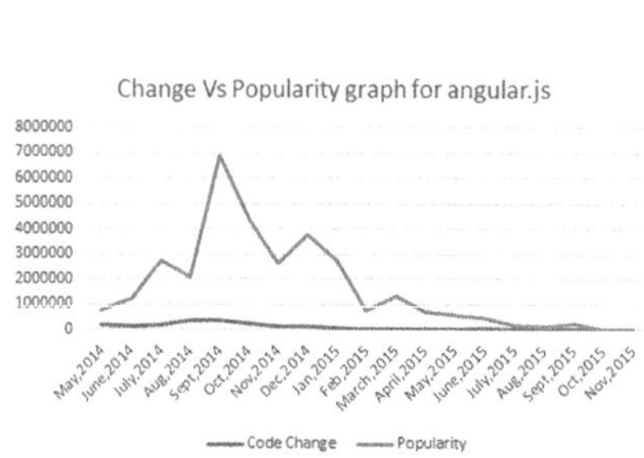

Fig. 2. Change vs. popularity graph for Angularjs

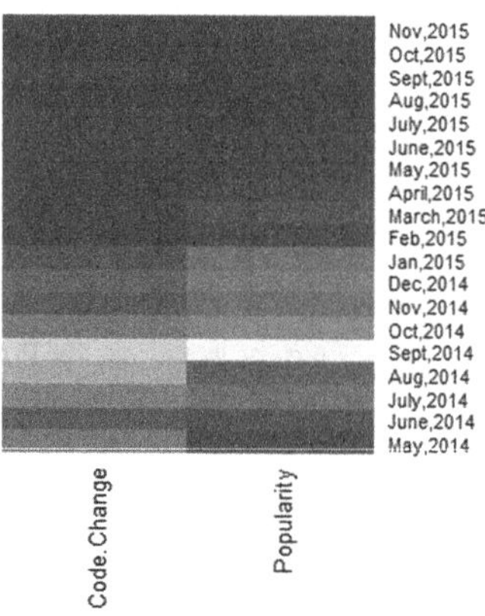

Fig. 3. Change vs. popularity graph for Angularjs

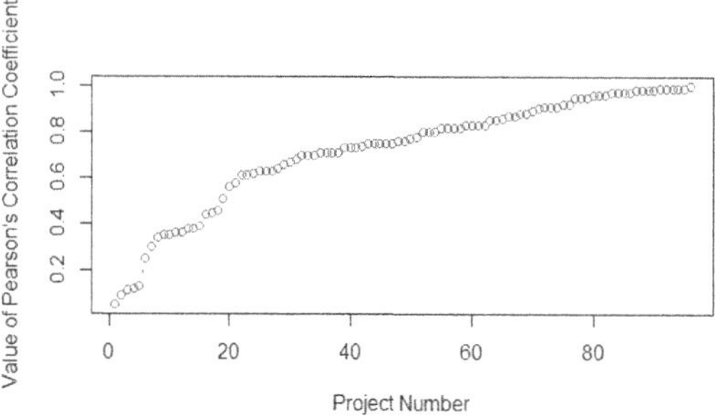

Fig. 4. Pearson's correlation coefficient for sampled projects

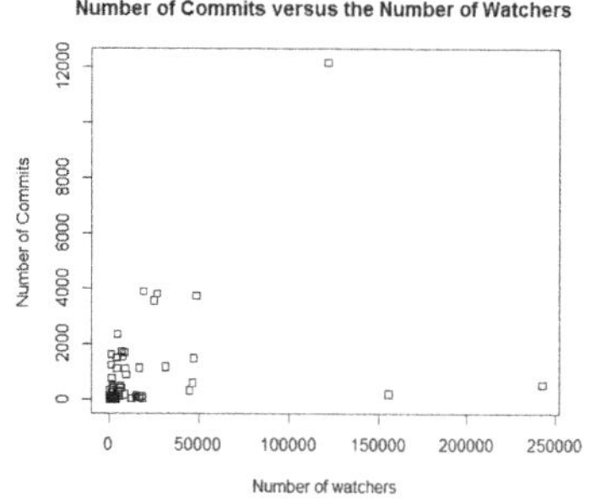

Fig. 5. Number of watchers vs. number of commits

showing weaker correlation, 23 projects have coefficient values between 0.3 and 0.7 showing moderate correlation, and 66 projects have values greater than 0.7 showing strong correlation.

When visualizing the data, it has been noticed that there were a few projects which had peak value for popularity and code changes at some point, whereas during the remaining points they had no activity (these projects have been filtered out as outliers) (Table 2).

In open source platforms like GitHub, watchers are used by developers to receive notifications on the status and progress of the projects, new discussions and events in the activity feed. The number of watchers can be assumed to be a proxy for the popularity of a project in GitHub. The Fig. 6 represents a plot correlating the project popularity with the number of watchers. Two facts can be observed from the figure. First, there is a strong correlation between the project popularity and the number of watchers (Pearson's correlation coefficient = 0.2006). Second, from the data sample that we have analyzed, it can be noted

Table 1. Popular (top-10 %) and very popular (top-1 %) systems

Programming language	No. of popular projects	No. of very popular projects
Java	2656	52
C++	1118	12
Javascript	339	12
Python	315	3

Table 2. Conclusion on correlation from coefficient values

Project name	r	t	Correlation
Gaia	0.8773773	7.5396	Strong correlation
Angularjs	0.7640059	4.8823	Strong correlation
cocos2d-x	0.3404814	1.4485	Moderate correlation
basemap	0.9972239	51.869	Strong correlation

that only a handful of projects have larger number of watchers than the number of pull requests.

Figure 7 represents a plot correlating the project popularity with the number of pull requests. It can be concluded from our analysis of the number of pull requests by the Pearson's method that there is a stronger correlation between the project popularity and the number of pull requests (Pearson's correlation coefficient = 0.8919) than that between the project popularity and the number of watchers. Second, it can be seen from Fig. 6 that the ratio of popularity change to the number of watchers is flatter with a slope of about 0.577 and from Fig. 7, the ratio of popularity change to the number of pull requests is elevated with a slope of about 1.0.

From the above evidence, it can be inferred that the number of pull requests is 73.2 % more important than the number of watchers to the change in project popularity for the 91 GitHub projects we have considered.

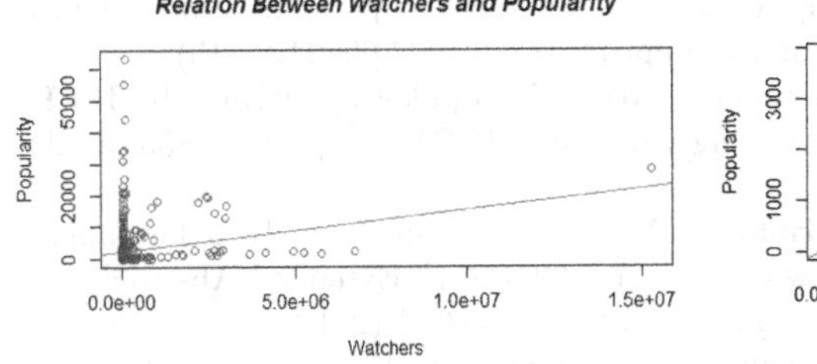

Fig. 6. Correlation between no. of watchers and project popularity

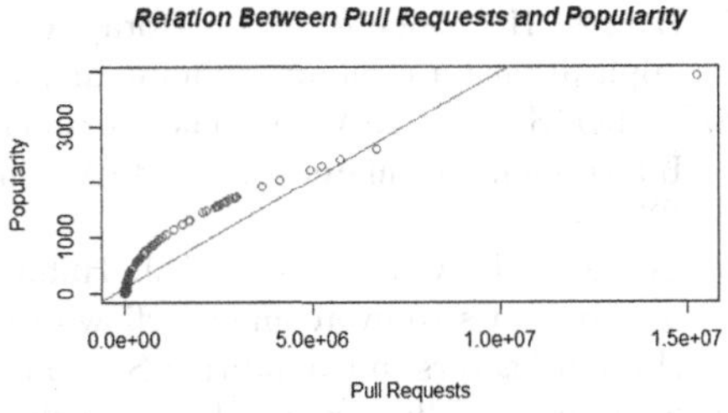

Fig. 7. Correlation between no. of pull requests and project popularity

5 Conclusion

In this project, we demonstrated the empirical analysis of our hypotheses on the data from 10 million GitHub repositories where we have filtered the projects on the basis of being highly and consistently popular with at least 1500 watchers increase each month and on the basis of the top 12 programming languages. Ultimately, we have used a total of 91 projects for our empirical study. We found that if the projects are consistently and highly popular, they attract consistent work from the developers' community and hence, the positive and strong correlation coefficient. When it comes to the second hypothesis, it was inferred that the number of pull requests has at least 50 % more contribution to the project popularity metric than the number of watchers. Finally, the number of pull requests is 73.2 % more important than the number of watchers for the project popularity.

References

1. LaToza, T.D., Towne, W.B., Van Der Hoek, A., Herbsleb, J.D.: Crowd development. In: 6th International Workshop on Cooperative and Human Aspects of Software Engineering (CHASE), pp. 85–88. IEEE (2013)
2. Yu, Y., Wang, H., Yin, G., Ling, C.X.: Who should review this pull-request: reviewer recommendation to expedite crowd collaboration. In: 2014 21st Asia-Pacific Software Engineering Conference, Jeju, pp. 335–342 (2014). doi:10.1109/APSEC.2014.57
3. Rahman, M.M., Roy, C.K.: An insight into the pull requests of GitHub. In: Proceedings of the 11th Working Conference on Mining Software Repositories, MSR, pp. 364–367 (2014)
4. Begel, A., Bosch, J., Storey, M.-A.: Social networking meets software development: perspectives from GitHub, MSDN, stack exchange, and topcoder. IEEE Softw. **30**(1), 52–66 (2013)
5. Dabbish, L., Stuart, C., Tsay, J., Herbsleb, J.: Social coding in GitHub: transparency and collaboration in an open software repository. In: Proceedings of the ACM Conference on Computer Supported Cooperative Work, pp. 1277–1286. ACM (2012)
6. Gousios, G.: The GHTorent dataset and tool suite. In: Proceedings of the 10th Working Conference on Mining Software Repositories, pp. 233–236. IEEE Press (2013)
7. Borges, H., Valente, M.T., Hora, A., Coelho, J.: On the popularity of GitHub applications: a preliminary note. arXiv preprint arXiv:1507.00604 (2015)
8. Weber, S., Luo, J.: What makes an open source code popular on GitHub? In: IEEE International Conference on Data Mining Workshop (ICDMW), pp. 851–855. IEEE (2014)
9. Lee, M.J., Ferwerda, B., Choi, J., Hahn, J., Moon, J.Y., Kim, J.: GitHub developers use rockstars to overcome overflow of news. In: CHI 2013 Extended Abstracts on Human Factors in Computing Systems, pp. 133–138. ACM (2013)
10. Sheoran, J., Blincoe, K., Kalliamvakou, E., Damian, D., Ell, J.: Understanding watchers on GitHub. In: Proceedings of the 11th Working Conference on Mining Software Repositories, MSR, pp. 336–339 (2014)

11. Weicheng, Y., Beijun, S., Ben, X.: Mining GitHub: why commit stops - exploring the relationship between developer's commit pattern and file version evolution. In: Proceedings of the 20th Asia-Pacific Software Engineering Conference (APSEC), APSEC 2013, pp. 165–169 (2013)
12. Bissyandé, T. F., Thung, F., Lo, D., Jiang, L., Réveillere, L.: Popularity, interoperability, and impact of programming languages in 100,000 open source projects. In: 2013 IEEE 37th Annual Computer Software and Applications Conference (COMPSAC), pp. 303–312. IEEE (2013)
13. McDonald, N., Goggins, S.: Performance and participation in open source software on GitHub. In: CHI 2013 Extended Abstracts on Human Factors in Computing Systems, CHI EA 2013, pp. 139–144 (2013)
14. Aggarwal, K., Hindle, A., Stroulia, E.: Co-evolution of project documentation and popularity within GitHub. In: Proceedings of the 11th Working Conference on Mining Software Repositories, pp. 360–363. ACM (2014)

Hidden Space Neighbourhood Component Analysis for Cancer Classification

Li Zhang(✉), Xiaojuan Huang, Bangjun Wang, Fanzhang Li, and Zhao Zhang

School of Computer Science and Technology & Joint International Research Laboratory of Machine Learning and Neuromorphic Computing, Soochow University, Suzhou 215006, Jiangsu, China
{zhangliml,wangbangjun,lfzh,cszzhang}@suda.edu.cn, 20154227006@stu.suda.edu.cn

Abstract. Neighbourhood component analysis (NCA) is a method for learning a distance metric which can maximize the classification performance of the K nearest neighbour (KNN) classifier. However, NCA suffers from the small size sample problem that the number of samples is much less than the number of features. To remedy this, this paper proposes a hidden space neighbourhood components analysis (HSNCA), which is a nonlinear extension of NCA. HSNCA first maps the data in the original space into a feature space by a set of nonlinear mapping functions, and then performs NCA in the feature space. Notably, the number of samples is equal to the number of features in the feature space. Thus, HSNCA can avoid the small size sample problem. Experimental results on DNA array datasets show that HSNCA is feasibility and efficiency.

Keywords: Neighbourhood components analysis · Nonlinear mapping · Small size sample problem · Feature space · Nearest neighbour

1 Introduction

At present, more and more attention has been paid to cancer classification based on gene expression. Many methods have been applied to cancer classification, such as support vector machines(SVMs) [1,2], and K nearest neighbor (KNN) [3]. It is well known, the KNN classifier is a simple and efficient classification method [4], whose classification performance greatly depends on both the number of neighbors and the distance measurement. Usually, KNN employs the Euclidean distance as a distance measurement. Unfortunately, the Euclidean distance measurement treats all features equally and does not consider the classification information. Thus, the resulted nearest neighbors may be not the real ones which are useful for classification. To incorporate the classification information, some distance metric learning methods have been proposed for improving the classification performance of KNN [5–11].

These metric learning methods try to construct a Mahalanobis distance over the input space and use it instead of the Euclidean distance. In fact, the goal

A. Hirose et al. (Eds.): ICONIP 2016, Part IV, LNCS 9950, pp. 44–51, 2016.
DOI: 10.1007/978-3-319-46681-1_6

of these methods is to find a linear transformation to map the original data into the feature space in which the Euclidean distance is still used. From the view of transformation, linear discriminant analysis (LDA), also called Fisher discriminant analysis (FDA) can be taken as one of metric learning methods. Although principle component analysis (PCA) can find a linear transformation, it ignores the classification information. Thus, PCA can not be taken as a metric learning method. However, LDA is not designed for KNN specially. Similar metric learning methods, such as Xing's method [9], maximally collapsing metric learning (MCML)[7], information-theoretic metric learning (ITML) [11] and relevant component analysis (RCA) [6] also do not leverage the full power of KNN classifier since these methods use all similar labeled inputs or side-information instead of K nearest neighbors.

This paper discusses learning a Mahalanobis distance metric for the KNN classifier and its application to cancer classification. The classical distance metric methods include large margin nearest neighbor (LMNN) [8], and neighborhood component analysis (NCA)[5]. LMNN learns the Mahalanobis distance by solving a semi-definite programming [8]. The idea behind LMNN is to separate data points from different classes by a large margin. Chopra et al. proposed a energy-based model for learning similarity metrics and applied it to face verification [10]. Neighborhood component analysis (NCA) was proposed for improve the classification performance KNN [5]. NCA attempts to maximize the leave-one-out classification accuracy on the training set. In addition, NCA can implement dimension reduction for data visualization and fast classification. Based on NCA, unsupervised NCA [12] was proposed for solving clustering problems, and neighborhood component feature selection method [13] was presented for high-dimensional data. To avoid the overfitting problem in NCA, Yang and Laaksonen developed a regularized neighborhood component analysis (RNCA) [14]. To implement nonlinear transformation, Qin et al. proposed a nonlinear neighborhood component analysis based on constructive neural networks [15]. To overcome the high computational cost, a fast NCA was presented in [16]. However, NCA and its most variants suffer from the small sample size problem that the number of samples is much less than the number of features.

To solve the small sample size problem in NCA, this paper proposes a hidden space neighbourhood components analysis (HSNCA), which is a nonlinear extension of NCA. HSNCA first maps the data in the original space into a feature space by a set of nonlinear hidden functions, and then performs NCA in the feature space. Notably, the number of samples is equal to the number of features in the feature space. Thus, HSNCA can avoid the small size sample problem. Experimental results on DNA array datasets show that HSNCA is feasibility and efficiency.

The rest of this paper is organized as follows. Section 2 proposes hidden space-based neighborhood component analysis. We perform experiments on gene expression data in Sect. 3 and conclude this paper in Sect. 4.

2 Hidden Space Neighbourhood Components Analysis

2.1 Algorithm Description

Zhang et al. proposed hidden space support vector machine (HSSVM) [17], which is an extension of support vector machine (SVM). HSSVM can adopt any nonlinear mapping functions which do not limit to the Mercer kernel function. Inspired by this idea, some algorithms have been extended into the hidden space such as PCA [18], LDA [19], and discriminant neighborhood embedding [20]. To solve the small sample size problem, we extend NCA into the hidden space, or NSNCA.

Given N training samples $\{\mathbf{x}_i, y_i\}_{i=1}^N$ with $\mathbf{x}_i \in \mathbb{R}^D$ and $y_i \in \{1, \cdots, C\}$ where D is the feature number and C is the class number, the goal of HSNCA is to find a distance metric that maximizes the performance of KNN. In the following, we describe the details of HSNCA.

First, we map the training samples into a hidden space by using a nonlinear hidden function $\varphi(\mathbf{x}_i)$. In [17,18,20], a possible nonlinear hidden function is given. Namely,

$$\varphi(\mathbf{x}_i) = [k(\mathbf{x}_i, \mathbf{x}_1), \ldots, k(\mathbf{x}_i, \mathbf{x}_N)]^T \tag{1}$$

where $k(\mathbf{x}_i, \mathbf{x}_j)$ is any symmetric kernel function and needs not obey Mercer's condition.

Let $\mathbf{z}_i = \varphi(\mathbf{x}_i)$, where $\mathbf{z}_i$ is the image of $\mathbf{x}_i$ in the hidden space. By doing so, we can have new training samples $\{\mathbf{z}_i, y_i\}_{i=1}^N$ with $\mathbf{z}_i \in \mathbb{R}^N$. Obviously, we simply carry out dimension reduction at a low cost when $N \ll d$.

Second, we directly perform NCA on $\{\mathbf{z}_i, y_i\}_{i=1}^N$ in the hidden space. Let $\mathbf{A}$ be the transformation matrix. The Mahalanobis distance metric in the hidden space can be defined as

$$d(\mathbf{z}_i, \mathbf{z}_j) = (\mathbf{A}\mathbf{z}_i - \mathbf{A}\mathbf{z}_j)^T(\mathbf{A}\mathbf{z}_i - \mathbf{A}\mathbf{z}_j) = (\mathbf{z}_i - \mathbf{z}_j)^T\mathbf{Q}(\mathbf{z}_i - \mathbf{z}_j) \tag{2}$$

where $\mathbf{Q} = \mathbf{A}^T\mathbf{A}$ is the metric. To obtain the transformation matrix $\mathbf{A}$, HSNCA is to optimize the following objective:

$$\max f(\mathbf{A}) = \sum_i \sum_{j \in C_i} p_{ij} \tag{3}$$

where $C_i = \{j | y_i = y_j\}$, and p_{ij} is the probability of $\mathbf{z}_i$ selecting $\mathbf{z}_j$ as its neighbor, which can be defined as:

$$p_{ij} = \frac{\exp(-\|\mathbf{A}\mathbf{z}_i - \mathbf{A}\mathbf{z}_j\|^2)}{\sum_{m \neq i} \exp(-\|\mathbf{A}\mathbf{z}_i - \mathbf{A}\mathbf{z}_m\|^2)} \tag{4}$$

The solution to (3) could be found by using a gradient based optimizer [5]. For the detailed introduction about solution, the reader is referred to [5].

2.2 Hidden Function

Here, we discuss hidden functions and their parameters. Generally, the linear kernel $k(\mathbf{z}_i, \mathbf{z}_j) = \mathbf{z}_i^T \mathbf{z}_j$ and the Gaussian radius basis function (RBF) kernel $k(\mathbf{z}_i, \mathbf{z}_j) = \exp(-\gamma \|\mathbf{z}_i - \mathbf{z}_j\|^2))$ could be used to construct a hidden function, where $\gamma > 0$ is the RBF kernel parameter [18,20]. Clearly, there is no parameter if the hidden function consists of linear kernels. For the Gaussian RBF kernel, however, we must determine γ in advance.

In [19,20], an n-fold cross-validation was used to find the optimal parameter γ. It is well-known that the cross-validation method has a high time complexity for multiple experiments on different parameter pairs. To avoid this problem, some model selection for the Gaussian RBF kernel has been discussed. We introduce two methods as follows and adopt them to determine γ in our experiments.

A median value method was adopted in [21]. Namely,

$$\gamma = \text{median}\left(\left\{\gamma_i | \gamma_i = \frac{1}{\|\mathbf{z}_i - \bar{\mathbf{z}}\|^2}, i = 1, \cdots, N\right\}\right) \tag{5}$$

where median(·) denotes the function taking the median value of the elements in the set, and $\bar{\mathbf{z}} = \frac{1}{N}\sum_{i=1}^{N} \mathbf{z}_i$ is the mean of all new training samples. The median value method avoids the procedure of long time parameter selection.

Xu et al. proposed a direct parameter setting formula for the kernel parameter [22]. The optimal γ is determined by maximizing the difference between the local and global structure measures. Given the distance rank on the given data set in increasing order, the local structure measure d_{near} is the $\lfloor(1-\alpha)(n(N-1))\rfloor$th distance and the global structure measure d_{far} is the $\lfloor\alpha(n(N-1))+1\rfloor$th distance, where $0 \le \alpha < 0.5$ and $n \le N$ is the number of selected samples. Then, the optimal γ is given by

$$\gamma = \frac{\ln d_{far}^2 - \ln d_{near}^2}{d_{far}^2 - d_{near}^2} \tag{6}$$

3 Experiments

3.1 Datasets

To validate the efficiency of HSNCA, we perform experiments on six gene microarray datasets, including Leukemia, Breast Cancer, Central Nervous System (CNS), Leukemia3, Lung Cancer, and Glioblastoma. The detail of these datasets is described as follows:

- The Leukemia dataset has 72 samples belonging to two classes, or ALL (Acute Lymphoblastic Leukemia) and AML (Acute Myeloid Leukemia). The training set consists of 38 bone marrow samples (27 ALL and 11 AML), over 7129 probes from 6817 human genes. In addition, 34 test samples is provided, with 20 ALL and 14 AML.

- The CNS dataset contains 60 patient samples, 21 are survivors (alive after treatment) and 39 are failures (succumbed to their disease). There are 7129 genes in the dataset. The training set consists of the first 10 survivors and 30 failures, the other 11 survivors and 9 failures are testing points.
- The Breast Cancer dataset has 97 samples belonging to two classes, or Relapse and Non-relapse. Each sample has 24481 genes. The training set contains 78 patient samples, 34 of which are from patients who had developed distance metastases within 5 years (labelled as "relapse"), the rest 44 samples are from patients who remained healthy from the disease after their initial diagnosis for interval of at least 5 years (labelled as "non-relapse"). Correspondingly, there are 12 relapse and 7 non-relapse samples in the test set.
- In the Leukemia3 dataset, there are two subsets for training and test, respectively. Training data contains 57 leukemia samples (20 ALL, 17 MLL and 20 AML). Testing data contains 4 ALL, 3 MLL and 8 AML samples. In the above publication, it mentioned only 3 AML testing samples.
- The Lung Cancer dataset consists of 203 snap-frozen lung tumors and normal lung. The 203 speciments include 139 samples of lung adenocarcinomas (labelled as ADEN), 21 samples of squamous cell lung carcinomas (labelled as SQUA), 20 samples of pulmonary carcinoids (labelled as COID), 6 samples of small-cell lung carcinomas (labelled as SCLC) and 17 normal lung samples (labelled as NORMAL). Each sample is described by 12600 genes. The first 20, 10, 4, 10 and 10 samples in the five classes are taken as the training samples, respectively. The rest is treated as the test ones.
- In the Glioblastoma dataset, there are 50 samples with 12625 features and 4 diagnostic classes (classic gliomas (CG), classic oligodendrogliomas (CO), nonclassic gliomas (NG) and nonclassic oligodendrogliomas (NO)) on the basis of DNA gene expression signatures. The first 10, 5, 10, and 10 samples in the CG, CO, NG and NO are taken as the training samples, respectively. The rest is treated as the test ones.

The first five datasets can be downloaded from http://datam.i2r.a-star.edu.sg/datasets/krbd/index.html, and the last dataset from http://www.biolab.si/supp/bi-cancer/projections/info/glioblastoma.htm. All data here are normalized such that each gene has mean 0 and variance 1 (Table 1).

Table 1. Description of five datasets

Dataset	# class	# gene	# training	# test
Leukemia	2 (ALL/AML)	7129	38	34
Breast cancer	2 (Relapse/non-relapse)	24481	78	19
Central nervous system	2 (Survivor/failure)	7129	40	20
Leukemia3	3 (ALL/MLL/AML)	12582	57	15
Lung cancer	5 (ADEN/SQUA/COID/SCLC/NORMAL)	12600	54	149
Glioblastoma	4 (CG/CO/NG/NO)	12625	35	15

3.2 Experimental Setting and Results

We compare three methods, including KNN, NCA and HSNCA. For the KNN classifier, K varies from 1 to 10. The gene expression data has such a high dimension that NCA can not directly deal with it. Thus, some dimension reduction method must be performed such as PCA before NCA is carried out. However, NSNCA does not need this kind of data processing. In NSNCA, we adopt the Gaussian RBF kernel to construct the hidden function. The parameter γ is determined by (5) and (6), respectively. Correspondingly, we use HSNCA1 and HSNCA2 to represent the two different parameter selection methods. For HSNCA2, let $\alpha = 0.1$ which follows the setting in [22], and $n = N$ for the small size sample problems.

Both NCA and HSNCA can perform dimension reduction. Thus, we can observe the variation of classification performance vs. projection dimension obtained by these two methods. We show that curves of the classification error vs. both projection dimension and nearest neighbor number on the Leukemia dataset in Fig. 1. By using PCA, the training data of the Leukemia dataset only consists of 23 features. Therefore the projection dimension is no greater than 23, see Fig. 1(a). For KNN, the curve does not change because all features are used. Clearly, HSNCA2 obtains the best classification error when the projection dimension is around 15. Although both NCA and NSNCA1 can achieve the same performance level as KNN, they are worse than KNN on most projection dimensions. From Fig. 1(b), we can see that the classification performance of these methods decrease as increasing nearest neighbor number. The main reason is that the number of training sample is too small. Obviously, NSNCA2 is much better than other methods.

Experimental results on six datasets are listed in Table 2, where the figures in the parenthesis are the nearest neighbor number K and the projection dimension corresponding to the given classification errors. Note that NCA does not obtain result on the Breast Cancer dataset for out of memory when PCA was performed on 24,481 features. It is possible to ruin some discriminant information for the

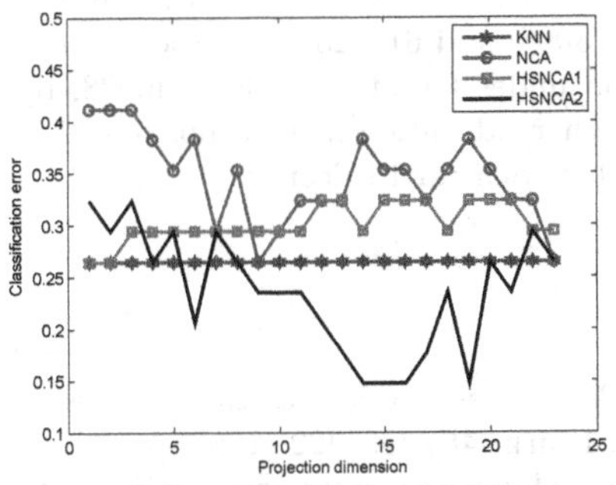

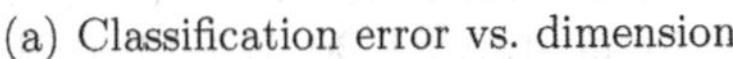

(a) Classification error vs. dimension

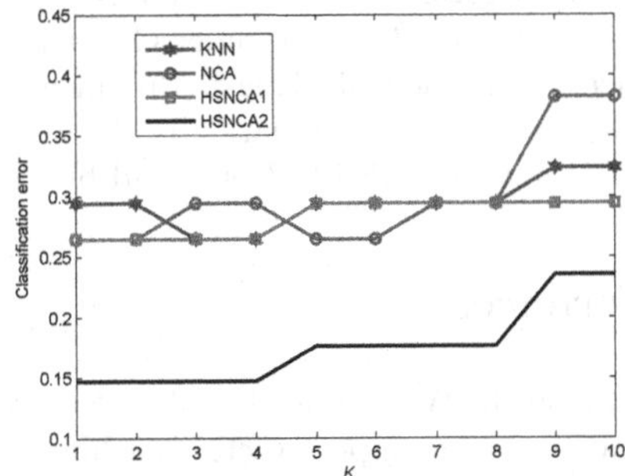

(b) Classification error vs. neighbor number

Fig. 1. Classification error vs. projection dimension (a) and vs. nearest neighbor number (b) on the Leukemia dataset

Table 2. Comparison of classification error rate (%) on six datasets

Dataset	*K*NN	NCA	HSNCA1	NSNCA2
Leukemia	26.47 (3,7129)*	26.47 (1,9)	26.47 (1,1)	**14.71** (1,14)
Breast cancer	36.84 (1,24481)	-	26.32 (1,10)	**21.05** (3,17)
CNS	40.00 (3,7129)	45.00 (3,11)	45.00 (1,2)	**35.00** (9,8)
Leukemia3	**0.00** (3,12582)	33.33 (7,7)	**0.00** (1,1)	**0.00** (1,12)
Lung cancer	**7.38** (3,12600)	23.49 (5,22)	8.72 (3,8)	**7.38** (3,5)
Glioblastoma	10.34 (1,12625)	27.59 (1,1)	**0.00** (1,1)	**0.00** (7,1)

Note: The figures in the parenthesis are the nearest neighbor number K and the projection dimension corresponding to the best classification errors.

PCA+NCA method, which can be supported by the results on the Leukemia3, Lung Cancer and Glioblastoma datasets. Contrarily, HSNCA enhances the discriminant information on three datasets, Leukemia, Breast Cancer and Glioblastoma. In addition, the parameter selection (6) is much better than (5) for the hidden function.

4 Conclusions

This paper develops HSNCA by introducing NCA into the hidden space and applies it to cancer classification based on gene expression data which is a classical small sample size problem. HSNCA greatly reduces the features by mapping the original data into a feature space for cancer classification. In the feature space, NCA can be carried out directly without small sample size problems. In addition, we provide two selection methods for hidden function parameter. Experimental results on DNA array datasets show that HSNCA is much better than PCA+NCA.

Acknowledgments. This work was supported in part by the National Natural Science Foundation of China under Grant Nos. 61373093, and 61402310, by the Natural Science Foundation of Jiangsu Province of China under Grant No. BK20140008, by the Natural Science Foundation of the Jiangsu Higher Education Institutions of China under Grant No. 13KJA520001, and by the Soochow Scholar Project.

References

1. Guyon, I., Weston, J., Barnhill, S., Vapink, V.: Gene selection for cancer classification using support vector machines. Mach. Learn. **46**, 389–422 (2002)
2. Li, J.T., Jia, Y.M., Li, W.L.: Adaptive huberized support vector machine and its application to microarray classification. Neural Comput. Appl. **20**, 123–132 (2011)
3. Li, L., Weinberg, C.-R., Darden, T.-A., Pedersen, L.-G.: Gene selection for sample classification based on gene expression data: study of sensitivity to choice of parameters of the GA/KNN method. Bioinformatics **17**, 1131–1142 (2001)

4. Cover, T., Hart, P.: Nearest neighbor pattern classification. IEEE Trans. Inf. Theor. **IT–13**, 21–27 (1967)
5. Goldberger, J., Roweis, S., Hinton, G., Salakhutdinov, R.: Neighbourhood components analysis. In: Advances in Neural Information Processing Systems, vol. 17, pp. 513–520. MIT Press (2004)
6. Shental, N., Hertz, T., Weinshall, D., Pavel, M.: Adjustment learning and relevant component analysis. In: Proceedings of 7th European Conference on Computer Vision, London, UK, pp. 776–792 (2002)
7. Globerson, A., Roweis, S.T.: Metric learning by collapsing classes. In: Advances in Neural Information Processing Systems, vol. 18 (2005)
8. Weinberger, K.Q., Saul, L.K.: Distance metric learning for large margin nearest neighbor classification. J. Mach. Learn. Res. **10**, 207–244 (2009)
9. Xing, E.P., Ng, A.Y., Jordan, M.I., Russell, S.: Distance metric learning, with application to clustering with side-information. In: Advances in Neural Information Processing Systems, vol. 14, pp. 521–528. MIT Press, Cambridge (2002)
10. Chopra, S., Hadsell, R., LeCunGoldberger, Y.: Learning a similiarty metric discriminatively, with application to face verification. In: Proceedings of IEEE Conference on Computer Vision and Pattern Recognition, San Diego, CA, pp. 349C–356 (2005)
11. Davis, J.V., Kulis, B., Jain, P., Sra, S., Dhillon, I.S.: Information-theoretic metric learning. In: Proceedings of 24th International Conference on Machine Learning, pp. 209–216. ACM, New York (2007)
12. Qin, C., Song, S., Huang, G., Zhu, L.: Unsupervised neighborhood component analysis for clustering. Neurocomputing **168**, 609–617 (2015)
13. Yang, W., Wang, K., Zuo, W.: Neighborhood component feature selection for high-dimensional data. J. Comput. **7**(1), 161–168 (2012)
14. Yang, Z., Laaksonen, J.: Regularized neighborhood component analysis. In: Ersbøll, B.K., Pedersen, K.S. (eds.) SCIA 2007. LNCS, vol. 4522, pp. 253–262. Springer, Heidelberg (2007)
15. Qin, C., Song, S., Huang, G.: Non-linear neighborhood component analysis based on constructive neural networks. In: Proceedings of 2014 IEEE International Conference on Systems, Man and Cybernetics, pp. 1997–2002. IEEE (2014)
16. Yang, W., Wang, K., Zuo, W.: Fast neighborhood component analysis. Neurocomputing **83**(6), 31–37 (2012)
17. Zhang, L., Zhou, W.D., Jiao, L.C.: Hidden space support vector machines. IEEE Trans. Neural Netw. **15**(6), 1424–1434 (2004)
18. Zhou, W., Zhang, L., Jiao, L.: Hidden space principal component analysis. In: Ng, W.-K., Kitsuregawa, M., Li, J., Chang, K. (eds.) PAKDD 2006. LNCS (LNAI), vol. 3918, pp. 801–805. Springer, Heidelberg (2006)
19. Zhang, L., Zhou, W.D., Chang, P.-C.: Generalized nonlinear discriminant analysis and its small sample size problems. Neurocomputing **74**, 568–574 (2011)
20. Ding, C., Zhang, L., Wang, B.J.: Hidden space discriminant neighborhood embedding. In: Proceedings of 2014 International Joint Conference on Neural Networks, pp. 271–277. IEEE (2014)
21. Zhang, L., Zhou, W.-D., Chang, P.-C., Liu, J., Yan, Z., Wang, T., Li, F.-Z.: Kernel sparse representation-based classifier. IEEE Trans. Sig. Process. **60**, 1684–1695 (2012)
22. Xu, Z., Dai, M., Meng, D.: Fast and efficient strategies for model selection of gaussian support vector machine. IEEE Trans. Syst. Man Cybern. - Part B: Cybern. **39**(5), 1292–1307 (2009)

Prediction of Bank Telemarketing with Co-training of Mixture-of-Experts and MLP

Jae-Min Yu and Sung-Bae Cho(✉)

Department of Computer Science, Yonsei University, Seoul, Republic of Korea
{yjam, sbcho}@yonsei.ac.kr

Abstract. Utilization of financial data becomes one of the important issues for user adaptive marketing on the bank service. The marketing is conducted based on personal information with various facts that affect a success (clients agree to accept financial instrument). Personal information can be collected continuously anytime if clients want to agree to use own information in case of opening an account in bank, but labeling all the data needs to pay a high cost. In this paper, focusing on this characteristics of financial data, we present a global-local co-training (GLCT) algorithm to utilize labeled and unlabeled data to construct better prediction model. We performed experiments using real-world data from Portuguese bank. Experiments show that GLCT performs well regardless of the ratio of initial labeled data. Through the series of iterating experiments, we obtained better results on various aspects.

Keywords: Bank telemarketing · Semi-supervised learning · Machine learning

1 Introduction

Marketing is a typical strategy to enhance business. Especially, telemarketing process mostly involves one or more calls. Making a call to contact every customer is expensive and time consuming. Therefore, it would be useful to call only customers who would be most willing to buy a certain product or service. One way to find these customers is to use a machine learning technique. This enables telemarketer to eliminate unnecessary contact to client through the evaluation of available customer information. Figure 1 shows the number of contacts performed for client in UCI bank marketing dataset. In the figure, x-axis means the number of contacts, and y-axis means the number of clients and the total number of telephone calls. The total number of calls is 124,943 from 45,198 clients. In order to contact clients, a telemarketer needs about 3 times more calls than the number of total clients. However, time-consuming contacts to clients do not always guarantee a good result. For this reasons, it is hard to predict the potential clients in telemarketing.

This research was supported by the MSIP (Ministry of Science, ICT and Future Planning), Korea, under the ITRC (Information Technology Research Center) support program (IITP-2016-R0992-15-1011) supervised by the IITP (Institute for Information & communications Technology Promotion).

A. Hirose et al. (Eds.): ICONIP 2016, Part IV, LNCS 9950, pp. 52–59, 2016.
DOI: 10.1007/978-3-319-46681-1_7

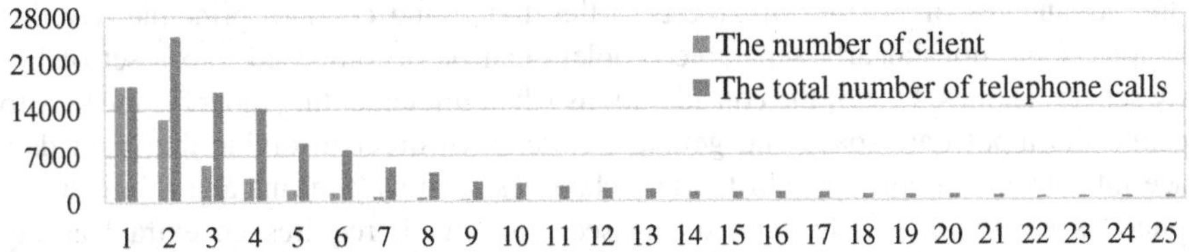

Fig. 1. The number of calls to clients in the UCI dataset

Many classification problems of financial field like prediction of telemarketing have complicated problem spaces according to data explosion and unforeseeable causal relationship of data. To solve this problem, Jacobs et al. presented mixture-of-experts model based on principle of divide and conquer [1]. Another issue of classification is about unlabeled data. In financial field, as unlabeled data have increased with time, unlabeled data is easier to get than labeled data [2]. These unlabeled data can be used to improve the performance of classification if the data can be used with labeled one.

In this paper, we present a global-local co-training (GLCT) method that can automatically predict the result of a phone call to sell long-term deposits by using co-training scheme [3]. Such GLCT is valuable to assist managers in prioritizing and selecting the next customers to be contacted during bank marketing. As a consequence, the time and costs of marketing would be reduced. To evaluate the usefulness of the proposed method for predicting bank marketing, we conducted experiments using real datasets collected from Portuguese retail bank.

2 Related Works

2.1 Co-training

Co-training is the multi-view semi-supervised learning approach proposed by Blum and Mitchell [4]. The algorithm works well if the feature set division of dataset satisfies two assumptions. First, each set of features is sufficient for classification. Second, the two feature sets of each instance are conditionally independent given the class. Co-training uses two learnable models with different feature sets. The models are trained independently and each model enlarges the opponent's training data set by predicting unlabeled instances iteratively. Despite the original co-training requires the conditional independencies between the two feature sets as views, it is not easy to separate feature sets nicely in real-world problems. To overcome this problem, Goldman and Zhou discovered that two models trained by different learning methods can be regarded as two different views [5].

2.2 Mixture-of-Experts

Mixture-of-experts models [1] consist of a set of experts, which model conditional probabilistic processes, and a gate which combines the probabilities of the experts. One of the advantages of mixture-of-experts is the divide and conquer principle. By this principle, some non-linearly separable problem can be decomposed into a set of linearly separable ones. One of the considerations when implementing model is the way to separate the input data space and generate experts. In order to divide the input data space into several subgroups, clustering techniques such as k-means algorithm can be applied as shown in [6]. In semi-supervised learning, it requires an extra learning method to deal with unlabeled data. The simplest way to train the mixture-of-experts model with both labeled and unlabeled data is to apply previous semi-supervised learning techniques such as self-training or co-training directly to the model.

2.3 Credit Scoring Using Machine Learning

Many researchers have demonstrated AI techniques such as artificial neural networks [7], decision tree [8], case-based reasoning [9], and support vector machine can be used as promising methods for credit scoring. In contrast with statistical models, AI techniques do not assume certain data distribution. These techniques automatically extract knowledge from training samples. According to previous studies, AI techniques are superior to statistical techniques in dealing with credit scoring problems, especially for non-linear pattern classification [10] (Table 1).

There is another attempt to solve financial problem by using machine learning. For example, Tsai performed that four different types of hybrid models are compared by combining classification and clustering methods with a real world dataset from a bank in Taiwan [12]. However, the approach of previous works did not consider characteristics of financial data including unlabeled data. The application of co-training approach to predicting bank telemarketing is a relatively new and unrevealed area.

Table 1. Related studies on prediction of financial data

Author	Problem	Data	Method
Li et al. (2006) [11]	Evaluation of consumer loans	Customer financial information (age, occupation, account)	SVM
Tsai et al. (2010) [12]	Credit rating	Customer credit information	EM + DT EM + NB
Milad et al. (2015) [13]	Risk assessment	Social lending information	Random forest

3 Global-Local Co-training for Prediction

3.1 Global-Local Co-training Algorithm

Figure 2 shows the procedure of the global-local co-training (GLCT) [14]. The main idea of the method is to use the global view and the local view simultaneously. Instead of dividing the feature set according to the views, we chose two different learning algorithms which generate both models.

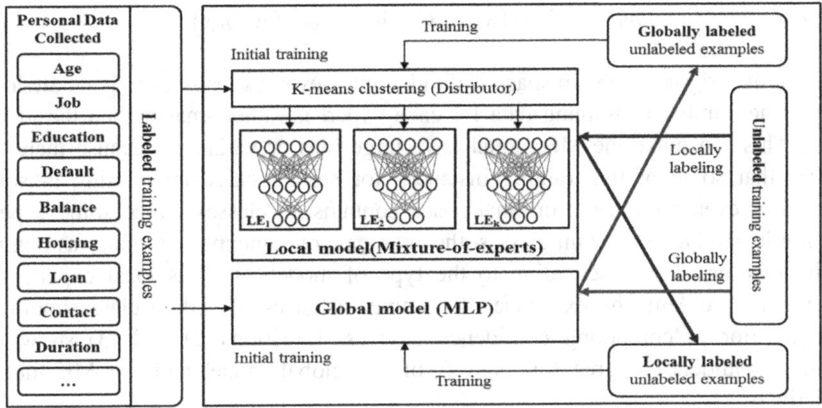

Fig. 2. The overall procedure of global-local co-training method

Formally, since the models can be imperfect and show poorer performance than the opponent in some conditions, the models should carefully predict labels, especially when there are not sufficient training data for local experts to generalize the entire problem. The algorithm compares the confidences from global model M_G and local model M_L to prevent degradations due to the imperfection. After initial training using personal data collected, the models can label instances for the opponents if only each model is more confident than the other one. The instances passed the first criteria are added to candidate sets U'_L or U'_G.

After obtaining the candidate sets, they are sorted with confidence degrees in descending order. At the next step, at most N instances for each class with higher confidence than the certain threshold e are chosen gradually from the top of the set to be labeled. By choosing instances with sufficient confidence, it is expected that the models would be improved for prediction at the next iteration. The chosen instances are added to the training data set for each model, training data set L_G for global model and training data set L_L for local model. The models are trained with newly extended labeled data set. If L_G and L_L do not change, the training ends. Afterwards, the algorithm iterates the training procedure until no additional data samples for L_L and L_G remained.

3.2 Measuring Confidence Degree

In order to obtain confidence degree of examples, we design the measure for confidences. Let *conf*(*w*,*x*) be the original confidence degree for input instance *x* defined by the corresponding model *w*. *conf*(*w*, *x*) can be changed by the type of model used to build the global or local model. In this paper, we implement both the global and the local models by using MLP. The output of MLP can be estimated as confidence degree, which can be simply defined as follows

$$conf(w, x) = \{w(x)|w(x)\,\text{is output of}\,w\,\text{for}\,x\} \tag{1}$$

Since the original problem space is divided into several subspaces to generate local experts, the number of training data for each expert is rather smaller than the global model. This can make the ME model have large bias or variance, because there are chances that some of the training instances for certain class are missing in some sub-regions even though the original space contains all classes. Since some experts have not learned about certain classes, they can generate unexpected values as outputs for the missing classes according to the type of models used as local experts. To prevent this problem, the existence of training instances of certain class should be checked prior to computing confidence degrees. Equations (2) and (3) show the modified confidence degree for class C of the global model and the ME model, respectively.

$$conf(M_G, \mathrm{x}, \mathrm{c}) = \begin{cases} M_{G,C}(x), & \textit{if}\, N_{M_{G,C}} > 0 \\ 0, & \textit{otherwise} \end{cases} \tag{2}$$

$$\begin{aligned} conf(M_L, \mathrm{x}, \mathrm{c}) &= \sum_{i=1}^{N} g_i conf(\mu i, x, c) \\ conf(\mu_i, \mathrm{x}, \mathrm{c}) &= \begin{cases} \mu_{i,C}(x), & \textit{if}\, N_{\mu,i,c} > 0 \\ 0, & \textit{otherwise} \end{cases} \end{aligned} \tag{3}$$

4 Experiments

This section presents the experiments conducted to evaluate the usefulness of the GLCT method on the prediction problems. The number of hidden nodes for each neural network is fixed to ten and learning rate is 0.3. There are two aspects of the experiments. One is to prove the usefulness of co-training based on unlabeled data. The other is that the proposed method shows better classification performance than other semi-supervised learning such as self-training.

4.1 Bank Telemarketing Data

This paper focuses on targeting through telemarketing phone calls to sell long-term deposits. Within a campaign, the human agents make phone calls to a list of clients to sell the deposit or, if the client calls the contact-center for any other reason, he is asked to subscribe the deposit. Thus, the result is a binary class (unsuccessful or successful contact). This paper considers real data collected from a Portuguese retail bank, from May 2008 to November 2010, in a total of 45,198 phone contacts. The training data is used for feature and model selection and includes all contacts executed up to April 2010. The test data is used for measuring the prediction capabilities of the GLCT method, including the most recent contacts in dataset, from May 2010 to November

2010, in total of 11,034. Table 2 shows a summary of used data sets. Table 3 describes the list of features with description.

Table 2. A summary of data sets used

Name	#Features	#Classes	#Samples
Personal data	14	2	45,198

4.2 Experimental Results

We conduct experiments to show the improvement of performance by using unlabeled data. For each experiment, the performances of models trained in three different conditions that are trained with all labeled data. All experiments were conducted ten times and results were averaged. In order to show that GLCT performs well regardless of the ratio of initial labeled data, we change the ratio of labeled instances to observe performances according to the amount of initial labeled data. We changed it from 5 % to 15 % as Table 4. Figure 3 shows the result of the test on dataset. As the result, the model trained with both labeled and unlabeled data showed better performance than the model without co-training. This result implied that GLCT can be used as the training method for the mixture-of-experts model in semi-supervised approach.

Finally, we compared GLCT with an alternative method to train the mixture-of-experts model to show the superiority of GLCT. In this experiment, the self-training algorithm [23] was used as the alternative technique to train the mixture-of-experts. As mentioned in mixture of experts section of related works, it is one of the conventional semi-supervised learning approach to train mixture-of-experts. Figure 4 shows the GLCT produces significantly better performance than self-training algorithm. In the beginning of initial iteration, there is quite a few difference in terms of accuracy.

Table 3. Attributes and description of dataset

Attribute	Description (data type)
Age	Age of client
Job	Type of job
Marital	Marital status (categorical: divorced, married, single, unknown)
Education	Education level of client
Default	Has credit in default? (categorical: no, yes, unknown)
Balance	Balance of account
Housing	Has housing loan? (categorical: no, yes, unknown)
Loan	Has personal loan? (categorical: no, yes, unknown)
Contact	Contact communication type
Duration	Last contact duration, in seconds
Campaign	Number of contacts performed during this campaign
Pdays	Number of days that passed by after the client was last contacted
Previous	Number of contacts performed before this campaign
P outcome	Outcome of the previous marketing campaign

Table 4. The number of labeled and unlabeled data according to the ratio

Algorithm	No. of labeled data	No. of unlabeled data
Mixture of local expert	1,708	32,456
Co-training based ME	1,708	32,456
Mixture of local expert	3,416	30,748
Co-training based ME	3,416	30,748
Mixture of local expert	5,124	29,040
Co-training based ME	5,124	29,040

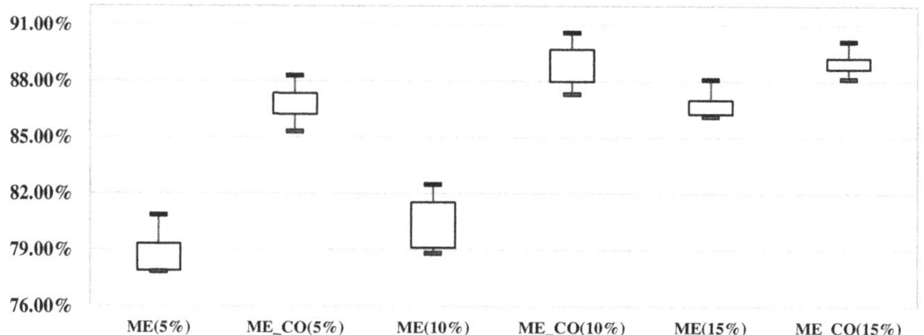

Fig. 3. The comparative results of model performance

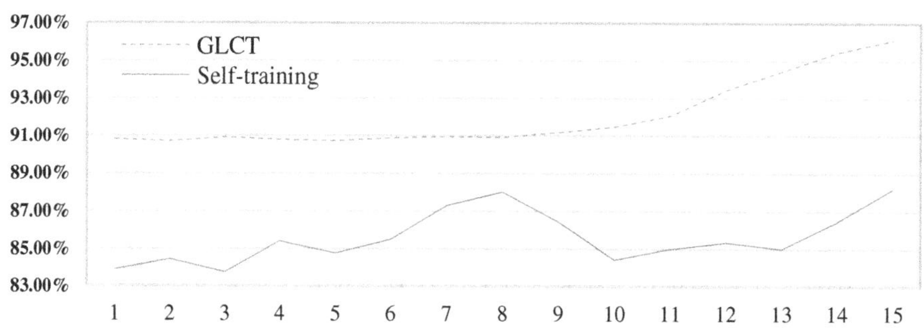

Fig. 4. Change of accuracies in iteration for two methods

Although iterations increased, self-training cannot overtake GLCT's on accuracies. Also, in case of self-training, variation of accuracy of classification is high regardless of increase or decrease. This reveals that absence of global model can degrade the performance of the mixture-of-experts.

5 Conclusion

In this paper, we have proposed a prediction method of bank marketing using the global local co-training with both labeled and unlabeled data. Local model (Mixture-of-experts) is a variant of divide-and-conquer paradigm, and it is proper to solve complicated problem such as bank marketing from unrelated information. Also, one of the characteristics of financial data is that a large-scale unlabeled data is generated from a lot of people in the world. To utilize the unlabeled data, we try to use global-local co-training method for predicting possible future clients in bank telemarketing using real dataset from Portuguese bank. The experiment was based on the ratio of labeled examples. As the result, we confirm the feasibility of this model in financial field. Through the series of iterating experiments, we obtained quantitative results which can be evaluated on various aspects. In the future, we need to validate the accuracy and effectiveness by comparing to other competitive methods.

References

1. Jacobs, R., Jordan, M., Nowlan, S., Hinton, G.: Adaptive mixture local experts. Neural Comput. **3**(4), 79–87 (1991)
2. Duda, R.O., Hart, P.E.: Pattern Classification and Scene Analysis. Wiley, Chichester (1993)
3. Wang, W., Zhou, Z.-H.: Analyzing co-training style algorithms. In: Proceedings of the 18th European Conference on Machine Learning, pp. 454–465 (2007)
4. Blum, A., Mitchell, T.: Combining labeled and unlabeled data with co-training. In: Proceedings of the 11th Annual Conference on Computational Learning Theory, pp. 92–100 (1998)
5. Goldman, S., Zhou, Y.: Enhancing supervised learning with unlabeled data. In: Proceedings of the 17th International Conference on Machine Learning, pp. 327–334 (2000)
6. Kuncheva, L.I.: Clustering-and-selection model for classifier combination. In: Proceedings of the 4th International Conference on Knowledge-Based Intelligent Engineering Systems and Allied Technologies, vol. 1, pp. 185–188 (2000)
7. West, D.: Neural network credit scoring models. Comput. Oper. Res. **27**, 11–12 (2000)
8. Hung, C., Chen, J.H.: A selective ensemble based on expected probabilities for bankruptcy prediction. Expert Syst. Appl. **36**(3), 5297–5303 (2009)
9. Shin, K.S., Han, I.: A case-based approach using inductive indexing for corporate bond rating. Decis. Support Syst. **32**(1), 41–52 (2001)
10. Huang, Z., Chen, H., Hsu, C.J., Chen, W.H., Wu, S.S.: Credit rating analysis with support vector machines and neural networks: a market comparative study. Decis. Support Syst. **37** (4), 543–558 (2004)
11. Li, S.T., Shiue, W., Huang, M.H.: The evaluation of consumer loans using support vector machines. Expert Syst. Appl. **30**(4), 772–782 (2006)
12. Tsai, C.F., Chen, M.L.: Credit rating by hybrid machine learning techniques. Appl. Soft Comput. **10**(2), 374–380 (2010)
13. Milad, M., Vural, A.: Risk assessment in social lending via random forests. Expert Syst. Appl. **42**(10), 4621–4631 (2015)
14. Yoon, J.-W., Cho, S.-B.: Global/local hybrid learning of mixture-of-experts from labeled and unlabeled data. In: Corchado, E., Kurzyński, M., Woźniak, M. (eds.) HAIS 2011, Part I. LNCS, vol. 6678, pp. 452–459. Springer, Heidelberg (2011)

Prioritising Security Tests on Large-Scale and Distributed Software Development Projects by Using Self-organised Maps

Marcos Alvares[1(✉)], Fernando Buarque de Lima Neto[1,2], and Tshilidzi Marwala[1,2]

[1] University of Johannesburg, Johannesburg, South Africa
marcos.alvares@gmail.com
[2] University of Pernambuco, Recife, Brazil

Abstract. Large-scale and distributed software development initiatives demand a systematic testing process in order to prevent failures. Significant amount of resources are usually allocated on testing. Like any development and designing task, testing activities have to be prioritised in order to efficiently validate the produced code. By using source code complexity measurement, Computational Intelligence and Image Processing techniques, this research presents a new approach to prioritise testing efforts on large-scale and distributed software projects. The proposed technique was validated by automatically highlighting sensitive code within the Linux device drivers source code base. Our algorithm was able to classify 3,077 from 35,091 procedures as critical code to be tested. We argue that the approach is general enough to prioritise test tasks of most critical large-scale and distributed developed software such as: Operating Systems, Enterprise Resource Planning and Content Management systems.

1 Introduction

Literature shows that source code complexity impacts the likelihood of software failures [1,3,13,16]. From a Software Engineering perspective, this relation is very plausible once source code complexity is directly related to several aspects of software quality, such as: simplicity, modularity and conciseness.

Although code complexity techniques can be a reliable resource for assisting security analysts to predict vulnerable spots and prioritise testing efforts, is not enough once each development team has its own ability and capacity to tackle complexity. A piece of software can be considered structurally complex but, at the same time, and well maintained by a team that can handle such level of complexity and effectively is skillful enough to avoid programming failures.

As a possible solution to the above mentioned challenge, this research encompasses a new approach to systematically prioritise large-scale software testing efforts not only based on source code information but also considering development team characteristics. Information about source code complexity and issues

A. Hirose et al. (Eds.): ICONIP 2016, Part IV, LNCS 9950, pp. 60–69, 2016.
DOI: 10.1007/978-3-319-46681-1_8

historical data were combined in order to extract information about which development teams should be tested first and increase the likelihood to uncover software issues.

Furthermore, experimental results of a large scale prioritising task at the Linux Device Drivers code base is provided on 257 development teams which were categorised. The development teams clusters were generated by using *Kohonen Maps* (also known as *Self-Organised Maps*) [7–9].

This new approach aims to assist development teams leaders to prioritise tests in order to seek for software vulnerabilities and increase efficiency of *bug hunting* initiatives, specially on large code databases maintained by large communities.

This article is structured in 4 sections. Section 2 presents some relevant references and the necessary background required for fully understanding of the proposed technique. Section 3 describes the proposed approach to clustering development teams and Sect. 4 shows a real test case by using Linux device drivers source code. Finally, Sect. 5 summarises findings, comments about future work and new perspectives.

2 Theoretical Background

This section provides the necessary background knowledge and references to fully understand the proposed technique described within Section

2.1 Cyclomatic Complexity

Cyclomatic complexity is a software metric developed by Thomas J. McCabe in 1976 [11] which quantitatively presents a numeric score for representing the structural complexity in a computer program. It directly measure the number of different possible flows through a source code [4,6,15].

This metric uses the program control flow graph characteristics to generate information about complexity. Nodes represents basic units and edges represents software flows. Each possible path within a control flow graph represents a different possibility of execution.

Cyclomatic complexity can be calculated by the follow equation:

$$F(E, P) = E - N + 2P \tag{1}$$

where E represents the total amount of edges in the graph, N represents the number of nodes and P represents *connected components* [2].

For instance, Fig. 1 represents the control flow of a very simple and hypothetical software.

The complexity of this graph calculated by: 5 − 5 + (2x1) = 2. This number means that this software has 2 possible execution flows.

Arthur H. Watson and Thomas J. McCabe proposed a successful test strategy based on this metric called *Basis Path Testing* [17] which states that each

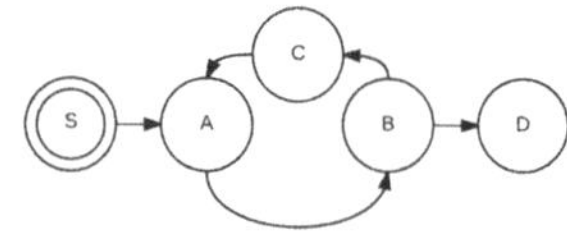

Fig. 1. Example of control flow graph

possible execution path within the program has to have a test case. This metric has been reviewed and successfully applied along the years to analyse source code [10,14].

For this research, *Cyclomatic Complexity* was used to measure the structural complexity of a source code.

2.2 Self-organising Maps

Self-Organising maps (SOM) is a specific type of neural network proposed by Teuvo Kohonen [8]. SOM network is an unsupervised learning technique which produces a low dimensionality map of a high dimensionality input space. This technique is useful for identify similarity patterns in complex data [5,12].

A SOM is composed by a matrix where each node has a two dimensional position (x, y) and a weight vector with the same dimensionality of the inputted data.

Groups of data can be visualised each of them represented by a different colour. Algorithm 1 describes the most popular training algorithm for SOM networks.

Algorithm 1. SOM Training Algorithm

```
Randomise weights for each node
repeat
  Choose a random element inside the input data
  for Each node inside the map do
    Calculate the Euclidean distance between the current node and the selected
    element
  end for
  Select the node with the smallest distance (Best Matching Unit BMU)
  Update all neighbours around the BMU to looks similar to the selected input
  element
until Number of iterations is done
return the trained map
```

3 Contribution

Development teams can be involved collaboratively in the development process of a large-scale software. Hierarchically, each team can be responsible for a section of the code. This section presents an approach to cluster development teams by their similarities including software complexity and issues introduced in order to guide testing efforts.

Usually, large scale software has multiple teams responsible for different functionalities. For instance, a hypothetical modern Operating System can be maintained by several development teams each of them looking after a different functionality, such as: file system modules, device drivers, scheduling, networking and security. This hypothetical scenario suits most open source projects which have distributed development environment.

The data used within experiments described in this article is structured in 5 main entities: team, member, basic unit, *metadata* and issue. Software development teams are responsible for basic units (*e.g.* objects, function or methods) and have team members. Each basic unit has *metadata* information (size, complexity and used types) and can be associated to issues.

In line with that, a first step is to collect information about which team is responsible for which part of the code. These entities and their relationships are going to compose a basic element of a input data together with information about each basic unit. Specific characteristics were targeted for each piece of code which can be collected from any Revision Control System[1], such as:

- Basic units identification (*e.g.* function signature);
- Code size (number of basic units and number of lines);
- Creation date for each file;
- Cyclomatic complexity for each basic unit;

The above mentioned information should be put together with data extracted from documentation and bug management system tools[2], such as:

- Team name: development team identifier;
- Issues found along a previously defined period;
- Number of changes for each issue over each basic unit;

Our aim is to automatically find a group of development teams responsible for low complexity code but which presents a high incidence score of software issues found along a certain period of time. For this task Self-Organised Map was used to clustering and visualising the desired pattern. The outcome of this process is a trained network which can be used to classify new input patterns.

Once all groups are identified, it is necessary to analyse each group properties. Sometimes some groups causes distortions at the generated map, for instance, development teams who are responsible for another teams (and are also indirectly responsible for their code) causes a distortion on the output map. Most of time will be necessary to filter these outliers and re-training the network in order to get an output map which correctly segment the input data set.

The last step is manually determine which group presents similar characteristics to what we are trying to spot. The proposed process can be described as the following steps:

[1] Such as: Git, Subversion, Mercurial and CVS.
[2] Such as: Redmine, Bugzilla, Atlassian Jira and Github.

1. Collect information about development teams;
2. Collect information about which team is responsible for which code;
3. Collect information about each basic unit within the analysed code;
4. Calculate complexity of each basic unit;
5. Define an analysis *Time Window*;
6. Collect all issues found inside this *Time Window*;
7. Create an input data set;
8. Train a SOM network with input data set;
9. Analyse each generated cluster characteristics;
10. Filter "*outliers*" inputs from the resulting map;
11. Repeat the two previous 3 steps n times until you get a map with good "*contrast*";
12. Segment the map and identify groups;
13. Select group which fits to the desired pattern;

The resulting output of the whole process is a set of development teams responsible for maintaining source code with low complexity and have introduced relevant amount of issues along the time specified in the *Time Window* parameter. This result may indicate where testing and training resources should be applied first in order to increase source code reliability and bug hunting initiatives.

4 Experiments

This section validates the proposed process by analysing the entire Linux device drivers source code. Linux is one of the most popular Open Source projects maintained following a distributed development model. The entire Linux project has 1,266 development teams of which 879 are related to device drivers source code base set[3].

This non-centralised development approach makes Linux one of the most portable platforms, used by a significant variety of electronic devices, such as: laptops, televisions, phones, tablets and smart watches. Most of the Linux source code is written in C programming language although it also have small extensions in Assembly Language.

Linux kernel version 3.16.1 was chosen to carry out the experiment described along this section. Table 1 shows specific characteristics for device drivers portion of Linux source code.

An important input data for the proposed method is a set of found issues along a certain amount of time. In line with this requirement, issues detected along the past 5 years were collected from the Linux issue tracking system[4]. According to it, the device drivers source code had 1,058 Bugs registered during the past 5 years. All these issues were manually registered and most of them have a "*revision id*" referencing which code fixes the described programming mistake.

[3] https://www.kernel.org/doc/linux/MAINTAINERS.

[4] https://bugzilla.kernel.org/.

Table 1. Linux device drivers source code characteristics

Characteristic	Value(#)
Teams	879
Maintainers	1,045
Functions	361,668
Files	10,666
Lines of code	8,474,075

A "*revision id*" is an unique identification used by the Version Control System to represent a change into the code. By using these knowledge base systems it was possible to automatically extract all required information for performing the proposed analysis by correlating the above mentioned maintainers file, information extracted from the issue tracker system and the source code by itself. Summarising all information extracted from these systems were:

- Information about C files;
- Information about functions;
- Complexity of functions;
- Information about issues;
- Information about development Teams;
- Relation between teams and files;
- Relation between functions and files;
- Relation between functions and issues;

The data set used as input was assembled by using these informations and used for training the self-organised map. Each row from the input data set represents one development team characteristics composed by three main informations:

- Complexity: represents the sum of the Cyclomatic Complexity for functions;
- Changes: represents the amount of code changes necessary to fix all issues found along a previously specified amount of time;
- Size/Defective: represents the proportion between the total amount of functions and the total amount of functions affected by issues.

From 879 development teams responsible for device drivers only 254 had issues registered during the past 5 years. All data series (Complexity, Changes and Size/Defective) were normalised to values between 0 and 1 in order to optimise the training process.

The used self-organising map was configured to use a 50x50 topology and 0.1 as learning ratio. Figure 2 shows an output map generated by presenting the above described input data set over 200,000 iterations.

It was possible to spot 6 groups. Table 2 shows each identified group and the amount of development teams inside each recognised cluster.

92.5 % of all development teams belong to groups 1 and 3 (blue and black regions). This means that the map does not provide a good "*contrast*" although useful information can be observed, such as:

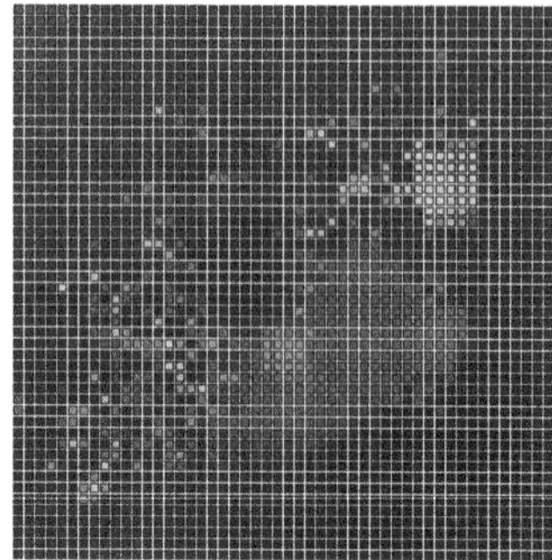

Fig. 2. SOM output map after 200,000 iterations.

Table 2. Development team groups found by the SOM algorithm

Group ID	Number of teams (#)
1	20
2	1
3	215
4	6
5	3
6	7
7	2

1. Groups 2, 5 and 7 are composed by general development teams[5];
2. Group 4 and 6 is composed by teams with small amount of vulnerabilities and high complexity source code.

Groups 2, 4, 5, 6 and 7 do not fit our targeted pattern. These teams have high contrasting features when compared with the rest of the input data set. These small amount of teams are affecting the clustering process. In order to sort out this issue, a second input data set was created containing all development teams from groups 1 and 3. This new input data set was used to generate a new map.

Additionally, the second data set was filtered for development teams responsible for maintaining source code bases smaller or equal than 800 functions. As part of future work, we are planning to manually test code changes produced by the resulting selected development teams. Then, specifically for this experiment, we are interested in a reasonable amount of code suitable for manual tests. After both selective processes, the second data set contains 192 teams to be clustered.

Figure 3 presents the SOM output map for the filtered input data set using the same learning rate and network topology over 200,000 iterations.

We could identify 10 different groups within this new map. This map presents a much more useful image once development teams are more well distributed

[5] Teams that contains other teams, like: Network Drivers or DRM Drivers.

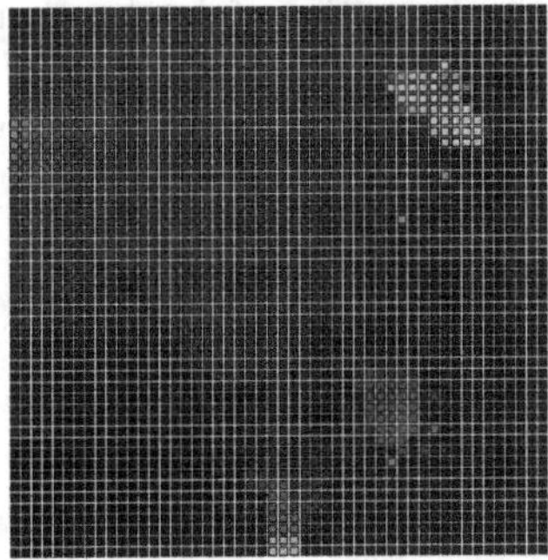

Fig. 3. SOM output map after 200,000 iterations using a filtered input data set.

Table 3. Groups produced by the filtered input data

Group ID	Number of teams (#)
1	41
2	1
3	1
4	7
5	2
6	25
7	103
8	4
9	1
10	7

presented into clusters. Table 3 shows the number of development teams within each identified cluster.

Groups 5 and 4 contain 9 development teams with characteristics close to what we are trying to spot (*e.g.* low complexity, high amount of issues and small code base). Table 4 presents a selected set of development teams and their characteristics. This information spots development teams with higher probability to introduce programming mistake in future.

The proposed approach could spot development teams responsible for popular device drivers (*e.g.* "*8169 10/100/1000 GIGABIT ETHERNET*" and "*AGP-GART*") which issues can have a significant impact over millions of users. Another interesting finding is the "*ASUS NOTEBOOKS AND EEEPC ACPI*" which is distributed within equipment used by social projects of digital inclusion[6].

Issues on such critical software can compromise services hosted in such devices. This results are encouraging as they illustrate how the proposed approach can help to prioritise testing efforts.

6 http://www.asus.com/About_ASUS/Corporate_Social_Responsibility/.

Table 4. Final selection of development teams

Team	Complexity	Changes(#)	Size/defective
8169 10/100/1000 GIGABIT ETHERNET	160	97	5.84
AGPGART	119	59	11.72
MARVELL GIGABIT ETHERNET	100	38	20.76
ASUS NOTEBOOKS AND EEEPC ACPI	137	36	8.9
ACER WMI LAPTOP EXTRAS	78	23	15.71
PCMCIA SUBSYSTEM	71	19	61.08
IPMI SUBSYSTEM	25	14	56.16
HIBERNATION	18	13	60
SUSPEND TO RAM	18	13	60

5 Conclusion

This paper presented a method to prioritise test efforts on large-scale distributed software development projects by using Self-Organised Maps. The proposed technique is based on source code complexity and historical data of programming issues and returns information about human tolerance to complexity instead of metrics purely based on the observed source-code.

Although the proposed technique was successfully tested in an open-source environment, it can be extended and applied to any software collaboratively developed by multiple teams. This is specially relevant for software which presents architecture based on modules or plug-ins.

Experiment were able to automatically select 9 critical development teams (from 254 teams). This new approach can swiftly help communities and companies to optimise resources during testing and validation stage. Another benefit is pointing out development teams that need to be trained in order to reduce the amount of issues in future. The proposed technique can also be used also as part of *bug hunting* initiatives.

References

1. Basili, V., Briand, L., Melo, W.: A validation of object-oriented design metrics as quality indicators. IEEE Trans. Softw. Eng. **22**(10), 751–761 (1996)
2. Bollobás, B.: Modern Graph Theory. Graduate Texts in Mathematics, vol. 184. Springer, Heidelberg (1998)
3. Cataldo, M., de Souza, C.: Exploring the impact of API complexity on failure-proneness. In: 9th International Conference on Global Software Engineering (2014)
4. Dibble, C., Gestwicki, P.: Refactoring code to increase readability and maintainability: a case study. J. Comput. Sci. Coll. **30**(1), 41–51 (2014)
5. Hammami, I., Mercier, G., Hamouda, A.: The Kohonen map for credal classification of large multispectral images. In: 2014 IEEE Geoscience and Remote Sensing Symposium, pp. 3706–3709. IEEE, July 2014

6. Henderson-Sellers, B., Tegarden, D.: A critical re-examination of cyclomatic complexity measures. In: Lee, M., Barta, B.-Z., Juliff, P. (eds.) Software Quality and Productivity. IFIP Advances in Information and Communication Technology, pp. 328–335. Springer, Heidelberg (1994)
7. Kohonen, T.: Self-organized formation of topologically correct feature maps. Biol. Cybern. **43**, 59–69 (1982)
8. Kohonen, T.: The self-organizing map. Proc. IEEE **78**(9), 1464–1480 (1990)
9. Lihong, M., Mingguang, W., Jun, J.: Joint investigation of cases using self-organized map network. In: 2011 International Conference on Electronics, Communications and Control (ICECC), pp. 1520–1523. IEEE, September 2011
10. Mccabe, T.: Cyclomatic complexity and the year 2000. IEEE Softw. **13**(3), 115–117 (1996)
11. McCabe, T.: A complexity measure. IEEE Trans. Softw. Eng. **4**, 308–320 (1976)
12. Mota, R.L.M., Shiguemori, E.H., Ramos, A.C.B.: Application of self-organizing maps at change detection in Amazon forest. In: 11th International Conference on Information Technology: New Generations, pp. 371–376. IEEE, April 2014
13. Nagappan, N., Ball, T., Zeller, A.: Mining metrics to predict component failures. In: 28th International Conference on Software Engineering, pp. 452–461 (2006)
14. Sarwar, S., Muhammd, M.: Cyclomatic complexity: the nesting problem. In: 2013 Eighth International Conference on Digital Information Management (ICDIM) (2013)
15. Shepperd, M.: A critique of cyclomatic complexity as a software metric. Softw. Eng. J. **3**(2), 30 (1988)
16. Viega, J., McGraw, G.: Building Secure Software: How to Avoid Security Problems the Right Way. Addison-Wesley Press, Melbourne (2001)
17. Watson, A., McCabe, T., Wallace, D.: Structured testing: a testing methodology using the cyclomatic complexity metric. NIST Spec. Publ. **500**(235), 1–114 (1996)

Android Malware Detection Method Based on Function Call Graphs

Yuxin Ding(✉), Siyi Zhu, and Xiaoling Xia

Key Laboratory of Network Oriented Intelligent Computation,
Department of Computer Sciences and Technology,
Harbin Institute of Technology Shenzhen Graduate School, Shenzhen, China
yxding@hitsz.edu.cn, {szhu_hitsz,xxia_hitsz}@sina.com

Abstract. With the rapid development of mobile Internet, mobile devices have been widely used in people's daily life, which has made mobile platforms a prime target for malware attack. In this paper we study on Android malware detection method. We propose the method how to extract the structural features of android application from its function call graph, and then use the structure features to build classifier to classify malware. The experiment results show that structural features can effectively improve the performance of malware detection methods .

Keywords: Android malware · Static detection · Machine learning

1 Introduction

With the rapid development of mobile internet, mobile devices have become unnecessary tools for people communication. A recent report from IDC shows that there are 470 millions of smartphones sold in 2015 in China, and more than eighty percent of these devices employ Android platform. Unfortunately, such popularity also attracts the malware developers, which has made mobile platforms a prime target for attack.

The Android market is the fastest growing mobile application platform. However, different from Apple's App store which may check the security of each available application manually by software security experts, there is lack of app security checking process before the applications are published on the Android market. The openness of the Android platform attracts both benign and malicious developers [1]. Furthermore, Android allows installing third-party applications that may increase the spread of Android malware. In this paper we study on Android malware detection method. We extract the structural features of android app from function call graphs, and use machine learning method to detect malware of Android platform.

2 Related Work

In our study, we focus on the static malware detection method. Static methods detect malware by analyzing the contents of each applications. The static analysis mechanism can reduce the cost and improve the performance. In the previous works, the following features are extracted to detect Android malware:

A. Hirose et al. (Eds.): ICONIP 2016, Part IV, LNCS 9950, pp. 70–77, 2016.
DOI: 10.1007/978-3-319-46681-1_9

Requested permission: In [2–4], researchers have developed several permission based approaches to detect whether the application is potentially malicious. In [2], the authors propose that Android malware often request a certain permission or a certain combination of two or three permissions. In [3] the authors propose that Android malware often requests a critical permission that is rarely requested. In [4], four probabilistic generative models are used to detect potentially Android malware.

Imported package: Zhou et al. [6] use package information imported by applications to detect Android malware.

API calls: Kim et al. [5] and Zhou et al. [6] also consider API calls to analyze the Android applications.

The above mentioned works employ the permissions or the usage of specific API functions to detect malware. These features are easy to circumvent, for example using kernel-based exploits or API-level rewriting [7]. Graph based features offer a robust representation for malware and have been successfully used to detect malware targeting Windows systems. Inspired by graph based detection methods, in this study we try to use the structural features of android app from function call graphs to detect malware of Android platform.

3 Extraction of Structural Features of Android Apps

Permission features and Android API features have been widely used to detect malware on Android platform. However, these features are separated features, which cannot represent the dependency relations between features. Therefore, the methods based on these features are not robust to fight against malware obfuscation technologies. To solve this issue, we use function graph to represent the behaviors of an Android application. We use the static analysis tool Androguard to generate the function call graph. The function call graph is a directed graph in which each node represents a function call, and an edge from a node v_1 to a node v_2 indicates a call from the function represented by v_1 to the function represented by v_2. Figure 1 shows an example of the function call graph of an Android application. A function call is represented as the combination of Class Name, Method Name, and Descriptor.

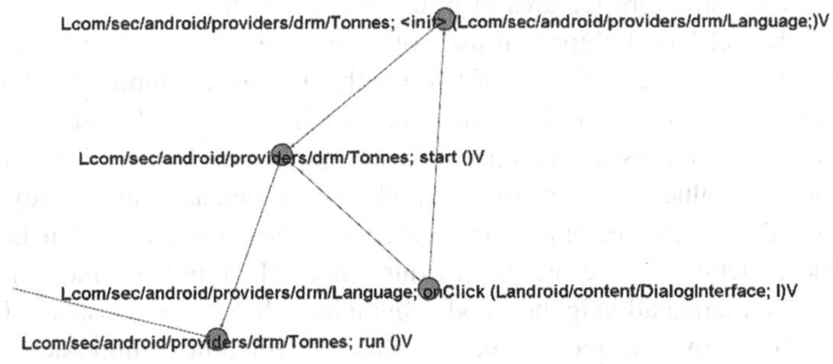

Fig. 1. Example of a function call graph

Android platform provides thousands of function calls. Analyzing the relations of all possible functions call is a labor-intensive work, in addition, most function calls are common functions which cannot bring security problems. In our work we choose 11 class packages as sensitive class packages, and we only consider function calls in sensitive class packages. The sensitive class packages we choose are as follows: android.accounts, android.app, android.bluetooth, android.content, android.location, android.media, android.net, android.nfc, android.provider, android.telecom, and android.telephony. These 11 packages cover the most sensitive resources, such as Bluetooth, user's location information, Internet network, SMS short message, telephone number and so on.

After obtaining the function call graph, we extract the feature of each node. The feature of each node is encoded as a bit vector of length 11, which is represented as $l(v_i)$. In a bit vector, each bit represents a sensitive class package. Table 1 gives an example of a feature vector of a function. The fourth bit represents the android.content package. The value of the fourth bit is one, which means the node calls one or several functions contained in android.content package. In this way, we assign a node of the graph a node feature vector.

Table 1. A feature vector of a function

Bit	1	2	3	4	5	6	7	8	9	10	11
Value	0	0	0	1	0	0	0	0	0	0	0

Next, we construct the structure feature of the function call graph. Our goal is to calculate the similarity of two function call graphs using structure feature. In general, we can measure the similarity of two graphs using graph isomorphism algorithm, subgraph isomorphism algorithm or maximum common subgraph algorithm. However, it has been proved that graph isomorphism problem and maximum common subgraph problem are NP complete problem. It is hard to find an efficiency algorithm which can be applied in our problem. The papers [8–11] proposed the graph kernel based algorithm to measure the similarity of two graphs. The graph kernel based algorithm is a high efficiency algorithm for graph matching, which can run in time linear in the number of nodes. In our work we use the neighborhood hash graph kernel proposed in [8] to construct the structure features of a function call graph.

The graph kernel based algorithm use both the node information and topological information to measure the similarity of two graphs. It uses a compression function to encode the two types of information into one vector called as the structure feature vector. In fact, the compression function is a decomposition kernel. The kernel generates a single hash value (a vector) for each node (also called as central node), which is calculated over the feature vectors of the neighbors of the central node. The hash value represents the structure feature around a central node. In this way, the graph based algorithm can enumerate all neighborhood subgraphs in linear time, instead of running an isomorphism test over all pairs of neighborhoods. Different compression functions have been proposed to encode the structure features of a graph. In our work we choose the compression function $f(v)$ proposed in [8], which is shown as Eq. (1).

In Eq. (1), v represents a central node, v_i is its neighbor node, $l(v_i)$ represents the bit vector of node v_i. $\oplus$ represents a bit-wise XOR on the binary vectors and *ROT* denotes a single-bit rotation to the left. The computation can be finished in constant time for each node, and the computation complexity is $\theta(\text{md})$ where d is the maximum out-degree and m is the length of the bit vector. $f(v)$ is also a bit vector which has the same length as $l(v_i)$. $f(v)$ represents the structure feature of node v. $F(G) = \{f(v_1), \ldots, f(v_n)\}$ denotes the structure feature set of graph G.

$$f(v) = ROT_1(l(v)) \oplus (l(v_1^{adj}) \oplus \ldots, \oplus l(v_n^{adj})) \quad (1)$$

To evaluate the similarity of two function call graphs, G_1 and G_2, a neighborhood hash graph kernel function is designed to evaluate the count of common identical substructures in two graphs. Usually the kernel value shows the size of the intersection of the structure feature sets $F(G_1)$ and $F(G_2)$. The neighborhood hash graph kernel function used in our work is shown in Eq. (2), where $\delta(f(v_1), f(v_2))$ is 1 when $f(v_1)$ is equal to $f(v_2)$, otherwise 0.

$$k(G_1, G_2) = \sum_{v_1 \in G_1} \sum_{v_2 \in G_2} \delta(f(v_1), f(v_2)) \quad (2)$$

4 Malicious Code Detection Model

In the detection model, we create three feature extraction modules which are used to extract different features. Our goal is to evaluate the performance of different features for malware detection. We extract three kinds of features: permission feature, function call feature and structure feature of function call graphs.

We decompile APK file and extract permissions used by an Android application from the configure file AndroidManifest.xml. An Android application sample is represented as a permission vector in which each dimension represents a permission.

To obtain the API calls, we first obtain the smali file (a Dalvik byte code file) by decompiling an APK file. By searching the byte code instruction "invoke", we can find all API calls used by an android application. Due to the large number of API calls, we need to choose API calls having strong classification ability as the features. We build a large dataset which contains 2000 benign samples and 2000 malicious samples. We choose 1200 API calls using the feature selection method based on entropy [12]. Finally, an Android sample is represented as an API vector with 1200 dimensions.

After obtaining the feature vectors of sample, we can use machine learning methods to build detection models. For permission features and API call features, we use the machine learning algorithms, SVM, decision tree, KNN and random forest to build the malware classification models, respectively. For the structure features, we use the KNN algorithm to build the classification model.

The architecture of the detection model is shown in Fig. 2.

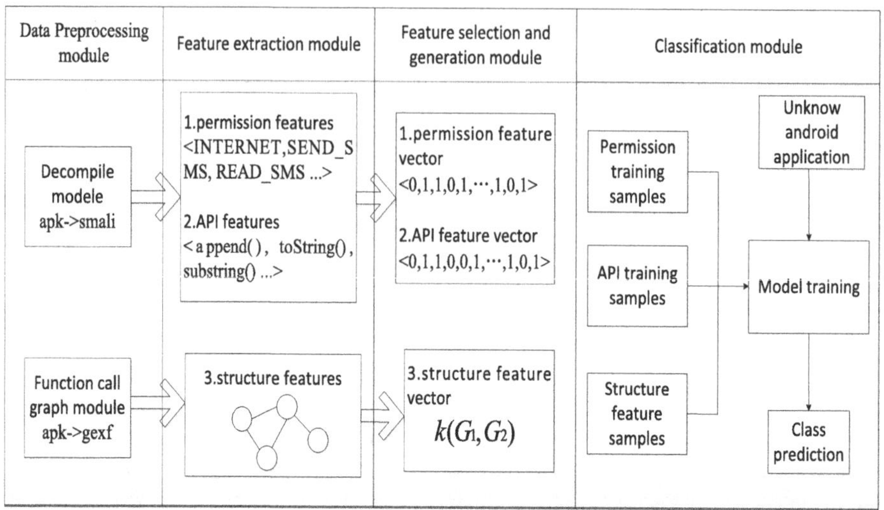

Fig. 2. System architecture

5 Experiments

5.1 Data Set Description

The experimental data consists of 3226 benign and 2149 malicious android applications. The malicious applications come from the Genome project [13] which provides a benchmark malware dataset. The benign applications come from both the official Google Play store as well as popular third-party markets. To demonstrate the efficiency of the structure features, we also extract the permission features and API features, and compare the performance of the classifiers trained using different features. The data partition is shown in Table 2.

Table 2. Experimental dataset description

	Training samples	Testing samples
Benign files	1613	1613
Malicious files	1074	1074

The performance of the classifiers is evaluated by three criteria: accuracy, the false negative rate (FNR), and the false positive rate (FPR). We define the following measurements. In our work we repeat each experiment five times, and each time we randomly select training and testing samples shown as Table 2. The average accuracy, average FPR and average FNR are calculated to evaluate the final performance of the proposed model.

The true positives (TP) value is the number of malicious files classified as malicious files.
The true negatives (TN) value is the number of benign files classified as benign files.
The false positives (FP) value is the number of benign files classified as malicious files.
The false negatives (FN) value is the number of malicious files classified as benign files.
Accuracy = (TP + TN)/(TP + TN + FP + FN)
FPR = FP/(TN + FP)
FNR = FN/(TP + FN)

5.2 Experimental Results

Results for detection methods based on permission features:

To select useful permission features, we build a dataset which contains 2000 benign samples and 2000 malicious samples. We test different number of permissions as features according to the feature selection methods proposed in [12]. Finally, we select 80 permissions as features of android applications. The experimental results for permission features are shown in Table 3.

From Table 3 we can see the four machine learning algorithms have the similar performance. The accuracy of the four algorithms is greater than 92.3 % and lower than 93.8. Compared to other algorithms, the random forest algorithm has the better performance, which has a high accuracy, a low false positive rate and false negative rate.

Table 3. Experimenatal results for permission features

	Accuracy	FNR	FPR
SVM	92.4 %	4.8 %	8.6 %
Random forest	93.8 %	2.8 %	8.3 %
Decision tree	93.1 %	3.7 %	8.8 %
KNN	92.8 %	4.1 %	8.2 %

Results for detection methods based on Android API features:

To select useful API features for classification, we also use the above mentioned dataset for API feature selection. The feature selection method is the same as the feature selection method in [12]. We test different number of API features. Finally, we select 1200 APIs as features of android applications. The experimental results for permission features are shown in Table 4.

From Table 4 we can see the four machine learning algorithms also have the similar performance. The accuracy of the four algorithms is greater than 94.0 % and lower than 95.4. Compared to other algorithms, the random forest algorithm has the better performance, which has a high accuracy, a low false positive rate and false negative rate.

Table 4. Experimenatal results for API features

	Accuracy	FNR	FPR
SVM	94.1 %	2.5 %	8.0 %
Random forest	95.5 %	1.3 %	6.6 %
Decision tree	94.6 %	2.1 %	7.5 %
KNN	94.1 %	2.1 %	8.1 %

From Tables 3 and 4, we can see the performance of API features is better than that of the permission features. In average the accuracy of the detection methods based on API features is about 1.5 % higher than that of the methods based on permission features, and the false positive rate of the API based methods is about 1.5 % lower that of the permission based methods.

Results for detection methods based on structure features:

We use the K-nearest neighbor algorithm to build the detection model based on structure features. In our experiments, we set the parameter K as 3, 7, and 11, respectively. The distance between a testing sample and each training sample is defined as Eq. (2) which represents the similarity between two graphs. The experimental results are shown in Table 5. From Table 5, we can see we get the best performance when K is equal to seven.

From Tables 3, 4, and 5, we can see the whole performance of the method based on structure features is better than the permission based methods and the API based methods. In average the accuracy of the structure based method is about 2.3 and 3.8 % higher than that of the API based methods and the permission based methods, respectively, and the false positive rate is about 2.8 and 4.3 lower than that of the API based methods and the permission based methods, respectively.

Table 5. Experimenatal results for structure features

	Accuracy	FNR	FPR
KNN (K = 3)	96.3 %	0.84 %	5.4 %
KNN (K = 7)	97.5 %	0.84 %	3.7 %
KNN (K = 11)	96.8 %	0.8 %	4.3 %

6 Conclusion

Malware often attack a target by executing API calls. So we can detect malware by evaluating the similarity of API calls of applications. In this paper, a feature extraction method based on function call graph is proposed. In this method, the structure information of function calls is extracted and is used to find similar API call subgraph between known malicious code and the detected sample. The experimental results that structure features of API calls are efficiency for malware detection. Compared with the permission features and API features, structure feature based malware detection methods have a high detection accuracy and low false detection rate.

Acknowledgment. This work was partially supported by Scientific Research Foundation in Shenzhen (Grant No. JCYJ20140627163809422, JCYJ20160525163756635), Guangdong Natural Science Foundation (Grant No. 2016A030313664) and Key Laboratory of Network Oriented Intelligent Computation (Shenzhen).

References

1. Enck, W., Octeau, D., Mcdaniel, P.: A study of android application security. In: Proceedings of the 20th Usenix Security Conference, p. 21 (2011)
2. Enck, W., Ongtang, M., McDaniel, P.: On lightweight mobile phone application certification. In: Proceedings of 16th ACM Conference on Computer and Communications Security, pp. 235–245 (2009)
3. Sarma, B.P., Li, N., Gates, C., Potharaju, R., Nita-Rotaru, C., Molloy, I.: Android permissions: a perspective combining risks and benefits. In: Proceedings of 17th ACM Symposium Access Control Models Technologies, pp. 13–22 (2012)
4. Peng, H., Gates, C., Sarma, B., Li, N., Qi, Y., Potharaju, R., Nita-Rotaru, C., Molloy, I.: Using probabilistic generative models for ranking risks of Android apps. In: Proceedings of ACM Conference on Computer and Communications Security, pp. 241–252 (2012)
5. Kim, S., Cho, J.I., Myeong, H.W., Lee, D.H.: A study on static analysis model of mobile application for privacy protection. In: Park, J.J., Chao, H.-C., Obaidat, M.S., Kim, J. (eds.) Computer Science and Convergence. Lecture Notes in Electrical Engineering, vol. 114, pp. 529–540. Springer, Heidelberg (2012)
6. Zhou, Y., Wang, Z., Zhou, W., Jiang, X.: Hey, you, get off of my market: detecting malicious apps in official and alternative Android markets. In: Proceedings of the 19th Annual Network & Distributed System Security Symposium, February 2012
7. Hao, H., Singh, V., Du, W.: On the effectiveness of api-level access control using bytecode rewriting in android. In: Proceedings of the ACM SIGSAC Symposium on Information, Computer and Communications Security (2013)
8. Kashima, H.: A linear-time graph kernel. In: Proceedings of the Ninth IEEE International Conference on Data Mining, pp. 179–188 (2009)
9. Gärtner, T., Lloyd, J.W., Flach, P.A.: Kernels and distances for structured data. Mach. Learn. **57**(3), 205–232 (2004)
10. Schölkopf, B., Platt, J., Hofmann, T.: Fast computation of graph kernels. In: Advances in Neural Information Processing Systems 19 NIPS, pp. 1449–1456 (2006, 2007)
11. Shervashidze, N., Schweitzer, P., Leeuwen, E.J.V., et al.: Weisfeiler-Lehman graph kernels. J. Mach. Learn. Res. **1**(3), 1–48 (2011)
12. Ding, Y., Yuan, X., Zhou, D., Dong, L., An, Z.: Feature representation and selection in malicious code detection methods based on static system calls. Comput. Secur. **30**, 514–524 (2011)
13. Malware Geome Project. http://www.malgenomeproject.org

Proposal of Singular-Unit Restoration by Focusing on the Spatial Continuity of Topographical Statistics in Spectral Domain

Kazuhide Ichikawa and Akira Hirose(✉)

Department of Electrical and Electronic Engineering,
The University of Tokyo, 7-3-1 Hongo, Bunkyo-ku, Tokyo 113-8654, Japan
k_ichikawa@eis.t.u-tokyo.ac.jp, ahirose@ee.t.u-tokyo.ac.jp
http://www.eis.t.u-tokyo.ac.jp/

Abstract. An interferogram which interferometric synthetic aperture radar (InSAR) acquires includes singular points (SPs), which cause an unwrapping error. It is very important to remove the SP. We propose a filtering technique in order to eliminate the distortion around a SP. In this proposed filter, a complex-valued neural network (CVNN) learns the continuous changes of topographical statistics in the spectral domain. CVNN predicts the spectrum around a singular unit (SU), i.e., the four pixels constituting a SP, to restore the SU. The proposed method is so effective in the removal of the distortion at the SU that it allows us to generate a highly accurate digital elevation model (DEM).

Keywords: Interferometric synthetic aperture radar · Singular point · Complex-valued neural network · Spectral domain

1 Introduction

Synthetic Aperture Radar (SAR) is a radar which generates high resolution images regardless of the pysical size of its antenna. SAR observes various targets such as terrain, disasters and glacier by actively radiating radio waves to receive scattered waves day and right by penetrating clouds weather. SAR has been investigated because it can obtain extensive information which optical sensors cannot acquire. Microwave has transmission and scattering characteristics different from those of visible and infrared light.

We make a study of interferometric SAR (InSAR) which focuses on the phase information. InSAR observes a target from two places. Obtained two results are called "master" and "slave." We make a SAR interferogram by multiplying master and conjugated slave. Each pixel of the interferogram has a wrapped phase value in the range of $(-\pi, \pi]$. It is possible to generate digital elevation model (DEM) by unwrapping this wrapped phase. The unwrapping process is to integrate the phase difference between neighboring pixels. Since a height map is conservative, the unwrapping can be easily conducted.

A. Hirose et al. (Eds.): ICONIP 2016, Part IV, LNCS 9950, pp. 78–85, 2016.
DOI: 10.1007/978-3-319-46681-1_10

However, an actual interferogram has massive rotation points, which is called singular points (SPs). The SPs cause an unwrapping error. We cannot uniquely decide elevation values when SPs exist so that the reliability of the generated DEM should be low. In order to solve this problem, various unwrapping methods [1,3,4,11,12,14–16] and local co-registration techniques [2,9,10] as well as many filtering methods [5,8,13,17] were proposed. These filters are effective in the removal of the SPs to a certain extent, but their process often smears dense fringes and/or faint features.

SPs originates not from the distortion of one pixel but from that of a set of neighboring pixels. The minimum unit constituting an SP is a set of four pixels, and we named the pixels a "singular unit (SU)." We have aimed to establish a more effective filtering method by restoring the distortion of this unit.

In this paper, we propose a filtering technique in the spectral domain which represents the topographic information. We have complex-valued neural network (CVNN) learn the continuous changes of topographical statistics to predict the spectrum that does not include the singularity, resulting in the restoration of the SU. We confirm that it is possible to preserve detailed fringes, which are smeared by conventional filters, to eliminate SPs, and to generate more accurate DEM by using the proposed complex-valued neural-network-based filtering technique in spectral domain.

2 SU Restoration by Using Compex-Valued Neural Network

2.1 Principle of the Proposed Filter

In this section, we propose a filtering method to restore a SU. A CVNN learns the changes in topographical statistics in the spectral domain. It predicts an appropriate local-window spectrum without SP at the SU position by paying attention to its continuous changes when the local window approaches the SU.

The phase value acquired from InSAR should continuously change because landscape is a conservative field. However, the distortion at SU arises under the influence of the interference in a single pixel and/or noise. We calculate ideal values which is free from distortion nature by using proper topographic information around SU. We have CVNN learn the continuous change of topographical statistics and predict the ideal spectrum. We inversely transform the ideal spectrum to real space to restore the SU.

Figure 1 shows the processing flow of the proposed filtering method. We call this filter CVNN-prediction filter. First of all, we detect an SU in a phase interferogram to set an $L \times L$-pixel window, at the center of which the SU exists. We prepare a set of $M \times M$-sized N local-window spectra $\boldsymbol{Z}_K \equiv [z_K(u,v), z_{K-1}(u,v), \cdots, z_{K-N+2}(u,v)]^T$ for local windows aligned upward away from the SU window by K pixels, $K-1$ pixels,$\cdots$, and $K-N+2$ pixels, respectively. The spectra are used as teacher-input signals. The output teacher signal is the next spectrum $z_{k-N+1}(u,v)$. Then we shift the teacher input

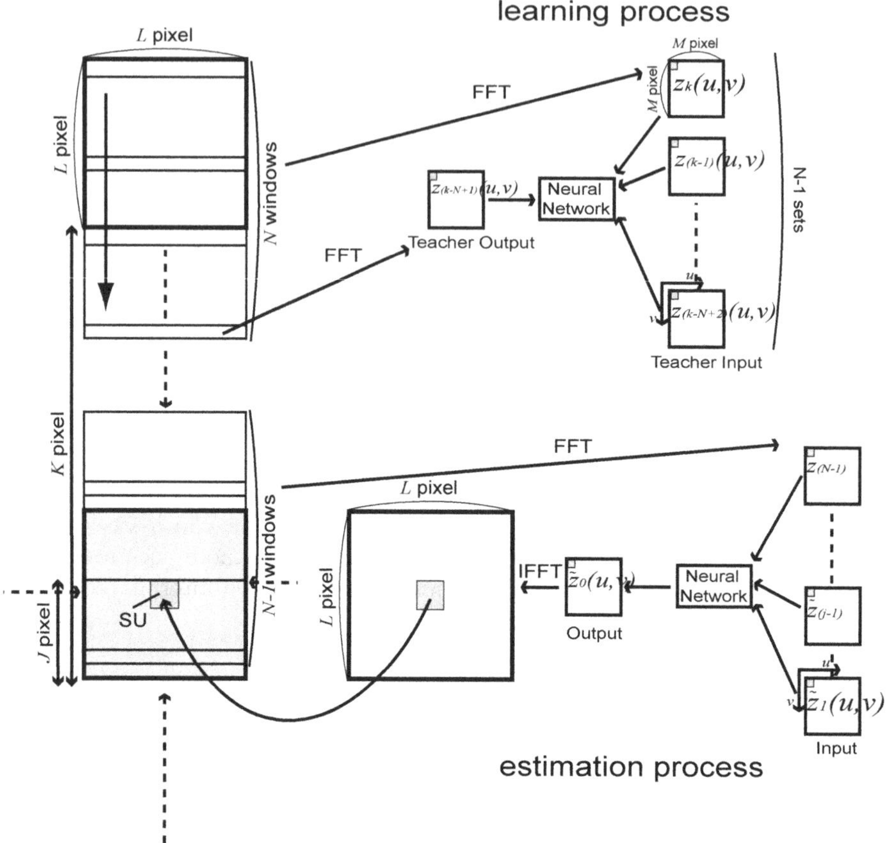

Fig. 1. The processing flow in CVNN-prediction filter.

and output spectra by 1 pixel downward to obtain $\boldsymbol{Z}_{K-1}$ to conduct the next learning step. A step may include multiple weight updates. We repeat this shift and learning process until the bottom of the last window in the set comes down to the position J-pixels upward away from the SU.

The estimation process starts after the learning process is finished. First, we input $\boldsymbol{Z}_{J+N-2} \equiv [z_{J+N-2}(u,v), z_{J+N-3}(u,v), \cdots, z_J(u,v)]^T$ to the learned CVNN to estimate $\tilde{z}_{J-1}(u,v)$. Next, using the estimated value, we input $\boldsymbol{Z}_{J+N-3} \equiv [z_{J+N-3}(u,v), z_{J+N-4}(u,v), \cdots, \tilde{z}_{(J-1)}(u,v)]^T$ to the CVNN and calculate $\tilde{z}_{J-2}(u,v)$. This series of operations is continued until $\tilde{z}_0(u,v)$ is predicted. We also predict $\tilde{z}_0(u,v)$ from the data right-hand side, left-hand side and the bottom by calculating the neighboring values in the same way. Then, we average the four predicted spectra to obtain the final predicted spectrum. The corresponding $L \times L$-pixel signal in the real space is computed by two-dimentional inverse Fourier transform. Finally, we restore the SU by replacing the pixel values of the SU by the central four pixel values. Many adjacent pairs of positive

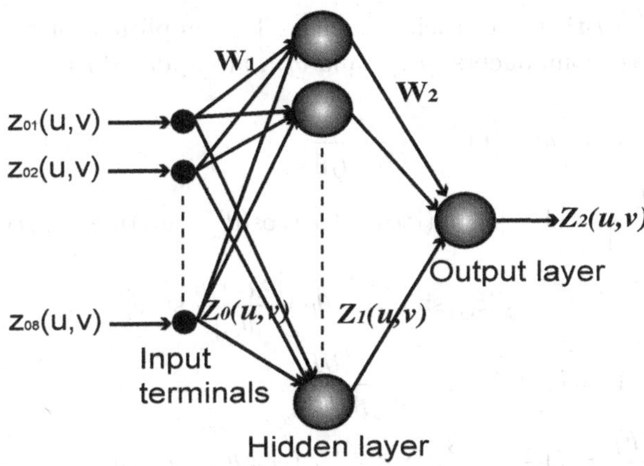

Fig. 2. Layered complex-valued neural network.

and negative SPs approach each other to disappear in their combination when a series of the operations is conducted to all of the SUs in the interferogram.

In this proposed CVNN-prediction filter, repetition of update in the learning process requires a large calculation cost. Then we focus on the continuous transition of the topographic information. That is, by stopping the update of weights at certain times in a window set, we have CVNN advance to the next learning window set regardless of whether the convergence is completed or not. Such learning process enables us to reduce the calculation cost and to predict the phase statistics that change continuously.

2.2 Layered Complex-Valued Neural Network

In this section, we describe the construction and the dynamics of CVNN used in the proposed method. Figure 2 shows the structure of the CVNN, which has input terminals, a hidden layer and an output layer. The connection weights w_{lji} between i-th input and j-th neuron in the l-th layer can be described with the amplitude and phase components, $|w_{lji}|$ and θ_{lji}. The internal state of each neuron θ_{lji} is expressed as

$$u_{lj} \equiv \sum_i w_{lji} z_{(l-1)i} = \sum_i |w_{lji}||z_{(l-1)i}| e^{j(\theta_{lji} + \theta_{(l-1)i})} \tag{1}$$

where z_{lj} denotes output signal of j-th neuron in l-th layer, and is generated by a nonlinear function as [7]

$$z_{lj} = f_{\mathrm{ap}}(u_{lj}) = \tanh(|u_{lj}|) e^{j \arg(u_{lj})} \tag{2}$$

We update the connection weights by using the steepest descent method to minimize an error function $E_l \equiv |\boldsymbol{z}_l - \hat{\boldsymbol{z}}_l|^2/2$ where $\boldsymbol{z}_l$ is the output signals and

$\hat{\boldsymbol{z}}_l$ stands for the l-th layer teacher signals. The amplitude of the connection weights and phase components, $|w_{lji}|$ and θ_{lji}, are updated as

$$|w_{lji}|(r+1) = |w_{lji}|(r) - K_1 \frac{\partial E_l}{\partial |w_{lji}|} \tag{3}$$

$$\begin{aligned}\frac{\partial E_l}{\partial |w_{lji}|} &= (1-|z_{lj}|^2)(|z_{lj}| - |\hat{z}_{lj}|\cos(\theta_{lj} - \hat{\theta}_{lj}))|z_{(l-1)i}|\cos\theta_{lji}^{\mathrm{rot}} \\ &\quad -|z_{lj}||\hat{z}_{lj}|\sin(\theta_{lj} - \hat{\theta}_{lj})\frac{|z_{(l-1)i}|}{|u_{lj}|}\sin\theta_{lji}^{\mathrm{rot}}\end{aligned} \tag{4}$$

$$\theta_{lji}(r+1) = \theta_{lji}(r) - K_2 \frac{1}{|w_{lji}|}\frac{\partial E_l}{\partial \theta_{lji}} \tag{5}$$

$$\begin{aligned}\frac{1}{|w_{lji}|}\frac{\partial E_l}{\partial \theta_{lji}} &= (1-|z_{lj}|^2)(|z_{lj}| - |\hat{z}_{lj}|\cos(\theta_{lj} - \hat{\theta}_{lj}))|z_{(l-1)i}|\sin\theta_{lji}^{\mathrm{rot}} \\ &\quad +|z_{lj}||\hat{z}_{lj}|\sin(\theta_{lj} - \hat{\theta}_{lj})\frac{|z_{(l-1)i}|}{|u_{lj}|}\cos\theta_{lji}^{\mathrm{rot}}\end{aligned} \tag{6}$$

where r represents the number of the weight updating, $\theta_{lji}^{\mathrm{rot}} \equiv \theta_{lj} - \theta_{(l-1)i} - \theta_{lji}$, and K_1 and K_2 are the constants that determine the learning speed. The teacher signals in the hidden layer $\hat{\boldsymbol{z}}_1$ are obtained through the backpropagation of the teacher signal $\hat{z}_2$.

3 Experimental Results

We compare the proposed CVNN-prediction filter with conventional filters, namely, Goldstein-Werner (GW) filter [6] and CMRF filter [17]. The experiment is conducted for the interferogram of Mt.Fuji observed by ALOS-2 (Advanced Land Observing Satellite-2) of Japan Aerospace Exploration Agency (JAXA). Figure 3 shows the interferogram of the skirts of Mt. Fuji, the size of which is 256 × 256 pixel. Figure 3(a) shows the raw phase interferogram, while Fig. 3(d) presents the interferogram filtered with GW filter, Fig. 3(g) with CMRF filter and Fig. 3(j) with CVNN-prediction filter. Figure 3(b), (e), (h) and (k) represent distributions of singular points included in (a), (d), (g) and (j), respectively.

Figure 3(d) looks smooth phase, but the fringes are smeared and the fine topographic information is lost. The density of the SP remains high in the region near the summit of Mt. Fuji (e.g. (20, 220)). Compared with Fig. 3(d), (g) has the fine fringes but the number of the SPs remains large in the summit. On the contrary, in Fig. 3(j), the fine information is preserved and the density of the SP becomes much lower than that of the conventional filters. The number of the SPs are 1,952 in Fig. 3(b), 329 in Fig. 3(e), 297 in Fig. 3(h), 71 in Fig. 3(k). The proposed filter is more effective than the two conventional filters at least in the viewpoint of the SP removal.

In order to evaluate the performance of the proposed filter, we also generate DEMs from the phase interferograms with unwrapping by using the minimum cost network flow method [1]. Figure 3(c) shows real terrain, and Fig. 3(f), (i)

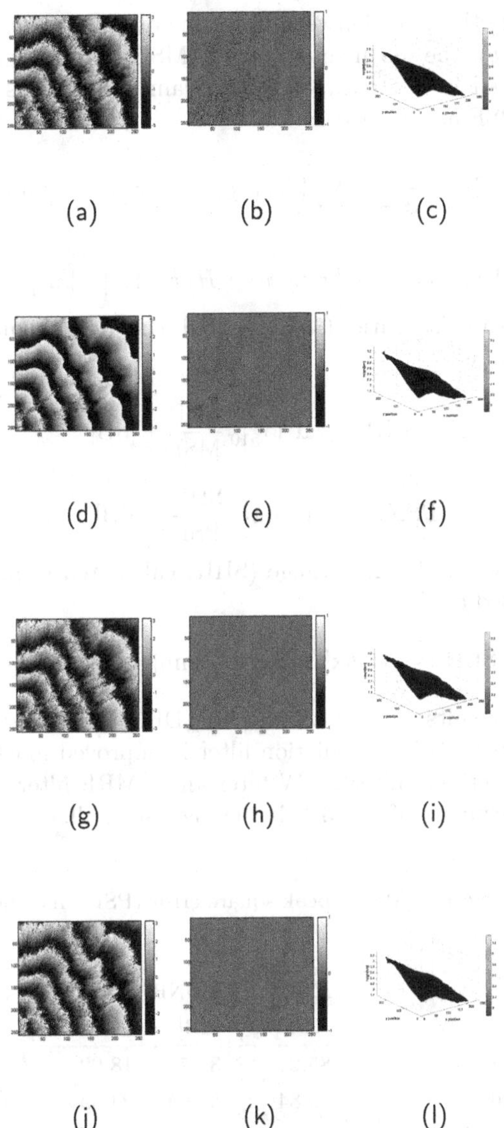

Fig. 3. (a) Raw interferogram, (b) its distribution of SPs, (c) real terrain, (d) interferogram filtered with GW filter, (e) its distribution of SPs, (f) DEM generated from (d), (g) interferogram filtered with CMRF filter, (h) its distribution of SPs, (i) DEM generated from (g), (j) interferogram filtered with CVNN prediction filter, (k) its distribution of SPs and (l) DEM generated from (j).

and (l) show DEMs generated from Fig. 3(d), (g) and (j), respectively. We apply the conical weighted averaging filter, the radius of which is 5 pixel, to the DEMs after unwrapping. We evaluate these DEMs except for the unprocessed edges.

In order to evaluate the error between the value of DEM, $H(x, y)$, and the real terrain data, $\hat{H}(x, y)$, the mean square error (MSE) and the peak square error (PSE) are calculated in the entire interferogram, where S is the area of the image. MSE and PSE are defined as

$$\mathrm{MSE} = \frac{1}{S}\sum_{x}\sum_{y}\{\hat{H}(x,y) - H(x,y)\}^2 \quad [\mathrm{m}^2] \tag{7}$$

$$\mathrm{PSE} = \max[\{\hat{H}(x,y) - H(x,y)\}^2] \quad [\mathrm{m}^2] \tag{8}$$

The mean signal-to-noise ratio (MSNR) and the peak signal-to-noise ratio (PSNR) are defined as

$$\mathrm{MSNR} = 10\log_{10}\frac{\mathrm{SHR}}{\mathrm{MSE}} \quad [\mathrm{dB}] \tag{9}$$

$$\mathrm{PSNR} = 10\log_{10}\frac{\mathrm{SHR}}{\mathrm{PSE}} \quad [\mathrm{dB}] \tag{10}$$

where SHR is the squared height-range (SHR) calculated from the real terrain data. This is denoted as

$$\mathrm{SHR} = \{\max[\hat{H}(x,y)] - \min[\hat{H}(x,y)]\}^2 \tag{11}$$

Table 1 shows the results. In the MSNR, the DEM generated from the interferogram filtered with CVNN-prediction filter is improved much more than the ones with the conventional filters, GW filter and CMRF filter, as well as in the PSNR. Thus the accuracy of Fig. 3(l) is the highest of all.

Table 1. Mean square error (MSE), peak square error (PSE) and their signal-to-noise (SN) ratios.

Method	MSE	MSNR	PSE	PSNR
	$\times 10$ [m^2]	[dB]	$\times 10^3$ [m^2]	[dB]
GW filter	85.21	33.78	18.02	20.52
CMRF filter	69.84	34.64	24.33	19.22
CVNN-prediction filter	66.88	34.83	12.10	22.25

4 Conclusion

We have proposed CVNN prediction filter that restores the distortion of SU by using the topographic statistics. We succeeded in generating more accurate DEM by using the proposed filter than the conventional methods.

References

1. Costantini, M.: A novel phase unwrapping method based on network programming. IEEE Trans. Geosci. Remote Sens. **36**(3), 813–821 (1998)
2. Danudirdjo, D., Hirose, A.: Local subpixel coregistration of interferometric synthetic aperture radar images based on fractal models. IEEE Trans. Geosci. Remote Sens. **51**(7), 4292–4301 (2013)
3. Danudirdjo, D., Hirose, A.: Anisotropic phase unwrapping for synthetic aperture radar interferometry. IEEE Trans. Geosci. Remote Sens. **53**(7), 4116–4126 (2015)
4. Danudirdjo, D., Hirose, A.: InSAR image regularization and DEM error correction with fractal surface scattering model. IEEE Trans. Geosci. Remote Sens. **53**(3), 1427–1439 (2015)
5. Goldstein, R.M., Zebker, H.A., Werner, C.L.: Satellite radar interferometry: two-dimensional phase unwrapping. Radio Sci. **23**, 713–720 (1988)
6. Goldstein, R.M., Werner, C.L.: Radar interferogram filtering for geophysical applications. Geophys. Res. Lett. **25**(21), 4035–4038 (1998)
7. Hirose, A.: Complex-Valued Neural Networks, 2nd edn. Springer, Heidelberg (2012)
8. Lee, J.S., Papathanassiou, K., Ainsworth, T., Grunes, M., Reigber, A.: A new technique for phase noise filtering of SAR interferometric phase images. IEEE Trans. Geosci. Remote Sens. **36**(5), 1456–1465 (1998)
9. Natsuaki, R., Hirose, A.: SPEC method - a fine co-registration method for SAR interferogram. IEEE Trans. Geosci. Remote Sens. **49**(1), 28–37 (2011)
10. Natsuaki, R., Hirose, A.: InSAR local co-registration method assisted by shape-from-shading. IEEE J. Select. Top. Appl. Earth Obs. Remote Sens. **6**(2), 953–959 (2013)
11. Oshiyama, G., Hirose, A.: Distortion reduction in singularity-spreading phase unwrapping with pseudo-continuous spreading and self-clustering active localization. IEEE J. Select. Top. Appl. Earth Obs. Remote Sens. **8**(8), 3846–3858 (2015)
12. Pritt, M., Shipman, J.: Least-squares two-dimensional phase unwrapping using FFT's. IEEE Trans. Geosci. Remote Sens. **32**(3), 706–708 (1994)
13. Suksmono, A.B., Hirose, A.: Interferometric SAR image restoration using Monte-Carlo metropolis method. IEEE Trans. Sig. Process. **50**(2), 290–298 (2002)
14. Suksmono, A.B., Hirose, A.: Progressive transform-based phase unwrapping utilizing a recursive structure. IEICE Trans. Commun. **E89–B**(3), 929–936 (2006)
15. Trouvé, E., Nicolas, J., Maître, H.: Improving phase unwrapping techniques by the use of local frequency estimates. IEEE Trans. Geosci. Remote Sens. **36**(6), 1963–1972 (1998)
16. Yamaki, R., Hirose, A.: Singularity-spreading phase unwrapping. IEEE Trans. Geosci. Remote Sens. **45**(10), 3240–3251 (2007)
17. Yamaki, R., Hirose, A.: Singular unit restoration in interferograms based on complex-valued Markov random field model for phase unwrapping. IEEE Geosci. Remote Sens. Lett. **6**(1), 18–22 (2009)

Inferring Users' Gender from Interests: A Tag Embedding Approach

Peisong Zhu[1], Tieyun Qian[1(✉)], Ming Zhong[1], and Xuhui Li[2]

[1] State Key Laboratory of Software Engineering, Wuhan University, Wuhan, China
{zhups24,qty,clock}@whu.edu.cn
[2] School of Information Management, Wuhan University, Wuhan, China
lixuhui@whu.edu.cn

Abstract. This paper studies the problem of gender prediction of users in social media using their interest tags. The challenge is that the tag feature vector is extremely sparse and short, i.e., less than 10 tags for each user. We present a novel conceptual class based method which enriches and centralizes the feature space. We first identify the discriminating tags based on the tag distribution. We then build the initial conceptual class by taking the advantage of the *generalization* and *specification* operations on these tags. For example, "Kobe" is a specialized instance of "basketball". Finally, we model class expansion as a problem of computing the similarity between one tag and a set of tags in one conceptual class in the embedding space.

We conduct extensive experiments on a real dataset from Sina Weibo. Results demonstrate that our proposed method significantly enhances the quality of the feature space and improves the performance of gender classification. Its accuracy reaches 82.25 % while that for the original tag vector is only 62.75 %.

Keywords: Gender classification · Users's interests · Conceptual class

1 Introduction

Existing approaches to gender classification rely heavily on the lexical [3,4,13], syntactic [11], or stylistic features [7–9]. These features need deep analysis of the contents of microblogs, which are hard to access and also time consuming to analyze. More importantly, a great number of users do not post microblogs at all. For this kind of users, it is impossible to extract any of the above features. Fortunately, the users in social media have other information besides their posts. In this paper, we propose to infer the users' gender using their interest tags.

This task is challenging since the tag feature space is extremely short and sparse. Each user is allowed to use a few tags, e.g., at most 10 tags in contrast to the 140 characters for each Chinese microblog. Moreover, the tags are usually diverse. For instance, users may use either "Red Devil" or "Manchester United" for the same team. Treating the tag set of each user as a document for classification produces poor results. This paper thus proposes a novel conceptual

A. Hirose et al. (Eds.): ICONIP 2016, Part IV, LNCS 9950, pp. 86–94, 2016.
DOI: 10.1007/978-3-319-46681-1_11

class (CC for short) based method to deal with this problem. A CC is a set of semantically similar or related tags [15]. It can also contains subconcepts or instances. We argue that the use of conceptual classes helps condense the tag feature space and improves the classification. The key idea is to effectively construct and expand the conceptual classes. To this end, we first select the distinguishing tags like "game" and "Beijing girl". We then build the initial conceptual classes by applying *generalization* and *specification* operations to these tags. Finally we expand the conceptual class by selecting the most similar tags where the similarity is defined on the embedding tag space. We conduct experiments on a real data set from Sina Weibo. Results demonstrate that our approach is very effective.

The most related work to ours is the work in [3], which also applied the CCs to gender classification. However, they only used two conceptual classes, one for male and the other for female. In contrast, our CCs contain 24 fine-grained classes. More importantly, the construction and expansion of CCs are entirely different. The CCs in [3] are built on the top of syntactic analysis on the contents. In contrast, we construct CCs based on the generalization and specification operations on discriminating tags. Finally, We expand the CCs using word embedding rather than the mutual information method used in [3]. Our method is independent of any class labels and thus can explore the tremendous unlabeled data in social media.

2 Related Work

2.1 Feature Set

Almost all existing studies for gender classification use the content information. Among which, the word or character n-grams are the most widely used features [3–5,7,12]. There are also a number of stylistic features extracted from the content, including ratio of punctuation, capital letters, unique words [7], slang words [9], word or sentence length [7,9], conceptual class [3], and the POS sequence [11]. While previous studies show that the performance of gender classification can be enhanced by using the above features, a main hinder is that they mainly exist in the content texts and need complex feature extraction procedure. In this paper, we propose to use the interest tags to infer the users' gender.

2.2 Classification Method

A number of machine learning approaches have been explored to solve the problem of gender prediction, for instance, SVMs [5,11–13], decision trees [1,5], Naïve Bayes [1,9,11,16], the maximum entropy learner [7], the Winnow algorithm [4,14], and logistic regression [2,3]. The classification method is not the focus of our research. Any supervised learning approaches can be used. In our study, we choose to use logistic regression as our classifier since it is not sensitive to the parameter settings as SVM.

3 A Conceptual Class Based Method

3.1 Building Initial Conceptual Classes

We first select the most distinguishing 100 tags from a public data set[1]. Among them 58,501 users specify their gender (not verified), 37,669 male and 20,832 female, respectively. Table 1 shows the top six discriminating tags and their counts in male and female.

Table 1. The distribution of sample discriminating tags

Interest tag	In male	In female
High-healed shoes	1	52
Beijing girl	1	47
Cancer boy	43	1
Dilettante female	0	40
Pro Evolution Soccer	40	0
LotionSPA	0	34

We find that these tags are quite gender specific and also consistent with our common sense knowledge. For example, males are fond of strenuous exercise like football while females often pay attention to beauty. However, these tags are too narrow to have a high coverage. On the other hand, we find general concepts are often used by a large number of users. For example, the count of football for male and female is 1217 and 97, respectively. This inspires us to use the discriminating tags as the first seed, and then generalize or specify the seed to form a conceptual class. Doing *generalization* is to find the superclass of a lower-level tag and doing *specification* is to find a subclass or an instance of a upper-level tag. The operations are manually done with the guidance of WikiTaxonomy and Baidu Baike.

Each conceptual class has two tags after two operations. We further choose another discriminating tag which is the counterpart of the lower level tag in CC. For example, we add "Yao Ming" into the CC containing "basketball" and "Kobe". The rationale behind this is to capture other aspects of the conceptual class. For those which are difficult to do two operations, we just discard them.

Finally we define 24 conceptual classes in total, each containing 3 tags. The detail is given in Table 2. We do not add more tags as we hope to have the least manual intervention and leave the expansion of CCs as an automatic procedure.

3.2 Expanding Conceptual Classes

The key point in conceptual class expansion is to find similar tags using the word embedding technique, where both the tag and conceptual class are represented as an embedded vector.

[1] http://xunren.thuir.org/share_EPSN.

Table 2. The 24 initial conceptual classes

Kobe, YaoMing, basketball	La Liga, PES, football	SEO, outsourcing, Internet startup
Lavida, Schumacher, car	Antique, Chuang-tzu, sinology	Prosecutor criminal police law
Naruto, dota, game	Whitening, anti-wrinkleskin care	Planning, marketing, financial planning
Olay, cosmetics, Estee Lauder	Hair, matching, dress up	Ping/xie, Chibi Maruko Chan, comic
Man, indoorsman, uncle	Couple, parent-offspring, family	Beijing girl, indoors woman, woman
Physics, mathematics, science	Yoga, Belly Dance, women fitness	Model, anchorwoman, office Lady
Romantic, emotional, sensitivity	Founder, consultant, social status	Boxing, golf, sporting event
Patriotic, Taiwan Strait, politics	Java, Android, IT technology	Prodigal, TV play, pink complex

Computing the Embedded Tag Vectors. We use a commonly used method Skip-Gram in the Word2Vec toolkit [10] for tag embedding. Note that although the user's tags are separate words, they indeed are related to each other. For instance, users who use "NBA" as a tag often use "Kobe" too. Furthermore, since there are at most ten tags for one user, we can set the window size to 10 to get each tag's context.

After embedding, each tag t_l is represented as a n-dimension vector, i.e., $\overrightarrow{t_l} = \overrightarrow{v_{l1}, ..., v_{ln}}$, where n is a predefined parameter. Following the practice in [10], we set n to 100, and use the hierarchical softmax for approximation.

Expanding Conceptual Classes. Given m tags in a conceptual class $CC = \{t_1, ..., t_m\}$, each tag t_j is a n-dimension vector $\overrightarrow{v_{t_j}} = \overrightarrow{v_{j,1}, ..., v_{j,n}}$, we present the following three strategies to represent a conceptual class as a feature vector.

1. the average of tag vectors in CC (AVG): $\overrightarrow{CC} = \overrightarrow{v_1, ..., v_n}$, where $v_i = \frac{\sum_{j=1}^{j=m} v_{j,i}}{m}$.
2. the dot product of tag vectors in CC (DPT): $\overrightarrow{CC} = \overrightarrow{v_1, ..., v_n}$, where $v_i = \prod_{j=1}^{j=m} v_{j,i}$.
3. the most similar tag vector in CC (MST): $\overrightarrow{CC} = \overrightarrow{t_k}$, $sim(t, t_k) = \max_{j=1}^{j=m} sim(t, v_{j,i})$

The similarity between a tag t and CC is defined as the cosine similarity of two vectors. We can then select the most similar tags to supplement CC. The expansion is performed in an iterative way controlled by two parameters k and K. In each iteration, top K most similar tags are added into the CC, and the procedure will repeat k times.

3.3 Condensing Tag Space

We use the expanded conceptual class to condense the tag space for both the training and test data. We do this by treating each conceptual class as a pseudo word. For each tag in the class, if it appears in a user's tag features, we then add the feature vector of the conceptual class to the tag feature vector. The rationale is to use the conceptual class to represent every tag it contains. Finally, we pool the user's feature vector by averaging all its embedded tag feature vectors as well as the feature vectors of the matched conceptual classes. Suppose a user u has m tags $T_u = \{t_1, ..., t_m\}$ in his/her profile which belong to n different conceptual classes $CC_u = \{CC_1, ..., CC_n\}$, the final feature vector of a user u can be defined as: $\overrightarrow{f_u} = \frac{\sum_{i=1}^{i=m} \overrightarrow{t_i} \sum_{j=1}^{j=n} \overrightarrow{CC_j}}{m+n}$.

4 Experimental Evaluation

We randomly select 500 users from the public data and manually check their genders. We build a corpus with 200 female and 200 male users for experiment. We further crawl 85212 friends of these users. The tags of the users and their friends are used to train the embedded tag vectors. We also try other corpora but they are worse than this one. All our experiments use the logistic regression classifier with L2 regularization [6]. The results are averaged over 5-fold cross validations. We report the accuracy as the evaluation metric.

4.1 Baselines

We have the following three baselines.

(1) Frequency-based representation of content vector (FRC): each word in a user's microblogs is represented as x:y, where x is a word id and y its frequency. This is a commonly used representation in gender classification [3,5,7].
(2) Frequency-based representation of tag vector (FRT): similar to FRC except that x is a tag.
(3) Distributed representation of tag vector (DRT): each tag is represented as a vector in the embedded space. Note that no existing study merely used FRT or DRT as features for gender classification. We propose these two baselines to examine how the embedded tag space differs from the original one, and show how CCs further improve the performance upon the embedding technique.

4.2 Effects of Parameters *k* and *K*

Here we show how parameters k and K affect the expansion of the conceptual classes in Table 3.

Table 3. A sample of expanded CC

Initial CC	Exp(panded) 3 k = 1, K = 3	Exp(panded) 6 k = 2, K = 3
Java Android IT technology	Python programmer mobile Internet	Python programmer mobile Internet software programming cloud computing

From Table 3, we observe that in each iteration, the top 3 ($K = 3$) similar tags are added to the conceptual class, and then this expanded CC (in addition to the initial CC) is used as the seed set for next iteration. More importantly, it can be seen that the expanded tags are highly correlated with those in the initial class, This clearly demonstrates that, with the guide of conceptual classes, the proposed tag expansion approach is restricted in a narrow field and thus capable of finding very similar or related tags.

We then show their effects on classification accuracy. We vary k from 1 to 5 stepped by one and K = 3,5,8,10. The results are shown in Table 4. The highlighted cells have the best performances for each line.

Table 4. Accuracy (%) of classification by using different parameters

	k = 1	k = 2	k = 3	k = 4	k = 5
K = 3	81.00	80.75	81.25	**82.25**	81.25
K = 5	81.25	**81.50**	81.00	80.75	80.50
K = 8	81.25	**81.50**	80.50	81.25	81.25
K = 10	**81.75**	80.75	81.00	80.75	80.50

From Table 4, it can be seen that the best performances for all the lines are reached when 10 to 16 tags are added into the conceptual class. Furthermore, we find a stair-like shape in the table. This indicates that the performance is enhanced much faster for the large K than those for the smaller ones, which is intuitive since a smaller K means a fewer number of tags are added and thus the performance does not change much. However, a small K means a strict condition for selection and thus tags with high quality will be chosen. This explains why the best performance, i.e., 82.25 % in the table is reached at $K = 3$.

4.3 Effects of Three Expansion Strategies

We show how the expansion strategies affect the performance of gender classification in Table 5.

Table 5. Classification accuracy (%) by using three expansion strategies ($k = 1..5$, $K = 10$)

Strategy	k = 1	k = 2	k = 3	k = 4	k = 5
AVG	81.75	80.75	81.00	80.75	80.50
DPT	78.50	79.50	78.75	78.25	78.50
MST	79.75	78.25	77.50	76.00	74.75

We can see that the AVG strategy is the best among the three strategies in all cases. We also find that MST outperforms DPT at the first point and the situation converses later. This is because MST can select the most similar tag in each iteration. However, the nearest point in the conceptual class may not well represent this class. When more tags are filled into the class according to the similarity to only one point, they may become less relevant and thus bring down the performance. In the following we will use the AVG strategy.

4.4 Comparison with Baselines

We compare our proposed approach with three baselines in Table 6.

Table 6. Comparison with baselines

Methods	Accuracy(%)
FRC	77.25
FRT	62.75
DRT	78.75
Initial CC (ICC)	79.75
Expanded CC (ECC)	82.25

We make the following important observations.

(1) In the original space, the frequency based tag representation (FRT) is the worst. It is much worse than the frequency based word representation (FRC). This is understandable because there are only at most ten tags in a user's profile while there are a large number of words in his/her microblogs.
(2) After tag embedding, the distributed tag representation (DRT) significantly improves the performance of the original FRT, and it also outperforms FRC. This shows that tag embedding alleviates the sparsity problem by borrowing information from other users' tags.
(3) The initial conceptual class (ICC) beats DRT. Note that the only difference between ICC and DRT is that ICC adds the concept class as a pseudo word to the tag vector. This infers that the CCs concentrate the feature vector toward the direction of the CC and further condense the tag space.

(4) The expanded conceptual class (ECC) reaches the best accuracy among all methods. This shows that our expanding approach is effective to find the most similar tags. It also demonstrates that when sufficiently related tags are added, CCs have very positive effects on centralizing the embedded tag space of DRT, which helps improve the accuracy from 78.75 % to 82.25 %.

5 Conclusion and Discussions

We proposed a novel conceptual class based framework to predict users' gender using their interest tags. The key idea is to build a set of initial CCs using generalization and specification operations and to expand the CCs in the embedded tag space. Results demonstrate our method significantly improves the performance.

Future directions: (1) The generalization and specification operations currently are manually done on distinguishing tags. A more automatic method will be to compute a score based on their frequency in corpus and the distance in taxonomy. (2) While our conceptual classes are gender specific, we hope the framework can be applied to other demographics like age and profession.

Acknowledgment. The work described in this paper has been supported in part by the NSFC Projects (61272275, 61572376, 61272110), the Wuhan Science and Technology Bureau "Chenguang Jihua" (2014072704011250).

References

1. Alowibdi, J.S., Buy, U.A., Yu, P.: Empirical evaluation of profile characteristics for gender classification on twitter. In: Proceedings of ICMLA, pp. 365–369 (2013)
2. Bamman, D., Eisenstein, J., Schnoebelen, T.: Gender identity and lexical variation in social media. J. Sociolinguistics **18**, 135–160 (2014)
3. Bergsma, S., Durme, B.V.: Using conceptual class attributes to characterize social media users. In: Proceedings of ACL, pp. 710–720 (2013)
4. Burger, J.D., Henderson, J., Kim, G., Zarrella, G.: Discriminating gender on Twitter. In: Proceedings of EMNLP, pp. 1301–1309 (2011)
5. Cheng, N., Chen, X., Chandramouli, R., Subbalakshmi, K.P.: Gender identification from e-mails. In: Proceedings of CIDM, pp. 154–158 (2009)
6. Fan, R.E., Chang, K.W., Hsieh, C.J., Wang, X.R., Lin, C.J.: Liblinear: a library for large linear classification. J. Mach. Learn. Res. **9**, 1871–1874 (2008)
7. Filippova, K.: User demographics and language in an implicit social network. In: Proceedings of EMNLP-CoNLL, pp. 1478–1488 (2012)
8. Garera, N., Yarowsky, D.: Modeling latent biographic attributes in conversational genres. In: Proceedings of ACL and IJCNLP, pp. 710–718 (2009)
9. Goswami, S., Sarkar, S., Rustagi, M.: Stylometric analysis of bloggers age and gender. In: Proceedings of ICWSM, pp. 214–217 (2009)
10. Mikolov, T., Sutskever, I., Chen, K., Corrado, G.S., Dean., J.: Distributed representations of words and phrases and their compositionality. In: Proceedings of NIPS (2013)

11. Mukherjee, A., Liu, B.: Improving gender classification of blog authors. In: Proceedings of EMNLP, pp. 207–217 (2010)
12. Peersman, C., Daelemans, W., Vaerenbergh, L.V.: Predicting age and gender in online social networks. In: Proceedings of SMUC, pp. 37–44 (2011)
13. Rao, D., Yarowsky, D., Shreevats, A., Gupta, M.: Classifying latent user attributes in twitter. In: Proceedings of SMUC, pp. 37–44 (2010)
14. Schler, J., Koppel, M., Argamon, S., Pennebaker, J.W.: Effects of age and gender on blogging. In: Proceedings of AAAI Spring Symposium on Computational Approaches for Analyzing Weblogs, pp. 199–205 (2005)
15. Sun, X., Xiao, Y., Wang, H., Wang, W.: On conceptual labeling of a bag of words. In: Proceedings of IJCAI, pp. 1326–1332 (2015)
16. Tang, C., Ross, K., Saxena, N., Chen, R.: What's in a name: a study of names, gender inference, and gender behavior in facebook. In: Proceedings of SNSMW (2011)

Fast Color Quantization via Fuzzy Clustering

László Szilágyi[1,2(✉)], Gellért Dénesi[1], and Călin Enăchescu[3]

[1] Faculty of Technical and Human Science of Tîrgu Mureş, Sapientia - Hungarian Science University of Transylvania, Tîrgu Mureş, Romania
lalo@ms.sapientia.ro

[2] Department of Control Engineering and Information Technology, Budapest University of Technology and Economics, Budapest, Hungary

[3] Department of Informatics, Petru Maior University of Tîrgu Mureş, Tîrgu Mureş, Romania

Abstract. This comparative study employs several modified versions of the fuzzy c-means algorithm in image color reduction, with the aim of assessing their accuracy and efficiency. To assure equal chances for all algorithms, a common framework was established that preprocesses input images in terms of a preliminary color quantization, extraction of histogram and selection of frequently occurring colors of the image. Selected colors were fed to clustering by studied c-means algorithm variants. Besides the conventional fuzzy c-means (FCM) algorithm, the so-called generalized improved partition FCM algorithm, and several versions of the generalized suppressed FCM were considered. Accuracy was assessed by the average color difference between input and output images, while efficiency tests monitored the total runtime. All modified algorithms were found more accurate, and some suppressed models also faster than FCM.

Keywords: Color quantization · Fuzzy clustering · Improved partition

1 Introduction

Despite of the constant progress, storage space and communication bandwidth is still limited, so it is important to retain the meaning of the data as much as possible while reducing its size. This requirement recently led to the development of several optimal color quantization methods, which employed self-organizing maps [1], batch neural gas [2], fish swarm algorithm [3], generic roughness measure [4], c-means clustering [5,6], hierarchical clustering [7], Gaussian mixture models [8], and combined soft computing techniques [9].

Earlier we introduced a histogram-based formulation of the problem applicable for various c-means clustering algorithms [6], color reduction being achieved in three steps. The preprocessing step performed an initial rigid color quantization, reducing the possible colors from $2^{24} \approx 16.77$ millions to $52^3 \approx 140$ thousands, by rounding the three intensity values of a pixel (situated in the range $\{0, 1, \ldots 255\}$ to the closest multiple of 5. This is followed by a color

A. Hirose et al. (Eds.): ICONIP 2016, Part IV, LNCS 9950, pp. 95–103, 2016.
DOI: 10.1007/978-3-319-46681-1_12

selection, which separates frequent quantized colors from rare ones via histogram thresholding. The clustering step applies the chosen version of c-means algorithm to the selected set of frequent colors, and performs fast histogram-based clustering. Finally a label is provided for each quantized color existing in the image, and the output image is created according to the labels of each pixel's quantized color. The output image contains a reduced number of colors and is suitable for palette based representation. There are two main parameters in the framework: (i) the number of clusters (or final colors) c, varying between 2 and 256; (ii) the percentage P (ranging between 80 % and 100 %) of pixels fed to clustering, which controls the histogram thresholding during preprocessing [6].

This paper provides a comparative study of various fuzzy c-means algorithms with improved partition, based on their application for color reduction within the above mentioned framework. Algorithms involved in this study include generalized improved partition fuzzy c-means (GIFP-FCM) [10,11], suppressed fuzzy c-means (s-FCM) [12], and various generalized versions of s-FCM [13]. Numerical analysis will be performed to reveal the accuracy and efficiency of the tested algorithms.

2 Employed Clustering Algorithms

The fuzzy c-means (FCM) algorithm and all others derived from it, which are involved in this study, partition a set of object data $\mathbf{X} = \{\mathbf{x}_1, \mathbf{x}_2, \ldots, \mathbf{x}_n\}$ into a predefined number of c clusters, based on the alternative optimization of a quadratic objective function. FCM minimizes

$$J_{\text{FCM}} = \sum_{i=1}^{c} \sum_{k=1}^{n} u_{ik}^m ||\boldsymbol{x}_k - \boldsymbol{v}_i||_{\mathbf{A}}^2 = \sum_{i=1}^{c} \sum_{k=1}^{n} u_{ik}^m d_{ik}^2, \tag{1}$$

where $\mathbf{v}_i$ represents the prototype or centroid of cluster i ($i = 1 \ldots c$), $u_{ik} \in [0, 1]$ is the fuzzy membership function showing the degree to which vector $\boldsymbol{x}_k$ belongs to cluster i, $m > 1$ is the fuzzyfication parameter (exponent), and d_{ik} represents the distance (any inner product norm defined by a symmetrical positive definite matrix $\mathbf{A}$) between $\boldsymbol{x}_k$ and $\boldsymbol{v}_i$. All algorithms in this study use probabilistic partition, meaning that the fuzzy memberships assigned to any input vector $\mathbf{x}_k$ with respect to clusters satisfy the probability constraint $\sum_{i=1}^{c} u_{ik} = 1$. The minimization of the objective function J_{FCM} is achieved by alternately applying the optimization of J_{FCM} over $\{u_{ik}\}$ with $\mathbf{v}_i$ fixed, $i = 1 \ldots c$, and the optimization of J_{FCM} over $\{\boldsymbol{v}_i\}$ with u_{ik} fixed, $i = 1 \ldots c$, $k = 1 \ldots n$ [14]. In each loop, the optimal values are deduced from the zero gradient conditions and Lagrange multipliers, and obtained as follows:

$$u_{ik}^{\star} = \frac{d_{ik}^{-2/(m-1)}}{\sum_{j=1}^{c} d_{jk}^{-2/(m-1)}} \quad \begin{array}{l} \forall i = 1 \ldots c, \\ \forall k = 1 \ldots n \end{array}, \tag{2}$$

$$\boldsymbol{v}_i^{\star} = \frac{\sum_{k=1}^{n} u_{ik}^m \boldsymbol{x}_k}{\sum_{k=1}^{n} u_{ik}^m} \quad \forall i = 1 \ldots c. \tag{3}$$

According to the alternating optimization (AO) scheme of FCM, Eqs. (2) and (3) are alternately applied, until cluster prototypes stabilize. The hard c-means (HCM) is a special case of FCM, which uses $m = 1$, and thus the memberships are obtained by the winner-takes-all rule: fuzzy membership functions in this case will be chosen from the set $\{0, 1\}$, while maintaining the probabilistic partition.

Fuzzy membership functions produced by FCM using Eq. (2) can be highly multimodal, especially in case of high numbers of clusters. Mathematically this means that for any chosen input vector $\boldsymbol{x}_k$, its distances measured from cluster prototypes $\boldsymbol{v}_1, \boldsymbol{v}_2, \ldots, \boldsymbol{v}_c$ are denoted by $d_{1k}, d_{2k}, \ldots, d_{ck}$. Let us sort these distances in increasing order, and let us denote them by $d_{s_1k}, d_{s_2k}, \ldots, d_{s_ck}$, meaning that d_{s_1k} is the minimum and d_{s_ck} is the maximum. In cases when two or more of the shortest distances are almost equal, the fuzzy memberships of vector $\boldsymbol{x}_k$ with respect to distant classes, namely $u_{s_ck}, u_{s_{c-1}k}, u_{s_{c-2}k}, \ldots$ have local maxima, instead of being close to zero. In other words, close to the boundary between two or more clusters, the fuzzy memberships with respect to all distant classes are relatively high. This phenomenon is being treated by the algorithms derived from FCM that are involved in this study.

FCM with Improved Partition. The notion of improved partition was introduced by Höppner and Klawonn [10] and later generalized by Zhu et al. [11]: FCM with generalized improved partition (GIFP-FCM) optimizes the objective function

$$J_{\text{GIFP}} = \sum_{i=1}^{c} \sum_{k=1}^{n} \mu_{ik}^m d_{ik}^2 + \sum_{k=1}^{n} a_k \sum_{i=1}^{c} \mu_{ik}(1 - \mu_{ik}^{m-1}), \quad (4)$$

whose the partition update formula

$$\mu_{ik}^{\star} = \frac{(d_{ik}^2 - a_k)^{-1/(m-1)}}{\sum_{j=1}^{c} (d_{jk}^2 - a_k)^{-1/(m-1)}} \qquad \begin{matrix} \forall i = 1 \ldots c, \\ \forall k = 1 \ldots n \end{matrix} . \quad (5)$$

while the cluster prototype update formula is the same as in FCM, but it uses the fuzzy membership functions μ_{ik}. Equation (5) explains us the difference between the behavior of GIFP-FCM and FCM: for any input vector $\boldsymbol{x}_k$, the square of its distances measured from all cluster prototypes are virtually reduced by a the same positive value a_k. The proposed context dependent formula for the choice of a_k is: $a_k = \omega \min\{d_{ik}^2, i = 1 \ldots c\}$, with $\omega \in [0.9, 0.99]$, thus keeping the square of all distorted distances positive. Using $\omega = 1$ would reduce GIFP-FCM to HCM, while $\omega = 0$ leads to FCM.

Suppression of the FCM Partition. The suppressed fuzzy c-means (s-FCM) algorithm [12] had the declared goal of reducing the execution time of FCM by improving its convergence speed, while preserving its good partition quality. The s-FCM algorithm does not minimize J_{FCM}. It is not introduced as an algorithm that minimizes any objective function. Instead of that, it manipulates with the AO scheme of FCM, by inserting an extra computational step in each iteration, placed between the partition update formula (2) and prototype update formula

(3). This new step deforms the partition (fuzzy membership functions) according to the following rule:

$$\mu_{ik} = \begin{cases} 1-\alpha+\alpha u_{ik} & \text{if } i = \arg\max_j\{u_{jk}\} \\ \alpha u_{ik} & \text{otherwise} \end{cases}, \tag{6}$$

where μ_{ik} ($i = 1 \dots c$, $k = 1 \dots n$) represents the fuzzy memberships obtained after suppression. During the iterations of s-FCM, these suppressed membership values μ_{ik} will replace u_{ik} in Eq. (3). Parameter $\alpha \in [0,1]$ controls the strength of suppression applied to the FCM partition. Suppression rate $\alpha = 1$ makes s-FCM identical with FCM, while $\alpha = 0$ with HCM. Any other values of α lead to algorithms that differ from all above mentioned ones.

Earlier we showed that suppressing the FCM partition is mathematically equivalent with a certain reduction of the distances d_{iw_k} to d'_{iw_k} with $w_k = \arg\min_j\{d_{ik}, j = 1 \dots c\}$, for any $k = 1 \dots n$, without changing distances d_{ik} with $i \neq w_k$ [13]. The degree of context dependent distance reduction was characterized by a so-called quasi learning rate η, whose formula was defined as $\eta \stackrel{\text{def}}{\equiv} 1 - d'_{iw_k}/d_{iw_k}$ and evaluated as $\eta = 1 - [1 + \frac{1-\alpha}{\alpha u_{w_k k}}]^{(1-m)/2}$, with $u_{w_k k} = \max\{u_{ik}, i = 1 \dots c\}$ being the largest fuzzy membership function value for the input vector $\mathbf{x}_k$. This quasi learning rate η enabled us to introduce several generalized suppression rules for the FCM partition. These generalization rules consider the suppression rate context sensitive and thus assign a dedicated α_k suppression rate to each input vector $\boldsymbol{x}_k$ in each iteration. In the following we enumerate some families of previously defined suppression rules: (1) Constant learning rate $\eta = \theta$, with $\theta \in [0,1]$ fixed; (2) Learning rate defined as a function of largest fuzzy membership value: $\eta = f(u_{w_k k})$; (3) Direct formula between largest fuzzy membership value before and after suppression: $\mu_{w_k k} = f(u_{w_k k})$. Suppression rules included in this study are exhibited in Table 1. Further details on suppression rules of the FCM partition can be found in [13].

Table 1. Proposed generalized suppression rules, where u_w and μ_w are the abbreviation of $u_{w_k k}$ and $\mu_{w_k k}$, respectively, and $q = 2/(1-m)$. In all cases, the suppression parameter varied in $\{0.1, 0.2 \dots 0.9\}$

Name	Definition	Suppression formula
gs$_\theta$-FCM	$\eta \stackrel{\text{def}}{\equiv} \theta$	$\alpha_k = [1 - u_w + u_w(1-\theta)^q]^{-1}$
gs$_\beta$-FCM	$\eta \stackrel{\text{def}}{\equiv} 1 - u_w^{\frac{\beta}{1-\beta}}$	$\alpha_k = \left[1 + u_w\left(u_w^{\frac{q\beta}{1-\beta}} - 1\right)\right]^{-1}$
gs$_\kappa$-FCM	$\eta \stackrel{\text{def}}{\equiv} \frac{1}{2} + \frac{2\kappa-1}{2}\sin(\pi u_w)$	$\alpha_k = \left[1 - u_w + u_w\left(\frac{1}{2} - \frac{2\kappa-1}{2}\sin(\pi u_w)\right)^q\right]^{-1}$
gs$_\tau$-FCM	$\mu_w \stackrel{\text{def}}{\equiv} \frac{u_w+\tau}{u_w\tau}$	$\alpha_k = \frac{1-\tau}{1+u_w\tau}$
gs$_\sigma$-FCM	$\mu_w \stackrel{\text{def}}{\equiv} u_w^\sigma$	$\alpha_k = \frac{1-u_w^\sigma}{1-u_w}$
gs$_\xi$-FCM	$\mu_w \stackrel{\text{def}}{\equiv} \left(\sin\frac{\pi u_w}{2}\right)^\xi$	$\alpha_k = \frac{1-\left(\sin\frac{\pi u_w}{2}\right)^\xi}{1-u_w}$

3 Results and Discussion

Quantitative characterization of accuracy was provided by the averaged color difference (ACD) between the original and the final color reduced image, employing the formula: $\text{ACD} = \frac{1}{n}\sum_{k=1}^{n} ||\boldsymbol{x}_k - \boldsymbol{v}_{\lambda_k}||$, where $\mathbf{x}_k$ is the color vector of pixel k, and λ_k is the label given to pixel k. Twenty-five color photographs (800×600 pixels, 8-bit color depth) were fed to algorithms presented in Sect. 2, each containing 3–30 thousand colors after initial color quantization. All algorithms were run at 40 different settings generated by parameters $c \in \{16, 24, 32, 48, 64, 96, 128, 256\}$ and $P \in \{80\,\%, 90\,\%, 95\,\%, 98\,\%, 100\,\%\}$. Benchmark indices like ACD, total runtime, and number of optimization cycles, were recorded. The upper part of Table 2 exhibits the results of the 1000 accuracy competitions (25 images $\times$ 40 scenarios) of nine selected algorithms. Although GIFP-FCM ($\omega = 0.9$) is ranked first, in best 2, and in best 3 the most times, s-FCM (at $\alpha = 0.8$) is the one that performs the most times better than FCM. The lower part of Table 2 shows the results of the 1000 efficiency competitions of the algorithms. HCM won almost all efficiency competitions, but it was excluded from this table as we were more interested in the competition of fuzzy ones, which also produced better quality. This way s-FCM ($\alpha = 0.5$) had top ranking the most times, but gs-FCM ($\tau = 0.5$) was found the one performing the most times quicker than FCM.

Table 3 lists ACD value statistics – average and standard deviation values obtained for the 25 images – relating about the outcome of the tested algorithms, involving pixel inclusion rate of 100 % and cluster number growing from 16 to 256. In most cases, GIFP-FCM ($\omega = 0.9$) is the best scoring tested algorithm. In case of very reduced number of colors, some suppressed FCM clustering models can compete with GIFP-FCM. HCM has the worst (but comparable) results out

Table 2. Overall accuracy and efficiency ranking based on 1000 runs (25 images $\times$ 40 scenarios). Figures indicate number of top ranked performances out of 1000

Accuracy	FCM	GIFP-FCM		s-FCM		gs-FCM			HCM
		$\omega = 0.9$	$\omega = 0.99$	$\alpha = 0.5$	$\alpha = 0.8$	$\xi = 0.9$	$\tau = 0.5$	$\theta = 0.5$	
Best	47	**458**	36	61	243	86	16	43	10
In top 2	117	**586**	160	177	410	247	118	165	20
In top 3	153	**677**	223	330	545	419	280	327	46
Better than FCM		838	701	797	**932**	860	779	805	280
Efficiency	FCM	GIFP-FCM		s-FCM		gs-FCM			
		$\omega = 0.9$	$\omega = 0.99$	$\alpha = 0.5$	$\alpha = 0.8$	$\xi = 0.9$	$\tau = 0.5$	$\theta = 0.5$	
Best	10	99	190	**357**	13	4	257	70	
In top 2	31	158	323	**663**	39	22	604	160	
In top 3	44	257	554	**840**	76	73	797	359	
Faster than FCM		340	698	803	323	417	**835**	610	

Table 3. Average ACD values of 25 images obtained at $P = 100\,\%$, for various values of the cluster number. For each scenario, benchmark results for HCM, FCM, and the best performing c-means algorithm is given.

Reduced colors c	HCM benchmark (mean ± SD)	FCM benchmark (mean ± SD)	Best benchmark (mean ± SD)	Best performing algorithm
16	8.601 ± 1.882	8.376 ± 1.825	8.004 ± 1.763	gs-FCM$_\xi$ at $\xi = 0.9$
24	7.337 ± 1.539	7.057 ± 1.531	6.472 ± 1.405	s-FCM at $\alpha = 0.8$
32	6.498 ± 1.489	6.255 ± 1.453	5.861 ± 1.363	GIFP-FCM at $\omega = 0.9$
48	5.622 ± 1.238	5.484 ± 1.219	5.177 ± 1.146	GIFP-FCM at $\omega = 0.9$
64	5.191 ± 1.220	4.968 ± 1.192	4.655 ± 1.118	GIFP-FCM at $\omega = 0.9$
96	4.583 ± 1.028	4.360 ± 1.024	4.121 ± 0.975	GIFP-FCM at $\omega = 0.9$
128	4.190 ± 0.973	4.030 ± 0.964	3.806 ± 0.917	GIFP-FCM at $\omega = 0.9$
192	3.808 ± 0.841	3.620 ± 0.835	3.368 ± 0.779	GIFP-FCM at $\omega = 0.9$
256	3.556 ± 0.776	3.388 ± 0.768	3.154 ± 0.722	GIFP-FCM at $\omega = 0.9$

of all tested algorithms, and most algorithms involving improved or suppressed partitions have their optimal parameter to perform better than FCM.

Figure 1 shows an example of test image used as input, and three output images created by the GIFP-FCM algorithm, at $P = 98\,\%$ pixel inclusion rate. Sixteen colors proved definitely too few to produce acceptable images, although every relevant detail is visible here as well. The output image that uses $c = 32$ colors is considerably of better quality, while those obtained with $c = 64 \ldots 256$ clusters hard to distinguish from the original one unless viewed strongly magnified. These images suggest that FCM-based algorithms are able to extract acceptable palettes of color images. Figure 2 exhibits several examples of reduced color images of good (or at least acceptable) quality, which use only 16 to 96 colors. For each item denoted by (a)–(d), the upper image is the full-color original, while the bottom one is the reduced color image. Here again we can see, that most color photos can be reproduced in good quality in 256 or less colors.

Figure 3 plots the variation of the ACD value against the number of reduced colors c, for nine selected images, at constant $P = 98\,\%$, when employing GIFP-FCM at $\omega = 0.9$. The higher number of clusters always resulted in better approximation quality. However, the same quality level (e.g. ACD = 6 units) is reached at $c = 16$ colors in case of Image 3, and at $c = 128$ colors in case of Image 4.

Fig. 1. Color reduction results of GIFP-FCM ($\omega = 0.9$): original image and images with a reduced number of color, at pixel selection rate $P = 98\,\%$. (Color figure online)

Fig. 2. Some images in original (upper row) and after color reduction (bottom row): (a) 16 colors; (b) 24 colors; (c) 48 colors; (d) 96 colors. (Color figure online)

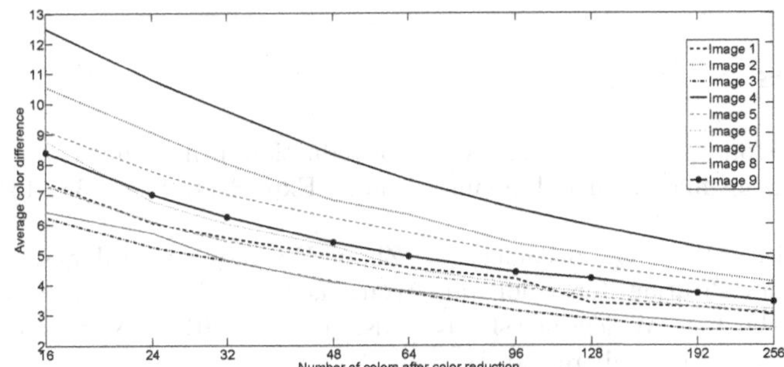

Fig. 3. ACD vs. number of clusters, at fixed pixel selection rate $P = 90\,\%$.

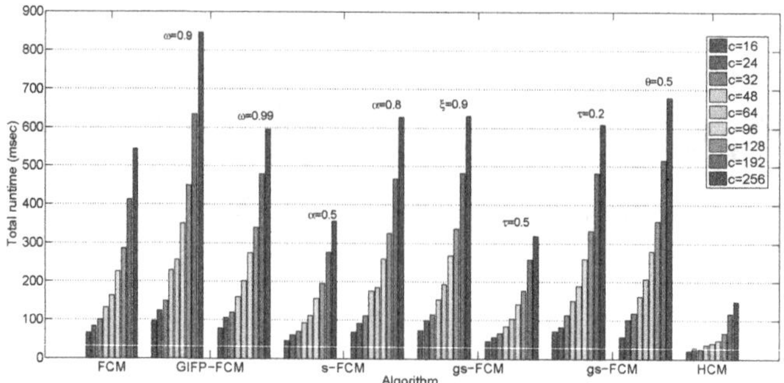

Fig. 4. Total runtime averaged over fifteen test images, plotted against applied algorithm and cluster number, at fixed pixel selection rate $P = 98\,\%$

Fig. 4 presents the average runtime of each tested algorithm, obtained in case of various cluster numbers c, and pixel inclusion rate $P = 98\,\%$. Averaging was performed over the twenty-five test images. All efficiency tests were performed on a PC with i7 processor running at 3.4 GHz frequency. Algorithms were allowed to perform up to 200 iterations unless convergence was reached. HCM was stopped after the first iteration that produced no change in labels, while all others used convergence threshold set to $\varepsilon = 0.1 \times c$ intensity units.

4 Conclusions

This paper involved several c-means clustering algorithm variants in a color quantization problem and provided a comparison via accuracy and efficiency benchmark. Improved and suppressed partitions both proved of accuracy than conventional FCM, and many of their variants were also found more efficient. We recommend using GIFP-FCM at $\omega = 0.9$ due to its best accuracy. In applications where time is also important, the generalized suppressed FCM of type τ could also be a fine choice, fixing the parameter around $\tau = 0.5$.

References

1. Rasti, J., Monadjemi, A., Vafaei, A.: Color reduction using a multi-stage Kohonen self-organizing map with redundant features. Exp. Syst. Appl. **38**, 13188–13197 (2011)
2. Celebi, M.E., Wen, Q., Schaefer, G., Zhou, H.: Batch neural gas with deterministic initialization for color quantization. In: Bolc, L., Tadeusiewicz, R., Chmielewski, L.J., Wojciechowski, K. (eds.) ICCVG 2012. LNCS, vol. 7594, pp. 48–54. Springer, Heidelberg (2012)
3. El-Said, S.A.: Image quantization using improved artificial fish swarm algorithm. Soft Comput. **19**, 2667–2679 (2015)
4. Yue, X.D., Miao, D.Q., Cao, L.B., Wu, Q., Chen, Y.F.: An efficient color quantization based on generic roughness measure. Patt. Recogn. **47**, 1777–1789 (2014)

5. Celebi, M.E.: Improving the performance of k-means in color quantization. Image Vis. Comput. **29**, 260–271 (2011)
6. Szilágyi, L., Dénesi, G., Szilágyi, S.M.: Fast color reduction using approximative c-means clustering models. In: IEEE International Conference on Fuzzy Systems (FUZZ-IEEE), pp. 194–201 (2014)
7. Celebi, M.E., Wen, Q., Hwang, S.: An effective real-time color quantization method based on divisive hierarchical clustering. J. Real-Time Imag. Process. **10**, 329–344 (2015)
8. Zeng, S., Huang, R., Kang, Z.: Image retrieval using spatiograms of colors quantized by Gaussian Mixture models. Neurocomputing **171**, 673–684 (2016)
9. Schaefer, G.: Soft computing-based colour quantisation. EURASIP J. Imag. Video Process. **2014**(8), 1–9 (2014)
10. Höppner, F., Klawonn, F.: Improved fuzzy partition for fuzzy regression models. Int. J. Approx. Reason. **5**, 599–613 (2003)
11. Zhu, L., Chung, F.L., Wang, S.: Generalized fuzzy c-means clustering algorithm with improved fuzzy partition. IEEE Trans. Syst. Man Cybern. B. **39**, 578–591 (2009)
12. Fan, J.L., Zhen, W.Z., Xie, W.X.: Suppressed fuzzy c-means clustering algorithm. Patt. Recogn. Lett. **24**, 1607–1612 (2003)
13. Szilágyi, L., Szilágyi, S.M.: Generalization rules for the suppressed fuzzy c-means clustering algorithm. Neurocomputing **139**, 298–309 (2014)
14. Bezdek, J.C.: Pattern Recognition with Fuzzy Objective Function Algorithms. Plenum, New York (1981)

Extended Dependency-Based Word Embeddings for Aspect Extraction

Xin Wang(✉), Yuanchao Liu, Chengjie Sun, Ming Liu, and Xiaolong Wang

School of Computer Science and Technology, Harbin Institute of Technology, Harbin 150001, China
{xwang,lyc,cjsun,mliu,wangxl}@insun.hit.edu.cn

Abstract. Extracting aspects from opinion reviews is an essential task of fine-grained sentiment analysis. In this paper, we introduce outer product of dependency-based word vectors and specialized features as representation of words. With such extended embeddings composed in recurrent neural networks, we make use of advantages of both word embeddings and traditional features. Evaluated on SemEval 2014 task 4 dataset, the proposed method outperform existing recurrent models based methods, achieving a result comparable with the state-of-the-art method. It shows that it is an effective way to achieve better extraction performance by improving word representations.

Keywords: Aspect extraction · Sequence labelling · Sentiment analysis · Word embeddings · Representation learning

1 Introduction

Identifying aspects in opinion reviews is the fundamental task for fine-grained sentiment analysis, which makes it possible to comprehensively analyse the advantages and shortcomings of products or services. It is also important to other sentiment-related such as opinion summarization. Thus, it has attracted increasing research interest in recent years.

When discussing 'aspects', researchers refer to one specific part of products or services instead of the whole. As shown in Table 1, 'battery life' is considered as an aspect, while 'laptop' is not.

Table 1. An example review tagged with BIO tagging schema.

Awesome	laptop	with	good	battery	life
O	O	O	O	B	I

Like most extraction tasks in fine-grained opinion mining, aspect extraction is usually formulated as a sequence labelling problem. Conventional BIO tagging

A. Hirose et al. (Eds.): ICONIP 2016, Part IV, LNCS 9950, pp. 104–111, 2016.
DOI: 10.1007/978-3-319-46681-1_13

scheme [11] is used in the tagging process. As shown in Table 1, 'B' is for the word at the beginning of an aspect, 'I' is for other tokens in the aspect, and 'O' is for those outside the aspect-related phrases.

In the subtask of aspect term extraction of SemEval 2014 task 4, the state-of-the-art systems leverage hand-crafted features. DLIREC system make use of syntactic features and resources including wordnet and opinion lexicon [12]. While IHS R&D Belarus system consider semantic-related features and named entities during extraction [1]. Such methods using traditional features represent tokens with one-hot vectors. It lead to a problem of data sparsity. To ease such problem, many features describe commonalities of words are introduced. For instance, Part-Of-Speech (POS) features describe syntactical generality of words, while word taxonomy features represent semantic commonalities of synonymous.

An alternative way to identify aspect terms in reviews is learning to label each token represented as dense vector with compositional models in an end to end manner. Liu *et al.* leverage word embedding to represent words, and capture the interaction between word with simple recurrent neural network (SRNN) and long short-term memory (LSTM) [8]. However, the capacity of word embedding describing uniqueness of tokens are limited. For example, when extracting aspect terms, words describe the whole product (e.g. 'laptop', 'machine' and 'notebook') should be labelled with distinguished tag with those who describe a specific part (such as 'battery' or 'screen'). Nevertheless, word vector of these words tend to be similar for such tokens sharing similar syntactic and semantic usages. Moreover, word embeddings used in previous works are trained based on natural surrounding context. For instance, representations of 'appetizer' and 'barbecued' are similar, while they take on different roles in review expressions. Such embeddings are limited in reflecting the functional characteristic in the token-level tagging tasks.

In this paper, we introduce dependency-based word embeddings to aspect term extraction to encode the structure information of context, and use outer product of task-specific traditional features and word vectors to get extended word representations to capture both commonness and speciality of word. Such representations are then composed in a recurrent neural network sequence labeller. Evaluated on SemEval 2014 task 4 dataset, the proposed method outperform existing recurrent-based models using traditional word vector, achieving a result comparable with the state-of-the-art method. It shows the effectiveness of our perspective to improve word representations. Moreover, one of our settings without using any language-depend features, resources or tools still gets satisfied result, which could be implement easily in other languages.

2 Extending Dependency-Based Word Embeddings

Hand-crafted features bear a data sparsity problem, while contextual-based word embeddings are limited in leveraging specialized features. Incorporating these two is an effective way to improve extraction performance. Different from previous work modifying neural models [8], we focus on the perspective of improving word representations to get better initial bias and final results.

2.1 Dependency-Based Word Embeddings

Dependency relation is an important feature to capture function of sepecific word in aspect extraction [12]. However, widely used word embeddings trained based on context window (bag-of-words context embeddings) are limited in encoding such information. These embeddings focus on topical similarities instead of functional similarities. For example, in bag-of-words (BoW) context embeddings trained on English Wikipedia corpus with *word2vec* skip-gram model [9], the top 10 similar words of 'appetizer' contain 'barbecued', which bears a different function although in a same topic. This is confusing to a neural token-level sequence labeller.

Levy and Goldberg transfer the concept of *context* from natural position adjacency to direct dependency relation, and introduce dependency-based word embeddings [7]. Figure 1 shows an example of different context of word 'food'. Original skip-gram model makes use of *bag-of-words context*, namely, the surrounding words in a window ('the', 'is' and 'always' here in a 2-word window). While dependency-based embeddings consider words with dependency relation as *syntactic context*. In the example in Fig. 1, 'the' and 'good' are treat as *syntactic context* in the process of dependency-based embeddings learning.

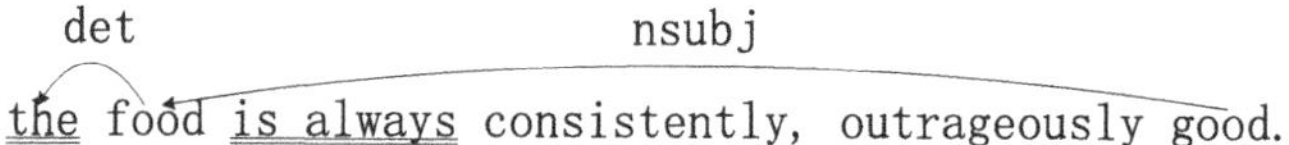

Fig. 1. An example of different context of word 'food'

Bag-of-words contexts encode domain similarities, while syntactic context induce functional similarities. For instance, in BoW embeddings, 'appetizer' and 'barbecued' tend to be similar. While in dependency-based embeddings, the most similar words of 'appetizer' are all co-hyponyms like 'omelette'. This is smooth for token-level sequence labelling tasks. Words with same function tending to share same tag in aspect term extraction have similar representations.

2.2 Extending Word Embeddings

BoW embeddings or dependency-based embeddings encode the syntactic and semantic similarities, which has a functional overlap with generalized features, such as, *part-of-speech* or *semantic category* of word. But there are features describing uniqueness of specific word remaining omitted, such as *word frequency* that count the occurrence of specific words or *gazetteer feature* which is the name list of aspect terms and directly describes whether a specific word could be a part of an aspect. With such features, a classifiers could easily tell 'laptop' is not an aspect term, though it's embedding is similar to 'screen'.

Integrating these *specialized features* with word embeddings bring more complete information to the classifier. However, simply connecting the dense embeddings and the sparse features confuses the models. Liu *et al.* learn abstract representation through compositional model and connect traditional features in classifier layer [8], but the incongruity of dense and sparse is still exist in the connection. While Toh and Wang convert the dense embeddings to few sparse binary tags of clusters [12]. However, there is nonnegligible loss of information during the clustering.

Yu *et al.* [13] make use of outer-product of word embeddings e and binary vector of traditional features f as the compound embeddings c.

$$c = f \otimes e \tag{1}$$

Not restricted to discrete value, in this work, we extend the traditional features vector to more general form. As shown in the left part of Fig. 2, both continuous values between [0–1] and discrete value (0 or 1) are leveraged. In this way, both generalized and specialized information are covered.

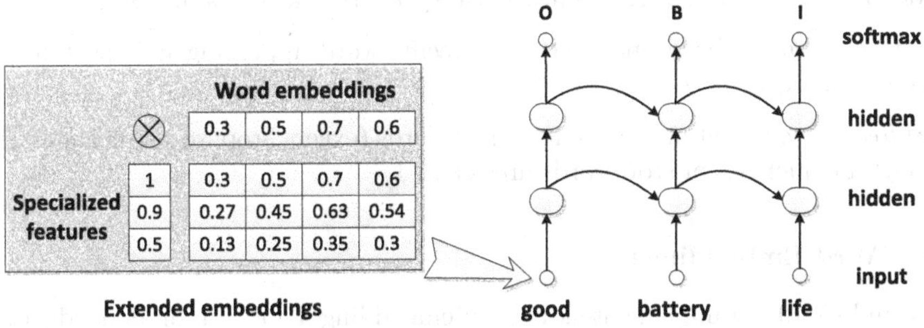

Fig. 2. Recurrent neural network based sequence labeller with extended embeddings.

2.3 Recurrent Neural Network Sequence Labeller

Recurrent neural networks (RNNs) have been proven to be effective compositional models in recent works [5,6]. RNNs compose the current input with previous hidden states and get the current hidden activation. In this way, recurrent structures encode the information of context [3]. There are several variants of RNN, such as LSTM and bi-directional RNN, have been proven to be effective in different NLP tasks. However, in the task of aspect term extraction, the target entities tend to be short and easy to cover with simple uni-directional structure [8]. Thus, in this work, we leverage the deep form (with multiple layers) of uni-directional Elman network.

The illustration of the tagging process is shown in Fig. 2. We (1) takes compound embeddings of each word as input of each time step, (2) generates the abstract representations in hidden layers by composing current input and previous hidden activation, and (3) finally gets the BIO tags through a softmax layer.

3 Experiment Settings

3.1 Data Set and Metric

The training and test data in SemEval-2014 task 4: aspect-based sentiment analysis evaluation campaign [10] is used to evaluate our methods. Specifically, we focus on the subtask of aspect term extraction on the domain of *Laptop* and *Restaurant*. Each domain has about 3000 sentences in training set and 800 sentences in test set.

Only if one extracted chunk exactly match the gold standard annotation, we consider it as a correct term. Then the standard precision, recall and F_1 value are computed to evaluate the effectiveness of the model.

3.2 Included Features

As described in Subsect. 2.2, we introduce specialized features of words.

gazetteer feature: The proportion of a specific word appearing in aspect term out of its all occurrence.

word frequency: Count of occurrences of all words (except stop words) normalized to [0–1]. Frequencies of stop words are set to 1.

3.3 Word Embeddings

Levy and Goldberg implement skip-gram embedding algorithm on dependency-based syntactic contexts. A 300-dimensional embeddings are trained on English Wikipedia [7]. We leverage this **Dependency Embedding** in our method and extend it with specialized features, while other two embeddings are used in the comparison. We tune the input embeddings during training to get more suitable representations.

SENNA embeddings trained by Collobert *et al.* [2] and **Google embeddings** learning using word2vec [9] are leveraged in the comparison.

3.4 Training Parameters

The recurrent structures are implemented with RNNLIB toolkit [4]. We carry out early stopping on validation set during training. Training process stops when there's no better result after 30 epochs. Experimentally, we set the learning rate to 0.005 and the momentum to 0.7 during stochastic gradient descent.

3.5 Baselines

HIS_RD and **DLIREC** are top systems in SemEval 2014 task 4 aspect term extraction subtask [1,12]. Both system leverage hand-craft features and decide the label of the tokens through CRF models.

SENNA Only represent for Elman network only using SENNA embeddings as input of each time step corresponding to each word context in the sentence. The structure are described in the right part of Fig. 2. While **SENNA+Softmax** represent for Elman network sequence labeller with traditional features extending the hidden activations in softmax layer. Results of these two system are reported in [8].

SENNA+Connection takes the connection of SENNA embeddings and specialized features as the input of each time step of neural labeller, which is a simple intuitive implementation of features incorporation.

4 Results and Discussion

4.1 Comparison with Baselines

Table 2 shows the F_1 score comparison of the proposed methods and baseline systems. Besides the standard test set, we also conduct 10-fold cross validation. Experiments are implemented on both domains of 'laptop' and 'restaurant'.

Table 2. F_1 score comparison with baseline systems.

	Cross validation		Test	
	Laptop	Restaurant	Laptop	Restaurant
HIS_RD	-	-	74.55	79.62
DLIREC	-	-	73.78	**84.01**
SENNA Only	76.97	80.08	73.86	79.89
SENNA+Softmax	78.22	81.52	73.70	81.36
SENNA + connection	77.14	80.93	73.19	80.41
Extended dependency	**79.35**	**83.21**	**75.45**	82.12

On the standard test set, our system achieves a new state-of-the-art performance in 'laptop' domain and outperforms previous connection based feature fusion methods in 'restaurant' domain. Compared with other strategies of integrating features and embeddings, such as connection in input layer or softmax layer, the proposed method achieve a significant improvement (according to the two-sided paired t-test with a confidence level of $\alpha = 0.05$). The results shows the effectiveness of our strategy of improving word representations.

4.2 Comparison of Embeddings

To further analyse the source of improvement, we compare different embeddings. Both result before and after fine-tuning are shown in Table 3. Extended group shows the result of extended embeddings which is the out-product of specialized features and original embeddings.

Table 3. Comparison of different embeddings

	Laptop		Restaurant	
	−tune	+tune	−tune	+tune
SENNA	60.85	73.86	75.78	79.89
Google	67.91	72.91	74.73	79.54
Dependency	68.69	74.52	76.98	81.34
Extended SENNA	70.13	74.67	76.58	81.55
Extended Google	71.78	75.33	77.37	**82.22**
Extended dependency	**72.80**	**75.45**	**77.65**	82.12

By comparing the extended group and original group, it is easy to find that the extending strategy is quite effective, especially on the untuned embeddings. Integrating features such as gazetteer of aspect terms bring task specific information to the model and achieve better initial bias as well as final results. Extended SENNA outperforming input connection and softmax connection ($F1$ value of these two shown in Table 2). This is due to the fact that the out-product strategy is free from heterogenicity of simple connection and balances the dense and sparse information with redundancy bits.

It is noticeable that dependency-based embeddings outperform BoW embeddings. The reason of such phenomenon is that dependency-based embeddings focus more on the semantic and syntactic function of word, which is critical for the task of identifying role of specific word in the context. While BoW embeddings encode topical information, and lead to lack of discrimination of word functions in the same product review domain.

4.3 Other Discussion

It is worth noting that the setting of *Extended Google* is language-independent. The embeddings is unsupervised trained and the feature is extracted without using any language-dependent resources or tools. The result still outperform baselines, which indicate our method has a promising potential of transferring to other language.

We test the strategy with different number of hidden layers. Elman network with 1–3 layers perform similarly. While the degradation in performance appears when the number of hidden layers continue to grow. It can be inferred that the shallow form of RNN already has enough capacity to handle short entity like aspect term in this task.

5 Conclusions

In this paper, we have explored to extract aspect terms with improved word embeddings. Dependency-based embeddings encode function of word and out-product of dense embeddings and specialized feature capture task-specific information. Evaluated on public dataset, the proposed method outperform existing feature integrating strategy, achieving a result comparable with the state-of-the-art method. An language-independent setting also achieve comparable result, showing the potential of proposed method implemented in other language.

Acknowledgments. This work is supported by the projects of China Postdoctoral Science Special Foundation (No. 2014T70340), National Natural Science Foundation of China (No. 61300114 and No. 61572151), and Specialized Research Fund for the Doctoral Program of Higher Education (No. 20132302120047).

References

1. Chernyshevich, M.: IHS R&D belarus: Cross-domain extraction of product features using conditional random fields. In: Proceedings of the 8th International Workshop on Semantic Evaluation (SemEval 2014), pp. 309–313 (2014)
2. Collobert, R., Weston, J., Bottou, L., Karlen, M., Kavukcuoglu, K., Kuksa, P.: Natural language processing (almost) from scratch. J. Mach. Learn. Res. **12**, 2493–2537 (2011)
3. Elman, J.L.: Finding structure in time. Cogn. Sci. **14**(2), 179–211 (1990)
4. Graves, A.: RNNLIB: a recurrent neural network library for sequence learning problems (2010). http://sourceforge.net/projects/rnnl
5. Irsoy, O., Cardie, C.: Opinion mining with deep recurrent neural networks. In: Proceedings of the 2014 Conference on Empirical Methods in Natural Language Processing (EMNLP), pp. 720–728 (2014)
6. İrsoy, O., Cardie, C.: Modeling compositionality with multiplicative recurrent neural networks. In: International Conference on Learning Representations (ICLR) (2015)
7. Levy, O., Goldberg, Y.: Dependency-based word embeddings. In: ACL (2), pp. 302–308 (2014)
8. Liu, P., Joty, S., Meng, H.: Fine-grained opinion mining with recurrent neural networks and word embeddings. In: Conference on Empirical Methods in Natural Language Processing (EMNLP 2015) (2015)
9. Mikolov, T., Sutskever, I., Chen, K., Corrado, G.S., Dean, J.: Distributed representations of words and phrases and their compositionality. In: Advances in Neural Information Processing Systems, pp. 3111–3119 (2013)
10. Pontiki, M., Galanis, D., Pavlopoulos, J., Papageorgiou, H., Androutsopoulos, I., Manandhar, S.: Semeval-2014 task 4: aspect based sentiment analysis. In: Proceedings of the 8th International Workshop on Semantic Evaluation (SemEval 2014), pp. 27–35 (2014)
11. Ramshaw, L.A., Marcus, M.P.: Text chunking using transformation-based learning. In: Third Workshop on Very Large Corpora (1995)
12. Toh, Z., Wang, W.: DLIREC: aspect term extraction and term polarity classification system. In: Proceedings of the 8th International Workshop on Semantic Evaluation (SemEval 2014), pp. 235–240 (2014)
13. Yu, M., Gormley, M.R., Dredze, M.: Combining word embeddings and feature embeddings for fine-grained relation extraction. In: North American Chapter of the Association for Computational Linguistics (NAACL) (2015)

Topological Order Discovery via Deep Knowledge Tracing

Jiani Zhang[1,2(✉)] and Irwin King[1,2]

[1] Shenzhen Key Laboratory of Rich Media Big Data Analytics and Applications, Shenzhen Research Institute, The Chinese University of Hong Kong, Shenzhen, China
{jnzhang,king}@cse.cuhk.edu.hk

[2] Department of Computer Science and Engineering, The Chinese University of Hong Kong, Shatin, N.T., Hong Kong

Abstract. The goal of discovering topological order of skills is to generate a sequence of skills satisfying all prerequisite requirements. Very few previous studies have examined this task from knowledge tracing perspective. In this paper, we introduce a new task of discovering topological order of skills using students' exercise performance and explore the utility of Deep Knowledge Tracing (DKT) to solve this task. The learned topological results can be used to improve students' learning efficiency by providing students with personalized learning paths and predicting students' future exercise performance. Experimental results demonstrate that our method is effective to generate reasonable topological order of skills.

Keywords: Knowledge tracing · Topological order · Recurrent neural networks

1 Introduction

Online education platforms have gained great popularity in recent years. Companies like Coursera and edX have attracted millions of students to enroll diversified online courses. However, these Massive Open Online Courses (MOOCs) are contributed by different institutions without an integrated structure. Moreover, in these online environments students often lack personalized instruction that helps them study more efficiently.

In order to give personalized instruction, we need to evaluate what a student knows and does not know in advance. Knowledge Tracing (KT) is such a task of modeling students' latent skills over time based on past study performance, where study performance is a sequence of exercises with correct or incorrect responses. Usually, we model students' mastery of skills as latent variables and students' performance as observed variables. As shown in Fig. 1, each exercise requires an underlying skill to answer the exercise correctly. Different exercises can map to the same skill, e.g., $Ex.1$ and $Ex.2$ both map to $SkillA$. If a student answers an exercise correctly, then the probability of his mastery of the

A. Hirose et al. (Eds.): ICONIP 2016, Part IV, LNCS 9950, pp. 112–119, 2016.
DOI: 10.1007/978-3-319-46681-1_14

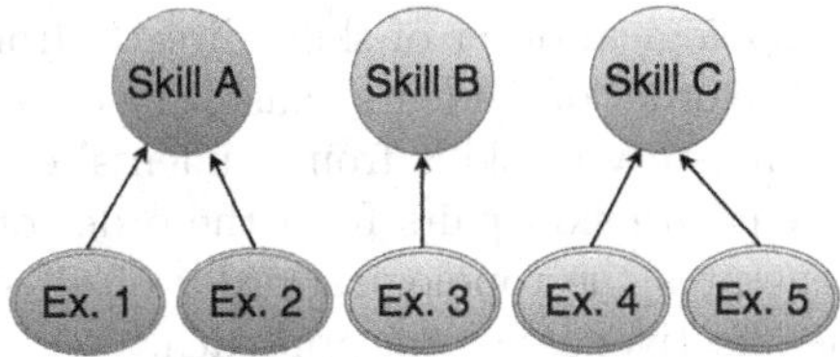

Fig. 1. Sequence of one student's exercise performance. Green nodes represent exercises with correct answers, while gray nodes represent exercises with incorrect answers. Blue skills are in *known* state and gray ones are in *unknown* state. (Color figure online)

underlying skill will increase. Otherwise, the probability will decrease. The most promising application of accurately evaluating students' underlying skills is to help determine which exercise is most suitable to give to students.

However, in previous KT studies, underlying skills are treated as isolated and independent individuals, which is unrealistic in reality. For example, when learning arithmetic operations, a student would not master *subtraction* before this student mastered *addition*. Among skills, there always exists a topological order which is an optimal sequence for students to learn skills one after another. As shown in Fig. 2, $SkillB$ is a prerequisite of $SkillC$. If a student has not mastered $SkillB$, then the probability of correctly answering $Ex.4$, whose underlying skill is $SkillC$, will decrease. This observation makes it possible to discover topological order of skills from students' exercise performance. The challenge of this task is that (1) the input data is only a sequence of binary responses to student's answers, (2) it is inherently difficult to represent human learning process by numerical simulations, and (3) there is no explicit description of skills in the observation data.

In this paper, we propose a rule-based method to discover topological order of skills from limited students' performance data with the aid of Deep Knowledge Tracing (DKT). DKT is a newly proposed method [1] to solve the KT problem using Recurrent Neural Networks (RNNs). One big advantage of deep learning methods is their capability of learning feature representation from large-scale datasets, where domain knowledge and structure can be discarded [2]. Since it

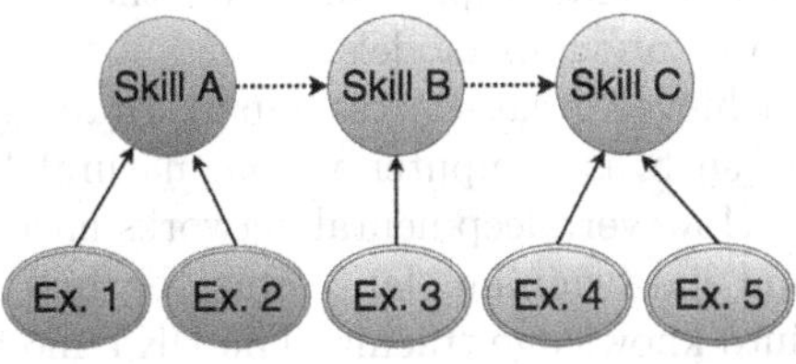

Fig. 2. Adding topological order of skills to predict students' future performance. For example, directed edge (A, B) indicates that $SkillA$ must be completed before $SkillB$ may be attempted and the same as directed edge (B, C).

is difficult to generate topological order of skills directly from students' performance data, we can use the order of students' mastering of skills as a bridge. Our method infers students' mastery of skills from students' exercise performance, and then discovers prerequisite skill pairs from the order of students' mastery of skills, and finally generates the topological order of all skills. Experimental results demonstrate the effectiveness of our approach.

We summarize three main contributions in our paper.

1. This is the first paper to introduce the task of discovering topological order of skills from students' exercise performance.
2. We propose a new method to discover topological order of skills utilizing Deep Knowledge Tracing.
3. Experimental results demonstrate the effectiveness of our method.

2 Related Work

Knowledge Tracing (KT) has increasingly received attention both in the psychology and computer science domain in the past two decades.

The dominant method of KT is Bayesian Knowledge Tracing (BKT), which was first introduced in 1994 [3] to implement mastery learning. Mastery learning maintains that a level of mastery must be achieved in prerequisite knowledge before proceeding to learn subsequent topics. BKT can be easily described with two learning parameters and two performance parameters. Many following variations raised by integrating personalization study [4,5], exercise diversity [6] and other information into Bayesian framework.

The BKT model and its extensions are the most popular models in Intelligent Tutoring Systems due to their strong interpretation properties of evaluating students latent knowledge state. However, several strong assumptions proposed in the first paper [3] have not improved yet in the follow-up study. Assumptions such as bnginary knowledge state representation, no forgetting mechanism and single skill modeling are all the causes that make BKT inflexibility and unrealistic.

Recently, Chris Piech et al. proposed a Deep Knowledge Tracing (DKT) [1] method that used Recurrent Neural Networks (RNNs) [7,8] to trace student's knowledge, which achieved great improvement on the prediction accuracy of students' performance over previous models.

Deep Learning has achieved great success in pattern recognition and machine learning domains [2,9], such as computer vision, natural language processing and speech recognition. However, deep neural networks have not attracted much attention in educational data mining. DKT was the first model to integrate deep learning models into knowledge tracing. The DKT model implemented the simplest one-hidden-layer RNNs, but it demonstrated a stunning improvement over the mainstay, i.e., BKT [10]. The prediction accuracy of students' future performance increased more than 15 %.

3 Topological Order Discovery Model

The goal of topological order discovery can be achieved by two main steps. The first step is to obtain students' mastery of skills from their exercise performance using DKT. The second step is to discover topological relationship between skills from students' mastery order of skills by two predefined rules.

3.1 Deep Knowledge Tracing Model

The DKT model [1] was primarily designed to predict the probability of answering the next exercise correctly. However, the actual output of this model is the probability of answering all the exercises correctly. From student's current exercise performance, we can easily find the exercise with the highest or lowest correct probability in the next time stamp within this model.

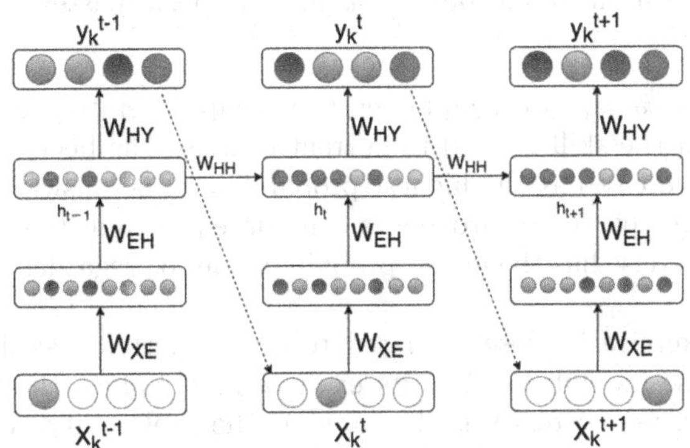

Fig. 3. Deep knowledge tracing model

As shown in Fig. 3, the framework for DKT is Recurrent Neural Networks (RNNs), where at each time t the input $\mathbf{x_k^t}$ from student k is an exercise tuple, i.e., $\mathbf{x_k^t} = \{q_k^t, a_k^t\}$. Exercises which require the same skill to answer correctly are labeled with the same label index in the preprocessing. Suppose there are N unique skills, the input $q_k^t \in \mathbb{R}^N$ is a one-hot encode presentation where only the corresponding index equals to 1 and others are all 0s. Accordingly, a_k^t is also $\in \mathbb{R}^N$ and preprocessed to one-hot encoding with the corresponding exercise index equals to 1 if answered correctly. The combined input for RNNs, i.e., $\mathbf{x_k^t}$, is $\in \mathbb{R}^{2N}$.

The output $\mathbf{y_k^t} \in \mathbb{R}^N$ is the probability that student k will answer each exercise correctly in the next time stamp. Evaluation function is the negative log likelihood. Let $\delta(q_k^{t+1})$ denote which exercise to be answered at time stamp $t+1$ and ℓ be cross entropy. The total loss for an exercise sequence is

$$\text{Loss} = \sum_t \ell((\mathbf{y_k^t})^T \delta(q_k^{t+1}), a_k^{t+1}) \tag{1}$$

Then we can use standard backpropagation algorithm [11] to train the model by computing gradient descents of the loss function.

3.2 Topological Order Discovery

After training the DKT model well, we can acquire the probability of answering all exercises correctly in the next time stamp. The relationship between two skills, e.g., skill A and B, can be divided into four different categories.

1. If B has been mastered, then A has also been mastered. $\Rightarrow$ 'master B $\rightarrow$ master A'.
2. If B has not been mastered, then A has not been mastered either. $\Rightarrow$ '$\neg$ master B $\rightarrow$ $\neg$ master A'.
3. If B has been mastered, but A has not been mastered yet. $\Rightarrow$ 'master B $\rightarrow$ $\neg$ master A'.
4. If B has not been mastered, but A has already been mastered. $\Rightarrow$ '$\neg$ master B $\rightarrow$ master A'.

Among these four cases, case 1 and case 2 can help generate prerequisite pairs. In other words, if the skill related to current exercise has been mastered, then the skill whose exercise has the highest probability to be answered correctly in the next time stamp can be regarded as a candidate prerequisite skill. Otherwise, the skill whose exercise has the lowest probability can be regarded as a candidate prerequisite skill.

Since each exercise is closely related to one underlying skill in the DKT model, we can use its skill label as its exercise label. Thus, the probability of answering an exercise correctly is the same as the probability of mastering its latent skill. We propose the following method to discover topological order of skills from students' exercise performance.

Firstly, generate partial order pairs. Since students always answer exercises with the same skill in sequence, we only need to care about whether this student has mastered the underlying skill at the last time stamp. If the probability of answering the input exercise is larger than 0.5, then it infers that the student has mastered this skill and vice versa. We apply two rules to generate partial order pairs: (1) if the current skill has been mastered, then the skill with the highest output probability can be regarded as a prerequisite, and (2) if the current skill has not been mastered, then the skill with the lowest output probability can be considered as a prerequisite.

Secondly, remove redundant links between skills. The most useful rule is that if '$skillA \rightarrow skillB$', '$skillB \rightarrow skillC$' and '$skillA \rightarrow skillC$', then '$skillA \rightarrow skillB \rightarrow skillC$' is enough. Pruning is an essential part for generating the final topological order of skills.

Last, use topological sorting algorithm to generate one topological order of skills. We apply Kahn's algorithm [12] to generate a topological order from the directed acyclic graph discovered in the previous step.

4 Experiments

In this section, we conduct experiments on real-world datasets to evaluate the effectiveness of our method.

4.1 Datasets

ASSISTment Dataset: This dataset was gathered in year 2009–2010 from the ASSISTments platform.[1] There are two partitions, one with labeled skills and the other without labeled skills. In our experiments, we train the model using the 'skill builder' one, which is a large, standard benchmark in KT topic. In this dataset, there are more than 4,000 students having answered over 446,000 exercises along with 111 unique labeled skills. Two assumptions are imposed on this dataset: (1) each exercise maps to only one skill, and (2) exercises with the same skill are preprocessed to have the same exercise tags.

In order to evaluate the plausibility of extracted topological order of skills, a hierarchical skill graph and its topological order in arithmetic subject was created based on empirical knowledge. Figure 4 demonstrates the ground truth among eight example skills.

4.2 DKT Model Results

First we train the DKT model on the ASSISTment dataset until the AUC value attains 0.86. In the process of training DKT, all one-hot encoding exercises and

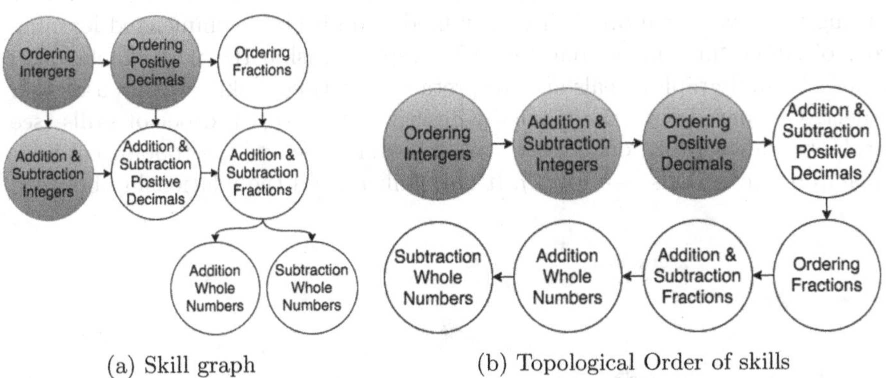

(a) Skill graph (b) Topological Order of skills

Fig. 4. A partial manually-labeled skill graph and its topological order in Arithmetic subject from the ASSISTment dataset. Each node represents a specific skill. Each arrow represents prerequisite relationship. Red nodes denote that the student has mastered these skills and green nodes can be the next highly-recommended skills for this student to learn. (Color figure online)

[1] ASSISTment dataset: https://sites.google.com/site/assistmentsdata/home/assistment-2009-2010-data.

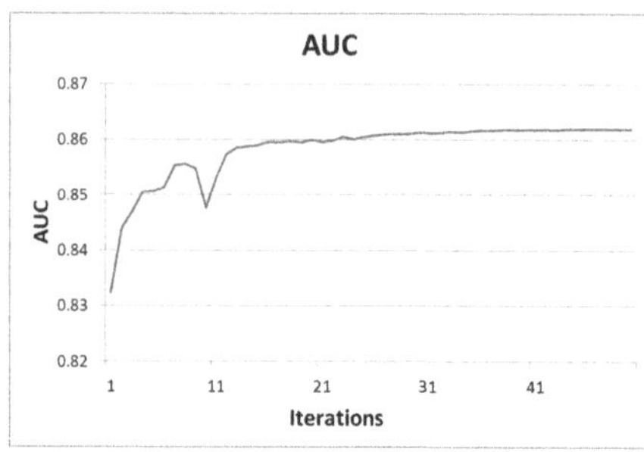

Fig. 5. The AUC in testing time over 50 iterations

Fig. 6. Total training time over 50 iterations

responses are mapped to a fixed-size dense space. We tried several values of the space size and finally set it to be 100.

Furthermore, we consistently use hidden dimensionality of 100 and only one hidden layer. More hidden layers can instead lead to overfitting issue and decrease the AUC value.

Figure 5 gives the AUC value on the testing data within 50 iterations. We can see that it converges very fast and reaches 0.86 after the $23rd$ iteration. The time for training each iteration requires nearly 10 min, which is shown in Fig. 6.

4.3 Topological Order Discovery Result

The prediction accuracy on the ASSISTment dataset guarantees the feasibility of using the topological order discovery model to obtain recommended learning order of latent skills in the dataset. After replacing skill IDs with labeled skill names, the skill graph reveals an interpretable ordering of skills within a certain subject. Compared with the manually-described topological order of skills (see Fig. 4), we can see that our model can accurately discover the topological relationship between skills (see Fig. 7). It only fails to find the relationship between

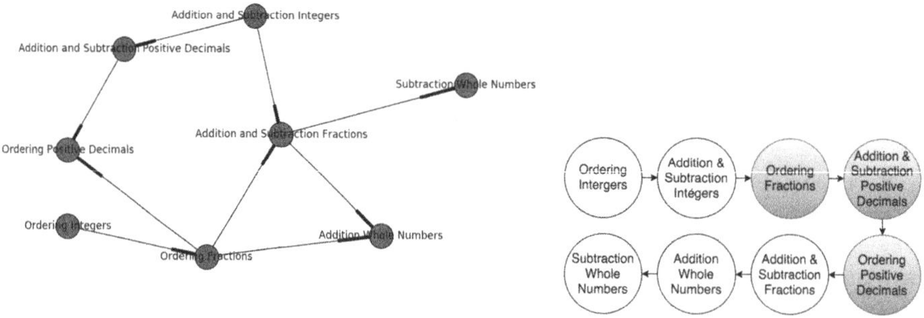

Fig. 7. Experimental results: example skill graph and topological order in ASSISTment dataset

two skills 'Ordering Fractions' and 'Ordering Positive Decimals', which actually have no clear prerequisite relationship between these two skills in reality.

Another interesting observation is that strongly associated skills described in the skill graph occurred far apart in the input exercise sequences. For example, '*Ordering Integers*' is a prerequisite of '*Ordering Fractions*'. However, even though these two skills do not appear in a sequence, '*Ordering Integers*' always appears to be a prerequisite of '*Ordering Fractions*' in our experiments.

5 Future Work

The method of topological order discovery at this stage is generated using some predefined rules. In the next step, we desire to create an end-to-end training model to learn topological order and integrate the topological information into the RNNs framework to improve the prediction accuracy of student performance.

Acknowledgement. The work described in this paper was partially supported by the Research Grants Council of the Hong Kong Special Administrative Region, China (No. CUHK 14208815 of the General Research Fund), and 2015 Microsoft Research Asia Collaborative Research Program (Project No. FY16-RES-THEME-005).

References

1. Piech, C., Spencer, J., Huang, J., Ganguli, S., Sahami, M., Guibas,L., Sohl-Dickstein, J.: Deep knowledge tracing. In: NIPS (2015)
2. LeCun, Y., Bengio, Y., Hinton, G.: Deep learning. Nature **521**(7553), 436–444 (2015)
3. Corbett, A.T., Anderson, J.R.: Knowledge tracing: modeling the acquisition of procedural knowledge. User Model. User-Adap. Interact. **4**(4), 253–278 (1994)
4. Pardos, Z.A., Heffernan, N.T.: Modeling individualization in a Bayesian networks implementation of knowledge tracing. In: De Bra, P., Kobsa, A., Chin, D. (eds.) UMAP 2010. LNCS, vol. 6075, pp. 255–266. Springer, Heidelberg (2010)
5. Yudelson, M.V., Koedinger, K.R., Gordon, G.J.: Individualized Bayesian knowledge tracing models. In: Lane, H.C., Yacef, K., Mostow, J., Pavlik, P. (eds.) AIED 2013. LNCS, vol. 7926, pp. 171–180. Springer, Heidelberg (2013)
6. Pardos, Z.A., Heffernan, N.T.: KT-IDEM: introducing item difficulty to the knowledge tracing model. In: Konstan, J.A., Conejo, R., Marzo, J.L., Oliver, N. (eds.) UMAP 2011. LNCS, vol. 6787, pp. 243–254. Springer, Heidelberg (2011)
7. Graves, A.: Generating sequences with recurrent neural networks. arXiv preprint arXiv:1308.0850 (2013)
8. Hochreiter, S., Schmidhuber, J.: Long short-term memory. Neural Comput. **9**(8), 17351780 (1997)
9. Schmidhuber, J.: Deep learning in neural networks: an overview. Neural Netw. **61**, 85117 (2015)
10. Khajah, M., Lindsey, R.V., Mozer, M.C.: How deep is knowledge tracing? arXiv preprint arXiv:1604.02416 (2016)
11. Hecht-Nielsen, R.: Theory of the backpropagation neural network. In: International Joint Conference on Neural Networks, pp. 593–605. IEEE (1989)
12. Kahn, A.B.: Topological sorting of large networks. Commun. ACM **5**(11), 558–562 (1962)

PTR: Phrase-Based Topical Ranking for Automatic Keyphrase Extraction in Scientific Publications

Minmei Wang[1,2], Bo Zhao[1,2], and Yihua Huang[1,2(✉)]

[1] The National Key Laboratory for Novel Software Technology, Department of Computer Science and Technology, Nanjing University, Nanjing, China
Wang.minmei1211@gmail.com, chawbhoppi@smail.nju.edu.cn, yhuang@nju.edu.cn
[2] Collaborative Innovation Center of Novel Software Technology and Industrialization, Nanjing, China

Abstract. Automatic keyphrase extraction plays an important role for many information retrieval (IR) and natural language processing (NLP) tasks. Motivated by the facts that phrases have more semantic information than single words and a document consists of multiple semantic topics, we present PTR, a phrase-based topical ranking method for keyphrase extraction in scientific publications. Candidate keyphrases are divided into different topics by LDA and used as vertices in a phrase-based graph of the topic. We then decompose PageRank into multiple weighted-PageRank to rank phrases for each topic. Keyphrases are finally generated by selecting candidates according to their overall scores on all related topics. Experimental results show that PTR has good performance on several datasets.

Keywords: Automatic keyphrase extraction · LDA · PageRank

1 Introduction

Keyphrases play an important role in information retrieval (IR) and natural language processing (NLP) tasks [1]. Keyphrases make readers rapidly understand a document and use the understood information for further tasks especially in scientific publications. With the sharp increase of the scientific publications, automatic keyphrase extraction has become urgent and essential.

Automatic keyphrase extraction concerns the automatic selection of important and topical phrases from the body of a document. There are two kinds of methods: supervised methods and unsupervised methods [2]. As training data is hard to procure on supervised methods, most researchers concentrate on unsupervised methods. Existing unsupervised methods can be categorized into four groups: graph-based ranking methods [3], topic-based clustering methods [4], simultaneous learning methods [5] and language modeling methods [6].

A. Hirose et al. (Eds.): ICONIP 2016, Part IV, LNCS 9950, pp. 120–128, 2016.
DOI: 10.1007/978-3-319-46681-1_15

In our work here, we focus on extracting keyphrases in scientific publications. Scientific publications have professional expression and clear semantics compared to other corpora. Many algorithms that have good performance are based on the graph-based ranking methods, the basic one is TextRank [3]. Following algorithms improve TextRank. TopicRank [7] makes some improvements by building graph with topics (a set of candidate keyphrases) rather than words. However, TextRank neglects topic information in the document. TopicRank brings in topic information by clustering phrases into topics through a Hierarchical Agglomerative Clustering (HAC) algorithm. But it only uses single document's information and each topic is of the same importance, which make TopicRank not cover major topics of the document perfectly.

In this paper, we propose an unsupervised method called PTR (Phrase-Based Topical Ranking). The core of our algorithm is that we rank candidate keyphrases by considering phrases graph rather than words graph and deal with a document by its semantic topics. With the fact that a document contains multiple semantic topics, we firstly find latent topics of documents and correlations of phrases with topics by LDA [8] for every document. Then we construct phrase-based graph and decompose PageRank into multiple weighted-PageRank on different topics. Finally, we choose top candidate keyphrases according to their scores on all topics.

Our method has several advantages over traditional graph-based methods. Intuitively, regarding phrases as the subject rather than words conforms to the fact that most keyphrases are combinations of words. Simultaneously, introducing the document's topics can make extracted keyphrases catch major topics and cover all topics. In addition, most other graph-based ranking methods like TextRank [3] use a window size S to measure the proximity distance between two words. This treatment need introduce an additional parameter and neglect co-occurrence information beyond the window size. In our method, we propose an automatic method to measure proximity distance between two phrases in a document by using their offset position.

2 Methodology

We propose a phrase-based topical ranking (PTR) method to automatically extract keyphrases from scientific publications. The process of our PTR method is illustrated in Fig. 1. The overall algorithm is divided into four steps. First, each document is pre-processed and extracted candidate keyphrases. We then acquire the topics of documents and candidate keyphrases by LDA model. Third, we construct weighted phrase-based graph and perform weighted-PageRank algorithm on document's related topics to rank candidates. Finally, we sum up scores on all related topics for candidates and select keyphrases with higher scores.

2.1 Pre-processing

Scientific publications corpus is pre-processed by segmentation, removal of stop words, part-of-speech (POS) tagging and candidate keyphrases selection. The

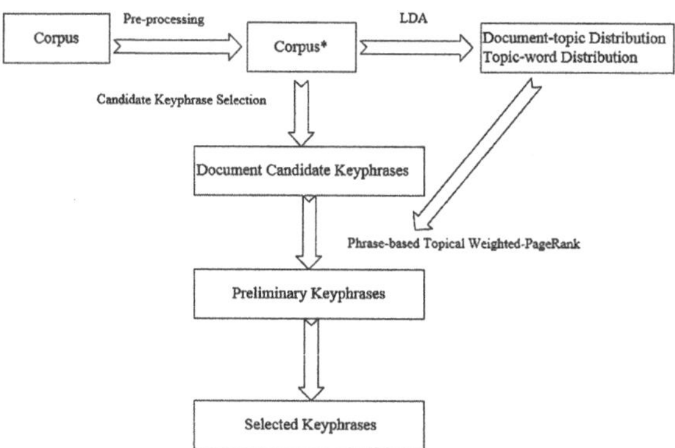

Fig. 1. The process of the proposed method.

first two stages are the basic steps that are widely used in NLP tasks. In the POS tagging stage, every document in corpus is annotated. The widely used POS tagging tool is Stanford Parser. We find noun phrases that satisfy pre-defined pattern are most likely to be keyphrases from previous researches [4,9]. We then extract noun phrases with same pattern as follows:

$$pattern = (JJ)^*(NN|NNS|NNP|NNPS)^+ \tag{1}$$

where JJ indicates adjectives, and various forms of nouns are represented using NN, NNS, NNP and $NNPS$. The pattern represents zero or more adjectives followed by one or more nouns. We regard this noun phrases as candidate keyphrases. In this stage, candidates containing one word in the long papers whose frequency of occurrence fall below a minimum threshold are removed.

2.2 Topic Model Construction

A document is usually composed of multiple semantic topics. LDA is a three-level hierarchical Bayesian model which can find hidden topic information for large collections of discrete data such as text corpora, we use LDA model for topic interpreter. In LDA, each word w of a document d is regarded to be generated by first sampling a topic z for a document from topic distribution ϑ^d, then sampling a word from the topic z's distribution φ^z. ϑ^d and φ^z are separately drawn from conjugate Dirichlet priors α and β. In LDA, the probability of word w for a given document d is represented as follows:

$$p(w|d,\alpha,\beta) = \sum_{z=1}^{K} p(w|z,\beta)p(z|d,\alpha) \tag{2}$$

When the model converges, the distributions of words is stable. We obtain the topic distributions of each word w, namely $p(w|z)$ and the topic distributions of a document, namely $p(z|d)$.

To run weighted-PageRank on a graph, we have to acquire the topics of documents and the correlations of candidate keyphrases with the topics by using LDA. For every topic, we define the correlation of candidate keyphrases (phrases) r_i and topic z as follows:

$$c(r_i|z) = \frac{\sum_{w \epsilon r_i} p(w|z)}{N(r_i)} \tag{3}$$

where $w \epsilon r_i$ represents word w is the containing words of r_i and $p(w|z)$ represents the probability of w belongs to topic z and $N(r_i)$ represents the number of words in phrase r_i. The $p(w|z)$ and $c(r|z)$ will be used in the next step.

2.3 Phrase-Based Topical Weighted-PageRank

Topical weighted-PageRank is a ranking algorithm derived from PageRank. PageRank is a well-known ranking algorithm that uses link information to assign global importance scores to web pages. In the typical PageRank algorithm, graph is usually directed and vertex nodes represent web pages. The basic idea of PageRank is that a vertex is important if there are many other important vertices pointing to it. The simplified equation to compute the importance of vertex v_i is as follows:

$$PR(v_i) = \frac{(1-\lambda)}{N} + \lambda \sum_{v_j \epsilon s(v_i)} \frac{PR(v_j)}{L(v_j)} \tag{4}$$

where N is the total number of vertices, $s(v_i)$ is the set of vertices link to v_i, $L(v_j)$ is the number of outbound links on vertex v_j.

Our phrase-based topical weighted-PageRank is undirected and based on three considerations. (1) Every document has multiple semantic topics. When running PageRank, we consider graphs on every document's topics. (2) Phrases are more suitable to the theme of a document than single words, thus we build graphs by phrases. (3) Every phrase has different probability that belongs to a topic, thus the initial weight of each vertex in the graph is different.

Phrase-based topical weighted-PageRank is divided into two stages: building weighted phrases graph and using weighted-PageRank. The aim of first stage is to build a weighted graph on each document's related topics. The second stage is to use topical weighted-PageRank algorithm to rank candidates on each document's related topics.

In the building weighted phrases graph stage, we use document's candidate keyphrases array *ckArray*, document's topic distributions array *dtArray* and a multidimensional array of topics's words distributions named TW where $TW(i)$ is an array of topic $z's$ word distributions to build graph. In this stage, we first sort array *dtArray* in descending order and pick topics with nonzero value as related topics. Then for chosen related topics of a document, we use the referred Eqs. (3) to compute the correlation of phrase r_i and topic z. Finally, we compute the proximity distance between every two phrases in each topic. We propose a method to automatically measure the proximity distance between two phrases. Our method bases on the fact that there are strong relations between two phrases which appear closely in the document. According to this fact, the equation to compute the proximity distance between phrase r_i and r_j is as follows:

$$dist(r_i, r_j) = \sum_{d_i \epsilon pos(r_i)} \sum_{d_j \epsilon pos(r_j)} \frac{1}{|d_i - d_j|} \tag{5}$$

where d_i and d_j respectively represent the location offset of phrase r_i and r_j. After the building graph stage, we get a weighted graph for every document's related topic.

When weighted phrases graph is built, we use weighted-PageRank algorithm to rank candidates on document's related topics. For each document's related topic, we obtain a graph containing a nonzero candidate keyphrases array r, a corresponding correlation array c built by Eqs. (3) and a multidimensional array $dist$ of proximity distance between every two candidate keyphrases from the last stage. In weighted-PageRank, vertices represent candidate keyphrases in array r, the edge between two vertices represents the corresponding proximity distance in $dist$. The equation to compute the importance of vertices is as follows:

$$PR_z(r_i) = (1 - \lambda)c_z(r_i) + \lambda \sum_{r_j \epsilon s(r_i)} \frac{dist(r_i, r_j)}{O(r_j)} PR_z(r_j) \tag{6}$$

where $c_z(r_i)$ is the initial weight of phrase r_i(correlation of phrase r_i and topic z), $dist(r_i, r_j)$ is the proximity distance between phrases r_i and r_j, and $O(r_j)$ represents sum weight of outbound links on vertex r_j. We gain the ranking score of candidate r_i on topic z, namely $PR_z(r_i)$.

2.4 Keyphrase Selection

After obtaining candidates ranking scores on each topic, we begin to rank candidates on all related topics as the final scores. We define the final scores of a candidate keyphrase as follows:

$$s(r_i) = \sum_{z=1}^{K} PR_z(r_i) * p(z|d) \tag{7}$$

where $p(z|d)$ is the probability of document d belongs to topic d, $PR_z(r_i)$ represents the ranking score of r_i on topic z. $s(r_i)$ is the final ranking score of candidate keyphrase r_i. We finally choose the top candidate keyphrases according to their scores in descending order.

Figure 2 is an example. In Fig. 2, every circle in the graph represents a candidate, different circle color represents different topic. The line between two circles is the relation between two candidate keyphrases, the thicker of the line is, the stronger relation two circles have. Document is composed of many candidate keyphrases (circles), each topic owns the subset of candidates in the document

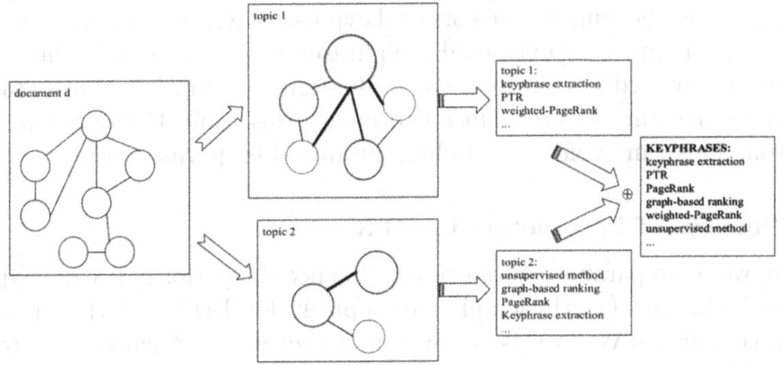

Fig. 2. An example of phrase-based topical ranking method for automatic keyphrase extraction.

because some candidate keyphrases are semantically dead in the topic. For each topic, different size of circle represents the relationship with the topic. The bigger of the circle is, the stronger relation the phrase and the topic have. After running weighted-PageRank algorithm for each topic, candidate keyphrases are sorted on two topics. We choose top candidate keyphrases according to their final scores.

3 Experiments

3.1 Datasets

We employ three standard scientific publication datasets for evaluating the algorithm. The first dataset named SemEval was built by Kim [10]. This dataset consists of 284 ACM full articles which is divided into a trial set containing 40 documents, a training set containing 144 documents and a test set containing 100 documents. The second dataset was built by Krapivin [11]. This dataset contains 2304 ACM full articles. The third dataset was built by Hulth [12] that contains 2000 abstracts which is divided into a trial set containing 500 abstracts, a training set containing 1000 abstracts and a test set containing 500 abstracts. For corpora consistency, we choose test set for SemEval dataset, a subset contains 400 articles of Krapivin dataset coincides with other methods, and 500 test sets for Hulth dataset. For SemEval and Krapivin dataset, the threshold in the pre-processing stage is 80 and 500 respectively.

3.2 Evaluation

To evaluate the effectiveness of the algorithm, we select a standard evaluation metric. The metric is precision/recall/F-measure represented as follows:

$$p = \frac{C_{correct}}{C_{extract}}, r = \frac{C_{correct}}{C_{standard}}, f = \frac{2pr}{p+r}$$

where $C_{correct}$ is the number of correct keyphrases extracted by the method, $C_{extract}$ is the number of extracted keyphrases and $C_{standard}$ is the number of annotated standard keyphrases. For evaluation, the words in both standard and extracted keyphrases are reduced to base forms using Porter Stemmer for comparison. This can avoid mismatching produced by phrases form.

3.3 Influences of Parameters to PTR

There are two main parameters that may influence the performance of keyphrase extraction including: (1) the number of topic K for LDA, (2) the number of extracted keyphrases W. In this subsection, we set different parameters to evaluate the influence of every parameter. Except the parameter under investigation, we set other parameters to the following values: W = 10, K = 50 for both three datasets.

We demonstrate the influence of the number of topics K of LDA model on three datasets in Table 1, we set K equals 30, 50 and 100 respectively. Table 1 shows the results when W is set to 10. We observe that performance does not change much as the number of topics varies. The results on three datasets indicate that learning latent topics from LDA model is appropriate and stable.

Table 1. Influence of number of topic K when the number of selected keyphrases W is 10 on three datasets

K	SemEval			Krapivin			Hulth		
	Precision	Recall	F-measure	Precision	Recall	F-measure	Precision	Recall	F-measure
30	18.6	13.3	15.5	21.0	39.3	27.4	26.8	31.4	28.9
50	19.0	13.5	15.8	20.1	37.6	26.2	27.5	32.2	29.7
100	18.1	12.8	15.0	21.7	40.9	28.4	26.2	30.5	28.1

We observe the influence of extracted keyphrases number W when W ranges from 5 to 15 in Table 2. Theoretically speaking, when W increases, there will be more correct keyphrases to be selected. But the increase speed of W is usually faster than the speed of correct keyphrases, so recall value will be larger while precision value will be smaller. We can see results are in good agreement with the theory from Table 2. F-measure value is a comprehensive indicator that considers both precision and recall value. It also influenced by the number of annotation keyphrases. According to our statistics on three datasets, the average numbers of annotation keyphrases for SemEval, Krapivin and Hulth are 15.13, 6.37 and 9.8 respectively. If the number is larger than the range of W, the F-measure value will be approximately stable. This is the reason why F-measure value on SemEval dataset is approximately stable. As for Krapivin dataset, the average number is 6.37, so when the size of w is between 5 and 10, we achieve good results. As the size of W is larger, more useless phrases are extracted thus the result gets worse. The average number of Hulth dataset is 9.8, so the result is better with the W which is between 10 and 15 than the result with the W which is between 5 and 10 (Table 2).

Table 2. Influence of the number of W when the number of topics is 50 on three datasets

W	SemEval			Krapivin			Hulth		
	Precision	Recall	F-measure	Precision	Recall	F-measure	Precision	Recall	F-measure
5	24.0	8.7	12.7	30.0	28.3	27.2	27.8	16.7	20.5
10	19.0	13.5	15.8	20.1	37.6	26.2	27.5	32.2	29.7
15	15.4	16.5	15.9	15.3	42.4	22.6	27.3	42.6	33.3

3.4 Results of Comparing with Baseline Methods

For comparison purpose, we use two basic baselines. The first is TextRank. TextRank ranks words in the document by introducing PageRank to text and select keyphrases containing high ranked words. The second is TopicRank. TopicRank introduces topic information for each document. For all the methods, we choose number of selected keyphrases as 10. F-measure comprehensively considers precision and recall, we choose F-measure for comparison metric. Table 3 shows the comparison results of PTR with TextRank and TopicRank method. From the experiments on these datasets, our PTR method's results are the highest compared to baseline methods. We draw the conclusion that our method outperforms baseline methods on several datasets.

Table 3. The comparison F-measure results of PTR with baseline methods on three datasets when the number of selected keyphrases W is 10

Method	SemEval	Krapivin	Hulth
TextRank	5.6	12.4	12.7
TopicRank	12.1	-	27.9
PTR	**15.8**	**27.4**	**29.7**

4 Conclusion

We propose an unsupervised automatic keyphrase extraction method called PTR in scientific publications. The introduction of weighted graph and topic information makes PTR achieve promising performance on several datasets. In future work, we consider introducing more proper method to measure distance between two phrases. Specifically, when we get two phrases in a document, we should use more information to judge the distance between them and thus rank their importance better.

Acknowledgments. This work was supported by China NSF Grants (No. 61572250 and No. 61223003) and Jiangsu Province Industry Support Program (BE2014131).

References

1. Nguyen, T.D., Kan, M.-Y.: Keyphrase extraction in scientific publications. In: Goh, D.H.-L., Cao, T.H., Sølvberg, I.T., Rasmussen, E. (eds.) ICADL 2007. LNCS, vol. 4822, pp. 317–326. Springer, Heidelberg (2007)
2. Hasan, K.S., Ng, V.: Automatic keyphrase extraction: a survey of the state of the art. In: Proceedings of Association for Computational Linguistics (ACL). Association for Computational Linguistics, Baltimore, Maryland (2014)
3. Mihalcea, R., Tarau P.: TextRank: bringing order into texts. Association for Computational Linguistics (2004)
4. Liu, Z., Huang, W., Zheng, Y., et al.: Automatic keyphrase extraction via topic decomposition. In: Proceedings of Conference on Empirical Methods in Natural Language Processing, pp. 366–376. Association for Computational Linguistics (2010)
5. Wan, X., Yang, J., Xiao, J.: Towards an iterative reinforcement approach for simultaneous document summarization and keyword extraction. In: Annual Meeting-Association for Computational Linguistics. vol. 45, no. 1, p. 552 (2007)
6. Tomokiyo, T., Hurst, M.: A language model approach to keyphrase extraction. In: Proceedings of ACL 2003 Workshop on Multiword Expressions: Analysis, Acquisition and Treatment, vol. 18, pp. 33–40. Association for Computational Linguistics (2003)
7. Bougouin, A., Boudin, F., Topicrank, D.B.: Graph-based topic ranking for keyphrase extraction. In: International Joint Conference on Natural Language Processing (IJCNLP), pp. 543–551 (2013)
8. Blei, D.M., Ng, A.Y., Jordan, M.I.: Latent Dirichlet allocation. J. Mach. Learn. Res. **3**, 993–1022 (2003)
9. Barker, K., Cornacchia, N.: Using noun phrase heads to extract document keyphrases. In: Hamilton, H.J. (ed.) Canadian AI 2000. LNCS (LNAI), vol. 1822, pp. 40–52. Springer, Heidelberg (2000)
10. Kim, S.N., Medelyan, O., Kan, M.Y. Semeval- task 5: automatic keyphrase extraction from scientific articles. In: Proceedings of 5th International Workshop on Semantic Evaluation, pp. 21–26. Association for Computational Linguistics (2010)
11. Krapivin, M., Autaeu, A., Marchese, M.: Large dataset for keyphrases extraction (2009)
12. Hulth, A.: Improved automatic keyword extraction given more linguistic knowledge. In: Proceedings of conference on Empirical Methods in Natural Language Processing, pp. 216–223. Association for Computational Linguistics (2003)

Neural Network Based Association Rule Mining from Uncertain Data

Sameen Mansha[1(✉)], Zaheer Babar[1], Faisal Kamiran[1], and Asim Karim[1,2]

[1] Information Technology University of Punjab, Lahore, Pakistan
{sameen.mansha,zaheer.babar,faisal.kamiran,asim.karim}@itu.edu.pk,
asim.karim@lums.edu.pk

[2] Lahore University of Management Sciences, Lahore, Pakistan

Abstract. In data mining, the U-Apriori algorithm is typically used for Association Rule Mining (ARM) from uncertain data. However, it takes too much time in finding frequent itemsets from large datasets. This paper proposes a novel algorithm based on Self-Organizing Map (SOM) clustering for ARM from uncertain data. It supports the feasibility of neural network for generating frequent itemsets and association rules effectively. We take transactions in which itemsets are associated with probabilities of occurrence. Each transaction is converted to an input vector under a *probabilistic framework*. SOM is employed to train these input vectors and visualize the relationship between the items in a database. Distance map based on the weights of winning neurons and support count of items is used as a criteria to prune data space. As shown in our experiments, the proposed SOM is a promising alternative to typical mining algorithms for ARM from uncertain data.

Keywords: Frequent itemset mining · Uncertain data · Self organizing map

1 Introduction

Association rule mining (ARM) was introduced in 1993 and it has been applied extensively in numerous applications since then. The process of forming association rules can be divided into two activities: *(i) frequent itemsets finding* and *(ii) association rule generation*. For example in market basket analysis, a dataset contains a number of items and transactions. Each transaction contains a list of items a customer has purchased. The dataset is analyzed to discover frequent items and associations among items that satisfy certain given constraints, e.g., support and confidence thresholds. Besides market-basket analysis analysis, frequent itemsets mining is also a core component in other domains, such as sequential-pattern mining and graph mining.

In many applications data is inherently noisy, a number of indirect data collection methodologies have lead to the proliferation of uncertain data such as data collected by sensors or in satellite images. In uncertain databases, the

A. Hirose et al. (Eds.): ICONIP 2016, Part IV, LNCS 9950, pp. 129–136, 2016.
DOI: 10.1007/978-3-319-46681-1_16

existence of each element is captured by a likelihood measure or probability that indicates their presence. Such databases are much more complex because of the additional challenges of representing the probabilistic information. Many researchers have studied the problem of frequent pattern mining from uncertain data. Chui et al. first solved the problem and took uncertainty of items into account by computing their expected support in their U-Apriori algorithm [7]. Afterwards, they proposed a probabilistic filter to prune candidates early [6]. To avoid candidate generation, UF-growth algorithm generates FP-trees to compute expected support of items [13]. A sampling based algorithm to find frequent patterns in streaming probabilistic data was proposed [18].

On the other hand, there has been some exploration on the topic concerning ARM employing neural networks. Artificial Neural networks (ANNs) is a biological inspired technology which has been used successfully for data mining problems such as pattern recognition, classification and prediction [8]. In 2000, a hopfield network was used to discover frequent pattern itemsets based on the usage of neural network for combinatorial optimization problems. For FIM, a two dimensional array of order $n \times p$ neurons with its corresponding energy function was considered sufficient given a set of n transactions and p items [10]. A cogency inspired measure for producing rules CRAM was introduced. Cogency is the probability of the assumed facts being true if the conclusion is true. CARM takes items as symbols and was efficient for dealing with infrequent items due to its congency-inspired approach. However, only one item consequent association rules were generated [14].

There are a few works directly related to the application of employing a Self Organizing Map (SOM) for ARM. They are used for various purposes in information processing and pattern recognition [11]. SOM was first used for clustering in a method proposed for mining association rules from a database of online recommendation [5]. SOM was applied for ARM on transactions based on the fact that SOM makes the clustering a two-level approach. First the data was clustered with a SOM and then SOM was clustered to decrease computational load [17]. A list of principles was defined to employ a self-organizing map for frequent itemset mining. A novel way of traversing data space was introduced if dependency among patterns accumulated in each item is considered for FIM [3]. A probabilistic method was proposed to identify and extract the support of all possible itemsets from the weight matrix of a trained map [1]. Knowledge embedded in SOM is decoded through the training of associative memory based on correlation matrix memory. Itemset support was estimated from the weight matrix of a trained associative memory to generate association rules [2]. Existing methods for FIM are not directly applicable to uncertain data. To the best of our knowledge no research articles exist to use SOM for FIM from uncertain data. Our main contribution is that we are first one to tackle following problem:

We preprocess transactions under a probabilistic framework to employ self-organizing map for ARM. To reduce overall time complexity, data for ARM is chosen based on distance map of clusters considering support count of items.

Table 1. Uncertain transaction database

Transaction ID	Milk	Bread	Chips
A	0.2	0.7	0
B	0.5	0	1

2 Association Rule Mining in Uncertain Data

Association rule analysis is commonly used for market basket databases for the analysis of customer behavior. Consider the case of a department store with the goal of maximizing sales. Customers can be analyzed to find sets of items that are all purchased by a large group of customers. Table 1 shows such customer information. Here customer A purchases milk 20 % and bread 70 % of his visits. Customer B buys milk in 50 % of times and chips every time she visits the stores. The supermarket uses a database that represents each customer as a single uncertain transaction. The existence of each item is captured by a likelihood measure or probability that indicates their presence in the transactions. A probability value in such a dataset might be derived based on historical data statistics. We assume that these transactions are mutually independent, i.e., the decision of customer A has no influence on customer B. This assumption is justifiable in many real world applications and independence between items is often assumed in the literature [4,7].

An uncertain dataset D consists of d transactions $D = \{t_x|1 = 1,\ldots,d\}$ where each transaction belongs to a set of items $I = \{i_k|1 = 1,\ldots,m\}$. When a set of items, called itemset contains k items, it is referred as k-itemset. Each item i_k in t_x is associated with a non-zero probability $P(i_k \in t_x) \in (0,1)$, which indicates the likelihood that item i_k is present in transaction t_x. For example probability of existence of bread in t_A is $P(Bread \in t_A) = 0.7$. Note that customer B always buys chips so $P(Chips \in t_B) = 1$. Under a *probabilistic framework*, for lossless data interpretation purposes *possible worlds* model is applied. It is assumed that each probability derives two possible worlds (W) per transaction. In one possible world W_1, item i_k is present in transaction t_d; in another world W_2, item i_k does not exist in t_i. We do not know which world is real but each possible world W is associated with $P(W)$ that expresses the probability of W being the true world. Let $P(W_1)$ be the probability of world W_1 being the real world, then we have $P(W_1) = P(i_k \in t_x)$ and $P(W_2) = 1 - P(i_k \in t_x)$. This concept is also applicable if number of items are increased in t_x. For example, let item i_m be another item in t_x with probability $P(i_m \in t_x)$. The probability of world in which t_x contains both items i_k and i_m is $P(i_k \in t_x).P(i_m \in t_x)$. Let $A_{z,x}$ contains set of items that xth transaction, i.e., t_x contains in the world $P(W_Z)$ and $d = |D|$ is the total number of transactions then probability of W_z can be calculated using (1):

$$P(W) = \prod_{x=1}^{d}\Big(\prod_{i \in A_{z,x}} P(i \in t_x). \prod_{j \notin A_{z,x}} (1 - P(j \in t_x))\Big) \tag{1}$$

W1

	M	*B*	*C*
tA	✗	✗	✗
tB	✗	✗	✓

W2

	M	*B*	*C*
tA	✓	✗	✗
tB	✗	✗	✓

W3

	M	*B*	*C*
tA	✗	✓	✗
tB	✗	✗	✓

W4

	M	*B*	*C*
tA	✓	✓	✗
tB	✗	✗	✓

W5

	M	*B*	*C*
tA	✓	✗	✗
tB	✗	✓	✓

W6

	M	*B*	*C*
tA	✓	✗	✗
tB	✗	✓	✓

W7

	M	*B*	*C*
tA	✓	✓	✗
tB	✗	✓	✓

W8

	M	*B*	*C*
tA	✗	✗	✗
tB	✗	✓	✓

Fig. 1. 8 Possible Worlds derived from dataset given in Table 1, $d = 2$ transactions and $u = 3$ uncertain items (milk and bread). Note that chips is a certain item.

The probability of world (W_2) in Fig. 1 is $P(Milk \in t_A) * (1 - P(Bread \in t_A)) * P(Chips \in t_B) * (1 - P(Milk \in t_B)) = 0.2 * 0.3 * 1 * 0.5 = 0.3$. To find frequent patterns, support count *sup* is calculated based on the total number of times, an item appears in database. The number of possible worlds of a database increases exponentially with both the number of transactions and number of uncertain items belonging to transactions. There are 2^u possible worlds where u is the total number of uncertain items that occur in all transactions of D. Fortunately, support count can be calculated without enumerating all possible worlds and finding the support of items in it. Equation (2) defines the support of k-itemset *sup*(i) as the sum of the product of the existential probability of each item from $i \in I$ in all transactions d:

$$sup(i) = \sum_{x=1}^{d} \prod_{i \in I} P(i \in t_x) \tag{2}$$

For example, the expected support of itemset milk is the sum of its expected support in t_A and t_B : $0.2 + 0.5 = 0.7$. An itemset i is frequent if and only if its expected support is higher or equal to a user specified minimum expected support threshold called $minSup$. Given pattern A and B (with $A \cap B = \varnothing$), if pattern AB is frequent, then A is also frequent (called the anti-monotonicity property). Also, $A \Rightarrow B, A \subset I, B \subset I$ is an *association rule* if the following conditions hold:

$$sup(AB) \geq minSup \tag{3}$$

$$conf(A \Rightarrow B) \geq minConf \tag{4}$$

where $\frac{sup(AB)}{sup(A)}$, denoted by $conf(A \Rightarrow B)$, is the confidence of $A \Rightarrow B$, and $minConf \in [0,1]$ is the *confidence threshold*. To verify (4), the values of *sup*(A) and *sup*(AB) have to be calculated first using (2). Rules that satisfy both $minSup$ threshold and a $minConf$ threshold are called association rules.

3 Association Rule Mining on SOM Clusters

Self-organizing map (SOM) is an unsupervised neural network algorithm. Generally SOM consists of a layer of units, called neurons that adapt themselves to a population of input patterns. The adaption is required through an iterative procedure which processes input patterns. Upon presentation of each input pattern, neuron closer to it and all its neighbors are moved closer to input pattern. After a significant number of iterations all neurons move into the area of high concentration of input patterns.

In this paper, we purpose using SOM for association rule mining. All transactions in D can be captured using an adjacency matrix X of size $d \times m$ where $d = |D|$ is the total number of transactions and $m = |I|$ is the number of items in D. The zth transaction is represented by a feature vector x_z comprising of existential probabilities of items in transaction t_z. Moreover x_{zk}, is set equal to existential probability $P(i_k \in t_z)$ of item i_k in transaction t_z. This matrix is used for identifying frequent patterns as discussed next. Adjacency matrix X can be created when transactions are extracted from database. To train input vectors $X = [x_1, x_2, \ldots, x_D]$, a SOM or GHSOM neural network can be initialized to generate map units [11,12]. Two users that buy similar items are likely to fall in one cluster. Collection of such users will form a cluster that influence each other strongly and buy similar items. In SOM, each cluster is represented by a neuron, and the geometric closeness of two neurons show the relative strength of association between respective transactions. Let $Y = \{y_1, y_2, \ldots, y_n\}$ be the set of neurons corresponding to the clusters to be discovered. The weight matrix of the SOM is given by $W = [w_1, w_2, \ldots, w_n]$ where w_i is the weight vector of size m of the ith neuron. Algorithm 1 shows the steps required in training a SOM for identifying dependent items. The SOM is initialized by randomly assigning weights in the interval $[-1, +1]$. Subsequently, features vectors $x \in X$ are fed randomly and repeatedly through the SOM. The process for training involves stepping through several training iteration until the patterns in dataset are learnt by SOM. For each input pattern x one neuron y' will "win" (which means that w'_y is the weight vector more similar to x). This is determined via the distance function $d(x, w_y)$. We use the Euclidean distance for this purpose. The process of determining winning neurons, better known as Best Matching Unit (BMU) selection leads the training. This winning neuron y' will have its weights adjusted so that it will have a stronger response to the input the next time it sees it (which means that the distance between x and w_y will be smaller). The neighborhood function $h(y')$ returns the neurons neighboring y'. For each neighboring neuron y, the corresponding weight w_y is updated such that it is brought closer to x. The learning rate parameter $\eta > 0$ controls the amount of update in each iteration. Furthermore, the update for the winning neuron is stronger than that for its neighbors. This is defined by the Gaussian function $g(y, y')$ centered on y'. The learning rate is reduced with the number of iterations to smooth out the convergence of the SOM. Convergence occurs when change in the weight matrix W from one iteration to the next becomes acceptably small. As different neurons win for different patterns, their ability to recognize that particular pattern will increase.

Algorithm 1. Self-Organizing Map for Clustering Uncertain Transactions

Input: Adjacency matrix $X = \{x_1, x_2, \ldots, x_d\}$ based on uncertain transaction database

Output: An array of clusters which shows specific cluster assigned to each transaction, Y= $\{y_1, y_2, \ldots y_n\}$

begin
  Initialize the weights of each neuron W= $\{w_1, w_2, \ldots w_n\}$ randomly.
  repeat select $x \in$ X randomly
    determine winning neuron y' such that $y' = \arg\min_{y \in Y} d(x, w_y)$
    for all $y \in h(y') \subset Y$
      $w_y = w_y + \eta(t)(x - w_y)g(y, y')$
    decrease η by a small amount
  until termination condition is true
end

4 Experimentation and Analysis

To visualize the formation of clusters and relationship with items, seven randomly generated uncertain transactions are given in Table 2. Probability and support of each item is shown in corresponding cell, e.g., $P(i_4 \in t_A) = 0.4$ and $sup(i_4) = 4.26$. Figure 2 represents output of our SOM when applied on data given in Table 2. We discover 4 clusters that are arranged in a 2×2 grid of cells. Each cell in the grid is identified by its 2-D coordinates with values starting from 0. For example, the bottom left and top right cells are identified as (0, 0) and (2, 2) respectively. Every time a transaction is placed on grid, its corresponding cell is marked with a pattern as shown in Fig. 2. In Table 2, $clusterID$ is assigned to each transaction, which represents the specific coordinates of cell on grid where transaction is placed. For example t_B is placed on cluster (0, 1). If more than one neurons lie in the same cluster they have similar components in their weight factor. This is shown with the background color of each grid cell indicating the normalized sum of the distances between the corresponding neurons weight and those of its neighbors. A darker background color of a cell signifies that the corresponding transaction is closer to its neighbors w.r.t. support count of items. For example transactions t_C and t_D are both located on same cluster (1, 0) and have similar probabilities for item representation. They contain non-zero probabilities for items: i_4,i_5 and i_7; and null values for i_1,i_2 and i_6. It shows the influence of one customer's choice on another or alternatively some items were bought together. If values of any item sets are greater than $minsup$ in these columns, then they are considered as potential candidates for generating association rules. Let $minsup$ be 1, then i_5 will be removed from the map unit because its support count is 0.9. If the support of any 1-itemset is less than minimum support count, then the support count of any k-itemset which contains this itemset will also be less than minimum support count. Remaining itemsets will be passed to U-Apriori to generate association rules. Candidate set generation is very tedious so structured map units can be used to choose data for ARM based on distance map shown in Fig. 2. Our method provides same

Table 2. Characteristics of uncertain transaction database

Transaction ID	P(i_1)	P(i_2)	P(i_3)	P(i_4)	P(i_5)	P(i_6)	P(i_7)	*ClusterID*
t_A	0	0.3	0.0045	0.4	0	1	0.5	(0,0)
t_B	0.00004	0	0	0	0	0	0	(0,1)
t_C	0	0	0.1	0.76	0.23	0	0.0045	(1,0)
t_D	0	0	0	1	0.4	0	0.0012	(1,0)
t_E	0	0.8	0	0.19	0.28	0	0.91	(1,1)
t_F	0	0.004	0.06	0.91	0.00002	0.000004	0.33354	(1,1)
t_G	0.0034	1	1	1	0.005	0.00012	1	(1,1)
SupportCount	0	2.104	1.1445	4.26	0.915	1	2.74924	

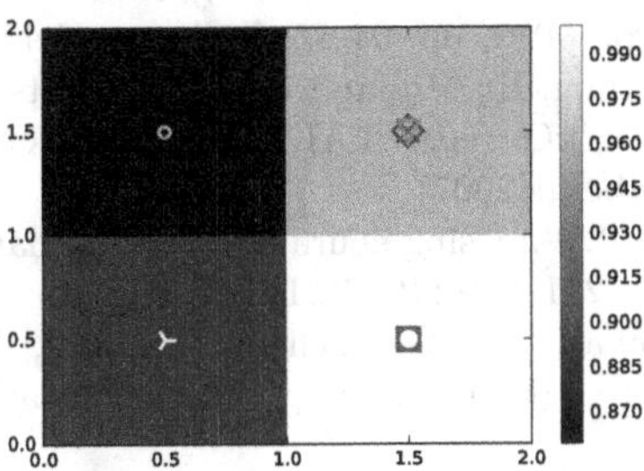

Fig. 2. Map formed after training a SOM

patterns as U-Apriori in less time. In this paper, our implementation of U-Apriori and SOM is based on MiniSom [16] and SPMF [9] tools.

5 Conclusion and Future Work

This work outlines the use of neural networks for association rule mining from uncertain data. U-Apriori is an efficient utility for FIM when database is relatively small. We explore how self-organizing map could be employed for association rule mining to reduce time complexity in large databases. We cluster transactions and consider items with atleast minimum support count to generate rules. This paper has opened up many venues for further research. We plan to apply SOM in such a way that its weight matrix can directly be used for ARM and providing input data to U-Apriori is not required. Our approach returns an estimate of whether an item is frequent or not with no indication of how good this estimation is. To judge the quality of uncertain items, we further plan to consider probability distribution of support, information about the confidence of support [4] and probabilistic association rule generation [15].

References

1. Baez-Monroy, V., Keefe, S.O.: The identification and extraction of itemset support defined by the weight matrix of a self-organising map. In: International Joint Conference on Neural Network (IJCNN), pp. 3518–3525. IEEE (2006)

2. Baez-Monroy, V.O., O'Keefe, S.: An associative memory for association rule mining. In: International Joint Conference on Neural Network (IJCNN), pp. 2227–2232. IEEE (2007)
3. Baez-Monroy, V.O., O'Keefe, S.: Principles of employing a self-organizing map as a frequent itemset miner. In: Duch, W., Kacprzyk, J., Oja, E., Zadrożny, S. (eds.) ICANN 2005. LNCS, vol. 3696, pp. 363–370. Springer, Heidelberg (2005)
4. Bernecker, T., Kriegel, H.-P., Renz, M., Verhein, F., Zuefle, A.: Probabilistic frequent itemset mining in uncertain databases. In: International Conference on Knowledge Discovery and Data Mining, pp. 119–128. ACM (2009)
5. Changchien, S.W., Lu, T.-C.: Mining association rules procedure to support on-line recommendation by customers and products fragmentation. Expert Syst. Appl. **20**, 325–335 (2001)
6. Chui, C.-K., Kao, B.: A decremental approach for mining frequent itemsets from uncertain data. In: Washio, T., Suzuki, E., Ting, K.M., Inokuchi, A. (eds.) PAKDD 2008. LNCS (LNAI), vol. 5012, pp. 64–75. Springer, Heidelberg (2008)
7. Chui, C.-K., Kao, B., Hung, E.: Mining frequent itemsets from uncertain data. In: Zhou, Z.-H., Li, H., Yang, Q. (eds.) PAKDD 2007. LNCS (LNAI), vol. 4426, pp. 47–58. Springer, Heidelberg (2007)
8. Craven, M.W., Shavlik, J.W.: Using neural network for data mining. Future Gener. Comput. Syst. **13**(2–3), 211–229 (1997). Elsevier Press
9. Fournier-Viger, P., Gomariz, A., Gueniche, T., Soltani, A., Wu, C., Tseng, V.S.: SPMF: a Java open-source pattern mining library. J. Mach. Learn. Res. **15**, 3389–3393 (2014)
10. Gaber, K., Bahi, M., El-Ghazawi, T.: Parallel mining of association rules with a hopfield type neural network. In: 12th IEEE International Conference on Tools with Artificial Intelligence, p. 0090. IEEE (2000)
11. Kohonen, T. (ed.): Self-Organizing Maps. Springer-Verlag New York Inc., Secaucus (1997)
12. Kohonen, T., Somervuo, P.: Self-organizing maps of symbol strings. Neurocomputing **21**(1), 19–30 (1998). Elsevier
13. Leung, C.K.S., Carmichael, C.L., Hao, B.: Efficient mining of frequent patterns from uncertain data. In: 7th IEEE International Conference on Data Mining Workshops (ICDMW), pp. 489–494. IEEE (2007)
14. Soltani, A., Akbarzadeh-T, M.-R.: Confabulation-inspired association rule mining for rare and frequent itemsets. IEEE Trans. Neural Netw. Learn. Syst. **25**(11), 2053–2064 (2014)
15. Sun, L., Cheng, R., Cheung, D.W., Cheng, J.: Mining uncertain data with probabilistic guarantees. In: 16th ACM SIGKDD International Conference on Knowledge Discovery and Data Mining, pp. 273–282. ACM (2010)
16. Vettigli, G.: MiniSom: minimalistic and NumPy Based Implementation of the Self Organizing Maps. https://github.com/JustGlowing/minisom
17. Yang, S., Zhang, Y.: Self-organizing feature map based data mining. In: Yin, F.-L., Wang, J., Guo, C. (eds.) ISNN 2004. LNCS, vol. 3173, pp. 193–198. Springer, Heidelberg (2004)
18. Zhang, Q., Li, F., Yi, K.: Finding frequent items in probabilistic data. In: ACM SIGMOD International Conference on Management of Data, pp. 819–832. ACM (2008)

Analysis and Knowledge Discovery by Means of Self-Organizing Maps for Gaia Data Releases

Marco Antonio Álvarez[1(✉)], Carlos Dafonte[1], Daniel Garabato[1], and Minia Manteiga[2]

[1] Department de Tecnologías de la Información y las Comunicaciones, Universidade da Coruña (UDC), Elviña, 15071 A Coruña, Spain
{marco.antonio.agonzalez,dafonte,daniel.garabato}@udc.es

[2] Department de Ciencias de la Navegación y de la Tierra, Universidade da Coruña (UDC), Paseo de Ronda 51, 15011 A Coruña, Spain
manteiga@udc.es

Abstract. A billion stars: this is the approximate amount of visible objects estimated to be observed by the Gaia satellite, representing roughly 1 % of the objects in the Galaxy. It constitutes the biggest amount of data gathered to date: by the end of the mission, the data archive will exceed 1 Petabyte. Now, in order to process this data, the Gaia mission conceived the Data Processing and Analysis Consortium, which will apply data mining techniques such as Self-Organizing Maps. This paper shows a useful technique for source clustering, focusing on the development of an advanced visualization tool based on this technique.

Keywords: Gaia mission · European Space Agency · Data mining · Artificial Intelligence · Self-Organizing Maps visualizations

1 Introduction

Gaia is one of the key missions of the European Space Agency (ESA), which will conduct a census of the Milky Way with unprecedented accuracy. It was successfully launched on December 19th, 2013. During the scientific operation Gaia will observe all visible objects with magnitudes $6 < G < 20$, an estimated one billion objects, representing approximately 1 % of the objects in the Galaxy (the largest amount achieved to date). Thus, it can develop a detailed 3D map that will allow us to answer several open questions about the composition, formation, and evolution of the Milky Way. Gaia will observe not only stars, but all types of astronomical objects with apparent brightness within the limits of the satellite, either galactic (asteroids, comets), or extragalactic (other galaxies, supernovas, quasars).

In order to analyze the data from the Gaia mission, the European Space Agency organized the Data Processing and Analysis Consortium (DPAC) which is composed of hundreds of scientists and engineers. DPAC is divided into nine

A. Hirose et al. (Eds.): ICONIP 2016, Part IV, LNCS 9950, pp. 137–144, 2016.
DOI: 10.1007/978-3-319-46681-1_17

Coordination Units (CUs). The present work is dedicated to algorithm development in CU8, which is responsible for source classification and astrophysical parameters (AP) estimation, as well as in CU9, with data mining and visualization developments, both responsible for providing the platforms for computer processing and data visualization.

Several methods will be applied to the observed objects, in order to obtain a wide variety of relevant information for analysis and data classification. Since the Gaia mission is the first astronomical mission that will unbiasedly observe the entire sky down to magnitude 20, a large number of outliers are expected to be found. This paper explains how these types of objects are to be classified, by means of an unsupervised learning technique that is known as Self-Organizing Maps.

In order to provide a user-friendly environment for this technique that is readily available to the scientists community, we developed a useful interface for analysis and data visualization that is supposed to become operational in the summer of 2016, when the first Gaia Data Release will become public.

2 Self-Organizing Maps

Self-Organizing Maps (SOM) come from the branch of competitive neural networks. They were proposed by Kohonen in 1988 [8]. Since then, they remain the quintessential unsupervised ANN. In fact, to date there have been published over 5,000 articles related to SOMs.

These maps are obtained by projecting a multidimensional continuous input space into a discrete two-dimensional output space. That is, the input dataset is projected onto a set of neurons, which are arranged topologically in a lattice. Generally, a 2D lattice is used for simplicity of calculation and subsequent visualization, but a lattice in 3 or more dimensions can be specified.

Neurons are the main components, which are trained by means of competitive learning; in that sense, each time that an input feeds the network all neurons compete between them to represent this element, but there is only one winner, the best representative neuron, which is calculated through a distance measure. In our case, euclidean distance proved to be the best in result terms, also in computational time.

This technique has the advantage of projecting the dataset into a two (or three)-dimensional grid, where each neuron is related to its neighbourhood, so that the grid preserves the input topology. This topological preservation facilitates the data exploration and highlights properties of the data at hand [7]. Thus, a SOM calculates a reduced set of prototypes and a topological relationship between them. Additionally, it can be quickly trained and it has been demonstrated to be very robust in the presence of noise. SOMs have not yet been widely applied to the field of astronomy, but an introduction to their possible applications can be found in [6].

Due to the vast amount of data to be handled, the processing time is an important aspect to keep in mind. This is why we developed the algorithm using distributed computing through Hadoop and Spark frameworks. Moreover, our algorithm has already been integrated into the SAGA's pipeline [10] at CNES in France with the goal of being available for scientific work. In addition, we developed a version of the algorithm which includes symbolic data [3].

3 Classification Tool

As it was commented before, the amount of data to be processed is enormous, in fact most of the tasks related with Gaia will be processed on a cluster; as a result, we developed the algorithm of the SOM in such a way that it can be executed on a distributed system. Previous works in our laboratory show that we developed and successfully tested a Hadoop [11] cluster version of the algorithm [4,5,9] for CU8, but now in CU9 it has been decided to migrate to Spark [12]. We therefore developed a new version of the algorithm to be deployed on this latter architecture.

By means of SOMs we would be able to analyze and classify the outliers provided by the satellite, but the main objective of our work is to provide a useful tool, which allows the scientific community to perform data analysis and classifications. This tool must allow users to request for complex analysis, in terms of computational time and the amount of data involved, in an easy way and through a friendly interface. Taking this into account we need at least two different modules, the user interaction module and the computation module, in a Client-Server architecture.

The users interact with the first module, the client, which is responsible for data visualizations and for getting information through communication messages with the second module, the server, that implements all the logic for data processing.

The server module has to be able to communicate with the cluster in order to request for the trainings and to retrieve the related files. We decided to implement a Rest (Representational State Transfer) service using Spring; for communications between server and clients we decide to use JSON (JavaScript Object Notation).

The client application has to be powerful, useful, complete, compatible with other Gaia tools, and easy to access and use. We decided to develop and implement a Web application, working with Apache Tapestry because of its potential and, to provide useful features for plots and statistics, JavaScript, 3js and SVG libraries were really helpful. We also integrated SAMP [2] (Simple Application Messaging Protocol) to allow data exchange with other Gaia tools, as described below.

4 Features

Once the application is finished, several visualization tools are available to unveil the datas physical nature and distribution. Users can visualize different representations of the entire SOM as well as some graphics for each neuron individually, studying as such the data from different levels.

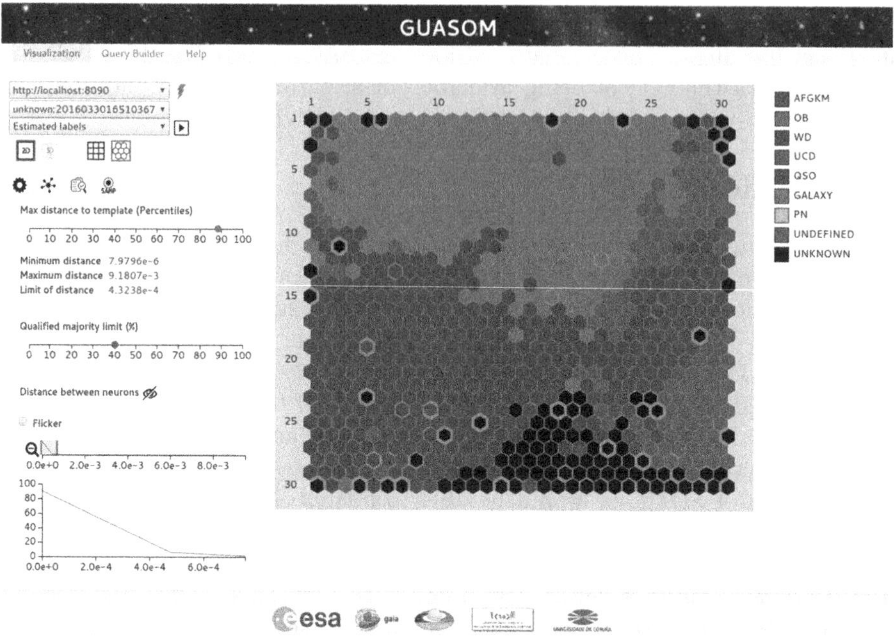

Fig. 1. Capture of the application with Gaia Labels visualization and the feature of distance between neurons activated

As we can see in Fig. 1, the interface has two defined areas: the left panel, where some controls and options are available, and the rest of the space, which is used for the SOM and neuron visualizations.

There is a useful option for representing the distance of each neuron to its neighbours. This option draws the lines between neurons with different width accordingly to the distance between them. More distance means more width because these neurons are less similar, giving the impression of a wall between them (1).

Classical SOM Visualizations

- *U-Matrix* (2a), which displays the distances among clusters. A dark color corresponds to a large distance and thus a gap between the clusters in the input space, whereas a light color between the clusters means that they are close to each other in the input space. Light areas can be thought of as dense regions in the input space, while dark areas correspond to more sparse ones.
- *Hits* (2b), which shows the number of observations falling in each neuron, represented by a color in a gray scale, so that a dark color indicates that the neuron contains few observations, while a light color indicates a high number of observations. This display is helpful to visualize the data density in each region of the SOM network and, hence, of the input space.

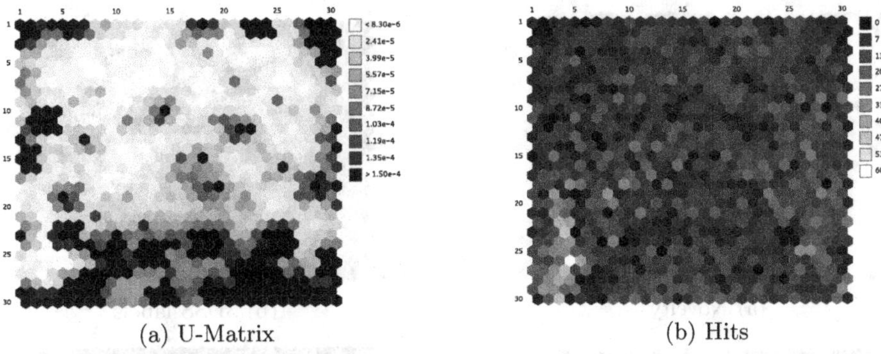

(a) U-Matrix (b) Hits

Fig. 2. Classic views

Specialized SOM Visualizations. These are more specialized visualizations, oriented towards exploring different features related with Astrophysics and its particular parameters. At this point it is convenient to mention that each neuron or cluster has a representative, called prototype, which is a virtual pattern that better represents or resembles the set of input patterns belonging to such a cluster. The neurons are labelled by comparing their prototype to different templates, which were obtained from different libraries such as SDSS Outliers Library spectra.

- *Novelty* (3a). This SOM plot shows the novelty of the neurons. A dark color corresponds to the more novel neurons, calculated by means of the distance between template and prototype. Less distance means less novelty because the observations of this neuron are quite similar to the prototype, and the prototype is well known.

  Item *Simbad labels* (3b). This visualization shows the representative class for each neuron. In this case, SOM clusters receive a color in function of the most frequent SIMBAD [1] identification. In the figure, black clusters do not have objects identified in SIMBAD, a grey color is assigned to clusters with a similar frequency among two or more classes, or with no classes greater or equals to the qualified majority limit. This value can be specified by the user.
- *Color distribution* (3c), which allows to evaluate the network organization according to the stars' colors, which are directly linked to the temperatures. Color distribution $G_{BP} - G_{RP}$ is obtained by subtracting the magnitudes corresponding to the integrated flux respectively of RP and BP spectra.
- *Gaia labels* (1), which represents the distribution of astronomical object classes obtained with Gaia photometric simulations. The color assigned to each cluster was set in function of the predominant class of the objects belonging to it.
- *Category distribution* (3d). This visualization shows the distribution of a unique category, selected by the user, on the map, allowing to determine which neurons have objects of this category.

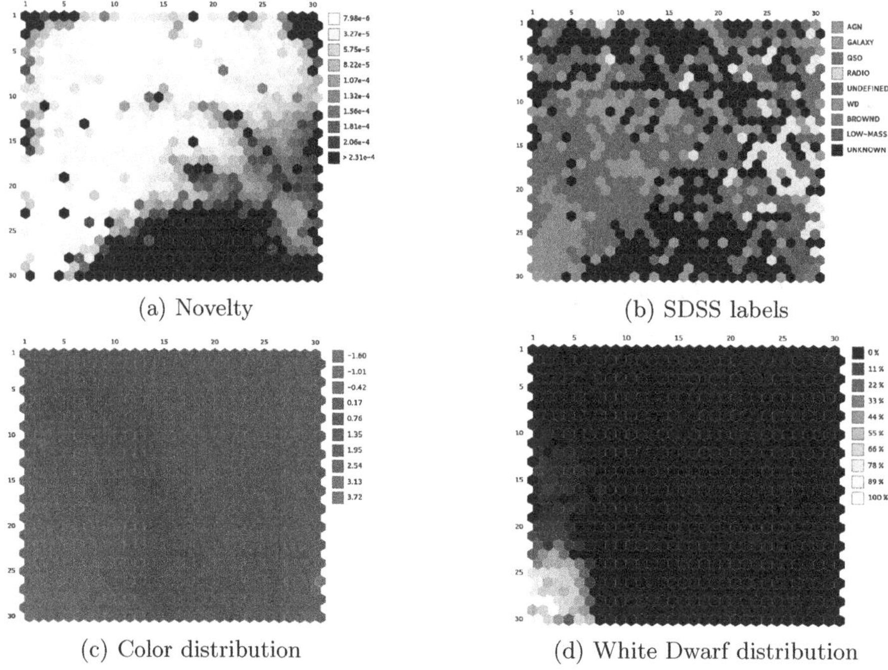

(a) Novelty

(b) SDSS labels

(c) Color distribution

(d) White Dwarf distribution

Fig. 3. Specialized views

These visualizations provide a remarkable improvement in detection of isolated areas, simplifying the identification of weird objects and similar ones.

An extra option that we have developed is the possibility to represent some of these graphs in 3D, allowing the user to visualize some combinations of the previous visualizations. For instance, a user could visualize a combination of the Umatrix and the Gaia Labels representations (4c).

For each neuron, a user could visualize the graphic of the prototype with the matched template (the template that is more similar to the prototype), in order to analyse their similarity, the distribution of the different type of objects which belongs to this neuron (4b) and, in addition, some extra information which is shown in the interface.

Another useful feature is the cross match section, where a user could select any of the observations belonging to a selected neuron and search for this observation on different external data bases, such as Simbad or SkyServer, based on its astronomical coordinates.

Finally, in order to make our application compatible with other tools such as Aladin, Topcat, Vaex, etc., we integrate the SAMP protocol, allowing users to select a neuron or a set of neurons and send their information to others. At the same time, our application could receive messages coming from other tools, giving the user the possibility to download or highlight this information on a specific SOM visualization (4a).

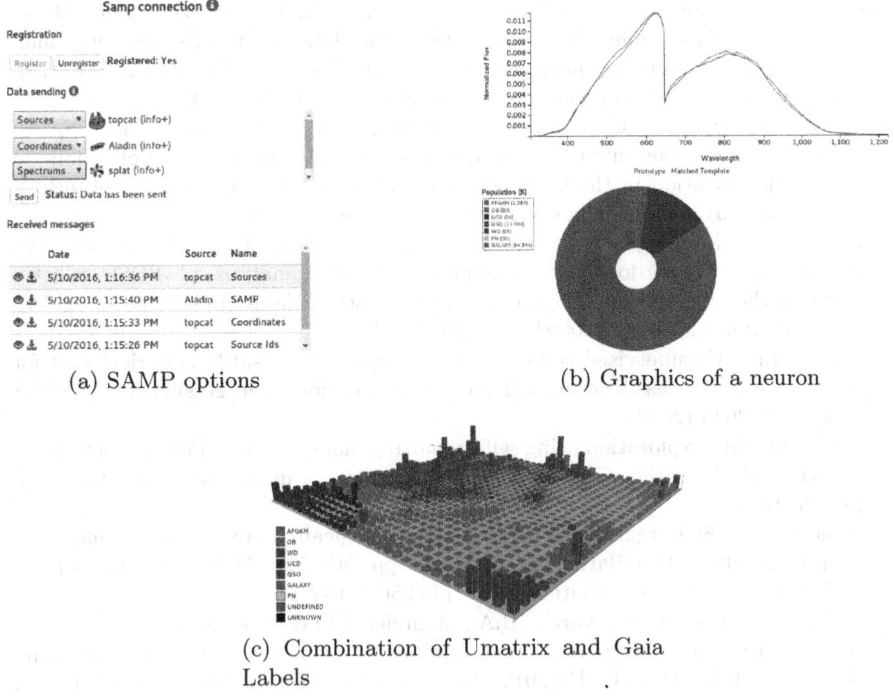

(a) SAMP options

(b) Graphics of a neuron

(c) Combination of Umatrix and Gaia Labels

Fig. 4. Special actions

5 Conclusion

This application was developed for the future data releases of the Gaia mission. We have already presented it in several meetings of Gaia, causing great expectancy, and we believe that the features of this application are really helpful for the desired purpose.

Due to the amount of data that Gaia will provide, data mining techniques are widely important for the treatment of these data. We are aware of the processing time required for the analysis, and have been optimizing our algorithm and our application so as to minimize the processing and the response time. In future works, we plan to include GPU processing in order to reduce the times even more.

We developed this tool in order to make this Artificial Intelligence technique, Self-Organizing Maps, more accessible to the astrophysical and astronomical communities, providing a useful interface for the analysis of Gaia data.

References

1. SIMBAD Astronomical Database. http://simbad.u-strasbg.fr/simbad/
2. Simple Application Messaging Protocol. http://www.ivoa.net/documents/SAMP/

3. del Coso, C., Fustes, D., Dafonte, C., Nóvoa, F.J., Rodríguez-Pedreira, J.M., Arcay, B.: Mixing numerical and categorical data in a self-organizing map by means of frequency neurons. Appl. Soft Comput. **36**, 246–254 (2015). http://www.sciencedirect.com/science/article/pii/S1568494615004512
4. Fustes, D., Dafonte, C., Arcay, B., Manteiga, M., Smith, K., Vallenari, A., Luri, X.: SOM ensemble for unsupervised outlier analysis. Application to outlier identification in the Gaia astronomical survey. Expert Syst. Appl. **40**(5), 1530–1541 (2013). http://dx.doi.org/10.1016/j.eswa.2012.08.069
5. Fustes, D., Manteiga, M., Dafonte, C., Arcay, B., Ulla, A., Smith, K., Borrachero, R., Sordo, R.: An approach to the analysis of SDSS spectro scopic outliers based on self-organizing maps. Astron. Astrophys. **559**, A7 (2013). http://dx.doi.org/10.1051/0004-6361/201321445
6. Geach, J.E.: Unsupervised self-organized mapping: a versatile empirical tool for object selection, classification and redshift estimation in large surveys. MNRAS **419**, 2633–2645 (2012)
7. Kaski, S.: Data exploration using self-organizing maps. In: Acta Polytechnica Scandinavica, Mathematics, Computing and Management in Engineering Series (82), March, 1997
8. Kohonen, T.: Self-organized formation of topologically correct feature maps. In: Neurocomputing: Foundations of Research, pp. 509–521. MIT Press, Cambridge (1988). http://dl.acm.org/citation.cfm?id=65669.104428
9. Ordóñez, D., Dafonte, C., Varela, B.A., Manteiga, M.: HSC: a multi-resolution clustering strategy in self-organizing maps applied to astronomical observations. Appl. Soft Comput. **12**(1), 204–215 (2012). http://dx.doi.org/10.1016/j.asoc.2011.08.052
10. Valette, V., Amsif, K.: CNES Gaia Data Processing Centre, a complex operation plan. In: 12th International Conference on Space Operations, June, 2012. http://www.spaceops2012.org/proceedings/documents/id1291264-Paper-001.pdf
11. White, T.: Hadoop: The Definitive Guide, 1st edn. O'Reilly Media Inc., Sebastopol (2009)
12. Wills, J., Owen, S., Laserson, U., Ryza, S.: Advanced Analytics with Spark: Patterns for Learning from Data at Scale, 1st edn. O'Reilly Media Inc., Sebastopol (2015)

Computational and Cognitive Neurosciences

The Impact of Adaptive Regularization of the Demand Predictor on a Multistage Supply Chain Simulation

Fumiaki Saitoh(✉)

Department of Industrial and Systems Engineering, Aoyama Gakuin University, 5-10-1, Fuchinobe, Chuo-ku, Sagamihara, Kanagawa, Japan
saitoh@ise.aoyama.ac.jp

Abstract. The supply chain is difficult to control, which is representative of the bullwhip effect. Its behavior under the influence of the bullwhip effect is complex, and the cost and risk are increased. This study provides an application of online learning that is effective in large-scale data processing in a supply chain simulation. Because quality of solutions and agility are required in the management of the supply chain, we have adopted adaptive regularization learning. This is excellent from the viewpoint of speed and generalization of convergence and can be expected to stabilize supply chain behavior. In addition, because it is an online learning algorithm for evaluation of the bullwhip effect by computer simulation, it is easily applied to large-scale data from the viewpoint of the amount of calculation and memory size. The effectiveness of our approach was confirmed.

Keywords: Online learning · Adaptive regularization · Supply chain · Inventory simulation · Bullwhip effect · Demand forecasting

1 Introduction

Supply chain management (SCM) is one of the most important factors in modern corporate management. The supply chain is the total process from production and procurement to the product's arrival to customers. It is a multi-stage system composed of many entities such as wholesalers, retailers, logistics bases, makers, and factories. In the inventory management of the supply chain, inventory costs are increased when the order quantity is too large. Conversely, if the order is too small, sales opportunity loss and the reduction of customer satisfaction will occur. It is difficult to find the optimum order in this trade-off in a supply chain with complex behavior. Recently, various related studies have been conducted.

The bullwhip effect can be cited as one of the causes of behavior complication in the supply chain. The bullwhip effect is a phenomenon in which demand fluctuation increases as it moves to the upper stream [11]. Increasing demand fluctuation will cause a large problem for corporate management and will attract cost and risk. Recently, various related studies have been conducted. Lead-time,

A. Hirose et al. (Eds.): ICONIP 2016, Part IV, LNCS 9950, pp. 147–155, 2016.
DOI: 10.1007/978-3-319-46681-1_18

batch order, price fluctuations, and demand prediction accuracy are considered the causes of the bullwhip effect.

In this research, we aim to suppress the bullwhip effect and stabilize the supply chain by improving the accuracy of demand forecasting. The bullwhip effect hardly occurs if perfect synchronization and the sharing of all information can be achieved between the entities constituting the supply chain. However, where this is impractical, each entity is requested to make decisions based on the data. Along with the improved accuracy of the demand forecast for each entity, this can be expected to reduce the impact of the bullwhip effect.

Almost all research that aims to reduce the bullwhip effect is based on a discussion of information-sharing or supply chain structure. Generally, in these simulation models, the moving average method and exponential smoothing method are used as demand predictors. However, because these are not based on learning or estimation, the forecast has low accuracy reliability. In recent years, statistical models and machine learning have been adopted as demand predictors in the supply chain model [1–3]. Demands are predicted by fitting neural networks or time-series regression models to the stored demand data from past trading. These models are assumed offline learning, and it is rare to apply the framework of online learning in these models.

This study provides an application of an online learning-based regression model of demand forecast in the inventory simulation. We expect stabilization of the supply chain and improvement of prediction accuracy. In this study, we apply online regression to the demand prediction model based on the simple models that are widely used in supply chain simulation. We consider speed of convergence and generalization performance, and adopted adaptive regularization as the learning algorithm. The effectiveness of our approach is confirmed through the evaluation of the behavior of the bullwhip effect in the inventory simulation using artificial data.

2 Bullwhip Effect and Inventory Simulation

Here, we provide an outline of the supply chain model utilized in this study. The supply chain model utilized in our simulation is composed of multiple companies that are deployed in series (see Fig. 1). We verify the behavior of the multi-stage inventory simulation by using this model. This model assumes that information about inventory levels and final demand is not shared between entities, the situation is the most unstable, and the bullwhip effect is likely to occur. The mathematical symbols used in the simulation are as below:

- t : Index representing the time series.
- i : Index representing the entities; the numbers $i = 1, 2, ..., n$ are allocated in order from the downstream.
- L_i : Lead time. Products that were ordered at the end of period t will arrive at the beginning of period $t + Li + 1$.
- q_t^i : Order quantity of the i-th entity for the $i + 1$-th entity, in period t.

y_t^i : Target inventory levels, that is, the target value for adjusting the inventory levels.
D_t : Final demand, that is, the demand of the most downstream in the supply chain, which corresponds to customer demand.
$\hat{d}_t^i$: Demand prediction of the i-th entity for orders received from one step downstream.

In this study, the trading of each entity is based on the periodic ordering system, represented by the following equation. Predicted demand value $\hat{d}_t^i$ in the t-th time period is calculated by:

$$\hat{d}_t^i = \begin{cases} f(D_{t-j}, ..., D_{t-p}), i = 1 \\ f(q_{t-j}^{i-1}, ..., q_{t-p}^{i-1}), \ \ otherwise \end{cases} \tag{1}$$

where $f()$ is the prediction model; moving average and exponential smoothing are most widely used in this function. D_t is final demand in the t-th time period, w is time width of the moving average, and $i(1, 2, ..., n)$ identifies the number of stages. Order quantity q_t^i for the i-th entity is calculated by Eqs. 2 and 3. These are determined based on the prediction results for the amount of demand and the inventory levels.

$$q_t^i = \begin{cases} y_{t+1}^i - y_t^i + D_t, \ \ i = 1 \\ y_{t+1}^i - y_t^i + q_t^{i-1}, otherwise \end{cases} \tag{2}$$

$$y_t^i = L_i \hat{d}_t + x\sqrt{L_i}\sigma, \tag{3}$$

where y_t^i is target stock amount, L_i is lead time, and σ is standard deviation of a demand-forecasting error.

The bullwhip effect is a phenomenon in which the variance of the order quantity of each entity becomes larger than the variance of the final demand. For the above model, the magnitude of the bullwhip effect is defined by the following equation:

$$R_{t+1}^i = \frac{Var[q_t^i]}{Var[D_t]}, \tag{4}$$

This is the variance ratio between the demand of the i-th entity and final demand. Here, we can grasp the magnitude of the bullwhip effect by comparing each value.

3 Appling Adaptive Regularization Models to Inventory Simulation

3.1 Demand Prediction Model in Inventory Simulation

Demand forecast plays a critical role in cost management and risk hedging related to supply chain inventory. In the typical supply chain simulation model, the moving average method is used for online-based demand forecasting to evaluate the bullwhip effect [4–6]. However, since the moving average has a delay in

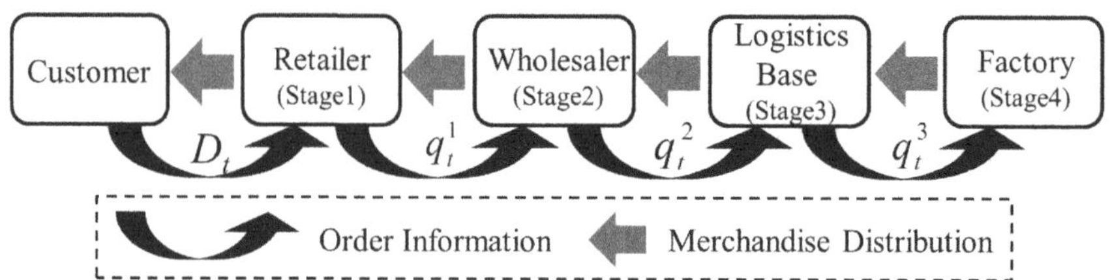

Fig. 1. The schematic diagram of multi-stage supply chain model (information decenterlized inventory simulation)

the trend of the time series data and is only simple noise reduction technique, the prediction accuracy is insufficient in practice.

Recently, various learning models are beginning to be applied to similar tasks, but it is rare to apply the framework of online learning. Since data that can be effectively utilized for management have become large scale in recent years, online learning plays an important role from the perspective of memory size and computational cost. Therefore, in this study, we apply online learning for estimation of the demand predictor. This ensures prediction accuracy without wasting memory and amount of calculation as well as the moving-average method through the adoption of the online learning regression predictor for past data.

The demand prediction model utilized in the present study is an auto-regression (AR) model, which can be expressed as follows:

$$\hat{d}_t^i = \boldsymbol{\mu}_{i,t}{}^{\mathrm{T}} \boldsymbol{x}_{i,t} \tag{5}$$

where $\boldsymbol{x}_{i,t} = (x_{t-1}^i, x_{t-2}^i, ..., x_{t-w}^i, 1)^{\mathrm{T}}$ denotes time-series data and the constant term in the width of time interval w, and $\boldsymbol{\mu}_{i,t}$ denotes the weight vector of the coefficients. T expresses the transpose of a vector. The weight vector $\boldsymbol{\mu}_{i,t}$ corresponding to the coefficient and the constant term is updated in the online learning for this task.

3.2 Demand Prediction Using Adaptive Regularization of Weight Vectors

The stochastic gradient method is widely used in online learning, but has some problems in its application to supply chain tasks. First, it has a large calculation cost and requires much time to reach convergence. Second, it is difficult to set the update width parameter. Finally, the convergence solution is unstable. Since a high generalization performance and rapid response are required in supply chain management, we consider the general stochastic gradient method to be unsuitable for inventory simulation.

Adaptive regularization of weight vectors (AROW) [7,8] is a robust learning algorithm that can be expected to avoid these problems. We apply the framework of AROW in a linear time series regression model as a learning algorithm for the demand predictor. E in the following equation is minimized by updating the variance-covariance matrix Σ and the weight vector μ.

$$E = l_2(y, \boldsymbol{\mu}^{\mathrm{T}}\boldsymbol{x}) + \frac{1}{2}\boldsymbol{x}^{\mathrm{T}}\boldsymbol{\Sigma}\boldsymbol{x} + \lambda D_{KL}(N(\boldsymbol{\mu}, \boldsymbol{\Sigma}||N(\tilde{\boldsymbol{\mu}}, \tilde{\boldsymbol{\Sigma}})) \tag{6}$$

where $\tilde{\boldsymbol{\mu}}$ and $\tilde{\boldsymbol{\Sigma}}$ denote the current solution for $\boldsymbol{\mu}$ and $\boldsymbol{\Sigma}$, and the first term on the right side denotes l_2 loss, defined as follows:

$$l_2(y, \boldsymbol{\mu}^{\mathrm{T}}\boldsymbol{x}) = \frac{1}{2}(\boldsymbol{\mu}^{\mathrm{T}}\boldsymbol{x} - y)^2 \tag{7}$$

where the second term on the right side, $\frac{1}{2}\boldsymbol{x}^{\mathrm{T}}\boldsymbol{\Sigma}\boldsymbol{x}$, denotes the regularization term. The third term on the right side represents the Kullback Leibler (KL) divergence, defined by the following equation.

$$D_{KL}(p||q) = \int p(x) \log \frac{p(x)}{q(x)} dx \tag{8}$$

KL divergence denotes distance from density p to density q. This term contributes to the adjustment of the amount of change during an update in accordance with the reliability of the estimation.

When applied to learning of the autoregressive model with adaptive regularization based on the above E, the following updated equations are obtained:

$$\boldsymbol{\mu}_{i,t} \leftarrow \boldsymbol{\mu}_{i,t-1} + \frac{y^i_{t-1} - \boldsymbol{x}^{\mathrm{T}}_{i,t-1}\boldsymbol{\mu}_{i,t-1}}{\boldsymbol{x}_{i,t-1}\boldsymbol{\Sigma}^i_{t-1}\boldsymbol{x}_{i,t-1} + \gamma} \tag{9}$$

$$\boldsymbol{\Sigma}^i_t \leftarrow \boldsymbol{\Sigma}^i_{t-1} + \frac{\boldsymbol{\Sigma}^i_{t-1}\boldsymbol{x}_{i,t-1}\boldsymbol{x}^{\mathrm{T}}_{i,t-1}\boldsymbol{\Sigma}^i_{t-1}}{\boldsymbol{x}_{i,t-1}\boldsymbol{\Sigma}^i_{t-1}\boldsymbol{x}_{i,t-1} + \gamma} \tag{10}$$

where $\gamma > 0$ denotes a regularization parameter. The weight vector is initialized by uniform random numbers, and the variance-covariance matrix is initialized as the diagonal matrix; all diagonal components are 1. As the number of time steps proceeds, the weight vector and the variance-covariance matrix are updated sequentially, and the demand for the next period is predicted with the learning results.

4 Experimental

Here, we confirm the effectiveness of our approach through computer simulation.

4.1 Experimental Settings

The structure of the inventory simulation for the experiment is a serial type of supply chain model, and is composed of retail, wholesale, logistics bases, and a factory. The relationships of the entities are distributed in an information-decentralized model, and information about final demand cannot be shared between entities. Adaptive regularization, the stochastic gradient method, and the moving average method are each applied to the task, and we evaluate the magnitude of the bullwhip effect and prediction accuracy. The magnitude of the bullwhip effect is evaluated using an index of Eq. 4 described above.

In the experiment, we have been using artificial data as final demand. The data were created using the following formula, taking into account the properties of variation in demand for the product:

$$D(t) = T(t) + C(t) + S(t) + I(t) \tag{11}$$

where trend is expressed as $T(t) = at$, cyclical variation is expressed as $C(t) = \rho_c \cos(k_c t)$, seasonal variation is expressed as $S(t) = \rho_s \sin(k_s t)$, and irregular variation is expressed as $I(t) \sim N(0, \sigma^2)$.

Table 1. Parameter settings for the experiment.

For artificial data		For supply chain model		For prediction model	
D_1	50	Service parameter: z	1.65	γ	10
ρ_c	5	Number of periods	500	w	10
ρ_s	1	Lead time: L_1	1		
k_c	0.02	Lead time: L_2	1		
k_s	0.01	Lead time: L_3	1		
a	0.1	Lead time: L_4	1		
σ	0.5				

The initial value of the prediction model parameters and the data for each enforcement are initialized by uniform random numbers, and the results of 100 trials are evaluated. One trial is composed of 500 steps in the experiment. The magnitude of the bullwhip effect in each setting of prediction is compared in the early steps, in the middle steps, and in the later steps. The parameters utilized in the simulation model and the data creation are shown in Table 1.

4.2 Experimental Results

First, we evaluate the magnitude of the bullwhip effect comparatively. Each vertical axis in Figs. 2(a–c) represents the evaluation index of the bullwhip effect. Figures 2(a–c) correspond to the results of the moving average method, stochastic gradient descent (SGD), and our approach (using AROW), respectively. Figures 2(a–c) correspond to auto-regression by adaptive regularization, the auto-regressive model using the stochastic gradient method, and the moving-average method, respectively. The dark gray bars, bright gray bars, and white bars correspond to the initial step ($20 < t < 100$), the middle step ($150 < t < 250$), and the late step ($350 < t < 450$), respectively. Each horizontal axis represents the

Table 2. Prediction accuracy.

	$t = 50$	$t = 100$	$t = 200$	$t = 500$
Stage.1: MA	4.02	5.03	6.75	5.85
Stage.1: SGD	227.86	212.76	210.83	240.01
Stage.1: AROW	0.64	0.58	0.52	0.53
Stage.2: MA	3.86	5.27	6.89	5.24
Stage.2: SGD	224.54	202.51	207.14	228.32
Stage.2: AROW	19.16	7.14	7.81	5.71
Stage.3: MA	3.69	5.38	7.24	4.93
Stage.3: SGD	237.31	204.66	218.26	229.88
Stage.3: AROW	45.20	17.52	17.35	14.11
Stage.4: MA	3.51	5.50	7.57	4.62
Stage.4: SGD	229.29	182.43	204.28	202.65
Stage.4: AROW	74.45	29.85	20.90	16.59

index of the stage in the supply chain, and, error bars in the figure represent standard deviations of the magnitude of bullwhip effect.

Next, we evaluate the prediction accuracy of each approach. Table 2 shows the results of the prediction accuracy. Each value indicates the mean absolute error, and the rows in the table represent the correspondence between object of comparison approaches to the stage of the supply chain. The columns correspond to the time transition.

4.3 Discussion

As shown in Figs. 2(a–c), it can be confirmed that the evaluation index for the bullwhip effect is higher for the moving average and stochastic gradient methods in the upstream of the supply chain. On the other hand, in our approach using AROW, suppression of the bullwhip effect can be confirmed. This result indicates that the improved prediction accuracy of online learning based on regularization contributes to the stability of the system. In this simulation model based on the periodic ordering system, the behavior of the first right-side term of Eq. 3 affects the bullwhip effect. Therefore, the index R is considered to be stabilized by our approach.

Next, we focus on the transition of prediction accuracy. The prediction accuracy of the stochastic gradient method in the early period is extremely low, and its slow convergence can be confirmed. Further, even in the latter period, the result using the stochastic gradient method was inferior to our approach. Considering the application to the practical supply chain, it was confirmed that the convergence speed of parameters is an important factor. There is no great difference in the prediction error of the moving average with our approach. However, because the moving average has delay between predictive value and real value,

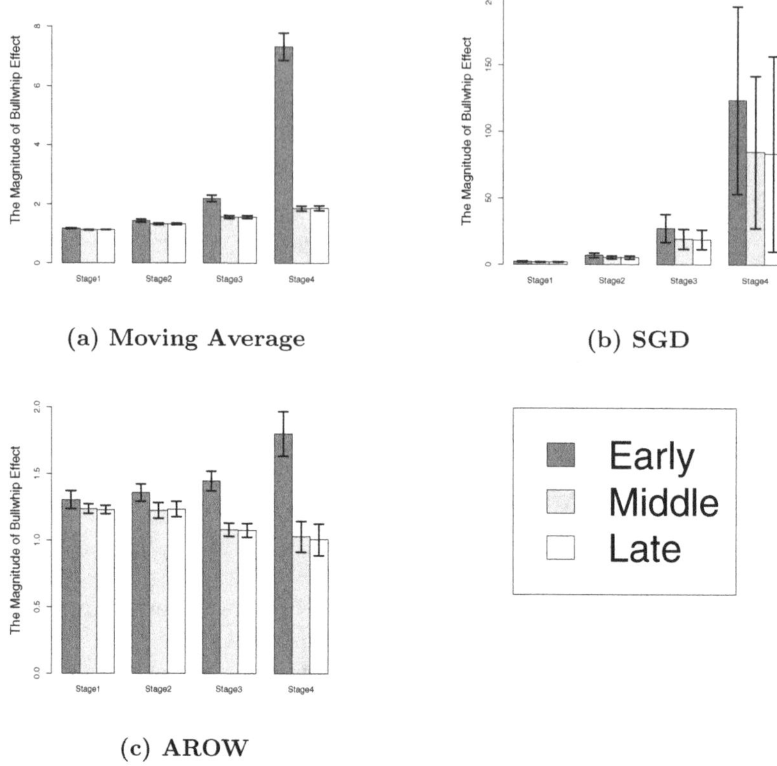

(a) Moving Average (b) SGD

(c) AROW

Fig. 2. Magnitude of the bullwhip effect. Under the situation where the bullwhip effect is not suppressed, the index of magnitude is increased in the upstream stages.

and is not a prediction that accurately captures the trend, it is believed that our approach is able to suppress the bullwhip effect.

5 Conclusion

For the purpose of online estimation and stabilization of the stock simulation, we applied AROW to the demand prediction model. The moving average method and the auto-regressive model based on the stochastic gradient method were used for inventory simulation as a comparison. Through the evaluation of experimental results, we confirmed the validity of our approach from the viewpoints of stability, convergence speed, and prediction accuracy. Further, when adopting online learning, it was confirmed that adaptive regularization is useful.

It has been found that convergence speed and stability of the prediction model's learning affect the degree of the bullwhip effect. The amount of data that can be handled is constrained by memory and processing load, and thus online learning is becoming important. Since supply chains are complicated by recent

ICT developments, we believe that online learning will contribute to a quick response to changes in the market. We are planning an application of a nonlinear predictive model and sophisticated simulation considering the structure of the supply chain as future work [9,10].

Acknowledgments. This work was supported by JSPS KAKENHI Grant-in-Aig for Young Scientists (B) Numbers 15K1625.

References

1. Carbonneau, R., Laframboise, K., Vahidov, R.: Application of machine learning techniques for supply chain demand forecasting. Eur. J. Oper. Res. **184**, 1140–1154 (2008)
2. Jaipuri, S., Mahapatr, S.S.: An improved demand forecasting method to reduce bullwhip effect in supply chains. Expert Syst. Appl. **41**, 2395–2408 (2014)
3. Fu, D., Ionescu, C., Aghezzaf, E., Keyser, R.D.: Quantifying and mitigating the bullwhip effect in a benchmark supply chain system by an extended prediction self-adaptive control ordering policy. Comput. Ind. Eng. **81**, 46–57 (2015)
4. Zhang, X.: The impact of forecasting methods on the bullwhip effect. Int. J. Prod. Econ. **88**, 15–27 (2004)
5. Chinh, N.Q., Zhengping, L., Siew, T.P., Xianshun, C., Soon, O.Y.: An agent-based simulation to quantify and analyze bullwhip effects in supply chains. In: IEEE International Conference on Systems, Man, and Cybernetics, pp. 4528–4532 (2013)
6. Chen, F., Ryan, J.K., Simchi-Levi, D.: The impact of exponential smoothing forecasts on the bullwhip effect. Nav. Res. Logistics **47**, 269–286 (2000)
7. Crammer, K., Kulesza, A., Dredze, M.: Adaptive regularization of weight vectors. In: Neural Information Processing Systems (NIPS) (2009)
8. Sugiyama, M.: Introduction to Statistical Machine Learning, 1st edn. Morgan Kaufmann, Burlington (2015)
9. Chaharsooghi, S.K., Heydari, J., Zegordi, S.H.: A reinforcement learning model for supply chain ordering management: an application to the beer game. Decis. Support Syst. **45**, 949–959 (2008)
10. Saitoh, F., Utani, A.: Coordinated rule acquisition of decision making on supply chain by exploitation-oriented reinforcement learning. In: Mladenov, V., Koprinkova-Hristova, P., Palm, G., Villa, A.E.P., Appollini, B., Kasabov, N. (eds.) ICANN 2013. LNCS, vol. 8131, pp. 537–544. Springer, Heidelberg (2013)
11. Wang, X., Disney, S.M.: The bullwhip effect: progress. Trends Dir. Eur. J. Oper. Res. **250**(3), 691–701 (2016)
12. Kristianto, Y., Helo, P., Jiao, J., Sandhu, M.: Adaptive fuzzy vendor managed inventory control for mitigating the bullwhip effect in supply chains. Eur. J. Oper. Res. **216**, 346–355 (2012)

The Effect of Reward Information on Perceptual Decision-Making

Devu Mahesan, Manisha Chawla, and Krishna P. Miyapuram(✉)

Center for Cognitive Science, Indian Institute of Technology Gandhinagar, Gandhinagar 382355, Gujarat, India
{devu.mahesan,manisha.chawla,kprasad}@iitgn.ac.in

Abstract. Decision making can be treated as a two-step process involving sensory information and valuation of various options. However, the integration of value and sensory information at a neural level is still unclear. We used electroencephalography (EEG) to investigate the effect of reward information on perceptual decision making using two- alternative discriminating task. The reward information was signalled before the appearance of the stimuli. Our findings suggest that economic value acts as a top-down influence early in the decision epoch possibly shifting the evaluation criteria to a more favourable outcome.

1 Introduction

Investigations into the neural mechanism of decision making has been mainly conducted by two groups of neuroscientists, either perceptual or value based. Identifying the neural activity which is responsible for the integration of both is a major challenge faced by system and cognitive neuroscientists.

Previous studies have used single trial analysis of EEG to build a timing diagram to study the neural correlates of perceptual decision making [1]. A late and an early component have been identified, where the early component arises 150 ms and the latter 300 ms post stimulus in an object recognition task [2]. The early component characterizes the speed of low level visual processing and feature discrimination [3] and the late component is claimed to represent the recognition process [4]. Philiastides and Sajda found evidence for these two components in a face vs. car discrimination task, the earliest component correlated to the psychophysical function and the latter represented the psychometric function. The appearance of the earliest component was consistent throughout; however with the decrease in the evidence, a systematic shifting of the later component forward in time was observed [5]. Thus, temporal components are evolved over time, beginning from an early visual processing till processing the accumulated evidence. Significant activation was observed in brain areas related to early visual processing for the early component, whereas activation in the lateral occipital cortex corresponded to the late component.

This research was supported by grant from Cognitive Science research Initiative, Department of Science and Technology, India (SR/CSRI/70/2014).

A. Hirose et al. (Eds.): ICONIP 2016, Part IV, LNCS 9950, pp. 156–163, 2016.
DOI: 10.1007/978-3-319-46681-1_19

Significant activation in areas such as anterior cingulate cortex (ACC) and dorsolateral prefrontal cortex (DPLFC) during difficult task indicated a top down influence of attention related to the task.

A host of imaging studies considering the integration of perceptual and value based decision making suggests that signals originating in the extrastriate visual cortex, regarding the perceptual signals are passed on to lateral intraparietal cortex (LIP), frontal eye field and superior colliculus for decision making [6].

However, when both the sensory as well as value information has to be integrated at the same time to make choices, it remains unknown how temporal components unfold and how the presence of reward information affects the appearance of these components. We used electroencephalography (EEG) to study how reward affects perceptual decision making using a two-alternative discrimination task. The reward information was provided before the discrimination task; hence they had to integrate both the information to come up with a decision. As the reward information do not convey any orientation detail and is uninformative, and the difficulty level of the Gabor patches are not manipulated, the task has been reduced to a simple perceptual decision making. This would help us to isolate the reward component in the decision making task.

2 Materials and Methods

2.1 Participants

Twelve healthy participants (2 women, 10 men), with normal or corrected-to-normal vision, were recruited from students of Indian Institute of Technology, Gandhinagar for the experiment. The mean age of the participants were 22.06 with an age range of 18–27 years. 15 of them were right-handed and 1 left-handed and reported no history of neurologic or neuropsychiatric disorders. Written consent was obtained from all the participants, which informed them of the conditions of the study, allowing them to stop the experiment at any time. The ethics committee of Indian Institute of Gandhinagar approved of the experiment and each participant received Rs. 120 for his/her contribution to the study. Each participant completed the experiment in one session within 35 min.

2.2 Stimuli

The experiment was programmed using E prime software which was adapted to EGI's Net-Station version 5.2. The displays were generated by a Dell computer of screen resolution 1920 × 1080. Responses were collected by button press on the keyboard..

Gabor patches were generated using online Gabor patch generator (http://www.cogsci.nl/gabor-generator), where two sets of Gabor patches (a frequency of 0.10 cycles/pixels enclosed within a Gaussian envelope of standard deviation of 60 pixels) were generated. In one half of the trials the patch which was oriented 5° in the clock wise direction, and in the other half, the patches were oriented 5° in the anticlockwise direction. The RGB values of the background colour were fixed at 128, 128, 128, where Colour 1 was 255, 255, 255 and Colour 2 was 0, 0, 0. If the participants responded correctly, a cash register sound was heard, which lasted for 1 s.

2.3 Experimental Design

Reward information and the orientation of the Gabor patches were manipulated at two levels in this study. On each trial, the Gabor patch was oriented either 5° clockwise (right from the vertical meridian) or 5° anticlockwise (left from the vertical meridian). Reward information was provided by arrows pointing to left and right, which could have a higher value towards the right and lower value towards the left or higher value towards left and lower value towards right. The block consists of four conditions: High reward and Gabor patch towards the right, high reward towards right and Gabor patch towards left, low reward towards right and Gabor patch towards right and low reward and Gabor patch towards left. A control group of balanced rewards was not considered as previous experiments shows that choice is not affected by the magnitude of the reward but by motion coherence in such conditions [7]. Each condition has 100 trials each, hence four hundred trials in all in a block. All the trials within the block were completely randomized.

2.4 Experimental Procedure

The participants were given a two-alternative forced-choice perceptual decision making task where they were asked to differentiate the orientation of the Gabor patches that is the rotation of the Gabor patches from its mean rotation.

Each participant took part in a behavioral and EEG training session which included one block of twelve trials. The experiment was consisted of one block of 400 trials each, 100 trials per condition. Figure 1 shows the sequence of events in a typical trial. Trials began with the appearance of a fixation cross on the screen which was jittered across 200 ms to 500 ms. Then the reward information appeared on the screen for 500 ms. Two types of reward information were possible. One where the higher value (twenty) is towards the right and the lower value (ten) is towards the left and another where the higher value (twenty) is towards the left and the lower value (ten) is towards the left. The participants were asked to judge the orientation of the Gabor patch which stayed on the screen for 150 ms and indicate their choice (right or left) when the question mark appears on the screen by pressing the right arrow key for Gabor patch oriented to right or the left arrow key for Gabor patch oriented to left. Participants were to respond within an assigned deadline of 1500 ms after the onset of the question mark, at the end of which a visual and auditory feedback was received which stayed on the screen for 1000 ms. Feedback indicated whether the response was correct, if so, then the amount of money received by the participant. If the choice was incorrect or if the participant could not provide a response within the deadline, a feedback of zero point was shown on the screen. If the participant could respond correctly within the deadline, they heard a cash register sound and earned either 20 or 10 according to the reward information provided to them. Incorrect responses earned no points and was followed by no auditory feedback.

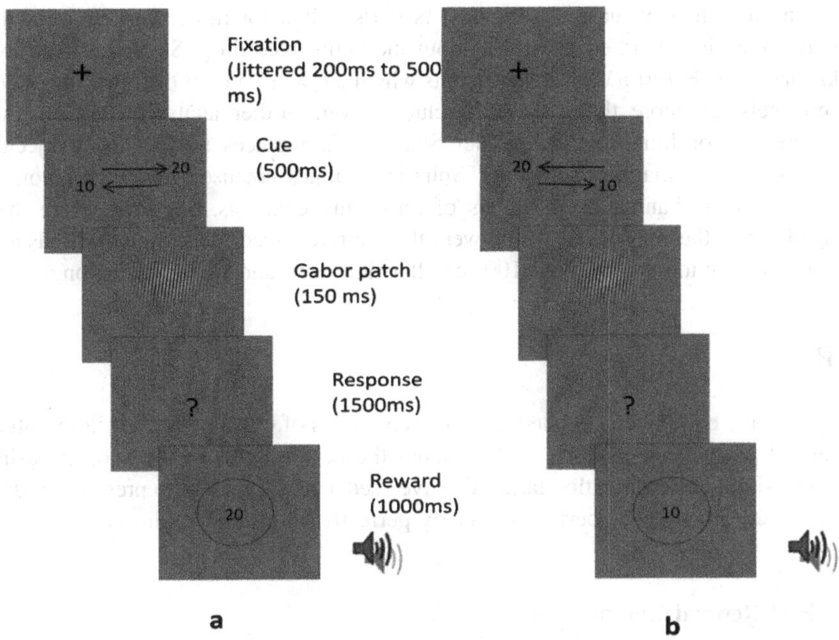

Fig. 1. Perceptual decision making task. On each trial, the participant viewed a Gabor patch and responded indicating the orientation of the Gabor patch (left or right). The reward information cues the magnitude of reward for the correct response; the arrow pointing to twenty denotes a value twice that of the arrow pointing towards 10. The two possible reward segments are shown in a and b. An incorrect answer would lead to zero reward points.

3 Analysis

Electroencephalography (EEG) was recorded using Electrical Geodesics, Inc.'s 128-channel Geodesic Sensor Net. Scalp voltages were gathered continuously from electrodes which were placed on the scalp of the participants throughout the duration of the experiment. The recording of the EEG signal occurred at a 0.1–30 Hz band-pass filter and at a sampling rate of 250 s/s. Coded trigger pulse were send to EEG system by E prime software where stimuli was presented to mark the stimuli and reward information onset. All signals were recorded relative to a vertex reference electrode. The EEG data was processed offline using Net Station 5 Geodesic EEG software version 5.2. The first step to the process of deriving ERPs was filtering. The raw data was subjected to a low pass filter of 30 Hz and high pass filter of 0.1 Hz. The continuous EEG was segmented as: reward locked, where the reward information was locked in place. EEG data was segmented into 1100 ms long epoch starting from 100 ms prior to reward onset. The 100-ms pre reward served as a baseline. Next, Net Stations artifact detection tool was run on each of the segmented file separately, which automatically detects eye blinks, eye movements and marks bad channels in the input file. For a channel to be classified as bad channel, the amplitude of the given segment for that particular channel should be more than 200 μV between its minimum and maximum amplitude values. If a channel is marked bad for more than 20 % of the segments, then it is marked bad throughout the entire recording. Segments with eye blinks that are > ± 140 μV, eye movements which are > ± 55 μV or if the number of bad channels are more than 10, are excluded from further analysis. Bad channel replacement algorithm provided in Net Station 5.2, replaces signals from rejected electrodes in the remaining segments. Spherical splines are used in interpolation of signals of a bad channel from signals of remaining channels. Segments were then averaged across the conditions, which were then re-referenced to average mastoids and a baseline correction was applied 100 ms after time zero and was 100 ms long.

4 Results

To identify the components related to neural correlates of effect of reward information on perceptual decision making, we measured the performance of the subjects using EEG on a simple discrimination task. We have used a 10–20 system to present our data post analysis. All subjects performed nearly perfectly on all the conditions.

4.1 Post Reward Components

In the interval between the onset of the reward information and just before the appearance of the Gabor patch we identified the following components post reward: P100, N100, P200, and N200. The temporal onset of the components was consistent across both post reward and post stimuli. The early components P100 and N100 peaked around 118 ms at O2 and 126 ms at F3 respectively. The late components P200 and N200 peaked around 200 ms at F3 and 199 ms at T6 respectively (Fig. 2).

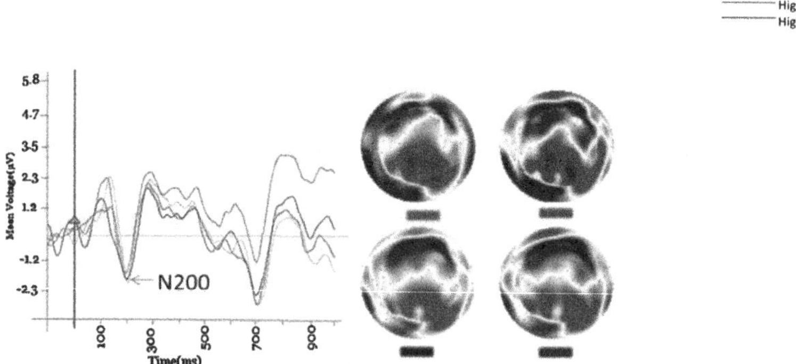

Fig. 2. N200 post reward peaking around 200 ms in F3. A negative going component peaked around 200 ms in the frontal lobe representing a top-down influence on the categorization task

4.2 Post Stimulus Components

In the interval between the appearance of the Gabor patch and just before the earliest response we identified the following components: P100, P200 and N200. The early component P100 peaked around 198 ms at F3. The late components P200 and N200 peaked at 198 ms at F3 and 200 ms at O2.

4.3 Early and Late Components

Researches investigating mechanisms involved in visual processing has been able to distinguish lower level encoding processing from higher level behavior relevant processes. The P100 and N100 components are the early perceptual task independent processes, which are involved in perceptual analysis where feature extraction is performed by the extrastriate visual cortex as it confirmed by the occurrence of P100 in the occipital cortex. Under the influence of spatial attention, attention is directed to the frame for further processing of the reward. Absence of N100 post stimulus indicates that N200 of the reward related component represents a top-down influence on categorization of the Gabor patch.

The components P200 and N200 represents the task relevant late components, which indexes mechanisms for higher level feature detections and stages of item encoding as shown by the presence of P200 [8]. The appearance of N200 represents the process of categorization for task relevant targets [8].

5 Discussion

Our research has used a two-alternative forced choice paradigm to study the effect of reward information on perceptual decision making. The sensory component of the task was manipulating the orientation of the Gabor patch, to which a value element was added by altering the reward associated with each correct choice. We have further investigated the temporal evolution of the components and the effect of reward on them. Figure 3 reviews our findings in the form of a temporal diagram. From previous EEG studies, we know that that the early component (≈170 ms) is related to events such as early perception [1]. Appearance of P100 has been implicated to early

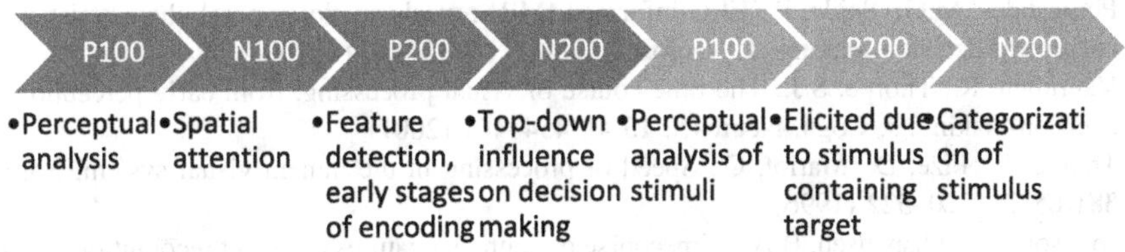

Fig. 3. Timing diagram of ERP components. Diagram summarizing temporal evolution of various components, starting from an early perceptual analysis to the late categorization tasks. Reward influences categorization task early in the decision epoch via a top-down influence on the task

perceptual analysis which is believed to have its sources from the extrastriate visual cortex [8]. Thus P100 can be considered as an early component arising due to early perception. Elicitation of N100 in the frontal lobe confirms our belief that information which is processed by the occipital cortex is conveyed to the dorsolateral frontal cortex [9]. Thus the presence of these components represents a perceptual event and do not directly relate to the task in hand.

As now there are evidences regarding long term representation and categorization of objects by the ventral stream of visual processing [1], the appearance of P200 and N200 post stimulus can be directly linked to categorization, task dependent tasks. Many a studies shows activation in LOC, which is a part of the ventral stream during object perception and recognition [10–13]. Thus the late component is more closely related to the decision-related event.

However, it is interesting to note the absence of N100 post-stimulus, indicating the recruitment of top down influence on the categorization task. Especially the component appearing ≈200 ms after the onset of the reward information can be correlated to this top-down influence. Previous EEG studies have shown the N200 component is implicated for covert attention during object recognition [14] This component is thus more likely to correspond to the top-down influence on categorization of the Gabor patch than a mere bottom-up processing, by the virtue of the absence of N100.

As our stimuli are not dynamic, a relation can be made between the temporal nature of the neural correlates and the fundamental character of the neural processes. Previous work using single trial analysis has identified a early and late component of perceptual decision making at different difficulty levels [1, 5]. Here, we have tried to see how these components are affected when reward information is introduced. An additional evidence for the existence of the early and late component has been provided by our findings. However, we have not addressed the issue of localization of the neural activity.

Thus, we can construct a temporal diagram which can separate the effect of reward information on perceptual decision making in human beings. Our results show that reward information can play a crucial role in assigning attentional resources early in the decision epoch while making choices possibly by shifting the evaluation criteria to the more favorable outcome.

References

1. Philiastides, M.G., Sajda, P.: EEG-informed fMRI reveals spatiotemporal characteristics of perceptual decision making. J. Neurosci. **27**(48), 13082–13091 (2007)
2. Vanrullen, R., Thorpe, S.J.: The time course of visual processing: from early perception to decision-making. J. Cogn. Neurosci. **13**(4), 454–461 (2001)
3. Thorpe, S., Fize, D., Marlot, C.: Speed of processing in the human visual system. Nature **381**(6582), 520–522 (1996)
4. Johnson, J.S., Olshausen, B.A.: Timecourse of neural signatures of object recognition. J. Vis. **3**(7), 4 (2003)
5. Philiastides, M.G., Ratcliff, R., Sajda, P.: Neural representation of task difficulty and decision making during perceptual categorization: a timing diagram. J. Neurosci. **26**(35), 8965–8975 (2006)

6. Puce, A., Allison, T., Gore, J.C., McCarthy, G.: Face-sensitive regions in human extrastriate cortex studied by functional MRI. J. Neurophysiol. **74**(3), 1192–1199 (1995)
7. Rorie, A., Gao, J., McClelland, J.L., Newsome, W.T.: Integration of sensory and reward information during perceptual decision-making in lateral intraparietal cortex (LIP). PLoS ONE **5**(2), e9308 (2010)
8. Luck, S.J.: An Introduction to the Event-Related Potential Technique. MIT Press, Cambridge (2014)
9. Foxe, J.J., Simpson, G.V.: Flow of activation from V1 to frontal cortex in humans. Exp. Brain Res. **142**(1), 139–150 (2002)
10. Malach, R., Reppas, J.B., Benson, R.R., Kwong, K.K., Jiang, H., Kennedy, W.A., Ledden, P.J., Brady, T.J., Rosen, B.R., Tootell, R.B.: Object-related activity revealed by functional magnetic resonance imaging in human occipital cortex. Proc. Natl. Acad. Sci. **92**(18), 8135–8139 (1995)
11. Grill-Spector, K., Kushnir, T., Hendler, T., Malach, R.: The dynamics of object-selective activation correlate with recognition performance in humans. Nat. Neurosci. **3**(8), 837–843 (2000)
12. Grill-Spector, K., Malach, R.: fMR-adaptation: a tool for studying the functional properties of human cortical neurons. Acta Psychol. **107**(1), 293–321 (2001)
13. Grill-Spector, K., Knouf, N., Kanwisher, N.: The fusiform face area subserves face perception, not generic within-category identification. Nat. Neurosci. **7**(5), 555–562 (2004)
14. Luck, S.J., Hillyard, S.A.: Spatial filtering during visual search: evidence from human electrophysiology. J. Exp. Psychol. Hum. Percept. Perform. **20**(5), 1000 (1994)

Doubting What to Eat: A Computational Model for Food Choice Using Different Valuing Perspectives

Altaf H. Abro(✉) and Jan Treur

Behavioural Informatics Group, Vrije Universiteit Amsterdam,
De Boelelaan 1081, 1081 HV Amsterdam, The Netherlands
{a.h.abro, j.treur}@vu.nl

Abstract. In this paper a computational model for the decision making process of food choices is presented that takes into account a number of aspects on which a decision can be based, for example, a temptation triggered by the food itself, a desire for food triggered by being hungry, valuing by the expected basic satisfaction feeling, and valuing by the expected goal satisfaction feeling.

Keywords: Computational model · Food choice · Hebbian learning · Desire

1 Introduction

The effect of overweight and obesity has been studied extensively in recent years, as all over the world it is becoming a substantial problem that increases the risk of health problems, such as diabetes, high blood pressure, cardiovascular diseases and other [1, 2]. Moreover, obesity was found to be associated with an increased risk of morbidity and mortality [3, 4]. One of the recent studies, published in The Lancet [5] has described the current trends in obesity and made future predictions about continuation of these trends: 18 % of men and 21 % of women will be obese by 2025. Researchers from different fields such as neuroscience focus on eating behaviour:

> 'What we eat, when and how much, all are influenced by brain reward mechanisms that generate 'liking' and 'wanting' for foods. As a corollary, dysfunction in reward circuits might contribute to the recent rise of obesity and eating disorders' [6], p. 43

Also from a computational perspective attempts have been made to get an understanding of the decision making process on food choice; e.g., [7]. Computational models can play a vital role in providing support to lifestyle change. To develop smart, human-aware applications, modelling the dynamics and interaction of mental states underlying behavioural choices is becoming more and more important.

In this paper the aspects of eating behaviour and particularly making choices of food are explored computationally. For that purpose it has been taken into account:

- How is desire for food generated in relation to a body unbalance interpreted as being hungry and/or by any kind of sensed external stimulus. For example, certain types of food may be sensed and considered attractive by their cues.

A. Hirose et al. (Eds.): ICONIP 2016, Part IV, LNCS 9950, pp. 164–174, 2016.
DOI: 10.1007/978-3-319-46681-1_20

- How can a person learn to adopt a rational behaviour on the basis of his past experiences; for example, this can involve Hebbian learning [8, 9].
- How does experienced satisfaction play a role, and how can the choice of food be altered through the satisfaction level.
- How influential is the role of goals; for example, a person wants to lose weight (long term goal) and adopts as a short term goal to eat healthy and light food.

2 Background

In the past decade a number of theories and models have addressed different aspects that may cause overweight and obesity. This involves different fields including, psychology, social science, health science, and neuroscience [6, 10]. Researchers attempt to develop interventions which help people change their lifestyle to avoid obesity. The current paper focuses on a neurologically inspired computational model, for which in this section some neurological background is discussed.

From a neurological perspective a large number of studies have been conducted on food-related behaviour, and how food addiction is formed; for example, discussing how highly palatable food activates reward pathways lead to obesity [11–13]. Studies suggest that dysregulation of brain reward pathways may contribute to increased consumption of highly palatable foods, which leads to weight gain and obesity [14]. Another reason for addictive behaviour formation is excessive eating over long period, which involves reward circuitry, and persons become habitual in that behaviour [13]. Different brain regions are involved in food reward circuitry which leads addiction behaviour: a prefrontal region and the amygdale, and also the limbic system integrating amygdala with hypothalamus and septal nuclei [15–17].

Changing the eating behaviour is a quite challenging task, as persons often become habitual in it. Decision making about food choices usually involves considering various options and comparing them in order to make a reasonable choice out of them that can satisfy hunger or taste. Every available food option is coupled with an associated feeling related to prediction whether that option will provide satisfaction or not. It depends on reward mechanisms, as a form of valuing of the options. In making a choice for an option such valuing processes play a significant role.

A desire on the one hand triggers preparations for responses and on other hand generates the associated feelings via recursive body and as-if body loops [18–20], which can predict the consequences and satisfaction of particular options before taking action. This is done by evaluating the options using loops involving interaction between feeling and preparation. The connections from feelings to preparations can be either assumed static or adaptive. Adaptive connection strengths depend on earlier experiences [21, 22]. They can be learnt over time as a form of neural plasticity by changing the connections strengths [9, 23].

Neuroscientific literature shows that goals guiding an individual's behaviour are linked with activations in the prefrontal cortex. Such activations can inhibit the activation of subcortical structures (e.g., cerebellum and basal ganglia), associated with habitual behaviour [24, 25] and have the ability to change habitual behaviour [26–28].

Such control mechanisms can help persons to change their preferences or choice of food by keeping goals in mind; for example, they may start to look for low calories food instead of high calories food.

3 The Computational Model

In this section the proposed computational model is discussed, based on the literature described in Sect. 2. The model was designed by considering a number of aspects of eating behaviour, such as food desire generation and preparation for actions to fulfil the desire. As causes of a food desire either internal (feeling less energy, hungry), or external (stimuli) causes are considered. In relation to a generated desire, particular actions are considered and the extent to which they will provide a feeling of satisfaction. Moreover, goals are considered in this model. An overview of the model is depicted in Fig. 1; the concepts used are explained in Table 1. The cognitive model reflects the underlying neurological concepts at an abstracted level, and is based on a network of temporal-causal relationships [29].

The process of a desire generation is modelled in two ways, as shown in Fig. 1. First, a desire can be generated through the metabolic activity (energy level). This process is modelled by the connections from metabolic state ms_{ub} that leads to the bodily unbalance (for being hungry) represented by body state bs_{ub}. A person senses

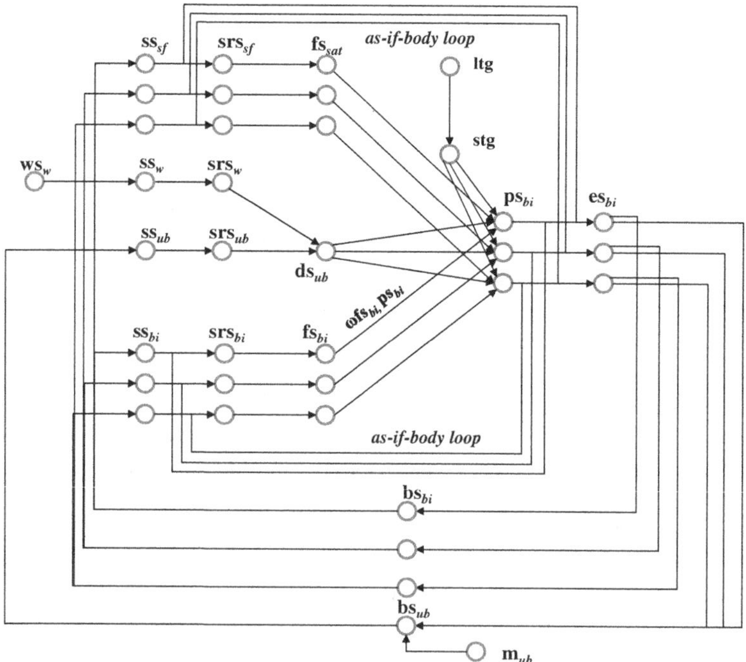

Fig. 1. Conceptual representation of the computational model

Table 1. Overview of the states of the proposed model (see also Fig. 1)

Formal	Informal name	Description
ws_w	World state *w*	This characterizes the current world situation which the person is facing the stimulus, in this case *w* is the food stimulus
ss_w	Sensor state for *w*	The person senses the world through the sensor state, providing sensory input
srs_w	Sensory representation state for *w*	Internal representation of sensory world information on *w*
m_{ub}	Metabolism for *ub*	Represents metabolic energy level: a low level while being inactive and a high level while being active
bs_{ub}	Body state of *ub*	A bodily state that represents the body state underlying being hungry; if a person is becoming hungry, the body unbalance *ub* increases, and after eating there will be lower or no unbalance
ss_{ub}	Sensor state for *ub*	The person senses the bodily unbalance state *ub*, providing sensory input
srs_{ub}	Sensory represent- ation of *ub*	Internal sensory representation for sensed bodily unbalance information on *ub*
ds_{ub}	desire for unbalance *ub*	Generating a desire based on sensory representation for a body unbalance *ub* (e.g., a state of being hungry)
ps_{bi}	Preparation for an action b_i	Preparation for an eating response
fs_{bi}	Feeling associated to body state b_i	A feeling state fs_{bi} for the eating action
srs_{bi}	Sensory representation of b_i	Internal sensory representation of body srs_b for *bi* in the brain. In this case here *b* represents the associated available food choices
ss_{bi}	Sensor state for b_i	The person senses the external body states or environment through the sensor state, providing sensory input.
ss_{sfbi}	Sensor state for satisfaction sf_{bi}	The person senses the external body states providing sensory input to the feelings of satisfaction.
srs_{sfbi}	Sensory representation of satisfaction sf_{bi}	Internal representation of the body aspects of feelings of satisfaction
fs_{sfbi}	Feeling for satisfaction sf_{bi}	Feeling of satisfaction; these are the feelings about the considered food choice, how much satisfactory it is.
es_{bi}	Effector state for action b_i	In the considered scenarios the action b_i is performed to eat food of particular choice for which a person is prepared, to reduce the body unbalance *ub*
bs_{bi}	Body state for b_i	This represents external body states related to eating that particular food.
ltg	Long term goal	This represents the long term goal, to lose weight, for example
stg	Short term goal	Short term goal refers to smaller incremental way of achieving long term goals for example start to eat healthy avoid from fast food etc.
$\omega_{fsbi,psbi}$	Learnt connections	These are the connections which can be learnt by Hebbian learning. This one models how the generated feeling affects the preparation for response bi.

the bodily unbalance via the internal sensor state ss_{ub} that enables the person to form the sensory body representation state srs_{ub} representing body unbalance in the brain. By this body unbalance the desire to eat something to reduce the body unbalance develops and the person prepares for the available food options. This is modelled by the connection from the desire state ds_{ub} to preparation states ps_{bi}.

Not only the desire but also the associated feelings have impact on the preparations; before performing any action a feeling state fs_{bi} for the (partly prepared) action is generated by a predictive *as-if-body loop* via the sensory representation state srs_{bi}.

In the considered scenario, the desire state ds_{ub} has an effect on a number of preparation states ps_{bi} for responses b_i which lead to sensory body representation srs_{bi} and to the feeling states fs_{bi}. Subsequently, the states fs_{bi} have strengthening impacts on the preparation state ps_{bi}, which in turn has an impact on feeling state, fs_{bi}, through srs_{bi} which makes the process recursive: an as-if body loop (e.g., [30]).

A desire, generated by body unbalance or by environmental factors or both, needs to be fulfilled by eating a particular selected food. A person takes action for a selected food option to reduce or eliminate the body unbalance. In this model the effect of an executed action b_i (by es_{bi}) is effectuated on the one hand through the connection from effector states es_{bi} to body states bs_{bi} that are bodily representations of direct valuations or satisfaction for the selected options. On the other hand the action effect is effectuated through the connection from effector states es_{bi} to body states bs_{ub}, this action effect represents reduction of the bodily unbalance bs_{ub} after eating.

Another process involves the expected satisfaction feeling of goal fulfilment after the particular action execution es_{bi} (eating particular food). Expected satisfaction feelings also have effect on making choices. For example, if a person takes fast food initially, after eating he may not feel satisfied as it was not healthy, so next time (s)he may prefer to go for healthy food, in order to feel more satisfaction.

The model as depicted in Fig. 1 shows the connection from body state bs_{bi} sensed by the person via sensor state ss_{sf} and represented by sensory representation state for goal fulfilment satisfaction srs_{sf}. This representation state srs_{sf} receives impact from the preparation state ps_{bi} via an *as-if-body loop* as well. The sensory information srs_{sf} then leads to a level of expected feelings of goal fulfilment satisfaction fs_{sf}. So on the basis of this satisfaction a person may continue to choose the same option (eat the same kind of food) again and again or he or she may change the option, e.g., from fast food to healthy food. In this way it will have impact on preparation states ps_{bi}.

Setting goals in life either long term or short term is also a common practice. Goals may relate to a healthy life style, behaviour change or other ambition. In this case long term goals can be considered to weight loss and short term goals can be considered in relation to taking a healthy diet. If a person does not have any goal about his or her health he or she may eat anything based on various personal characteristics/preference of wanting and liking. On the other hand, if a person wants to aim at being healthy then this may affect what is considered suitable to eat. In the considered scenario as a long term goal ltg it is taken that a person wants to lose weight and that leads to short term goal stg which is to take a healthy diet to achieve long term goal. A short term goal has a direct influence on the preparation states ps_{bi}. So by keeping goals in mind a person can change his eating behaviour and start giving preference to healthy food.

Another factor that is included in the model, is how the effect of the feeling on the preparation can be learnt over time based on past experiences. For this a Hebbian learning mechanism has been included by which such connections may automatically emerge or strengthen. The connections from feeling fs_{bi} to preparation ps_{bi} with weights $\omega_{fsbi,psbi}$ have been made adaptive in this way. The strengths $\omega_{fsbi,psbi}$ are adapted by the Hebbian learning principle, that connected neurons that are frequently activated simultaneously strengthen their connections over time [8, 9, 31].

In the example scenarios it has been shown that if a person starts changing eating behaviour than the particular connection gets more strength over time due to learning. Each time the person experiences a certain feeling for a prepared action, through Hebbian learning this strengthens the association between feeling and preparation. So the recursive process of predictive *as-if body* loops via feeling states fs_{bi} to the preparation states ps_{bi}, involves Hebbian learning in order to enable adaptivity.

The model description presented above (Fig. 1 and Table 1) represents a network of cognitive and affective states in a conceptual manner. The following elements are also considered to be given as part of such a conceptual representation:

- For each connection from state X to state Y a *weight* $\omega_{X.,Y}$ (a number between −1 and 1), for strength of impact; a negative weight is used for suppression
- For each state Y a *speed factor* η_Y (a positive value) for timing of impact
- For each state Y (a reference to) a *combination function* $\mathbf{c}_Y(...)$ used to aggregate multiple impacts from different states on one state Y

For a numerical representation of the model the states Y get activation values indicated by $Y(t)$: real numbers between 0 and 1 over time points t, where the time variable t ranges over the real numbers. The conceptual representation of the model (as shown in Fig. 1 and in Table 1) can be transformed in a systematic or even automated manner into a numerical representation as follows [29]:

- At each time point t state X connected to state Y has an *impact* on Y defined as

$$\mathbf{impact}_{X,Y}(t) = \omega_{X,Y}X(t)$$

where $\omega_{X,Y}$ is the weight of the connection from X to Y

- The *aggregated impact* of multiple states X_i on Y at t is determined using a *combination function* $\mathbf{c}_Y(..)$:

$$\begin{aligned}\mathbf{aggimpact}_Y(t) &= \mathbf{c}_Y(\mathbf{impact}_{X1,Y}(t), \ldots, \mathbf{impact}_{Xk,Y}(t)) \\ &= \mathbf{c}_Y(\omega_{X1,Y}X_1(t), \ldots, \omega_{Xk,Y}X_k(t))\end{aligned}$$

Where X_i are the states with connections to state Y

- The effect of $\mathbf{aggimpact}_Y(t)$ on Y is exerted over time gradually, depending on *speed factor* $\eta\eta_Y$:

$$\begin{aligned}Y(t + \Delta t) &= Y(t) + \eta_Y[\mathbf{aggimpact}_Y(t) - Y(t)]\Delta t \\ \text{or } \mathbf{d}Y(t)/\mathbf{d}t &= \eta_Y[\mathbf{aggimpact}_Y(t) - Y(t)]\end{aligned}$$

- Thus the following *difference* and *differential equation* for Y are obtained:

$$\begin{aligned}Y(t + \Delta t) &= Y(t) + \eta_Y[\mathbf{c}_Y(\omega_{X1,Y}X_1(t), \ldots, \omega_{Xk,Y}X_k(t)) - Y(t)]\Delta t \\ \mathbf{d}Y(t)/\mathbf{d}t &= \eta_Y[\mathbf{c}_Y(\omega_{X1,Y}X_1(t), \ldots, \omega_{Xk,Y}X_k(t)) - Y(t)]\end{aligned}$$

In the model considered here, for all states for the standard combination function the *advanced logistic sum combination function* **alogistic**$_{\sigma,\tau}$(...) is used [10]:

$$\mathbf{c}_Y(V_1,\ldots V_k) = \mathbf{alogistic}_{\sigma,\tau}(V_1,\ldots,V_k) = \left(\frac{1}{1+e^{-\sigma(v_1+\ldots+v_k-\tau)}} - \frac{1}{e^{\sigma\tau}}\right)(1+e^{-\sigma\tau})$$

(Here σ is a *steepness* parameter and τ a *threshold* parameter. The advanced logistic sum combination function has the property that activation levels 0 are mapped to 0 and it keeps values below 1. For example, for the prepration state ps_{bi} the model is numerically represented in difference equation form as

$$\mathbf{aggimpact}_{psbi}(t) =$$
$$\mathbf{alogistic}_{\sigma,\tau}(\omega_{srsub,psbi}srs_{ub}(t), \omega_{fsbi,psbi}fs_{bi}(t), \omega_{fssat,psbi}fs_{sat}(t), \omega_{stg,psbi}stg_{,}(t))$$
$$ps_{bi}(t+\Delta t) = ps_{bi}(t) + \eta_{psbi}[\mathbf{aggimpact}_{psbi}(t) - ps_{bi}(t)]\Delta t$$

The numerical representation of Hebbian learning is:

$$\omega_{fsbi,psbi}(t+\Delta t) = \omega_{fsbi,psbi}(t) + [\eta fs_{bi}(t)ps_{bi}(t)(1-\omega_{fsbi,psbi}(t)) - \zeta\omega_{fsbi,psbi}(t)]\Delta t$$

4 Simulation Results

The computational model has been used to conduct a number of simulation experiments according to different scenarios performed using the Matlab environment. In this section one is described in detail. The following scenario is used:

> James is facing overweight leading to an obesity problem, and has difficulties in everyday life routine regarding his diet (dietary pattern). To overcome this problem, he set goals to lose weight by eating healthy food. Here food is categorized in three categories, light food (very low calories), healthy food (low calories), and fast food (high calories). So, he has three available food options to choose from, each of which gives some extent of satisfaction as expected from that food. As he wants to lose weight, he will want to go for healthy or light food. In daily routine, when he feels hungry he looks around to have healthy food to eat.

The parameter values used are shown in Tables 2 and 3.

In the scenario the strength of the weights $\omega_{fsbi,psbi}$ of the connections from feelings to the considered preparations change over time through the Hebbian learning mechanism. Initial values of the activation levels for all states have been chosen 0 except the metabolic activity which depends on the scenario, and activation values for goals either are 0 or higher, also depending on the scenario. The simulation is executed for some scenarios for 180 time points and for other scenarios simulations for 1500 time points; the time step $\Delta t = 0.1$. Learning rate $\eta = 0.016$ and extinction rate $\zeta = 0.000015$ have been used when Hebbian learning is used. Details of the values for parameters used in the simulation are given in Table 2 (threshold **τ** and steepness σ) and in Table 3 (connection weights). In the scenario depicted in Fig. 2, the simulation shows five times hunger.

Table 2. Values of threshhold, steepness and update speed

State	τ	σ	State	τ	σ	State	τ	σ
bs_{ub}	0.5	8	fs_{b1}	0.16	2	ss_{sf1}	0.14	2.4
ss_{ub}	0.14	2.4	fs_{b2}	0.16	2	ss_{sf2}	0.14	2.4
srs_{ub}	0.14	2.4	fs_{b3}	0.16	2	ss_{sf3}	0.14	2.4
ds_{ub}	0.5	7	es_{b1}	0.14	2.4	srs_{sf1}	0.14	2.4
ps_{b1}	0.5	8	es_{b2}	0.14	2.4	srs_{sf2}	0.14	2.4
ps_{b2}	0.5	8	es_{b3}	0.14	2.4	srs_{sf3}	0.14	2.4
ps_{b3}	0.5	8	bs_{b1}	0.14	2.4	fs_{sat1}	0.16	2.4
srs_{b1}	0.14	2.4	bs_{b2}	0.14	2.4	fs_{sat2}	0.16	2.4
srs_{b2}	0.14	2.4	bs_{b3}	0.14	2.4	fs_{sat3}	0.16	2.4
srs_{b3}	0.14	2.4	ss_{b2}	0.14	2.4	stg	0.14	2.4
ss_{b1}	0.14	2.4	ss_{b3}	0.14	2.4			

Table 3. Values of parameters used: connection weights

Weight		Weight		Weight		Weight	
$\omega_{\mathrm{m}ub,\mathrm{bs}ub}$	1	$\omega_{\mathrm{srsb1,fs}b1}$	1	$\omega_{\mathrm{es}b1,\mathrm{bs}ub}$	-0.15	$\omega_{\mathrm{bs}b1,\mathrm{ss}sf1}$	1
$\omega_{\mathrm{bs}ub,\mathrm{ss}ub}$	1	$\omega_{\mathrm{srsb2,fs}b2}$	1	$\omega_{\mathrm{es}b2,\mathrm{bs}ub}$	-0.15	$\omega_{\mathrm{bs}b2,\mathrm{ss}sf2}$	1
$\omega_{\mathrm{ss}ub,\mathrm{srs}ub}$	1	$\omega_{\mathrm{srsb3,fs}b3}$	1	$\omega_{\mathrm{es}b3,\mathrm{bs}ub}$	-0.15	$\omega_{\mathrm{bs}b3,\mathrm{ss}sf3}$	1
$\omega_{\mathrm{srs}ub,\mathrm{ds}ub}$	1	$\omega_{\mathrm{fs}b1,\mathrm{psb1}}$	0.4	$\omega_{\mathrm{es}b1,\mathrm{bs}b1}$	1	$\omega_{\mathrm{ss}sf1,\mathrm{srs}sf1}$	1
$\omega_{\mathrm{ws}w,\mathrm{ss}w}$	1	$\omega_{\mathrm{fs}b2,\mathrm{psb2}}$	0.4	$\omega_{\mathrm{es}b2,\mathrm{bs}b2}$	1	$\omega_{\mathrm{ss}sf2,\mathrm{srs}sf2}$	1
$\omega_{\mathrm{ss}w,\mathrm{srs}w}$	1	$\omega_{\mathrm{fs}b3,\mathrm{psb3}}$	0.4	$\omega_{\mathrm{es}b3,\mathrm{bs}b3}$	1	$\omega_{\mathrm{ss}sf3,\mathrm{srs}sf3}$	1
$\omega_{\mathrm{srs}w,\mathrm{ds}ub}$	0.8	$\omega_{\mathrm{psb1,es}b1}$	1	$\omega_{\mathrm{bs}b1,\mathrm{ss}b1}$	1	$\omega_{\mathrm{ps}b1,\mathrm{srs}sf1}$	1
$\omega_{\mathrm{ds}ub,\mathrm{psb1}}$	1	$\omega_{\mathrm{psb2,es}b2}$	1	$\omega_{\mathrm{bs}b2,\mathrm{ss}b2}$	1	$\omega_{\mathrm{ps}b2,\mathrm{srs}sf2}$	1
$\omega_{\mathrm{ds}ub,\mathrm{psb2}}$	1	$\omega_{\mathrm{psb3,es}b3}$	1	$\omega_{\mathrm{bs}b3,\mathrm{ss}b3}$	1	$\omega_{\mathrm{ps}b3,\mathrm{srs}sf3}$	1
$\omega_{\mathrm{ds}ub,\mathrm{psb3}}$	1	$\omega_{\mathrm{stg,ps}b1}$	0	$\omega_{\mathrm{ss}b1,\mathrm{srs}b1}$	1	$\omega_{\mathrm{srs}sf1,\mathrm{fs}sat1}$	0
$\omega_{\mathrm{psb1,srs}b1}$	1	$\omega_{\mathrm{stg,ps}b2}$	0	$\omega_{\mathrm{ss}b2,\mathrm{srs}b2}$	1	$\omega_{\mathrm{srs}sf2,\mathrm{fs}sat2}$	0
$\omega_{\mathrm{psb2,srs}b2}$	1	$\omega_{\mathrm{stg,ps}b3}$	0	$\omega_{\mathrm{ss}b3,\mathrm{srs}b3}$	1	$\omega_{\mathrm{srs}sf3,\mathrm{fs}sat3}$	0
$\omega_{\mathrm{psb3,srs}b3}$	1	$\omega_{\mathrm{fs}sat1,\mathrm{ps}b1}$	0.3	$\omega_{\mathrm{fs}sat2,\mathrm{ps}b2}$	0.3	$\omega_{\mathrm{fs}sat3,\mathrm{ps}b3}$	0.3
$\omega_{\mathrm{ltg,stg}}$	0.8						

The role of the learning mechanism is illustrated, so the person can change his eating behaviour by learning through past experiences. When the person notices a body unbalance or hunger that triggers the desire to eat then initially the person may have equal tendency for all available options Fig. 2(b). But through the associated feelings valuation of the options takes place and the person can learn Fig. 2(e): the eating behaviour can be changed by adapting it slowly due to learning from his experiences. In this simulation Hebbian learning is used on the weights $\omega_{\mathrm{fs}bi,\mathrm{ps}bi}$ of the connections between associated feelings and preparations so these links are dynamic in this scenario; the weights $\omega_{\mathrm{es}bi,\mathrm{bs}bi}$ of the connections from execution states to body state are 0.3, 0.9, 0.6 respectively, these values are only for this simulation, change in these connections will make a change in the learning connections as well.

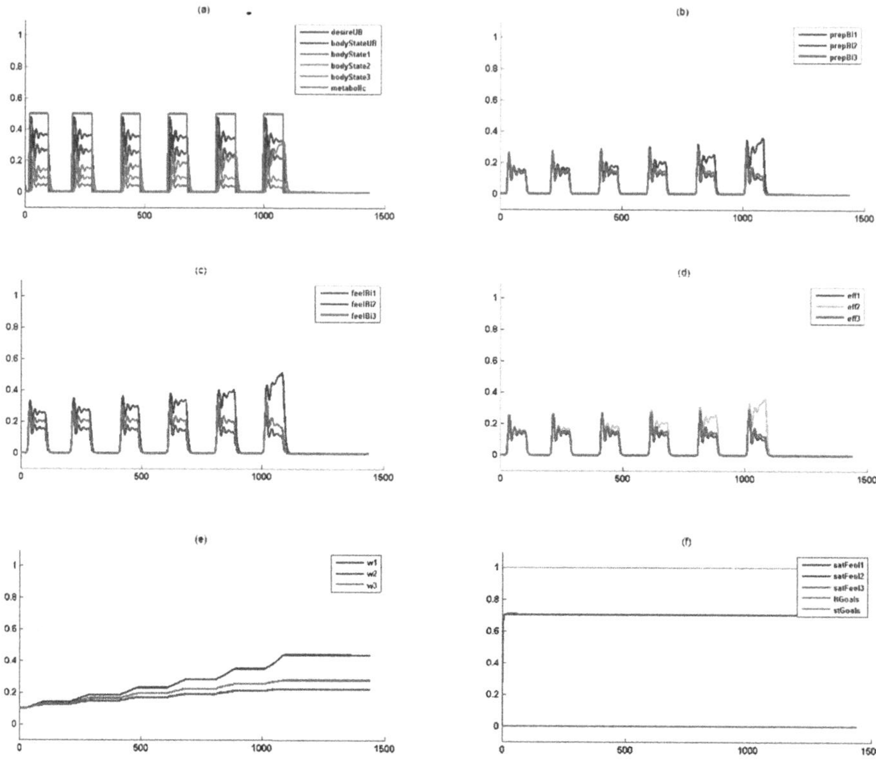

Fig. 2. Simulation results of five times body unbalance (hunger), with Hebbian learning

5 Conclusion

The model presented in this paper is a neurologically inspired computational model for making food choices by involving a number of internal and external factors. The focus of this paper is to formalise the dynamics and interaction of internal states which are involved in decision of making food choice. This model will help to understand the complex process from food desire generation to food choices that evolve based on internal prediction in combination with associated feelings for valuation of the available food options. The model also incorporates the role of goals and their influence on decision making on food choices. The simulation results suggest that this model is capable of learning of making food choices on the one hand through associated feelings to fulfil the desire, and on the other hand through levels of satisfaction in relation with various types of food and goals. The proposed computational model for making food choices can be used to develop human-aware intelligent systems that can help and support persons with overweight and obesity.

In future work, the model will be extended with food desire regulation strategies, and further focus will be on the social and environmental factors; in addition it may be further extended with more personal characteristics.

References

1. WHO: Global status report on noncommunicable diseases 2014. World Health organization (2014)
2. WHO: Obesity: preventing and managing the global epidemic. World Health Organization, Geneva (2000)
3. Stroebe, W.: Dieting, Overweight, and Obesity: Self-regulation in a Food-rich Environment. American Psychological Association, Washington, DC (2008)
4. McGee, D.L.: Body mass index and mortality: a meta-analysis based on person -level data from twenty-six observational studies. Ann. Epidemiol. **15**, 87–97 (2005)
5. Vos, T., Barber, R.M., Bell, B., et al.: Global, regional, and national incidence, prevalence, and years lived with disability for 301 acute and chronic diseases and injuries in 188 countries, 1990–2013: a systematic analysis for the Global Burden of Disease Study 2013. Lancet **386**(9995), 743–800 (2015)
6. Berridge, K.C., Ho, C.-Y., Jocelyn, M.R., Alexandra, G.D.: The tempted brain eats: pleasure and desire circuits in obesity and eating disorders. Brain Res. **1350**, 43–64 (2010)
7. Bosse, T., Hoogendoorn, M., Memon, Z.A., Treur, J., Umair, M.: A computational model for dynamics of desiring and feeling. Cogn. Syst. Res. **19–20**, 39–61 (2012)
8. Hebb, D.O.: The Organisation of Behaviour. Wiley, New York (1949)
9. Bi, G.Q., Poo, M.M.: Synaptic modification by correlated activity: Hebb's Postulate. Ann. Rev. Neurosci. **24**(1), 139–166 (2001)
10. Berridge, K.C., Terry, E.R., Aldridge, J.W.: Dissecting components of reward: 'liking', 'wanting', and learning. Curr. Opin. Pharmacol. **9**, 65–73 (2009)
11. Finlayson, G., King, N., Blundell, J.E.: Liking vs. wanting food: importance for human appetite control and weight regulation. Neurosci. Biobehav. Rev. **31**(7), 987–1002 (2007)
12. Berner, L.A., Avena, N.M., Hoebel, B.G.: Bingeing, self-restriction, and increased body weight in rats with limited access to a sweet-fat diet. Obesity **16**, 1998–2002 (2008)
13. Davis, C.: From passive overeating to 'food addiction': a spectrum of compulsion and severity. ISRN Obesity **2013**, 1–20 (2013)
14. Berthoud, H.R., Lenard, N.R., Shin, A.C.: Food reward, hyperphagia, and obesity. AJP: regulatory, integrative and comparative. Physiology **300**(6), R1266–R1277 (2011)
15. Elliott, R., Friston, K.J., Dolan, R.J.: Dissociable neural responses in human reward systems. J. Neurosci. **20**(16), 6159–6165 (2000)
16. Barbara, J.R., Erin, L.M., Liane, S.R.: Portion size of food affects energy intake in normal-weight and overweight men and women. Am. J. Clin. Nutr. **76**(6), 1207–1213 (2002)
17. Koob, G.F., Nora, D.V.: Neurocircuitry of addiction. Neuropsychopharmacology **35**(1), 217–238 (2010)
18. Damasio, A.R.: The Feeling of What Happens: Body and Emotion in the Making of Consciousness. Harcourt Brace, New York (1999)
19. Damasio, A.R.: Looking for Spinoza. Vintage books, London (2004)
20. Damasio, A.R.: Self Comes to Mind: Constructing the Conscious Brain. Pantheon books, New York (2010)
21. Damasio, A.R.: Descartes' Error: Emotion Reason and the Human Brain. Papermac, London (1994)
22. Damasio, A.R.: The somatic marker hypothesis and the possible functions of the prefrontal cortex. Philos. Trans. Roy. Soc. Lond. Ser. B, Biol. Sci. **351**(1346), 1413–1420 (1996)
23. Gerstner, W., Kistler, W.M.: Mathematical formulations of Hebbian learning. Biol. Cybern. **87**, 404–415 (2002)

24. Ashby, F.G., Turner, B.O., Horvitz, J.C.: Cortical and basal ganglia contributions to habit learning and automaticity. Trends Cogn. Sci. **14**, 208–215 (2010)
25. De Wit, S., et al.: Differential engagement of the ventromedial prefrontal cortex by goal-directed and habitual behaviour toward food pictures in humans. J. Neurosci. **29**, 11330–11338 (2009)
26. Mowrer, O.H.: Learning Theory and the Symbolic Processes. Wiley, New York (1960)
27. Watson, J.B.: Psychology as behaviourist views it. Psy. Review **20**, 158–177 (1913)
28. Webb, T.L., Paschal, S., Aleksandra, L.: Planning to break unwanted habits: habit strength moderates implementation intention effects on behaviour change. Br. J. Soc. Psychol. **48**(3), 507–523 (2009)
29. Treur, J.: Dynamic modeling based on a temporal–causal network modeling approach. Biol. Inspired Cogn. Architect. **16**, 131–168 (2016)
30. Damasio, A.R.: The Feeling of What Happens: Body Emotion and the Making of Consciousness. Vintage, London (2000)
31. Gerstner, W., Kistler, W.M.: Mathematical formulations of Hebbian learning. Biol. Cybern. **87**, 404–415 (2002)

A Novel Graph Regularized Sparse Linear Discriminant Analysis Model for EEG Emotion Recognition

Yang Li[1,2], Wenming Zheng[1(✉)], Zhen Cui[1(✉)], and Xiaoyan Zhou[3]

[1] Key Laboratory of Child Development and Learning Science of Ministry of Education, Research Center for Learning Science, Southeast University, Nanjing 210096, Jiangsu Province, People's Republic of China
{wenming_zheng,zhen.cui}@seu.edu.cn

[2] School of Information Science and Engineering, Southeast University, Nanjing 210096, Jiangsu Province, People's Republic of China

[3] School of Electronic and Information Engineering, Nanjing University of Information Science and Engineering Technology, Nanjing 210096, Jiangsu Province, People's Republic of China

Abstract. In this paper, a novel regression model, called graph regularized sparse linear discriminant analysis (GraphSLDA), is proposed to deal with EEG emotion recognition problem. GraphSLDA extends the conventional linear discriminant analysis (LDA) method by imposing a graph regularization and a sparse regularization on the transform matrix of LDA, such that it is able to simultaneously cope with sparse transform matrix learning while preserve the intrinsic manifold of the data samples. To cope with the EEG emotion recognition, we extract a set of frequency based EEG features to training the GraphSLDA model and also use it as EEG emotion classifier for testing EEG signals, in which we divide the raw EEG signals into five frequency bands, i.e., δ, θ, α, β, and γ. To evaluate the proposed GraphSLDA model, we conduct experiments on the SEED database. The experimental results show that the proposed algorithm GraphSLDA is superior to the classic baselines.

Keywords: EEG · Emotion recognition · Sparse LDA

1 Introduction

Human being has the most abundant emotions among the animals. Emotions are always accompanied with a man whatever he does. In daily life of human communications, emotional perception plays an important role [2]. Thus, it is useful to analyze the emotion of human beings in order to understand their intrinsic feelings and behaviors.

With the rapid development of pattern recognition technology, emotion analysis has become a very hot research topic in many fields such as affective computing, computer vision et al. Professor Picard R W from MIT first proposed

A. Hirose et al. (Eds.): ICONIP 2016, Part IV, LNCS 9950, pp. 175–182, 2016.
DOI: 10.1007/978-3-319-46681-1_21

the concept of affective computing [6]. Subsequently, Pang putted forward the concept of sentiment analysis in 2002 [4]. Sentiment analysis is becoming a hot research field of pattern recognition. Although there are a growing number of scholars to turn to sentiment analysis, there is still lack of dataset. What is more, most of the datasets contain only audio signal, visual signal, or audio and visual signals, whereas ignoring the relationship between human brain and emotion.

Brain is the nerve center of human, which has a complexly physiological structure and function. It has a great significance to study human brain. In 1992, the German neuropsychiatrist Berger H first placed electrodes on human scalp, and used these electrodes to record physiological signals of brain. Since then, the researches based on physiological signal had attracted numerous scientists and researchers in clinical medicine area and had been used in a wide range of applications. This kind of physiological signal is referred to as the electroencephalogram (EEG) signal. The acquisition of EEG is carried out by recording synchronous firing of neurons. EEG signal is captured through the electrodes placed at various regions of human scalp. The electrodes record the physiological signals produced by the brain cell activities. Since the EEG signal directly reflects the physiological activity of brain, it is very different from both visual and audio signals. On the contrary, it record a person's emotional activity truthfully. Therefore, the study of EEG signal may open a promising door to further uncover human emotion mechanism.

Recently, some researchers have explored the problem of EEG based human emotional analysis. As neurons of emotions are activated in a special region according to human cognition mechanism, the choice of channels associated with electrodes is often considered in the task of emotion recognition [3]. By further pre-defining some specific electrodes/channels, Petrantonakis et al. [5] proposed a EEG-based descriptor called higher order crossings (HOC). With the development of deep learning techniques, Zheng [10] employed Deep Belief Network(DBN) to try to extract high-level features of EEG emotion signals. Besides, EEG signals are also considered as one source to perform human emotion analysis of multi-modality [9]. For example, Soleymani M [7] tried to recover affective tags of video sequences by fusing multiple resources from EEG, pupillary response and gaze distance. The aforementioned methods have preliminarily investigated the problem of channel selection or feature extraction based on EEG signals for emotion recognition. As mentioned in these works, emotion recognition based on EEG signals just arouses researcher's attention, and need to be further explored due to its potential applications.

In this paper, we propose a novel EEG emotion recognition method, called graph regularized sparse linear discriminant analysis (GraphSLDA), to jointly perform channel selection and feature extraction. GraphSLDA extends the conventional linear discriminant analysis (LDA) method by imposing a graph regularization and a sparse regularization on the transform matrix of LDA. The sparse regularization is used to cope with channel choice by constraining the space of features of each entire channel with group sparsity, whereas the graph regularization is used to preserve intrinsic manifold structures in the process of data embedding to reduce over-fitting of trained models. To suppress those noises and further extract those representative frequency signals, we divide the raw EEG signals into five frequency bands, i.e., δ, θ, α, β, and γ, to feed into our proposed GraphSLDA model. To evaluate our method, we conduct experiments on the public and prevalent SEED database. The experimental results show that the proposed GraphSLDA model is more effective in the EEG based emotion recognition task.

2 GraphSLDA Model

2.1 From LDA to LSR

Linear discriminant analysis (LDA) is originally one of the most widely-used feature extraction method which aims to learn a transform matrix that maximizes the Fisher's discriminant criterion. Recently, Torre [8] indicated that from the view of regression, LDA is able to be reformulated as a weighted reduced rank regression (WRRR) optimization problem. More specifically, suppose that we have a training feature matrix $\mathbf{X} = [\mathbf{x}_1, \cdots, \mathbf{x}_N] \in \mathbb{R}^{d\times N}$ and its corresponding label matrix is denoted by $\mathbf{Y} = [\mathbf{y}_1, \cdots, \mathbf{y}_N] \in \mathbb{R}^{c\times N}$ whose i^{th} column $\mathbf{y}_i = [y_{i,1}, \cdots, y_{i,c}]^T$ has a binary entry 1 or 0, where d is the dimension of feature vectors, N is the number of the training samples, c is the class number, and 1 or 0 means whether the ith sample belongs to the jth class. Then, according to the work of Torre [8], the optimization problem of LDA can be formulated as follows:

$$\min_{\mathbf{A},\mathbf{B}} \|(\mathbf{Y}\mathbf{Y}^T)^{-1/2}(\mathbf{Y} - \mathbf{A}\mathbf{B}^T\mathbf{X})\|_F^2, \tag{1}$$

where $\mathbf{A} \in \mathbb{R}^{c\times r}$ and $\mathbf{B} \in \mathbb{R}^{d\times c}$ are the regression coefficient matrices and r is their rank.

In this paper, we will use two feasible operations to reduce the complexity of the novel LDA formulation in advance. Firstly, we assume that the numbers of training samples belonging to each class are approximately equal such that $\mathbf{Y}\mathbf{Y}^T$ can be replaced by an identity matrix. Secondly, we further fix coefficient matrix $\mathbf{A}$ as an identity matrix as well. Thus, we will obtain a least squares regression (LSR) based formulation of LDA as the following optimization problem:

$$\min_{\mathbf{B}} \|\mathbf{Y} - \mathbf{B}^T\mathbf{X})\|_F^2. \tag{2}$$

2.2 GraphSLDA

Given one subject, let the matrix $\mathbf{X}_k$ denote the features of training samples at the k-th ($k = 1, \cdots, K$) channel, where features are stacked column by column, then the EEG feature matrix of training data can be written as $\mathbf{X} = [\mathbf{X}_1^T, \cdots, \mathbf{X}_K^T]^T$. Thus Eq. (2) can be reformulated as follows,

$$\min_{\mathbf{B}_i} f_d(\mathbf{B}) = \|\mathbf{Y} - \sum_{i=1}^{K} \mathbf{B}_i^T\mathbf{X}_i\|_F^2, \tag{3}$$

where $\mathbf{B}_i$ is a block matrix of coefficients corresponding to the features $\mathbf{X}_i$ of the i-th channel. Different from the above LSR in Eq. (2), GraphSLDA will take into consideration local manifold structures of features as well as the contribution of different channels to EEG emotion recognition. Based on manifold assumption [1], given two feature vectors $\mathbf{x}_i$ and $\mathbf{x}_j$ close to each other in the original feature space, their embedding representations on a new transform space should also preserve their original adjacent relationship, which can benefit for improve the classification accuracy to some extent. Formally, let G denote the k nearest neighbor graph of N training samples, where each sample is regarded as a vertex, $\mathbf{W}$ be its corresponding weight matrix whose element $W_{i,j}$ associating with i-th sample and j-th sample takes the value of 1 or 0 according to the following rule:

$$W_{i,j} = \begin{cases} 1, \mathbf{x}_i \in \mathcal{N}(\mathbf{x}_j) \text{ or } \mathbf{x}_j \in \mathcal{N}(\mathbf{x}_i); \\ 0, \text{otherwise.} \end{cases} \tag{4}$$

Thus the manifold regularization term of regression coefficient can be written as:

$$f_m(\mathbf{B}) = \sum_{i=1}^{K} \mathrm{tr}(\mathbf{B}_i^T \mathbf{X}_i \mathbf{L} \mathbf{X}_i^T \mathbf{B}_i) = \mathrm{tr}(\mathbf{B}^T \mathbf{X} \mathbf{L} \mathbf{X}^T \mathbf{B}), \tag{5}$$

where $\mathbf{B} = [\mathbf{B}_1^T, \mathbf{B}_2^t, \cdots, \mathbf{B}_K^T]^T$, $\mathbf{L} = \mathbf{D} - \mathbf{W}$ is the Laplacian matrix, $\mathbf{D} = \mathrm{diag}(d_1, \cdots, d_N)$ and $d_i = \sum_1^N W_{i,j}$, and $\mathrm{tr}(\cdot)$ means the trace of a matrix.

To further do channel selection, we add a sparse selection constraint on the projecting matrix $\mathbf{B}_i$ by introducing group sparse term, formally,

$$f_c(\mathbf{B}) = \sum_{i=1}^{K} (\|\mathbf{B}_i\|_F + \|\mathbf{B}_i\|_F^2). \tag{6}$$

By combining the above three equations f_d, f_m, f_c to form the final objective function, i.e., the proposed GraphSLDA model,

$$\min_{\mathbf{B}} \|\mathbf{Y} - \sum_{i=1}^{K} \mathbf{B}_i^T \mathbf{X}_i\|_F^2 + \lambda \sum_{i=1}^{K} \mathrm{tr}(\mathbf{B}_i^T \mathbf{X}_i \mathbf{L} \mathbf{X}_i^T \mathbf{B}_i) + \mu \sum_{i=1}^{K} (\|\mathbf{B}_i\|_F + \|\mathbf{B}_i\|_F^2), \tag{7}$$

where λ and μ are the balance parameters corresponding to two regularization terms.

2.3 Optimization

To solve the objective model in Eq. (7), we use the inexact augmented Lagrange multiplier (ALM) strategy, which is often used to the problem of quadratic optimization. Concretely, we introduce an auxiliary variable $\mathbf{C}$ which is expected to be equal to $\mathbf{B}$. Thus, the problem of Eq. (7) can be converted to the optimization problem with a equation constraint as the following:

$$\begin{aligned} &\min_{\mathbf{B},\mathbf{C}} \|\mathbf{Y} - \sum_{i=1}^{K} \mathbf{C}_i^T \mathbf{X}_i\|_F^2 + \lambda \sum_{i=1}^{K} \mathrm{tr}(\mathbf{B}_i^T \mathbf{X}_i \mathbf{L} \mathbf{X}_i^T \mathbf{B}_i) + \mu (\sum_{i=1}^{K} \|\mathbf{B}_i\|_F + \sum_{i=1}^{K} \|\mathbf{C}_i\|_F^2), \\ &\text{s.t.} \quad \mathbf{C}_i = \mathbf{B}_i, i = 1, \cdots, K. \end{aligned} \tag{8}$$

Then the augmented Lagrange function can be expressed as:

$$L(\mathbf{B}_i, \mathbf{C}_i, \mathbf{T}_i) = \|\mathbf{Y} - \sum_{i=1}^{K} \mathbf{C}_i^T \mathbf{X}_i\|_F^2 + \sum_{i=1}^{K} \mathrm{tr}[\mathbf{T}_i^T(\mathbf{C}_i - \mathbf{B}_i)]$$
$$+\frac{\kappa}{2}\sum_{i=1}^{K} \|\mathbf{C}_i - \mathbf{B}_i\|_F^2 + \lambda \sum_{i=1}^{K} \mathrm{tr}(\mathbf{C}_i^T \mathbf{X}_i \mathbf{L} \mathbf{X}_i^T \mathbf{C}_i) + \mu(\sum_{i=1}^{K} \|\mathbf{B}_i\|_F + \sum_{i=1}^{K} \|\mathbf{C}_i\|_F^2), \quad (9)$$

where κ is a regularization parameter, and $\mathbf{T}_i$ is the Lagrange multiplier. To solve Eq. (9), we employ iterate optimization on the three variables, and the optimization steps are summarized as follows:

1. Fix $\mathbf{B}_i$ and $\mathbf{T}_i$, and Update $\mathbf{C}_i$: $\mathbf{C}_i = \arg\min_{\mathbf{C}_i} \|\mathbf{Y} - \sum_{i=1}^{K} \mathbf{C}_i^T \mathbf{X}_i\|_F^2 + \sum_{i=1}^{K} \mathrm{tr}[\mathbf{T}_i^T(\mathbf{C}_i - \mathbf{B}_i)] + \lambda \sum_{i=1}^{K} \mathrm{tr}(\mathbf{C}_i^T \mathbf{X}_i \mathbf{L} \mathbf{X}_i^T \mathbf{C}_i) + \frac{\kappa}{2} \sum_{i=1}^{K} \|\mathbf{C}_i - \mathbf{B}_i\|_F^2 + \lambda \sum_{i=1}^{K} \|\mathbf{C}_i\|_F^2$;
2. Fix $\mathbf{C}_i$ and $\mathbf{T}_i$, and Update $\mathbf{B}_i$: $\mathbf{B}_i = \arg\min_{\mathbf{B}_i} \frac{\mu}{\kappa}\|\mathbf{B}_i\|_F + \frac{1}{2}\|\mathbf{B}_i - (\mathbf{C}_i + \frac{1}{\kappa}\mathbf{T}_i)\|_F^2$;
3. Update $\mathbf{T}_i$ and κ: $\mathbf{T}_i = \mathbf{T}_i + \kappa(\mathbf{C}_i - \mathbf{B}_i)$ and $\kappa = \min(\rho\kappa, \kappa_{max})$;
4. Check Convergence: $\|\mathbf{C}_i - \mathbf{B}_i\|_\infty \leq \epsilon$.

In Step 1, by setting gradient of the objective function as 0, we can obtain a close-form solution of $\mathbf{C}_i$:

$$\mathbf{C} = [\frac{2\mathbf{X}(\lambda\mathbf{Y} + \mathbf{I})\mathbf{X}^T}{\kappa} + (1 + \frac{2\mu}{\kappa})\mathbf{I}]^{-1}(\frac{2\mathbf{X}\mathbf{Y}^T - \mathbf{T}}{\kappa} + \mathbf{B}), \quad (10)$$

where $\mathbf{C} = [\mathbf{C}_1^T, \cdots, \mathbf{C}_K^T]^T$ and $\mathbf{I}$ denotes the identity matrix. As the other steps are similar to Algorithm 1 of the previous work [11], we omit their solutions.

2.4 Testing

After obtaining the optimal solution $\hat{\mathbf{B}}_i$ using the above method, for a given testing sample $\mathbf{x}_t$, we can conveniently calculate its corresponding emotion label vector: $\mathbf{l}_t = \sum_{i=1}^{K} \hat{\mathbf{B}}_i^T(\mathbf{x}_t)_i$. Then, its emotion category can be determined by the following criterion:

$$\text{emotion\_label} = \arg\max_k\{\mathbf{l}_t(k)\}, \quad (11)$$

where $\mathbf{l}_t(k)$ indicates the k-th element of label vector $\mathbf{l}_t$.

3 Experiments

The proposed GraphSLDA is verified on the emotion dataset SEED [10], which contains fifteen subjects across different sessions. Following the same feature extraction step in [10], we use differential entropy (DE) of EEG signals as the features to feed into our model.

In our experiments, we use the leave one session out strategy so that other researchers can reproduce all results of our proposed method. That is to say, one session of a subject is used for testing while the remaining sessions of this subject are used for training, and we loop all sessions for testing and average all testing accuracies as the final performance of this subject. The mean accuracy of all subjects is shown in Tables 1 and 2. Here we compare the SVM based method, which is often used in the classification of EEG signals as a state-of-the-art method. Note that we don't compare the method of [10] due to its no reproduction in both coding and data partition. From these results, we can observe that our GraphSLDA achieves better accuracies than SVM at all cases except in the θ band. In addition, we can obtain two external observations from the two tables:

1) High frequency band carries more emotion information. As shown in Table 1, the performance is improved with the increase of frequency, i.e., $\gamma > \beta > \alpha > \theta > \delta$.
2) Combination of different frequency bands can benefit for emotion recognition. As shown in Table 2, the use of five frequency bands can reach the best result, which indicates that the five frequency bands are complementary with each other for reflecting human emotion.

To verify the effectiveness of two constraint terms in our model, we especially conduct the experiments with the cases of $\lambda = 0$ or $\mu = 0$, which means removing the manifold constraint or sparse constraint respectively. For this, we choose the third subject as the evaluation data, and the results are shown in Table 3. If only considering the sparse constraint, i.e., $f_d(\mathbf{B}) + f_c(\mathbf{B})$, the improvement is about 4.8 % points compared to the original model LSR. After adding the manifold structure preservation term, we can further promote the emotion recognition accuracy more than 4.2 %. Thus it demonstrates that the gains of our method come from the use of two constraint terms to some extent.

Besides, we depict the activity maps of brain electrodes for different bands according to the learnt weight matrix $\mathbf{B}$ in our proposed GraphSLDA model in Fig. 1. In the maps, the energy of each electrode is calculated by using the 2-norm operation on the corresponding row of the matrix $\mathbf{B}$. We can find that the areas of lateral temporal and front brain are heavily activated especially in α, β and γ frequency bands. That means these areas have more contribution to emotion expression, which also conforms to the cognition observation in psychology. For δ and θ bands, their maps seem have no regularity, which provides a possible explanation why their performances are still inferior to other three frequency bands.

We can find that in α, β and γ bands lateral temporal and front of the brain area have a relatively deeper color. That means these areas have more contribution to classification. And this is consistent with the study of psychology. From α and β bands, the maps could not find some regularity, it explained why the classification accuracies in α, β and γ bands are better than that in α and β bands (Tables 1 and 2).

Table 1. The mean accuracies of different frequency bands

Band	δ	θ	α	β	γ
GraphSLDA	**0.5673**	0.6116	**0.6985**	**0.7803**	**0.7884**
SVM	0.5604	**0.6323**	0.6707	0.7495	0.7597

Table 2. The mean accuracies of different combinations on frequency bands

Band	$\beta+\gamma$	$\alpha+\beta+\gamma$	$\theta+\alpha+\beta+\gamma$	$\delta+\theta+\alpha+\beta+\gamma$
GraphSLDA	**0.8263**	**0.8484**	**0.8628**	**0.8841**
SVM	0.7843	0.8097	0.8278	0.8523

Table 3. The performance of two constraints, i.e., sparse and manifold regularization terms.

Model	$f_d(\mathbf{B})$	$f_d(\mathbf{B})+f_c(\mathbf{B})$	$f_d(\mathbf{B})+f_c(\mathbf{B})+f_m(\mathbf{B})$
GraphSLDA	0.6871	0.7353	0.7777

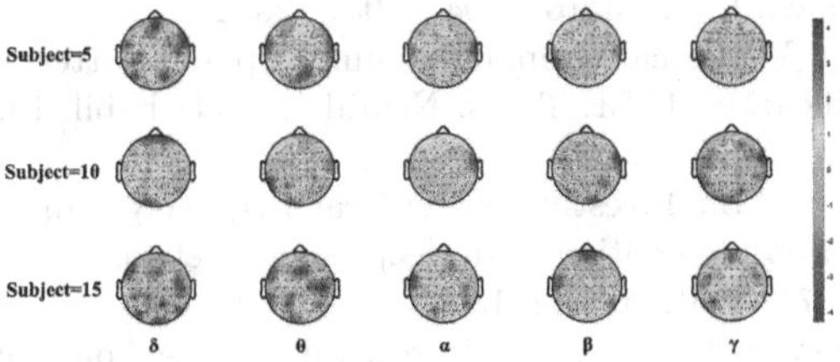

Fig. 1. Examples of brain electrodes activity maps. Deeper color represents larger energy in the learnt model **B**.

4 Conclusion

In this paper, we proposed a novel model called graph regularized sparse linear discriminant analysis (GraphSLDA) to deal with EEG emotion recognition. In GraphSLDA, two regularization terms, group sparse term and manifold preserving term, are introduced into the linear discriminative analysis model. As observed from the experiments, we can benefit from three folds: (1) adaptively implement channel selection by using sparse term; (2) reduce overfitting for small dataset by using manifold embedding; and (3) meanwhile learn more discriminative projecting space by using LDA. The comparison experiments also demonstrate that our proposed GraphSLDA is more effective.

Acknowledgement. This work was supported by the National Basic Research Program of China under Grant 2015CB351704, the National Natural Science Foundation of China (NSFC) under Grants 61231002 and 61572009, the Natural Science Foundation of Jiangsu Province under Grant BK20130020.

References

1. Belkin, M., Niyogi, P.: Laplacian eigenmaps and spectral techniques for embedding and clustering. In: NIPS, vol. 14, pp. 585–591 (2001)
2. Chibelushi, C.C., Bourel, F.: Facial expression recognition: a brief tutorial overview. CVonline: On-Line Compendium Comput. Vis. **9** (2003)
3. Li, M., Lu, B.L.: Emotion classification based on gamma-band EEG. In: Annual International Conference of the IEEE Engineering in Medicine and Biology Society, EMBC 2009, pp. 1223–1226. IEEE (2009)
4. Pang, B., Lee, L., Vaithyanathan, S.: Thumbs up?: sentiment classification using machine learning techniques. In: Proceedings of the ACL-02 Conference on Empirical Methods in Natural Language Processing, vol. 10, pp. 79–86. Association for Computational Linguistics (2002)
5. Petrantonakis, P.C., Hadjileontiadis, L.J.: Emotion recognition from EEG using higher order crossings. IEEE Trans. Inf. Technol. Biomed. **14**(2), 186–197 (2010)
6. Picard, R.W., Picard, R.: Affective Computing, vol. 252. MIT Press, Cambridge (1997)
7. Soleymani, M., Pantic, M., Pun, T.: Multimodal emotion recognition in response to videos. IEEE Trans. Affect. Comput. **3**(2), 211–223 (2012)
8. De la Torre, F.: A least-squares framework for component analysis. IEEE Trans. Pattern Anal. Mach. Intell. **34**(6), 1041–1055 (2012)
9. Wang, H., Zheng, W.: Local temporal common spatial patterns for robust single-trial EEG classification. IEEE Trans. Neural Syst. Rehabil. Eng. **16**(2), 131–139 (2008)
10. Zheng, W.L., Lu, B.L.: Investigating critical frequency bands and channels for EEG-based emotion recognition with deep neural networks. IEEE Trans. Auton. Mental Develop. **7**(3), 162–175 (2015)
11. Zheng, W.: Multi-view facial expression recognition based on group sparse reduced-rank regression. IEEE Trans. Affect. Comput. **5**(1), 71–85 (2014)

Information Maximization in a Feedforward Network Replicates the Stimulus Preference of the Medial Geniculate and the Auditory Cortex

Takuma Tanaka(✉)

The Center for Data Science Education and Research, Shiga University, 1-1-1 Banba, Hikone, Shiga 522-8522, Japan
tanaka.takuma@gmail.com

Abstract. Central auditory neurons exhibit a preference for complex features, such as frequency modulation and pitch. This study shows that the stimulus preference for these features can be replicated by a network model trained to maximize information transmission from input to output. The network contains three layers: input, first-output, and second-output. The first-output-layer neurons exhibit auditory-nerve neuron-like preferences, and the second-output-layer neurons exhibit a stimulus preference similar to that of cochlear nucleus, medial geniculate, and auditory cortical neurons. The features detected by the second-output-layer neurons reflect the statistical properties of the sounds used as input.

Keywords: Information maximization · Auditory information processing · Auditory cortex · Pitch selectivity · Frequency modulation selectivity

1 Introduction

Neurons in the auditory system of the brain can detect multiple dimensions of the complex features of sounds. For example, human speech is composed of a series of combinations of features such as phoneme, tone, stress, length, and prosody. The precise timing and temporal variation of these features convey critical information about the content of speech. To correctly comprehend human speech, an auditory system must be able to detect the features precisely and encode these features in the firing patterns of neurons.

The neural encoding of sensory inputs has been intensively studied both experimentally and theoretically. Experiments on auditory information processing have revealed that auditory-nerve neurons respond to sine wave-like tones [1] and that central auditory neurons encode complex features such as sound intensity and pitch. More than half of medial geniculate neurons in cats exhibit non-monotonic rate–sound-intensity functions [2], which implies that these neurons can be interpreted as intensity-coding neurons. Bendor and Wang reported

A. Hirose et al. (Eds.): ICONIP 2016, Part IV, LNCS 9950, pp. 183–190, 2016.
DOI: 10.1007/978-3-319-46681-1_22

that neurons in the auditory cortex of marmoset monkeys responded to both pure tones and missing fundamental harmonic complex sounds and called them pitch-selective neurons [3]. This type of high-level feature selectivity is thought to have emerged as a result of the integration of simpler low-level feature selectivity, such as that exhibited by auditory-nerve neurons, in a circuit with hierarchical structure. Hierarchical structure has been widely used in the theoretical modeling of visual information processing [4–7]. Theoretical studies of visual processing have demonstrated that stimulus selectivity to complex features such as the boundary between two gratings emerges in a generative model of natural scenes [8,9] and in network models that maximize the amount of information conveyed by the output [7,10]. These studies suggest that feature representation in the auditory system can be understood on the basis of a similar framework. In fact, a previous model of auditory feature representation [11] showed that units representing complex auditory features emerged in a generative model of sounds. However, the "spikes" in this model were generated by the maximum likelihood estimation, and therefore whether neurons can perform such a computation remains unclear. Moreover, complex auditory feature detection has not been treated in terms of maximizing information transmission.

Therefore, this study examines auditory feature detection in a feedforward network of rate-coding neuron models using an algorithm based on the information maximization principle. The network consists of three layers: input, first-output, and second-output. Short waveforms from a natural sounds dataset and a human speech corpus are provided to the input layer. The first-output-layer neurons respond to the wavelet-like waveforms. The second-output-layer neurons encode more complex features, such as pitch, tone intensity, and upward and downward frequency modulation. The selectivity of these model neurons for these complex features is comparable to that of experimentally reported cochlear nucleus, medial geniculate, and auditory cortical neurons. These results suggest that the central auditory neurons can be understood in terms of information maximization and that an extended network model based on the information maximization principle could replicate the more complex feature detection of the auditory cortices.

2 Model

The Pittsburgh natural sounds dataset [12] was used as the natural-sound input, and the Priority Areas "Spoken Dialogue" Simulated Spoken Dialogue Corpus (PASD) was used as the human-speech input. The former was down-sampled to 11 kHz, and the latter was down-sampled to 8 kHz. The input time series was shifted and scaled to have zero mean and unit variance. Consecutive samples ($N = 200$) were randomly chosen from the input time series to be used as input to the network at each time step. The network model and learning algorithm described previously [7] were used. The network consists of input, first-output, and second-output layers, each containing N model neurons. The value of the i-th input at time step t is $x_i(t)$. The states of the neurons in the first- and second-output layers at time t are determined by

$$u_i(t) = f\left(\sum_{j=1}^{N} V_{ij} x_j(t)\right), \tag{1}$$

and

$$z_i(t) = f\left(\sum_{j=1}^{N} W_{ij}(|u_j(t)| - \overline{|u_j|})\right), \tag{2}$$

respectively, where V_{ij} is the connection weight from the input neuron j to the first-output-layer neuron i, W_{ij} is the connection weight from the first-output-layer neuron j to the second-output-layer neuron i, $\overline{|u_j|}$ is the average of $|u_j(t)|$ over time, and

$$f(x) = 2 \arctan \tanh \frac{x}{2} \tag{3}$$

is the activation function. The integration time constant τ was set to 10^4 steps in all simulations. The weights V_{ij} were updated once every 1000 steps. The updates of the matrices $\mathbf{V}$ and $\mathbf{W}$ were performed 10^4 times with $\epsilon = \epsilon_0$, 1.9×10^5 times with $\epsilon = 10\epsilon_0$, and 8×10^5 times with $\epsilon = \epsilon_0$ using the Newton method described in [7], where $\epsilon_0 = 10^{-6}$ for natural-sound input and $\epsilon_0 = 10^{-5}$ for human-speech input.

3 Results

3.1 First-Output-Layer Neurons

The input and first-output layer can be regarded as a network performing independent component analysis [6]. Similar to a previous study of independent component analysis [13], the first-output-layer neurons exhibit selectivity to Gabor wavelet-like sound waveforms. Figure 1A shows the column vectors of $\mathbf{V}^{-1}$, sound waveforms that the first-output-layer neurons are selective for, of the network in which natural sounds are used as the input. They have Gabor function-like shapes with different frequencies, amplitudes, phases, dispersions, and center

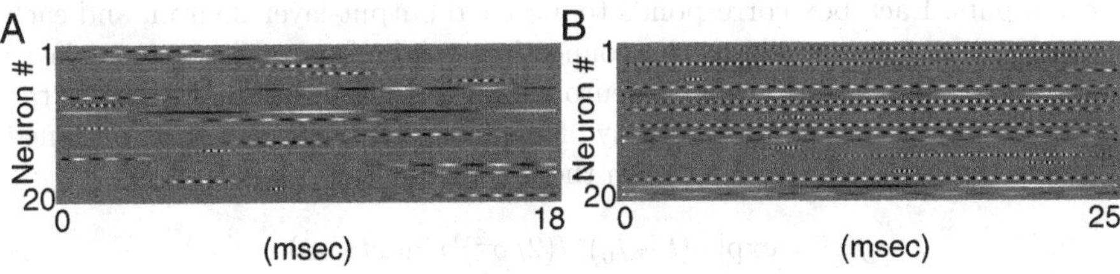

Fig. 1. Preferred sound waveform of 20 first-output-layer neurons in networks with (A) natural-sound input and (B) human-speech input.

positions. Figure 1B shows the selectivity of the first-output-layer neurons with human-speech input. The waves in Fig. 1B resemble sine waves without amplitude modulation because the majority of human speech is composed of vowels. Selectivity to these sine waves with and without amplitude modulation is similar to the stimulus preference of auditory-nerve neurons [1, 13].

The response of these neurons to continuous sine waves is shown in Fig. 2. The horizontal and vertical axes represent the frequency and amplitude of sine waves, respectively. The density represents the maximal value of the first-output-layer neuron output obtained by varying the phases of the sine waves. Almost all neurons respond to a small-amplitude tone with the preferred frequency and to a wide range of tones if the amplitude is increased. This figure shows that the wavelet-like connection weights result in unimodal, V-shaped tuning curves with only one preferred frequency [14].

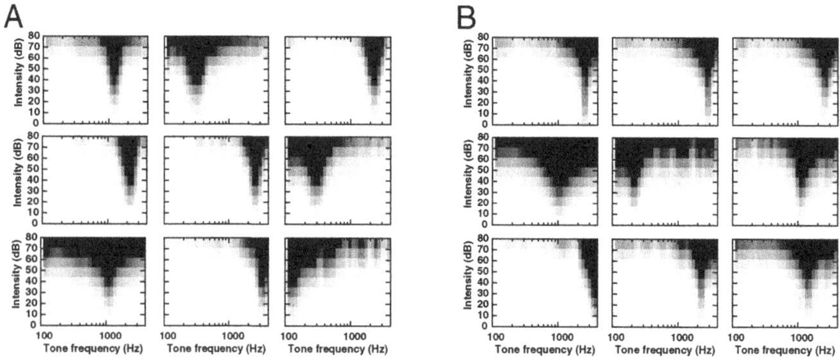

Fig. 2. Tuning curve of first-output-layer neurons in networks with (A) natural-sound input and (B) human-speech input.

3.2 Second-Output-Layer Neurons

Figure 3 shows the connection weights from the first-output-layer neurons to the second-output-layer neurons in the networks with natural-sound and human-speech inputs. Each box corresponds to a second-output-layer neuron, and each line in the box corresponds to the connection weight from a first-output-layer neuron to the second-output-layer neuron. The vertical position of a line in the box represents the preferred frequency of the first-output-layer neuron obtained by fitting the row vector of $\mathbf{V}^{-1}$ with the Gabor function, that is, ω of

$$g(t) = \exp[-(t - t_0)^2/(2\pi\sigma^2)] \sin(\omega t + \phi). \quad (4)$$

The horizontal position and the length of the line represent t_0 and σ, respectively. The color represents the value of the connection weight W_{ij}, with red and blue

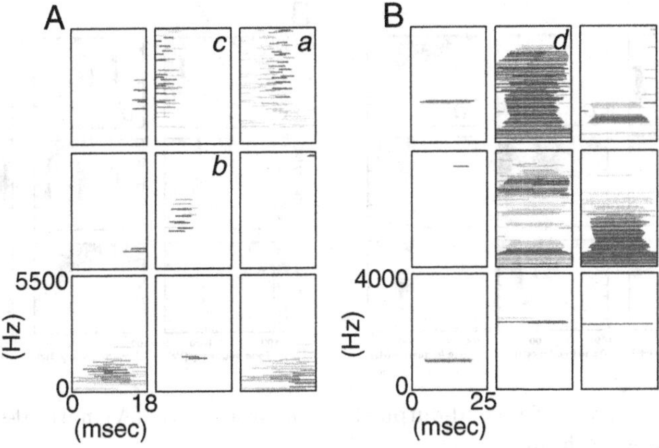

Fig. 3. Connection weights from first- to second-output-layer neurons for the networks with (A) natural-sound input and (B) human-speech input. The second-output-layer neurons are represented as boxes. Connections from the first-output-layer neurons correspond to the horizontal lines, whose vertical position, horizontal position, and length represent ω, t_0, and σ of the fitted Gabor function, respectively. Red and blue indicate positive and negative connection weights, respectively. (Color figure online)

indicating positive (excitatory) and negative (inhibitory) connection weights, respectively.

The stimulus selectivities exhibited by the first-output-layer neurons in the networks with natural-sound and human-speech inputs are substantially different from each other. Because the difference in the stimulus preference of the first-output-layer neurons affects the stimulus preference of the second-output-layer neurons, Figs. 3A and B show completely different types of stimulus preferences in these two networks. Therefore, the results of the two networks with different inputs are presented separately.

Natural Sounds

Frequency-Modulation Selectivity. Second-output-layer neuron *a* in Fig. 3A receives positive connection weights from first-output-layer neurons selective for low-frequency tones in the earlier half and for high-frequency tones in the latter half, while it receives negative connection weights from neurons selective for high-frequency tones in the earlier half and for low-frequency tones in the latter half. Therefore, this neuron is selective for a frequency change from a low tone to a high tone. Selectivity to frequency modulation is found in more than half of the neurons in the cochlear nucleus [15]. Neuron *a* appears to correspond to these cochlear nucleus neurons responsive to frequency modulation.

Intensity Tuning. Figure 4A shows the tuning curves of the second-output-layer-neurons trained with natural-sound input. These tuning curves are much more variable than those of the first-output-layer neurons (Fig. 2A). Tuning curve *b*

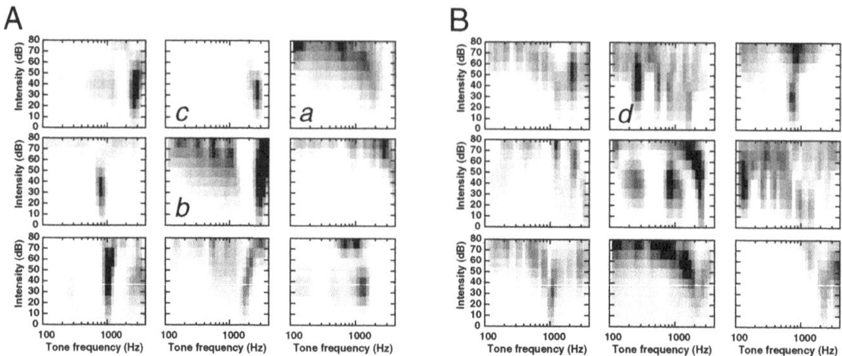

Fig. 4. Tuning curves of second-output-layer neurons with (A) natural-sound input and (B) human-speech input.

in Fig. 4A has two preferred frequencies; this suggests that this neuron receives strong positive connection weights from two first-output-layer neurons with different frequency preferences, which is evident in Fig. 3A (*b*). This is similar to auditory cortical neurons classified as double U-shaped tuning curves [16]. The tuning curve *c* (Fig. 3A) is similar to circumscribed neurons in the auditory cortex [16], which do not respond to increased amplitude of sine waves at any frequency. Therefore, this neuron is selective for a tone with a specific frequency and intensity level. Physiologically, this type of neuron is reported to compose approximately 20 % of the neurons in a cat's auditory cortex [16].

Human Speech

Pitch Selectivity. Figure 3B shows the connections from first- to second-output-layer neurons in the network with human-speech input. Some of these neurons receive positive inputs from first-output-layer neurons with preferred frequencies that are multiples of a value (neuron *d*). That is, these neurons respond to the pitch of a tone, and, consequently, have intensive multimodal tuning curves (Fig. 4, neuron *d*). This type of stimulus preference has been reported in auditory cortical neurons [16]. Because the present learning algorithm maximizes the information transmission from input to output, this result suggests that pitch selectivity in the auditory cortex emerges to encode human and other animal voices efficiently. Indeed, human voices are primarily composed of tones with a principal frequency and its higher harmonics. Owing to this statistical property of human voices, second-output-layer neurons acquire selectivity to tones with frequencies f, $2f$, $3f$, ... during training.

4 Discussion

This study presented the properties of model neurons in networks trained to maximize the amount of information transmitted to the output layers, using natural sounds and human speech as inputs. The first-output-layer neurons respond to wavelet-like stimuli and have unimodal tuning curves. This property is consistent with the experimentally reported properties of auditory-nerve neurons [1]. It is also consistent with previous theoretical studies [11,13], showing that wavelet-like functions are information-efficient in encoding natural sounds. The properties of the second-output-layer neurons are affected by the type of input. Natural sounds, which contain abundant abrupt changes in pitch, favor the emergence of second-output-layer neurons selective for pitch change. In contrast, human voices are dominated by continuous waves with higher harmonics, making pitch selectivity advantageous in encoding information. These properties are also consistent with previous experimental and theoretical results [3,11,15].

The training algorithm of the present model is based on the information maximization principle, that is, the information conveyed by the output neurons is maximized during training. The fact that neurons in the two output layers exhibit stimulus preferences similar to the cochlear nucleus, medial geniculate, and auditory cortical neurons suggests that the neurons in the central nervous system have evolved to encode as much information as possible by forming an information-efficient circuit. This is corroborated by previous studies which showed that the properties of simple and complex cells in the primary visual cortex can be replicated by the information maximization model [7,17]. If, as suggested by the present model, the sensory information processing in the central nervous system can be understood in terms of information maximization, a model with a larger number of layers would replicate and predict the properties of neurons in the higher sensory cortices.

Acknowledgments. This work was supported by JSPS KAKENHI Grant Numbers 15H04266 and 16K16123.

References

1. de Boer, E., de Jongh, H.R.: On cochlear encoding: potentialities and limitations of the reverse-correlation technique. J. Acoust. Soc. Am. **63**, 115–135 (1978)
2. Rouiller, E., de Ribaupierre, Y., Morel, A., de Ribaupierre, F.: Intensity functions of single unit responses to tone in the medial geniculate body of cat. Hear. Res. **11**, 235–247 (1983)
3. Bendor, D., Wang, X.: The neuronal representation of pitch in primate auditory cortex. Nature **436**, 1161–1165 (2005)
4. Fukushima, K.: Neocognitron: a self-organizing neural network model for a mechanism of pattern recognition unaffected by shift in position. Biol. Cybern. **36**, 193–202 (1980)
5. Felleman, D.J., Van Essen, D.C.: Distributed hierarchical processing in the primate cerebral cortex. Cereb. Cortex **1**, 1–47 (1991)
6. Bell, A.J., Sejnowski, T.J.: The "independent components" of natural scenes are edge filters. Vis. Res. **37**, 3327–3338 (1997)
7. Tanaka, T., Nakamura, K.: Information maximization principle explains the emergence of complex cell-like neurons. Front. Comput. Neurosci. **7**, 165 (2013)

8. Karklin, Y., Lewicki, M.S.: A hierarchical Bayesian model for learning nonlinear statistical regularities in nonstationary natural signals. Neural Comput. **17**, 397–423 (2005)
9. Karklin, Y., Lewicki, M.S.: Emergence of complex cell properties by learning to generalize in natural scenes. Nature **457**, 83–86 (2009)
10. Tanaka, T., Aoyagi, T., Kaneko, T.: Replicating receptive fields of simple and complex cells in primary visual cortex in a neuronal network model with temporal and population sparseness and reliability. Neural Comput. **24**, 2700–2725 (2012)
11. Karklin, Y., Ekanadham, C., Simoncelli, E.P.: Hierarchical spike coding of sound. In: Advances in Neural Information Processing Systems, pp. 3032–3040 (2012)
12. Smith, E.C., Lewicki, M.S.: Efficient auditory coding. Nature **439**, 978–982 (2006)
13. Lewicki, M.S.: Efficient coding of natural sounds. Nature Neurosci. **5**, 356–363 (2002)
14. Kiang, N.Y.S., Sachs, M.B., Peake, W.T.: Shapes of tuning curves for single auditory-nerve fibers. J. Acoust. Soc. Am. **42**, 1341–1342 (1967)
15. Britt, R., Starr, A.: Synaptic events and discharge patterns of cochlear nucleus cells II. Frequency-modulated tones. J. Neurophysiol. **39**, 179–194 (1976)
16. Sutter, M.L.: Shapes and level tolerances of frequency tuning curves in primary auditory cortex: quantitative measures and population codes. J. Neurophysiol. **84**, 1012–1025 (2000)
17. Bell, A.J., Sejnowski, T.J.: An information-maximization approach to blind separation and blind deconvolution. Neural Comput. **7**, 1129–1159 (1995)

A Simple Visual Model Accounts for Drift Illusion and Reveals Illusory Patterns

Daiki Nakamura(✉) and Shunji Satoh

Graduate School of Information Systems,
The University of Electro-Communications, Tokyo 182-8585, Japan
daiki@hi.is.uec.ac.jp, shun@is.uec.ac.jp

Abstract. Computational models of vision should not only be able to reproduce experimentally obtained results; such models should also be able to predict the input–output properties of vision. We assess whether a simple computational model of neurons in the Middle Temporal (MT) visual area proposed by the authors can account for illusory perception of "rotating drift patterns," by which humans perceive illusory rotation (clockwise or counterclockwise) depending on the background luminance. Moreover, to predict whether a pattern causes visual illusion or not, we generate an enormous set of possible visual patterns as inputs to the MT model: $8^8 = 16,777,216$, possible input patterns. Numerical quantities of model outputs by computer simulation for 8^8 inputs were used to estimate human illusory perception. Using psychophysical experiments, we show that the model prediction is consistent with human perception.

Keywords: Visual illusion · Lucas–Kanade method · MT · Computational model

1 Introduction

Estimating the outside world's motion from retinal inputs (optical flows) is an important visual function, but motion estimation by humans is not perfectly accurate; humans perceive visual illusions. From a functional viewpoint, we can regard illusory motions as estimation failures of optical flows caused by an imperfect estimator in the brain. Three conventional approaches to reveal vision properties using illusory images are (i) discovery of novel illusory static or video images by human experts, (ii) quantitative examination of illusory perception by psychological and physiological experiments, and (iii) development of physiologically plausible models to reproduce illusory perception as an input–output relation of those computational models.

Humans perceive illusory rotation when observing Fraser–Wilcox (FW) stimuli like those in the left column of Fig. 1a [1, 2]. Humans also perceive illusory rotation when FW stimuli disappear [3]. The direction of illusory rotation (clockwise/counterclockwise) depends on the background luminance of afterimages. Humans perceive clockwise rotation when a prior and a post stimuli resemble those shown respectively in the upper left and upper center panels of Fig. 1a, for which the background luminance is bright. In contrast, humans perceive counterclockwise rotation if the background luminance is dark. The circular–periodic designs of prior patterns are

A. Hirose et al. (Eds.): ICONIP 2016, Part IV, LNCS 9950, pp. 191–198, 2016.
DOI: 10.1007/978-3-319-46681-1_23

identical, as shown in Fig. 1a. Herein, we designate the illusory rotation when a Fraser–Wilcox stimuli disappears as "drift illusion."

Here, a simple question confronts us: can we discover novel illusory patterns with no human effort? Assuming that the prior stimuli of Fig. 1 include eight luminance values in one period (a circular sector of 45°), and that a luminance value is represented by a number of 8 digits, then because the number of possible patterns is $8^8 = 16,777,216$, psychological experiments using human subjects are unsuitable to classify the 16 million patterns as illusory ones or not. Almost 400 days would be necessary for one person to classify 16 million patterns if the person were forced to classify stimuli within 2 s/pattern without any break. However, an accurate computational model reflecting human perception can classify all those patterns in 4 days if 20 ms per input pattern can simulate human perception. We can regard a computational model of a visual system as an indefatigable virtual vision. This approach would be useful to validate the plausibility of models. If model prediction were not consistent with human perception, we would be able to infer that the model must be improved.

The explanations in this paper demonstrate that (iv) a simple model for optical-flow estimation, which is NOT based on brain research, is sufficient to reproduce some properties of vision and illusion, and (v) other illusory patterns besides FW pattern can be extracted from numerous possible inputs by numerical simulation of the simple model, which is regarded as an indefatigable human subject. First, we examine whether our computational model reproduces the luminance dependency of drift illusion as presented in Fig. 1a. Second, we prepare possible patterns composed of circular periodic patterns (Fig. 4 shows examples). Then we obtain model predictions for these patterns by numerical simulation and comparison to psychological experimentally obtained results. Finally, after comparing the model predictions with human

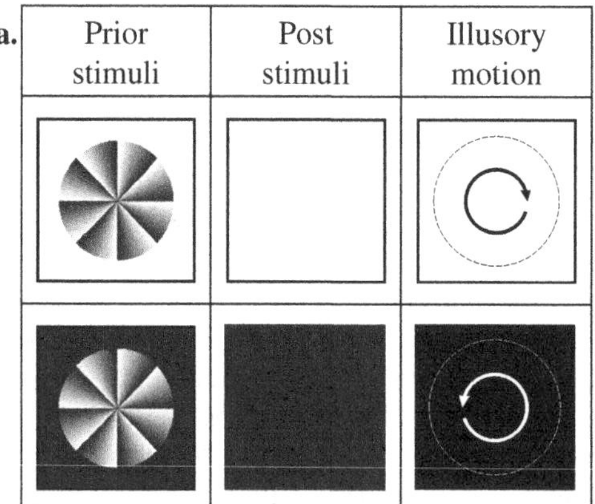

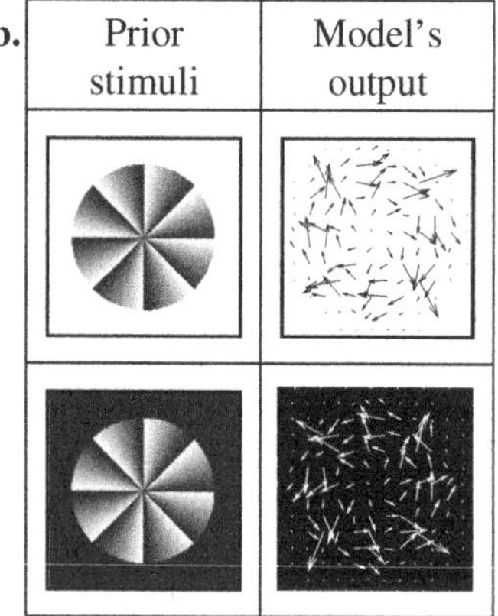

Fig. 1. a. Examples of drift illusion. The illusory rotation direction depends on the background luminance. Humans perceive clockwise illusory rotation when the background luminance is bright (top), but perceive counterclockwise illusory rotation for the dark background (bottom). b. Output vectors (optical flows, estimated perception of motion) obtained from our computational model. Clockwise rotation vectors are obtained with relative background luminance of 1.0 (top). Counterclockwise rotation vectors appear with relative background luminance of 0.0 (bottom).

perceptions obtained from psychological experiment, we evaluated the plausibility of our computational model.

2 Reproduce Rotational Illusion Dependent on Background Luminance

Humans perceive illusory rotation in a post stimuli followed by prior stimuli (Fig. 1a). The illusory rotation direction varies according to the background luminance [3]. We examined whether our computational model can reproduce drift illusion dependent on background luminance.

2.1 Computational Model: Modified Lucas–Kanade Method [4]

We proposed that a modified method to estimate optical flow from video images can be a model of neurons in the Middle Temporal (MT) visual area in which the outside world's motions embedded in retinal images are decoded as neural outputs of the MT area [5]. Our model, based on Lucas–Kanade (LK) method, adopts multi-resolution processing and a modification to avoiding division by zero. We have demonstrated that the model accounts precisely for MT responses [5].

We refer to two-dimensional optical flows at time t as outputs of MT neurons at time t. Estimated optical flows $\hat{\boldsymbol{v}}(x,y,t) = (\hat{v}_x(x,y,t)\ \hat{v}_y(x,y,t))^{\mathrm{T}}$ are calculated using the following formula.

$$\hat{\boldsymbol{v}}(x,y,t) = \left\{ \begin{pmatrix} S_{xx}(x,y,t) & S_{xy}(x,y,t) \\ S_{xy}(x,y,t) & S_{yy}(x,y,t) \end{pmatrix} + \varepsilon^2 E \right\}^{-1} \begin{pmatrix} -S_{xt}(x,y,t) \\ -S_{yt}(x,y,t) \end{pmatrix} \tag{1}$$

$$\left(S_{ij}(x,y,t) \stackrel{\text{def}}{=} w(x,y) * \left\{ \frac{\partial I(x,y,t)}{\partial i} \frac{\partial I(x,y,t)}{\partial j} \right\} \right) \tag{2}$$

In that equation, $I(x,y,t)$ represents the relative luminance of input image in the (x,y) coordinate system, E is an identity matrix, $*$ is the convolution operator, and ε^2 is a parameter to avoid zero division. The $w(x,y)$ is a Gaussian window with variance (window size) σ^2; parameter $\varepsilon^2 = 1.0 \times 10^{-5}$. The multi-resolution scheme is realized by application of a set of various values of window size, but we fixed it as $\sigma^2 = 11/6$ in this article for simple discussion and rapid simulation. LK method was derived under the following assumptions: (1) temporal changes of a texture are caused only by an objective motion, (2) a change in luminance can be expressed by first-order approximation of the Taylor expansion, and (3) optical flows are constant in the window $w(x,y)$. Because disappearing patterns like drift illusion violate the assumption described above, the estimate $\hat{\boldsymbol{v}}$ would not be correct. It might correspond to human illusion.

2.2 Numerical Simulation: Rotational Directions and the Rotational Strength

We examined whether our model (Eq. 1) can account for drift illusion depending on the background luminance.

The input image size is 400 × 400 pixels. The circular pattern diameter is 300 pixels. Inputs are gray scale images of which the luminance ranged from 0.0 (darkest, black) to 1.0 (brightest, white). We show estimated optical flow vectors, which were down-sampled for visualization (Fig. 1b). In Fig. 1b, clockwise rotation vectors appeared when the relative background luminance was 1.0 (Fig. 1b, top). Counter-clockwise rotation vectors appeared when the relative background luminance was 0.0 (Fig. 1b, bottom).

To quantitate the outputs, spatially averaged rotation $\bar{R}$ was evaluated using the following equation.

$$\bar{R} = \frac{1}{|S|}\int_S \mathrm{rot}_{2D}\hat{\boldsymbol{v}}(x,y,t)dS = \frac{1}{|S|}\int_S \frac{\partial \hat{v}_y(x,y,t)}{\partial x} - \frac{\partial \hat{v}_x(x,y,t)}{\partial y} dS \quad (3)$$

Therein, S denotes the area of circular patterns. $\bar{R} > 0$ coincides with a counter-clockwise rotation. Figure 2 shows the rotation $\bar{R}$ obtained from our model with respect to background luminance. The smallest negative value of $\bar{R}$, clockwise rotation, was obtained at maximum background luminance ($I = 1.0$). In contrast, the largest positive value for counterclockwise rotation was obtained at minimum relative luminance ($I = 0.0$). The magnitude of rotation was zero at background luminance $I = 0.5$.

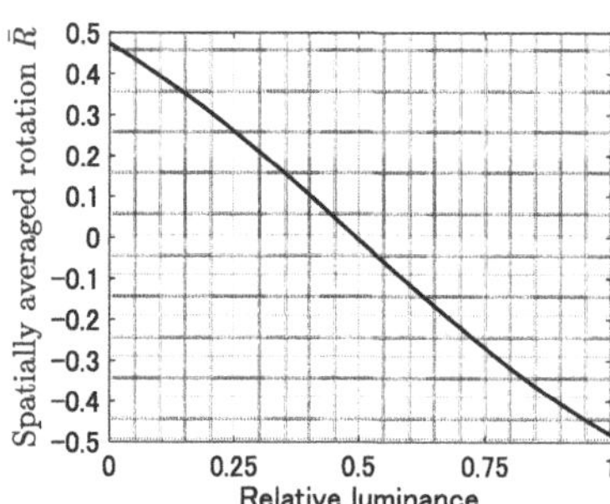

Fig. 2. Rotations of model outputs with respect to the relative luminance. The smallest negative value for clockwise rotation was obtained at maximum background luminance ($I = 1.0$). Similarly, the largest positive value for counterclockwise rotation was obtained at minimum relative luminance ($I = 0.0$).

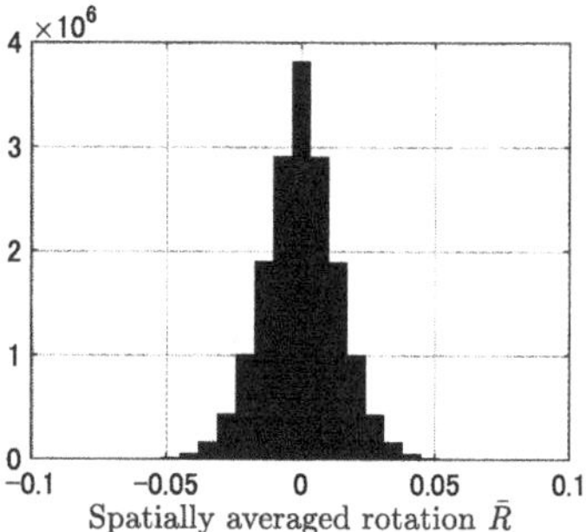

Fig. 3. Histogram of spatially averaged rotation $\bar{R}$ for 16,777,216 stimulus. Almost all stimuli result in little rotation $\bar{R} \simeq 0$, while some stimuli have clockwise or counterclockwise rotation.

2.3 Discussion

The results presented in the previous section indicate that the model successfully accounts for the background dependence of the human illusory perception for the drift illusion. The model (improved LK method) assumes that "(1) temporal changes of a texture are caused only by an objective motion." In other words, it does not presume suddenly disappearing objects such as in the case of the drift illusion. Although our model's outputs for drift illusion are meaningless from an engineering viewpoint, it is interesting that these rotating vectors representing optical flows are consistent with human perception.

From Eq. 1, we ascertained that the temporal derivative term $\partial I(x, y, t)/\partial t$ affects rotational direction and rotational strength. Because background luminance affects the temporal derivatives which are luminance differences between a prior- and post-stimulus, the background luminance directly affects the direction of optical flows.

3 Model Predictions and Psychological Experiments

We can evaluate the correlation between human perception and model prediction using prior/post images with white background.

3.1 Circular Stimulus

As portrayed in Fig. 4, a prior stimulus is composed of circular sectors of 45°. A luminance pattern in a circular sector comprises eight gray levels: a combination of $I \in \{0/7, 1/7, 2/7, \cdots, 7/7\}$. The number of possible patterns is $8^8 = 16,777,216$.

3.2 Selection of Stimuli for Psychological Experiment

We obtained the rotation $\bar{R}$ for all possible stimuli of a white background. Figure 3 shows a histogram of rotation $\bar{R}$. Figure 3 indicates that almost all stimuli have small rotation $\bar{R} \simeq 0$, although some stimuli cause a clockwise or counterclockwise rotation vector. This result implies that almost all stimuli would not be illusory patterns, but some patterns with large $|\bar{R}|$ might cause human illusions. The simulation time was less than 60 h (dual processor of Xeon E5-2630 v2 2.6 GHz, 32 GB RAM).

For psychological experiments, 33 patterns were chosen randomly from the 16 million patterns so that the rotation $\bar{R}$ of selected 33 patterns were distributed as uniformly as possible, and that a selected pattern contains both black and white ($I = 0.0$ and $I = 1.0$). Real values of Fig. 4 signify $\bar{R}$ s from −0.0281 to 0.0368.

3.3 Methods

Human subjects were seated in a dark room with their head resting on a chin rest fixed 1 m from the display. At the center of a gamma-corrected CRT monitor (GDM-F520;

Sony Corp.) with a refresh rate of 85 Hz, 33 selected stimuli were displayed. The display resolution was 1024 × 768 pixels. The screen visual angle was 22.0 × 16.6°. The circular stimulus diameter was 13.0° (300 pixels). The maximum luminance (white; $I = 1.0$) was $81.3\,\mathrm{cd/m^2}$.

The 33 prior stimuli in Fig. 4 were displayed randomly. Each stimulus was displayed 10 times. Post stimuli were uniformly white. Prior stimuli were presented for 1500 ms. Subsequently, prior stimuli disappeared and post stimuli (uniform white) were displayed. Then, subjects were forced to report, as soon as possible, the direction

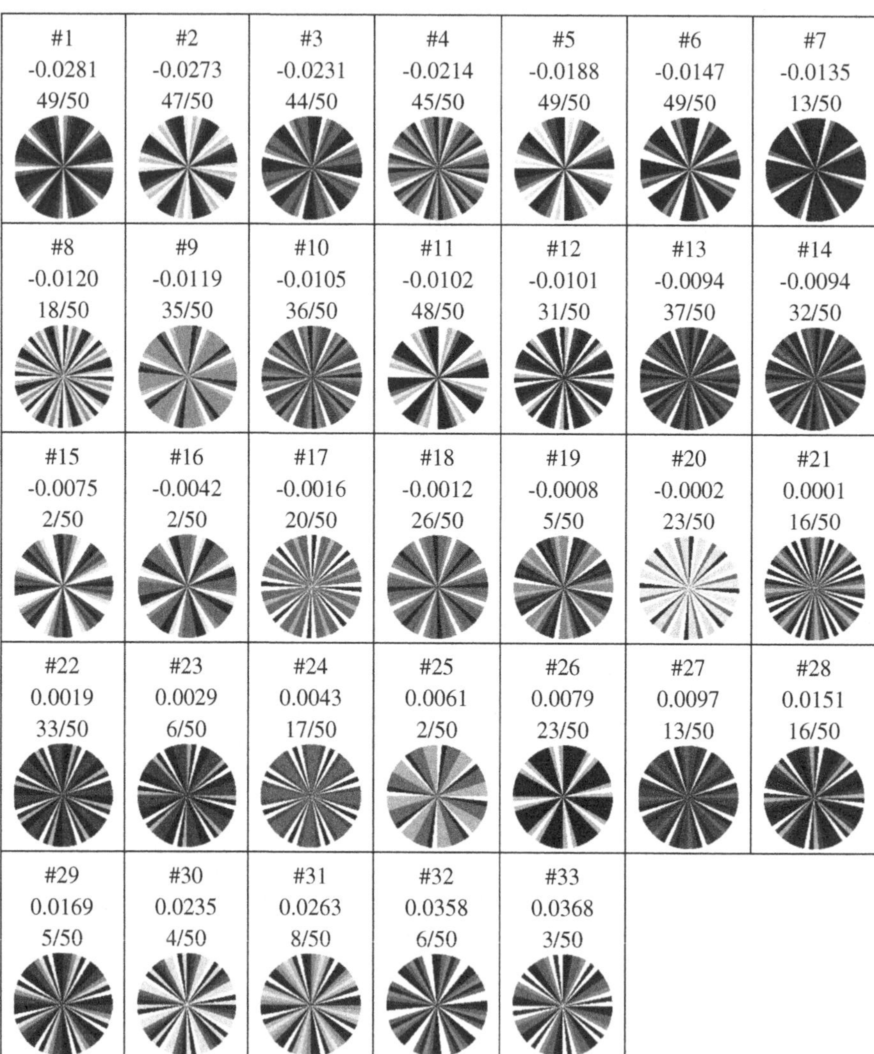

Fig. 4. Stimuli used in psychological experiments. #1–#33 are the indexes of stimuli. Negative and positive real values are spatially averaged rotation $\bar{R}$. Fractional numbers are the probability of human judgment to clockwise rotation of perception for 50 trials.

of rotation after the disappearance prior stimuli (either clockwise or counterclockwise; 2AFC) with a rotary device (PowerMate NA16029; Griffin Technology). The participants were five naïve subjects (23–24 years old).

This study was approved by the ethical committee of the University of Electro-Communications.

3.4 Correlation Between Model Estimation and Psychological Experiment

The fractional number in Fig. 4 is the probability of human judgment for "clockwise" rotation. For example, 49/50 of #1 means that humans tend to perceive clockwise illusory rotation, and 3/50 of #33 perceive counterclockwise rotating illusion.

Next, we compared model predictions with human responses. We adopt the following formula to transform rotation $\bar{R}$ into the stochastic judgment of clockwise motion $\Pr(CW)$

$$\Pr(CW) = \frac{1}{2}\left(1 - \mathrm{erf}\left(\frac{\bar{R}}{s\sqrt{2}}\right)\right). \tag{4}$$

Therein, erf() is the error function; s is a positive parameter. We assumed that the chance level corresponds to circumstances in which $\bar{R} = 0$ and $\Pr(CW) = 0.5$. The free parameter s of Eq. 4 was determined by application of a nonlinear fitting of the model function $\Pr(CW)$ to 33 data of human judgment. The best parameter was $s = 0.05$.

Figure 5 presents a scatter plot of model judgment and human judgment. If the model prediction was perfectly correct, markers in Fig. 5 should be on the diagonal line. Unfortunately, the model prediction was not perfect, but we can see a positive correlation between them. The correlation coefficient was 0.75. Model prediction was not perfect. However, we successfully obtained illusory patterns besides the FW pattern as shown in #1 and #33 of Fig. 4.

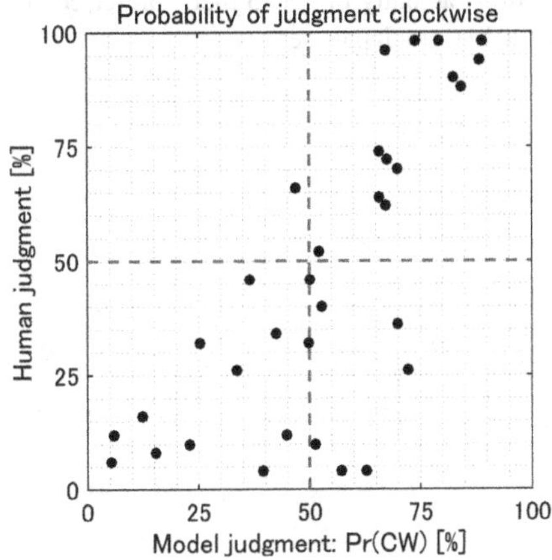

Fig. 5. Scatter plot of model judgment and human judgment.

4 Conclusions

We first showed that our computational model reproduced illusory rotation that is dependent on background luminance. Then, we strove to discover other drift illusion patterns beside a well-known illusory pattern. Numerical simulations showed positive correlation between human perception and model prediction, and revealed some novel illusory patterns.

The model, however, is not perfect. Some stimuli which had large rotational optical flow vectors did not induce illusion, which implies that we should improve our computational model. There is a room for improvement regarding the model's judgment function using the symmetric error function.

As described in this paper, we restricted ourselves to consideration of circular patterns for the convenience of limiting the simulation time. In principle, it is possible to examine all possible two-dimensional patterns. We believe that this approach will help to generate completely new illusory patterns, and facilitate the evaluation of various vision models.

Acknowledgements. This work was partially supported by JSPS KAKENHI (24500371 and 16K00204) and NIJC Riken, JAPAN.

References

1. Fraser, A., Wilcox, K.J.: Perception of illusory movement. Nature **281**(5732), 565–566 (1979)
2. Faubert, J., Herbert, A.M.: The peripheral drift illusion: a motion illusion in the visual periphery. Perception **28**, 617–621 (1999)
3. Hayashi, Y., Ishii, S., Urakubo, H.: A computational model of afterimage rotation in the peripheral drift illusion based on retinal ON/OFF responses. PLoS ONE **9**, e115464 (2014)
4. Lucas, B.D., Kanade, T.: An iterative image registration technique with an application to stereo vision. In: Proceedings of Imaging Understanding Workshop, pp. 121–130 (1981)
5. Nakamura, D., Satoh, S.: A novel computational concept of the physiological properties of MT neurons - Do MT neurons actually prefer to their 'preferred speeds'? - IEICE Technical report, **113**(500), 41–46 (2014). (in Japanese)

An Internal Model of the Human Hand Affects Recognition of Graspable Tools

Masazumi Katayama(✉) and Yusuke Akimaru

Department of Human and Artificial Intelligent Systems,
Graduate School of Engineering, University of Fukui,
3-9-1 Bunkyo, Fukui-shi, Fukui 910-8507, Japan
katayama@h.his.u-fukui.ac.jp

Abstract. In this study, we validated a plausibility of a hypothesis that in the human brain an internal simulation of grasping contributes to tool recognition. Such an internal simulation must be performed by utilizing internal models of the human hand. An internal model corresponding to a geometrically transformed hand shape was retrained by an experimental paradigm we built. The retrained internal model of the dominant hand affected cognitive judgments of object size of tools used by the dominant hand and however did not influence these of tools used by the non-dominant hand. While, those results in the training condition of the non-dominant hand showed the reverse tendency of the former results. The above results indicate the plausibility of the hypothesis.

Keywords: Graspable tool · Recognition · Internal model · Human hand

1 Introduction

When you feel a thirst, you can immediately find a drinking cup even in the complicated environment, although there are many kinds of cups. Thus, an object concept of a tool should be maintained by universally representing its tool in the brain. Traditionally, it has been considered that the object concept is expressed based on symbols of declarative memories, visual features and so on. However, it is recently supported that the object concept is also related to sensorimotor experiences of tool use (e.g., [1,2]). Even if you are verbally explained about an unknown tool, you can not really understand its tool. You are able to really know about its tool by actually repeating tool use.

While, it is biologically ascertained that internal models of the human body are represented in the brain. For example, an inverse dynamics model of a monkey's eye is represented in the cerebellum [3] and an internal model of the human arm is acquired in the cerebellum [4]. The presence of an internal simulation of action has been also pointed out, because the motion-related cortical areas partly activated even in motion imagery (e.g., [5,6]). A patient that the parietal cortex was partly damaged became impossible to take into account physical

A. Hirose et al. (Eds.): ICONIP 2016, Part IV, LNCS 9950, pp. 199–207, 2016.
DOI: 10.1007/978-3-319-46681-1_24

constraints in motion imagery of hand movements [6]. In an observation task, moreover, although the motion-related cortical areas partly activate when viewing graspable tools, the activations in these areas were smaller than the above case when viewing objects of the other categories [7]. From this point of view, we have proposed a hypothesis that in the human brain an internal simulation of grasping a tool contributes to tool recognition: judging whether we can grasp a target object is useful to recognize as its tool. Such an internal simulation can be realized by utilizing internal models of the human hand. In order to validate the plausibility of the hypothesis, by using an experimental paradigm to retrain the internal mode of the transformed hand shape, we investigated a relationship between the trained internal model and cognitive judgments of tool size [8].

2 Methods

Fifteen participants joined in the first experiment (right-handed, aged 18–24) and fifteen participants joined in the second experiment (right-handed, aged 18–22). Each participant was tested by the Edinburgh Handedness Inventory. They were completely naive with regard to the specific purpose and jointed this experiment after signing an informed consent agreement.

An experimental system has built in a dark room and constructed by a finger motion measurement device (CyberGlove, Right and left hand types, CyberGlove Systems Inc.), two three dimensional motion measurement devices (FASTRAK, Polhemus; OPTOTRAK3020, Northern Digital Inc.), a mirror, a display (XL2720T, BenQ Inc.), an experimental chair using an ergonomically designed car seat (RECARO GmbH & Co. KG.), and a chin rest to fix the head (see Fig. 1). Finger lengths between joints of the hand were measured and a hand and forearm shaped using the measured lengths were displayed on a monitor. Participants were able to see the screen with the mirror placed in front of them. The position, size, orientation and hand shape of the displayed hand were adjusted to

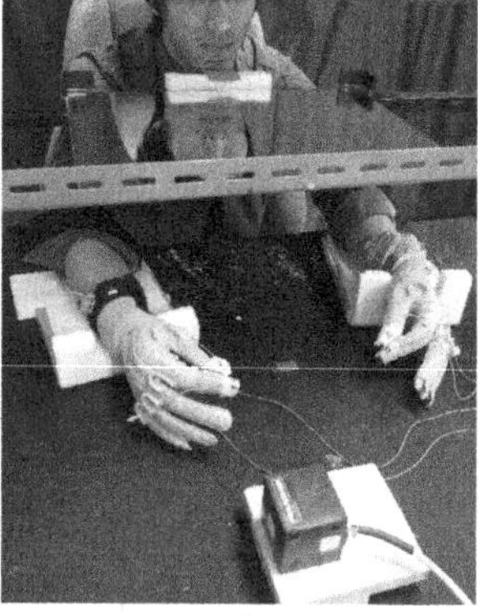

Fig. 1. Experimental setup.

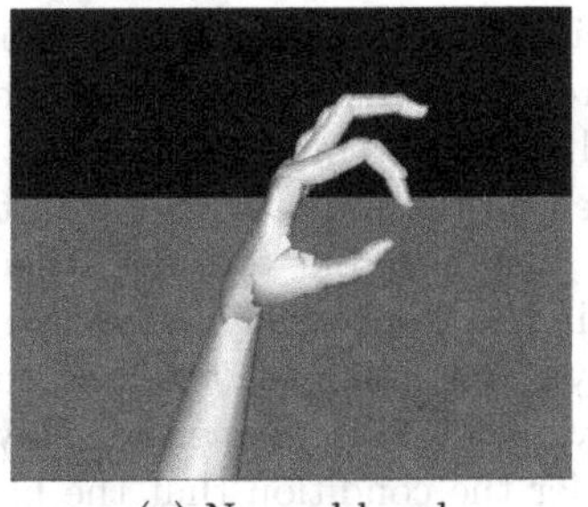

(a) Normal hand

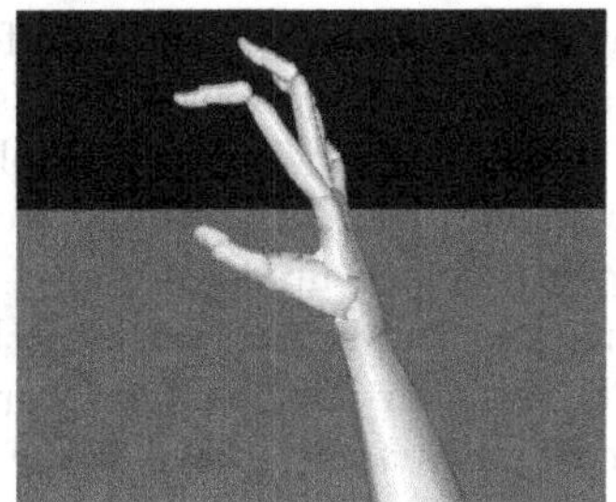

(b) Geometrically transformed hand

Fig. 2. Displayed hand.

look like their own hand. Infrared light-emitting diodes (IR LED markers) were attached to each fingertip in order to accurately measure the fingertip positions by OPTOTRAK with a sampling frequency of 200 Hz. Joint angles of the right and left hands of each participant were measured by two CyberGloves, respectively, and the hand position and orientation ware measured by FASTRAK with a sampling frequency of 60 Hz. The displayed hand and forearm could be moved in synchronization with the participant's hand movement.

Fig. 3. Screen during training.

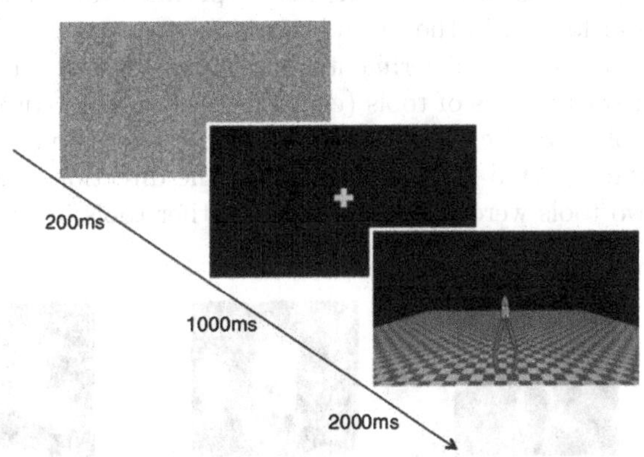

Fig. 4. Screens during measurement.

A basic idea in an experimental paradigm we have built is to examine a relationship between cognitive process for graspable tools and a retrained internal model of the human hand. The shape of the participant's hand was geometrically transformed and the hand was displayed on a monitor. In the transformed hand, the length between CM and MP joints of the thumb and the lengths between MP and PIP joints of the other fingers were lengthened to 1.8 times, respectively (see Fig. 2).

Before the experiments, in order for a participant to feel the displayed hand like one's hand, synchronous tactile stimuli were simultaneously given to both the participant's hand and the displayed hand for a few minutes. Then, two small circles and the normal hand were displayed on the monitor and a participant repeatedly executed the finger movement task that put the fingertips of the thumb and the index finger to the small circles, respectively, as shown in Fig. 3. The positions of the circles were randomly changed in each trial and they repeated 20 trials as a set. After the trials, only the two circles were displayed and they performed the same task 20 times under the condition that the hand was not displayed. The errors were detected from the differences between the measured grip apertures and the premeasured correct widths. The training were continued until the average of the errors became less than 8 mm. When the training did not finish, the training finished after 5 sets. In the training, they trained also with the contralateral hand alternately. After the training, we examined two kinds of measurements described below. After a break for a few minutes, two small circles and the geometrically transformed hand were displayed, the same participant repeatedly executed the task, and the errors of the grip apertures were detected in the same way. In the training, they trained also with the contralateral untransformed hand alternately. After the training, the same measurements were executed. In order to maintain the learning effect, the training of a set was executed between the measurements. The right-hand was geometrically transformed in the first experiment and The left-hand was geometrically transformed in the second experiment.

Before the experiments, participants answered a questionnaire about images of thirty kinds of tools (e.g., Fig. 5): the hand (right and/or left) that uses each tool, frequency in tool use, grip type and so on. Base on the results, two tools were selected from the tools that the direction of grip was given, and moreover two tools were selected from the other tools in each participant. An image was

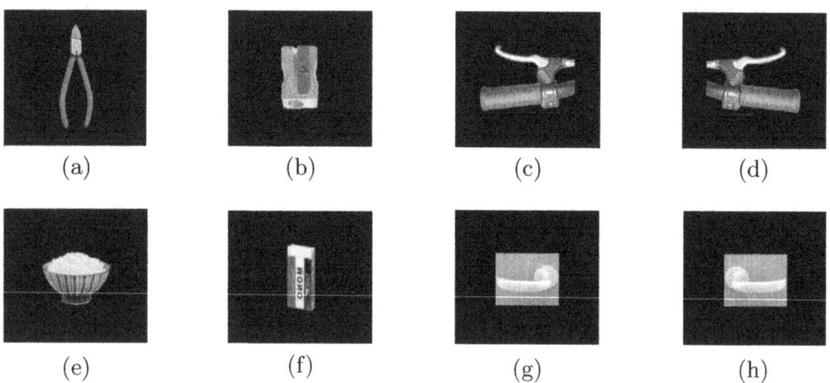

Fig. 5. Examples of tools. a: a nipper, b: a pencil sharpener, c and d: handles of a bicycle, e: a rice bowl, f: a rubber eraser, g and h: door knobs. The images of c, d, g and h are the tools that each grip direction is given.

randomly selected from within different ten image sizes of its selected tool and the image was displayed as shown in Fig. 4. Participants were instructed to answer whether they recognized as its tool with regard to object size: If they felt too small or large as its tool, they answered "No". Because 10 trials in each image size were executed, the number of the measurements was 100 trials per each tool (10×10 trials). Moreover, as another measurement, participants were instructed to answer a verbal estimation of the apparent size of a displayed tool image, using a 10-point scale in which 1 corresponded to the size of a 1 yen coin and 10 corresponded to the size of a compact disk (CD) (see [9]). The image display was the same as the above measurement. The number of the measurements was 10 trials per each tool.

3 Results

The results of cognitive judgment for tool size were shown in Figs. 6 and 7. The lower and upper thresholds were detected from the two intersection points that the probability become 0.5, by interpolating between the data points by a sigmoid function. It is shown that the object size between the lower and upper thresholds is recognized as its tool. In a right handle of the transformed right-hand condition, a larger object becomes to be recognized as its tool, as shown in Fig. 6(b), although the result in the case of a left handle shown in Fig. 6(a) does not change. While, in the case of the transformed left-hand condition shown in Fig. 7, these results show the reverse tendency of the above results. Here, in order to examine the cognitive tool sizes for all tools, the change rate of these thresholds was calculated as follows:

$$\text{Change rate} = \frac{\text{CTH2} - \text{CTH1}}{\text{CTH1}} \times 100, \quad (\%) \tag{1}$$

where CTH1 and CTH2 express the centers between the lower and upper thresholds in the normal hand condition and the transformed hand condition, respectively. The average values of the change rates were shown in Figs. 8 and 9. In the

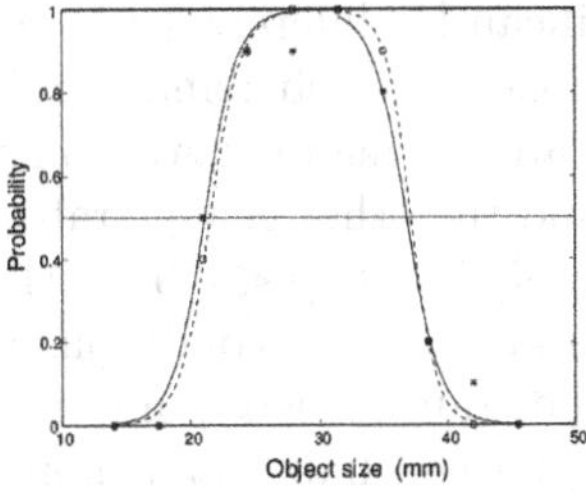

(a) a left handle of a bicycle

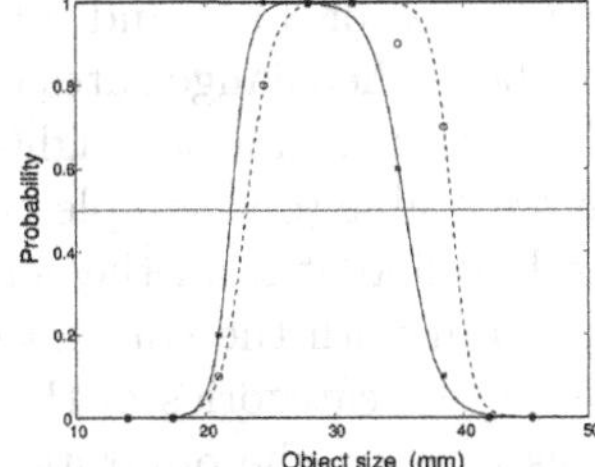

(b) a right handle of a bicycle

Fig. 6. Typical results of cognitive judgment for tool size that were measured under the transformed right-hand condition. The solid lines stand for the results in the norman hand condition and the dashed lines express the results in the transformed right-hand condition. The vertical axis is the probability that answered "Yes".

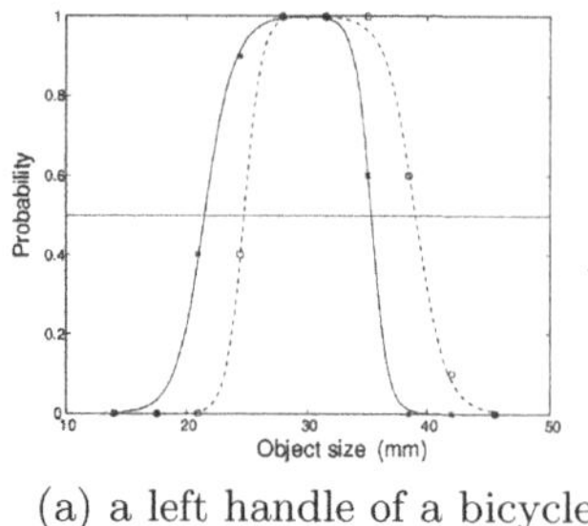

(a) a left handle of a bicycle

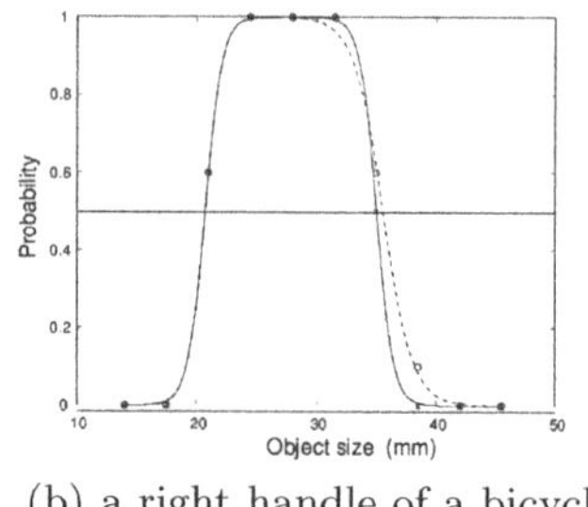

(b) a right handle of a bicycle

Fig. 7. Typical results of cognitive judgment for tool size under the transformed left-hand condition.

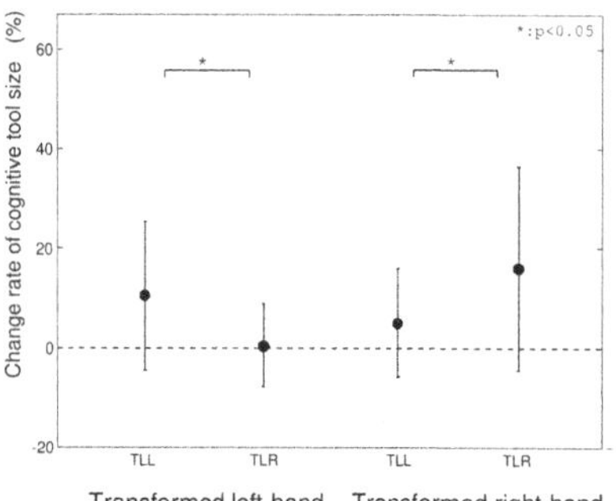

Fig. 8. Change rates of cognitive tool size in both of the conditions. TLR and TLL stand for the tools used by the right hand and the left hand, respectively. Vertical lines are standard deviations.

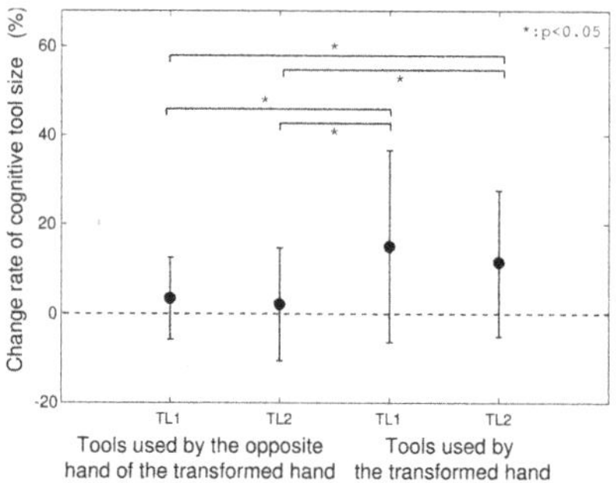

Fig. 9. Change rates of cognitive tool size for tool categories. TL2 and TL1 stand for the tools that each grip direction is given and the other tools, respectively. Vertical lines are standard deviations.

transformed right-hand condition of Fig. 8, the change rates of TLR and TLL are significantly different ($p < 0.05$) and in the transformed left-hand condition the change rates of TLR and TLL are also significantly different ($p < 0.05$). Moreover, both the change rates of TLR of the transformed left-hand condition and TLL of the transformed right-hand condition are not significantly different from the zero value (one sample t-test, $p > 0.05$) and the other change rates are significantly different from the zero value (one sample t-test, $p < 0.05$). All the change rates in both the transformed right-hand and left-hand conditions were divided into two categories of the tools used by the transformed hand and the other tools used by the opposite hand of the transformed hand, and moreover each category was divided into TL1 and TL2. Figure 9 shows each average value of the change rates of the four cases. Four combinations of the change rates between the two categories are significantly different (Tukey-Kramer method, $p < 0.05$). The change rates of the tools used by the transformed hand are significantly different from the zero value (one sample t-test, $p < 0.05$), although these of the other tools used by the opposite hand of the transformed hand are not significantly different from the zero value (one sample t-test, $p > 0.05$). Here, although these results depend on the difference of the object sizes of the tool categories, the object sizes are not significantly different ($p > 0.05$).

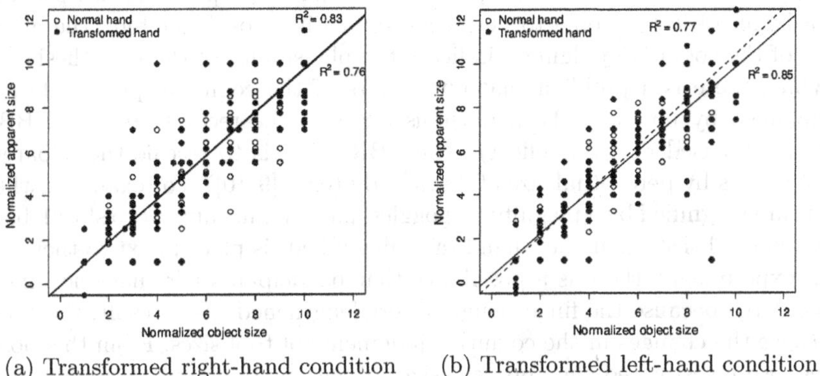

(a) Transformed right-hand condition (b) Transformed left-hand condition

Fig. 10. Normalized apparent size. R^2 is a coefficient of determination of a linear regression analysis.

Figure 10 shows the normalized apparent sizes calculated as follows:

$$\mathrm{NAS_i} = \frac{10 - 1}{\mathrm{AS_{max}} - \mathrm{AS_{min}}}(\mathrm{AS_i} - \mathrm{AS_{min}}) + 1. \quad (2)$$

Here, $\mathrm{NAS_i}$ and $\mathrm{AS_i}$ stand for an ith normalized apparent size and an ith apparent size, respectively, and $\mathrm{AS_{max}}$ and $\mathrm{AS_{min}}$ express the maximum and the minimum values of the apparent sizes measured with each tool, respectively. It is noted that $\mathrm{AS_{max}}$ and $\mathrm{AS_{min}}$ in the transformed hand condition were the values of the same tool as the normal hand condition. The difference of two linear regression lines in the normal and transformed hand conditions, were statistically tested by an analysis of covariance. The difference of the slopes is not significant in both the cases of Figs. 10(a) and (b) ($p > 0.05$), and also the difference of the average values is not significant in both the cases of these fugues ($p > 0.05$).

4 Discussion

In the transformed hand condition, although participants were unable to skillfully execute the finger movement task before the training, they gradually became possible to accurately perform the task without viewing the displayed hand. Here, note that human brain should acquire an internal model to relate fingertip positions and joint angles through the training. In the basis of the hypothesis, the internal model of the right hand should be utilized in an internal simulation of grasping tools used by the right hand and it is not used in case of tools used by the left hand. Thus, the hypothesis predicts that the retrained internal model of the right hand influences the cognitive judgments of tools used by the right hand and it does not affect the cognitive judgments of tools used by the left hand. Moreover, because the displayed hand was geometrically transformed to become possible to grasp larger objects, the hypothesis also predicts that a larger object become to be recognized as its tool. Therefore, the above results of the cognitive judgment indicate the plausibility of the hypothesis.

While, there is a problem that our results of the cognitive judgments can be explained by not only the hypothesis but also the recently reported BBR effect (body-based rescaling effect). The BBR effect is to rescale the apparent size of objects by perceptual size of one's body (e.g., [9,10]). For example, when objects are magnified by a magnifying goggles, participants appear to shrink back to near-normal size when one's hand (also magnified) is placed next to them [9]. In our experiments, there is a possibility that participants felt that one's hand became large because the finger lengths were lengthened. As a result, the effect may cause the changes in the cognitive judgments of tool sizes. From this point of view, we investigated the apparent sizes of various tools in the same way as Linkenauger et al. [9]. If the changes in the cognitive judgment were caused by the BBR affect, the apparent sizes in the transformed hand condition should become smaller than those in the normal hand condition. As shown in Fig. 10, however, both the apparent sizes of the normal and transformed hand conditions are not different. Thus, the results of Fig. 10 show that the BBR effect did not arise under the transformed hand condition in our experimental paradigm. From the above considerations, we demonstrated the plausibility of the hypothesis that an internal model of the human hand contributes to tool recognition.

Acknowledgments. This research was partially supported by MEXT KAKENHI (C) No. 15K00200.

References

1. Katayama, M., Kawato, M.: A neural network model integrating visual information, somatosensory information and motor command. J. Robot. Soc. Japan **8**, 757–765 (1990). in Japanese
2. Borghi, A.M.: Object concepts and action. In: Grounding Cognition: The Role of Perception and Action in Memory, Language, and Thinking, pp. 2–34. Cambridge University Press, Cambridge (2005)
3. Shidara, M., Kawano, K., Gomi, H., Kawato, M.: Inverse-dynamics model eye movement control by purkinje cells in the cerebellum. Nature **365**, 50–52 (1993)
4. Imamizu, H., Miyauchi, S., Tamada, T., Sasaki, Y., Takino, R., Putz, B., Yoshioka, T., Kawato, M.: Human cerebellar activity reflecting an acquired internal model of a new tool. Nature **403**(6766), 192–195 (2000)
5. Jeannerod, M.: The representing brain: neural correlates of motor intention and imagery. Behav. Brain Sci. **17**, 187–245 (1994)
6. Sirigu, A., Duhamel, J.R., Cohen, L., Pillon, B., Dubois, B., Agid, Y.: The mental representation of hand movements after parietal cortex damage. Science **273**(5281), 1564–1568 (1996)

7. Chao, L.L., Martin, A.: Representation of manipulable man-made objects in the dorsal stream. NeuroImage **12**, 478–484 (2000)
8. Katayama, M., Kurisu, T.: Human object recognition based on internal models of the human hand. In: Yamaguchi, Y. (ed.) Advances in Cognitive Neurodynamics (III), pp. 591–598. Springer, Heidelberg (2013)
9. Linkenauger, S.A., Ramenzoni, V., Proffitt, D.: Illusory shrinkage and growth: bady based scaling affects the perception of size. Psychol. Sci. **21**(9), 1318–1325 (2010)
10. van der Hoort, B., Guterstam, A., Ehrsson, H.H.: Being barbie: the size of one's own body determines the perceived size of the world. PLoS ONE **6**(5), 1–10 (2011)

Perceptual Representation of Material Quality: Adaptation to BRDF-Morphing Images

K. Kudou(✉) and K. Sakai

Department of Computer Science, University of Tsukuba, 1-1-1 Tennodai, Tsukuba 305-8753, Japan
kudou@cvs.cs.tsukuba.ac.jp, sakai@cs.tsukuba.ac.jp
http://www.cvs.cs.tsukuba.ac.jp/

Abstract. Perception of the material quality of a surface depends on its reflectance properties. Recent physiological studies reported the neural selectivity to glossy surfaces in the Inferior Temporal cortical areas [e.g., 1]. In the present study, we examine the hypothesis that basis neurons are selective to typical materials, and that the combinations of their responses are representative of a variety of natural materials. To assess the hypotheses, we performed a psychological experiment based on adaptation. If adaptation to a specific material is observed, the presence of neurons that are selective to the specific material is predicted. We performed adaptation tests with six typical material qualities including gloss, matte, metal and wood. We observed the adaptation to certain materials but not to some other materials. This result indicates the presence of basis neurons that are selective to materials, which is fundamentally important for understanding cortical representation of surface materials.

Keywords: Vision · Visual cortex · Adaptation · Psychophysics · Material quality · Surface reflectance properties

1 Introduction

Perception of material quality depends on three factors: shape, illumination, and surface reflectance properties. Of those, surface reflectance properties are the most important factor for the material perception. Recent studies have focused on the perception of glossiness [2, 3]. A physiological study reported the presence of neurons that are selective to glossy surfaces in the Inferior Temporal (IT) cortex of the macaque monkey [1]. A psychophysical study has reported the human adaptation to glossy surfaces [4]. However, the real world presents a variety of materials (matte, metal, wood, etc.) not limited to glossiness. We hypothesized the presence of *basis neurons* that are selective to surface materials fundamental in the cortical representation. Cortical neurons in the visual areas have been suggested to represent arbitrary features by the linear combinations of a few fundamental features [5, 6]. We expect the presence of basis neurons that represent typical materials, and the linear combination of their responses for the material perception in general. In the present study, we psychologically investigated this hypothesis. If adaptation to a specific material is observed, then the presence of neurons that are selective to this material is suggested. We observed adaptation not

A. Hirose et al. (Eds.): ICONIP 2016, Part IV, LNCS 9950, pp. 208–212, 2016.
DOI: 10.1007/978-3-319-46681-1_25

only to gloss, but also to other typical materials. Results demonstrate the presence of basis neurons that are selective to fundamental materials.

2 Methods

2.1 Procedure

The experimental procedure is presented in Fig. 1. After we had subjects adapt to materials A and B on the left and right sides, test stimuli were presented on the screen. We asked subjects to judge which test stimulus is more similar to the material quality of the reference stimulus. For instance, we set the material A on the left and B on the right in the adaptation stimuli and have subjects adapt to them. After adaptation, when two test stimuli of the same surface that have a 50–50 mix of A and B are presented, the right test stimulus is expected to look more like A and the left more like B. We obtain a psychometric function by repeating the trials in which test stimuli vary in the mixture ratio of A and B. The difference in the perceptual equality (P.E.) between non-adaptation (i.e., trials without adaptation) and adaptation is the magnitude of adaptation. In this study, we tested six materials and obtained six psychometric functions. Eight observers participated in this experiment, six males and two females in early twenties of their age. The experiment was approved by the ethics committee of the institute at University of Tsukuba.

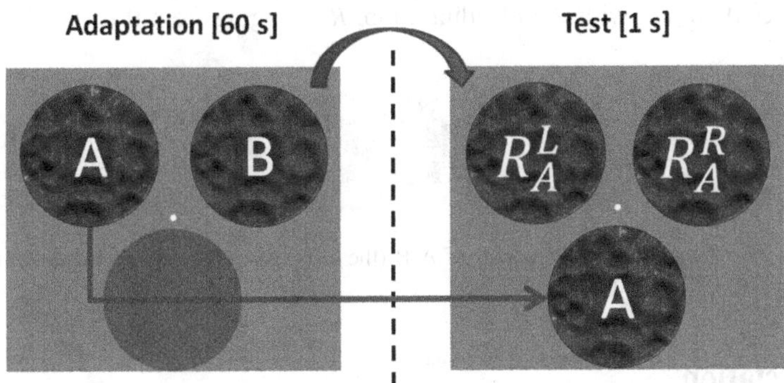

Fig. 1. Experimental procedure. Following the adaptation stimuli (left), three test stimuli (right) were presented. Participants were asked to judge which stimulus on the top most closely resembles the reference (bottom) in terms of material quality. Participants repeated the trial 24 times with the interval of 6 s.

2.2 Stimuli

2.2.1 Selection of BRDF

We selected characteristic materials for the experiment. Surface reflectance properties can be described by a Bidirectional Reflectance Distribution Function (BRDF). BRDF is a function that defines how incident light is reflected from one point on a surface.

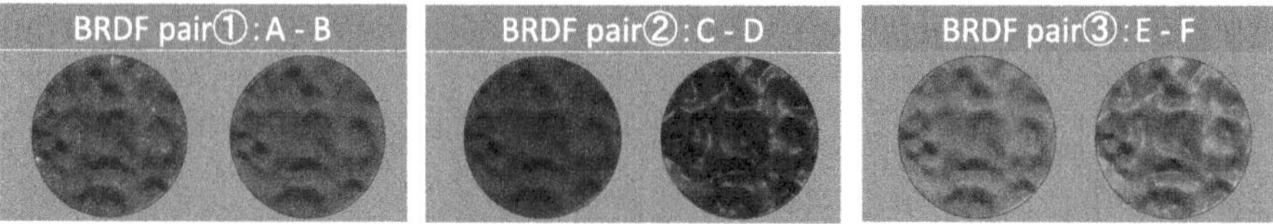

Fig. 2. Rendered stimuli of the selected BRDF pairs. We selected three BRDF pairs for which each pair is distant in *c–d* space and which are visually as dissimilar in appearance as possible. PBRT (Physically Based Rendering Toolkit) [9] is used to render the stimuli. For the illumination environment, we used Circus_Backstage from the sIBL Archive [10] by which subjects can easily perceive the material quality of the stimuli. We used three types of random terrains.

We selected three material pairs from MERL/MIT BRDF dataset such that each pair is distant on *c–d* space [7] and is visually as dissimilar in appearance as possible. Examples of the rendered stimulus pairs with distinct BRDFs are presented in Fig. 2.

2.2.2 Morphing BRDF

For test stimuli, we generated a set of stimulus with an arbitrary mixing ratio of A and B. Morphing each selected BRDF pair, we can generate a set of stimuli with continuously changing surface reflectance properties. BRDF Morphing was done by α blend [8].

$$MorphingBRDF = R_A \times A + (1 - R_A) \times B, \tag{1}$$

where R_A s the ratio of the BRDF of A; A and B are paired. Figure 3 shows examples of the rendered stimuli with the blending ratio, R_A, ranging from 0 to 1.

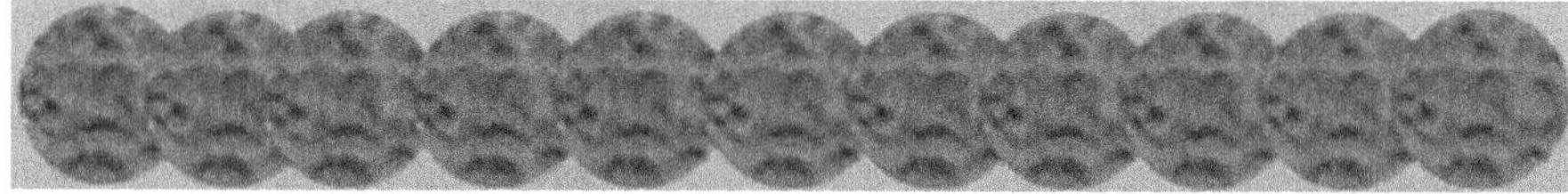

Fig. 3. A series of morphing for the pair of A-B (the leftmost in Fig. 2). R_A increases from left to right (R_A = 0.0, 0.1,…, 1.0).

3 Expectation

Figure 4 presents expected psychometric functions in which the ratio of the answer to the left is shown as a function of the difference in R_A of the pair of test stimuli. Without adaptation (blue dotted line), the P.E. of the obtained psychometric function is zero. With adaptation (red solid line), the P.E. of the obtained psychometric function is expected to shift in the right direction because, with adaptation to A, the answer ratio to the left decreases. We alter the reference to B and repeat the experiment in order to test the adaptation to B. Although psychometric functions are flipped, that with adaptation (red solid line) is expected to shift in the right direction as well.

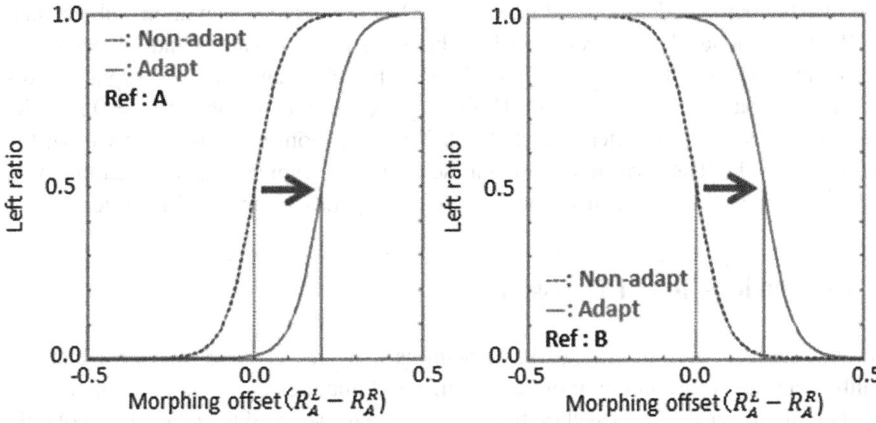

Fig. 4. Translation of the psychometric function is expected after adaptation. The left panel shows the expected psychometric function for reference A. If adaptation is observed, the psychometric function shifts in the right direction (from blue solid line (without adaptation) to red solid line (with adaptation)). The right panel shows the expectation for reference B in which we observe the shift in the right direction as well. (Color figure online)

4 Results

We performed adaptation experiments for all three BRDF pairs. The results for BRDF pair A-B are presented in Fig. 5. As might be expected, the obtained psychometric functions shifted in the right direction, indicating that adaptation to A is evoked (Fig. 5, left). In addition, the psychometric function for the reference B shifted right. This fact also indicates that adaptation to B is evoked (Fig. 5, right). We obtained similar results

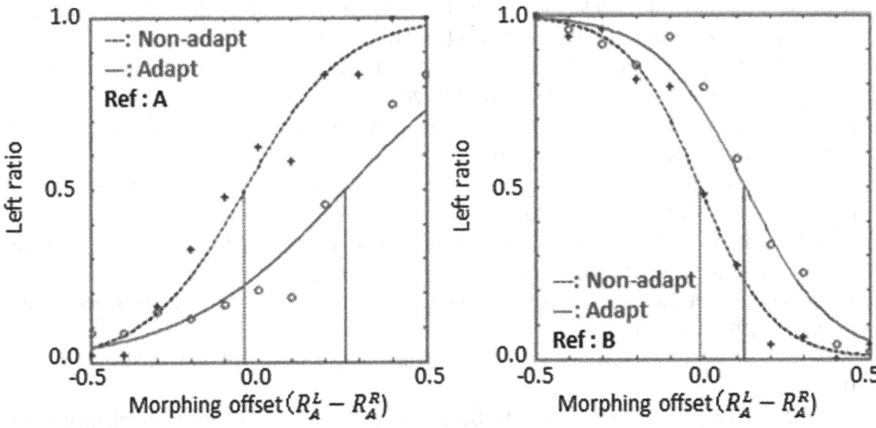

Fig. 5. Results for the adaptation to BRDF pair A-B. The mean ratio of the answer to the left among eight subjects is shown as a function of the morphing offset. The left and right panels present the results for references A and B, respectively.

for all BRDF pairs, and conducted three-way ANOVA (factor: adaptation, subject, and BRDF) to elucidate the difference in P.E. between non-adaptation and adaptation to each of material qualities. The analysis showed that the magnitude of adaptation differed in material quality ($p \leq 0.01$). However, irrespective of subject ($p = 0.28$), the adaptation effect was significant ($p \leq 0.01$). No interaction was found in terms of the subject ($p > 0.1$). These results show that adaptation to typical material qualities was evoked, and that the extent of adaptation effects depends on material qualities.

5 Conclusions and Discussion

We hypothesized the presence of basis neurons that are selective to typical material qualities, and the representation of various material qualities by linear combinations of their neural responses. Our psychophysical experiments showed the adaptation not only to gloss but also to matte, metal, and wood. These results suggest the presence of neurons that are selective to a variety of material qualities. The magnitude of adaptation varied among the material qualities, which may support the idea that some fundamental material qualities are represented by basis neurons, and the other qualities by their linear combinations.

Acknowledgements. This work was supported by a grant-in-aid from JSPS (KAKENHI 26280047) and a grant-in-aid for Scientific Research on Innovative Areas, "Shitsukan" (No. 251 35704) from MEXT, Japan.

References

1. Nishio, A., Goda, N., Komatsu, H.: Neural selectivity and representation of gloss in the monkey inferior temporal cortex. J. Neurosci. **32**(31), 10780–10793 (2012)
2. Kim, J., Marlow, P.J., Anderson, B.L.: The perception of gloss depends on highlight congruence with surface shading. J. Vis. **11**(9), 1–19 (2011)
3. Sakai, K., Meiji, R., Abe, T.: Facilitatory mechanisms of specular highlights in the perception of depth. Vis. Res. **115**, 188–198 (2015)
4. Motoyoshi, I., Nishida, S., Sharan, L., Adelson, E.H.: Image statistics and the perception of surface qualities. Nature **447**, 206–209 (2007)
5. Hatori, Y., Mashita, T., Sakai, K.: Sparse coding generates curvature selectivity in V4 neurons. JOSA **33**(4), 527–537 (2016)
6. Olshausen, B.A., Field, D.J.: Emergence of simple-cell receptive field properties by learning a sparse code for natural images. Nature **381**, 607–609 (1996)
7. Ferwerda, J.A., Pellacini, F.: A psychophysically based model of surface gloss perception. In: SPIE 4299, 8 Jun 2001
8. Woo, M., Neider, J., Davis, T.: OpenGL Programming Guide, 3rd edn. Addison-Wesley, Boston (2000)
9. Pharr, M., Humphreys, G.: Physically Based Rendering from Theory to Implementation. Morgan Kaufmann, Burlington (2010)
10. sIBL Archive. http://www.hdrlabs.com/sibl/archive.html

GPU-Accelerated Simulations of an Electric Stimulus and Neural Activities in Electrolocation

Kazuhisa Fujita[1,2(✉)] and Yoshiki Kashimori[2]

[1] National Institute of Technology, Tsuyama Collage, 654-1 Numa, Tsuyama, Okayama 708-8506, Japan
k-z@nerve.pc.uec.ac.jp
[2] University of Electro-Communications, 1-5-1 Chofugaoka, Chofu, Tokyo 182-8585, Japan

Abstract. To understand mechanism of information processing by a neural network, it is important to well know a sensory stimulus. However, it is hard to examine details of a real stimulus received by an animal. Furthermore, it is too hard to simultaneously measure a received stimulus and neural activities of a neural system. We have studied the electrosensory system of an electric fish in electrolocation. It is also difficult to measure the electric stimulus received by an electric fish in the real environment and neural activities evoked by the electric stimulus. To address this issue, we have applied computational simulation. We developed the simulation software accelerated by a GPU to calculate various electric stimuli and neural activities of the electrosensory system using a GPU. This paper describes comparison of computation time between CPUs and a GPU in calculation of the electric field and the neural activities.

Keywords: GPGPU · CUDA · Acceleration · Electrolocation

1 Introduction

An electric fish can recognize object parameters, such as material, size, distance and shape, in complete darkness. The ability to recognize these object parameters is provided by the electrosensory system of the fish. The fish generates an electric field using its electric organ. An object around the fish distorts the electric field and make an electric image on fish's body surface. The fish can extract the object parameters from the electric image on the body surface using the electrosensory system. However, it is not well known what features of the electric image represent the object parameters and how the nerve system extracts the object's features from the electric image. Furthermore, it is difficult to simultaneously measure an electric image in the real environment and neural activities of the nerve system. To address this issue, we have made a numerical models for calculation of the electric field and the neural activities. However, the computational costs of these models are high. Thus, to accelerate to calculate many types of electric stimuli and neural activities evoked by the stimuli, we developed

A. Hirose et al. (Eds.): ICONIP 2016, Part IV, LNCS 9950, pp. 213–220, 2016.
DOI: 10.1007/978-3-319-46681-1_26

the simulation software to calculate various electric stimuli and neural activities accelerated by a GPU. This paper describes comparisons of computation time between CPU and GPU in calculation of the electric field and neural activities.

In recent years, GPGPU (General Purpose computing of GPU) is used for scientific numerical applications. A GPU is a computation unit which is specialized in the calculation for drawing 3D graphics. A GPU has massive simple computational units and delivers good performance in parallel computation. Taking this advantage, a GPU is often used for parallel computing of scientific computation. There are two popular GPGPU environments, CUDA of nVidia and the OpenCL of OpenCL Working Group. In particular, CUDA is used for scientific and technological calculation because of ease of development. In the present study, we used the CUDA in software development for GPGPU.

GPGPU is widely used in the field of neuroscience. In particular, it has been used to challenge to real-time simulation of neural activities. Yamazaki and Igarashi have succeeded in real-time simulation of a large-scale model of the cerebellum using a GPU [7]. Beyeler et al. have calculated the neural activities of the cortical network by a GPU and implements the navigation of the robot by the calculated activities [1]. It has been also used in the calculation of deep neural network that has become a hot topic in recent years.

In the present study, in order to reveal details of the electrical stimulation and the activity of the electrical receptors and neurons of the electrosensory nucleus using the computer simulation, we tried to accelerate simulation of electrical stimulation and of neural activities using a GPU. We report the effect of a GPU on computation time of the simulations.

2 Electrolocation

In this study, we dealt with a electric stimulus and electrosensory system of a weakly electric fish for electrical localization. The weakly electric fish generates an electric field around it using a electric organ in its tail. If there is an object that has different electric properties to water around the fish, the electric field is modulated and an electric image is made on the body surface of the fish. The electric receptors of fish's body surface always monitor the state of the electric image. The fish can detect object's properties such as distance, size, shape, resistance, and capacitance from the electric image. An electric image received by the fish is converted to neural activity by electro receptors on fishs skin and is projected to the higher electrosensory nucleus. The higher nucleus extracts the object's features.

3 Methods

3.1 Models

A Model of an Electric Field. In this subsection, we denote the method of the calculation of an electrical field generated by a weakly electric fish. The fish creates an electric field around itself using an electric organ. In the real environment, it is difficult to measure the potential of the electric field generated by the fish at all locations and the electrical stimulus received by the fish. Heiligenberg [5], Hoshimiya et al. [6] and Fujita and Kashimori [4] have developed the numerical model to calculate the electric field and have investigated the electrical stimulus received by the weakly electric fish on its skin. We used the model of Fujita and Kashimori [4] but the capacitance of the object was not considered for simplification in the present study.

Figure 1a shows the model for calculating the electric field generated by the fish. The space of the water tank including the fish is divided by the square cells. We modeled the electric properties of the material that occupied the space between the $(m1, n)$th and (m, n)th squares as the equivalent circuit illustrated in Fig. 1b. The center of each square represented a node of the electric network. The current flowing from the $(m1, n)$th node to the (m, n)th node, $I(m1, n; m, n)$, was described by the following formula:

$$I(m-1, n; m, n) = Y(m-1, n; m, n)(V(m-1, n) - V(m, n)), \tag{1}$$

where $Y(\cdot)$ is admittance between nodes and $V(m, n)$ is the potential of the (m, n)th node. In this study, we did not consider capacitance of the material thus the admittance between the nodes is regarded as inverse of the resistance.

The (m, n)th node received the currents flowing from the four neighbors shown in Fig. 1b. The conservation law of the current to the (m, n)th node is given by

$$I(m-1, n; m, n) + I(m+1, n; m, n) + I(m, n-1; m, n) + I(m, n+1; m, n) = 0. \tag{2}$$

From Eqs. 1 and 2, the potential of the (m, n)th node, $V(m, n)$, yields

$$\begin{aligned} V(m, n) =(&\sum_{X=m-1,m+1} Y(X, n; m, n)V(X, n) \\ &+ \sum_{X=n-1,n+1} Y(m, X; m, n)V(m, X))/(\sum_{X=m-1,m+1} Y(X, n; m, n) \\ &+ \sum_{X=n-1,n+1} Y(m, X; m, n)). \end{aligned} \tag{3}$$

To determine the potentials of the nodes, it is necessary to repeatedly calculate this equation until the values of the potentials converge.

In this study, the water tank is a square where each side is $\sqrt{N}$. N is the number of nodes and is equal to 2^n. All nodes in the water tank have same resistance $R_{\text{water}} = 3.8 \times 10^2\,\Omega$. The potential of boundary of the water tank assuming that extends infinitely water was zero. There is a dipole modeled on the electric organ of the weak electric fish in the water tank. The dipole is located at $(3\sqrt{N}/8, 3\sqrt{N}/8)$ and $(5\sqrt{N}/8, 5\sqrt{N}/8)$. The potential of the dipole is $V(3\sqrt{N}/8, 3\sqrt{N}/8) = 100$ mV and $V(5\sqrt{N}/8, 5\sqrt{N}/8) = -100$ mV. In this study, to measure calculation time of various methods, the number of calculations of the potential was 1000000 times.

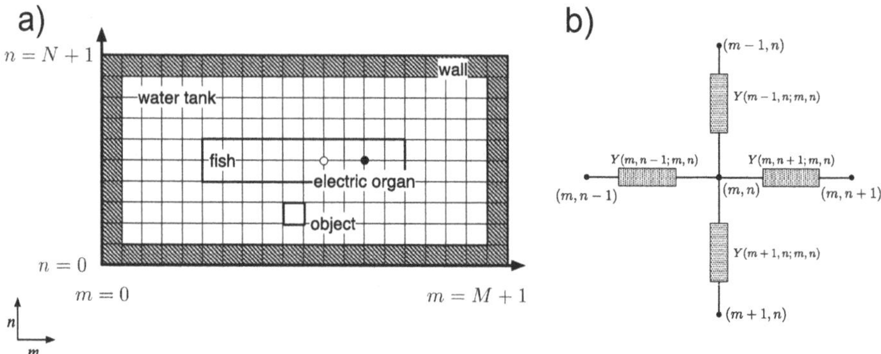

Fig. 1. A model for calculating the electric field

A Model of an Electroreceptor. An electric stimulus is encoded as spikes by an electroreceptor afferent. In this study, we used the simple model proposed by Brandman and Nelson [2] as an electroreceptor afferent. The voltage between inner and outer skin, $v(t)$, is denoted by

$$v(t) = ci(t), \tag{4}$$

where c is resistance of skin and $i(t)$ is the electric current through the skin. The electric current, $i(t)$, is denoted by

$$i(t) = I \sin(2\pi t/T), \tag{5}$$

where I and T is an amplitude and a period of the sinusoidal current, respectively. The sine function represents a sinusoidal wave of electric organ discharge. The afferent nerve of the electroreceptor fires when the voltage, $v(t)$, is more than the threshold θ. Thus, the spike of the afferent nerve, $s(t)$, is indicated by

$$s(t) = \begin{cases} 1 & \text{if } v(t) > \theta(t) \\ 0 & \text{otherwise} \end{cases}. \tag{6}$$

When the afferent nerve fires, $\theta(t + \Delta t) = \theta(t) + b$ to consider the adaptation of the threshold. Because $\theta(t)$ recover with time, θ yields

$$\frac{d\theta(t)}{dt} = -b/a, \tag{7}$$

where a is a time constant.

We set the parameters: $c = 1\,\mathrm{G}\,\Omega$, $I = 1\mathrm{nA}$, $a = 5\,\mathrm{ms}$, and $b = 1$ shown in the paper of Brandman and Nelson [2]. In this study, we did not consider mutual connections between the receptors. In other words, the receptors independently respond to electric stimuli.

A Neuronal Model. We applied the leaky integrate and fire (LIF) models as a neuron in electrosensory nuclei. This model is the simplest neuronal model because the dynamics of the membrane potential is described by the first order differential equation. In the LIF model, the membrane potential is denoted by

$$\tau \frac{dV(t)}{dt} = -V(t) + V_0 + RI(t), \tag{8}$$

where τ is the time constant, R is resistance of the membrane, and $I(t)$ is an input current. A spike of the neuron, $s(t)$, is indicated by

$$s(t) = \begin{cases} 1 & \text{if } V(t) > \theta \\ 0 & \text{otherwise} \end{cases}, \tag{9}$$

where, θ is threshold.

We set the parameters: $RI(t) = 20\,\text{mV}$, $\tau = 1\,\text{ms}$, $V_0 = -70\,\text{mV}$, and $\theta = -55\,\text{mV}$. In this study, we did not consider mutual connections between the neurons.

3.2 Implementation

The flow of processing by CPUs is depicted in Fig. 2a. “Calculate potential” in Fig. 2 means the process of the calculation of the potential of the electric field and membrane potentials of neurons. Parameters and variables are sequential initialized. The calculation of the potentials of the nodes is executed one by one. We parallelized the process in the box of Fig. 2a using the OpenMP. The flow of processing by a GPU is shown in Fig. 2b. In the GPU computing, the initializing and computation processes are parallelized.

The activities of electroreceptors and neurons were calculated using the fourth-order Runge-Kutta method. Step size was 0.025 ms.

All of the programs were developed and executed on Debian 8.4 which is one of Linux distortions. The program codes for CPUs were developed by C++ with OpenMP. The codes for CPUs were compiled using GCC4.9.2. The program codes for GPU computing were developed using the CUDA toolkit 7.5 that is the CUDA development environment. All programs were compiled with the -O2 option that is the optimization option. The variable types of the all programs were double-precision floating-points.

To compare calculation time, we used three computers. The Computer1 is equipped with two Xeon E5-2640v3 CPUs (2.60 GHz, 8 cores), 32 GB of RAM, and a Tesla K20C that is a board specialized for GPU numerical computing. The Computer1 can use up to 32 threads. The Computer2 is equipped with the Core i7 2700k CPU (3.5 GHz, 4 cores), 8 GB of RAM, and GTX750Ti. GTX750Ti is not a fast GPU in the current products but we used it because the product is low price and the low power consumption. The architecture of Computer2 in a computer used is old. The type of the memory is DDR3 and PCIe is Gen2. Therefore, the bandwidth of data transfer is low while the clock of the CPU is highest. Computer3 has one Xeon E5-2640v3 CPU with 16 GB RAM. The parallelized programs by OpenMP were executed using 32, 8, and 16 threads on the computer1, the computer2, and the computer3, respectively.

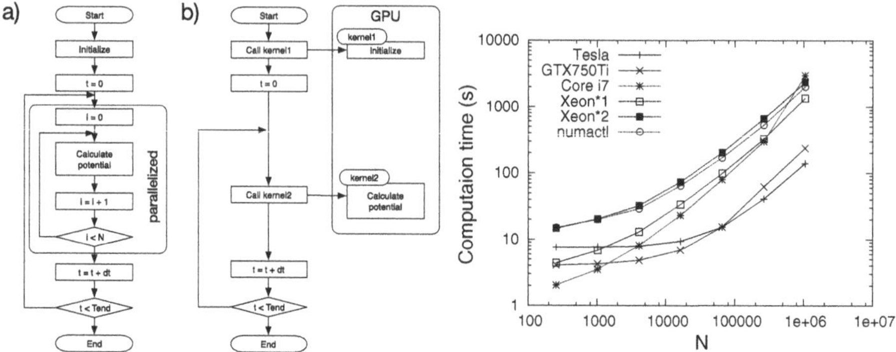

Fig. 2. Flow chart

Fig. 3. Computation time of calculating an electric field

4 Results

We examined computation times of the electric field and the activities of electroreceptor afferents and higher nucleus. Here we explain the meanings of the labels of lines in the graphs. "Tesla" represents the computation time of the Tesla K20c mounted on the Computer1. "GTX750Ti" represents the computation time of the GTX750Ti mounted on the Computer2. "Xeon*2" represents the computation time of two Xeon E5-2640 v3 CPUs mounted on the Computer1. "Core i7" represents the computation time of Core i7 2700k mounted on the Computer2. "Xeon*1" represents the computation time of one Xeon E5-2640 v3 CPU mounted on the Computer3. "numactl" represents the computation time of represents the computation time of the two Xeon E5-2640 v3 CPUs mounted on the Computer1 with the numactl command. numactl is a command in order to effectively use the NUMA when software without optimization of NUMA is executed. In all calculations by the CUDA, the number of threads is 128 if there is no particular mention. In each of the calculations, we set N with $2^8 = 256$, $2^{10} = 1024$, $2^{12} = 4096$, $2^{14} = 16384$, $2^{16} = 65536$, $2^{18} = 262144$, and $2^{20} = 1048576$.

Figure 3 denotes the computation times of calculation of the electric field. The computation times of the GPUs is faster than that of the CPUs for $N \geq 2^{12}$. Furthermore, increase of the calculation time of the GPU was moderate compared to that of the CPUs. In this calculation, the difference in calculation time between the "Tesla" and "GTX750Ti" was small. The computation times of the CPUs are interesting. In most cases, computation times depend on the clock frequency not but the number of threads of a CPU. However, the difference is gradually shrinking. When the number of the nodes is 2^{20}, the computation time of "Core i7" that has the highest clock frequency became the slowest. The long computation of "Xeon*2" might be cause by the cost of thread creation. numactl did not improve the computation time of "Xeon*2".

Figure 4 represents the computation time of calculating the activities of receptor afferents. For $N \leq 2^{12}$, the computation times of the CPUs were smaller speed than that of the GPUs. However, the computation times of the GPUs is shorter than that of the CPUs when the N is large number. In particular, the computation time of "Tesla" gently increased. On the other hand, it did not show a large difference in computation time of "Xeon*2" and that of "GTX750Ti". When N is 2^{20}, the computation time of "Xeon*2" increased rapidly. However, it was improved by numactl. Thus, it seems that the data transfer between memory and a CPU was a bottleneck because the program did not effectively use the NUMA.

Figure 5 represents the computation times of neural activities. The results displayed the almost same tendency with the calculation of the electroreceptor's activities.

Calculation times of the electric field and neural activities have different features. The different features derives from difference of constitution of computations. The simulation of the electric field consists of only simple calculations, such as addition, subtraction, multiplication, and division while the simulation of the neuronal activities consists of the simple calculations and the if statement.

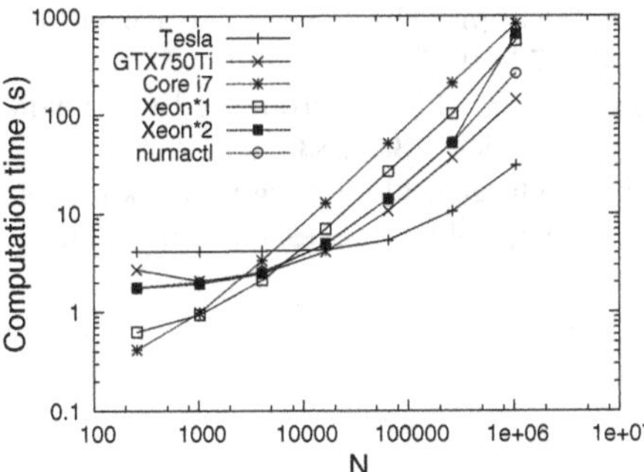

Fig. 4. Computation time of calculating receptors' responses

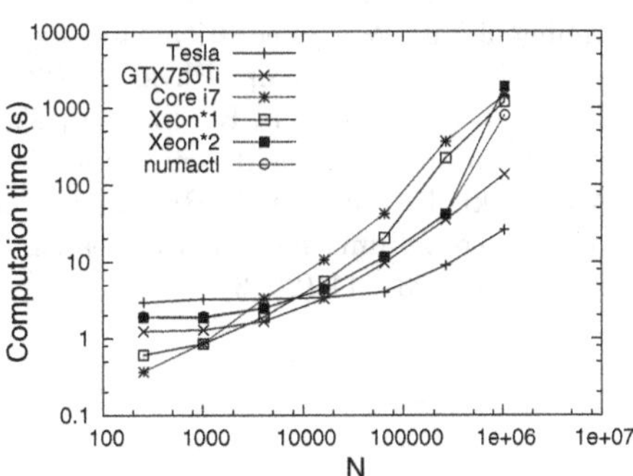

Fig. 5. Execution time of calculating neural activities

5 Conclusion

We developed simulation programs of the electric field and activities of electroreceptors and neurons for parallel processing by CPUs and a GPU. From verification of computation times, it was found that GPU-accelerated programs were faster than programs parallelized by CPUs. CUDA development environment is easy to develop a software for a GPU and the software is effectively fast. Thus, it may be better to use CUDA if the development cost slightly increases. When we use multiple CPUs, it is necessary to take particular care for the computer architecture in order to effectively accelerate a program.

In the future work, we will develop GPU-accelerated simulation software of the higher nucleus using the detailed neuronal model proposed by Doiron [3]. In addition, we will use the single precision floating point number in our programs because a NVIDIA's GPU generally has larger number of processors for single precision.

Acknowledgment. This work was supported by JSPS KAKENHI Grant Number 15K07146.

References

1. Beyeler, M., Oros, N., Dutt, N., Krichmar, J.L.: A GPU-accelerated cortical neural network model for visually guided robot navigation. Neural Netw. Official J. Int. Neural Netw. Soc. **72**, 75–87 (2015)
2. Brandman, R., Nelson, M.E.: A simple model of long-term spike train regularization. Neural Comput. **14**(7), 1575–1597 (2002)
3. Doiron, B., Laing, C., Longtin, A., Maler, L.: Ghostbursting: a novel neuronal burst mechanism. J. Comput. Neurosci. **12**(1), 5–25 (2002)
4. Fujita, K., Kashimori, Y.: Modeling the electric image produced by objects with complex impedance in weakly electric fish. Biol. Cybern. **103**(2), 105–118 (2010)
5. Heiligenberg, W.: Theoretical and experimental approaches to spatial aspects of electrolocation. J. Comp. Physiol. **103**(3), 247–272 (1975)
6. Hoshimiya, N., Shogen, K., Matsuo, T., Chichibu, S.: The apteronotus EOD field: waveform and EOD field simulation. J. Comp. Physiol. **135**, 283–290 (1980)
7. Yamazaki, T., Igarashi, J.: Realtime cerebellum: a large-scale spiking network model of the cerebellum that runs in realtime using a graphics processing unit. Neural Netw. **47**, 103–111 (2013)

Analysis of Similarity and Differences in Brain Activities Between Perception and Production of Facial Expressions Using EEG Data and the NeuCube Spiking Neural Network Architecture

Hideaki Kawano[1(✉)], Akinori Seo[1], Zohreh Gholami Doborjeh[2], Nikola Kasabov[2], and Maryam Gholami Doborjeh[2]

[1] Faculty of Engineering, Kyushu Institute of Technology, Kitakyushu 804-8550, Japan
kawano@ecs.kyutech.ac.jp

[2] Knowledge Engineering and Discovery Research Institute, Auckland University of Technology, Auckland 1142, New Zealand

Abstract. This paper is a feasibility study of using the NeuCube spiking neural network (SNN) architecture for modeling EEG brain data related to perceiving versus mimicking facial expressions. It is demonstrated that the proposed model can be used to study the similarity and differences between corresponding brain activities as complex spatio-temporal patterns. Two SNN models are created for each of the 7 basic emotions for a group of Japanese subjects, one when subjects are perceiving an emotional face and another, when the same subjects are mimicking this emotion. The evolved connectivity in the two models are then subtracted to study the differences. Analysis of the models trained on the collected EEG data shows greatest similarity in sadness, and least similarity in happiness and fear, where differences in the T6 EEG channel area were observed. The study, being based on the well-known mirror neuron concept in the brain, is the first to analyze and visualize similarity and differences as evolved spatio-temporal patterns in a brain-like SNN model.

Keywords: Mirror neuron system · Facial expression · EEG data · NeuCube · Spiking neural network (SNN)

1 Introduction

Facial expression is a fundamental tool in human communication. Understanding the facial expression effects on a third person is of a crucial importance to develop a comprehensible communication. Neuropsychological studies reported that communications through facial expressions are highly related to the Mirror Neuron System (MNS). MNS principle has been introduced in 1990s by Rizzolatti when he discovered similar areas of the brain became activated when a monkey performed an action and when a monkey observed the same action

A. Hirose et al. (Eds.): ICONIP 2016, Part IV, LNCS 9950, pp. 221–227, 2016.
DOI: 10.1007/978-3-319-46681-1_27

performed by another [1]. The MNS in human was also confirmed by an experiment using functional magnetic resonance imaging (fMRI) data [2]. Different facial expressions of emotion have different effects on the human brain activity. The brain processes of perceiving an emotional facial expression and mimicking expression of the same emotion are spatio-temporal processes. The analysis of collected Spatio-Temporal Brain Data (STBD) related to these processes could reveal personal characteristics or abnormalities that would lead to a better understanding of the brain processes related to the MNS. This can be achieved only if the models created from the STBD can capture both spatio and temporal components from this data. Despite of the rich literature on the problem, such models still do not exist.

Recently, a brain-inspired Spiking Neural Network (SNN) architecture, called NeuCube [4–6], has been proposed to capture both the time and the space characteristics of STBD, such as EEG, fMRI, DTI, etc. In contrast to traditional statistical analysis methods that deal with static vector-based data, the NeuCube has been successfully shown to be a rich platform for STBD mapping, learning, classification and visualization [7–9].

In this paper, the NeuCube was used to model EEG data recorded during a facial expression task (both perceiving and mimicking) to investigate the brain activity patterns elicited from 7 kinds of emotional faces (anger, contempt, disgust, fear, happiness, sadness, and surprise) in terms of similarity and differences. The models allow for a detail understanding on the problem.

2 The NeuCube Spiking Neural Network Architecture

The NeuCube architecture [4] consists of: an input encoding module; a 3D recurrent SNN reservoir/cube (SNNc); an evolving SNN classifier. The encoding module converts continuous data streams into discrete spike trains. As one implementation, a Threshold Based Representation (TBR) algorithm is used for encoding. The NeuCube is trained in two learning stages. The first stage is unsupervised learning based on spike-timing-dependent synaptic plasticity (STDP) learning [10] in the SNNc. The STDP learning is applied to adjust the connection weights in the SNNc according to the spatiotemporal relations between input data variables. The second stage is a supervised learning that aims at learning the class information associated with each training sample. The dynamic evolving SNNs (deSNNs) [11] is employed as an output classifier. In this study, the NeuCube is used for modelling, learning, and visualization of the case study EEG data corresponding to different facial expressions.

3 The Case Study STBD: EEG Data from Facial Expression

Eleven male Japanese participants, including 9 right-handed and 2 left-handed, aged between 22 and 25 years old (M = 23.2, SD = 1.2), participated in the case

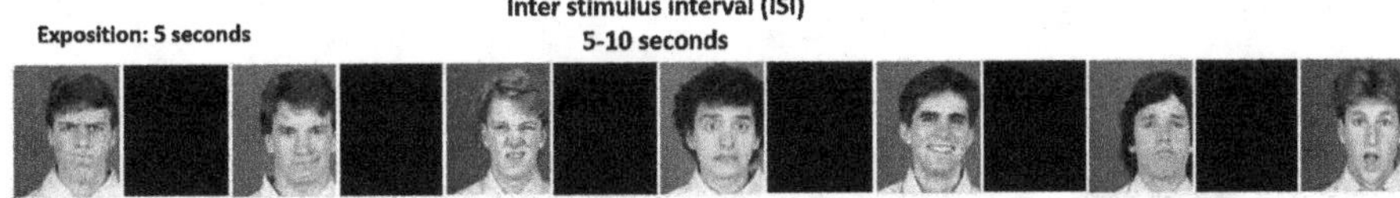

Fig. 1. The facial expression-related task: the order of emotion expressions is: anger, contempt, disgust, fear, happiness, sad, and surprise. Each subject watched 56 images during an experiment.

study of the facial expression task. As facial stimuli, JACFEE collection [12] was used, consisting of 56 colour photographs of 56 different individuals. Each individual illustrates one of the seven different emotions, i.e. anger, contempt, disgust, fear, happiness, sadness, surprise. The collection is equally divided into male and female populations (28 males, 28 female).

During the experiments, subjects were wearing EEG headset (Emotive EPOC+) which consists of 14 electrodes with the sampling rate of 128 Hz and the bandwidth is between 0.2 and 45 Hz.

The EEG data was recorded while the subjects were performing two different facial expression tasks. During the first presentation, subjects were instructed to perceive different facial expression images shown on a screen, and in the second presentation they were asked to mimic the facial expression images.

Each facial expression image was exposed for 5 s followed by randomly 5 to 10 s inter stimulus interval (ISI) as shown in Fig. 1.

4 Analysis of the Spatiotemporal Connectivity in a Trained SNNc of a NeuCube Model

A 3D brain-like SNNc is created to map the Talairach brain template of 1471 spiking neurons [13,14]. The spatiotemporal data of EEG channels were encoded into spike trains and entered to the SNNc via 14 input neurons which spatial locations in the SNNc correspond to the 10–20 system location of the same channels on the sculp. The SNNc is initialized with the use of the "small world" connectivity [4].

The following NeuCube parameter values were used in the simulations: TBR: 0.5, small word connectivity distance: 2.5, STDP rate: 0.01.

During the unsupervised STDP learning, the SNNc connectivity evolves with respect to the spike transmission between neurons. Stronger neuronal connection between two neurons means stronger information (spikes) exchanged between them. Figure 2 illustrates the trained SNNc with EEG data of perceiving and mimicking the 7 different facial expressions. It also shows the differences between the SNNc connectivity of perceiving versus mimicking, which was obtained after the two corresponding models were subtracted.

It can be seen from Fig. 2 that when a SNNc was trained on the EEG data related to facial expressions of both perceiving and mimicking conditions, similar

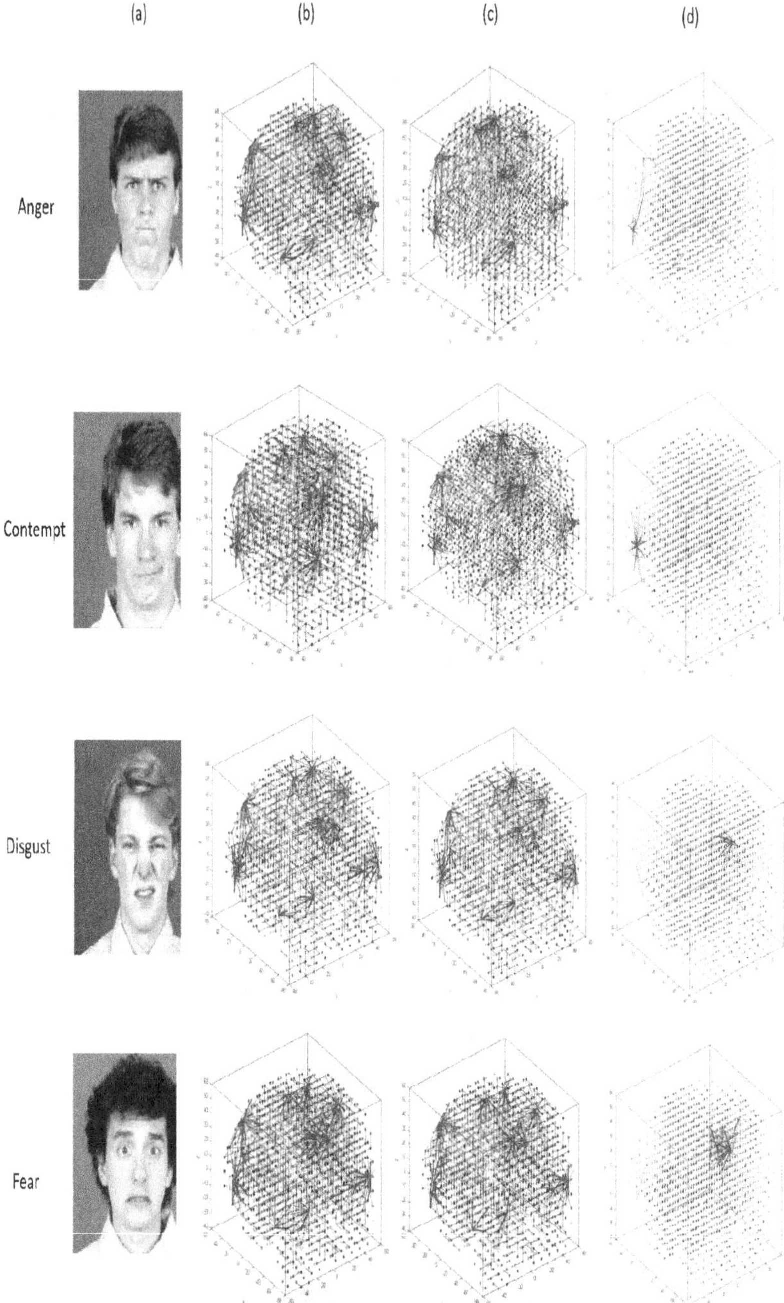

Fig. 2. (a) Exposing emotional facial expressions on a screen; (b) Connectivity of a SNNc trained on EEG data related to perceiving the facial expression images by a group of subjects; (c) Connectivity of a SNNc trained on EEG data related to mimicking the facial expressions; (d) Subtraction of the SNNc models from (a) and (b) to visualize, study and understand differences between perceiving and mimicking emotions

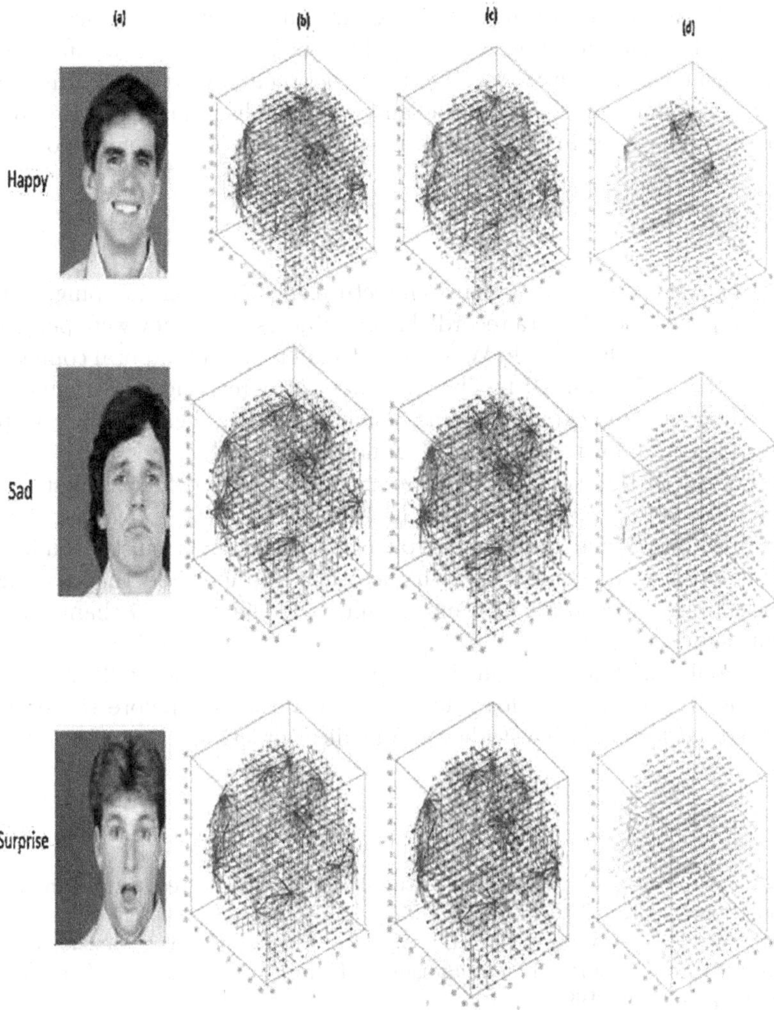

Fig. 2. (*continued*)

neuronal connections were evoked in the SNNc reflecting similar cortical activities. Particularly, greater similarity can be observed in the right hemisphere of the SNNc for anger, contempt, sadness, surprise. This finding proves a neurological fact that this emotional information is usually processed across specific domains of the right hemisphere of the brain [15]. It also reflects the MNS principle in facial expression of emotion. Among the all presented emotional faces, some of them can be considered as dominant emotions if the brain activity patterns of perceiving and mimicking that emotions have a high level of similarity. This similarity is mostly observed for sadness.

Some differences between perceiving and mimicking emotions are also observed. It is seen from Fig. 2(d) that those neurons located around the T7 EEG channel represent the most differences between perceiving and mimicking facial expressions in anger, contempt, and less in sadness and surprise. We can also observe differences in the T6 area for fear, disgust and happy emotions.

5 Conclusion

In this paper we used the NeuCube architecture of SNN [4] for mapping, learning and visualization of EEG data recorded from subjects when they were performing a facial expression-related task. We observed similar spatiotemporal connectivity created in the SNNc trained by the EEG data of perceiving a particular facial expression versus mimicking the same facial expression. This finding can prove the principle of the mirror neurons in human brain.

Making use of the NeuCube SNN architecture allowed for the first time to discover the level of variation in the brain activity patterns against different facial expressions. We identified that role of mirror neurons can be dominant in sadness emotion when compared with the other emotions and that the biggest differences were recorded for fear and happiness in the T6 EEG channel area of the right hemisphere.

This is only the first study in this respect. Further studies will require more subject data to be collected for a more models developed before the proposed method is used for cognitive studies and medical practice.

References

1. Gallese, V., Fadiga, L., Fogassi, L., Rizzolatti, G.: Action recognition in the premotor cortex. Brain **119**, 593–609 (1996)
2. Iacoboni, M., Woods, R.P., Brass, M., Bekkering, H., Mazziotta, J.C., Rizzolatti, G.: Cortical mechanisms of human imitation. Science **286** (5449), 2526–2528 (1999)
3. Binkofski, F., Buccino, G., Posse, S., Seitz, R.J., Rizzolatti, G., Freund, H.-J.: A fronto-parietal circuit for object manipulation in man: evidence from an fMRI-study. Eur. J. Neurosci. **11**, 3276–3286 (1999)
4. Kasabov, N.: NeuCube: a spiking neural network architecture for mapping, learning and understanding of spatio-temporal brain data. Neural Netw. **52**, 62–76 (2014)
5. Tu, E., Kasabov, N., Yang, J.: Mapping temporal variables Into the NeuCube for improved pattern recognition, predictive modelling and understanding of stream data. IEEE Trans. Neural Netw. Learn. Syst, 1–13 (2016)
6. Kasabov, N., Scott, N., Tu, E., Marks, S., Sengupta, N., Capecci, E.: Evolving spatio-temporal data machines based on the NeuCube neuromorphic framework: design methodology and selected applications, Neural Netw. (2016, to appear)
7. Doborjeh, M.G., Capecci, E., Kasabov, N.: Classification and segmentation of fMRI spatio-temporal brain data with a Neucube evolving spiking neural network model. In: IEEE SSCI, U.S.A, Orlando (2014)

8. Doborjeh, M.G., Wang, G., Kasabov, N., Kydd, R., Russell, B.R.: A Neucube spiking neural network model for the study of dynamic brain activities during a GO/NO_GO task: a case study on using EEG data of healthy Vs addiction vs treated subjects. IEEE Trans. Biomed. Eng. (2016, to appear)
9. Maryam Gholami, D., Nikola, K.: Dynamic 3D clustering of spatio-temporal brain data in the NeuCube spiking neural network architecture on a case study of fMRI data. In: Neural Information Processing, Istanbul (2015)
10. Song, S., Miller, K.D., Abbott, L.F.: Competitive Hebbian learning through spike-timing-dependent synaptic plasticity. Nature Neurosci. **3**(9), 919–926 (2000)
11. Kasabov, N., Dhoble, K., Nuntalid, N., Indiveri, G.: Dynamic evolving spiking neural networks for on-line spatio-and spectro-temporal pattern recognition. Neural Netw. **41**, 188–201 (2013)
12. Matsumoto, D., Ekman, P.: Japanese, Caucasian facial expressions of emotion (JACFEE) [Slides]. Intercultural and Emotion Research Laboratory, Department of Psychology, San Francisco State University, San Francisco (1988)
13. Talairach, J., Tournoux, P.: Co-planar Stereotaxic Atlas of the Human Brain. 3-Dimensional Proportional System: An Approach to Cerebral Imaging. Thieme Medical Publishers, New York (1988)
14. Koessler, L., Maillard, L., Benhadid, A., Vignal, J.P., Felblinger, J., Vespignani, H., Braun, M.: Automated cortical projection of EEG sensors: anatomical correlation via the international 10–10 system. Neuroimage **46**(1), 64–72 (2009)
15. Alfano, K.M., Cimino, C.R.: Alteration of expected hemispheric asymmetries: valence and arousal effects in neuropsychological models of emotion. Brain Cogn. **66**, 213–220 (2008)

Self and Non-self Discrimination Mechanism Based on Predictive Learning with Estimation of Uncertainty

Ryoichi Nakajo[1], Maasa Takahashi[1], Shingo Murata[2], Hiroaki Arie[2], and Tetsuya Ogata[1(✉)]

[1] Department of Intermedia Art and Science, Waseda University, Tokyo, Japan
{nakajo,takahashi}@idr.ias.sci.waseda.ac.jp, ogata@waseda.jp
[2] Department of Modern Mechanical Engineering, Waseda University, Tokyo, Japan
{murata,arie}@sugano.mech.waseda.ac.jp

Abstract. In this paper, we propose a model that can explain the mechanism of self and non-self discrimination. Infants gradually develop their abilities for self–other cognition through interaction with the environment. Predictive learning has been widely used to explain the mechanism of infants' development. We hypothesized that infants' cognitive abilities are developed through predictive learning and the uncertainty estimation of their sensory-motor inputs. We chose a stochastic continuous time recurrent neural network, which is a dynamical neural network model, to predict uncertainties as variances. From the perspective of cognitive developmental robotics, a predictive learning experiment with a robot was performed. The results indicate that training made the robot predict the regions related to its body more easily. We confirmed that self and non-self cognitive abilities might be acquired through predictive learning with uncertainty estimation.

Keywords: Self/non-self cognition · Cognitive developmental robotics · Recurrent neural network

1 Introduction

Infants gradually acquire their cognitive abilities by interacting with the environment. Self and non-self discrimination is the foundation for self–other cognition, which is crucial for their emotional development, such as development of a theory of mind [7]. In particular development of motor skills, infants initially move their bodies randomly [9]. Three-month-old infants frequently look at their hands [10], and six-month-old infants start to reach for surrounding objects [2]. Although the targets of infants' learning appear to shift from their own bodies to external objects over the course of their development, the mechanism underlying self and non-self discrimination is yet unclear.

Cognitive developmental robotics [1] is an approach that could explain infants' developmental processes. This approach is based on the constructive

A. Hirose et al. (Eds.): ICONIP 2016, Part IV, LNCS 9950, pp. 228–235, 2016.
DOI: 10.1007/978-3-319-46681-1_28

method and aims to build a computational model by repeatedly proposing hypotheses that are tested by computational implementation and verification with an agent in a real environment. Predictive processing [3] has attracted attention as a scheme for explaining the computational model of the brain. The scheme uses an internal model to produce and predict sensory-motor inputs, and learns to minimize the error between the predictions of the sensory-motor inputs and the actual feedback it receives.

Nagai and colleagues evaluated the predictabilities of the self–other correspondence by using Hebbian learning for imitative interactions [5]. Nishide and colleagues proposed a computational model that allowed a robot to learn sensory-motor mapping to predict a region of the robot's own body preferentially [6]. In this work, variables such as the time window and biases used to update the learnable parameters related to the prediction of the external objects were designed by the experimenter. However, designed parameters may be unsuitable for an agent to achieve self and non-self cognition; thus, a computational model that can automatically predict the predictability with its own scheme is required.

In this study, we propose a possible model to explain the mechanism of self and non-self discrimination. We hypothesized that infants' cognitive abilities for self and non-self discrimination are developed through predictive learning via their sensory-motor inputs and assessment of their uncertainties, which is the opposite of assessing predictabilities. To evaluate our hypothesis with respect to cognitive developmental robotics, we carried out experiments using a small humanoid robot implementing a neurodynamical model called the stochastic continuous time recurrent neural network (S-CTRNN) [4]. This model can predict not only sensory-motor inputs, but also their uncertainties as variances. In the experiments, the robot moves its arm randomly, and then predicts the positions of its right hand and external objects. From the results of this robot experiment, we confirmed that the proposed model might allow the robot to distinguish a region of its body from non-self objects.

2 Model

For a robot to achieve self and non-self discrimination, we use S-CTRNN. S-CTRNN assumes that the ith dimensional target state at time step t of the sth sequence ($\hat{y}_{t,i}^{(s)}$) has Gaussian noise with a mean of zero and a variance of $v_{t,i}^{(s)}$. Under this assumption, S-CTRNN can predict and generate the sensory-motor inputs and the uncertainties of the next state from the current inputs. The uncertainties are determined as variances in the S-CTRNN model. The scheme for S-CTRNN is shown in Fig. 1.

In S-CTRNN, the internal state of the ith neural unit at time step t ($1 \leq t$) of the sth sequence ($u_{t,i}^{(s)}$) is

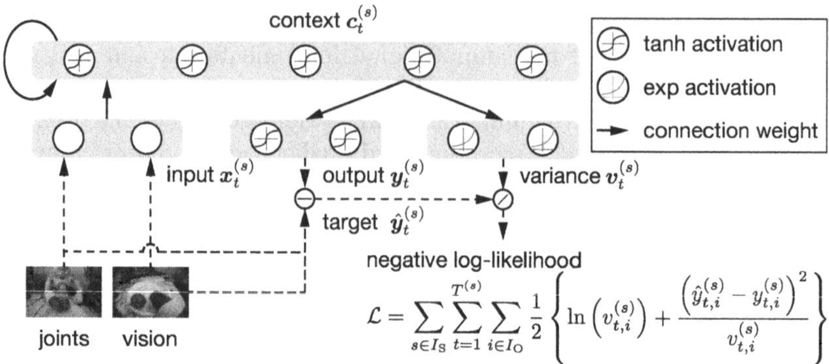

Fig. 1. Schematic of S-CTRNN. S-CTRNN predicts and generates the next sensory-motor inputs and their uncertainties from the current state, $\boldsymbol{x}_t$. The context, $\boldsymbol{c}_t^{(s)}$, and the output, $\boldsymbol{y}_t^{(s)}$, are activated by a logistic function, and the variance, $\boldsymbol{v}_t^{(s)}$, is activated by $\exp(\cdot)$. The learnable parameters in the network are optimized through minimizing the negative log likelihood determined along with Gaussian distribution via the gradient descent method with backpropagation through time.

$$u_{t,i}^{(s)} = \begin{cases} \left(1 - \dfrac{1}{\tau_i}\right) u_{t-1,i}^{(s)} + \dfrac{1}{\tau_i}\left({}^t\boldsymbol{w}_i \cdot [\boldsymbol{x}_t^{(s)}; \boldsymbol{c}_{t-1}^{(s)}]\right) + b_i & (i \in I_{\mathrm{C}}), \\ {}^t\boldsymbol{w}_i \cdot \boldsymbol{c}_t^{(s)} + b_i & (i \in I_{\mathrm{O}} \cup I_{\mathrm{V}}), \end{cases} \tag{1}$$

where I_{C}, I_{O}, and I_{V} are the index sets of the respective neural units, τ_i is the time constant value of the ith leaky integrator context unit, $\boldsymbol{w}_i$ is the connection weight vector to the ith unit, b_i is the bias given to the ith unit, $\boldsymbol{x}_t^{(s)}$ is the input vector at time step t of the sth sequence, and $\boldsymbol{c}_t^{(s)}$ is a vector of the activation value in the context layer at time step t of the sth sequence. The learnable parameters in the network consist of the connection weights, $\boldsymbol{w}_i$, the biases, b_i, and the initial internal states of the sth sequence, $(u_{0,i}^{(s)})$. These parameters are optimized to minimize the negative log likelihood and are updated through the gradient descent method with backpropagation through time [8].

3 Experiment

3.1 Task Design

We designed an experiment in which a small humanoid robot called NAO learned the mapping between sensory-motor inputs via S-CTRNN. Here, the robot corresponds to an infant. In this experiment, the robot randomly moved its right arm, and observed the right hand and an external object. The robot's right arm movements were generated based on its joint angles. These joint angles were updated every 0.1 s during the movements, and the next joint angles were determined

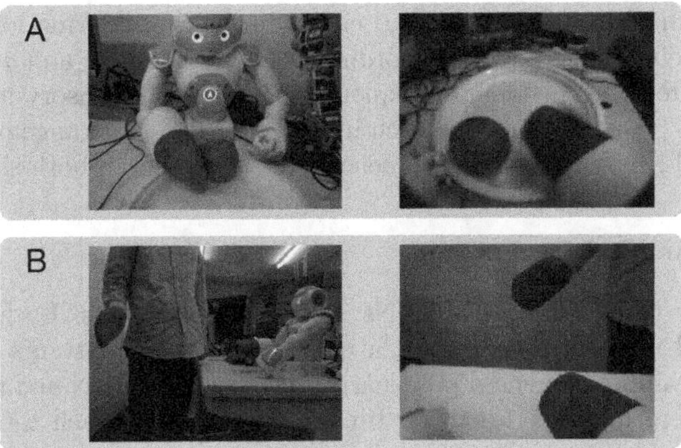

Fig. 2. Task design: in (A) and (B), overviews of experiments are shown on the left, and images from the robot's camera are shown on the right. (A) The robot moves its right arm randomly and can hit the red ball on the plate (*ball-observation*). (B) During the robot's arm action, the human experimenter wears a red glove on their right hand and moves it in front of the robot (*hands-observation*). In the *hands-observation* situation, the robot's hand cannot touch the experimenter's hand. (Color figure online)

by the current joint angles and random values. The robot's joint angles were adjusted when a camera mounted on the robot's mouth could not see its right hand. The observed external objects were a ball on a plate and the human experimenter's hand. When observing the ball (*ball-observation*), the robot could hit the ball with its right hand. The ball hit by the robot rolled to the robot because the plate leaned to the robot. When observing the human experimenter's hand (*hands-observation*), the experimenter was about 50 cm from the robot, and the robot could not touch the experimenter's hand. Thus, the robot could sometimes interact with the external object during its actions in the *ball-observation* situation, whereas it could not in the *hands-observation* situation. In addition, in the *ball-observation* situation, the robot's arm movements were sometimes disturbed by collisions with the ball and the plate on which the ball was set. Therefore, the arm movements were different during the two situations, and the amplitude in *hands-observation* was broader than that in *ball-observation*. These two situations are shown in Fig. 2.

For object recognition, the robot wore a green attachment and the red ball was put on the plate in *ball-observation*, and the experimenter wore a red glove on their right hand in *hands-observation*. In both situations, the robot began its arm actions from the initial position, and its neck joints were fixed because the camera was mounted on the robot's mouth. During the actions, the robot recorded its arm joint angles and the sensory inputs obtained from images captured by its camera at 10 fps for 20 s. Each recorded training sequence was composed of the joint angles of the right arm of the robot, and sensory inputs from the camera images. The robot's arm had four degrees of freedom. From the camera images, the two-dimensional centroids of the green and red sensory inputs were extracted.

Therefore, each training sequence had eight-dimensional information consisting of four robot joint angles and four coordinates for the colors (x and y coordinates for green and red). In the training sequences, the recorded sensory-motor inputs were scaled to $[-0.8, 0.8]$ because each input had a different range of values. To train the S-CTRNN, 20 training sequences were recorded in both situations.

3.2 Experimental Evaluation

We separately trained the S-CTRNNs in the two situations. Each S-CTRNN had 35 neural units in the context layer, and the time constants were set as 7.5. Twenty training sequences were learned in each S-CTRNN and its learnable parameters were updated 1,000,000 times. After training each S-CTRNN, we confirmed the passages of the uncertainties of the sensory inputs for the robot's hand and the external objects. In addition, we performed a forward propagation to predict the next state and the uncertainties of the sensory-motor inputs. The predicted uncertainties were evaluated by comparing the similarities between the two trajectories of the internal states in the variance layer. The similarities of the sth sequence were measured by calculating the correlate coefficient matrix, $R^{(s)}$, as

$$R^{(s)} = \left(r_{ij}^{(s)}\right) \quad (i \in I_{\mathrm{V}} \wedge j \in I_{\mathrm{V}}), \tag{2}$$

$$r_{ij}^{(s)} = \frac{{}^{t}\left(\Delta \boldsymbol{u}_i^{(s)}\right) \cdot \left(\Delta \boldsymbol{u}_j^{(s)}\right)}{\|\Delta \boldsymbol{u}_i^{(s)}\|_2 \ \|\Delta \boldsymbol{u}_j^{(s)}\|_2} \quad (i \in I_{\mathrm{V}} \wedge j \in I_{\mathrm{V}}), \tag{3}$$

$$\Delta \boldsymbol{u}_i^{(s)} = \left\{u_{t+1,i}^{(s)} - u_{t,i}^{(s)}\right\}_{t=1}^{T^{(s)}-1} \quad (i \in I_{\mathrm{V}}), \tag{4}$$

where $T^{(s)}$ is the time length of the sth sequence. We also observed the development of the relationship between the predictions of the robot's motor commands and the sensory inputs by comparing the correlation coefficient matrices early in the training with those later on.

4 Results and Discussion

Passages of the predicted variances during training of each S-CTRNN are shown in Fig. 3. In Fig. 3, the variance of sensory inputs consisting of the robot's hand and the external objects are accumulated for the $x - y$ coordinates of centroids, all time steps, and all sequences. The variances of the position of the robot's hand are lower than that of the position of the external objects later in the training, even though both variances early in the training are similar. This may be because the prediction of the position of the robot's hand is incidental to that of its arm joints. The predicted variances for the position of the robot's right hand in the *hands-observation* situation are about 10^2 times larger than those in the *ball-observation* situation. This difference might arise from the robot's arm

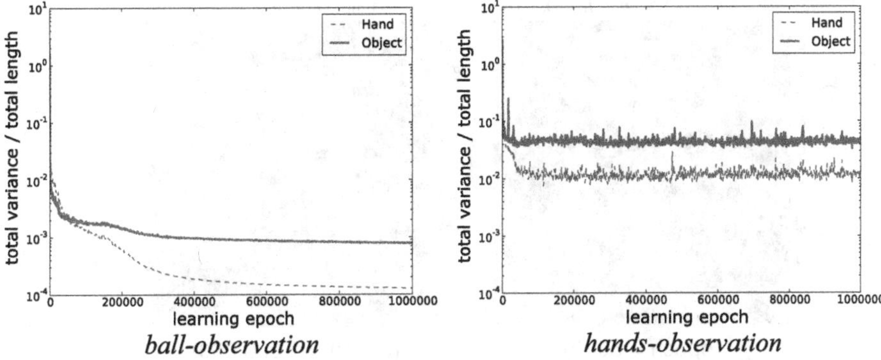

Fig. 3. Passages of the predicted variances for *ball-observation* (left) and *hands-observation* (right). Red solid and green dotted lines indicate the variances of the positions of the robot's right hand and that of the external objects, respectively. (Color figure online)

movements in these two situations. Unlike in *hands-observation*, the amplitude of the robot's arm movements are decreased because of collisions with the ball and the plate in *ball-observation*. The smaller amplitude of the movements may make prediction of the position of the robot's right hand easier.

The correlation coefficient matrices at training epochs 5000 and 1,000,000 given by Eqs. 2, 3 and 4 are shown in Fig. 4. In both the *ball-observation* and *hands-observation* situations, the correlations between the movements of the robot's arm and the position of its hand are higher than those between the arm movements and the position of the external objects at epoch 1,000,000, although those at epoch 5000 are non-uniform. Furthermore, comparing the matrices of *ball-observation* and *hands-observation* at epoch 1,000,000, shows that the correlations between the robot's motor command and the sensory inputs of the external objects are closer to those of the robot's hand in the *ball-observation* situation.

The results in Fig. 3 suggest that the model allows the robot to distinguish the region of its body from the non-self object. In addition, the distinction between the self and non-self may be acquired gradually through the predictive learning of sensory-motor inputs. The robot would initially identify self with non-self, and suddenly notice the boundary between them. The difference between *ball-observation* and *hands-observation* correlations at epoch 1,000,000 might arise from the contingency between the behaviors of the robot and the external objects (Fig. 4). This was because the robot could perceive its own joint angles, which affected the trajectories of the ball in *ball-observation* situation, even though the robot received no information about the effects of the trajectories of the experimenter's hand in the *hands-observation* situation.

The model proposed by Nishide and colleagues required the additional parameters designed by the experimenter to modify the prediction of the sensory inputs of the external objects [6]. Therefore, the relationship between the predictabilities of the robot's motor commands and the sensory inputs depended on

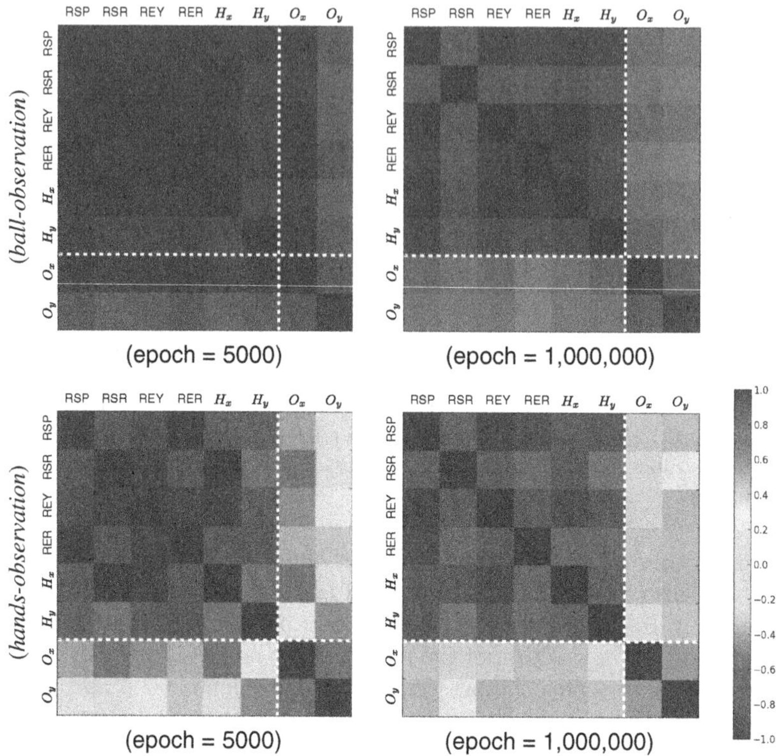

Fig. 4. Correlation coefficient matrices for the internal state in the variance layer at training epochs 5000 (right) and 1,000,000 (left) for *ball-observation* (top) and *hands-observation* (bottom). RSP, RSR, REY, and RER indicate the robot's four joint angles of right shoulder pitch, shoulder roll, elbow yaw, and elbow roll, respectively. H_x and H_y show the position of the robot's right hand. Similarly, O_x and O_y show the position of the external objects in the respective situations.

these designed parameters. Furthermore, the predictabilities were evaluated by the prediction error observed from the outside of the model. In our model, the scheme in which the model can predict the uncertainties of the sensory inputs allows the robot to distinguish the region of its body from the external objects without the designed parameters. Thus, the same model can learn to predict the uncertainties of the sensory inputs of the different external objects.

5 Conclusion

We developed a model that could explain the mechanism of self and non-self discrimination. We used the S-CTRNN to predict the uncertainty as the variance in its scheme for our proposed model. To evaluate the model from the perspective of cognitive developmental robotics, we conducted an experiment using a robot, and trained the S-CTRNN to predict the next state and their uncertainties of the sensory-motor inputs. Two situations were used for this experiment: one in which the robot could interact with the external object, and the other in which the robot could not interact with the object. The results showed that the uncertainties of the sensory inputs related to the robot's own body became lower than the unrelated inputs. Furthermore, the results also indicated that the differences in contingencies between the robot's own body and the external objects affected the uncertainties of the external objects. The differences between the estimated uncertainties and the relationships between the robot's joint angles and sensory inputs could be used to distinguish self from non-self.

Although we used two separate experimental situations, an infants' developmental progress will include both situations such as the interruption of an object's motion by their parents. Therefore, in future work, we intend to use our proposed model to investigate this scenario. Furthermore, the experimental tasks will be extended to interactions with parents, such as turn-taking and imitation, which are crucial to developing infants' social skills.

Acknowledgements. This work was supported by JSPS KAKENHI Grant Numbers 24119003, 15H01710, and 16H05878.

References

1. Asada, M., Hosoda, K., Kuniyoshi, Y., Ishiguro, H., Inui, T., Yoshikawa, Y., Ogino, M., Yoshida, C.: Cognitive developmental robotics: a survey. IEEE Trans. Auton. Ment. Dev. **1**(1), 12–34 (2009)
2. Berthier, N.E., Keen, R.: Development of reaching in infancy. Exp. Brain Res. **169**(4), 507–518 (2006)
3. Clark, A.: Whatever next? Predictive brains, situated agents, and the future of cognitive science. Behav. Brain Sci. **36**(3), 181–204 (2013)
4. Murata, S., Namikawa, J., Arie, H., Sugano, S., Tani, J.: Learning to reproduce fluctuating time series by inferring their time-dependent stochastic properties: application in robot learning via tutoring. IEEE Trans. Auton. Ment. Dev. **5**, 298–310 (2013)
5. Nagai, Y., Kawai, Y., Asada, M.: Emergence of mirror neuron system: immature vision leads to self-other correspondence. In: Proceedings of the 1st Joint IEEE International Conference on Development and Learning and on Epigenetic Robotics (2011)
6. Nishide, S., Nobuta, H., Okuno, H.G., Ogata, T.: Preferential training of neurodynamical model based on predictability of target dynamics. Adv. Robot. **29**(9), 587–596 (2015)
7. Premack, D., Woodruff, G.: Does the chimpanzee have a theory of mind? Behav. Brain Sci. **1**, 515–526 (1978)
8. Rumelhart, D.E., Hinton, G.E., Williams, R.J.: Learning internal representations by error propagation. In: Parallel Distributed Processing: Explorations in the Microstructure of Cognition, Chap. 8, pp. 318–362. MIT Press, Cambridge (1986)
9. Thelen, E.: Rhythmical stereotypies in normal human infants. Anim. Behav. **27**(3), 699–715 (1996)
10. White, B.L.: The New First Three Years of Life. Touchstone, New York (1995)

A Framework for Ontology Based Management of Neural Network as a Service

Erich Schikuta[1(✉)], Abdelkader Magdy[1], and A. Baith Mohamed[2]

[1] University of Vienna, Vienna, Austria
{erich.schikuta,shaabana52}@univie.ac.at
[2] Arab Academy for Science and Technology and Maritime Transport, Alexandria, Egypt
baithmm@hotmail.com

Abstract. Neural networks proved extremely feasible for problems which are hard to solve by conventional computational algorithms due to excessive computational demand, as NP-hard problems, or even lack of a deterministic solution approach. In this paper we present a management framework for neural network objects based on ontology knowledge for the cloud-based neural network simulator N2Sky, which delivers neural network resources as a service on a world-wide basis. Core of this framework is the Neural Network Query Engine, N2Query, which allows users to specify their problem statements in form of natural language queries. It delivers a list of ranked N2Sky resources in return, providing solutions to these problems. The search algorithm applies a mapping process between a domain specific problem ontology and solution ontology.

Keywords: Neural network as a service · Virtual organization · Semantic description · Cloud computing

1 Introduction

Virtual organizations [4] share a wide variety of geographically distributed computational resources (such as workstations, clusters and supercomputers), storage systems, databases, libraries and special purpose scientific instruments to present them as a unified integrated resource transparently. In the course of our research we designed and developed N2Sky [12], a virtual organisation for the computational intelligence (CI) community, providing access to neural network resources and enabling infrastructures to foster federated cloud resources. N2Sky aroused strong interest even beyond the CI community[1]. This endeavour of providing an environment for the access to practically unlimited resources faces one specific challenge: We propose a centralized registry approach collecting all semantic knowledge of neural network objects by semantic web technologies.

In this paper we present a novel semantic management framework for neural network resources. Thus, the paper is structured as follows: The key components for realizing Neural Network as a Service principle are presented in Sect. 2.

[1] http://cacm.acm.org/news/171642-neural-nets-now-available-in-the-cloud/.

A. Hirose et al. (Eds.): ICONIP 2016, Part IV, LNCS 9950, pp. 236–243, 2016.
DOI: 10.1007/978-3-319-46681-1_29

Section 3 proposes the semantic management architecture of the novel N2Query system. The implementation decisions and a use case study for our proposed system are presented in Sect. 4. Finally, the paper concludes with a summary and an outlook to our future work.

2 Towards Neural Network as a Service

We aim for an infrastructure enabling large-scale applications and transparent access to "high-end" resources from the desktop, providing a uniform "look & feel" to a wide range of resources and location independence of computational resources as well as data. A first attempt was N2Grid [11], which fostered the resources of the Grid infrastructure. Overcoming the deficiency of the Grid the cloud computing paradigm provides access to large amounts of computing power by aggregating resources based on quality of service levels [15], both hardware and softwares, on the one hand offering them as a single system view and on the other hand, more importantly, supporting modern business models. cloud computing has been evolved from technologies like cluster computing and grid computing and has given rise to sky computing [7]; an architectural concept that denotes federated cloud computing. Sky computing is an emerging computing model where resources from multiple clouds are leveraged to create s large scale distributed infrastructure. Key elements to reach our overall goal of a cooperative problem solution environment for the computational intelligence community are on the one hand N2Sky, a VO for provisioning, and on the other hand ViNNSL, a domain specific language for the description of neural network resources. N2Sky and ViNNSL provide the baseline for N2Query, which enables the user to specify his/her query in form of a natural language description of the problem statement and delivers a list of ranked N2Sky resource-URIs, which provide solutions for these problems. Thus, we aim for delivering a domain specific query interface to the user resembling a "Neural Network Google".

2.1 N2Sky

N2Sky is an artificial neural network provisioning environment facilitating the users to create, train, evaluate neural networks fostering different types of resources from clouds of different affinity, e.g. computational power, disk space, networks etc. The vision of N2Sky foresees a huge number of neural network objects stored all over the Internet which deliver problem solution capabilities at will of the members of the N2Sky virtual organisation. Neural network objects are on the one hand generic neural network paradigms which can be instantiated by learning mechanisms solving a specific problem on a given training data set and on the other hand already trained neural networks for given application problems. The number of these neural network objects is expected to be very large and continuously growing. Another characteristic is that these neural network objects are distributed on a world wide scale on the Internet administratively under the umbrella of the N2Sky virtual organization on participating resource nodes.

2.2 ViNNSL

To describe and identify neural network objects in N2Sky we developed ViNNSL (Vienna Neural Network Specification Language) [1], an XML-based domain specific language which allows to specify neural network objects in a standardized way by attributing them with semantic information. Originally it was developed as communication framework to support service-oriented architecture based neural network environments. Thus, ViNNSL is capable to describe the static structure, the training and execution phase neural network objects in a distributed infrastructure, as grids and clouds. In its last extension [10] it supports semantic information too, describing the usage scenario of a network objects for given problem domain. Thus, ViNNSL allows dynamic resource-to-resource communication regarding the semantics of neural network resources, which supports both structural information and semantic information, as describing the usage of a network objects for given problems by a natural language approach.

3 Semantic Management Architecture of Neural Network Resources

In the effort to find an existing neural network paradigm or already trained object we applied the technique of ontology alignment. An ontology defines a set of representational primitives to model a domain of knowledge. The representational primitives are typically classes, attributes or relationships. These contain information about their meaning and constraints on their logically consistent application. So, an ontology represents the knowledge of a specific domain. To combine knowledge of different domains leads to ontology alignment, where one ontology is mapped to another. Hereby three ontology combination paradigms can be distinguished, ontology linking, ontology mapping, and ontology importing [5].

3.1 Solution and Problem Ontology Management

For our problem we apply ontology linking, where individuals from distinct ontologies are coupled with links. The concept is as follows: We administer basically two ontologies, a problem ontology and a solution ontology:

- The problem ontology consists of a hierarchical organisation of typical neural network application problems, as classification, optimization, approximation, storage, pattern restoration, cluster analysis, feature extraction etc. In the ontology hierarchy these main domains are finer distinguished till the single problem specifications in the leave nodes.
- The solution ontology stores all known N2Sky neural network objects organized according to their paradigm, as perceptron, multi-layer backpropagation, self-organizing maps (Kohonen cards), recurrent networks (Elman, Jordan, etc.), cellular neural networks, etc. The idea is that paradigm families are appropriate for specific solution mechanisms. Here the ontology delivers a fine grained structure finally giving the neural network objects (trained neural networks for a specific problem) as leaves.

Figure 1 shows an example of our implemented problem-solution ontology.

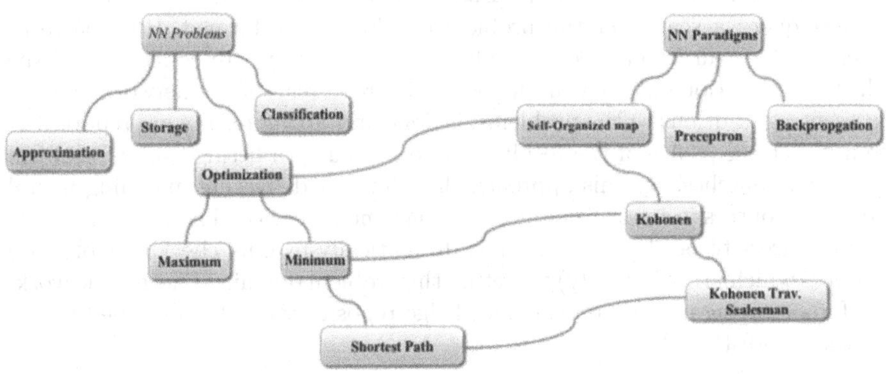

Fig. 1. Problem-solution hierarchal ontology structures

Now, we define a mapping of problem ontology nodes, describing a specific problem, to solution ontology nodes, denoting network objects which deliver a solution for this problem. Links can be defined not only between leaves of the hierarchies but also between internal nodes. So a node specifying a more general problem in the problem ontology can link to a subtree in the solution hierarchy identifying a set of similar neural network objects delivering solutions for related problems. Following the link from the problem specification in the problem ontology delivers the URI of an available Kohonen neural network object for execution in the solution ontology [9]. If the pattern "travelling salesman" would have not showed up in the problem ontology at least the higher level link from "optimization⇒minimum" would result the set of all Kohonen networks as possible solution candidates. This mapping between problems to solutions can be done on the one hand by N2Sky administrators manually, and on the other hand by an automatic mapping during insertion of a new network objects based on the ViNNSL semantic information of this object. This workflow for integrating new neural network resources into the knowledge repository of N2Sky can be generically described by the following algorithm A1:

A1/1 N2Sky resource provider (*RP*) attributes its Neural Network Resource (*NNR*) with a ViNNSL Description (*VD*) specifying structural and semantic information.

A1/2 *RP* sends *VD* together with *NNR* or URI of *NNR* to N2SKy knowledge repository.

A1/3 *VD* is integrated into Problem Ontology (*PO*) according to problem domain.

A1/4 *NNR* or its URI is integrated into Solution Ontology (*SO*) according to network paradigm.

A1/5 Link between *VD* insertion node in *PO* and *NNR* insertion node in *SO* is created.

3.2 Querying the Ontologies

The search algorithm is as follows: Based on the natural language keywords of the user query a scan over the problem ontology is performed. Hits, patterns matching the scan, resembled by nodes in the hierarchy, are collected and the links to the solution ontology are followed. There, a scan of the network objects, representing solutions to the problems, is done and fitting results are reported to the user. The sequence of the results can be guided by a fitting rank of problem to solution matches. By this approach the effort for delivering matching neural network resources is centralized in the management service. The number of network resources to be checked is pruned dramatically by only checking (solution) resources which are (obviously) targeting the problem domain. The generic workflow for executing a query on the knowledge repository can be described by the following algorithm A2:

A2/1 User describes in natural language his/her Problem Description PD using N2Query interface.
A2/2 Cognitive representation of the problem description delivering a synset SS (set of cognitive synonyms).
A2/3 Problem description is classified according to SS in PO delivering set of PO nodes.
A2/4 Links from PO nodes to respective SO nodes are followed.
A2/5 SO nodes are starting points of tree search delivering URIs of possible solution candidates.
A2/6 VD of solution candidates are analyzed and ranked according to match with PD.
A2/7 Ranked list is reported to user.

4 N2Query Implementation and Use Case

The knowledge repository of the N2Query system implementation is realized by ontology technique using semantic web languages to represent and implement problem-solution ontologies in form of hierarchical structure of classes and sub-classes nodes. The user describes his/her query in form of natural language description of the problem statement. WordNet package [8] is used to understand the semantic description of user query. The problem-solution ontologies are realized by using RDF and OWL semantic web languages [14]. Apache Jena [6] is applied to write RDF and OWL statements in a Java framework. The next challenge is the storage of the realized ontologies on the internet. Actually, huge storage repositories for RDF data have been developed, which store the RDF triples in a relational database (RDB) [2]. We used Sesame framework [13], for creating, parsing, storing and querying RDF information [3]. Sesame supports two query languages: SPARQL and SeRQL [13]. In our work SPARQL language is applied for the problem to solution mapping process between the two ontologies. Figure 2 shows a sample query of SPARQL code applied on the problem

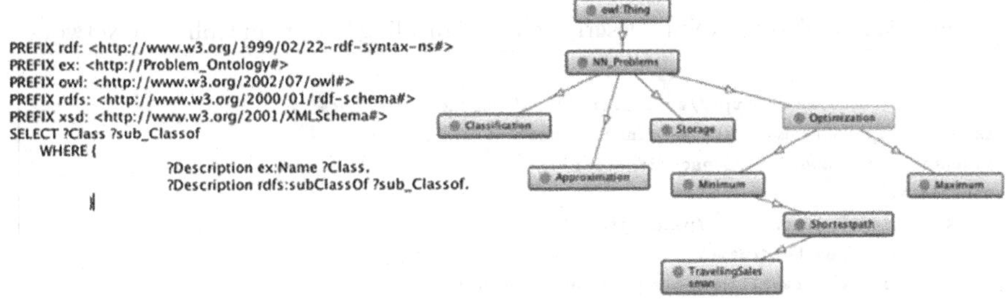
PREFIX rdf: <http://www.w3.org/1999/02/22-rdf-syntax-ns#>
PREFIX ex: <http://Problem_Ontology#>
PREFIX owl: <http://www.w3.org/2002/07/owl#>
PREFIX rdfs: <http://www.w3.org/2000/01/rdf-schema#>
PREFIX xsd: <http://www.w3.org/2001/XMLSchema#>
SELECT ?Class ?sub_Classof
WHERE {
?Description ex:Name ?Class.
?Description rdfs:subClassOf ?sub_Classof.
}
owl:Thing
NN_Problems
Classification
Storage
Optimization
Approximation
Minimum
Maximum
Shortestpath
TravellingSalesman

Fig. 2. SPARQL query (left), the hierarchical description of problem ontology (right)

ontology and shows the hierarchical structure of this ontology in form of classes and sub-classes using Protégé Framework[2].

4.1 Use Case: Travelling Salesman Problem

In the following use case we will highlight the use of the different semantic tools for the execution of the two workflows presented above, A1 and A2. The focus of the use case will be on a NN resource, a parallelized Kohonen network, providing a solution to the well-known Travelling Salesman problem [9]. We start with the integration phase A1. The provider of the NN resource, the Kohonen network, uses the ViNNSL language for the description of the problem and paradigm domain in step A1/1. Hereby the `paradigm` and `problem domain` tags of ViNNSL are used, as shown in the sketch of the respective XML code 1.1. In step A1/3 the description of the NN resource, Optimization → Minimum → ShortestPath → TravelingSalesman, is integrated into the problem Ontology by the RDF statements as shown in Fig. 3. Analogously the NN resource is integrated into the Solution Ontology accordingly to its paradigm family in step A1/4, Self-Organized Map → Kohonen → Kohonen Traveling Salesman. In step A1/5 an appropriate link from problem to Solution Ontology is created pointing from problem description to the respective physical NN resource, see Fig. 1. Now in step A2/1, querying the N2Query tool by a natural language phrase like, "Minimization of path length between cities", the N2Query system tries to recognise the semantic representation of this problem by using WordNet package (step A2/2), and receives in the synnet set a phrase like "Traveling Salesman". In step A2/3 the SPARQL Query algorithm classifies the user query into one of problem classifications in problem domain and, following the link to the Solution Ontology (step A2/4), delivers by a subtree traversal (A2/5) the possible URIs of the well trained NN objects. Matching the ViNNSL descriptions in step A2/6 the system produces and reports a ranked list of qualified solution URIs to the user's problem (A2/7).

[2] http://protege.stanford.edu/.

Listing 1.1. Sketch of ViNNSL Description for Traveling Salesman Kohonen Network

```
<?xml version="1.1" enconding="UTF-8"?>
<description xmlns:xsi="http://www.w3.org/2001/XMLSchema-instance"
  xsi:noNamespaceSchemaLocation="vinnsl20.xsd">
  <identifier>8765445678123desc</identier>
  <metadata>
     <paradigm>Optimization</paradigm>
     <name>Kohonen Net</name>
     <description>Travelling Salesman</description>
  </metadata>
  <problemDomain>
     <applicationField>Optimization</applicationField>
     <applicationField>Minimum</applicationField>
     <applicationField>Shortestpath</applicationField>
     <applicationField>TravelingSalesman</applicationField>
  </problemDomain>
  <executionEnvironment>
     <parallel>
        <software> ... </software>
        <hardware> ... </hardware>
     </parallel>
  </executionEnvironment>
  <structure> ... </structure>
  <data> ... </data>
</description>
```

```
<rdf:RDF
    xmlns:rdf="http://www.w3.org/1999/02/22-rdf-syntax-ns#"
    xmlns:ex="http://Problem_Ontology/"
    xmlns:owl="http://www.w3.org/2002/07/owl#"
    .
  <rdf:Description rdf:about="http://Problem_Ontology/NN_Problems">
    <rdf:type rdf:resource="http://www.w3.org/2002/07/owl#Class"/>
    .
  <rdf:Description rdf:about="http://Problem_Ontology/Approximation">
    <ex:Des>approximation, parataxis, approach, approach</ex:Des>
    <ex:Name>Approximation</ex:Name>
    <rdfs:subClassOf rdf:resource="http://Problem_Ontology/NN_Problems"/>
    <rdf:type rdf:resource="http://www.w3.org/2002/07/owl#Class"/>
  </rdf:Description>
    .
    .
  <rdf:Description rdf:about="http://Problem_Ontology/Optimization">
    <ex:Des>optimize, optimization, estimate, estimation, approximation, optimum</ex:Des>
    <ex:Name>Optimization</ex:Name>
    <rdfs:subClassOf rdf:resource="http://Problem_Ontology/NN_Problems"/>
    <rdf:type rdf:resource="http://www.w3.org/2002/07/owl#Class"/>
  </rdf:Description>
</rdf:RDF>
```

Fig. 3. RDF code of problem ontology

5 Conclusion and Future Work

We presented a new technology of problem solving techniques using our proposed N2Query system by allowing users to describe problems in natural language description. N2Query delivers the solution fostering semantic techniques. We have outlined the implementation process of problem-solution ontologies using semantic web languages and other tools. N2Query will be part of the next version of N2Sky, which will also bring a comprehensive redesign of N2Sky's architecture using micro-services, docker technique and dynamic cloud deployment.

References

1. Beran, P.P., Vinek, E., Schikuta, E., Weishäupl, T.: ViNNSL - the vienna neural network specification language. In: Proceedings of the International Joint Conference on Neural Networks, IJCNN 2008, pp. 1872–1879. IEEE, June 2008
2. Bönström, V., Hinze, A., Schweppe, H.: Storing rdf as a graph. In: Web Congress, 2003. Proceedings. First Latin American. pp. 27–36. IEEE (2003)
3. Broekstra, Jeen, Kampman, Arjohn, van Harmelen, Frank: Sesame: A Generic Architecture for Storing and Querying RDF and RDF Schema. In: Horrocks, Ian, Hendler, James (eds.) ISWC 2002. LNCS, vol. 2342, pp. 54–68 Springer, Heidelberg (2002)
4. Foster, I., Kesselman, C., Tuecke, S.: The Anatomy of the Grid: Enabling Scalable Virtual Organizations. Int. J. Supercomputer Applications pp. 15(3) (2001)
5. Homola, M., Serafini, L.: Towards formal comparison of ontology linking, mapping and importing. Proc. DL **10**, 291–302 (2010)
6. Horridge, M., Bechhofer, S.: The owl api: A java api for owl ontologies. Semantic Web **2**(1), 11–21 (2011)
7. Keahey, K., Tsugawa, M., Matsunaga, A., Fortes, J.: Sky computing. Internet Computing, IEEE **13**(5), 43–51 (2009)
8. Princeton University: WordNet: A Lexical Database for English. https://wordnet.princeton.edu
9. Schabauer, H., Schikuta, E., Weishäupl, T.: Solving Very Large Traveling Salesman Problems by SOM Parallelization on Cluster Architectures. In: Sixth International Conference on Parallel and Distributed Computing Applications and Technologies (PDCAT05). pp. 954–958. IEEE Computer Society, Dalian, China (2005)
10. Schikuta, E., Huqqani, A., Kopica, T.: Semantic extensions to the Vienna Neural Network Specification Language. In: 2015 International Joint Conference on Neural Networks (IJCNN). pp. 1–8 July 2015
11. Schikuta, E., Weishaupl, T.: N2Grid: Neural Networks in the Grid. In: Neural Networks, 2004. Proceedings. 2004 IEEE International Joint Conference on. vol. 2, pp. 1409–1414 vol. 2 July 2004
12. Schikuta, E., Mann, E.: N2Sky - Neural Networks as Services in the Clouds. In: International Joint Conference on Neural Networks. IEEE, USA (2013)
13. Sesame: Sesame. http://rdf4j.org
14. Staab, S., Studer, R.: Handbook on ontologies. Springer Science & Business Media (2013)
15. Vinek, E., Beran, P.P., Schikuta, E.: Classification and Composition of QoS Attributes in Distributed, Heterogeneous Systems. In: Cluster, Cloud and Grid Computing (CCGrid), 2011 11th IEEE/ACM International Symposium on. pp. 424–433 May 2011

Computational Model of the Cerebellum and the Basal Ganglia for Interval Timing Learning

Ohki Katakura[1(✉)] and Tadashi Yamazaki[1,2]

[1] Graduate School of Informatics and Engineering, The University of Electro-Communications, Tokyo, Japan
o.katakura@uec.ac.jp

[2] Artificial Intelligence Research Center, National Institute of Advanced Industrial Science and Technology, Tokyo, Japan

Abstract. In temporal information processing, both the cerebellum and the basal ganglia play essential roles. In particular, for interval timing learning, the cerebellum exhibits temporally localized activity around the onset of the unconditioned stimulus, whereas the basal ganglia represents the passage of time by their ramping-up activity from the onset of the conditioned stimulus to that of the unconditioned stimulus. We present a unified computational model of the cerebellum and the basal ganglia for the interval timing learning task. We report that our model reproduces the localized activity in the cerebellum and the gradual increase of the activity in the basal ganglia. These results suggest that the cerebellum and the basal ganglia play different roles in temporal information processing.

Keywords: Computational model · Cerebellum · Basal ganglia · Interval timing · Relay hypothesis

1 Introduction

Time, one of the most important concept for our lives, is obviously processed in our brain. Neural mechanisms of time perception are still unclear, but many lines of studies have suggested that at least, the cerebellum and the basal ganglia play important roles in temporal information processing [1,2]. The cerebellum processes time shorter than 1 s for precise motor control and the basal ganglia processes time longer than 1 s for cognitive tasks [1,2]. In time interval learning tasks, both the cerebellar nucleus and the thalamus (the output stages of the cerebellum and the basal ganglia, respectively) represent the learned time by their peak activities [3–5]. Activity of the cerebellar nucleus increases just before the represented timing and shows a narrow peak at the timing [3,6]. On the other hand, in the thalamus, the activity is ramped-up from the beginning towards the instructed timing [4,5].

A. Hirose et al. (Eds.): ICONIP 2016, Part IV, LNCS 9950, pp. 244–251, 2016.
DOI: 10.1007/978-3-319-46681-1_30

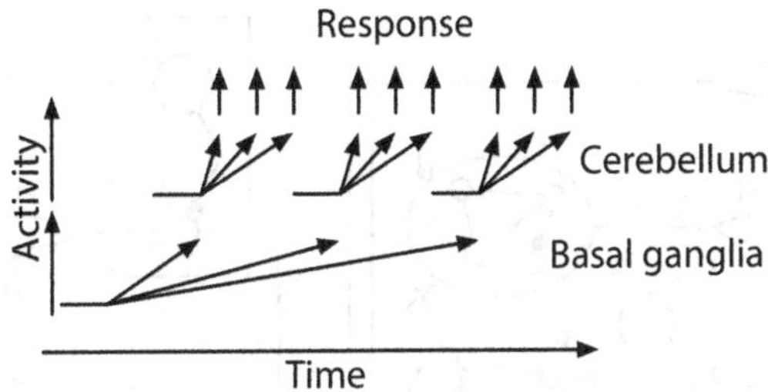

Fig. 1. Illustration of a relay hypothesis [7]. The basal ganglia represents long and rough learned interval. Then the cerebellum activated by ramped-up activity of the basal ganglia via the cerebral cortex, control precision of represented timing. These synergistic works realize long and precise time representation.

From the different representation of time by the cerebellum and the basal ganglia, Tanaka has proposed a relay hypothesis [7]. In this hypothesis, the basal ganglia is in charge of relatively longer time interval for rough estimation of the learned interval. When the activity reaches a threshold, it activates the cerebellum via the cerebral cortex to emit the motor response with precise timing (Fig. 1). A relay hypothesis proposes that the cerebellum and the basal ganglia work synergistically to represent a long time passage with a fine temporal precision.

In this work, we built a neural network model composed of the cerebellum and the basal ganglia. We modeled them according to their known anatomical structures [8–10], and connected them via the cerebral cortex (Cbx). The overall structure is illustrated in Fig. 2. We assumed that the granular layer (GL) of the cerebellum and the network composed of the external capsule of globus pallidus (GPe) and the subthalamic nucleus (STN) in the basal ganglia represent time by their population activities as internal clocks [11,12].

We succeeded to reproduce the distinct activity patterns of the basal ganglia and the cerebellum in a stimulated interval timing learning task. These results suggest that the basal ganglia and the cerebellum play different roles in temporal information processing and support the relay hypothesis.

2 Methods

2.1 Model

Figure 2 illustrates the schematic of our model. The model is composed of the cerebral cortex (Cbx), the striatum (Str), the external globus pallidus (GPe), the internal globus pallidus (GPi), the subthalamic nucleus (STN), the thalamus (Tha), the pontine nucleus (PN), the granular layer (GL), the Purkinje cell layer (Pkj), the cerebellar nucleus (CN). Each component can contain multiple neurons. Let $K = K' = \{\text{Cbx}, \text{Str}, \text{GPi}, \text{Tha}, \text{GPe}, \text{STN}, \text{PN}, \text{GL}, \text{Pkj}, \text{CN}\}$. The model is built based on internal clock models [11,12].

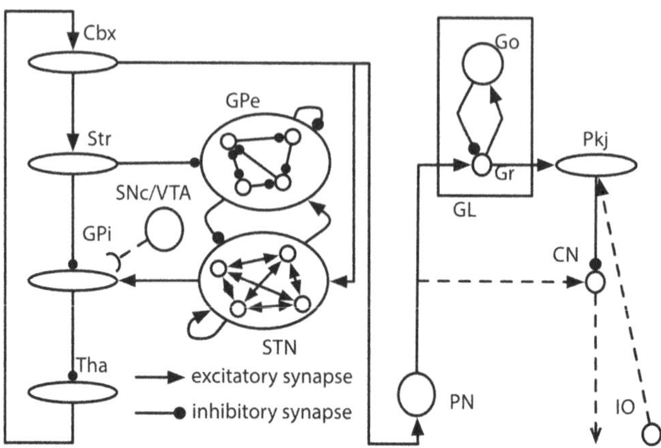

Fig. 2. Schematic circuit of the model. The model is composed of the cerebral cortex (Cbx), the striatum (Str), the external globus pallidus (GPe), the internal globus pallidus (GPi), the subthalamic nucleus (STN), the thalamus (Tha), the pontine nucleus (PN), the granular layer (GL) which is the recurrent network of the granule and the Golgi cells (Gr, Go), the Purkinje cell layer (Pkj), and the cerebellar nucleus (CN). The substantia nigra pars compacta (SNc), the ventral tegmental area (VTA), and the inferior olive (IO) are sources of instruction signals. We modeled the recurrent network of GPe and STN in the basal ganglia and GL in the cerebellum as internal clocks. External signals are injected to the Cbx, and the cerebellum and the basal ganglia receive the signals from Cbx.

Unless stated otherwise below, all neurons obey the same equation. Let $v_{ki}(t)$ be the internal state of neuron i in $k \in K$, which is calculated as

$$\tau_k \dot{v}_{ki} = -v_{ki}(t) + I_k + \sum_{k' \in K'} \sum_j w_{ki \leftarrow k'j} z_{k'j}(t), \tag{1}$$

where τ_k is the time constant, I_k is an external input, $w_{ki \leftarrow k'j}$ is the synaptic weight from a presynaptic neuron j in $k' \in K'$. Each weight is set at one of positive (for excitatory synapse), zero (no synaptic connection), or negative (for inhibitory synapse) values depending on the connectivity. $z_{ki}(t)$ is the output activity of the neuron calculated as

$$z_{ki}(t) = \begin{cases} 1 & v_{ki}(t) > 1, \\ 0 & v_{ki}(t) < 0, \\ v_{ki}(t) & \text{otherwise}. \end{cases} \tag{2}$$

Here we consider only one representative neuron in $k \in$ {Cbx, Str, GPi, Tha, PN, Pkj, CN}, all but GPe, STN, GL.

For $k \in \{\text{GPe}, \text{STN}, \text{GL}\}$, we use the other activation function as

$$z_{ki}(t) = \begin{cases} v_{ki}(t) & v_{ki}(t) > 0, \\ 0 & \text{otherwise}. \end{cases} \tag{3}$$

For GL, the cerebellar granular layer composed of the granule cells and the Golgi cells, we use the following equation for internal states.

$$v_{\mathrm{GL}i}(t) = I_{\mathrm{GL}i}(t) - \sum_j w_{\mathrm{GL}i \leftarrow \mathrm{GL}j} \sum_{s=1}^{t} \exp\left(-\frac{t-s}{\tau_{\mathrm{GL}}}\right) z_{\mathrm{GL}j}(s-1), \qquad (4)$$

where $I_{\mathrm{GL}i}(t)$ is the external input from PN, $w_{\mathrm{GL}i \leftarrow \mathrm{GL}j}$ is the synaptic weight of recurrent connection from neuron j, τ_{GL} is the time constant.

All the parameter values are set by hand.

2.2 Plasticity

We assume synaptic plasticity on the synapses from GL to Pkj and those from STN to GPi. For each trial in a simulated task described below, we update synaptic weight $w_{ki \leftarrow k'j}$ as follows.

$$w_{ki \leftarrow k'j} \leftarrow \begin{cases} c_{\mathrm{LTD}} w_{ki \leftarrow k'j} & z_{k'j}(t_{\mathrm{learn}}) > 0, \\ c_{\mathrm{LTP}} w_{ki \leftarrow k'j} & \text{otherwise}, \end{cases} \qquad (5)$$

where $(k, k') \in \{(k, k') | k = \mathrm{Pkj}, k' = \mathrm{GL}\} \cup \{(k, k') | k = \mathrm{GPi}, k' = \mathrm{STN}\}$, t_{learn} is the onset of unconditioned stimulus, $c_{\mathrm{LTD}} < 1$ and $c_{\mathrm{LTD}} > 1$ are constants.

Previous studies have mainly suggested striatal dopaminergic plasticity in the basal ganglia [8]. On the other hand, previous studies have also suggested dopamine receptors in the globus pallidus, and Mamad *et al.* have demonstrated dopaminergic impact to the pallido-subthalamic neurotransmission [13]. Thus, to simplify our model, we assumed that dopaminergic input to GPi modifies the synaptic weight of subthalamo-pallidal pathway (i.e. connections from STN to GPi) and modeled such plasticity as a simple perceptron.

2.3 Task

We carried out simulation of an interval timing learning task. In the task, a brief conditioned stimulus (CS) is fed to Cbx. The CS activates the network persistently due to the cortico-basal ganglia-thalamo-cortical positive feedback loop. After a certain time period from the CS onset, which is chosen from 400 ms, 800 ms, and 1200 ms, another brief unconditioned stimulus (US) is given to the network, which induces the plasticity. We did not model the US itself and the pathway feeding the US to the network explicitly. Rather we assumed that the US played a functional role for learning.

3 Results

Figure 3 shows raster plots of active states of GL, GPe, and STN. In response to a brief CS, the neurons repeat to become active and inactive randomly due to the random recurrent connections. Because different neurons become active with different temporal patterns, the population of active neurons is uniquely determined for each time step. In other words, the gradual change of the population of active neurons can represent the passage of time. The dynamics is qualitatively similar to internal clock models [11,12].

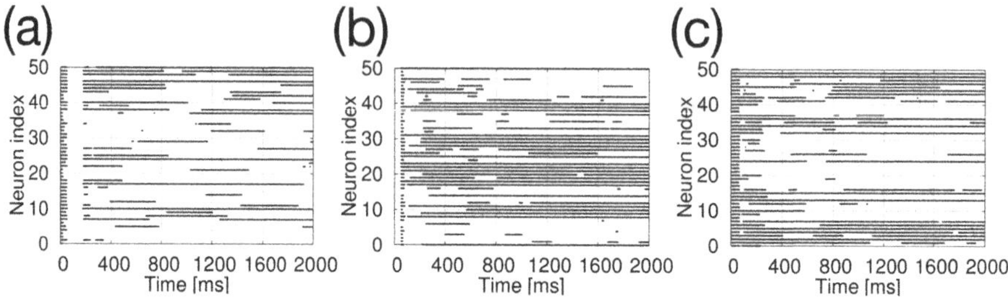

Fig. 3. Raster plots of active states ($z_{ki}(t) > 0$) of the first 50 neurons of (a) GL, (b) GPe, and (c) STN. Every neuron becomes active and inactive alternatively, and the populations of active neurons are unique at each time point.

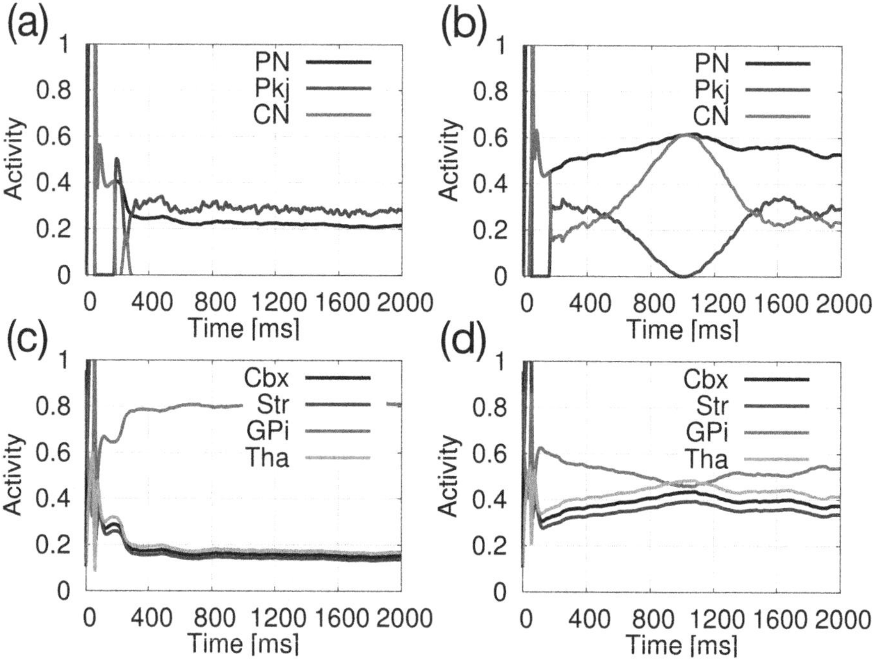

Fig. 4. Activities of the cerebellum (a, b) and the basal ganglia (c, d) before (a, c) and after (b, d) learning of an interval of 1000 ms. Before training seen in (a, c), all neurons shown in exhibited temporally constant activity. After learning seen in (b, d), their activities tended to modulate slowly in time to represent the learned timing. Different neuron types are represented in grayscale.

Figure 4 shows the activities of the cerebellum and the basal ganglia before and after learning of an instructed time interval at 1000 ms. Before learning, all neurons in both the cerebellum (Fig. 4a) and the basal ganglia (Fig. 4c) exhibit strong transient activity in response to a brief CS. After that, the activity is sustained to a certain level due to the cortico-basal ganglia-thalamo-cortical positive feedback loop.

After learning, Pkj in the cerebellum decreases the activity around 1000 ms, which is the instructed timing by the US. Because of the disinhibition, CN increases the activity around the same timing, suggesting the success of learning a time interval of 1000 ms. On the other hand, Cbx, Str, and Tha shows a gradual increase of their activities starting from the CS onset until 1000 ms with the same rate, whereas GPi the opposite activity pattern, suggesting that the basal ganglia also learned the same time interval.

Finally, we varied the time interval to learn and set it at one of 400, 800, and 1200 ms (Fig. 5). For all intervals, the present model successfully learns the instructed interval.

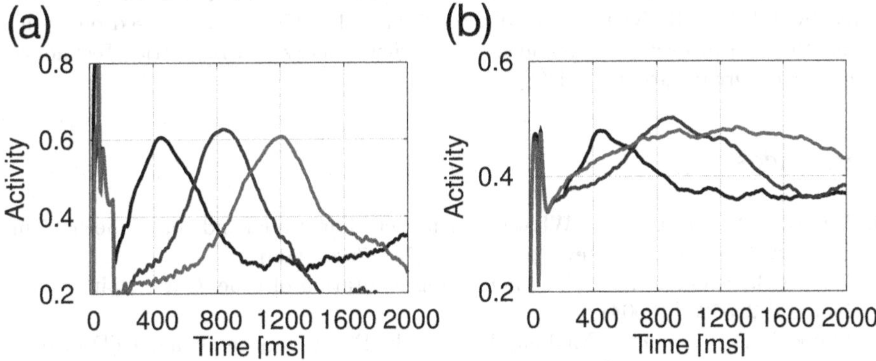

Fig. 5. Activities of (a) CN and (b) Tha after learning 400 (black), 800 (dark gray), and 1200 ms (gray) intervals. Activity of CN increased just before the learned timing and showed narrow peaks. On the other hand, that of Tha constantly increased from the onset to the learned timing and showed broad peaks.

4 Discussion

The present study proposed an unified model of the cerebellum and the basal ganglia for interval timing learning tasks. These two brain areas exhibited distinct temporal profiles during delay periods. Specifically, the cerebellum started to become active just before the US onset, whereas the basal ganglia exhibited a ramp-up activity starting from the CS offset towards the US onset.

The present model assumed that the recurrent networks composed of the GL with granule cells and Golgi cells in the cerebellum and GPe and STN in the basal ganglia act as an internal clock, respectively. Due to the recurrent inhibitory network, the population of active neurons in these networks change gradually in time, and so a certain population is uniquely determined for each time step from the CS onset. Thus, our model represents the passage of time by the gradual change of the active neuron population in time [11,12].

The difference between the temporal profile of neuron activity in the cerebellum and the basal ganglia may come from the excitatory-excitatory connections and inhibitory-inhibitory connections within STN and GPe, respectively. These connections could balance the activities of excitatory STN neurons and inhibitory GPe neurons, resulting in a gradual increase of the total neural activity.

The present result provides a theoretical framework for a relay hypothesis [7], which proposes that the basal ganglia provides a rough estimation of the instructed timing, whereas the cerebellum controls the timing of actual responses. This mechanism would be important to represent a long time passage with a fine temporal resolution.

Acknowledgment. We would like to thank Professor Masaki Tanaka at Hokkaido University for fruitful discussions on his relay hypothesis. Part of this work was supported by JSPS KAKENHI Grant Number 26119511. This paper is based on results obtained from a project commissioned by the New Energy and Industrial Technology Development Organization (NEDO).

References

1. Buhusi, C.V., Meck, W.H.: What makes us tick? Functional and neural mechanisms of interval timing. Nat. Rev. Neurosci. **6**, 755–765 (2005)
2. Ivry, R.B., Spencer, R.M.: The neural representation of time. Curr. Opin. Neurobiol. **14**, 225–232 (2004)
3. Mauk, M.D., Garcia, S., Medina, J.F., Steele, P.M.: Does cerebellar LTD mediate motor learning? Toward a resolution without a smoking gun. Neuron **20**, 359–362 (1998)
4. Tanaka, M.: Cognitive signals in the primate motor thalamus predict saccade timing. J. Neurosci. **27**, 12109–12118 (2007)
5. Tanaka, M., Kunimatsu, J.: Contribution of the central thalamus to the generation of volitional saccades. Eur. J. Neurosci. **33**, 2046–2057 (2011)
6. McCormick, D.A., Thompson, R.F.: Neuronal responses of the rabbit cerebellum during acquisition and performance of a classically conditioned nictitating membrane-eyelid response. J. Neurosci. **4**, 2811–2822 (1984)
7. Tanaka, M., Kunimatsu, J., Ohmae, S.: Neural representation of time. Brain Nerve **65**, 941–948 (2013)
8. Kandel, E.R., Schwartz, J.H., Jessell, T.M., Siegelbaum, S.A., Hudspeth, A.J.: Principles of Neural Science, 5th edn. McGraw-Hill Companies Inc., New York (2013)
9. Kita, H., Tachibana, Y., Nambu, A., Chiken, S.: Balance of monosynaptic excitatory and disynaptic inhibitory responses of the globus pallidus induced after stimulation of the subthalamic nucleus in the monkey. J. Neurosci. **25**, 8611–8619 (2005)

10. Mauk, M.D., Donegan, N.H.: A model of Pavlovian eyelid conditioning based on the synaptic organization of the cerebellum. Learn. Memory **3**, 130–158 (1997)
11. Yamazaki, T., Tanaka, S.: Neural modeling of an internal clock. Neural Comput. **17**, 1032–1058 (2005)
12. Yamazaki, T., Tanaka, S.: A neural network model for trace conditioning. J. Neural Syst. **15**, 23–30 (2005)
13. Mamad, O., Delaville, C., Benjelloun, W., Benazzouz, A.: Dopaminergic control of the globus pallidus through activation of D2 receptors and its impact on the electrical activity of subthalamic nucleus and substantia nigra reticulata neurons. PLoS ONE **10**(3), 1–16 (2015)

Bihemispheric Cerebellar Spiking Network Model to Simulate Acute VOR Motor Learning

Keiichiro Inagaki(✉) and Yutaka Hirata

Department of Robotic Science and Technology, College of Engineering, Chubu University, Kasugai, Japan
{kay, yutaka}@isc.chubu.ac.jp

Abstract. The vestibuloocular reflex (VOR) is an adaptive control system. The cerebellar flocculus is intimately involved in the VOR adaptive motor control. The cerebellar flocculus has bihemispheric architecture and the several lines of unilateral lesion study indicated that each cerebellar hemisphere plays different roles in the leftward and rightward eye movement control and learning. However, roles of bihemispheric cerebellar architecture underlying the VOR motor learning have not been fully understood. Here we configure an anatomically/physiologically plausible bihemispheric cerebellar neuronal network model composed of spiking neurons as a platform to unveil roles and capacities of bihemispheric cerebellar architecture in the VOR motor learning.

Keywords: Eye movement · Spike timing dependent plasticity · Computer simulation · Cerebellar lesion

1 Introduction

The cerebellum plays an essential role in motor learning, namely acquisition of new motor skills, maintenance and improvement of the acquired motor performances throughout lifetime. The vestibuloocular reflex (VOR), which maintains stable vision during head motion by counter-rotating the eyes in the orbit, has been one of the most popular model systems to investigate the role of the cerebellum in biological adaptive control. Adaptive control of the VOR is postulated to occur in the cerebellar flocculus because ablation of the flocculus precludes learning, and excludes learned memories [1]. It was reported that the unilateral lesion of the cerebellum caused initiation of ipsi-lateral smooth pursuit deficits [2]. It was also reported that the unilateral lesion of the cerebellum caused large deficit of learning changes of ipsi-lateral VOR, but small deficit in contra-lateral VOR [3]. These lines of evidence indicate that each cerebellar hemisphere plays different roles in the leftward and rightward eye movement control and learning. However, roles and capacities of bihemispheric cerebellar architecture underlying the VOR motor learning have not been fully understood. To date, several computational models for VOR motor learning have been proposed (e.g. [4, 5, 6, 7, 8]). These models, however, described the cerebellar neuronal circuitry without distinguishing two hemispheres.

A. Hirose et al. (Eds.): ICONIP 2016, Part IV, LNCS 9950, pp. 252–258, 2016.
DOI: 10.1007/978-3-319-46681-1_31

Here we configured an anatomically and physiologically plausible bihemispheric cerebellar neuronal network model composed of spiking neurons, and evaluated the bihemispheric structure on VOR motor learning, and consequence of unilateral lesion of the cerebellum.

2 Cerebellar Neuronal Network Model

Figure 1 illustrates the structure of the proposed bihemispheric cerebellar neuronal network model. The model was constructed by extending our previous single hemisphere model whose details are described elsewhere [5, 6]. We constructed and assembled left and right cerebellar hemisphere models and integrated their outputs. The proposed model consists of 9 subsystems, two of which are left and right cerebellar flocculus neuronal networks. Each subsystem is described as a transfer function whose characteristics are determined based upon experimental data in squirrel monkeys [9]. $G_{preFL}^{ecopy}(s)$, $G_{preFL}^{visual}(s)$, and $G_{preFL}^{vestib}(s)$ represent characteristics of pre-flocculus pathways each of which processes efference copy, visual, and vestibular signal, respectively. $G_{io}(s)$ represents characteristics of another pre-floccular pathway which processes the visual error signal (retinal slip) to generate Purkinje cell complex spikes. $G_{postFL}(s)$ describes characteristics of the post-floccular pathway which converts Purkinje cell simple spike outputs to a part of motor commands to move the eyes. $G_{nonFL}^{visual}(s)$ and $G_{nonFL}^{vestib}(s)$ describe characteristics of the non-floccular visual pathway and the non-floccular vestibular pathway, respectively.

The subsystem for the cerebellar flocculus neuronal network was constructed based on the known anatomical and physiological evidence [10–13]. The model consists of 20 Purkinje, 900 Golgi, 60 basket/stellate, and 10000 granule cells (Fig. 1 Cerebellar flocculus). Synaptic connections among these cell types are as follows. Each granule cell receives 6 excitatory mossy fiber inputs and 3 inhibitory Golgi cell inputs. Here, mossy fibers transfers vestibular, visual (retinal slip), and efference copy. In our model, 9 types of granule cells are prepared by setting different fractions of vestibular, visual and efference copy signals conveyed via mossy fiber inputs, basing upon physiological evidence in monkeys [14]. Each Golgi cell receives 20 excitatory mossy fibers and 100 granule cell inputs. Each basket/stellate cell receives excitatory synaptic inputs from 250 granule cells. Each Purkinje cell receives excitatory inputs from 10000 granule cells and a single climbing fiber as well as inhibitory inputs from 10 basket/stellate cells. Note that, the phase of input signals (mossy fiber and climbing fiber activities) for the two cerebellar hemispheres are 180 degrees sifted each other. The cerebellar LTD [15, 16] and LTP [17, 18] are also implemented as changes in synaptic weight between parallel fibers and Purkinje cells. LTD and LTP at synapses between parallel fibers and Purkinje cells are also implemented as a spike timing dependent plasticity. In the model, LTD and LTP are evoked based on the spike timing of granule cell (parallel fiber) and climbing fiber. With the learning rule, the synaptic weight between a granule cell and a Purkinje cell is updated so that it is decreased by LTD when a granule cell spike coincides with the firing of a climbing fiber, or increased by LTP when a granule cell fires a spike without a firing of a climbing fiber. Note that the synaptic weights are

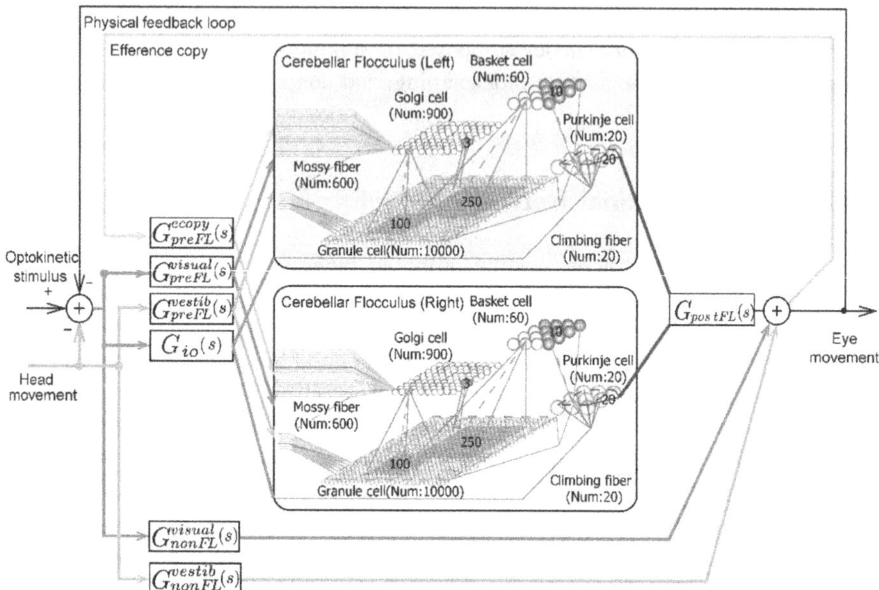

Fig. 1. VOR system model with bihemispheric cerebellar neuronal network and 7 subsystems each of which represents a different anatomical pathway processing different modality of signal. In the cerebellar neuronal network model, flocculus is explicitly described by spike neuron models to evaluate the cerebellar motor learning in terms of the spike timing dependent plasticity. The organization of the flocculus neuronal network and each of synaptic connections are determined by the anatomical and physiological evidence. The input signal for the left cerebellar hemisphere has opposite sign with that for the right cerebellar hemisphere [22].

restricted to be positive. More detailed description of the cerebellar flocculus model and the learning rule are described in [5, 6].

In scaling down the cerebellar cortical neuronal network to a computationally feasible dimension, it is suggested that the convergence/divergence ratios are more important than the cell ratios. Consequently, the model was scaled down, maintaining the convergence/divergence ratios as much as possible.

3 Model Simulation

We simulated three major experimental paradigms; VORe, VORs, and VORd. All paradigms consist of sinusoidal head rotation and/or optokinetic stimulus (OKS) (frequency: 0.5 Hz, amplitude: 40 deg/sec). In VORs, the head rotates in phase with the OKS, while in VORe the head and OKS rotate 180 degrees out of phase. In our simulation, continuous application of VORe or VORs was utilized for gain up (high gain) training or gain down (low gain) training, respectively, as in the animal behavioral experiments [9, 19]. We simulated high gain training and low gain training for 2 h. To quantify the newly acquired VOR memory in each condition, we simulated

VORd every 30 min for 20 stimulus cycles and measured the VOR gain. In VORd, the head was rotated without OKS. The VOR gain in the dark (g) was estimated by using the following linear regression:

$$e(t) = gh \cdot (t - \tau) + dc + \varepsilon(t) \quad (1)$$

where $h(t)$, $e(t)$, dc, and $\epsilon(t)$ are head velocity, eye velocity, a dc component and residual, respectively. τ is the latency between head velocity and eye velocity. We also simulated unilateral cerebellar lesion after 2 h VOR motor learning (VORe or VORs) to evaluate roles of each hemisphere in VOR motor learning.

4 Results

First, we investigate if the proposed model can reproduce VOR motor learning as changes in VOR gain, and modulation of Purkinje cell simple spikes during the VORd. Figure 2 summarizes the learning curves of VOR gain during high and low gain training. The trend of learning changes are comparable to those demonstrated experimentally in monkey [19] and goldfish [20]. Quantitatively, the VOR gain changed from 0.80 to 1.04 (+0.24) after 2-hour high gain training, and from 0.80 to 0.67 (–0.13) after 2-hour low gain training.

Purkinje cell simple spikes change firing modulation during VORd after VOR motor learning [9]. Figure 3 summarized adaptive changes in firing of Purkinje cell simple spikes in an experiment using squirrel monkey (top) and in the model

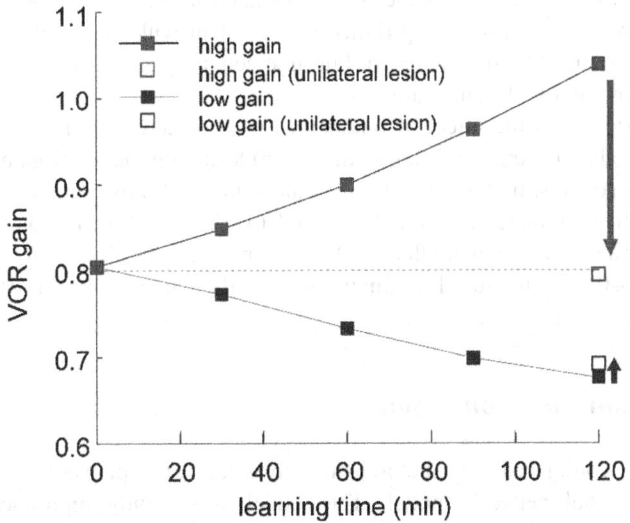

Fig. 2. Learning curves of VOR gain during high gain training (red) and low gain training (blue) in the model simulation. Effect of unilateral (left hemisphere) ablation post 2 h VOR motor learning is indicated as open squares. (Color figure online)

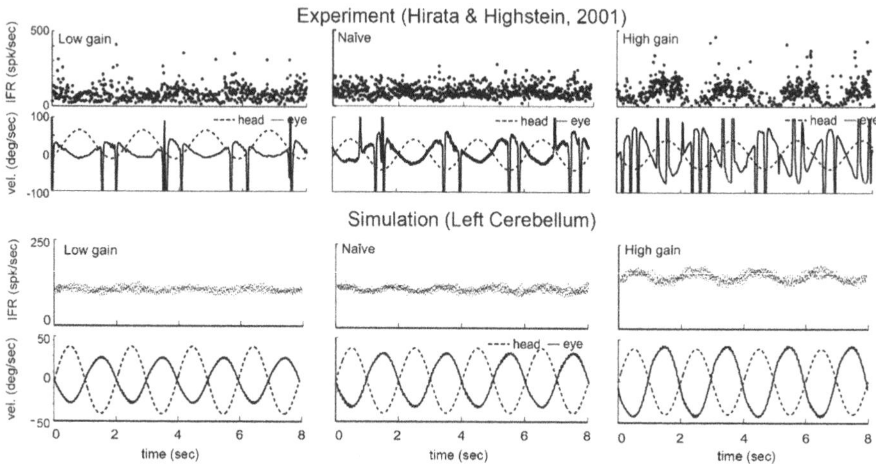

Fig. 3. Response of Purkinje cell simple spikes during VORd in an experiment using squirrel monkey (top) and simulation (bottom). Left: post low gain training, Middle: pre VOR motor learning, and Right: post high gain training. Solid line and dotted line indicate eye velocity and head velocity, respectively. The firing of Purkinje cell simple spikes is depicted as instantaneous firing rate.

simulation (bottom). Before any training (naïve condition), Purkinje cell simple spikes during VORd hardly modulated (Fig. 3, top-middle). Purkinje cell simple spikes changed firing modulation that would account for changes in VOR gain [9]. Namely, Purkinje cell simple spikes slightly modulated out of phase with head rotation after low gain training (Fig. 3, top-left), while they modulated in phase with the head rotation after high gain training (Fig. 3, top-right). In consistent with the monkey experiment, our model reproduced these changes in Purkinje cell simple spikes modulation during VORd after low and high gain trainings.

We also simulated the effect of unilateral lesion of cerebellum after VOR motor learning. The effect of unilateral (left hemisphere) lesion on the VOR gain is shown in Fig. 2 as open squares. In our simulation, all the gain learned by high gain training was abolished with the unilateral lesion (from 1.04 to 0.79 (–0.25)). In contrast, the unilateral lesion caused very small effect on learned memory after low gain training (from 0.67 to 0.69 (+0.02)). It caused minimal effect on the naïve VOR gain (from 0.80 to 0.72. data not shown).

5 Discussion and Conclusion

Presently we configured an anatomically/physiologically plausible bihemispheric cerebellar neuronal network model. The model successfully reproduced adaptive changes in VOR gains in both high and low gain training paradigms. The model demonstrated a significant reduction of learned VOR gain after high gain training, while minimal impairment of learned gain was observed after low gain training, suggesting different roles of each cerebellar hemisphere in high and low gain VOR motor learning and memory retention. Here, high and low gain learning might have occurred separately in each cerebellar hemisphere, and currently we are evaluating this possibility with reproduction of asymmetrical gain changes after unilateral lesion using the constructed model.

It was reported that an acutely acquired VOR gain decayed quickly, and decay time constants were significantly faster after gain increase training than after gain decrease training [19], indicating a necessity of different learning and memory mechanisms for high and low gain training. Our simulation results imply that the different learning and memory mechanisms for high and low gain training might be originated from bihemispheric cerebellar architectures.

Finally, our results also assure the usefulness of the model in evaluation of signal processing in cerebellar cortex, and interaction of bilateral cerebellar hemispheres. In the future study, we employ the model to evaluate learning mechanisms and roles of synaptic plasticity in direction selective and other context specific VOR motor learning [20, 21].

Acknowledgements. This work was supported in part by JSPS KAKENHI Grant-In-Aid for Scientific Research (B) (24300115 and 16H02901, YH) and Grant-in-Aid for Young Scientists (B) (15K16086, KI).

References

1. Nagao, S., Kitazawa, H.: Effects of reversible shutdown of the monkey flocculus on the retention of adaptation of the horizontal vestibulo-ocular reflex. Neuroscience **118**(2), 563–570 (2003)
2. Staube, A., Scheuerer, W., Eggert, T.: Unilateral cerebellar lesions affect initiation of ipsilateral smooth pursuit eye movements in humans. Ann. Neurol. **42**, 891–898 (1997)
3. Ito, M., Jastreboff, P.J., Miyashita, Y.: Specific effects of unilateral lesions in the flocculus upon eye movements in albino rabbits. Exp. Brain Res. **45**(1–2), 233–242 (1982)
4. Tabata, H., Yamamoto, K., Kawato, M.: Computational study on monkey VOR adaptation and smooth pursuit based on the parallel control-pathway theory. J. Neurophysiol. **87**, 2176–2189 (2002)
5. Inagaki, K., Hirata, Y.: The model of vestibuloocular reflex explicitly describing cerebellar neuronal network model. Inst. Electron. Inf. Commun. Eng. **J94-D**(5), 1293–1304 (2007)
6. Inagaki, K., Kobayashi, S., Hirata, Y.: Analysis of frequency selective vestibuloocular reflex motor learning using cerebellar spiking neuron network mode. Inst. Electron. Inf. Commun. Eng. **J94-D**(5), 919–928 (2011)
7. D'Angelo, E., Mapelli, L., Casellato, C., Garrido, J.A., Luque, N., Monaco, J., Prestori, F., Pedrocchi, A., Ros, E.: Distributed circuit plasticity: new clues for the cerebellar mechanisms of learning. Cerebellum **15**(2), 1–13 (2015)
8. Yamazaki, T., Nagao, S., Lennon, W., Tanaka, S.: Modeling memory consolidation during posttraining periods in cerebellovestibular learning. Proc. Natl. Acad. Sci. U.S.A. **112**, 3456–3541 (2015)
9. Hirata, Y., Highstein, S.M.: Acute adaptation of the vestibuloocular reflex: signal processing by floccular and ventral parafloccular Purkinje cells. J. Neurophysiol. **85**, 2267–2288 (2001)

10. Eccles, J.C., Ito, M., Szentagothai, J.: The Cerebellum as a Neuronal Machine. Springer, Heidelberg (1967)
11. Marr, D.: A theory of cerebellar cortex. J. Physiol. **202**, 437–470 (1969)
12. Albus, J.S.: A theory of cerebellar function. Math. Biosci. **10**, 25–61 (1972)
13. Ito, M.: The Cerebellum and Neural Control. Raven Press, New York (1984)
14. Lisberger, S.G., Fuchs, A.F.: Role of primate flocculus during rapid behavioral modification of vestibuloocular reflex. II. Mossy fiber firing patterns during horizontal head rotation and eye movement. J. Neurophysiol. **41**, 764–777 (1978)
15. Ito, M.: Long-term depression. Annu. Rev. Neurosci. **12**, 85–102 (1989)
16. Ito, M.: The Cerebellum: Brain for an Implicit Self. Financial Press, Upper Saddle River (2012)
17. Hirano, T.: Depression and potentiation of the synaptic transmission between a granule cell and a Purkinje cell in rat cerebellar culture. Neurosci. Lett. **119**, 141–144 (1990)
18. Sakurai, M.: Synaptic modification of parallel fibre - Purkinje cell transmission in in virto guinea-pig cerebellar slices. J. Physiol. **394**, 463–480 (1987)
19. Kuki, Y., Hirata, Y., Blazquez, P.M., Heiney, S.A., Highstein, S.M.: Memory retention of vestibuloocular reflex motor learning in squirrel monkeys. NeuroReport **15**(6), 1007–1011 (2004)
20. Yoshikawa, A., Hirata, Y.: Different mechanisms for gain-up and gain-down vestibuloocular reflex motor learning revealed by directional differential learning tasks. Inst. Electron. Inf. Commun. Eng. **J92-D**(1), 176–185 (2009)
21. Hirata, Y., Lockard, J.M., Highstein, S.M.: Capacity of vertical VOR adaptation in squirrel monkey. J. Neurophysiol. **88**, 3194–3207 (2002)
22. Purves, D., Augustine, G.J., Fitzpatrick, D., Katz, L.C., LaMantia, A.S., McNamara, J.O., Williams, S.M.: Neuroscience, 2nd edn. Sinauer Associates Inc., Sunderland (2004)

Theory and Algorithms

Modeling the Propensity Score with Statistical Learning

Kenshi Uchihashi and Atsunori Kanemura(✉)

National Institute of Advanced Industrial Science and Technology (AIST), Tsukuba 305-8568, Japan
atsu-kan@aist.go.jp

Abstract. The progress of the ICT technology has produced datasources that continuously generate datasets with different features and possibly with partial missing values. Such heterogeneity can be mended by integrating several processing blocks, but a unified method to extract conclusions from such heterogeneous datasets would bring consistent results with lower complexity. This paper proposes a flexible propensity score estimation method based on statistical learning for classification, and compared its performance against classical generalized linear methods.

Keywords: Propensity scores · Missing value estimation · Observational studies · Statistical learning · Deep learning

1 Introduction

Extracting scientific conclusions from data requires unbiased estimation, which may be difficult for modern datasets that have been produced from recent information communication systems. This is because these datasets may come from different sources (heterogeneity) and/or may have missing values because of the different costs in measurement or the nature of data acquisition designs. Complete-case analysis, which ignores data samples with missing variables and uses only fully observed ones, allows statistical and machine learning methods to run on such data with missing values, but, as a result, conclusions drawn are often biased. Moreover, the removal of incomplete cases reduces the sample efficiency, making conclusions variable. Therefore, we need a method that extracts unbiased conclusions from biased datasets.

Typical solutions to cope with the issues of heterogeneity and missingness include pre-processing heuristics such as mean-value imputation and the use of flexible models. Statistical theory offers ways to cope with these issues [1,2]; and among useful techniques in the theory we use matching by the propensity score and describe how to use statistical learning in estimating the propensity score. The model assumes heterogeneity in data; i.e., datasets have both continuous and discrete variables.

A. Hirose et al. (Eds.): ICONIP 2016, Part IV, LNCS 9950, pp. 261–269, 2016.
DOI: 10.1007/978-3-319-46681-1_32

In this paper, we adopt nonlinear models from statistical machine learning for estimating propensity scores, instead of widely-utilized generalized linear models, and validate them using several real datasets. Nonlinear models have richer representation of data and are expected to better fit the propensity score function. In particular, we evaluate the gradient boosting decision tree (GBDT) and the convolutional neural network (CNN) and compare their performance with the classical logistic regression and also a random baseline method. The experimental results show that the accuracy of matching measured by t-values and the Kullback-Leibler (KL) divergence is better for the nonlinear methods than the (generalized) linear one.

2 Statistical Analysis Under Missing Values

Here we describe how to analyze datasets with a bias in missing value mechanisms and how to estimate the outcome with reduced bias with a bias adjustment technique based on the propensity score.

Notation. We denote covariates by $\boldsymbol{x}$, the outcome of interest by y, and the missingness indicator by $r \in \{0, 1\}$. A data point (data sample) consists of these three types of variables. The outcome y is also called *target* and indicator r is called *treat* in this paper.

Control and Matching. In scientific experiments, we collect measurement samples both in a treatment condition ($r = 1$) and a control condition ($r = 0$) and compare the differences between the two conditions. Here, we (implicitly) make a *matching* between measurements from two groups, keeping the covariates (variable of no-interest) $\boldsymbol{x}$ the same between the two groups while changing the condition of interest. We are interested in quantifying the difference of the outcome y between the two groups.

2.1 Missing Mechanisms

The mechanism generating missing values are classified into three categories, namely, MCAR (missing completely at random), MAR (missing at random), MNAR (missing not at random) [1]. This paper assumes MAR behind the all datasets we analyze. MCAR indicates missing values take place statistically independently ("randomly") for individual occurences; that is, MCAR means r is statistically independent from $\boldsymbol{x}$ and y, or $p(r|\boldsymbol{x}, y) = p(r)$. In this case complete-case analysis yields no bias and we can safely ignore incomplete samples but the sample efficiency decreases. MAR assumes missingness r depends only on covariates $\boldsymbol{x}$ and not on outcome y, i.e., $p(r|\boldsymbol{x}, y) = p(r|\boldsymbol{x})$. In MAR, if we control the covariates, we have the same missing probabilities; that is, $p(r|\boldsymbol{x}_1, y_1) = p(r|\boldsymbol{x}_2, y_2)$ for $\boldsymbol{x}_1 = \boldsymbol{x}_2$ (and y_1 and y_2 can be different from each other) and we can estimate r only from $\boldsymbol{x}$. MNAR is the most general case and we cannot simplify the missing mechanism from $p(r|\boldsymbol{x}, y)$, which means even if we control the covariates $\boldsymbol{x}$, r and y are not independent.

In MAR, we can explain r only by the observable quantities, $\boldsymbol{x}$; that is, r and y are conditionally independent given $\boldsymbol{x}$:

$$p(r|\boldsymbol{x}, y) = p(r|\boldsymbol{x}). \tag{1}$$

This equation says that whether a missing value takes place or not is completely explainable only from the observations, $\boldsymbol{x}$, and the missing outcome y is not required. This property allow us to analyze the entire dataset (including partially missing cases) without bias by the use of the propensity score [3], as described below. We assume the missing mechanism of the datasets we use in this paper is MAR.

2.2 The Propensity Score

In MAR, although complete-case analysis generally gives biased conclusions, we can draw unbiased conclusions by matching based on the covariates. This is because if the values of the covariates are the same between two samples, one of these samples can be considered as treatment and the other as control. When the dimensionality of the covariates is not extremely low, it is difficult to perform matching but using a *balancing score* can reduce the problem to the matching of one-dimensional variable [4]. A balancing score b summarizes the covariates to keep the information on missing values, and defined as a function of $\boldsymbol{x}$ as

$$p(r|\boldsymbol{x}, b(\boldsymbol{x})) = p(r|b(\boldsymbol{x})). \tag{2}$$

Conditioning only on the balancing score suffices to have conditional independence with r, the missing value indicator.

A well-known and widely utilized example is the *propensity score*, which is the probability of having a missing value given the covariates $\boldsymbol{x}$ and defined as

$$e(\boldsymbol{x}) = p(r = 1|\boldsymbol{x}). \tag{3}$$

The propensity score is a balancing score [4] and thus can be considered as a one-dimentional summary of the information on the missing values that the covariates have. Conditioning on the propensity score e, we can regard the missing value takes place at random.

The standard estimation model for the propensity score is logistic regression [3], which is a generalized linear model. Although logistic regression is suited for analyzing the contribution of each covariate and for interpreting p-values, it is not as flexible as highly nonlinear machine learning models and in nonlinear settings logistic regression yield degraded performance compared to them [5]. Therefore, for the purposes of this paper, accurate estimation of the propensity score, logistic regression is not the optimal choice.

In this paper, we apply two of popular statistical machine learning methods, GBDT and CNN, and compare with logistic regression in the matching accuracy defined as the distance between the covariates.

2.3 Matching

If the values of the propensity score is the same, the probability distribution of the covariates are the same [6]. Therefore, if we make a matching based on the propensity score proximity from two groups (one with $r = 0$ and the other with $r = 1$), it is expected that the missing value is close to the value of that variable from another group. We employ the greedy matching method, where the two samples closest in the propensity score sense are made as a pair.

2.4 Propensity Score Estimation with CNN

CNN is a neural network that consists of stacked convolution and pooling layers [7]. Although CNN has initially been popular in image recognition fields since it is invariant under geometric warping of input variables, it is not widely used in other fields including time series analysis and natural language processing with high accuracy [8,9]. CNN's computational complexity is much lower that that of general deep networks because CNN learns the filters (convolution kernels) whose dimentionality is much lower than that of the input variables. Thanks to the low dimentionality of network parameters (weights), training CNNs is faster and robust to overfitting since gradient divergence is less likely. Still, CNN has nonlinearity to the input and more flexible than linear models.

The convolution layer extracts local features by filtering (convolution) with the invariance in translation and/or size change of input. The target of learning is the filter weights. The pooling layer sub-samples the output from its preceding convolution layer to reduce the dimentionality and become robust to local changes in the input. As a pooling mechanism, max pooling or average pooling are popular and known to produce satisfactory results. CNN's stack has alternating convolution and pooling layers to achieve nonlinear feature extraction before the final layer where full connection is used to integrate these features for classification.

3 Experiments

3.1 Datasets

We performed experiments using three real-world datasets, `lalonde`, `lindner`, and `ACTG175`. The `lalonde` dataset is derived from the treated group in the National Supported Work Demonstration (NSW) and the comparison sample from the Current Population Survey (CPS). This dataset consists of 10 variables. The `lindner` dataset contain data on 996 patients treated at the Lindner Center, Christ Hospital, Cincinnati in 1997. The patients received a Percutaneous Coronary Intervention (PCI). This dataset consists of 10 variables. The AIDS Clinical Trials Group Protocol 175, which also known as the `ACTG175` dataset, contains 2139 HIV-infected subjects. In this dataset, four different treatments are randomized subjects. In this paper, we consider two groups: ZDV monotherapy and the other three treatments. The missing value of this dataset is the CD4 count at 96 ± 5 weeks, which is the post-baseline between these two treatment groups. This dataset consists of 27 variables.

3.2 Estimating the Propensity Score

We estimated the propensity score by using three different methods, namely, logistic regression (LogisticR), GBDT, and CNN. The input to a learning model is the set of the covariates in a dataset, and the target output is a dummy variable `treat`, which indicates the assignment to treat or control conditions. The estimated value for a sample is the probability of that sample belonging to the treat condition.

The missing value indicator r is stored as the variable `treat` in the `lalonde` and `ACTG175` datasets, and the `lindner` dataset has `abcix` as r. The outcome variable in `lalonde` is `re78` and that in `lindner` is `lifepres` but for `ACTG175` the original outcome `cd496` has missing values and we make a new missing value indicator.

The hyperparameters for each model were selected as those that yielded highest five-fold cross validation accuracy. For CNN the selected filter size for the convolution layer was 3×1 and for the pooling layer max pooing with 2×1 was selected. The activation function was ReLU. Learning was performed for 20 epochs where the batch size was 50.

3.3 Matching with Propensity Scores

We performed greedy matching based on the estimated propensity scores. To have quantitatively evaluate how successful matching was, i.e., how the covariates became closer after matching, we employed two measures, t-values and Kullback-Leibler (KL) divergence. The baseline matching algorithm was random matching.

Evaluation by t-Values. We compared t-values between the missing group and the observable group after matching. The t value measures the distance between two groups after correcting with mean and variance values and is defined as follows:

$$t = \frac{\overline{x}_1 - \overline{x}_0}{\sqrt{s_1^2/N_1 + s_0^2/N_0}}, \tag{4}$$

where $\overline{x}_r$ and s_r^2 ($r \in \{0,1\}$) are the mean and the variance, respectively, for the covariates $\boldsymbol{x}_r$ for each group. The smaller t-value is, the closer the two distributions are.

Evaluation by KL Divergence. We calculate and compare the KL divergence between the two groups after matching. The KL divergence between two probability densities p and q is defined as

$$D_{\mathrm{KL}}(p||q) = \int_{-\infty}^{\infty} p(x) \ln \frac{p(x)}{q(x)} dx. \tag{5}$$

Here, p and q are the densities for the missing group and the observable group, respectively. Then $D_{\mathrm{KL}}(p||q)$ measures how similar the covariates are between the two groups.

Simulation. We randomly divided a dataset into five parts, where four of them are used for training, the rest one is used for testing, and repeated this ten times with different partitions, producing 50 different training and test data pairs. A learning model was trained and tested using one of the data pairs. In the testing step, the estimated propensity scores were used for greedy matching and t-values and KL divergences were calculated.

Figures 1, 2, and 3 show the average t-values for each dataset with three learning methods. All the models have their highest peaks at around $t = 0$, but CNN has produced the highest one, which means matching with CNN was best among the compared four methods.

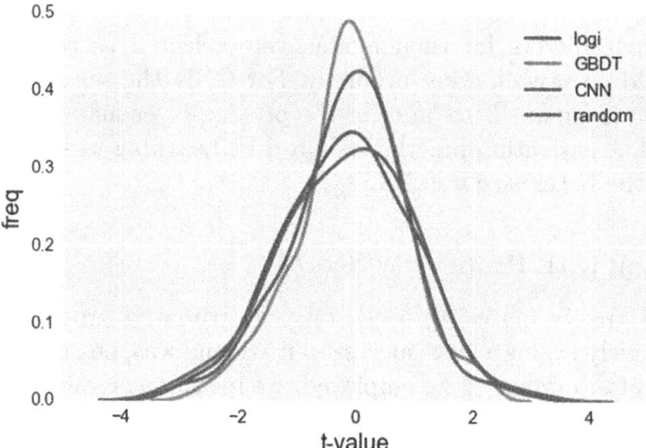

Fig. 1. t-value distributions for `lalonde` dataset

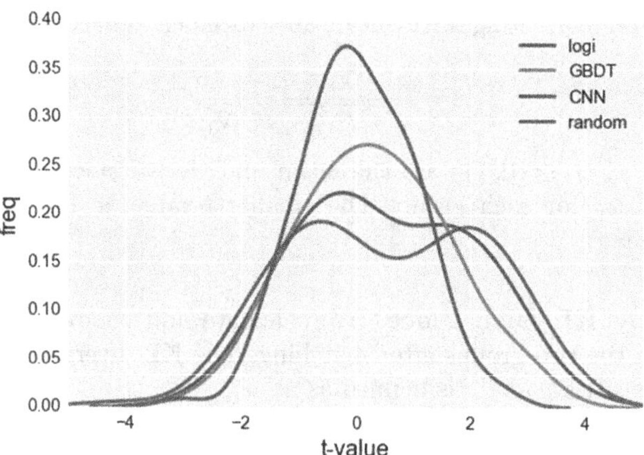

Fig. 2. t-value distributions for `lindner` dataset

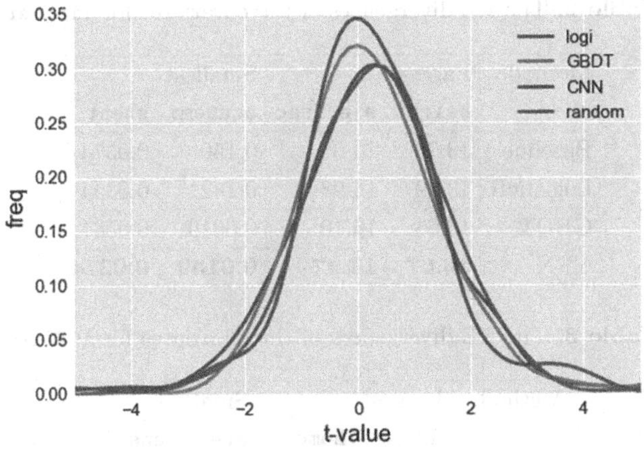

Fig. 3. t-value distributions for `ACTG175` dataset

Tables 1, 2, and 3 show the KL divergence between the covariates of the two groups. We approximate the continuous KL divergence defined in (5) by replacing p and q with their histogram representations. Among many covariates, four were selected based on the results of CNN; the covariates whose KL divergences were largest and 2nd largest and those smallest and 2nd smallest. For the largest-KL covariates CNN has produced far better results compared to the other methods and even for the smallest-KL covariates CNN was comparable.

The results shown above indicates the superiority of nonlinear models, in particular CNN, for estimating the propensity score. The reason behind this was two-fold. First is the complexity of the MAR mechanism. Some MAR are indeed linear and well modeled by logistic regression, but some are more complicated and require flexible models like CNN, which can cope with different missing patterns. Second is variable selection with max pooling. The covariates should be multi-dimensional to satisfy MAR [10] but there is no guarantee that the covariates are the minimum set of variables that matter; in other words, some covariates have nothing to do with missing value occurrences. Therefore the integration of variable selection into a learning model is advantageous and max pooling in CNN naturally incorporates it.

Table 1. The KL divergences of the covariates for `lalonde`

Method	Largest		Smallest	
	`u75`	`u74`	`nodegr`	`educ`
Baseline	0.0380	0.0216	0.103	16.54
LogisticR	0.0378	0.0181	0.125	**15.03**
GBDT	0.0113	**0.0178**	**0.0549**	15.05
CNN	**0.00578**	0.0211	0.0903	16.23

Table 2. The KL divergences of the covariates for `lindner`

Method	Largest		Smallest	
	`height`	`ejecfrac`	`acutemi`	`stent`
Baseline	13.70	14.62	0.150	0.0574
LogisticR	13.79	14.98	0.182	0.0394
GBDT	13.68	16.10	0.0619	0.0581
CNN	**13.67**	**14.47**	**0.0139**	**0.0274**

Table 3. The KL divergences of the covariates for `ACTG175`

Method	Largest		Smallest	
	`drugs`	`homo`	`age`	`cens`
Baseline	0.0171	0.0167	**2.73**	0.0110
LogisticR	0.0154	0.0151	3.00	0.0862
GBDT	0.0133	0.0202	3.22	0.0821
CNN	**0.0130**	**0.0144**	2.76	**0.0108**

4 Conclusion

We have demonstrated nonlinear statistical learning methods, GBDT and CNN, outperform the classical linear model, logistic regression, over all the three real-world datasets in the matching accuracy with propensity score estimation.

Acknowledgments. This study was supported in part by the New Energy and Industrial Technology Development Organization (NEDO), Japan, and by JSPS KAKENHI 26730130 and 15K12112.

References

1. Little, R.J.A., Rubin, D.B.: Statistical Analysis with Missing Data, 2nd edn. Wiley, Hoboken (2002)
2. Carpenter, J., Kenward, M.: Multiple Imputation and its Applications. Wiley, Hoboken (2013)
3. Guo, S., Fraser, M.W., Analysis, P.S.: Statistical Methods and Applications, 2nd edn. Sage, Thousand Oaks (2015)
4. Rosenbaum, P.R., Rubin, D.B.: The central role of the propensity score in observational studies for causal effects. Biometrika **70**(1), 41–55 (1983)
5. Bishop, C.M.: Pattern Recognition and Machine Learning. Springer, Heidelberg (2006)
6. Rosenbaum, P.R.: Design of Observational Studies. Springer, Heidelberg (2010)
7. LeCun, Y., Bottou, L., Benjio, Y., Haffner, P.: Gradient-based learning applied to document recognition. Proc. IEEE **86**(11), 2278–2324 (1998)
8. Sainath, T.N., Weiss, R.J., Wilson, K.W., Vinyals, O.: Learning the speech front-end with raw wave from CLDNNs. In: Proceedings of INTERSPEECH (2015)

9. Lei, T., Barzilay, R., Jaakkola, T.: Molding CNNs for text: non-linear, non-consecutive convolutions. In: Proceedings of Empirical Methods on Natural Language Processing (EMNLP) (2015)
10. Rubin, D.B.: Matched Sampling for Causal Effects. Cambridge University Press, Cambridge (2006)

Analysis of the DNN-kWTA Network Model with Drifts in the Offset Voltages of Threshold Logic Units

Ruibin Feng[1], Chi-Sing Leung[1(✉)], and John Sum[2]

[1] Department of Electronic Engineering,
City University of Hong Kong, Kowloon Tong, Hong Kong
rfeng4-c@my.cityu.edu.hk, eeleungc@cityu.edu.hk
[2] Institute of Technology Management,
National Chung Hsing University Taichung, Taichung, Taiwan
pfsum@dragon.nchu.edu.tw

Abstract. The structure of the dual neural network-based (DNN) k-winner-take-all (kWTA) model is much simpler than that of other kWTA models. Its convergence time and capability under the perfect condition were reported. However, in the circuit implementation, the threshold levels of the threshold logic units (TLUs) in the DNN-kWTA model may have some drifts. This paper analyzes the DNN-kWTA model under the imperfect condition, where there are some drifts in the threshold level. We show that given that the inputs are uniformly distributed in the range of $[0, 1]$, the probability that the DNN-kWTA model gives the correct output is greater than or equal to $(1 - 2\Delta)^n$, where Δ is the maximum drift level. Besides, we derive the formulas for the average convergent time and the variance of the convergent time under the drift situation.

Keywords: Winner-take-all · Dual neural network · Threshold logic unit · Convergence

1 Introduction

In the winner-take-all (WTA) problem, we aim at finding out the largest number from n numbers [1,2]. The WTA problem can be generalized to the kWTA problem, in which we need to find out the k largest inputs from n inputs. Among many models, the dual neural network (DNN)-based kWTA [3] is with the simplest structure. It contains $2n + 1$ connections only, while other kWTA models usually require n^2 connections. Recently, the stability was studied through in [3]. Besides, the bounds on its convergence time was studied in [4] and the analytical equations for the exact convergence time was derived in [5].

The DNN-kWTA model is suitable for hardware implementation because of its simple structure. However, the circuit implementation can never be perfect [6–9]. For instance, thermal noise [8,10] may be injected into the model. In [10],

A. Hirose et al. (Eds.): ICONIP 2016, Part IV, LNCS 9950, pp. 270–278, 2016.
DOI: 10.1007/978-3-319-46681-1_33

the dynamic behavior of the DNN-kWTA model under the noisy condition was reported. Another imperfect condition is that the offset voltages of the threshold logic units (TLUs) in the DNN-kWTA model may not be exactly equal to zero, i.e., there may be some drifts in the threshold level. Practically, we may not know the exact values of the drifts or the distribution of the drifts. We probably know the range of the drifts only.

This paper studies the behavior of the drifted DNN-kWTA model. We assume that we only know the maximum drift level, Δ. We first present a sufficient condition for checking whether the drifted model can produce the correct outputs or not. With this sufficient condition, we have an efficient method to study the successful rate of producing the correct outputs. Furthermore, for uniformly distributed inputs in the range of $[0, 1]$, we derive a formula to estimate that the probability that the model produces the correct outputs is greater than or equal to $(1-2\Delta)^n$. Besides, we derive some formulae to describe the convergence time under the drift situation. With the convergence time results, we can obtain the guideline of the sampling time for obtaining the kWTA result.

Section 2 presents the background information. Section 3 studies how the drift situation the performance (the chance to produce the correct outputs). Section 4 presents how the drift the convergence time. The paper is then concluded in Sect. 5.

2 Background

The DNN-kWTA model, shown in Fig. 1, has n input-output nodes and one hidden node. The n input-output receives the n inputs, denoted $\{u_1, \cdots, u_n\}$, and the signal y from the hidden node. The outputs of the n input-output nodes are denoted as $\{x_1, \cdots, x_n\}$. Its dynamics are governed by

$$\epsilon \frac{dy}{dt} = \sum_{i=1}^{n} x_i(t) - k, \tag{1}$$

$$x_i(t) = g(u_i - y), \text{ for } i = 1, \cdots, n, \tag{2}$$

where ϵ is the characteristic time of the circuit. For simplicity, we set $\epsilon = 1$. The function $g(s)$ is a threshold logic unit function, given as:

$$g(s) = \begin{cases} 1 \text{ if } s \geq 0, \\ 0 \text{ otherwise.} \end{cases} \tag{3}$$

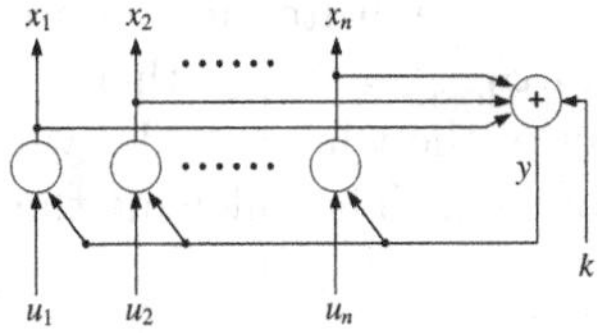

Fig. 1. Structure of a DNN-kWTA network.

Without loss of generality, we assume that the inputs are all distinct and bounded between 0 and 1. Let $\{u_{\kappa_1}, \cdots, u_{\kappa_n}\}$ be the sorted inputs in ascending order and $\{\kappa_1, \cdots, \kappa_n\}$ be the corresponding sorted index list. Furthermore, let $\{x_{\kappa_1}, \cdots, x_{\kappa_n}\}$ be the corresponding outputs. By (1) and (2), $y(t)$ converges in finite time. Define two index sets: $\mathcal{S}_{\text{loss}} = \{\kappa_1, \cdots, \kappa_{n-k}\}$ and $\mathcal{S}_{\text{win}} = \{\kappa_{n-k+1}, \cdots, \kappa_n\}$. Also, let y^* be the state value of y at equilibrium, where $y^* \in (u_{\kappa_{n-k}}, u_{\kappa_{n-k+1}}]$ and let x_i^* be the state value of x_i at equilibrium. For the DNN-kWTA model without drifts, at equilibrium, $x_i^* = 1$ if $i \in \mathcal{S}_{\text{win}}$, and $x_i^* = 0$ if $i \in \mathcal{S}_{\text{loss}}$. The above equation means that only the k outputs $\{x^*_{\kappa_{n-k+1}}, \cdots, x^*_{\kappa_n}\}$ with the k largest inputs are equal to 1. Other $n-k$ outputs are equal to 0.

3 DNN-*k*WTA with Drift

The drifts in the offset voltage of TLUs may cause the unknown behaviors in the outputs x_i's. Unfortunately, it is difficult to acquire any information about the exact drift values, except the range of the drifts. Denoting the drift of the ith node as δ_i, the network dynamics can be expressed as

$$\text{state equation } \frac{d\tilde{y}}{dt} = \sum_{i=1}^{n} \tilde{x}_i - k \tag{4}$$

$$\text{output equation } \tilde{x}_i = g(u_i + \delta_i - \tilde{y}), \tag{5}$$

where $\delta_i \in [-\Delta, \Delta]$ is the drift level, and Δ is the maximum drift level. We can interpret that we have a new set of inputs, namely the **drifted inputs**, given by

$$\tilde{u}_i = u_i + \delta_i. \tag{6}$$

Let $\{\tilde{u}_{\pi_1}, \cdots, \tilde{u}_{\pi_n}\}$ be the sorted list of $\tilde{u}_i$'s in ascending order, and $\{\pi_1, \cdots, \pi_n\}$ be the sorted index list. Furthermore, let $\{\tilde{x}_{\pi_1}, \cdots, \tilde{x}_{\pi_n}\}$ be the corresponding outputs. Define two new index sets, given by $\tilde{\mathcal{S}}_{\text{loss}} = \{\pi_1, \cdots, \pi_{n-k}\}$ and $\tilde{\mathcal{S}}_{\text{win}} = \{\pi_{n-k+1}, \cdots, \pi_n\}$. With this interpretation, one can follow the methods in [4] to prove that the drifted model converges within finite time too. That is, $\tilde{y}(t)$ converges to $\tilde{y}^*$ in finite time, where $\tilde{y}^* \in (\tilde{u}_{\pi_{n-k}}, \tilde{u}_{\pi_n-k+1}]$. For the drifted DNN-$k$WTA model, at equilibrium, we have $\tilde{x}_i^* = 1$ if $i \in \tilde{\mathcal{S}}_{\text{win}}$, and $\tilde{x}_i^* = 0$ if $i \in \tilde{\mathcal{S}}_{\text{loss}}$.

The above equation means that only the k outputs $\{\tilde{x}_{\pi_{n-k+1}}, \cdots, \tilde{x}_{\pi_n}\}$ are equal to 1. Other $n-k$ outputs are equal to 0. With the effect of drifts, there is one critical question: whether the drifted model produces the correct outputs. Clearly, if $\tilde{\mathcal{S}}_{\text{win}} = \mathcal{S}_{\text{win}}$ and $\tilde{\mathcal{S}}_{\text{loss}} = \mathcal{S}_{\text{loss}}$, then the set of the winner nodes under drift is equal to the set of the winner nodes without drift, i.e., the drifted model produces the correct outputs. The following theorem gives us the sufficient condition for producing the correct outputs.

Theorem 1. *Given the maximum drift level Δ, if $u_{\kappa_{n-k+1}} - u_{\kappa_{n-k}} > 2\Delta$, then the network produces the correct output.*

Proof. In the proof, we need to show that the winner set of the drifted model is equal to that of the model without drift, and that the loser set of the drifted model is equal to that of the model without drift. That means, we need to show

$$\tilde{x}^*_{\kappa_i} = \begin{cases} 1 \text{ for } i = n-k+1, \cdots, n, \\ 0 \text{ for } i = 1, \cdots, n-k. \end{cases} \tag{7}$$

For $i = 1, \cdots, n-k$, we have

$$\tilde{u}_{\kappa_i} = u_{\kappa_i} + \delta_{\kappa_i} \leq u_{\kappa_i} + \Delta \leq u_{\kappa_{n-k}} + \Delta. \tag{8}$$

For $i = n-k+1, \cdots, n$, we have

$$u_{\kappa_{n-k}} - \Delta \leq u_{\kappa_i} - \Delta \leq u_{\kappa_i} + \delta_{\kappa_i} = \tilde{u}_{\kappa_i}. \tag{9}$$

Now, if $u_{\kappa_{n-k+1}} - u_{\kappa_{n-k}} > 2\Delta$, then (8) and (9) becomes

$$u_{\kappa_i} + \delta_{\kappa_i} \leq u_{\kappa_{n-k}} + \Delta < u_{\kappa_{n-k+1}} - \Delta, \text{ for } i = 1, \cdots, n-k, \tag{10}$$

$$u_{\kappa_{n-k}} + \Delta < u_{\kappa_{n-k+1}} - \Delta \leq u_{\kappa_i} + \delta_{\kappa_i} \text{ for } i = n-k+1, \cdots, n. \tag{11}$$

We do not know the exact ordering in $\{u_{\kappa_1} + \delta_{\kappa_1}, \cdots, u_{\kappa_{n-k}} + \delta_{\kappa_{n-k}}\}$ and the exact ordering in $\{u_{\kappa_{n-k+1}} + \delta_{\kappa_{n-k+1}}, \cdots, u_{\kappa_n} + \delta_{\kappa_n}\}$ **Note that the two orderings are defined by π_i's rather than κ_i's.** From (10) and (11), $\{u_{\kappa_1} + \delta_{\kappa_1}, \cdots, u_{\kappa_{n-k}} + \delta_{\kappa_{n-k}}\}$ are the $n-k$ smallest drifted inputs, i.e., $\tilde{x}^*_{\kappa_i} = 0$ for $i = 1, \cdots, n-k$, and $\{u_{\kappa_{n-k+1}} + \delta_{\kappa_{n-k+1}}, \cdots u_{\kappa_n} + \delta_{\kappa_n}\}$ are the k largest drifted inputs from $\{\tilde{u}_1, \cdots, \tilde{u}_n\}$, i.e., $\tilde{x}^*_{\kappa_i} = 1$ for $i = n-k+1, \cdots, n$. The proof is complete. ∎

Theorem 1 gives us a fast way to study the performance of the network instead of simulating the neural dynamics. The advantage of using Theorem 1 is that we do not need to know the density function of the drifts. For example, when we have a number of sets of inputs, we can apply Theorem 1 to estimate the successful rate of producing correct output without simulating the dynamics and without knowing the density function of the drifts.

To illustration the application of Theorem 1, we generate a number of sets of inputs with the Beta distribution: $\text{Beta}_{a,b}(u) = \frac{\Gamma(a+b))}{\Gamma(a)\Gamma b} u^{a-1}(1-u)^{b-1}$, where $\Gamma(\cdot)$ is the Gamma function. The Beta distribution is a family of distribution functions defined on the interval between 0 and 1. We set a and b to 2. We consider three settings: $n = 5$ with $k = 3$, $n = 11$ with $k = 6$, and $n = 21$ with $k = 11$. For each setting, we generate 100,000 sets of inputs. We then added the drifts to the inputs. The drifts d_i's are discrete random variables, They are equal to $\pm\Delta$ with equal probability.

We then use Theorem 1 to measure the successful rate of the drifted DNN-kWTA model. We shows two successful rates. The "estimated successful rate" (ESR) is based on Theorem 1. The "actual successful rate" (ASR) is directly measured from of $\tilde{u}_i$'s. The results are summarized in Fig. 2. It can be seen that the ASR is greater than the ESR because Theorem 1 gives the sufficient

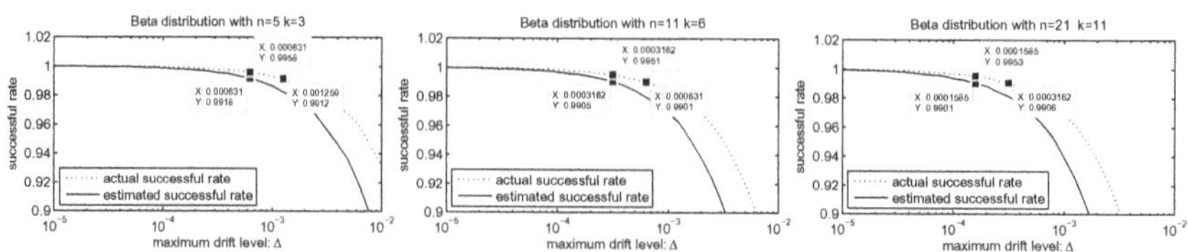

Fig. 2. The successful rate: the inputs are with Beta distribution.

condition. For instance, for $n = 11$ with $k = 6$, when Δ is equal to 0.0003162, the ASR is 0.9951, while the ESR is 0.9905. To achieve a high successful rate, saying 0.99, the maximum drift value should be small. For instance, for $n = 11$ with $k = 6$, the Δ from the ESR should not be smaller than 0.0003162, while the Δ from the ASR should not be smaller than 0.000631. Although the "actual successful rate" provides a better guideline, the advantages of our approach is that we do not need to know the actual density function of the drifts and the exact drift values.

Although Theorem 1 provides us an easy way to estimate the successful rate, we still need to generate a number of data sets. When the inputs are independently uniformly distributed in $[0, 1]$, we can have a formula to estimate the probability that the model produces the correct outputs. The formula is summarized in the following theorem.

Theorem 2. *Assume that the inputs $\{u_1, \cdots, u_n\}$ are independently uniformly distributed in $[0, 1]$. The probability that the network works properly is greater than $(1 - 2\Delta)^n$.*

Proof. Based on Theorem 1, we know $u_{\kappa_{n-k+1}} - u_{\kappa_{n-k}} > 2\Delta$ is the sufficient condition for the network generating correct outputs. Define $\phi = u_{\kappa_{n-k+1}} - u_{\kappa_{n-k}}$. According to order statistics theories [12], the probability density function is given by $\zeta(\phi) = n(1 - \phi)^{n-1}$. Hence, the probability of the sufficient condition is

$$\text{Prob}(u_{\kappa_{n-k+1}} - u_{\kappa_{n-k}} > 2\Delta) = \int_{2\Delta}^{1} n(1 - \phi)^{n-1} d\phi = (1 - 2\Delta)^n. \tag{12}$$

The proof is complete. ■

From probability theory, a non-unform distribution can be transformed into a uniform one. Besides, after the transformation, the ordering of the transformed inputs is identical to that of the original inputs. Hence Theorem 2 can be used to estimate the successful rate when we have a compressor to perform the transformation.

To illustration the application of Theorem 2, we use the same setting of n and k as in Theorem 1. For each setting, we generate 100,000 sets of inputs with uniform distribution. We then added the drifts to the inputs. The drifts d_i's are discrete random variables. They are equal to $\pm\Delta$ with equal probability.

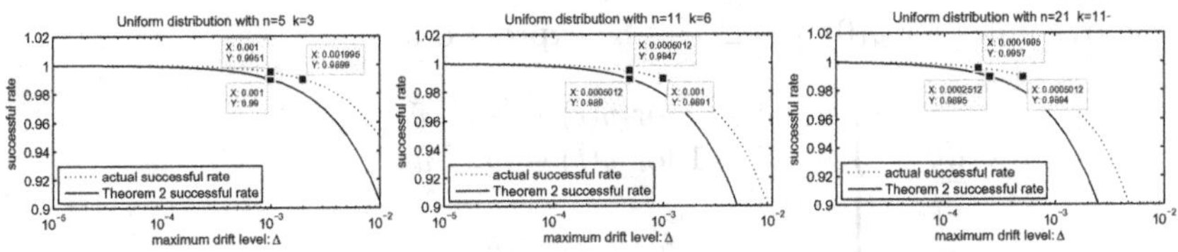

Fig. 3. The successful rate: the inputs are with uniform distribution.

We then use Theorem 2 to estimate the successful rate. We shows two successful rates. The "Theorem 2 successful rate" (ESR) is based on Eq. (12) in Theorem 2. The "actual successful rate" (ASR) is directly measured from the sorted list of $\{\tilde{u}_1, \cdots, \tilde{u}_n\}$. The results are summarized in Fig. 3. It can be seen that the ASR is greater than the ESR because Theorem 2 is based on the sufficient condition. To achieve a high successful rate of producing correct outputs, saying 0.99, the Δ should be small. For instance, for $n = 21$ with $k = 11$, the maximum drift from the ESR should not be greater than 0.000251, while the Δ from the AST should not be greater than 0.000501. Although the ASR provides a better guideline on the maximum drift level, the advantage of using Theorem 2 is that we do not need to generate a number of sets of inputs.

4 Convergence Time

This section analyzes the convergence time of the drifted DNN-kWTA model. With the convergence time result, we can obtain the guideline of the sampling duration for sampling the result of the kWTA process. We consider the starting condition of $y(0) = -\Delta$. First of all, we have an estimate on the position $\tilde{u}_{\pi_i}$, given by the following theorem.

Theorem 3. *Given the maximum drift level Δ, we have*

$$\tilde{u}_{\pi_i} \geq u_{\kappa_i} - \Delta, \textit{ and } \tilde{u}_{\pi_i} \leq u_{\kappa_i} + \Delta. \tag{13}$$

Proof. First at all, we have

$$\begin{aligned}
u_{\kappa_i} + \delta_{\kappa_i} &\geq u_{\kappa_i} - \Delta, \\
u_{\kappa_{i+1}} + \delta_{\kappa_{i+1}} &\geq u_{\kappa_{i+1}} - \Delta \geq u_{\kappa_i} - \Delta, \\
\vdots \quad \vdots \quad &\vdots \\
u_{\kappa_n} + \delta_{\kappa_n} &\geq u_{\kappa_n} - \Delta \geq u_{\kappa_{n-1}} - \Delta \geq \cdots\cdots \geq u_{\kappa_i} - \Delta.
\end{aligned}$$

Hence, there are at least $(n - i + 1)$ drifted inputs "$\tilde{u}_i = u_i + \delta_i$" 's are greater than $u_{\kappa_i} - \Delta$. That means, we have $\tilde{u}_{\pi_i} \geq u_{\kappa_i} - \Delta$. Similarly, we can prove that $\tilde{u}_{\pi_i} \leq u_{\kappa_i} + \Delta$. The proof is complete. ■

The dynamics for $y(0) = -\Delta$ can be expressed as

$$\frac{dy}{dt} = \begin{cases} n-k & \text{for } y(t) \in [-\Delta, \tilde{u}_{\pi_1}) \\ n-k-1 & \text{for } y(t) \in [\tilde{u}_{\pi_1}, \tilde{u}_{\pi_2}) \\ \cdots & \\ 1 & \text{for } y(t) \in [\tilde{u}_{\pi_{n-k}}, \tilde{u}_{\pi_{n-k-1}}) \end{cases}. \tag{14}$$

Hence the convergence time is given by

$$T_{-\Delta} = \frac{\tilde{u}_{\pi_1}+\Delta}{n-k} + \frac{\tilde{u}_{\pi_2}-\tilde{u}_{\pi_1}}{n-k-1} + \cdots + \frac{\tilde{u}_{\pi_{n-k-1}}-\tilde{u}_{\pi_{n-k-2}}}{2} + \frac{\tilde{u}_{\pi_{n-k}}-\tilde{u}_{\pi_{n-k-1}}}{1}. \tag{15}$$

Although we do not know the exact values of the drifts, based on Theorem 3, we can replace the sorted ordering indices of the drifted inputs with the original sorted ordering indices of the original inputs, given by

$$\begin{aligned} T'_{-\Delta} \leq & \frac{u_{\kappa_1}-\Delta+\Delta}{n-k} + \frac{u_{\kappa_1}-(u_{\kappa_1}-\Delta)}{n-k} + \frac{u_{\kappa_2}-u_{\kappa_1}}{n-k-1} \\ & + \frac{u_{\kappa_3}-u_{\kappa_2}}{n-k-2} + \cdots + \frac{u_{\kappa_{n-k-1}}-u_{\kappa_{n-k-2}}}{2} + \frac{u_{\kappa_{n-k}}-u_{\kappa_{n-k-1}}}{1} \end{aligned} \tag{16}$$

$$\leq \frac{u_{\kappa_1}-0}{n-k} + \frac{u_{\kappa_2}-u_{\kappa_1}}{n-k-1} \cdots + \frac{u_{\kappa_{n-k}}-u_{\kappa_{n-k-1}}}{1} + \frac{\tilde{u}_{\pi_{n-k}}-(u_{\kappa_{n-k}}-\Delta)}{1}. \tag{17}$$

From Theorem 3, as $\tilde{u}_{\pi_{n-k}} \leq u_{\kappa_{n-k}} + \Delta$, we have

$$T_{-\Delta} \leq T'_{-\Delta} = \sum_{i=1}^{n-k} \frac{u_{\kappa_i} - u_{\kappa_{i-1}}}{n-k-i+1} + 2\Delta, \text{ where } u_{\kappa_0} = -\Delta. \tag{18}$$

The above equation gives us a bound on the convergence time under the drift situation.

From the order statistics, if the inputs are uniformly distributed between 0 and 1, we can obtain the mean and variance of $T'_{-\Delta}$, given by

$$E[T'_{-\Delta}] = \frac{1}{n+1}\sum_{i=1}^{n-k}\frac{1}{i} + 2\Delta, \tag{19}$$

$$Var[T'_{-\Delta}] = \frac{1}{(n+1)(n+2)}\sum_{i=1}^{n-k}\frac{1}{i^2} - \frac{1}{(n+1)^2(n+2)}(\sum_{i=1}^{n-k}\frac{1}{i})^2. \tag{20}$$

To illustration the application of (19) and (20), an experiment was conducted with $n = \{20, 40, \cdots, 1000\}$, $k = 0.5n$ and $\epsilon = 1$. We consider two maximum drift values: $\Delta = \{0.0005, 0.001\}$. For each n, we generate 10,000 sets of inputs. We then measure the actual convergence time and obtain the statistics of the calculated values. Based on these, we can also construct the three sigma confident levels of the convergence time.

Apart from the measurement, we can use our expressions, (27) and (28), to obtain the bounds on the convergence time. The results are shown in Fig. 4.

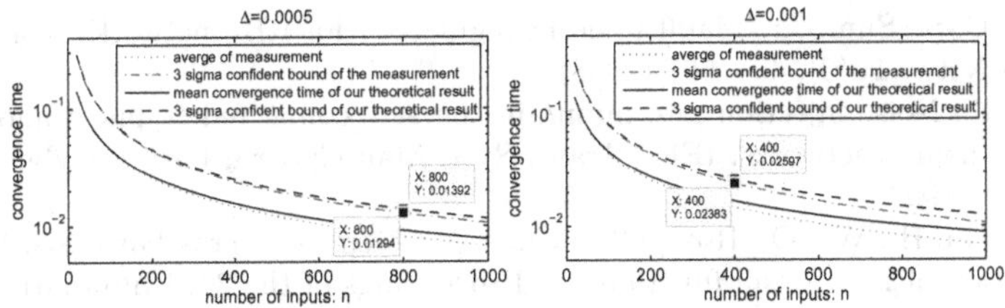

Fig. 4. Convergence time: the inputs are with uniform distribution.

From the curves, we can estimate the sampling duration of the drifted model, For instance, for n = 400 and $\Delta = 0.001$, with the three sigma confident level of the measurement result, the sampling duration is 0.02383. With the three sigma confident level of our theoretical result, the sampling duration can be reduced to 0.02597. Clearly, our theoretical result can give a accurate sampling duration. **Note that in our theoretical result we do need to generate a number of sets of inputs.** What we need is the maximum drift level only.

5 Conclusion

This paper theoretically studied the behaviors of DNN-kWTA model under the drift situation, where there are some drifts in the offset voltage of the TLUs. In our analysis, we only assume that the range of the drift is given. The sufficient condition (Theorem 1) for the network working properly was given. Based on it, the lower bound of successful rate is provided in Theorem 2. Finally we analyzed the convergence time of the drifted DNN-kWTA model (see (27) and (28)). Several experiments were carried to verify our theoretical results. One of the further directions is to study the convergence time from any initial point.

References

1. Lazzaro, J., Ryckebusch, S., Mahowald, M.A., Mead, M.A.: Winner-take-all networks of O(N) complexity. In: Advances in Neural Information Processing Systems, vol. 1, pp. 703–711 (1989)
2. Sum, J., Leung, C.-S., Tam, P., Young, G., Kan, W., Chan, L.W.: Analysis for a class of winner-take-all model. IEEE Trans. Neural Netw. **10**, 64–71 (1999)
3. Hu, X., Wang, J.: An improved dual neural network for solving a class of quadratic programming problems and its k-winners-take-all application. IEEE Trans. Neural Netw. **19**, 2022–2031 (2008)
4. Wang, J.: Analysis and design of a k-winners-take-all model with a single state variable and the heaviside step activation function. IEEE Trans. Neural Netw. **21**(9), 1496–1506 (2010)
5. Xiao, Y., Liu, Y., Leung, C.-S., Sum, J., Ho, K.: Analysis on the convergence time of dual neural network-based. IEEE Trans. Neural Netw. Learn. Syst. **23**(4), 676–682 (2012)

6. Leung, C.-S., Sum, J.: A fault tolerant regularizer for RBF networks. IEEE Trans. Neural Netw. **19**(3), 493–507 (2008)
7. Wang, L.: Noise injection into inputs in sparsely connected Hopfield and winner-take-all neural networks. IEEE Trans. Syst. Man Cybern. Part B Cybern. **27**(5), 868–870 (1997)
8. Hu, M., Li, H., Wu, Q., Rose, G.S., Chen, Y.: Memristor crossbar based hardware realization of BSB recall function. In: Proceedings of the 2012 International Joint Conference on Neural Networks (IJCNN) (2012)
9. He, J., Zhan, S., Chen, D., Geiger, R.L.: Analyses of static and dynamic random offset voltages in dynamic comparators. IEEE Trans. Circuits Syst. I **56**(5), 911–919 (2009)
10. Sum, J., Leung, C.-S., Ho, K.: Effect of input noise and output node stochastic on Wang's kWTA. IEEE Trans. Neural Netw. Learn. Syst. **24**(9), 1472–1478 (2013)
11. Bowling, S.R., Khasawneh, M.T., Kaewkuekool, S., Cho, B.R.: A logistic approximation to the cumulative normal distribution. J. Ind. Eng. Manag. **2**(1), 114–127 (2009)
12. Arnold, B.C., Balakrishnan, N., Nagaraja, H.N.: A First Course in Order Statistics (Classics in Applied Mathematics). SIAM (2008)

Efficient Numerical Simulation of Neuron Models with Spatial Structure on Graphics Processing Units

Tsukasa Tsuyuki[1(✉)], Yuki Yamamoto[2], and Tadashi Yamazaki[1,3]

[1] Graduate School of Informatics and Engineering, The University of Electro-Communications, Tokyo, Japan
t.tsuyuki@uec.ac.jp

[2] Faculty of Medicine, Tokyo Medical and Dental University, Tokyo, Japan

[3] RIKEN Brain Science Institute, Neuroinformatics Japan Center, Saitama, Japan

Abstract. Computer simulation of multi-compartment neuron models is difficult, because writing the computer program is tedious but complicated, and it requires sophisticated numerical methods to solve partial differential equations (PDEs) that describe the current flow in a neuron robustly. For this reason, dedicated simulation software such as NEURON and GENESIS have been used widely. However, these simulators do not support hardware acceleration using graphics processing units (GPUs). In this study, we implemented a conjugate gradient (CG) method to solve linear equations efficiently on a GPU in our own software. CG methods are known much faster and more efficient than the Gaussian elimination, when the matrix is huge and sparse. As a result, our software succeeded to carry out a simulation of Purkinje cells developed by De Schutter and Bower (1994) on a GPU. The GPU (Tesla K40c) version realized 3 times faster computation than that a single-threaded CPU version for 15 Purkinje cells.

Keywords: Computer simulation · Spatial model · Graphics processing units · Conjugate gradient method

1 Introduction

Integrate-and-fire model and Hodgkin-Huxley model are known as basic mathematical models that describe the change of membrane potential. In these models, the structure of neurons is neglected; a neuron is regarded as a material point. On the other hand, spatial neuron models consider the morphological structure such as soma, dendrites and axons. Spatial neurons are composed of equipotential segments called "compartments" connected with axial resistances. A change of membrane potential is described by a partial differential equation (PDE) called a cable equation. In general, a cable equation is discretized both in time (implicit) and space. Then, linear simultaneous equation are obtained, and we solve them to compute a membrane potential for each compartment and time step [1].

A. Hirose et al. (Eds.): ICONIP 2016, Part IV, LNCS 9950, pp. 279–285, 2016.
DOI: 10.1007/978-3-319-46681-1_34

Because writing a program to solve the equations is difficult, dedicated simulation software such as NEURON [2] and GENESIS [3] are used. However, they have two issues. First, they do not support modern computing accelerators such as graphics processing units (GPUs). They employ a data structure called array of structure (AoS), which typically compromises the parallel computing capability of accelerators. Second, they use the basic numerical method called Gaussian elimination to solve the equations. Gaussian elimination is known that having computational complexity of $O(N^3)$ in the case of N th-order dense matrices. Thus, the computational time increases in the order of N^3 as the size of the coefficient matrix N increases. Moreover, Gaussian elimination is difficult to compute in parallel on accelerators. On the other hand, there is a numerical method called a conjugate gradient (CG) method [4] to solve the equations. A CG method is a series of dot products and matrix-vector products. Also, the coefficient matrix does not change during iteration until convergence in a CG method. By these reasons, CG methods are known much faster and more efficient than Gaussian elimination, when the coefficient matrix is huge and sparse.

The purpose of this study is to develop our own software for numerical simulation of spatial models on GPUs. We used programming library called CUDA C [5] which is a kind of GPGPU technology and based on C language. We implemented the CG method on a GPU because the coefficient matrix obtained by discretizing a cable equation is sparse and symmetrical positive definite.

2 Methods

2.1 Cable Equation

Figure 1 illustrates how to "compartmentalize" a neuron model with spatial structure. A neuron is divided into many equipotential segments called "compartments" while connecting neighboring segments with a resistance (axial resistance). For example, a one-dimensional dendrite is compartmentalized by splitting it into many pieces of small compartments with the length of Δx. Each compartment is composed of the membrane capacitance and resistance, and the external current including synaptic current. Neighboring compartments are connected by an axial resistance (Fig. 2).

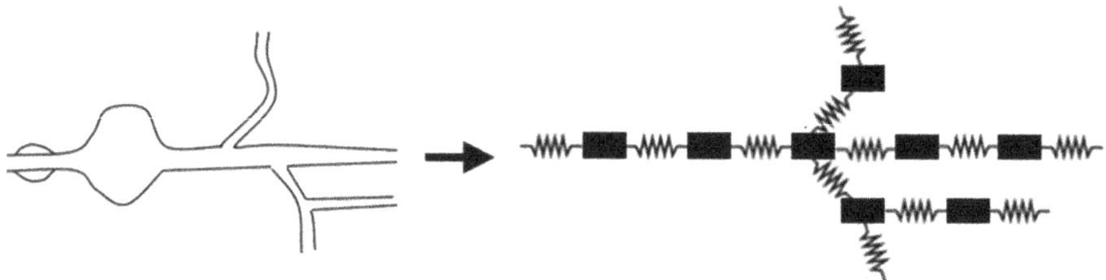

Fig. 1. Spatial discretization of a neuron model. Left: an exemplified model neuron. Right: the compartmental representation. A black square represents a compartment. Compartments are connected with axial resistances.

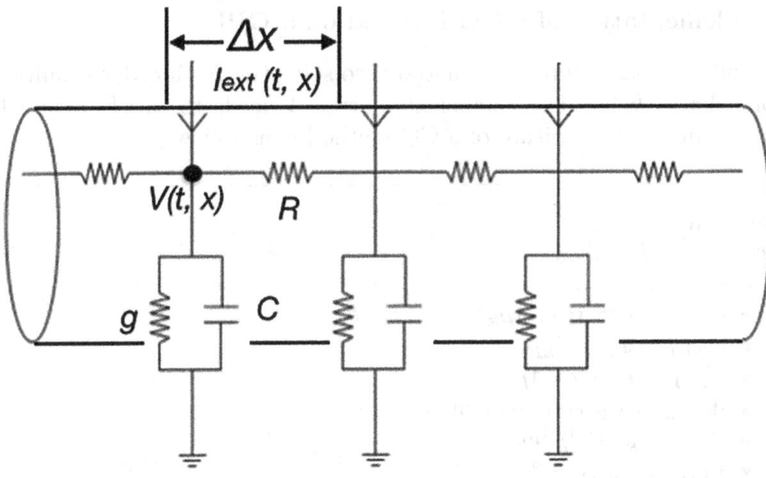

Fig. 2. A dendrite is regarded as a cable. It is spatially separated with multiple compartments with the same length Δx. Parameters are $V(t,x)$, the membrane potential in location x and time t; R, the axial resistance of a cable; g, the membrane conductance; C, the membrane capacitance; I_{ext}, the current injected externally.

A change of membrane potential on a cable is described by a partial differential equation (PDE) called a cable equation [1]:

$$C\frac{\partial V}{\partial t} = \frac{a}{2R}\frac{\partial^2 V}{\partial x^2} - gV + I_{\mathrm{ext}} \tag{1}$$

where C is the membrane capacitance, g is the leak conductance, R is the axial resistance, a is the surface area of a cable, I_{ext} is the external current including synaptic currents, $V = V(t,x)$ is the membrane potential in location x and time t. Equation (1) is discretized by taking the backward difference for the time (Δt) and the second-order central difference for the space (Δx), and Eq. 2 is obtained.

$$-\sigma V_{i-1}(t+\Delta t) + (1+2\sigma+\gamma)V_i(t+\Delta t) - \sigma V_{i+1}(t+\Delta t) = V_i(t) \tag{2}$$

where $V_i(t)$ is the membrane potential of compartment $i(=0,1,\cdots,N-1)$ and time t, $\sigma = a\Delta t/2RC(\Delta x)^2$, and $\gamma = g\Delta t/C$. Equation (2) is represented in a matrix form:

$$\boldsymbol{A}\boldsymbol{x} = \boldsymbol{b} \tag{3}$$

where

$$A = \begin{pmatrix} (1+2\sigma+\gamma) & -\sigma & 0 & \cdots \\ -\sigma & (1+2\sigma+\gamma) & -\sigma & \cdots \\ 0 & -\sigma & (1+2\sigma+\gamma) & \cdots \\ \vdots & \vdots & \vdots & \ddots \end{pmatrix}$$

$$\boldsymbol{x} = (\cdots, V_{i-1}(t+\Delta t), V_i(t+\Delta t), V_{i+1}(t+\Delta t), \cdots)^T$$
$$\boldsymbol{b} = (\cdots, V_{i-1}(t), V_i(t), V_{i+1}(t), \cdots)^T.$$

2.2 Implementation of a CG Method on a GPU

CG methods are basic numerical methods to solve the N-th order simultaneous equations $A\boldsymbol{x} = \boldsymbol{b}$ [4]. The coefficient matrix A needs to satisfy symmetrical positive definite. An algorithm of a CG method is as follows:

- give $\boldsymbol{x_0}$
- $\boldsymbol{r_0} = \boldsymbol{p_0} = \boldsymbol{b} - A\boldsymbol{x_0}$
- for $k = 0, 1, 2, ...$
  - $\alpha_k = ||\boldsymbol{r_k}||^2/(\boldsymbol{p_k}, A\boldsymbol{p_k})$
  - $\boldsymbol{x_{k+1}} = \boldsymbol{x_k} + \alpha_k \boldsymbol{p_k}$
  - $\boldsymbol{r_{k+1}} = \boldsymbol{r_k} - \alpha_k A\boldsymbol{p_k}$
  - if $||\boldsymbol{r_{k+1}}|| \leq \epsilon ||\boldsymbol{b}|| \rightarrow$ exit
  - $\beta_k = ||\boldsymbol{r_{k+1}}||^2/||\boldsymbol{r_k}||^2$
  - $\boldsymbol{p_{k+1}} = \boldsymbol{r_{k+1}} + \beta_k \boldsymbol{p_k}$

CG methods have two features. First, most of CG methods are composed of dot products and matrix-vector products. Second, the coefficient matrix does not change during iteration for convergence.

$$A = \begin{pmatrix} \mathbf{1} & 0 & \mathbf{3} & 0 & \mathbf{2} \\ 0 & \mathbf{1} & 0 & \mathbf{4} & \mathbf{6} \\ \mathbf{2} & \mathbf{5} & \mathbf{3} & 0 & 0 \\ 0 & 0 & \mathbf{7} & \mathbf{8} & 0 \\ 0 & \mathbf{1} & 0 & \mathbf{3} & \mathbf{9} \end{pmatrix}$$

val	1	3	2	1	4	6	2	5	3	7	8	1	3	9
col	0	2	4	1	3	4	0	1	2	2	3	1	3	4
row	0	3	6	9	11	14								

Fig. 3. CRS format (right) of an exemplified array (left). CRS format consists of three arrays. "val" array stores nonzero elements of the coefficient matrix A (size: the number of nonzero elements in A). "col" array stores column numbers of the nonzero elements (size: the number of nonzero elements in A). "row" array stores the positions of elements in val array that appear first in each row (size: $N + 1$).

Because neurons have tree-like morphological structure, most of the elements in matrix A is zero. These zero elements are not necessary. In this study, the sparse matrix was represented by the compressed row storage (CRS) format (Fig. 3). A dot product was implemented on one thread using shared memory. A matrix-vector product was implemented by using one thread per one row.

Moreover, a CG method was implemented by using a modern CUDA [5] technology called "Dynamic Parallelism", in which a kernel code can invoke other kernel codes. Specifically, a host code invokes a parent kernel code (one thread), which in turn invokes child kernel codes (many threads). Child kernel codes perform dot products and a matrix-vector product. A parent kernel code computes α_k, β_k and tests convergence. In this way, the entire algorithm of the

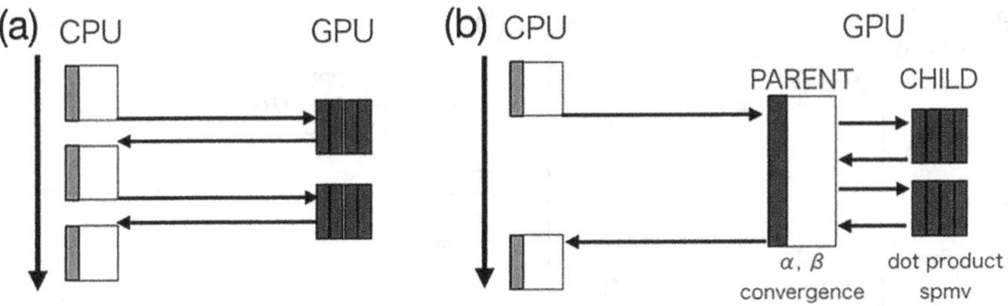

Fig. 4. Illustration of the flow of a CG method on a GPU without/with Dynamic Parallelism. (a) Without Dynamic Parallelism, communications between a CPU and a GPU must be made at every iteration for convergence. (b) With Dynamic Parallelism, a host code launches a parent kernel, which in turn launches child kernels. Child kernels compute dot products and a matrix-vector product. The parent kernel computes α_k, β_k and tests convergence.

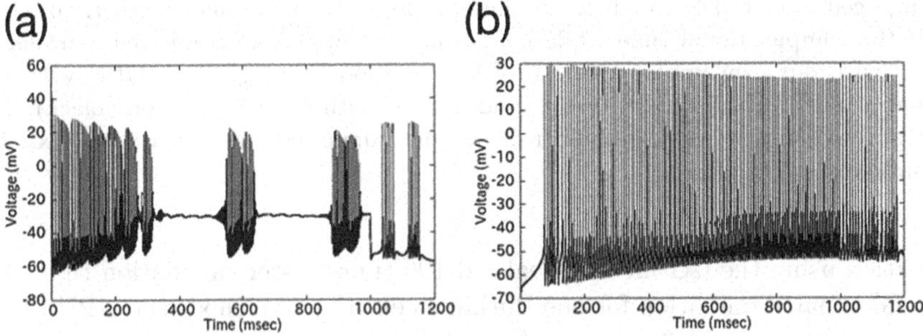

Fig. 5. Simulation results of the Purkinje cell model [6] with current injection of different amplitudes (a, 3.0nA; b, 0.5nA) by our software. The horizontal axis represents time (ms) and the vertical axis the membrane potential (mV). Our results match perfectly with the original ones calculated using GENESIS simulator [6].

CG method was implemented in a kernel code, which means that there is no need to return to the host code for each iteration (Fig. 4).

3 Results

We implemented simulation a program of Purkinje cells developed by De Schutter and Bower (1994) [6], which is composed of 1,600 compartments with 10 ion channels and 13 gate variables for each compartment. Figure 5 shows the membrane potential while varying the amplitude of the externally injected current to soma. These results matched almost perfectly with the results obtained using GENESIS simulator [6]. Thus, our software succeeded to carry out a simulation of Purkinje cells on a GPU.

We measured a computational time with temporal resolution of 20 microseconds for 5,000 iteration (= 100 milliseconds). First, in a single-threaded CPU, a

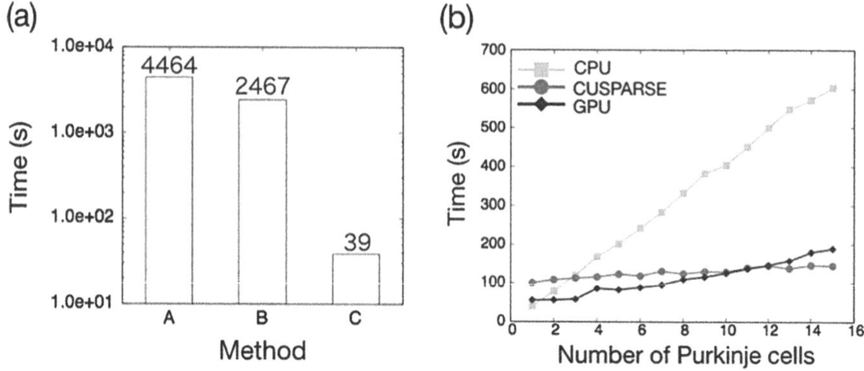

Fig. 6. (a) Comparison of computational time on a CPU with (A) Gaussian elimination, (B) CG method without CRS format, (C) CG method with CRS format. The vertical axis represents time (s) in a log scale spent to complete simulation of a single Purkinje cell model. The numbers on the bars indicate the actual number. (b) Scaling of the computational time while increasing the number of simulated neurons on a CPU with our own implementation of CG method (light gray), a GPU with CG method using NVIDIA library (gray), and a GPU with our CG method (black). The horizontal axis represents the number of neurons simulated, and the vertical axis the computational time (s).

simulation using the CG method realized 100 times faster calculation than that using Gaussian elimination for one Purkinje cell (Fig. 6a). Next, the GPU (Tesla K40c) version realized 3 times faster calculation than that a single-threaded CPU version for 15 Purkinje cells.

We also compared the performance of the CG method we implemented with these implemented using CUBLAS [7] and CUSPARSE [8] libraries developed by NVIDIA (Fig. 6b).

Figure 6a shows that a simulation using CRS format realized 80 times faster calculation than that not using it. This result means that the coefficient matrix obtained by discretizing a cable equation is a strong sparse matrix. That is why using CRS format accelerated the calculation. Figure 6b shows that the computational time increased as increasing the number of Purkinje cells, which suggests poor scaling property.

4 Discussion

To use the power of GPUs for efficient computer simulation of spatial neuron models, we have been developing our own software. A CG method combined with CRS format is a standard method in the field of general computational science, and owing to this technology, we achieved about 80 times improvement of the computational speed while keeping to obtain the same numerical results.

We conducted computer simulation of the cerebellar Purkinje cell model, which is composed of 1,600 compartments. We were able to reproduce the same numerical results with the previous ones. Typically, it is extremely difficult to reproduce numerical results of realistic, large-scale neuron models. This observation suggests that our software computes neuron dynamics quite accurately on a GPU.

Although the present version has demonstrated a good performance, further performance improvement will be expected as follows. First, in this study, we labeled compartments from soma to dendritic and axon terminals. The numbering, however, should be done in the opposite direction [9], which could make the connection matrix A more diagonal. Second, we employed CRS format for storing the matrix. CRS format makes the CG method more efficient, and this is appropriate for CPUs. However, in CRS format, coalesced memory access is no longer available, which is important for GPUs. For this purpose, we might use another format such as ELL [10].

Finally, our next target is to construct a network of spatial model neurons. Specifically, we will model the cerebellar granular layer network composed of granule cells and Golgi cells [11]. After the implementation, we will have a minimal functional circuit of the cerebellar cortex. This is our future work.

Acknowledgments. Part of this study was supported by JSPS KAKENHI Grant Number 26430009.

References

1. Koch, C., Segev, I.: Methods in Neuronal Modeling. MIT Press, Cambridge (1998)
2. Carnevale, N.T., Hines, M.L.: The NEURON Book. Cambridge (2005)
3. Bower, J.M., Beeman, D.: The Book of GENESIS: Exploring Realistic Neural Models with the GEneral NEural SImulation System, 2nd edn. Springer-Verlag, Heidelberg (1998)
4. Hestenes, M.R., Stiefel, E.: Methods of conjugate gradients for solving linear systems. J. Res. Nat. Bur. Stand. **49**, 409–436 (1952)
5. NVIDIA: CUDA C PROGRAMMING GUIDE (PG-02829-001_v7.5) (2015). http://docs.nvidia.com/cuda/pdf/CUDA_C_Programming_Guide.pdf
6. De Schutter, E., Bower, J.M.: An active membrane model of the cerebellar purkinje cell I. Simulation of current clamps in slice. J. Neurophys. **71**, 375–400 (1994)
7. NVIDIA: CUBLAS LIBRALY (DU-06702-001_v7.5) (2015). http://docs.nvidia.com/cuda/pdf/CUBLAS_Library.pdf
8. NVIDIA: CUSPARSE LIBRARY (DU-06709-001_v7.5). (2015) http://docs.nvidia.com/cuda/pdf/CUSPARSE_Library.pdf
9. Hines, M.: Efficient computation of branched nerve equations. Int. J. Bio-Med. Comput. **15**, 69–76 (1984)
10. Grimes, R., Kincaid, D., Young, D.: ITPACK 2.0 user's guide. Technical report CNA-150. U. Texas (1979)
11. Solinas, S., Nieus, T., D'Angelo, E.: A realistic large-scale model of the cerebellum granular layer predicts circuit spatio-temporal filtering properties. Front. Cell. Neurosci. **4**, 12 (2010)

A Scalable Patch-Based Approach for RGB-D Face Recognition

Nesrine Grati[1(✉)], Achraf Ben-Hamadou[2], and Mohamed Hammami[1]

[1] Multimedia InfoRmation Systems and Advanced Computing Laboratory (MIRACL), Sfax University, Road Sokra Km 3 BP 802, 3018 Sfax, Tunisia
grati.nesrine@gmail.com, mohamed.hammami@fss.rnu.tn
[2] Driving Assistance Research Center, Valeo Vision, 34 rue St-André Z.I. des Vignes, 93012 Bobigny, France
achraf.ben-hamadou@valeo.com

Abstract. This paper presents a novel approach for face recognition using low cost RGB-D cameras under challenging conditions. In particular, the proposed approach is based on salient points to extract local patches independently to the face pose. The classification is performed using a scalable sparse representation classification by an adaptive and dynamic dictionaries selection. The experimental results proved that the proposed algorithm achieves significant accuracy on three different RGB-D databases and competes with known approaches in the literature.

Keywords: Face recognition · RGB-D data · Sparse representation classification · Dynamic dictionary · Kinect

1 Introduction

Face recognition based on image and depth data has been substantially boosted these few years thanks to the ubiquity of consumer RGB-D cameras. Indeed, the performance of classical image-based face recognition methods are largely affected by imaging conditions. Conversely, 3D based approaches are less sensitive to lighting conditions and reliably describe the anatomical 3D face structure [6]. Nevertheless, these techniques still limited to bulky and costly 3D scanners. With the arrival of consumer RGB-D cameras (*e.g.*, MS Kinect, Asus Xtion Pro Live, *etc.*) simultaneously providing image and depth data, the new trend in the field is to take benefits of the complementary of these two modalities to improve face recognition systems.

This paper deals with face recognition from low-quality RGB-D data under arbitrary face pose variation. The proposed approach is totally independent from the face structure and semantic and just based on salient points detection to extract further local descriptor. The Sparse Representation Classification is introduced using adaptive and dynamic dictionaries selection which improves the recognition performances and speed-up the matching process.

The remaining of this paper is organized as follows. In Sect. 2 we summarize the works related to face recognition using RGB-D cameras and closely related to

A. Hirose et al. (Eds.): ICONIP 2016, Part IV, LNCS 9950, pp. 286–293, 2016.
DOI: 10.1007/978-3-319-46681-1_35

our contribution. Section 3 details the main contributions of the proposed face recognition approach. Then, Sect. 4 shows the experimental results to assess the performance of our approach on three different public databases and to compare with state-of-the-art systems. Finally, future works and some relevant perspectives are given in Sect. 5 to conclude this contribution.

2 Related Work

Human Face Analysis is an important in computer vision involving a wide range applications like of lip-reading [17], face recognition [4], speaker identification [16], and Human Machine Interaction [15], visual tracking [14], *etc.* With the arrival of the new RGB-D data, many works are developed to deal with the face recognition problem. In a recent work [12], the nose tip is manually detected and the facial scans are aligned with a generic face model using the Iterative Closest Point (ICP) algorithm to normalize the head orientation and generate a cononical frontal view for both image and depth data. Then, Sparse Representation Classification (SRC) is used separately on both modalities to perform the late fusion and face recognition. Conversely, Ciaccio *et al.* [4] used a large number of image sets in the gallery under different poses angles from a single RGB-D data to handle the pose variation issue. Also, the face pose is estimated via the detection and alignment of standard face landmarks (*e.g.*, eye corners, nose tip, *etc.*) in the images [19]. Each face is then represented using a set of extracted patches centered on the detected landmarks and described by a set of LBP descriptor, covariance of edge orientation, and pixel location and intensity derivative. The classification is then performed by computing distances between patch descriptors, inferring probabilities, and lately performing a Bayesian decision. In the same vein, [9] fits a 3D face model to the face data to generate a single 3D textured face model for each person in the gallery. The approach requires to estimate the pose for any new probe to be able to apply it to all 3D textured models in the gallery. This allows to generate 2D images by plan projection and then compute the LBP descriptor on the whole projected 2D images to perform the classification using an SRC algorithm. Goswami *et al.* [8] additionally combined image local descriptors to geometric attributes for computing face signatures. Indeed, face landmarks are first localized and projected in the depth maps to compute the Euclidean distances between them. Then, the random forest classifier is used for the identity classification. Other works require only accurate nose tip localization to align all the faces. The face pose is not directly estimated but still precise landmark estimation is necessary for alignment. For example, Xu *et al.* [5] propose a multimodal approach for face recognition under uncontrolled situations. A new local feature ELMDP (Enhanced Local Mixed Derivative Pattern), which is a mixed feature descriptor of different orders of local derivative patterns and local binary patterns, is proposed and it is respectively applied on 2D and depth image. For classification, Nearest Neighbor algorithm is used for the combined features with confidence weights. Boutelaa *et al.* [3] assessed the contribution depth information for different face analysis tasks like gender or ethnicity prediction. For face representation, four compact local descriptors are compared, namely, LBP, Local Phase Quantization (LPQ), Histogram of Oriented Gradients (HOG) and Binarized Statistical Image Features (BSIF). For face detection the nose tip is localized from the 3D model. After face region extraction and for each descriptor, the RGB and depth images are first divided into several local regions from which local histograms are extracted and then concatenated into an enhanced feature histogram used for classification using a support vector machine classifier (SVM) with a non-linear kernel.

A commonality of these approaches requires a pre-processing to precisely localize the face, estimate its pose, or even accurately localize face landmarks which is prone to error propagation in the sequential processing and a further dependency in the approach. In this work, our objective is to use both intensity and depth information using Sparse Representation Classifier to identify face. The hole face [5] and all facial blocks [3] are used to describe face with a local features, while in our approach only the most informative and discriminant regions defined around face salient points (*i.e.,* detected using SURF) are used. Differently to [9,12] where sparse representation is global as the whole face description is used as input, each local salient point description is considered separately in our approach. Besides, although considering all the training samples in the input dictionary for SRC, our algorithm is based on an adaptive and dynamic dictionary selection to speed-up the SRC and improve its performances.

3 Proposed Approach

As sketched in Fig. 1, the proposed approach involves online and offline phases sharing some processing blocks like raw data preprocessing (*i.e.,* median and bilateral filtering), face rough localization (*i.e.,* using Viola-Jones algorithm), and descriptor computing. The face region is cropped and resized to 96×96 pixels according to its mean depth to ensure a normalized spatial resolution. To get rid of face landmarks localization, we only consider salient image keypoints (*i.e.,* SURF) without any further semantic analysis and without loss of generality. In other words, we do not try to catch specific anatomical reference points on the face. That said, the repeatability of image feature points for face analysis was proven. In order to improve the performance of the SRC in terms of recognition performance and computational time and complexity, we introduce a Dynamic Dictionary Selection to ensure the scalability of the linear regression estimation in the SRC. Each extracted keypoint is classified separately with the SRC. In the following of the section, we detail our system while focusing of the main contributions.

3.1 Feature Keypoints and RGB-D Local Descriptors

We used the speed-up robust features (SURF) [1] to extract interest keypoints on the cropped face images. SRUF is a scale and in-plane rotation invariant feature. The number of extracted keypoints is variable and depends on face textures

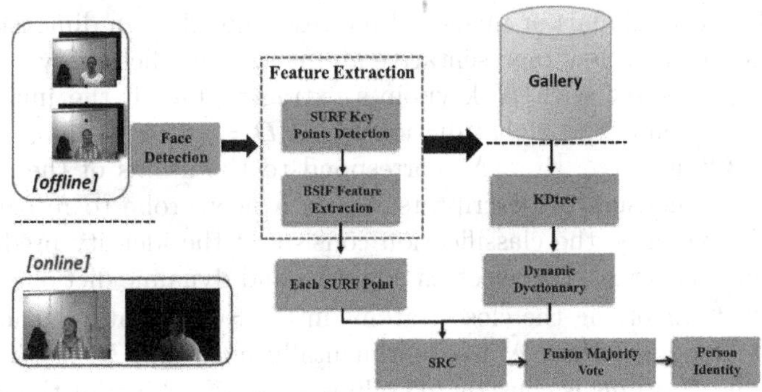

Fig. 1. The general flowchart of the proposed RGB-D face recognition system.

and also the position of the person in the frustum of the RGB-D camera. The keypoints coordinates are mapped on the depth crop using the sensor calibration parameters (see Fig. 2 for an example).

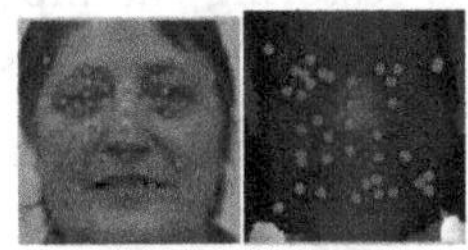

Fig. 2. Example of detected SURF points from RGB and depth images.

Around each keypoint, we crop from both image and depth data two patches of 20×20 pixels. The mapping between image and depth map can be ensured by the RGB-D sensor geometric calibration [2]. The LBP descriptor was widely used in the field of face analysis. However, the binary patterns are generic and hand-crafted. We believe that the binary patterns can be optimized for specific context and particularly for face analysis. Typically, BSIF [11] allows to learn linear filters to replace the LBP patterns. We followed [11] to learn specific filters w_i for image and depth data using independent component analysis (ICA).

After this processing and for a given face acquisition, we end up with two sets of keypoints described with both SURF and BSIF features computed respectively on image and depth data.

3.2 SRC with Dynamic Dictionary Selection

Sparse Representation Classification has attracted great attention in the past few years especially after its application firstly on face recognition [18]. The main objective of the SRC is to compute a new representation of an input signal in term of a linear combination of atoms in an complete dictionary. This is partly due to the fact that signals or images of interest, though high dimensional, can often be coded using a few representative atoms in some dictionary.

Given a gallery set with N keypoints extracted for all the images of all the persons, we define the global dictionary as $\boldsymbol{D} = [\boldsymbol{d}_1, \boldsymbol{d}_2, ..., \boldsymbol{d}_N] \in \mathbb{R}^{M \times N}$ where $\boldsymbol{d}_i \in \mathbb{R}^M$ and $i \in [1, ..., N]$ correspond to the atoms of the dictionary and M is the dimension of descriptors. Given a new probe to recognize with K extracted keypoints, the classification consists of the identity prediction for each keypoint separately. We select an adaptive and dynamic dictionaries noted by $\tilde{\boldsymbol{D}}_k \in \mathbb{R}^{M \times \tilde{N}}$ involving the closest atoms in the SURF feature space using a the fast KD-Tree algorithm. $\tilde{N}$ is experimentally fixed and $\tilde{N} \ll N$. The tree covering all the N atoms is constructed offline while searching for the $\tilde{N}$ nearest neighbors is relatively fast. This allows to substantially speed-up the SRC process and maintain a scalable classification. After the dynamic dictionaries selection, the SRC is performed for each keypoint k to represent its BSIF feature vector $\boldsymbol{y}_k$ as a linear combination of the corresponding $\tilde{\boldsymbol{D}}_k$ dictionary:

$$\boldsymbol{y}_k = \tilde{\boldsymbol{D}}_k \ \boldsymbol{x}_k + \epsilon_k \tag{1}$$

where ϵ_k captures noise and $\boldsymbol{x}_k \in \mathbb{R}^{\tilde{N}}$ is the sparse coefficient vector whose entries are zeros except the one restoring the input signal $\boldsymbol{y}_k$. In noisy conditions, however, coefficients associated with other identities may be non null. The estimation of the sparse coefficients $\boldsymbol{x}_k$ is formulated by a LASSO problem with an ℓ_1 minimization using [13]:

$$\min_{\boldsymbol{x}_k} \|\tilde{\boldsymbol{D}}_k \ \boldsymbol{x}_k - \boldsymbol{y}_k\|_2 + \lambda \|\boldsymbol{x}_k\|_1 \tag{2}$$

Finally, the identity associated to the j-th keypoint is classically given by the class generating the less residual. Afterward, the fusion of all the obtained identities is performed by a majority vote rule.

4 Experimental Results

4.1 The Databases

We evaluated the performance of our approach on three public RGB-D face databases; Biwi [7], Eurecom [10], and Curtinfaces [12] where for each face, image and depth data are acquired simultaneously. The Biwi database contains 24 sequence of 20 different persons where each person roles his face in different orientations within ±75in yaw, ±60in pitch, and ±50in roll. Eurecom Database is composed by 52 subjects, 14 females and 38 males. Each person has a set of 9 images in two different sessions. Each session contains 9 settings: neutral, smiling, open mouth, illumination variation, left end right profile, occlusion on the eyes, occlusion on the mouth, and finally occlusion with a white paper-sheet. CurtinFaces Database involves 52 subjects. Each subject has 97 images captured under different variations: combinations of 7 facial expression, 7 poses, 5 illuminations, and 2 occlusions. CurtinFaces database with low quality face models is more challenging in terms of variations of poses, and expression and illumination face models.

4.2 Experimental Protocol

We considered scenarios including pose, illumination and facial expression variations. We split each of the three databases into training and test sets. For BIWI database, in the gallery each person is represented by 30 images including different angles of pose and in the test set, 8 images are used separately for each pose. For EurecomKinect database, the first session set is selected for training and the second one for test. For CurtinFaces, 18 images containing only one kind of variations in illumination, pose or expression are selected for the training. The testing images include the rest non-occluded faces, there are a total of 6 different pose with 6 different expressions added to the neutral frontal views.

4.3 Experimental Results and Analysis

The first experiment is carried out to compare our results on BIWI database with those obtained by [9] under different yaw angles. The recognition performances for each modality separately and after fusion are reported in Fig. 3 in the right side, while we compare the fusion rates with [9] in the left side. Our approach outperforms [9] which is dependent on the precision of landmark points localization and uses only RGB information for face identification, especially we achieve an improvement of 13 % with ±30 yaw angle. For the second experiment we evaluate the proposed system against pose and expression variations simultaneously. The recognition rates are reported in Table 1. For an angle around ±30, our system achieves an average recognition rate of 96.91 % and 100 % for frontal view. With a yaw angle up to 60, depth performance decrease slightly which affects the fusion performance. An average rate of 92.25 % is achieved in our work which outperforms [4] (84.6 %).

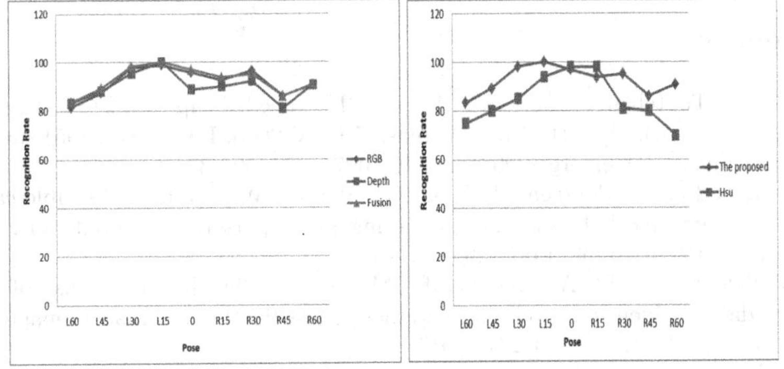

Fig. 3. Identification Rate under yaw angles variation on BIWI database.

Table 1. CurtinFace database performance

Pose	RGB	Depth	**Fusion**	Cov+LBP [4]
Frontal	100 %	100 %	**100 %**	N/A
Yaw ±30	95.77 %	84.41 %	**96.91 %**	94.2 %
Yaw ±60	85.5 %	78.07 %	**88.22 %**	84.6 %
Yaw ±90	83.7 %	51.08 %	**83.84 %**	75.0 %

We notice that generally image performances are better than those of depth data, this is caused by the noise and the low quality of depth sensor. Nether less, the fusion of the two modalities leads to even better performances.

For Eurecom database, we obtained an average recognition rate of 92.51 % for images and 77.30 % for depth data, and of 91.55 % after fusion, which is better than the recognition rate obtained in RISE [8] (*i.e.,* 89.0 % after fusion). It is noticeable that, the original depth data are used in our work as in [8], however in [3], new depth data are generated to enhance depth data quality.

5 Conclusion

This paper presents a scalable solution for RGB-D face recognition based on local descriptor extracted from face patches. These patches are defined around a set of salient points detected independently to the face structure and head pose. However, it is important to populate the gallery with different poses for every person which is doable by defining a precise protocol like the one in Biwi database. Also, we introduced the dynamic dictionaries selection for a scalable SRC classification to improved the recognition performances and make it faster. The experiments performed on the RGB-D databases demonstrate the effectiveness of the proposed algorithm on RGB-D face. Nevertheless, there are still room for improvement like enhancing the preprocessing of depth data. Also, we intend to improve the fusion by considering more advanced fusion strategies.

References

1. Bay, H., Tuytelaars, T., Gool, L.: SURF: speeded up robust features. In: Leonardis, A., Bischof, H., Pinz, A. (eds.) ECCV 2006. LNCS, vol. 3951, pp. 404–417. Springer, Heidelberg (2006). doi:10.1007/11744023_32
2. Ben-Hamadou, A., Soussen, C., Daul, C., Blondel, W., Wolf, D.: Flexible calibration of structured-light systems projecting point patterns. Comput. Vis. Image Underst. **117**(10), 1468–1481 (2013)
3. Boutellaa, E., Hadid, A., Bengherabi, M., Ait-Aoudia, S.: On the use of kinect depth data for identity, gender and ethnicity classification from facial images. Pattern Recogn. Lett. **68**, 270–277 (2015)
4. Ciaccio, C., Wen, L., Guo, G.: Face recognition robust to head pose changes based on the RGB-D sensor. In: 2013 IEEE Sixth International Conference on Biometrics: Theory, Applications and Systems (BTAS), pp. 1–6. IEEE (2013)

5. Dai, X., Yin, S., Ouyang, P., Liu, L., Wei, S.: A multi-modal 2D + 3D face recognition method with a novel local feature descriptor. In: 2015 IEEE Winter Conference on Applications of Computer Vision (WACV), pp. 657–662. IEEE (2015)
6. Faltemier, T.C., Bowyer, K.W., Flynn, P.J.: A region ensemble for 3-D face recognition. IEEE Trans. Inf. Forensics Secur. **3**(1), 62–73 (2008)
7. Fanelli, G., Dantone, M., Gall, J., Fossati, A., Van Gool, L.: Random forests for real time 3D face analysis. Int. J. Comput. Vis. **101**(3), 437–458 (2013)
8. Goswami, G., Vatsa, M., Singh, R.: RGB-D face recognition with texture and attribute features. IEEE Trans. Inf. Forensics Securi. **9**(10), 1629–1640 (2014)
9. Hsu, G.S.J., Liu, Y.L., Peng, H.C., Wu, P.X.: RGB-D-based face reconstruction and recognition. IEEE Trans. Inf. Forensics Secur. **9**(12), 2110–2118 (2014)
10. Huynh, T., Min, R., Dugelay, J.-L.: An efficient LBP-based descriptor for facial depth images applied to gender recognition using RGB-D face data. In: Park, J.-I., Kim, J. (eds.) ACCV 2012. LNCS, vol. 7728, pp. 133–145. Springer, Heidelberg (2013). doi:10.1007/978-3-642-37410-4_12
11. Kannala, J., Rahtu, E.: BSIF: Binarized statistical image features. In: 2012 21st International Conference on Pattern Recognition (ICPR), pp. 1363–1366. IEEE (2012)
12. Li, B.Y., Mian, A., Liu, W., Krishna, A.: Using kinect for face recognition under varying poses, expressions, illumination and disguise. In: 2013 IEEE Workshop on Applications of Computer Vision (WACV), pp. 186–192. IEEE (2013)
13. Mairal, J., Bach, F., Ponce, J., Sapiro, G.: Online learning for matrix factorization and sparse coding. J. Mach. Learn. Res. **11**, 19–60 (2010)
14. Rekik, A., Ben-Hamadou, A., Mahdi, W.: 3D face pose tracking using low quality depth cameras. In: VISAPP (2), pp. 223–228 (2013)
15. Rekik, A., Hamadou, A., Mahdi, W.: Human machine interaction via visual speech spotting. In: Battiato, S., Blanc-Talon, J., Gallo, G., Philips, W., Popescu, D., Scheunders, P. (eds.) ACIVS 2015. LNCS, vol. 9386, pp. 566–574. Springer, Heidelberg (2015)
16. Rekik, A., Ben-Hamadou, A., Mahdi, W.: Unified system for visual speech recognition and speaker identification. In: Battiato, S., et al. (eds.) ACIVS 2015. LNCS, vol. 9386, pp. 381–390. Springer, Heidelberg (2015). doi:10.1007/978-3-319-25903-1_33
17. Rekik, A., Ben-Hamadou, A., Mahdi, W.: An adaptive approach for lip-reading using image and depth data. Multimedia Tools Appl. **75**(14), 8609–8636 (2016)
18. Wright, J., Yang, A.Y., Ganesh, A., Sastry, S.S., Ma, Y.: Robust face recognition via sparse representation. IEEE Trans. Pattern Anal. Mach. Intell. **31**(2), 210–227 (2009)
19. Zhu, X., Ramanan, D.: Face detection, pose estimation, and landmark localization in the wild. In: 2012 IEEE Conference on Computer Vision and Pattern Recognition (CVPR), pp. 2879–2886. IEEE (2012)

Gaussian Processes Based Fusion of Multiple Data Sources for Automatic Identification of Geological Boundaries in Mining

Katherine L. Silversides(✉) and Arman Melkumyan

Australian Centre for Field Robotics, University of Sydney, Sydney, Australia
{katherine.silversides,arman.melkumyan}@sydney.edu.au

Abstract. Mining stratified ore deposits such as Banded Iron Formation (BIF) hosted iron ore deposits requires detailed knowledge of the location of orebody boundaries. In one Marra Mamba style deposit, the alluvial to bedded boundary only creates distinctive signatures when both the magnetic susceptibility logs and the downhole chemical assays are considered. Identifying where the ore to BIF boundary occurs with the NS3-NS4 stratigraphic boundary requires both natural gamma logs and chemical assays. These data sources have different downhole resolutions. This paper proposes a Gaussian Processes based method of probabilistically processing geophysical logs and chemical assays together. This method improves the classification of the alluvial to bedded boundary and allows the identification of concurring stratigraphic and mineralization boundaries. The results will help to automatically produce more accurate and objective geological models, significantly reducing the need for manual effort.

Keywords: Gaussian processes · Signal processing · Banded Iron Formation · Geophysical logging · Geochemical assay

1 Introduction

The commercial mining of stratified ore bodies requires sufficient understanding of the location and properties of the ore. This is achieved through modelling the properties of the data obtained through exploration drilling. Before this modelling can be completed the location of the relevant boundaries must first be determined. Exploration data can cover many thousands of holes. Manual processing of this data can require considerable time and is prone to inconsistencies and errors. Therefore an objective automatic method is desirable. The Banded Iron Formation (BIF) hosted iron ore deposits of the Hamersley Ranges of Western Australia are one type of stratified ore deposit. This paper studies a typical Marra Mamba type iron ore deposit where the BIF has been locally enriched to a higher iron content [1]. The data collected from the exploration holes is magnetic susceptibility, density, natural gamma and chemical assays.

Many machine learning techniques have been used to process geophysical logs [2], however the logs do not contain sufficient information to identify all boundaries. In the BIF-hosted deposits described here there are boundaries that require both geophysical

A. Hirose et al. (Eds.): ICONIP 2016, Part IV, LNCS 9950, pp. 294–301, 2016.
DOI: 10.1007/978-3-319-46681-1_36

logs and chemical assays to be completely and accurately defined. The different data sources cannot be directly processed together, as the geophysical logs and the chemical assays have downhole resolutions of 0.1 m and 2 m respectively. While the logs can be averaged to the 2 m intervals this involves a significant loss of data.

This paper investigates a signature classification method that combines geophysical logs with chemical assays for boundary classification. It will focus on two geological boundaries. The first boundary is between the alluvial and bedded material at the top of the ore deposit. The magnetic susceptibility contains a distinctive step-like feature at this boundary (Fig. 1A); however similar features can occur at other places in the log. The chemical assays of both aluminum trioxide (Al_2O_3) and titanium dioxide (TiO_2) change from higher values in the alluvial to lower in the bedded (Fig. 1B and C). Once again, similar features also occur in other places in the logs. Therefore, the alluvial to bedded boundary is only distinctive when both data sources are used. This boundary was identified in [3], and this paper extends this method to apply to different input data and different types of boundaries.

The second boundary is different, as it combines both a stratigraphic boundary (a change within the bedded units) and a mineralization boundary (a change from minable ore to waste). The NS3-NS4 shale bands occur at a transition between two BIF units, and this transition can be identified by their distinctive signature in the natural gamma logs (Fig. 2A), but not in the magnetic susceptibility or the assays. This signature has been previously identified using Gaussian Processes [4]. The mineralized BIF to waste BIF (ore to BIF) boundary at the base of the ore zone can occur at this stratigraphic boundary [5]. However, the presence of this second boundary cannot be determined from the geophysical logs and requires the use of the chemical assays. The ore to BIF boundary is characterized by a decrease in the iron (Fe) and an increase in the silicon dioxide (SiO_2) (Fig. 2B). For the purpose of this study a Fe grade cut-off was used to define ore and waste. Although other chemical assays can be considered, Fe is considered the key variable for determining iron ore from waste rock.

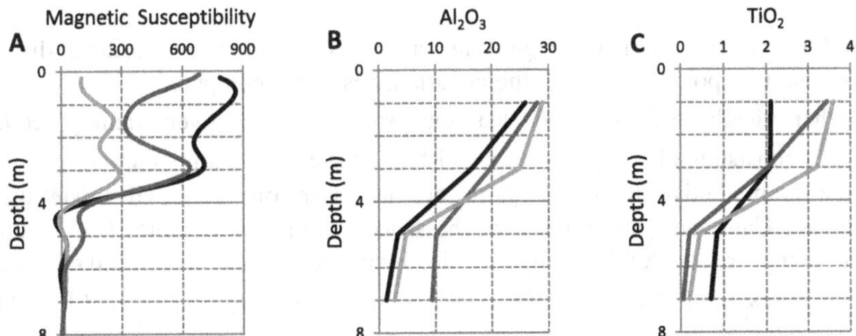

Fig. 1. The alluvial to bedded boundary creates signatures in the magnetic susceptibility logs (A), Al_2O_3 (B), and TiO_2 (C). Note the difference between the 0.1 m sampling for the magnetic susceptibility and the 2 m sampling for the assays.

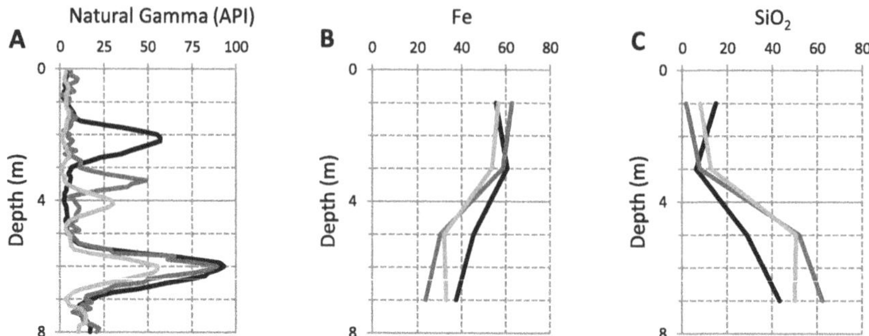

Fig. 2. The NS3-NS4 marker shales create signatures in the natural gamma (A), and the ore to BIF boundary create signatures in the Fe (B) and SiO_2 (C).

2 Background

Gaussian Processes (GPs) is a probabilistic method of modelling functions representing quantities of interest within a given set of data. Mathematically a Gaussian Process is an infinite collection of random variables, any finite number of which has a joint Gaussian distribution. Machine learning using GPs consists of training and inference. The training step optimizes the unknown hyperparameters to produce a probabilistic model that best represents the training library. These hyperparameters are then used to predict the values of the function of interest at new locations [6, 7].

The relationship between the inputs and outputs of the GP are partially described by the covariance function. The covariance function also defines the number and type of hyperparameters that are required for the model. This paper uses the single length-scale squared exponential covariance function, with the following analytic expression:

$$k(r) = \exp\left[-\frac{r^2}{2l^2}\right] \quad (1)$$

where l is the characteristic length-scale and r is the multi-dimensional distance between the two points for which the covariance is being computed.

Mathematically, the supervised learning problem uses a given training set $D = \{x_i, y_i\}_{i=1}^{N}$ consisting of N input point $x_i \in R^D$ and the corresponding outputs $y_i \in R$ to compute the predictive distribution $f(x_*)$ at a new test point x_*. A GP model uses a multivariate Gaussian distribution over the space of function variables $f(\mathbf{x})$ mapping input to output spaces. A GP is fully specified by its mean function $\mu(x)$ and covariance function $k(\mathbf{x}, \mathbf{x}')$, so $f(\mathbf{x}) \sim GP(\mu(\mathbf{x}), k(\mathbf{x}, \mathbf{x}'))$. Using $(X, \mathbf{f}, \mathbf{y}) = (\{x_i\}, \{f_i\}, \{y_i\})_{i=1}^{N}$ for the training set and $(X_*, \mathbf{f}_*, \mathbf{y}_*) = (\{x_{*i}\}, \{f_{*i}\}, \{y_{*i}\})_{i=1}^{N}$ for the testing points, the joint Gaussian distribution is:

$$\begin{bmatrix} y \\ \mathbf{f}_* \end{bmatrix} \sim N\left(\mu, \begin{bmatrix} K(X,X) + \sigma^2 I & K(X,X_*) \\ K(X_*,X) & K(X_*,X_*) \end{bmatrix}\right) \quad (2)$$

where $\mathbf{f}_* = f(x_*)$ is the predicted function without noise. In Eq. (2) N(μ,Σ) is a multivariate Gaussian distribution with mean μ and covariance Σ and K is the covariance matrix computed between all the points. The predictive distribution for new points can be obtained as $p(f_i|X_*, X, \mathbf{y}) = N(\boldsymbol{\mu}_*, \boldsymbol{\Sigma}_*)$ where

$$\boldsymbol{\mu}_* = K(X_*, X)\left[K(X,X) + \sigma^2 I\right]^{-1} y$$

$$\Sigma_* = K(X_*, X_*) - K(X_*, X)\left[K(X,X) + \sigma^2 I\right]^{-1} K(X, X_*) + \sigma^2 I. \tag{3}$$

The GP model is trained by maximizing the log of the marginal likelihood (lml) with respect to the hyper-parameters θ, where:

$$\log p(\mathbf{y}|\mathrm{X}, \boldsymbol{\theta}) = -\frac{1}{2}\mathrm{y}^{\mathrm{T}}[\mathrm{K}(\mathrm{X},\mathrm{X}) + \sigma^2 I]^{-1}\mathrm{y} - \frac{1}{2}\log|\mathrm{K}(\mathrm{X},\mathrm{X}) + \sigma^2 I| - \frac{\mathrm{N}}{2}\log 2\pi. \tag{4}$$

3 Method

3.1 Geophysical Data

The natural gamma logs were processed to identify the NS3-NS4 shale bands using the method described in [4]. A similar method was used to identify the alluvial to bedded boundary in the magnetic susceptibility logs. An addition pre-processing step using a low pass filter was applied to these logs, due to higher variability.

For each boundary an initial library was built containing both positive examples of the desired signature and negative examples of other signatures. Each example was an 8 m section of log where the measurements were acquired at 0.1 m intervals down the hole. The boundary was located at a set depth in the example, either 4 m (magnetic susceptibility) or 7 m (natural gamma). Each library had a corresponding output file assigning the positive examples a value of 1 and the negative examples an output of 0. While the magnetic susceptibility only required one library, the NS3-NS4 boundary required six libraries to cover variations in both signature location and amplitude.

A GP was used to learn the optimal hyperparameters from each library using a single length-scale squared exponential covariance function. The resulting GP models were used to classify all of the logs of the same type (i.e. magnetic susceptibility or natural gamma) from the site. The logs are much longer than the libraries; therefore each log was processed in 8 m sections using a sliding window with 0.1 m increments.

The GP provided both an output and a standard deviation (SD). The output represents the mean value of the probability distribution of the classification. This value describes how similar a section of log is to the signatures in the library. As outputs of 1 and 0 represent an exact match to the positive and negative library examples respectively, an output 0.5 or greater is considered a positive identification of the signature. The standard deviation provides a measure of the confidence for each GP output. Therefore, adding and subtracting the standard deviation from the GP output gives the range of the most probable classification values. When this range includes values on both sides of 0.5, the GP classification cannot be confidently assigned. Therefore a signature can be divided into certain and uncertain categories using the rules:

Certain signature:	Output − SD ≥ 0.5
Uncertain signature:	Output ≥ 0.5 and
	Output − SD < 0.5
Certain no signature:	Output + SD < 0.5
Uncertain no signature:	Output < 0.5 and
	Output + SD ≥ 0.5

Once the GP had classified all the logs, probabilistic criteria were applied to identify areas of the logs that were uncertain and needed to be included in the training library. This method identified signatures that were not already included in the library, to ensure that the full range of signatures was included. The GPs were then used to select the point in each log with the highest output and therefore the highest probability of being the boundary. The accuracies were determined using a manual geological interpretation. A GP boundary was considered correct if it was located within 4 m of the manually identified boundary. This allows for small differences in the boundary definitions as well as any error introduced into the manual interpretation due to the coarser 2 m resolution. If a boundary is not found in a hole, it is considered correct if the boundary is not present in the geological interpretation.

3.2 Chemical Assays

GPs were trained on chemical assays for both the alluvial to bedded boundary (Al_2O_3 and TiO_2) and the ore to BIF boundary (Fe and SiO_2) using a similar method to that described above. The chemical assays did not require pre-processing as they are averaged over larger intervals and do not have the small fluctuations observed in the geophysical data. The assays are sampled in 2 m intervals; therefore each 8 m example contained 4 data points for each input. Each example consisted of 8 data points in total, compared to 80 points for the magnetic susceptibility. The boundary was located in the middle of the example (4 m) for both boundary types.

The GP was trained using two length scales, one for each assay input. The ratio of the Al_2O_3 length scale to the TiO_2 length scale was restricted to within 15:1. This was guided by the relative magnitudes of the data. The Fe and SiO_2 have similar magnitudes; therefore the length scales were limited so that each one could not be more than double the value of the other. Without this linking the GP training can make one length scale very large, essentially eliminating any input from that assay type.

The trained GP was used to process all of the assays using an 8 m sliding window, for each hole selecting the depth with the highest probability of containing the boundary. The results were divided into the four categories described in the previous section and compared to the manual geological interpretation to determine the accuracy.

3.3 Combined Method

To locate a boundary using both the geophysical logs and the chemical assays, the two GPs trained for each boundary were used together. For each boundary the GPs were run independently down the data for the same hole. Although both GPs used 8 m example, due to the difference in the sampling size, there were 20 magnetic susceptibility or natural gamma GP outputs for every assay GP output. Therefore each assay output was combined with the highest magnetic susceptibility or natural gamma output for the 2 m that the assay covered. The outputs and standard deviations were combined using a weighted average. For the alluvial to bedded boundary the assay output was assigned double the weight of the magnetic susceptibility output as it contains two data sources, Al_2O_3 and TiO_2. Therefore Al_2O_3, TiO_2 and magnetic susceptibility had an equal impact on the output. However, when combining the NS3-NS4 and ore to BIF boundaries the NS3-NS4 natural gamma output was given double the weight of the Fe and SiO_2 assay output. This was because the NS3-NS4 signature is distinctive to this location only. The ore to BIF boundary, however, can occur at other location in the hole. Therefore the NS3-NS4 boundary was given more weight to ensure that the correct area was chosen. The highest combined output for each hole was selected as the boundary location. This was divided into categories and compared to the manual geological interpretation to determine accuracy as described above.

4 Results and Discussion

4.1 Alluvial to Bedded Boundary

The three methods described above for the magnetic susceptibility logs and Al_2O_3 and TiO_2 assays were used to process the same holes and their performances were analyzed (Table 1). For the magnetic susceptibility logs the certain holes had an accuracy of 73.9 %. The assays and combined method had improved accuracies of 90.9 % and 93.5 % respectively. When this is split into the signature and no signature groups, the magnetic susceptibility had the lowest accuracies for both, 72.3 % and 75.7 % respectively. In contrast, for the certain signatures the assays had an accuracy of 88.7 %

Table 1. Results for the alluvial to bedded boundary

		Magnetic susceptibility		Assays (Al_2O_3 and TiO_2)		Combined	
		No. of holes	Accuracy (%)	No. of holes	Accuracy (%)	No. of holes	Accuracy (%)
Certain holes	All	4699	73.9	4500	90.9	4050	93.5
	Signature	2537	72.3	3016	88.7	2217	94.0
	No sig.	2162	75.7	1484	95.6	1833	93.0
Uncertain holes	All	1235	42.3	1434	70.9	1884	70.6
	Signature	736	49.9	783	70.1	1213	83.5
	No sig.	499	31.1	651	71.9	671	47.2
% of holes certain		79.2		75.8		68.3	

and this improved to 94.0 % for the combined method. This improvement in accuracy would have a large impact on the accuracy of any surfaces modelled using the automatically identified boundaries.

The combined method had the lowest boundary detection rate, reflecting the elimination of the locations that only contain the assay or magnetic susceptibility signatures. Similarly the combined method has the lowest number of certain holes, reflecting the holes that only contain one of the signatures being classified as uncertain.

The magnetic susceptibility GP is overconfident, producing the highest number of certain holes, but the lowest accuracy. This may be caused by very similar magnetic susceptibility signatures that occur at other locations in the log and cannot be readily distinguished from those at the alluvial to bedded boundary.

The uncertain holes contain very variable results and require further manual investigation to determine the boundary location. The results are similar to those of the certain signatures, with magnetic susceptibility having the lowest accuracy, 42.3 %, and the assays and combined method achieving similar results with 70.9 % and 70.6 % respectively. The uncertain signatures also had similar patterns to the certain signatures, with the magnetic susceptibility, assays and combined results having accuracies of 49.9 %, 70.1 % and 83.5 % respectively.

4.2 NS3-NS4 and Ore to BIF Boundaries

The natural gamma logs and Fe and SiO_2 assays were processed together using the combined method described above and the accuracies were calculated (Table 2). Unlike the alluvial to bedded results, here the different data sources were not identifying the same boundary. The NS3-NS4 stratigraphic boundary is identified in the natural gamma logs and the ore to BIF boundary is identified in the assays. The combined method identifies where both boundaries occur together.

This method had an accuracy of 98.4 % for the certain holes. However, this decreases for the certain signatures to 84.2 %, as the combined GPs identified some NS3-NS4 signatures that are not at the ore to BIF boundary. This generally occurred where there was a very high GP output for the NS3-NS4 and an uncertain GP output for the assays. This method has a lower number of certain holes than the alluvial to bedded boundary, due to identifying two boundaries not a single boundary. There were many holes that contained only one of the two boundaries. This produced one high and one low GP output, which combined to make an uncertain result.

Table 2. Results for the NS3-NS4 and ore to BIF boundaries

		No. of holes	Accuracy (%)
Certain holes	All	3341	98.4
	Signature	183	84.2
	No signature	3158	99.3
Uncertain holes	All	2593	70.1
	Signature	508	45.5
	No signature	2085	76.1
Percent of holes certain		56.3	

Automatically identifying these boundaries is valuable as it reduces the weeks of manual labor required to process these logs, and provides objective inputs to the models used to predict the volumes and tonnages of minable ore.

5 Conclusions

The GP machine learning methods presented in this paper can be used to automatically classify different types of geological boundaries by fusing multiple downhole data sources with different resolutions. When identifying the alluvial to bedded boundary seperately, magnetic susceptibility and assays (Al_2O_3 and TiO_2) had accuracies of 72.3 % and 88.7 % respectively where a certain signature was identified. The combined GP method had an improved accuracy of 94.0 %, but identified fewer boundaries than the individual methods.

The NS3-NS4 shales and the Fe and SiO_2 assays identify different types of boundaries, stratigraphic and mineralization. Each data source only has information about one boundary. The combined method can identify the places where they occur together, with an accuracy of 98.4 % where the GP is certain and 84.2 % in the certain signatures. This combined method is therefore proposed as a suitable method for boundary classification from data sources with different resolutions. Automatically identifying these boundaries decreases the amount of manual effort required and provides an objective, repeatable output that can be used for mine modelling.

Acknowledgements. This work has been supported by the Australian Centre for Field Robotics and the Rio Tinto Centre for Mine Automation.

References

1. Thorne, S.W., Hagemann, S., Webb, A., Clout, J.: Banded iron formation-related iron ore deposits of the Hamersley Province, Western Australia. In: Hagemann, S., Rosiere, C., Gutzmer, J., Beukes N.J. (eds.) Banded Iron Formation-Related High Grade Iron Ore, Rev. Econ. Geol. **15**, 197–221 (2008)
2. Borsaru, M., Zhoua, B., Aizawa, T., Karashima, H., Hashimoto, T.: Automated lithology prediction from PGNAA and other geophysical logs. Appl. Radiat. Isotopes **64**, 272–282 (2006)
3. Silversides, K., Melkumyan, A.: Integration of downhole data sources with different resolution for improved boundary detection. In: 12th SEGJ International Symposium, Tokyo (2015)
4. Silversides, K., Melkumyan, A., Wyman, D.: Fusing gaussian processes and dynamic time warping for improved natural gamma signal classification. Math. Geosci. **48**, 187–210 (2016)
5. Silversides, K., Melkumyan, A., Hatherly, P., Wyman, D.: Boundary classification for automated geological modelling. In: 35th APCOM Symposium, pp. 133–120. AusIMM, Australia (2011)
6. Bishop, C.M.: Pattern Recognition and Machine Learning. Springer, Berlin (2006)
7. Rasmussen, C.E., Williams, C.K.I.: Gaussian Processes for Machine Learning. Springer Science+Business Media, LLC, Heidelberg (2006)

Speaker Detection in Audio Stream via Probabilistic Prediction Using Generalized GEBI

Koki Sakata, Shota Sakashita, Kazuya Matsuo, and Shuichi Kurogi(✉)

Kyushu Institute of Technology, Tobata, Kitakyushu, Fukuoka 804-8550, Japan
{sakata,sakashita}@kurolab.cntl.kyutech.ac.jp,
{matsuo,kuro}@cntl.kyutech.ac.jp
http://kurolab.cntl.kyutech.ac.jp/

Abstract. This paper presents a method of speaker detection using probabilistic prediction for avoiding the tuning of thresholds to detect a speaker in an audio stream. We introduce g-GEBI (generalized GEBI) as a generalization of BI (Bayesian Inference) and GEBI (Gibbs-distribution-based Extended BI) to execute iterative detection of a speaker in audio stream uttered by more than one speaker. Then, we show a method of probabilistic prediction in multiclass classification to classify the results of speaker detection. By means of numerical experiments using recorded real speech data, we examine the properties and the effectiveness of the present method. Especially, we show that g-GEBI and g-BI (generalized BI) are more effective than the conventional BI and GEBI in incremental speaker detection task.

Keywords: Probabilistic prediction · Speaker detection · Generalized Gibbs-distribution-based extended Bayesian inference

1 Introduction

This paper presents a method of speaker detection using probabilistic prediction. Here, from [1], as a branch of speaker recognition discipline, the speaker detection is the act of detecting a specific speaker in an audio stream, and the underlying theory encompasses segmentation, identification and verification of speakers. So far, we have developed a method of probabilistic prediction for text-prompted speaker verification for avoiding the tuning of the thresholds for acceptance and rejection [2]. Namely, in our previous studies [3,4], we have tuned the thresholds for speaker and text verification by the method of EER (equal error rate) for FAR (false acceptance rate) and FRR (false rejection rate) to be almost the same. However, the tuning is not so easy and it is practically desirable that FAR and FRR can be tuned depending on the risk or the security level of verification. To solve this problem, we have introduced probabilistic prediction which allows the users to decide on the level of risk they are prepared and to take appropriate

A. Hirose et al. (Eds.): ICONIP 2016, Part IV, LNCS 9950, pp. 302–311, 2016.
DOI: 10.1007/978-3-319-46681-1_37

action within a proper understanding of the uncertainties, as we can see in weather and climate forecasting [5].

From another point of view, our methods for speech recognition [2–4] employ multistep inference to reduce recognition error, where BI (Bayesian inference) is extended to GEBI (Gibbs-distribution-based extended BI) so that unregistered speakers can be rejected by thresholding the posterior probability derived by the inference. The GEBI has shown better performance in text-prompted speaker verification, where a single speaker utters an audio stream. However, different from the task of speaker verification, the speaker detection task assumes that there are more than one speaker in an audio stream. This indicates that the probability of a speaker at a time should decrease with the increase of time during an audio stream uttered by the other speaker. Since the GEBI does not have such a function, this article introduces a generalized GEBI which realizes BI, GEBI and their intermediate inference by tuning its parameters.

Here, note that our speech processing system employs competitive associative nets (CAN2s). The CAN2 is an artificial neural net for learning efficient piecewise linear approximation of nonlinear function [6], and we have shown that feature vectors of pole distribution extracted from piecewise linear predictive coefficients obtained by the bagging (bootstrap aggregating) version of the CAN2 reflect nonlinear and time-varying vocal tract of the speaker [7]. Although the most common way to characterize speech signal in the literature is short-time spectral analysis, such as Linear Prediction Coding (LPC) and Mel-Frequency Cepstrum Coefficients (MFCC) [8], the bagging CAN2 learns more precise information than LPC and MFCC (see [7] for details).

We formulate the method of probabilistic prediction for speaker detection using g-GEBI in Sect. 2, and show the experimental results and analysis in Sect. 3, followed by the conclusion in Sect. 4.

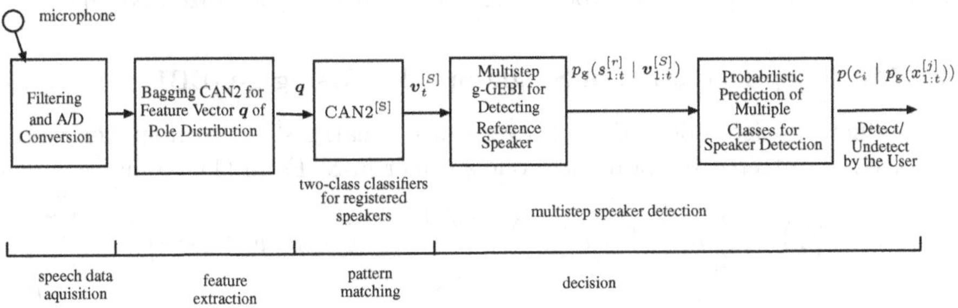

Fig. 1. Diagram of text-prompted speaker verification system using CAN2s

2 Probabilistic Prediction for Speaker Detection

Figure 1 shows an overview of the present speaker detection system using CAN2s. In the same way as general speaker recognition systems [8], it consists of four steps: speech data acquisition, feature extraction, pattern matching, and making a decision. In this research study, we use a feature vector of pole distribution obtained from a speech signal (see [7] for details).

2.1 Multistep Inference Using g-GEBI

Here, we formulate multistep inference using g-GEBI for speaker detection (see [4] for details of GEBI). In order to execute multistep inference of speaker detection in an audio stream, we assume that a sequence of spoken words $\mathrm{D}_{1:t} = \mathrm{D}_1\mathrm{D}_2 \cdots \mathrm{D}_t$ is uttered by a sequence of speakers $s_{1:t} = s_1 s_2 \cdots s_t$, where s_t denotes the speaker of the word x_t. Furthermore, let $\mathrm{LM}^{[S]}$ be a set of learning machines $\mathrm{LM}^{[s^{[i]}]}$ for registered speakers $s^{[i]} \in S = \{s^{[i]} \mid i \in I^{[S]} = \{1, 2, \cdots, |S|\}\}$, and each $\mathrm{LM}^{[s^{[i]}]}$ learns to predict a single-step verification of the speaker s_t at each time t by $v^{[s^{[i]}]} = 1$ for the acceptance and $v^{[s^{[i]}]} = -1$ for the rejection.

Now, for multistep speaker detection, let us suppose that the joint probability of the output vector $\boldsymbol{v}^{[S]} = (v^{s^{[1]}}, \cdots, v^{s^{[|S|]}})$ of all $\mathrm{LM}^{[s^{[i]}]}$ for a spoken word x of $s \in S$ is given by $p(\boldsymbol{b}^{[S]}|s) = \prod_{s^{[i]} \in S} p(b^{[s^{[i]}]}|s)$ for simplicity. Let $\boldsymbol{v}_{1:t}^{[S]} = \boldsymbol{v}_1^{[S]}\boldsymbol{v}_2^{[S]} \cdots \boldsymbol{v}_t^{[S]}$ be a sequence of $\boldsymbol{v}^{[S]}$ obtained from a speech of a speaker s, we calculate the following posterior probability for $t = 1, 2, \cdots$, recursively, as

$$p_{\mathrm{g}}\left(s \mid \boldsymbol{v}_{1:t}^{[S]}\right) = \frac{1}{Z_t} p_{\mathrm{g}}\left(s \mid \boldsymbol{v}_{1:t-1}^{[S]}\right)^{(t-t_0-1)^\alpha/(t-t_0)^\alpha} p\left(\boldsymbol{v}_t^{[S]} \mid s\right)^{\beta/(t-t_0)^\alpha}, \quad (1)$$

where α $(1 \geq \alpha \geq 0)$, β (≥ 0) and t_0 (≥ 0) are constants, and Z_t is an normalization constant for holding $\sum_{s \in S} p_{\mathrm{g}}\left(s \mid \boldsymbol{v}_{1:t}^{[S]}\right) = 1$. Note that the conventional (naive) BI is given for $\alpha = 0$ and $\beta = 1$, and GEBI for $\alpha = 1$ and $t_0 = 0$ (see [4] for details). Thus, we call the above inference g-GEBI (generalized GEBI).

2.2 Multistep Detection of Single Speaker Using g-GEBI

We conduct multistep detection of a reference speaker $s^{[r]} \in S$ in spoken word sequence $x_{1:T}$ uttered by more than one speaker in S. From (1), we have

$$p_{\mathrm{g}}\left(s^{[r]} \mid \boldsymbol{v}_{1:t}^{[S]}\right) = \frac{1}{Z_t} p_{\mathrm{g}}\left(s^{[r]} \mid \boldsymbol{v}_{1:t-1}^{[S]}\right)^{(t-t_0-1)^\alpha/(t-t_0)^\alpha} p\left(\boldsymbol{v}_t^{[S]} \mid s^{[r]}\right)^{\beta/(t-t_0)^\alpha}, \quad (2)$$

$$p_{\mathrm{g}}\left(\overline{s^{[r]}} \mid \boldsymbol{v}_{1:t}^{[S]}\right) = \frac{1}{Z_t} p_{\mathrm{g}}\left(\overline{s^{[r]}} \mid \boldsymbol{v}_{1:t-1}^{[S]}\right)^{(t-t_0-1)^\alpha/(t-t_0)^\alpha} p\left(\boldsymbol{v}_t^{[S]} \mid \overline{s^{[r]}}\right)^{\beta/(t-t_0)^\alpha}, \quad (3)$$

where $\overline{s^{[r]}} = S \backslash \{s^{[r]}\}$. For simpler expression, let us consider the logit of $p_{\mathrm{g}}\left(s^{[r]} \mid \boldsymbol{v}_{1:t}^{[S]}\right)$, or the log-odds (or the logarithm of odds) of the odds $\frac{p_{\mathrm{g}}\left(s^{[r]} \mid \boldsymbol{v}_{1:t}^{[S]}\right)}{p_{\mathrm{g}}\left(\overline{s^{[r]}} \mid \boldsymbol{v}_{1:t}^{[S]}\right)}$ given by

$$l_{1:t} = \log \left(\frac{p_{\mathrm{g}}\left(s^{[r]} \mid \boldsymbol{v}_{1:t}^{[S]}\right)}{p_{\mathrm{g}}\left(\overline{s^{[r]}} \mid \boldsymbol{v}_{1:t}^{[S]}\right)} \right) = \log \left(\frac{p_{\mathrm{g}}\left(s^{[r]} \mid \boldsymbol{v}_{1:t}^{[S]}\right)}{1 - p_{\mathrm{g}}\left(s^{[r]} \mid \boldsymbol{v}_{1:t}^{[S]}\right)} \right), \tag{4}$$

where the inverse transform is obtained by

$$p_{\mathrm{g}}\left(s^{[r]} \mid \boldsymbol{v}_{1:t}^{[S]}\right) = \frac{1}{1 + \exp\left(-l_{1:t}\right)} \tag{5}$$

Then, from the above Eqs. (2) and (3), we have

$$l_{1:t} = (t - t_0 - 1)^{\alpha}(t - t_0)^{-\alpha} l_{1:t-1} + \beta (t - t_0)^{-\alpha} r_t \tag{6}$$

where

$$r_t = \log \frac{p\left(\boldsymbol{v}_t^{[S]} \mid s^{[r]}\right)}{p\left(\boldsymbol{v}_t^{[S]} \mid \overline{s^{[r]}}\right)} \tag{7}$$

is the logarithm of the ratio of the likelihoods $p\left(\boldsymbol{v}_t^{[S]} \mid s^{[r]}\right)$ and $p\left(\boldsymbol{v}_t^{[S]} \mid \overline{s^{[r]}}\right)$.

2.3 Truncation of Logit and Probability

Now, let us suppose that the reference speaker $s^{[r]}$ to be detected and other speaker $s \in S \backslash \{s^{[r]}\}$ utter a sequence of words alternately, and t_i be a time until when a speaker has uttered and the other speaker utters from $t_i + 1$ to $t_i + \Delta_t$. Then, from (6) for $t = t_i + \Delta t$, we have

$$l_{1:t_i+\Delta t} = (t_i - t_0)^{\alpha}(t_i - t_0 + \Delta t)^{-\alpha} l_{1:t_i} + \beta (t_i - t_0 + \Delta t)^{-\alpha} \Delta t \left(\frac{1}{\Delta t} \sum_{k=t_i+1}^{t_i+\Delta t} r_k \right). \tag{8}$$

Here, r_k in the second term on the right hand side is supposed to have a population mean for a sequence of spoken words uttered by a single speaker, thus a sample mean $\left(\frac{1}{\Delta t}\sum_{k=t_i+1}^{t_i+\Delta t} r_k\right)$ is supposed to converge with the increase of Δt. On the other hand, the first term on the right hand side does not change for $\alpha = 0$ or the absolute value for $\alpha > 0$ decreases with the increase of Δt. As a result, for a reasonable length Δt of the sequence of words and appropriate α, β and t_0, we expect that $l_{1:t_i+\Delta t}$ is big enough for detecting $s^{[r]}$ and small enough not for detecting $s^{[r]}$ by a sequence of words uttered by $s^{[r]}$ and $s \in S \backslash \{s^{[r]}\}$, respectively. To have this done, we employ the truncation of $l_{1:t}$ with a lower bound $l_{\min}$ and an upper bound $l_{\max}$ for all t as

$$l_{1:t} := \begin{cases} l_{\min} & \text{if } l_{1:t} \leq l_{\min}, \\ l_{1:t} & \text{if } l_{\min} < l_{1:t} < l_{\max}, \\ l_{\max} & \text{if } l_{\max} \leq l_{1:t}. \end{cases} \tag{9}$$

In the experiments shown below, we use $p_{\max} = 0.99$ and $p_{\min} = 0.01$ which correspond to $l_{\max} = \log(p_{\max}/p_{\min}) \simeq 4.6$ and $l_{\min} = \log(p_{\min}/p_{\max}) = -l_{\max} \simeq -4.6$.

Incidentally, in the text-prompted speaker verification tasks [4], we have assumed $t_i = t_0 = 0$. Thus, the second term itself in (8) converges with the increase of Δt for $\alpha = 1$ (GEBI), while it does not converge when $\alpha = 0$ (BI), which seems to have contributed to the better performance of GEBI in that task (see [4] for details). However, in the speaker detection task, we have to tune α, β and t_0 as examined in the experiments shown below.

2.4 Probabilistic Prediction for Speaker Detection

As shown in [4], we introduce multiple classes to classify the results of speaker detection, and then introduce probabilistic prediction.

Multiclass Classification for Speaker Detection. Let us suppose that a sequence of spoken words consists of subsequences uttered by a reference speaker $s^{[r]}$ to be detected and the other speaker $s \neq s^{[r]}$, alternatively. Moreover, let us assume that the period from $t_i + 1$ to $t_i + t_{\text{tran}}$ is transient and the period from $t_i + t_{\text{tran}}$ to t_{i+1} is stationary in the sequence of speakers, where t_i is the time when a subsequence of a speaker is terminated and t_{tran} denotes the duration of the transient. Then, we define four classes as follows.

c_{rs}: Class of spoken words which are uttered by the reference speaker during a stationary period.
c_{rt}: Class of spoken words which are uttered by the reference speaker during a transient period.
c_{ot}: Class of spoken words which are uttered by a speaker other than the reference speaker during a transient period.
c_{os}: Class of spoken words which are uttered by a speaker other than the reference speaker during a stationary period.

In order to simplify the explanation, let $C = \{c_i \mid i \in I^{[C]})\}$ be the set of above classes, where $I^{[C]} = \{\text{rs}, \text{rt}, \text{ot}, \text{os}\}$.

Probabilistic Prediction in Multiclass Classification. Let $X^{[\text{test}]} = \{(x_{1:T}^{[j]}, c(x_{1:T}^{[j]})) \mid j \in I^{[\text{test}]}\}$ denote a test dataset (see Sect. 3 for details), where $x_{1:T}^{[j]}$ is the j-th sequence of spoken words in $X^{[\text{test}]}$ and $c(x_{1:T}^{[j]}) = c(x_1^{[j]})c(x_2^{[j]}) \cdots c(x_T^{[j]})$ indicates the sequence of target (correct) classes of $x_{1:T}^{[j]}$ to be classified. Let $p_{\text{g}}(x_{1:t}^{[j]})$ denote the g-GEBI probability $p_{\text{g}}\left(s^{[r]} \mid \boldsymbol{v}_{1:t}^{[S]}\right)$ obtained for $x_{1:t}^{[j]}$ by applying (6), (9) and (2) for $t = 1, 2, \cdots, T$, recursively. Namely, $p_{\text{g}}(x_{1:t}^{[j]})$ indicates the g-GEBI probability of the speaker s of $x_{1:t}^{[j]}$ at t being the reference speaker $s^{[r]}$. Then, from the conventional single step BI, we have

$$p\left(c_i \mid p_{\rm g}(x_{1:t}^{[j]})\right) = \frac{p\left(p_{\rm g}\left(x_{1:t}^{[j]}\right) \mid c_i\right) p(c_i)}{\sum_{c_l \in C} p\left(p_{\rm g}(x_{1:t}^{[j]}) \mid c_l\right) p(c_l)}, \tag{10}$$

where $p(c_i)$ indicates the prior probability of $c_i \in C$, and $p\left(p_{\rm g}(x_{1:t}^{[j]}) \;\middle|\; c_i\right)$ denotes the likelihood. As a strategy of executing probabilistic prediction without using thresholds, we predict the class of $x_t^{[j]}$ by

$$\hat{c}(x_t^{[j]}) = \operatorname*{argmax}_{c_i \in C} p\left(c_i \mid p_{\rm g}(x_{1:t+t_{\rm delay}}^{[j]})\right), \tag{11}$$

and provide the class $\hat{c}(x_t^{[j]})$ and the probability $p(\hat{c}(x_t^{[j]}) \mid p_{\rm g}(x_{1:t+t_{\rm delay}}^{[j]}))$ to the user or the decision maker. Here, for a reliable prediction, we use a positive delay $t_{\rm delay} > 0$ because $x_t^{[j]}$ and the following appropriate number of words of a speaker is expected to provide a stabler prediction of the class of $x_t^{[j]}$ than only $x_t^{[j]}$.

3 Experiments

3.1 Experimental Setting

We have recorded speech data sampled with 8 kHz of sampling rate and 16 bits of resolution in a silent room of our laboratory. They are from seven speakers (2 female and 5 male speakers): $\tilde{S} = \{\text{fHS}, \text{fMS}, \text{mKK}, \text{mKO}, \text{mMT}, \text{mNH}, \text{mYM}\}$ for ten Japanese spoken words (or digits) $D = \{$/zero/, /ichi/, /ni/, /san/, /yon/, /go/, /roku/, /nana/, /hachi/, /kyu/ $\}$. Here, note that to execute the experiments for unregistered speaker as shown below, we have employed LOOCV method, where we select a speaker $s^{[\text{unreg}]} \in \tilde{S}$ for unregistered speaker and the other speakers $s \in S = \tilde{S} \backslash s^{[\text{unreg}]}$ are used for registered speakers. For each speaker and each word, ten samples are recorded on different dates and times among two months. We denote each spoken word by $x = x_{s,d,n}$ for $s \in \tilde{S}$, $d \in D$ and $n \in N = \{1, 2, \cdots, 10\}$, and the given dataset by $X = (x_{s,d,n} | s \in \tilde{S}, d \in D, n \in N)$.

From the given dataset X, we have generated training and test sequences of spoken words $x_{1:T}^{[\text{train},j]}$ and $x_{1:T}^{[\text{test},j]}$ for $j = 1, 2, \cdots$, respectively. For each training and test sequence $x_{1:T}^{[j]} = x_{1:T}^{[\text{train},j]}$ and $x_{1:T}^{[j]} = x_{1:T}^{[\text{test},j]}$, the constituent word $x_t^{[j]}$ is selected from X as $x_t^{[j]} = x_{s,d,n} \in X$, where the word d and the sample number n are selected randomly from D and N, respectively. In order to determine s, we select the reference speaker $s^{[r]}$ randomly from $S = \tilde{S} \backslash \{s^{[\text{unreg}]}\}$ for each $x_{1:T}^{[j]}$. Furthermore, we divide each sequence $x_{1:T}^{[j]}$ into L_i-length subsequences as $x_{t_i+1:t_i+L_i}^{[j]}$ for $t_0 = 0$, $t_{i+1} = t_i + L_i$ and $i = 0, 1, 2, \cdots$, where L_i are set differently in each experiment as shown later. Then, the speaker of $x_{t_i+1:t_i+L_i}$ is selected randomly from $s \in \tilde{S} \backslash s^{[r]}$ for even $i = 2m$ ($m = 0, 1, 2, \cdots$), and set $s^{[r]}$ for odd $i = 2m+1$ ($m = 0, 1, 2, \cdots$). Thus, for multiclass classification shown in Sect. 2.4, the target class $c(x_t^{[j]})$ is given by

$$c(x_t^{[j]}) = \begin{cases} c_{\text{ot}} \text{ for } t = t_{2m} + k \text{ and } k = 1, 2, \cdots, t_{\text{tran}} \\ c_{\text{os}} \text{ for } t = t_{2m} + k \text{ and } k = t_{\text{tran}} + 1, \cdots, L_{2m} \\ c_{\text{rt}} \text{ for } t = t_{2m+1} + k \text{ and } k = 1, 2, \cdots, t_{\text{tran}} \\ c_{\text{rs}} \text{ for } t = t_{2m+1} + k \text{ and } k = t_{\text{tran}} + 1, \cdots, L_{2m+1} \end{cases} \tag{12}$$

where $m = 0, 1, 2, \cdots$. Finally, let $X^{[\text{train}]} = \{(x_{1:T}^{[\text{train},j]}, c(x_{1:T}^{[\text{train},j]})) \mid j \in I^{[\text{train}]}$ and $X^{[\text{test}]} = \{(x_{1:T}^{[\text{test},j]}, c(x_{1:T}^{[\text{test},j]})) \mid j \in I^{[\text{test}]}$ be the training and test dataset for the multiclass classification. Here, note that $x_t^{[\text{train},j]}$ in $x_{1:T}^{[\text{train},j]} \in X^{[\text{train}]}$ and $x_t^{[\text{test},j]}$ in $x_{1:T}^{[\text{test},j]} \in X^{[\text{test}]}$ are not the same but independent and identically distributed (i.i.d).

In order to evaluate the performance of the present method for unknown (untrained) data and unregistered speaker as a speaker other than the reference speaker, we employ a combination of LOOCV (leave-one-out cross-validation) and OOB (out-of-bag) estimate (see [4] for details). For the learning machines, we have used CAN2s for learning piecewise linear approximation of nonlinear functions (see [7] for details).

As a result of tuning α and β in (6), we show the result for $(\alpha, \beta) = (0, 1)$, $(1, 1)$, $(0, 0.25)$ and $(0.3, 1)$, to each of which we refer as BI, GEBI, g-BI (generalized BI) and g-GEBI, respectively. Since $t - t_0$ in (6) may increase to very large, we set $t_0 = 0$ at the initial time $t = 0$, and update $t_0 = t$ when $l_{1:t} = l_{\min}$ and $t - t_0 \geq 500$. We use $p(c_i) = 1/|C|$ for all c_i in (10). We have tuned $t_{\text{delay}} = 4$ for (11) and $t_{\text{tran}} = 3$ for (12).

3.2 Experimental Results and Analysis

Experimental Result of Probabilistic Prediction. In Fig. 2, we show the mean and the variance of $p_{\text{g}} = (x_{1:t}^{[j]})$ vs. t. For each test sequence $x_{1:t}^{[j]}$, the length of subsequences are L_i = 100, 20, 30, 15, 30, 10, 30, 5, 30, 20, 30, 15, 30, 10, 30, 5, 50 for $i = 0, 1, 2, \cdots, 16$, respectively, and the reference speaker $s^{[r]} \in S$ utters in the periods for $i = 2m + 1$ ($m = 0, 1, 2, \cdots$) after the other speaker $s \in S \backslash \{s^{[r]}\}$ utters in the periods for $i = 2m$. In Fig. 2(a), we can see that the variance of p_{g} for BI is very big, which seems inappropriate for speaker detection. In (b), we can see the effect of GEBI whose p_{g} is the geometric mean $\left(\prod_{k=0}^{t} p\left(v_k^{[S]} \mid s^{[r]}\right)\right)^{1/t}$ from the initial time to the current time t [4], and the result seems inappropriate for speaker detection. By means of tuning α and β to avoid the above disadvantages, we have the results of (c) g-BI and (d) g-GEBI.

In Fig. 3, we show the probability distribution of the classes $p\left(c_i \mid p_{\text{g}}(x_{1:t}^{[j]})\right)$ obtained by (10) for g-BI and g-GEBI, where we can see the boundary of the

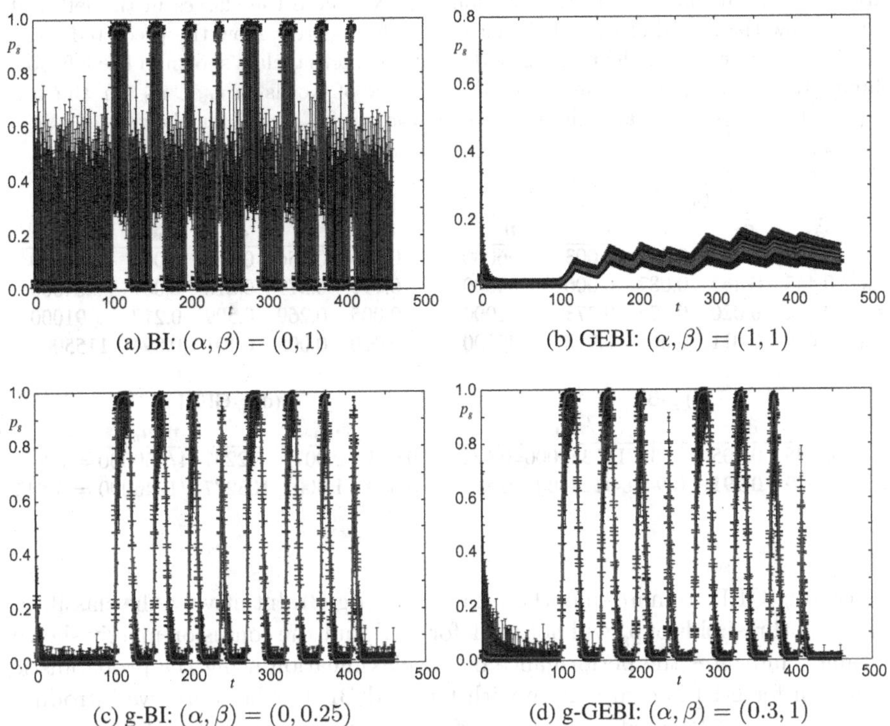

(a) BI: $(\alpha, \beta) = (0, 1)$ (b) GEBI: $(\alpha, \beta) = (1, 1)$

(c) g-BI: $(\alpha, \beta) = (0, 0.25)$ (d) g-GEBI: $(\alpha, \beta) = (0.3, 1)$

Fig. 2. Experimental result of the mean (blue line) and the error bars of $p_g = (x_{1:t}^{[j]})$ vs. t for (a) BI, (b) GEBI, (c) g-BI and (d) g-GEBI. The mean is obtained for each LOOCV datasets, so that there are seven blue lines in each picture although almost the same except for (b). The plus and minus error bars indicate RMS (root mean square) of positive and negative errors from the mean, respectively. (Color figure online)

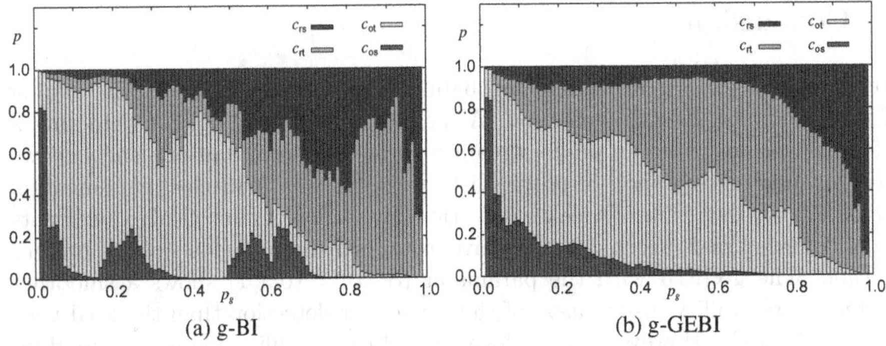

(a) g-BI (b) g-GEBI

Fig. 3. Experimental result of $p = p\left(c_i \mid p_g(x_{1:t}^{[j]})\right)$ vs. $p_g = (x_{1:t}^{[j]})$. The horizontal axis indicates p_g and the vertical length of a colored bar indicates the probability $p = p(c_i \mid p_g)$ of the class c_{rs} (blue), c_{rt} (green), c_{ot} (yellow), and c_{os} (red). (Color figure online)

Table 1. Experimental result of confusion matrix, where the classes in the leftmost column show the actual class and the classes in the top row show the predicted class. In each table, n indicates the number of data in each actual class shown in the leftmost column. In (c) and (d), n_{all} indicates the number of words in $c_{\mathrm{rs}} \cup c_{\mathrm{rt}}$ for the row of $c_{\mathrm{rd}} = (c_{\mathrm{rs}} \cup c_{\mathrm{rt}}) \cap \hat{c}_{\mathrm{d}}$, and the number of words in $c_{\mathrm{os}} \cup c_{\mathrm{ot}}$ for the row of $c_{\mathrm{od}} = (c_{\mathrm{os}} \cup c_{\mathrm{ot}}) \cap \hat{c}_{\mathrm{d}}$.

(a) g-BI

	$\hat{c}_{\mathrm{rs}}$	$\hat{c}_{\mathrm{rt}}$	$\hat{c}_{\mathrm{ot}}$	$\hat{c}_{\mathrm{os}}$	n
c_{rs}	0.752	0.200	0.045	0.003	266000
c_{rt}	0.445	0.462	0.087	0.006	84000
c_{ot}	0.031	0.020	0.675	0.273	91000
c_{os}	0.006	0.011	0.065	0.918	1155000

(b) g-GEBI

	$\hat{c}_{\mathrm{rs}}$	$\hat{c}_{\mathrm{rt}}$	$\hat{c}_{\mathrm{ot}}$	$\hat{c}_{\mathrm{os}}$	n
c_{rs}	0.784	0.156	0.056	0.004	266000
c_{rt}	0.197	0.579	0.218	0.007	84000
c_{ot}	0.005	0.269	0.509	0.217	91000
c_{os}	0.000	0.005	0.101	0.894	1155000

(c) g-BI

	$\hat{c}_{\mathrm{rs}}$	$\hat{c}_{\mathrm{os}}$	n/n_{all}
c_{rd}	0.995	0.005	238714/ 350000=0.682
c_{od}	0.009	0.991	1094884/1246000=0.879

(d) g-GEBI

	$\hat{c}_{\mathrm{rs}}$	$\hat{c}_{\mathrm{os}}$	n/n_{all}
c_{rd}	0.993	0.007	$226744/350000 = 0.648$
c_{od}	0.000	1.000	$1052772/1246000 = 0.845$

classes for g-GEBI is more smoother than that for g-BI. Intuitively, the classification using thresholds of p_{g} will not work for g-BI, but the confusion matrix shown in Table 1 indicates the performance of the classification using the probabilistic prediction for g-BI is competitive with for g-GEBI. Furthermore, we introduce the decision of "detected", "undetected" and "undecidable" for the predicted words in $\hat{c}_{\mathrm{rs}}$, $\hat{c}_{\mathrm{os}}$, and the other classes (i.e. transient $\hat{c}_{\mathrm{rt}}$ and $\hat{c}_{\mathrm{ot}}$), respectively. Then, for the words in decided class $\hat{c}_{\mathrm{d}} = \hat{c}_{\mathrm{rs}} \cup \hat{c}_{\mathrm{os}}$, we have the confusion matrix in Table 1(c) and (d), where $c_{\mathrm{rd}} = (c_{\mathrm{rs}} \cup c_{\mathrm{rt}}) \cap \hat{c}_{\mathrm{d}}$ and $c_{\mathrm{od}} = (c_{\mathrm{os}} \cup c_{\mathrm{ot}}) \cap \hat{c}_{\mathrm{d}}$. The result shows a very high performance of the classification for both g-BI and g-GEBI.

4 Conclusion

We have presented a method of probabilistic prediction to avoid the tuning of thresholds for speaker detection. After introducing g-GEBI for detecting a reference speaker from an audio stream uttered by more than one speaker, we have formulated a method of probabilistic prediction in multiclass classification for classifying the result of speaker detection. By means of numerical experiments using recorded real speech data, we have examined the properties of the present method. The g-GEBI with the parameter $(\alpha, \beta) = (0.3, 1)$ shows a smoother posterior probability distribution of the classes for detection than the g-BI with $(\alpha, \beta) = (0, 0.25)$. However, the performance of the classification is very good for both the g-GEBI and the g-BI. We would like to analyze this result in detail and other properties of the present method much more. Especially, the analysis for the cases where there are unregistered speakers may be harder but important for practical use.

References

1. Beigi, H.: Fundamentals of speaker recognition. Springer-Verlag New York Inc. (2011)
2. Kurogi, S., Sakashita, S., Takeguchi, S., Ueki, T., Matsuo, K.: Probabilistic prediction in multiclass classification derived for flexible text-prompted speaker verification. In: Arik, S., Huang, T., Lai, W.K., Liu, Q. (eds.) ICONIP 2015. LNCS, vol. 9489, pp. 216–225. Springer, Heidelberg (2015). doi:10.1007/978-3-319-26532-2_24
3. Kurogi, S., Ueki, T., Mizobe, Y., Nishida, T.: Text-prompted multistep speaker verification using Gibbs-distribution-based extended Bayesian inference for reducing verification errors. In: Lee, M., Hirose, A., Hou, Z.-G., Kil, R.M. (eds.) ICONIP 2013. LNCS, vol. 8228, pp. 184–192. Springer, Heidelberg (2013). doi:10.1007/978-3-642-42051-1_24
4. Kurogi, S., Ueki, T., Takeguchi, S., Mizobe, Y.: Properties of text-prompted multistep speaker verification using Gibbs-distribution-based extended Bayesian inference for rejecting unregistered speakers. In: Loo, C.K., Yap, K.S., Wong, K.W., Teoh, A., Huang, K. (eds.) ICONIP 2014. LNCS, vol. 8835, pp. 35–43. Springer, Heidelberg (2014). doi:10.1007/978-3-319-12640-1_5
5. Slingo, J., Palmer, T.: Uncertainty in weather and climate prediction. Phil. Trans. R. Soc. A **369**, 4751–4767 (2011)
6. Kurogi, S., Ueno, T., Sawa, M.: A batch learning method for competitive associative net and its application to function approximation. In: Proceedings of the SCI 2004, vol. V, pp. 24–28 (2004)
7. Kurogi, S., Mineishi, S., Sato, S.: An analysis of speaker recognition using bagging CAN2 and pole distribution of speech signals. In: Wong, K.W., Mendis, B.S.U., Bouzerdoum, A. (eds.) ICONIP 2010. LNCS, vol. 6443, pp. 363–370. Springer, Heidelberg (2010). doi:10.1007/978-3-642-17537-4_45
8. Campbell, J.P.: Speaker recognition: a tutorial. Proc. IEEE **85**(9), 1437–1462 (1997)

Probabilistic Prediction for Text-Prompted Speaker Verification Capable of Accepting Spoken Words with the Same Meaning but Different Pronunciations

Shota Sakashita, Satoshi Takeguchi, Kazuya Matsuo, and Shuichi Kurogi(✉)

Kyushu Institute of Technology, Tobata, Kitakyushu, Fukuoka 804-8550, Japan
{sakashita,takeguchi}@kurolab.cntl.kyutech.ac.jp,
{matsuo,kuro}@cntl.kyutech.ac.jp
http://kurolab.cntl.kyutech.ac.jp/

Abstract. So far, we have presented a method of probabilistic prediction using GEBI (Gibbs-distribution based Bayesian inference) for flexible text-prompted speaker verification. For more flexible and practical verification, this paper presents a method of verification capable of accepting spoken words with the same meaning but different pronunciations. For example, Japanese language has different pronunciations for a digit, such as /yon/ and /shi/ for 4, /nana/ and /shichi/ for 7, which are usually uttered via unintentional selection, and then it is a practical problem in speech verification of words involving digits, such as ID numbers. With several assumptions, we present a modification of GEBI for dealing with such words. By means of numerical experiments using recorded real speech data, we examine the properties of the present method and show the validity and the effectiveness.

Keywords: Probabilistic prediction · Text-prompted speaker verification · Gibbs-distribution-based extended Bayesian inference · Words with the same meaning but different pronunciations

1 Introduction

So far, we have presented a method of probabilistic prediction for text-prompted speaker verification [1]. Here, from [2], text-prompted speaker verification has been developed to combat spoofing from impostors and digit strings are often used to lower the complexity of processing. From another perspective, the method employs multistep inference using GEBI (Gibbs-distribution based extended Bayesian inference) for reducing single step verification error, and introduced the probabilistic prediction for flexible verification without using thresholds for verification. Here, note that, as shown in [3], we can see that the probabilistic prediction in weather and climate forecasting allows the users to take flexible and appropriate action within a proper understanding of the uncertainties. Thus, we are analyzing to develop such flexible system for text-prompted speaker verification.

A. Hirose et al. (Eds.): ICONIP 2016, Part IV, LNCS 9950, pp. 312–320, 2016.
DOI: 10.1007/978-3-319-46681-1_38

From the point of view of more flexible and practical verification, this paper presents a method of verification capable of accepting spoken words with the same meaning but different pronunciations. For example, Japanese language has different pronunciations for a digit, such as /yon/ and /shi/ for 4, /nana/ and /shichi/ for 7, which are usually uttered via unintentional selection. English language also has such words, such as /zero/ and /oh/ for 0. Thus, it is a practical problem in text-prompted speaker verification using spoken words consisting of digits, such as ID numbers.

Here, note that our speech processing system employs competitive associative nets (CAN2s). The CAN2 is an artificial neural net for learning efficient piecewise linear approximation of nonlinear function [4], and we have shown that feature vectors of pole distribution extracted from piecewise linear predictive coefficients obtained by the bagging (bootstrap aggregating) version of the CAN2 reflect nonlinear and time-varying vocal tract of the speaker [5]. Although the most common way to characterize speech signal in the literature is short-time spectral analysis, such as Linear Prediction Coding (LPC) and Mel-Frequency Cepstrum Coefficients (MFCC) [6], the bagging CAN2 learns more precise information than LPC and MFCC (see [5] for details).

We formulate the method of probabilistic prediction in Sect. 2, followed by showing experimental results and analysis in Sect. 3, and the conclusion in Sect. 4.

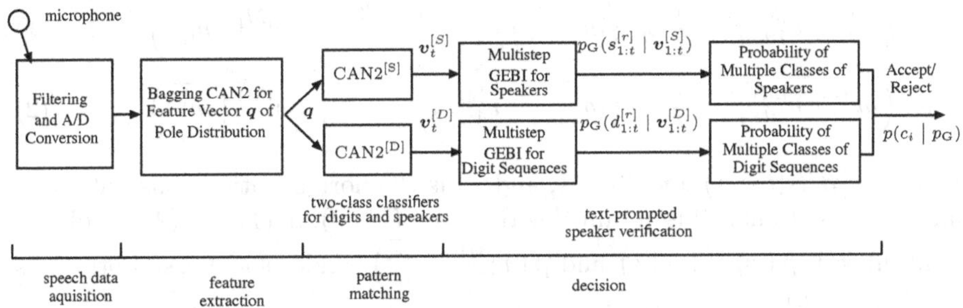

Fig. 1. Diagram of text-prompted speaker verification system using CAN2s

2 Probabilistic Prediction for Text-Prompted Speaker Verification

Figure 1 shows an overview of the present text-prompted speaker verification system using CAN2s. In the same way as general speaker recognition systems [6], it consists of four steps: speech data acquisition, feature extraction, pattern matching, and making a decision. In this research study, we use a feature vector of pole distribution obtained from a speech signal (see [5] for details).

2.1 Speaker and Text Verification of Words with Different Pronunciations

In order to achieve text-prompted speaker verification of spoken word involving a word with the same meaning but different pronunciations, let $S = \{s^{[i]} \mid i \in I^{[S]}\}$ be a set of speakers, $D^{[i]} = \{d^{[i:j]} \mid j \in I^{[D^{[i]}]}\}$ a set of words with the same meaning but different pronunciations, $D = \{d^{[i:j]} \mid i \in I^{[\widetilde{D}]}, j \in I^{[D^{[i]}]}\}$ a set of words whose meanings or pronunciations are different each other, and $\widetilde{D} = \{D^{[i]} \mid i \in I^{[\widetilde{D}]}\}$ a family of the sets $D^{[i]}$ indicating a family (set) of word meanings. Here, $I^{[S]}$, $I^{[\widetilde{D}]}$ and $I^{[D^{[i]}]}$ are index sets. Let $\mathrm{LM}^{[M]}$ for $M = S$ and D be a set of learning machines $\mathrm{LM}^{[m]}$ $(m \in I^{[M]})$, and each $\mathrm{LM}^{[m]}$ learns to predict a single-step verification as $v^{[m]} = 1$ for the acceptance of a speech segment of a speaker $m = s^{[i]}$ or a word $m = d^{[i:j]}$, and $v^{[m]} = 0$ for the rejection. Here, let us suppose that we have speech segments of spoken words obtained by some appropriate segmentation method, and this research focuses on multistep verification of spoken word sequences.

Let $\boldsymbol{v}_{1:T}^{[M]} = \boldsymbol{v}_1^{[M]} \boldsymbol{v}_2^{[M]} \cdots \boldsymbol{v}_T^{[M]}$ be an output sequence of $\mathrm{LM}^{[M]}$ for reference sequence $m_{1:T}^{[r]} = m_1^{[r]} m_2^{[r]} \cdots m_T^{[r]}$, where $m_t^{[r]} = s_t^{[r]} \in S$ and $m_t^{[r]} = D_t^{[r]} \in \widetilde{D}$ for reference speaker and reference word, respectively. Then, from GEBI (see [7] for details), we have posterior probability recursively for $t = 1, 2, \cdots, T$ as follows,

$$p_{\mathrm{G}}\left(m_{1:t}^{[r]} \mid \boldsymbol{v}_{1:t}^{[M]}\right) = \frac{1}{Z_t} p_{\mathrm{G}}\left(m_{1:t-1}^{[r]} \mid \boldsymbol{v}_{1:t-1}^{[M]}\right)^{\beta_t/\beta_{t-1}} p\left(\boldsymbol{v}_t^{[M]} \mid m_t^{[r]}\right)^{\beta_t}, \quad (1)$$

$$p_{\mathrm{G}}\left(\overline{m_{1:t}^{[r]}} \mid \boldsymbol{v}_{1:t}^{[M]}\right) = \frac{1}{Z_t} p_{\mathrm{G}}\left(\overline{m_{1:t-1}^{[r]}} \mid \boldsymbol{v}_{1:t-1}^{[M]}\right)^{\beta_t/\beta_{t-1}} p\left(\boldsymbol{v}_t^{[M]} \mid \overline{m_t^{[r]}}\right)^{\beta_t}. \quad (2)$$

where $\beta_t = \beta/t$ $(t \geq 1)$ and $\beta_0 = 1$, and Z_t is the normalization constant. Note that the conventional BI is obtained with $\beta_t = 1$ $(t \geq 0)$. In (1) and (2), we obtain the likelihood, $p\left(\boldsymbol{v}_t^{[M]} \mid m_t^{[r]}\right)$ and $p\left(\boldsymbol{v}_t^{[M]} \mid \overline{m_t^{[r]}}\right)$, under some assumptions as follows. For $m_t^{[r]} = s_t^{[r]}$, we have

$$p\left(\boldsymbol{v}_t^{[S]} \mid s_t^{[r]}\right) = \prod_{s^{[i]} \in S} p(v_t^{[s^{[i]}]} \mid s_t^{[r]}), \quad (3)$$

$$p\left(\boldsymbol{v}_t^{[S]} \mid \overline{s_t^{[r]}}\right) = \frac{\sum_{s \in S \backslash \{s_t^{[r]}\}} p(\boldsymbol{v}_t^{[s]} \mid s)\, p(s)}{\sum_{s \in S \backslash \{s_t^{[r]}\}} p(s)} = \frac{\sum_{s \in S \backslash \{s_t^{[r]}\}} \prod_{s^{[i]} \in S} p(v_t^{[s^{[i]}]} \mid s)}{|S| - 1}, \quad (4)$$

where we assume conditional independence and uniform prior probability of speakers as $p(s) = 1/|S|$. Similarly, for $m_{1:t}^{[r]} = D_{1:t}^{[r]}$, we have

$$p\left(\boldsymbol{v}_t^{[D]} \mid D_t^{[r]}\right) = \frac{\sum_{d\in D_t^{[r]}} p\left(\boldsymbol{v}_t^{[D]}, d\right)}{\sum_{d\in D_t^{[r]}} p(d)} = \frac{\sum_{d\in D_t^{[r]}} \prod_{d^{[i:j]}\in D} p\left(v_t^{[d^{[i:j]}]} \mid d\right)}{|D^{[r]}|} \tag{5}$$

$$p\left(\boldsymbol{v}_t^{[D]} \mid \overline{D_t^{[r]}}\right) = \frac{\sum_{d\in D\backslash D_t^{[r]}} p\left(\boldsymbol{v}_t^{[D]} \mid d\right) p(d)}{\sum_{d\in D\backslash D_t^{[r]}} p(d)} = \frac{\sum_{d\in D\backslash D_t^{[r]}} \prod_{d^{[i:j]}\in D} p\left(v_t^{[d^{[i:j]}]} \mid d\right)}{|D| - |D^{[r]}|} \tag{6}$$

where we assume conditional independence and uniform prior probability of words as $p(d) = 1/|D|$. The validity of the conditional independence assumption in the above equations affects the performance of the present method, while it depends on the performance of learning machines which produce $\boldsymbol{v}_t^{[S]}$ and $\boldsymbol{v}_t^{[D]}$.

2.2 Probabilistic Prediction for Speaker and Text Verification

As shown in [1], we define multiple classes to classify the verification results, and then execute probabilistic prediction for speaker and text verification.

Multiclass Classification for Speaker and Text Verification. For speaker verification, we use the following three classes, where we suppose all elements in each input and reference speaker sequence, respectively, consists of the same speaker;

$c_{+1}^{[S]}$ (Class of correct speakers): class of speakers satisfying $s_{1:T} = s_{1:T}^{[r]}$ $(\in S_{1:T})$ for the input $s_{1:T}$ and the reference $s_{1:T}^{[r]}$, where S is the set of registered speakers, and $S_{1:T}$ denotes the set of $s_{1:T}$ whose all elements s_t $(t = 1, 2, \cdots, T)$ are registered speaker $s_t \in S$.

$c_{-1}^{[S]}$ (Class of incorrect speakers): class of speakers satisfying $s_{1:T} \neq s_{1:T}^{[r]}$ for $s_{1:T}, s_{1:T}^{[r]} \in S_{1:T}$.

$c_0^{[S]}$ (Class of unregistered speakers): class of speakers satisfying $s_{1:T} \neq s_{1:T}^{[r]}$ for $s_{1:T} \notin S_{1:T}$.

For text (or word sequence) verification, we consider the following $N+1$ classes of $T(= mN)$-length word sequence consisting of m times of N-length subsequences:

$c_i^{[\widetilde{D}]}$ for $i = 0, 1, 2, \cdots, N$ (Class of word sequences with correct ratio being i/N): class of input and reference word sequences, or $d_{1:T}$ and $d_{1:T}^{[r]}$, respectively, each of them consists of m times of N-length subsequence, and i words in input and reference subsequences are the same.

In order to simplify the explanation, let $C^{[S]} = \{c_i^{[S]} \mid i \in I^{[C^{[S]}]})\}$ be the set of speaker verification classes, $C^{[\widetilde{D}]} = \{c_i^{[\widetilde{D}]} \mid i \in I^{[C^{[\widetilde{D}]}]}\}$ be the set of text verification classes, C denote $C^{[S]}$ or $C^{[\widetilde{D}]}$, and $I^{[C]}$ denote $I^{[C^{[S]}]} = \{-1, 0, 1\}$ or $I^{[C^{[\widetilde{D}]}]} = \{0, 1, 2, \cdots, N\}$.

Probabilistic Prediction in Multiclass Classification. Let $X^{[\text{test}]} = \{(x_{1:T}^{[j]}, c(x_{1:T}^{[j]})) \mid j \in I^{[\text{test}]}\}\}$ denote a test dataset (see Sect. 3 for details), where $x_{1:T}^{[j]}$ is the jth sequence of spoken words in $X^{[\text{test}]}$ and $c(x_{1:T}^{[j]}) = c(x_1^{[j]})c(x_2^{[j]}) \cdots c(x_T^{[j]})$ indicates the sequence of target (correct) classes of $x_{1:T}^{[j]}$ to be classified. Here, suppose that $x_t^{[j]}$ is spoken by $s(x_t^{[j]}) \in S$, the meaning is $d(x_t^{[j]}) \in \widetilde{D}$, and the class $c(x_t^{[j]})$ indicates a class in $C^{[S]}$ and $C^{[\widetilde{D}]}$ for verification of speaker and word, respectively.

Let $p_{\text{G}}(x_{1:T}^{[j]})$ denote the GEBI probability $p_{\text{G}}\left(m_{1:T}^{[r]} \mid \boldsymbol{v}_{1:T}^{[M]}\right)$ obtained for $x_{1:T}^{[j]}$ by applying (1) and (2) for $t = 1, 2, \cdots, T$, recursively. Then, from the conventional BI, we have

$$p\left(c_i \mid p_{\text{G}}(x_{1:T}^{[j]})\right) = \frac{p\left(p_{\text{G}}\left(x_{1:T}^{[j]}\right) \mid c_i\right) p(c_i)}{\sum_{c_l \in C} p\left(p_{\text{G}}(x_{1:T}^{[j]}) \mid c_l\right) p(c_l)}, \tag{7}$$

where $p\left(p_{\text{G}}(x_{1:T}^{[j]}) \mid c_i\right)$ denotes the likelihood of $p_{\text{G}}(x_{1:T}^{[j]})$ given c_i, and $p(c_i)$ indicates the prior probability of $c_i \in C$. We obtain the distribution of $p\left(p_{\text{G}}(x_{1:T}^{[j]}) \mid c_i\right)$ from training dataset $X^{[\text{train}]}$ and assume uniform prior $p(c_i) = 1/|C|$ in the experiments shown below.

With the probability $p\left(c_i \mid p_{\text{G}}(x_{1:T}^{[j]})\right)$ for $c_i \in C$, a user or a decision maker is expected to make flexible verification as shown in [1]. As a strategy of executing probabilistic prediction without using thresholds, we can predict the class of $x_t^{[j]}$ by

$$\hat{c}(x_T^{[j]}) = \operatorname*{argmax}_{c_i \in C} p\left(c_i \mid p_{\text{G}}(x_{1:T}^{[j]})\right), \tag{8}$$

and provide the class $\hat{c}(x_T^{[j]})$ and the probability $p(\hat{c}(x_t^{[j]}) \mid p_{\text{G}}(x_{1:T}^{[j]}))$ to the user or the decision maker. The average classification error (ACE) of $\hat{c}(x_T^{[j]})$ is given by

$$L_{\text{ACE}} = \frac{1}{n}\left[\sum_{j \in I^{[\text{test}]}} \mathbf{1}\{c(x_T^{[j]}) \neq \hat{c}(x_T^{[j]})\}\right] = \frac{1}{n}\left[\sum_{\{j | c(x_T^{[j]}) \neq \hat{c}(x_T^{[j]})\}} 1\right], \tag{9}$$

where $\mathbf{1}\{z\}$ indicates an indicator function, equal to 1 if z is true, and 0 if z is false, and $\{j | c(x_T^{[j]}) \neq \hat{c}(x_T^{[j]})\}$ indicates the set of indices satisfying $c(x_T^{[j]}) \neq \hat{c}(x_T^{[j]})$ for $j \in I^{[\text{test}]}$.

3 Experiments

3.1 Experimental Setting

We have recorded speech data sampled with 8 kHz of sampling rate and 16 bits of resolution in a silent room of our laboratory. They are from seven speakers (2 female and 5 mail speakers): $S = \{\mathrm{fHS}, \mathrm{fMS}, \mathrm{mKK}, \mathrm{mKO}, \mathrm{mMT}, \mathrm{mNH}, \mathrm{mYM}\}$. We use ten digits for a set of word meanings, or $\widetilde{D} = \{0, 1, 2, \cdots, 9\}$, and 14 digits (words) of Japanese pronunciations, where we usually and unintentionally use one of two different pronunciations for the digits $d^{[i:j]}$ for $i = 0, 4, 7, 9$. Namely, we usually use pronunciations as $d^{[i:0]}$ = /zero/, /ichi/, /ni/, /san/, /yon/, /go/, /roku/, /nana/, /hachi/, /kyu/ for the digits $0, 1, 2, \cdots, 9$, respectively, and sometimes unintentionally use $d^{[i:1]}$ = /rei/, /shi/, /shichi/, /ku/ for $i = 0, 4, 7, 9$, respectively. For each speaker and each digit, ten samples are recorded on different dates and times among two months. We denote each spoken digit by $x = x_{s,d,l}$ for $s \in S$, $d \in D = \{d^{[i:j]} \mid i = 0, 1, \cdots, 9; j = 0, 1\}$ and $l \in L = \{1, 2, \cdots, 10\}$, and given dataset by $X = (x_{s,d,l} | s \in S, d \in D, l \in L)$.

By meas of random selection from X, we have generated training dataset $X^{[\mathrm{train}]} = \{(x_{1:T}^{[j]}, c(x_{1:T}^{[j]})) \mid j \in I^{[\mathrm{train}]}\}$ for making the likelihood $p\left(p_{\mathrm{G}}(x_{1:T}^{[j]}) \,\middle|\, c_i\right)$ given in (7) and test dataset $X^{[\mathrm{test}]} = \{(x_{1:T}^{[j]}, c(x_{1:T}^{[j]})) \mid j \in I^{[\mathrm{test}]}\}$ for evaluating the performance of probabilistic prediction. A data $x_{1:T}^{[j]}$ indicates $T(=15)$-length spoken digits for $T = m \times N = 15$ with $m(=3)$ times of $N(=5)$-length digit sequences indicating some ID numbers for verification. Of course, $x_{1:T}^{[j]}$ in $X^{[\mathrm{train}]}$ and $X^{[\mathrm{test}]}$ are not the same but should be independent and identically distributed (i.i.d). To have this done, for each of training and test datasets, we have generated 1,000 data for each combination of 3 classes of speaker sequences involving correct, incorrect and unregistered speakers and 6 classes of digit sequences involving i/N correct digits for $i = 0, 1, \cdots, N = 5$.

We have employed OOB (out-of-bag) estimate to evaluate the performance of learning machines $\mathrm{LM}^{[M]}$, and LOOCV (leave-one-out cross-validation) method to evaluate the performance for testing spoken words uttered by unregistered speaker (see [7] for details). For the regression learning machines, we have used CAN2s for learning piecewise linear approximation of nonlinear functions (see [5] for details).

3.2 Experimental Results and Analysis

We have executed two experiments. The first one is to examine the effect of the addition of words with different pronunciations by means of using a default set given by $D_{\mathrm{def}} = \{d^{[i:0]} \mid i = 0, 1, \cdots, 9\}$ and augmented sets given by $D_{\mathrm{def}\oplus 4'7'} = D_{\mathrm{def}} \cup \{/\mathrm{shi}/, /\mathrm{shichi}/\}$ and $D_{\mathrm{def}\oplus 0'4'7'9'} = D_{\mathrm{def}} \cup \{/\mathrm{rei}/, /\mathrm{shi}/, /\mathrm{shichi}/, /\mathrm{kyu}/\}$. The 2nd one is to examine the effect of the combination of different pronunciations by means of comparing D_{def} with $D_{\mathrm{def}\ominus 7 \oplus 7'} = D_{\mathrm{def}} \backslash \{/\mathrm{nana}/\} \cup \{/\mathrm{shichi}/\}$ and $D_{\mathrm{def}\oplus 7'} = D_{\mathrm{def}} \cup \{/\mathrm{shichi}/\}$.

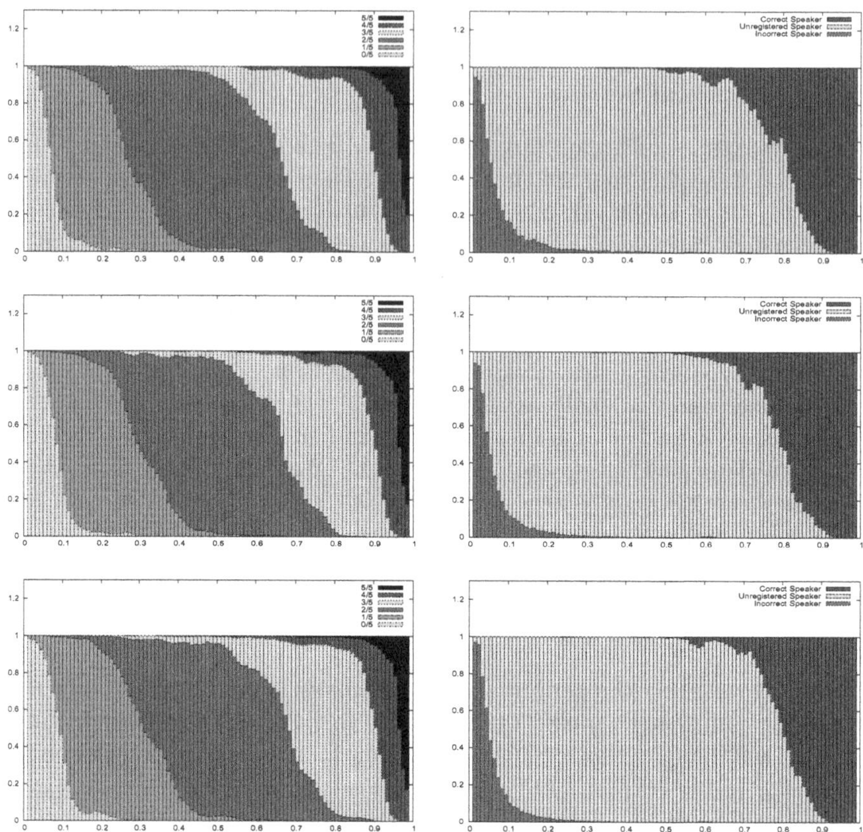

Fig. 2. Experimental result of posterior probability $p\left(c_i \mid p_{\mathrm{G}}(x_{1:T}^{[j]})\right)$ for spoken digits $x_{1:T}^{[j]}$ in D_{def} (top), $D_{\mathrm{def}\oplus 4'7'}$ (middle) and $D_{\mathrm{def}\oplus 0'4'7'9'}$ (bottom), respectively, and the classes c_i in $C^{[\tilde{D}]}$ and $C^{[S]}$ for verifying digits (left) and speakers (right), respectively. The horizontal axis indicates $p_{\mathrm{G}}(x_{1:T}^{[j]})$ and the vertical length of a colored bar indicates the probability of a class c_i corresponding to the color. (Color figure online)

The posterior probability for the first experiment is shown in Fig. 2, where we can see that the boundaries of the classes are smooth and it seems natural to estimate the class to be verified by (8) without using thresholds. This indicates a validity and an effectiveness of the probabilistic prediction using GEBI. Here, note that, as shown in [1], we could not have smooth boundaries by means of using the conventional Bayesian inference, or using $\beta_t = 1$ in (1) and (2), which is not shown in this paper due to the limit of space. We could not find out any clear difference of posteriors for D_{def}, $D_{\mathrm{def}\oplus 4'7'}$ and $D_{\mathrm{def}\oplus 0'4'7'9'}$ in Fig. 2.

In Table 1, we show an experimental result of ACE given by (9) of the probabilistic prediction obtained by (8). The columns of D_{def}, $D_{\mathrm{def}\oplus 4'7'}$ and $D_{\mathrm{def}\oplus 0'4'7'9'}$ show the effect of "pronunciation addition". Namely, we can see

Table 1. Experimental result of ACE. Class index $i = +1, 0, -1$ is of the classes $c_i^{[S]}$ in speaker verification, and $i = 5, 4, \cdots, 0$ is of $c_i^{[\tilde{D}]}$ in text verification.

			Pronunciation addition		Pronunciation combination	
	Class index i	D_{def}	$D_{\text{def}\oplus 4'7'}$	$D_{\text{def}\oplus 0'4'7'9'}$	$D_{\text{def}\ominus 7\oplus 7'}$	$D_{\text{def}\oplus 7'}$
Speaker verification	+1	0.010	0.004	0.004	0.010	0.007
	0	0.123	0.135	0.078	0.172	0.127
	−1	0.066	0.044	0.035	0.071	0.049
	Mean	0.066	0.061	0.039	0.085	0.061
Text verification	5	0.089	0.055	0.044	0.081	0.049
	4	0.431	0.662	0.659	0.473	0.666
	3	0.175	0.155	0.170	0.178	0.161
	2	0.125	0.152	0.137	0.119	0.124
	1	0.178	0.181	0.195	0.158	0.178
	0	0.097	0.075	0.107	0.091	0.076
	Mean	0.182	0.213	0.219	0.184	0.209

that, in speaker verification, the smallest ACE is obtained for $D_{\text{def}\oplus 0'4'7'9'}$, followed by in order for $D_{\text{def}\oplus 4'7'}$ and D_{def}. This is considered owing that there are much more features for speaker verification in $D_{\text{def}\oplus 0'4'7'9'}$ than in others. In text verification, the inverse order is obtained, i.e., the smallest ACE for D_{def}, followed by in order for $D_{\text{def}\oplus 4'7'}$ and $D_{\text{def}\oplus 0'4'7'9'}$. This is considered because the verification of larger number of spoken digits with the same meaning but different pronunciations is more difficult. Here, note that ACE for D_{def} is obtained by the same way as shown in our previous research study [1], but the present ACE is less than the one in [1]. This is because the previous and the present researches have been executed independently, and the parameter tuning of learning machines has not been optimized sufficiently in this research. Furthermore, the ACE for the class index $i = 4$ for text verification is very large because it is hard to discriminate the words in $c_4^{[\tilde{D}]}$ and $c_5^{[\tilde{D}]}$ as shown in [1]. However, this problem is solved in [1] by combining these classes into one class, which is easily introduced from the point of view of the present method or the probabilistic prediction using GEBI in multiclass classification derived for the verification (see [1] for details).

From the columns of D_{def}, $D_{\text{def}\ominus 7\oplus 7'}$ and $D_{\text{def}\oplus 7'}$ in Table 1, we can see that the mean ACE for D_{def} is smaller than that for $D_{\text{def}\ominus 7\oplus 7'}$ in both speaker and text verification. This indicates that the pronunciation /nana/ is better than the pronunciation /shichi/ for text-prompted speaker verification. This is supposed to be because the pronunciation /shichi/ is similar to the pronunciation /ichi/ in D_{def}, but the analysis has not been done, yet. The ACE of D_{def} and $D_{\text{def}\oplus 7'}$ has the same relationship as the effect of "pronunciation addition" explained above.

4 Conclusion

We have presented a method of probabilistic prediction for text-prompted speaker verification of spoken words involving a word with the same meaning and different pronunciations. We have formulated the method using GEBI to deal with such words. By means of numerical experiments using recorded real speech data, we have examined the properties to show the validity and the effectiveness of the present method. The performance of the present method evaluated by ACE is not satisfactory from the comparison with the results shown in [1], and we would like to improve it in our future research studies.

References

1. Kurogi, S., Sakashita, S., Takeguchi, S., Ueki, T., Matsuo, K.: Probabilistic prediction in multiclass classification derived for flexible text-prompted speaker verification. In: Arik, S., Huang, T., Lai, W.K., Liu, Q. (eds.) ICONIP 2015. LNCS, vol. 9489, pp. 216–225. Springer, Heidelberg (2015). doi:10.1007/978-3-319-26532-2_24
2. Beigi, H.: Fundamentals of Speaker Recognition. Springer-Verlag New York Inc., New York (2011)
3. Slingo, J., Palmer, T.: Uncertainty in weather and climate prediction. Phil. Trans. R. Soc. A **369**, 4751–4767 (2011)
4. Kurogi, S., Ueno, T., Sawa, M.: A batch learning method for competitive associative net and its application to function approximation. In: Proceedings of the SCI 2004, vol. V, pp. 24–28 (2004)
5. Kurogi, S., Mineishi, S., Sato, S.: An analysis of speaker recognition using bagging CAN2 and pole distribution of speech signals. In: Wong, K.W., Mendis, B.S.U., Bouzerdoum, A. (eds.) ICONIP 2010. LNCS, vol. 6443, pp. 363–370. Springer, Heidelberg (2010). doi:10.1007/978-3-642-17537-4_45
6. Campbell, J.P.: Speaker recognition: a tutorial. Proc. IEEE **85**(9), 1437–1462 (1997)
7. Kurogi, S., Ueki, T., Takeguchi, S., Mizobe, Y.: Properties of text-prompted multistep speaker verification using gibbs-distribution-based extended Bayesian inference for rejecting unregistered speakers. In: Loo, C.K., Yap, K.S., Wong, K.W., Teoh, A., Huang, K. (eds.) ICONIP 2014, Part II. LNCS, vol. 8835, pp. 35–43. Springer, Heidelberg (2014)

Segment-Level Probabilistic Sequence Kernel Based Support Vector Machines for Classification of Varying Length Patterns of Speech

Shikha Gupta[1], Veena Thenkanidiyoor[2], and Dileep Aroor Dinesh[1](✉)

[1] School of Computing and Electrical Engineering, Indian Institute of Technology Mandi, Mandi 175001, H.P., India
shikha_g@students.iitmandi.ac.in, addileep@iitmandi.ac.in

[2] Department of Computer Science and Engineering, National Institute of Technology Goa, Ponda 401403, Goa, India
veenat@nitgoa.ac.in

Abstract. In this work we propose the segment-level probabilistic sequence kernel (SLPSK) as dynamic kernel to be used in support vector machine (SVM) for classification of varying length patterns of long duration speech represented as sets of feature vectors. SLPSK is built upon a set of Gaussian basis functions, where half of the basis functions contain class specific information while the other half implicates the common characteristics of all the speech utterances of all classes. The proposed kernel is computed between the pair of examples, by partitioning the speech signal into fixed number of segments and then matching the corresponding segments. We study the performance of the SVM-based classifiers using the proposed SLPSK using different pooling technique for speech emotion recognition and speaker identification and compare with that of the SVM-based classifiers using other kernels for varying length patterns.

1 Introduction

Support vector machines (SVMs) being a kernel method for classification requires to design a suitable kernel as a measure of similarity between a pair of examples that are varying length in nature. The kernels designed for varying length patterns are referred to as dynamic kernels [1]. A speech utterance is subjected to short-time analysis that involves performing spectral analysis on each frame of about 20 ms duration. Each frame of speech is represented by a real valued feature vector. The duration of the utterances varies from one utterance to another. The number of frames also differs from one utterance to another. Hence, each speech utterance is represented as varying length pattern. These varying length patterns are either considered as sets of feature vectors or sequence of feature vectors depending on the tasks. In the tasks such as acoustic modeling of subword units of speech such as phonemes, triphones and syllables, duration of data is short and there is a need to model temporal dynamics. In such cases varying

A. Hirose et al. (Eds.): ICONIP 2016, Part IV, LNCS 9950, pp. 321–328, 2016.
DOI: 10.1007/978-3-319-46681-1_39

length speech patterns are represented as sequences of feature vectors. On the other hand, in the tasks such as speaker identification, spoken language identification, and speech emotion recognition the duration of speech data is long and preserving sequence information is not critical. In such cases these varying length speech patterns are represented as sets of feature vectors. Focus of this work is on constructing kernels for varying length patterns represented as sets of feature vectors. Fisher kernel using GMM-based likelihood score vectors [2], probabilistic sequence kernel [3], GMM supervector kernel [4], GMM-UBM mean interval kernel [5], GMM-based intermediate matching kernel [1], GMM-based pyramid match kernel [6] and example-specific density based matching kernel [7] are some of the state-of-the-art dynamic kernels for sets of feature vectors.

In this work we extend the probabilistic sequence kernel (PSK) [3] to include local information in matching the two speech utterances and to maintain temporal ordering of the feature vectors. PSK maps a set of feature vectors onto a high dimensional probabilistic score space. The probabilistic score space for a class is obtained by using the posterior probability of components of adapted GMM built for that class and posterior probability of component of class-independent Gaussian mixture model (CIGMM) to which the data of a class is adapted. PSK does not included temporal information while computing the kernel. In this work we propose segment-level probabilistic sequence kernel (SLPSK) as dynamic kernel for building SVM-based classifier for classification of speech signals represented as varying length sets of feature vectors. We propose to divide each speech signal into fixed number of segments. We propose to compute PSK of the local feature vectors of a particular segment from the two examples. Then the proposed SLPMK is computed as a combination of PSKs corresponding to all the segments. As the kernel is computed at segment level, it is expected to include more local information. Salient features of the proposed SLPSK are: (i) maintaining the temporal ordering of the feature vectors in a speech signal for some extent, and (ii) using the local information for matching between two speech utterances represented as sets of feature vectors. The effectiveness of the SVM-based classifiers using proposed SLPSK is studied for speech emotion recognition and speaker identification tasks.

The rest of paper is organized as follows. In Sect. 2, a brief description of probabilistic sequence kernel (PSK) for sets of feature vectors is presented. The proposed segment-level probabilistic sequence kernel (SLPSK) for sets of feature vectors is described in Sect. 3. This section also present the pooling techniques used in the construction of SLPMK. In Sect. 4 studies on speech emotion recognition and speaker identification tasks are presented. The conclusion is presented in Sect. 5.

2 Probabilistic Sequence Kernel for Sets of Feature Vectors

In this section we present probabilistic sequence kernel (PSK) constructed between pair of examples represented as sets of feature vectors. Let $\mathbf{X} = \{\mathbf{x}_1, \mathbf{x}_2, \ldots, \mathbf{x}_T\}$ be a set of local feature vectors. PSK [3] maps a sets of feature vectors onto a fixed dimensional probabilistic feature vector obtained using Gaussian mixture model (GMM). The PSK uses universal background model (UBM) with Q components and the class-specific GMMs obtained by adapting the UBM. The UBM, also called as class independent GMM (CIGMM), is a large GMM built using the training data of all the classes. A local feature vector $\mathbf{x}$ is represented in a higher dimensional feature space as a vector of responsibility terms of the $2Q$ components (Q from a class-specific adapted GMM and other Q from UBM), $\boldsymbol{\Psi}(\mathbf{x}) = [\gamma_1(\mathbf{x}), \gamma_2(\mathbf{x}), ..., \gamma_{2Q}(\mathbf{x})]^\top$. Since the element $\gamma_q(\mathbf{x})$ indicates the probabilistic alignment of $\mathbf{x}$ to the qth component, $\boldsymbol{\Psi}(\mathbf{x})$ is called the probabilistic alignment vector. Thus a probabilistic alignment vector includes the information specific to a class as well as the global information common to all the classes. A set of local feature vectors $\mathbf{X}$ is represented as a fixed dimensional vector $\boldsymbol{\Phi}_{\text{PSK}}(\mathbf{X})$, and is given by

$$\boldsymbol{\Phi}_{\text{PSK}}(\mathbf{X}) = \frac{1}{T}\sum_{t=1}^{T}\boldsymbol{\Psi}(\mathbf{x}_t) \tag{1}$$

Then, the PSK between two examples $\mathbf{X}_m = \{\mathbf{x}_{m1}, \mathbf{x}_{m2}, ..., \mathbf{x}_{mT_m}\}$ and $\mathbf{X}_n = \{\mathbf{x}_{n1}, \mathbf{x}_{n2}, ..., \mathbf{x}_{nT_n}\}$ is given as

$$K_{\text{PSK}}(\mathbf{X}_m, \mathbf{X}_n) = \boldsymbol{\Phi}_{\text{PSK}}(\mathbf{X}_m)^\top \mathbf{S}^{-1}\boldsymbol{\Phi}_{\text{PSK}}(\mathbf{X}_n) \tag{2}$$

where, $\mathbf{S}$ is the correlation matrix. The PSK in [3], does not include temporal ordering of the local feature vectors. In many speech application, including the temporal information helps to build a better classifier. Also in many applications, preserving local information also helps to build a better discriminative classifier [7]. In the following section, we propose segment-level PSK (SLPSK) to include local information as well as temporal information in the computation of PSK. It is seen from (1) that $\boldsymbol{\Phi}_{\text{PSK}}(\mathbf{X})$ is obtained by pooling all the probabilistic alignment vectors corresponding to each local feature vectors of $\mathbf{X}$ and taking their average. This is called average pooling [8]. In the next section we also propose to explore different pooling technique such as sum pooling [8] and max pooling [9] in the construction of $\boldsymbol{\Phi}_{\text{PSK}}(\mathbf{X})$.

3 Segment-Level Probabilistic Sequence Kernel

In this section, we propose segment-level PSK (SLPSK). In SLPSK, speech utterance represented as a set of feature vectors is divided into a fixed number of segments and then feature vectors of each segment is mapped onto probabilistic feature vector. SLPSK between a pair of speech utterances is computed by matching the corresponding segments.

Let $\mathbf{X}_m = \{\mathbf{x}_{m1}, \mathbf{x}_{m2}, \ldots, \mathbf{x}_{mT_m}\}$ and $\mathbf{X}_n = \{\mathbf{x}_{n1}, \mathbf{x}_{n2}, \ldots, \mathbf{x}_{nT_n}\}$ be the sets of feature vectors for two examples (speech utterances). Let N be the number of segments into which each utterance is divided. Let $\mathbf{X}_m^k = \{\mathbf{x}_{m1}^k, \mathbf{x}_{m2}^k, \ldots, \mathbf{x}_{mT_m^k}^k\}$ and $\mathbf{X}_n^k = \{\mathbf{x}_{n1}^k, \mathbf{x}_{n2}^k, \ldots, \mathbf{x}_{nT_n^k}^k\}$ be the subsets of local feature vectors of $\mathbf{X}_m$ and $\mathbf{X}_n$ belonging to k^{th} segment in their respective speech utterance. We propose to compute PSK between the two subsets of local feature vectors in the kth segment. The corresponding fixed dimensional vectors $\boldsymbol{\Phi}^k_{\text{PSK}}(\mathbf{X}^k_m)$ and $\boldsymbol{\Phi}^k_{\text{PSK}}(\mathbf{X}^k_n)$ are obtained by pooling their respective probabilistic alignment vectors. The different pooling techniques are presented in the end of this section. The segment-specific PSK between $\mathbf{X}^k_m$ and $\mathbf{X}^k_n$ is computed using

$$K^k_{\text{PSK}}(\mathbf{X}^k_m, \mathbf{X}^k_n) = \boldsymbol{\Phi}^k_{\text{PSK}}(\mathbf{X}^k_m)^\top \mathbf{S}_k^{-1} \boldsymbol{\Phi}^k_{\text{PSK}}(\mathbf{X}^k_n) \tag{3}$$

The correlation matrix $\mathbf{S}_k$ is defined as follows

$$\mathbf{S}_k = \frac{1}{M_k} \mathbf{R}_k^\top \mathbf{R}_k \tag{4}$$

where $\mathbf{R}_k$ is the matrix whose rows are the probabilistic alignment vectors for local feature vectors of kth segment and M_k is the total number of local feature vectors in k^{th} segment. The SLPSK for the $\mathbf{X}_m$ and $\mathbf{X}_n$ is then computed as combination of the segment-specific PSKs as follows:

$$K_{\text{SLPSK}}(\mathbf{X}_m, \mathbf{X}_n) = \sum_{k=1}^{N} K^k_{\text{PSK}}(\mathbf{X}^k_m, \mathbf{X}^k_n) \tag{5}$$

Since, PSK is a valid positive semidefinite kernel [3], the segment specific PSK is also a valid positive semidefinite kernel. Hence, the SLPSK is also a valid positive semidefinite kernel because the sum of valid positive semidefinite kernel is a valid positive semidefinite kernel.

Next, we discuss different pooling techniques used for pooling the probabilistic alignment vectors of sets of local feature vectors corresponding to each segments.

Pooling Techniques for Constructing $\boldsymbol{\Phi}_{\text{PSK}}(\mathbf{X})$:

Let $\mathbf{X}^k = \{\mathbf{x}^k_1, \mathbf{x}^k_2, \ldots, \mathbf{x}^k_{T^k}\}$ be the segment-level feature vectors corresponding the kth segment of an utterance. In this work we propose to explore 3 pooling techniques that are popular in the image domain [9]. They are:

(i) **Average Pooling**: In average pooling, $\boldsymbol{\Phi}_{\text{PSK}}(\mathbf{X}^k)$ is obtained by pooling all the probabilistic alignment vectors corresponding to each local feature vectors of $\mathbf{X}^k$ and taking their average. It is given as

$$\boldsymbol{\Phi}_{\text{PSK}}(\mathbf{X}^k) = \frac{1}{T^k} \sum_{t=1}^{T^k} \psi(\mathbf{x}^k_t) \tag{6}$$

(ii) **Sum Pooling**: In sum pooling, $\boldsymbol{\Phi}_{\text{PSK}}(\mathbf{X}^k)$ is obtained by adding all the probabilistic alignment vectors corresponding to each local feature vectors of $\mathbf{X}^k$. It is given by:

$$\boldsymbol{\Phi}_{\text{PSK}}(\mathbf{X}^k) = \sum_{t=1}^{T^k} \psi(\mathbf{x}^k_t) \tag{7}$$

The $\boldsymbol{\Phi}_{\text{PSK}}(\mathbf{X}^k) = [\Phi_1(\mathbf{X}^k), \Phi_2(\mathbf{X}^k), \ldots \Phi_{2\text{Q}}(\mathbf{X}^k)]^T$ is then normalized using sum normalization as suggested in [8]. The normalized q^{th} value of $\boldsymbol{\Phi}_{\text{PSK}}(\mathbf{X}^k)$ is given as:

$$\Phi_q(\mathbf{X}^k) = \frac{\Phi_q(\mathbf{X}^k)}{\sum_{j=1}^{2Q} \Phi_q(\mathbf{X}^k)} \tag{8}$$

(iii) **Max Pooling**: In max pooling, $\boldsymbol{\Phi}_{\text{PSK}}(\mathbf{X}^k)$ is obtained by taking maximum of each dimension of all probabilistic alignment vectors corresponding to each local feature vectors of $\mathbf{X}^k$. It is given by

$$\boldsymbol{\Phi}_{\text{PSK}}(\mathbf{X}^k) = \max(\psi(\mathbf{x}_1^k), \psi(\mathbf{x}_2^k), .., \psi(\mathbf{x}_t^k), .., \psi(\mathbf{x}_{T^k}^k)) \tag{9}$$

The $\boldsymbol{\Phi}_{\text{PSK}}(\mathbf{X}^k) = [\Phi_1(\mathbf{X}^k), \Phi_2(\mathbf{X}^k), \ldots \Phi_{2\text{Q}}(\mathbf{X}^k)]^T$ is then normalized using l_2 normalization as suggested in [9]. The normalized q^{th} value of $\boldsymbol{\Phi}_{\text{PSK}}(\mathbf{X}^k)$ is given as:

$$\Phi_q(\mathbf{X}^k) = \frac{\Phi_q(\mathbf{X}^k)}{||\boldsymbol{\Phi}_{\text{PSK}}(\mathbf{X}^k)||_2} \tag{10}$$

4 Experimental Studies

In this section, effectiveness of the proposed kernel is studied for speech emotion recognition and speaker identification tasks using SVM-based classifiers. Speech emotion recognition task involves automatically identifying the emotional state of a speaker from his/her voice. Speaker identification task involves identifying a speaker among a known set of speakers using a speech utterance produced by the speaker.

We have considered Mel frequency cepstral coefficients (MFCC) as features. The MFCC are the most successful and extensively used features for speaker recognition. A frame size of 20 ms and a shift of 10 ms are used for feature extraction from the speech signal of an utterance. Every frame is represented using a 39-dimensional feature vector. Here, the first 12 features are Mel frequency cepstral coefficients and the 13th feature is log energy. The remaining 26 features are the delta and acceleration coefficients. We consider, LIBSVM [10] tool to build the SVM-based classifiers. In this study, one-against-the-rest approach with class dependent kernel gram matrix is considered for 7-class and 4-class speech emotion recognition tasks and 122-class speaker identification task. The value of trade-off parameter, C in SVM is chosen empirically as 10^{-3}.

The Berlin emotional speech database (Emo-DB) [11] and the German FAU Aibo emotion corpus (FAU-AEC) [12] are used for studies on speech emotion recognition task. Emo-DB contains 494 utterances belonging to seven emotional categories with the number of utterances for the category given in parentheses: fear (55), disgust (38), happiness (64), boredom (79), neutral (78), sadness (53), and anger (127). These utterances correspond to ten sentences in German language uttered by five male and five female actors. We have considered 80 % of the utterances for training and the remaining for testing. In FAU-AEC, we have considered four super classes of emotions such as anger, emphatic, neutral, and motherese. We have considered an almost balanced subset of the corpus defined for these four classes by CEICES of the Network of Excellence HUMAINE funded by European Union [12]. We perform the classification at the chunk (speech utterance) level in the Aibo chunk set. The speaker-independent speech emotion recognition accuracy presented is the average classification accuracy along with 95 % confidence interval obtained for 3-fold stratified cross validation. The 3-fold cross validation is based on the three splits defined in Appendix A.2.10 of [12].

Speaker identification experiments are performed on the 2002 and 2003 NIST speaker recognition (SRE) corpora [13]. We have considered the 122 male speakers that are common to the 2002 and 2003 NIST SRE corpora. Training data for a speaker includes a total of about 3 min of speech from the single conversations in the training set of 2002 and 2003 NIST SRE corpora. The test data from the 2003 NIST SRE corpus is used for testing the speaker recognition systems. The speaker identification accuracy presented is the classification accuracy obtained for test examples. The training and test datasets as defined in the NIST SRE corpora are used in studies.

In our studies, the SVM-based classifiers using the SLPSK is built using different values for Q corresponding the number of Gaussian components and N correspond to the number of segmental division. The classification accuracies for the SVM-based classifier using SLPSK are given in Table 1 for speech emotion recognition and speaker identification tasks using different pooling techniques. The best performances are shown using bold phase. SLPSK computed using sum pooling is performed better for speaker identification task and SLPSK computed using max pooling performed better for speech emotion recognition tasks.

Table 1. Classification accuracy (in %) of the SVM-based classifiers with SLPSK for speech emotion recognition (SER) and speaker identification (Spk-ID) tasks using different pooling technique for the different values of Q and N.

Q	N	SVM using SLPSK with average poolong			SVM using SLPSK with sum pooling			SVM using SLPSK with max pooling		
		SER		Spk-ID	SER		Spk-ID	SER		Spk-ID
		Emo-DB	FAU-AEC		Emo-DB	FAU-AEC		Emo-DB	FAU-AEC	
256	1	85.40±0.28	65.05±0.10	83.12	90.20±0.29	65.42±0.09	81.23	88.00±0.28	65.68±0.13	81.34
	2	86.80±0.29	66.67±0.11	83.67	91.00±0.28	**66.29±0.13**	82.21	89.60±0.26	66.35±0.11	82.01
	4	83.80±0.29	65.23±0.13	81.22	88.40±0.28	65.16±0.11	82.12	88.20±0.28	65.19±0.10	80.98
512	1	89.00±0.27	64.46±0.09	88.63	91.20±0.31	64.61±0.12	88.63	89.60±0.30	65.12±0.11	87.32
	2	**91.18±0.26**	65.06±0.12	89.67	**92.60±0.29**	65.66±0.10	88.72	90.60±0.31	65.34±0.12	88.76
	4	89.40±0.29	63.63±0.11	86.34	89.80±0.30	64.06±0.09	86.43	88.80±0.28	63.70±0.10	86.24
1024	1	89.80±0.30	65.93±0.12	90.23	91.00±0.29	65.40±0.13	90.09	90.40±0.28	64.40±0.13	89.34
	2	88.60±0.28	**67.05±0.10**	**91.01**	90.20±0.27	66.28±0.12	**91.67**	**91.80±0.26**	**66.78±0.10**	**90.12**
	4	85.80±0.26	64.88±0.13	89.45	87.60±0.28	64.17±0.11	88.67	86.60±0.27	64.05±0.12	88.23

Table 2 compares the accuracies for speech emotion recognition and speaker identification tasks obtained using the GMM-based classifiers and SVM-based classifiers using the state-of-the-art dynamic kernels including PSK and the proposed SLPSK. Fisher kernel (FK) using GMM-based likelihood score vectors [2], GMM supervector kernel (GMMSVK) [4], GMM-UBM mean interval kernel (GUMIK) [5], GMM-based intermediate matching kernel (GMMIMK) [1],

GMM-based pyramid match kernel (GMMPMK) [6] and example-specific density based matching kernel (ESDMK) [7] are the state-of-the-art dynamic kernel based SVM classifiers considered for the study. In this study, the GMMs whose parameters are estimated using the maximum likelihood (ML) method (MLGMM) and by adapting the parameters of the UBM or CIGMM to the data of a class (adapted GMM) are considered to build GMM-based classifiers. The GMMs are built using the diagonal covariance matrices. The accuracies presented in Table 2 are the best accuracies observed among the GMM-based classifiers and SVM-based classifiers with dynamic kernels using different values for their parameters. The details of the experiments and the best values for the parameters can be found in [1,6,7]. It is seen that, the SVM-based classifiers using the proposed SLPSK perform significantly better than the GMM-based classifiers and SVM-based classifiers using state-of-the-art dynamic kernels for all the tasks. It is also seen that SLPMK-based SVM classifier performed significantly better than that of the PSK-based SVM classifiers. The better performance of the SVM-based classifier using the proposed kernel is mainly due to the capabilities of the SLPSK in capturing the local information better than the other dynamic kernels and also maintaining temporal information to some extent.

Table 2. Comparison of classification accuracy (CA) (in %) of the GMM-based classifiers and SVM-based classifiers using FK, PSK, GMMSVK, GUMIK, GMMIMK, GMMPMK and proposed ESDMK for speech emotion recognition (SER) task and speaker identification (Spk-ID) task. Here, CA95 %CI indicates average classification accuracy along with 95 % confidence interval. Q indicates the number of components considered in building GMM for each class or the number of components considered in building CIGMM or the number of virtual feature vectors considered. The pair (J, b) indicates values of J and b considered in constructing the pyramid. k indicates the number of neighbors considered in ESDMK. (Q, N) indicates the number of components considered in building GMM and the number of segments in SLPSK.

Classification model		SER				Spk-ID	
		Emo-DB		FAU-AEC			
		$Q/(J,b)/k/(Q,N)$	CA95%CI	$Q/(J,b)/k/(Q,N)$	CA95%CI	$Q/(J,b)/k/(Q,N)$	CA
MLGMM		32	66.81±0.44	128	60.00±0.13	64	76.50
Adapted GMM		512	79.48±0.31	1024	61.09±0.12	1024	83.08
SVM using	FK	256	87.05±0.24	512	61.54±0.11	512	88.54
	PSK	1024	87.46±0.23	512	62.54±0.13	1024	86.18
	GMMSVK	256	87.18±0.29	1024	59.78±0.19	512	87.93
	GUMIK	256	88.17±0.34	1024	60.66±0.10	512	90.31
	GMMIMK	512	85.62±0.29	1024	62.48±0.07	1024	88.54
	GMMPMK	(11,2)	88.65±0.23	(5,4)	64.73±0.16	(6,4)	90.26
	ESDMK	4	92.00±0.27	4	65.33±0.09	4	91.38
	SLPSK with average pooling	(512,2)	**91.18±0.26**	(1024,2)	**67.05±0.10**	(1024,2)	**91.01**
	SLPSK with sum pooling	(512,2)	**92.60±0.29**	(256,2)	**66.29±0.13**	(1024,2)	**91.67**
	SLPSK with max pooling	(1024,2)	**91.80±0.26**	(1024,2)	**66.78±0.10**	(1024,2)	**90.12**

5 Conclusion

In this paper, we proposed the segment-level probabilistic sequence kernels (SLPSKs) for classification of varying length patterns represented as sets of feature vectors using the SVM-based classifiers. The SLPSK is computed by partitioning the speech signal into finer segments and computing the pooled probabilistic alignment vector of corresponding segment and then matching the corresponding part using a probabilistic sequence kernel. The effectiveness of the proposed SLPSKs in building the SVM-based classifiers for classification of varying length patterns of long duration speech is demonstrated using studies on speech emotion recognition and speaker identification tasks. The performance of the SVM-based classifiers using the proposed SLPSK is significantly better than the GMM-based classifiers and that of the SVM-based classifiers using the state-of-the-art dynamic kernels. The proposed SLPSK can be used for classification of varying length patterns extracted from video, audio, music, and so on, represented as sets of feature vectors using SVM-based classifiers.

References

1. Dileep, A.D., Chandra Sekhar, C.: GMM-based intermediate matching kernel for classification of varying length patterns of long duration speech using support vector machines. IEEE Trans. Neural Netw. Learn. Syst. **25**(8), 1421–1432 (2014)
2. Smith, N., Gales, M., Niranjan, M.: Data-dependent kernels in SVM classification of speech patterns. Technical report CUED/F-INFENG/TR.387, Cambridge University Engineering Department, Trumpington Street, Cambridge, CB2 1PZ, U.K., April 2001
3. Lee, K-A., You, C.H., Li, H., Kinnunen, T.: A GMM-based probabilistic sequence kernel for speaker verification. In: Proceedings of INTERSPEECH, Antwerp, Belgium, pp. 294–297, August 2007
4. Campbell, W.M., Sturim, D.E., Reynolds, D.A.: Support vector machines using GMM supervectors for speaker verification. IEEE Signal Process. Lett. **13**(5), 308–311 (2006)
5. You, C.H., Lee, K.A., Li, H.: An SVM kernel with GMM-supervector based on the Bhattacharyya distance for speaker recognition. IEEE Signal Process. Lett. **16**(1), 49–52 (2009)
6. Dileep, A.D., Chandra Sekhar, C.: Speaker recognition using pyramid match kernel based support vector machines. Int. J. Speech Technol. **15**(3), 365–379 (2012)
7. Sachdev, A., Dileep, A.D., Thenkanidiyoor, V.: Example-specific density based matching kernel for classificationof varying length patterns of speech using support vector machines. In: Proceedings of ICONIP, Istanbul, Turkey, pp.177–184, November 2015
8. Yu, K., Lv, F., Huang, T., Wang, J., Yang, J., Gong, Y.: Locality-constrained linear coding for image classification. In: Proceedings of CVPR 2010, pp. 3360–3367. IEEE (2010)
9. Yang, J., Yu, K., Gong, Y., Huang, T.: Linear spatial pyramid matching using sparse coding for image classification. In: Proceedings of CVPR 2009, pp. 1794–1801. IEEE (2009)
10. Chang, C.-C., Lin, C.-J.: LIBSVM: a library for support vector machines. ACM Trans. Intell. Syst. Technol. (TIST) **2**(3), 27 (2011)
11. Burkhardt, F., Paeschke, A., Rolfes, M., Weiss, W.S.B.: A database of German emotional speech. In: Proceedings of INTERSPEECH, Lisbon, Portugal, pp. 1517–1520, September 2005
12. Steidl, S.: Automatic classification of emotion-related user states inspontaneous childern's speech. Ph.D. thesis, Der Technischen Fakultät der Universität Erlangen-Nürnberg, Germany (2009)
13. The NIST year 2003 speaker recognition evaluation plan (2003). http://www.itl.nist.gov/iad/mig/tests/sre/2003/

Attention Estimation for Input Switch in Scalable Multi-display Environments

Xingyuan Bu, Mingtao Pei(✉), and Yunde Jia

Beijing Laboratory of Intelligent Information Technology,
School of Computer Science, Beijing Institute of Technology,
Beijing 100081, People's Republic of China
{buxingyuan,peimt,jiayunde}@bit.edu.cn

Abstract. Multi-Display Environments (MDEs) have become commonplace in office desks for editing and displaying different tasks, such as coding, searching, reading, and video-communicating. In this paper, we present a method of automatic switch for routing one input (including mouse/keyboard, touch pad, joystick, etc.) to different displays in scalable MDEs based on the user attention estimation. We set up an MDE in our office desk, in which each display is equipped with a webcam to capture the user's face video for detecting if the user is looking at the display. We use Convolutional Neural Networks (CNNs) to learn the attention model from face videos with various poses, illuminations, and occlusions for achieving a high performance of attention estimation. Qualitative and quantitative experiments demonstrate the effectiveness and potential of the proposed approach. The results of the user study also shows that the participants deemed that the system is wonderful, useful, and friendly.

Keywords: Multi-display environment · Attention estimation · Input switch · Convolutional neural network

1 Introduction

Multi-display environments (MDEs) facilitate the efficiency of the users [1]. For instance, multiple displays in our office desk are used to edit and display different tasks, such as coding, searching, reading, and communicating. Over the past years, many approaches have been proposed to provide input switch in MDEs. Ashdown et al. [2] tracked user's head based on the 3D face model created by two cameras to switch the mouse pointer between monitors and use the mouse to move within each monitor, but this method lacks scalability and is not flexible in MDEs with more displays. Lander el al. [3] used head-mounted devices to estimate user's eye gaze to switch input, and allowed the user to move freely in MDE, but equipping the user with expensive and obtrusive devices is not user-friendly.

In this paper, we discuss an automatic switching technique for routing one input (including mouse and keyboard, touch pad, joystick, etc.) to different displays in scalable MDEs. We set up an MDE in our office desk, as shown in

A. Hirose et al. (Eds.): ICONIP 2016, Part IV, LNCS 9950, pp. 329–336, 2016.
DOI: 10.1007/978-3-319-46681-1_40

Fig. 1, in which each display is equipped with a webcam (on the top of the display frame) to capture the user's face video for detecting which display the user is looking at. Under this configuration, each webcam captures video image data independently, and then feed these data to the Convolutional Neural Networks (CNNs) to estimate the user's attention for automatically input switching. The independent webcam and attention estimation can make our approach exceeding in efficiency, scalability, user-friendliness. Dostal et al. [4] set up the same MDE in which web cameras are mounted on the displays to detect which display the user is attending. Their techniques were explored for techniques for visualizing display changes in multi-display environments.

Fig. 1. Our multi-display environment

Estimating user's attention is widely used in many fields [5,6], but still a challenging problem due to unpredictable user's mobility and complex condition [7]. Recent work in face recognition and head pose estimation [8,9] has demonstrated a considerable advantage of the deep Convolutional Neural Networks (CNNs) [10]. Instead of multiple CNNs, we use an integral deep tree architecture combining CNNs and traditional decision tree to estimate user's attention. The split node in the tree architecture splits the source sample space based on an intuitive idea that the preceding learning stages should make the data more distinguishable in subsequent learning stages [11]. Based on [11,12], we use perceptron as split node, then learn the split node and CNN parameters at the same time. The experiment results show the tree CNN outperforms primary CNN without significant increase in computational complexity, and achieves accurate input switch in multi-display environments.

2 Our Approach

Suppose there are N displays $D_i (i = 1, 2, \cdots, N)$ in the MDE. Let DC_i denotes the corresponding the webcam of the display D_i, and $I_{i,t}$ denotes the image captured by DC_i at time t. We detect face region $F_{i,t}$ in $I_{i,t}$ by the cascade detector [13] and feed the detected face region to the deep tree architecture for attention estimation. That is, at time t, we estimate whether the user is looking at the display D_i using the face region $F_{i,t}$, and a prediction score $P_{i,t}$ is computed for each display. And the input will be switched to the display with the highest prediction score.

2.1 Data Collection

We build an attention dataset called BIT-Attention Dataset which contains 525,000 face images. Faces in this dataset undergo large variances in view point, pose, illumination, occlusion and facial features as shown in Fig. 2. During data collecting process, a sign is shown on each display in turn. The sign is programmed to move over the display, and the participant is requested to track it only by changing head pose. This mechanism allows us to capture more data in the edge area where it is difficult to estimate user's attention. Images from DC_i are labeled as positive samples when the sign is shown on D_i (the user is looking at D_i), and the images from ones other than DC_i are labeled as negative samples.

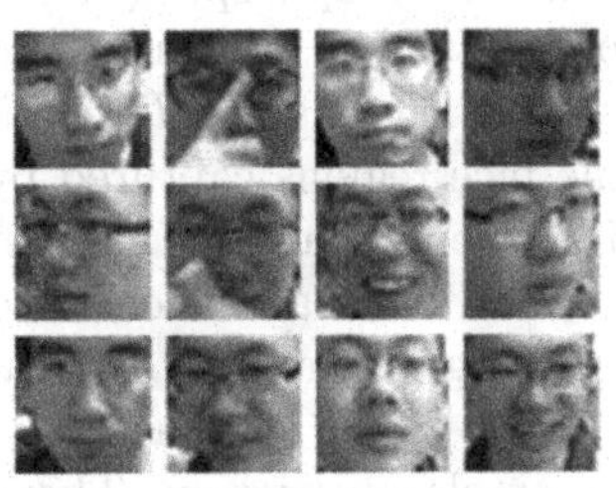

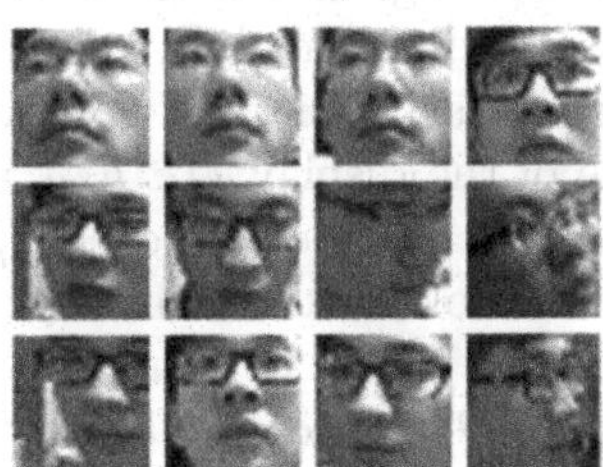

Fig. 2. The example images of our attention dataset

2.2 The Basic Architecture of Our CNN

Convolutional Neural Networks become increasingly popular in computer vision. CNN is an "end-to-end" system that it can learn an effective representation from raw input images. Therefore, we use CNN to estimate user attention. That is, the input to the CNN is the face region $F_{i,t}$ detected in $I_{i,t}$, the output of the CNN is the prediction score $P_{i,t}$, which indicates the probability that the user is looking at display D_i. Figure 3 illustrates the basic architecture of our CNN.

Our network architecture contains three stages, each stage consists of the convolutional layer, followed by the ReLU to prevent overfitting and accelerate fit the training set, and max pooling layer. The first two stages execute convolutions with 5×5 and 3×3 kernels, respectively, and strides are both 1. The max pooling

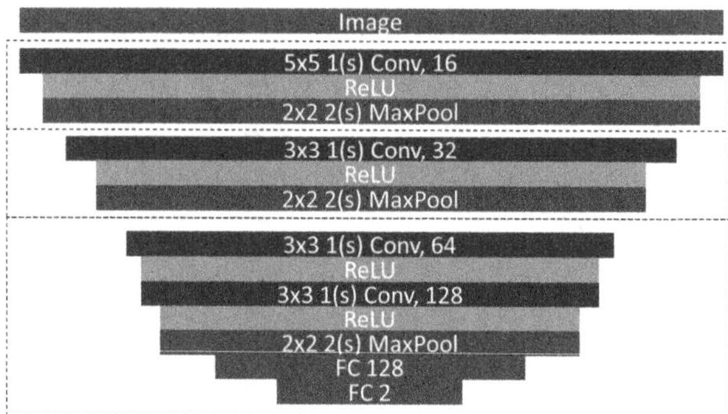

Fig. 3. The basic architecture of our convolutional neural network. It contains three stages indicated by dashed box. The first stage executes a convolution with 5×5 kernerls and 1 stride to produce 16 feature maps, and followed by ReLU and max pooling layer with 2 stride.

layer is employed to reduce the feature dimension at the end of each stage. The third stage contains two convolutional layers to capture more high-level features, then ends with two fully-connected layers, and last one is a 2-way SoftMax.

The challenge of attention estimation is the variant sample space caused by unconstrained user mobility and complex environments, which means there is a large intra-class variation. For example, suppose that the display D_i is on the left side of the display D_j, $F_{i,t1}$ and $F_{i,t2}$ are the face regions captured by the webcam DC_i at time t_1 and t_2, $F_{j,t1}$ is captured by the webcam DC_j at time t_1. If the user looks at the right edge of D_i at time t_1 and looks at the center of D_i at time t_2, $F_{i,t1}$ will be similar to the $F_{j,t1}$ since the right edge of D_i is near the left edge of D_j, but dissimilar to the $F_{i,t2}$ which has the same label as $F_{i,t1}$, thus causing intra-class variation is larger than between-class variation. Therefore, we employ some nodes to split samples into a series of subspaces. Samples with similar appearance will be assigned to the same subspace, and features will be learnt in the subspace to distinguish the similar samples.

Inspired by [11,12], we use perceptron as split node and insert perceptron in the joint of two stages as demonstrated in Fig. 4. Each node splits the sample space into two subspaces. Let S represent the source samples space and $S = \{S_1, S_2, \cdots, S_i, \cdots, S_n\}$, where S_i is a sample in the source space. The node splits source space into two spaces represented by S_+, S_- and the split function can be formulated as

$$\begin{cases} S_i \in S_+, D(S_i) \geq 0 \\ S_i \in S_-, D(S_i) < 0 \end{cases}, \tag{1}$$

and $D(X)$ is the split hyperplane:

$$D(X) = W^T \cdot X + b, \tag{2}$$

where W is the normal vector, b is the intercept of the hyperplane, X is a sample in the source space (i.e. the output of the current CNN stage). The hyperplane

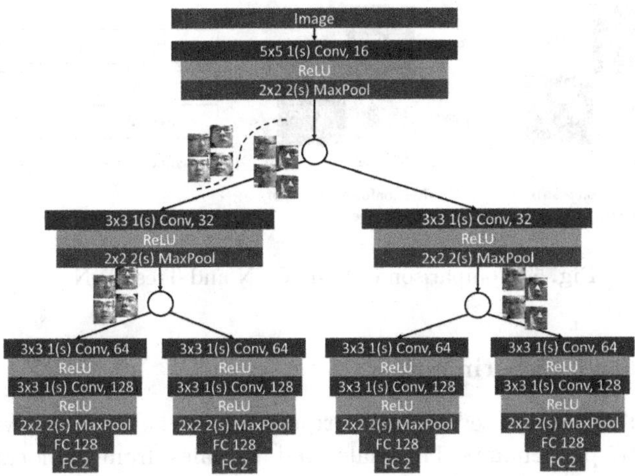

Fig. 4. Tree CNN is formed by the basic CNN and split nodes. In each joint of two stages, a split node is inserted to divide the basic CNN into two branches, so the CNN presents four branches in the final layer.

is optimized to assign samples with similar appearance to the same subspace by minimizing intra-class variation and maximizing between-class variation. In order to optimize the split node jointly with CNN in the same manner as [12], we define the loss as

$$Loss = \frac{\frac{1}{N}\sum\limits_{S_i \in S} D(S_i)^2}{(\frac{1}{N_+}\sum\limits_{S_i \in S_+} D(S_i) - \frac{1}{N_-}\sum\limits_{S_i \in S_-} D(S_i))^2}, \tag{3}$$

where N, N_+, N_- are the number of samples in the source space and subspaces.

3 Experiments

3.1 Quantitative Experiment

We compared the performance of the basic CNN and Tree CNN on our dataset. Ninety percent of the data is used for training, and the rest is for testing. The basic CNN and Tree CNN are both optimized by stochastic gradient descent with mini-batch because the loss of split node only works in batch case. The result is summarized in Fig. 5.

Tree CNN splits source sample space into serval subspaces according to representation of the sample. Although the Tree CNN structure is more complex than the basic CNN, there is no significant increase in computational complexity when testing. Given a sample, it will only choose one path from root to leaf, so the computational complexity is just like the basic CNN. Therefore, the performance improvement is brought by the combination of basic CNN and split node rather than more kernels in networks.

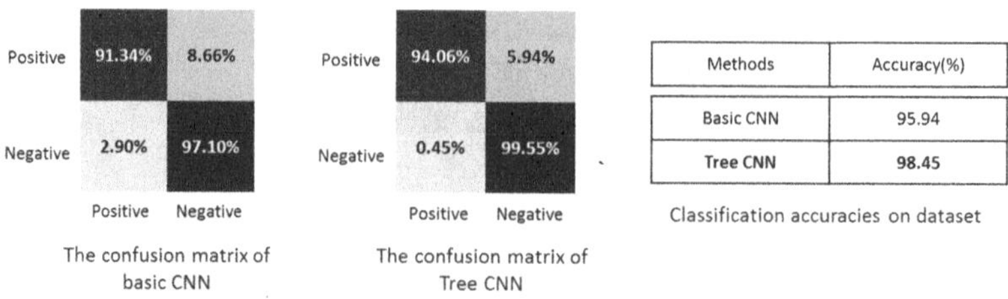

Fig. 5. Comparison of basic CNN and Tree CNN

3.2 Qualitative Experiment

User Study: We evaluated the user experience in the MDE shown in Fig. 1. We recruited 47 participants, 17 females and 32 males, from the local university to volunteer for our study, whose ages are from 18 to 23 years. The participants are asked to accomplish an input switch task with four methods: multiple mouse switching, shortcut key switching, mouse moving switching, and input switch by attention estimation.

With multiple mouses, user must choose the correct mouse paired to the display in which task happened. While with a single mouse, user could press shortcut key (like the keys F1 and F2) to switch input to the target display, or move the mouse from the current display to the target. With our method, the participant is just naturally looking at the display to switch the input.

The task is to click a series of 30×30 pixels red buttons which appear in different display continuously. The participants are asked to click the red button as soon as they can. At the beginning, one button arises in a random position of a random display, and it would disappear when being clicked, then a new button would arise again in another display, the participants need to switch input to the display and click the button. This process repeats ten times. After using all four methods to accomplish the task, the participants fill questionnaires with their fully experience about the rating of satisfaction, learnability and usability. Additionally, more subjective opinions like usefulness and friendliness are collected as well. The completion time of the four methods is also recorded.

Evaluation: The mean completion time is plotted in Fig. 6. We can see that the user spent much time to perform the task with multiple mouses (mean time = 2.58 s). With the shortcut key (mean time = 2.30 s), the user needs to take time to think about the relation between the display and the shortcut key. As the distance from the origin position of the mouse to the target position (red button) usually is quite large since there are many displays in the MDE, so the time cost just decreases little as the participants must move a mouse across long distance (mean time = 2.04 s).

Our method (mean time = 1.61 s) is faster than other traditional methods, since users just need to turn their face to the target display (the mouse would automatically route to the target display).

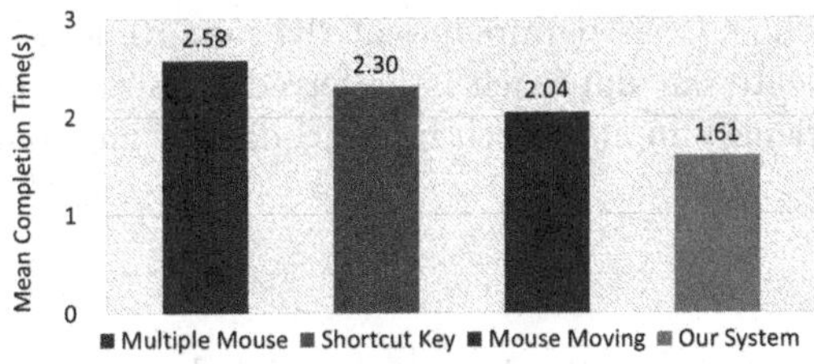

Fig. 6. Mean completion time

According to the Fitt's law [14]: $MT = a + b\log_2(\frac{D}{W} + 1)$, where MT is the average time to compete the movement, D is the distance from the starting point to the target, W is the width of the target, and a,b are the parameters of device. Our approach can reduce the D significantly, that is the reason why our method is faster than the others.

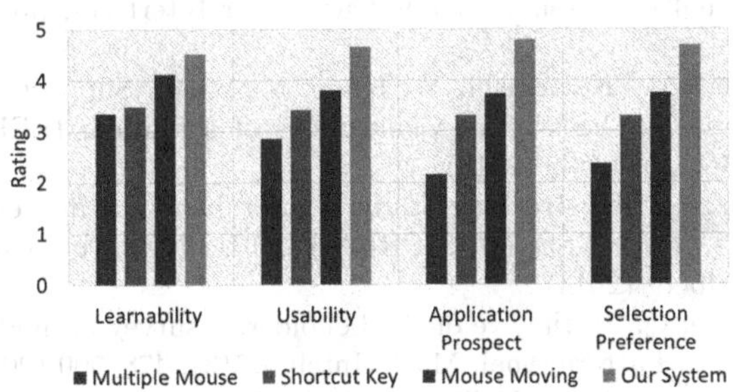

Fig. 7. Mean user ratings

Figure 7 shows the user ratings on a five-point low to high scale in our questionnaires, 1 = very unsatisfied, 5 = satisfied. We survey the satisfaction ratings of learnability and usability. Additionally, more subjective opinions like "How much application prospect do you think it has" and "How much preference do you have for it if you work in multi-display environments next time" are collected as well. The majority of participants are satisfied with our MDE, and they deemed that the system is wonderful, useful, and friendly.

4 Conclusion

We have presented an approach for automatically switching input in multi-display environments by estimating user's attention, which makes our system more user-friendly and free from space configuration. To achieve accurate and robust attention estimation, we employ a tree-like convolutional neural network which splits sample space into subspaces with similar appearance and learns the split node as well as CNN parameters at the same time. Finally, the results of experiments indicate our approach is preferred because it provides efficient, scalable and user-friendly input switch in multi-display environments.

References

1. Andrews, C., Endert, A., North, C.: Space to think: large high-resolution displays for sensemaking. In: Proceedings of the SIGCHI Conference on Human Factors in Computing Systems, pp. 55–64. ACM (2010)
2. Ashdown, M., Oka, K., Sato, Y.: Combining head tracking and mouse input for a gui on multiple monitors. In: CHI 2005 Extended Abstracts on Human Factors in Computing Systems, pp. 1188–1191. ACM (2005)
3. Lander, C., Gehring, S., Krüger, A., Boring, S., Bulling, A.: Gazeprojector: accurate gaze estimation and seamless gaze interaction across multiple displays. In: Proceedings of UIST (2015)
4. Dostal, J., Kristensson, P.O., Quigley, A.: Subtle gaze-dependent techniques for visualising display changes in multi-display environments. In: Proceedings of the 2013 International Conference on Intelligent User Interfaces, pp. 137–148. ACM (2013)
5. Das, D., Rashed, M., Kobayashi, Y., Kuno, Y., et al.: Supporting human-robot interaction based on the level of visual focus of attention. IEEE Trans. Hum.-Mach. Syst. **45**(6), 664–675 (2015)
6. Lu, W., Jia, Y.: An eye-tracking study of user behavior in web image search. In: Pham, D.-N., Park, S.-B. (eds.) PRICAI 2014. LNCS, vol. 8862, pp. 170–182. Springer, Heidelberg (2014)
7. Hansen, D.W., Ji, Q.: In the eye of the beholder: a survey of models for eyes and gaze. IEEE Trans. Pattern Anal. Mach. Intell. **32**(3), 478–500 (2010)
8. Asteriadis, S., Karpouzis, K., Kollias, S.: Visual focus of attention in non-calibrated environments using gaze estimation. Int. J. Comput. Vis. **107**(3), 293–316 (2014)
9. Dong, Z., Jia, S., Wu, T., Pei, M.: Face video retrieval via deep learning of binary hash representations. In: Thirtieth AAAI Conference on Artificial Intelligence (2016)
10. Krizhevsky, A., Sutskever, I., Hinton, G.E.: Imagenet classification with deep convolutional neural networks. In: Advances in Neural Information Processing Systems, pp. 1097–1105 (2012)
11. Bulo, S., Kontschieder, P.: Neural decision forests for semantic image labelling. In: Proceedings of the IEEE Conference on Computer Vision and Pattern Recognition, pp. 81–88 (2014)
12. Xiong, C., Zhao, X., Tang, D., Jayashree, K., Yan, S., Kim, T.K.: Conditional convolutional neural network for modality-aware face recognition. In: Proceedings of the IEEE International Conference on Computer Vision, pp. 3667–3675 (2015)
13. Viola, P., Jones, M.J.: Robust real-time face detection. Int. J. Comput. Vis. **57**(2), 137–154 (2004)
14. Hutchings, D.: An investigation of Fitts' law in a multiple-display environment. In: Proceedings of the SIGCHI Conference on Human Factors in Computing Systems, pp. 3181–3184. ACM (2012)

Deep Dictionary Learning vs Deep Belief Network vs Stacked Autoencoder: An Empirical Analysis

Vanika Singhal, Anupriya Gogna, and Angshul Majumdar(✉)

Indraprastha Institute of Information Technology, Delhi, India
{vanikas, anupriyag, angshul}@iiitd.ac.in

Abstract. A recent work introduced the concept of deep dictionary learning. The first level is a dictionary learning stage where the inputs are the training data and the outputs are the dictionary and learned coefficients. In subsequent levels of deep dictionary learning, the learned coefficients from the previous level acts as inputs. This is an unsupervised representation learning technique. In this work we empirically compare and contrast with similar deep representation learning techniques – deep belief network and stacked autoencoder. We delve into two aspects; the first one is the robustness of the learning tool in the presence of noise and the second one is the robustness with respect to variations in the number of training samples. The experiments have been carried out on several benchmark datasets. We find that the deep dictionary learning method is the most robust.

Keywords: Deep learning · Dictionary learning · Classification

1 Introduction

A typical neural network consists of an input layer where the samples are presented and an output layer with the targets (see Fig. 1). In between these two is the hidden or representation layer. If the representation is known, solving the network weights between the hidden layer and the output is straightforward. Therefore the main challenge in neural network learning is to learn the network weights between the input and the hidden layer. This forms the topic of 'representation learning'.

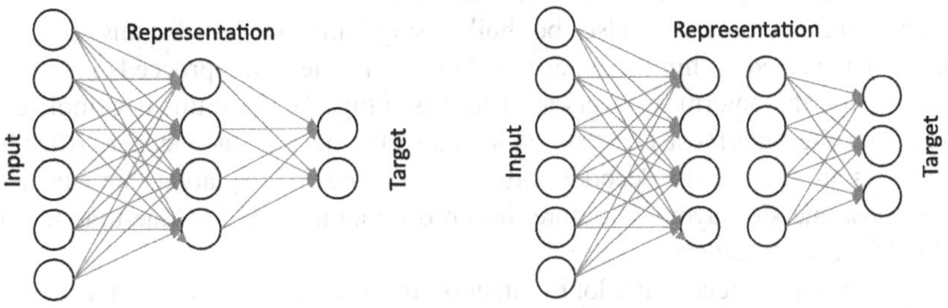

Fig. 1. Left – typical neural network. Right – segregated neural network.

A. Hirose et al. (Eds.): ICONIP 2016, Part IV, LNCS 9950, pp. 337–344, 2016.
DOI: 10.1007/978-3-319-46681-1_41

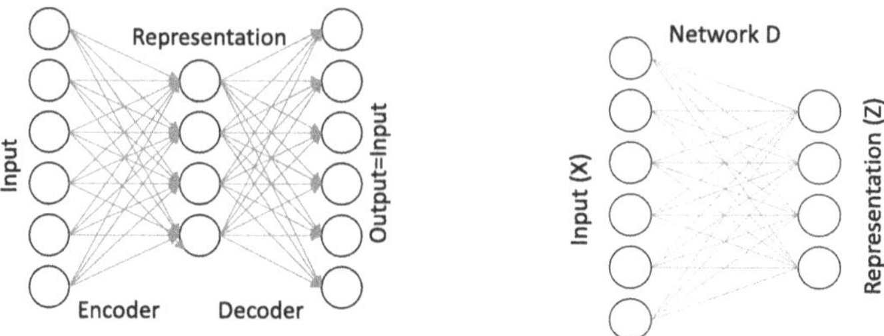

Fig. 2. Left – autoencoder; Right – restricted Boltzmann machine

There are two popular approaches to learn the representation – autoencoder and restricted Botlzmann machine. The architectures are shown in Fig. 2. An autoencoder learns the encoding and decoding weights between the input and itself – it is self supervised. The Euclidean cost function between the input and the decoded-encoder version of the input is minimized. This formulation makes the cost function amenable for gradient based optimization techniques. Usually the standard back propagation algorithm is used for learning these weights.

As the name suggests, the restricted Boltzmann machine (RBM) minimizes the Boltzmann cost function. Basically it tries to learn the network weights such that the similarity (in a probabilistic sense) between the representation and projection of the input is maximized. The usual limits of probability prevents degenerate solutions. As there is no output, the standard backpropagation algorithm cannot be used for RBM training; it is solved using contrastive divergence [1].

For RBM, once it is learnt, the targets are attached to its output, and fine-tuned by backpropagating errors. This leads to the complete neural network. For the autoencoder, after training, the decoder is removed and the target are attached after the encoder layer. The complete architecture is fine-tuned to form the neural network.

A single (hidden) layer neural network is relatively easy to train; therefore autoencoders or RBMs are hardly used for training such shallow neural networks. Such pretraining and fine-tuning is usually required for learning deep neural networks. Deeper architectures can be built by cascading RBMs. Depending on how they are trained, one can have two slightly different versions – deep Botlzmanm machine (DBM) or deep belief network (DBN). Once the deep architecture is learnt, the targets are attached to the deepest/final layer and fine-tuned with backpropagation. This completes the training for the deep neural network.

Deeper architectures can also be built using autoencoder. In this case, one autoencoder is nested within the other (see Fig. 3). The learning proceeds in a greedy fashion. At first the outermost layers are learnt (see Fig. 3). Once this is complete, the features from the outermost layer act as inputs to the nested autoencoder. After both autoencoders are trained, the decoder portion is removed and the targets are attached to the innermost encoder layer. As before, backpropagation is used to fine-tune the final neural network architecture.

Deep learning has received a lot of attention from academia and industry; in recent times deep learning enjoys widespread media coverage. Dictionary learning on the

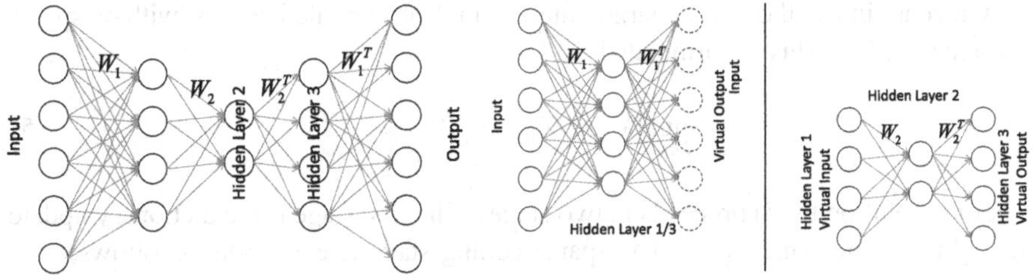

Fig. 3. Left – 2 layer stacked autoencoder. Right – greedy training.

other hand is popular only in academic circles. In dictionary learning, the objective is to learn a basis for representing the data. Since dictionary learning requires factorizing the data matrix into a dictionary and features; in earlier days it used to be called matrix factorization [2]. In recent years it has been popularized by the advent of K-SVD [3]; in the modern version one learns dictionaries such that the learned features are sparse.

So far all studies in dictionary learning employ a shallow (single layer) architecture. In a recent work [4], it was shown how deeper architectures can be built from dictionary learning. Since there is no published work on this topic, we will briefly introduce it in the following section. The main contribution of this work is to empirically compare deep dictionary learning (DDL) with SAE and DBN. We will study how the classification accuracies vary in the presence of noise in the data; and how they how perform when the training data is limited. The results will be presented in Sect. 3. The conclusions of this work is discussed in Sect. 4.

2 Deep Dictionary Learning

In dictionary learning one learns a basis/dictionary for expressing the data in terms of the coefficients. The basic formulation is as follows,

$$X = DZ \tag{1}$$

where D is the dictionary, Z are the coefficients and X is the training data (known).

The earliest methods [2, 4] solved the problem by formulating it as,

$$\min_{D,Z} \|X - DZ\|_F^2 \tag{2}$$

This was solved using the method of optimal directions [4] by alternately updating the dictionary (3) and the coefficients (4).

$$D_k \leftarrow \min_{D} \|X - DZ_{k-1}\|_F^2 \tag{3}$$

$$Z_k \leftarrow \min_{D,Z} \|X - D_k Z\|_F^2 \tag{4}$$

In recent times, there is a large interest in learning dictionaries with a sparse representation [3]. This is formulated as:

$$\min_{D,Z} \|X - DZ\|_F^2 \text{s.t.} \|Z\|_0 \leq \tau \tag{5}$$

As before, solution to (5) proceeds in two stages. The first stage is the dictionary update stage which is the same as (3). The sparse coding stage is expressed as follows,

$$Z_k \leftarrow \min_{Z} \|X - DZ\|_F^2 \text{s.t.} \|Z\|_0 \leq \tau \tag{6}$$

This is solved using by some greedy algorithm like orthogonal matching pursuit.

In deep dictionary learning, one learns multiple levels of dictionaries. The formulation for two levels is shown (5); it is easy to generalize for more levels.

$$X = D_1 D_2 Z \tag{7}$$

One might feel 'what is the requirement for learning multiple levels, if we can collapse the two levels into a single one as $D = D_1D_2$?'. One level dictionary learning (1) is a bi-linear problem, whereas two level dictionary learning (5) is a tri-linear problem. These are completely different problems and hence the coefficient obtained from (1) is not the same as the obtained from (5). Owing to the inherent non (bi/tri) linearity, dictionary learning is non-linear even without the introduction of activation functions. The learning problem for (5) is expressed as:

$$\min_{D_1, D_2, Z} \|X - D_1 D_2 Z\|_F^2 \text{ s.t.} \|Z\|_0 \leq \tau \tag{8}$$

Solving the tri-linear problem (8) is possible, but has not been studied before. On the other hand shallow dictionary learning (bi-linear) is a well studied problem. Unlike other basic building blocks of deep learning (such as autoencoder and RBM), dictionary learning enjoys several theoretical convergence guarantees [6–9]. Therefore, instead of solving the deep dictionary learning problem (8) directly, one would like to convert it single level dictionary learning problem in a greedy fashion. With the substitution $Z_1 = D_1Z$ (7) can be expressed as,

$$X = D_1 Z_1 \tag{9}$$

This boils down to a shallow dictionary learning problem for which there are many algorithms. In this work, we employ the block co-ordinate descent based techniques to solve (6). In the second stage, the former substitution leads to,

$$Z_1 = D_2 Z \tag{10}$$

This too is a shallow dictionary learning problem with sparsity coefficients. We already studied the solution for the same. Here we have shown the greedy learning paradigm for two levels. One can easily extend it to multiple levels.

2.1 Relationship with Neural Network

In the traditional interpretation of dictionary learning, it learns a basis (D) for representing (Z) the data (X). The columns of D are called 'atoms'. In [4], they look at dictionary learning in a different manner. Instead of interpreting the columns as atoms, one can think of them as connections between the input and the representation layer. To showcase the similarity, we have kept the color scheme intact in Fig. 4.

Unlike a neural network which is directed from the input to the representation, the dictionary learning kind of network points in the other direction – from representation to the input. This is what is called 'synthesis dictionary learning' in signal processing. The dictionary is learnt so that the features (along with the dictionary) can synthesize/generate the data. This establishes the connection between dictionary learning and neural network kind of representation learning. Building on that, one can build deeper architectures with dictionary learning. An example of two layer architecture is also shown in Fig. 4.

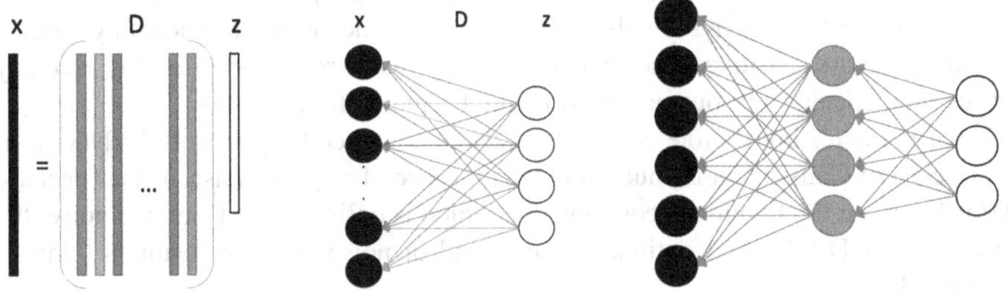

Fig. 4. Left – dictionary learning. middle – neural network interpretation. Right – deep dictionary learning.

3 Experimental Results

3.1 Datasets

We carried our experiments on several benchmarks datasets. These are the full MNIST dataset and the variations of MNIST; the images are of size 28 × 28. The full dataset has 50,000 training images and 10,000 test images. The variations datasets are more challenging than the more popular MNIST dataset primarily because they have fewer training samples (12,000) and larger number of test samples (50,000). This dataset was built for evaluating deep learning algorithms [10]. The variations are –

1. basic (smaller subset of MNIST)
2. basic-rot (smaller subset with random rotations)
3. bg-rand (smaller subset with uniformly distributed noise in background)
4. bg-img (smaller subset with random image background)
5. bg-img-rot (smaller subset with random image background plus rotation)

Comparison was performed between deep dictionary learning (DDL), deep belief network (DBN) and stacked autoencoder (SAE).

3.2 Evaluating Robustness with Respect to Noise

We evaluate the effects to two types of common additive noise – Gaussian noise and Impulse noise. We will study how the classification accuracy from different deep learning tools varies with the addition of noise. For Table 1, 10 % (standard deviation) Gaussian noise has been added both to the training and testing data. For impulse noise 10 % of the same samples have been corrupted by 1's or 0's.

Here all the representation learning tools are only used for feature extraction. The classifier used is a nearest neighbor classifier. This is because our objective is to understand how the feature extraction capacity of different tools vary with the addition of noise; we had to use the same classifier for all of them. More sophisticated parametric classifiers like neural network and support vector machine could also have been used, but in such tuned techniques it is difficult to gauge how much of the classification accuracy pertains to the feature extraction capability of the deep learning tool and how much of the accuracy is ascribed to the tuning of the classifier.

The results show that except for one dataset bg-rand which had background noise in the original data (therefore addition of noise did not change the characteristics of the dataset), our method always yields the best results. The other deep learning tools – DBN and SAE are sensitive to noise, in some cases (basic-rot, bg-img-rot) the accuracy reduces dramatically; but our proposed method remains fairly robust.

What is interesting to note is that the stacked autoencoder performs fairly well in the presence of Gaussian noise but not in the presence of impulse noise. This is because SAE is based on the Euclidean cost function which is optimum for Gaussian noise; the formulation of DBN is not optimal for any kind of noise and hence suffers (almost) equally in both cases.

Table 1. Variation of classification accuracy with noise

Name of dataset	10 % Gaussian noise			10 % Impulse noise		
	DDL	DBN	SAE	DDL	DBN	SAE
MNIST	**97.56**	97.34	90.86	**96.97**	96.38	58.01
basic	**97.78**	96.68	88.18	**95.07**	91.02	55.09
basic-rot	**86.92**	29.08	86.28	**84.98**	37.28	55.50
bg-rand	83.92	**85.27**	36.74	82.81	**83.02**	23.03
bg-img	**76.68**	55.68	70.06	**70.94**	52.17	52.70
bg-img-rot	**68.58**	30.43	68.69	**68.58**	35.51	52.93

3.3 Evaluating Robustness with Respect to Varying Number of Training Samples

In this sub-section we will see how the accuracy varies when the number of training samples vary. We test on two cases – full samples and first 66 % of training samples. The samples from each class are randomly distributed therefore each class has approximately equal distribution in the partial training sets. The first 'x' samples were taken to ensure reproducibility in research.

Table 2. Variation of classification accuracy with number of training samples

Name of dataset	12K training samples			8K training samples		
	DDL	DBN	SAE	DDL	DBN	SAE
MNIST	97.86	97.62	96.23	97.70	96.11	95.94
basic	97.8	96.98	93.32	95.29	85.35	92.4
basic-rot	90.83	86.92	93.35	87.80	81.86	92.21
bg-rand	89.43	88.89	37.90	87.40	82.77	37.29
bg-img	73.60	61.27	72.91	73.10	51.32	72.17
bg-img-rot	73.58	35.70	69.68	72.26	30.49	61.40

As before, we use a simple nearest neighbor classifier for these experiments. The logic remains the same as before. The results are shown in Table 2.

The result show that our proposed technique always yields the best results. The results are apparently obvious – as the number of training samples decrease there is a fall in the accuracy. However, what is interesting is that DBN is the worst hit; both SAE and our proposed DDL are hit by the reduction in the number of training samples, but the fall in accuracy is small. For DBN the fall in classification accuracy from is significantly larger.

4 Conclusion

Deep Belief Network (DBN) and Stacked Autoencoders (SAE) are time tested tools for representation learning. In this work we compare a new deep learning tool – deep dictionary learning (DDL) with DBN and SAE. There is no published work on DDL, hence we briefly introduce it. We show how dictionary learning can be interpreted as a neural network model. Once the architectural similarity is established we show how deeper structures can be built by greedy learning – each block requiring solving the well studied problem of dictionary learning.

This is the first work that pits deep learning tools against each other in two challenging practical scenarios – noise (Gaussian, Impulse) and reduced number of training samples. In the presence of noise, we find that the DDL performs better than the others in all situations (in general). The stacked autoencoder performs well in the presence of Gaussian noise but is hard hit when the noise is of impulsive nature. The DBN (which is neither optimally suited for any noise) performs equally bad in both cases.

When the number of training samples are reduced, all the deep learning tools perform worse. However, performance of our proposed DDL and SAE, degrade smoothly. But the accuracy for DBN drastically falls when the number of training samples reduce.

This work performs an empirical analysis of deep learning tools. At least from empirical analysis on the datasets used in this paper, we conclude that the newly developed deep dictionary learning method performs considerably better than the others and should be the preferred choice in such scenarios.

References

1. Sutskever, I., Tieleman, T.: On the convergence properties of contrastive divergence. In: AISTATS (2010)
2. Lee, D.D., Seung, H.S.: Learning the parts of objects by non-negative matrix factorization. Nature **401**(6755), 788–791 (1999)
3. Rubinstein, R., Bruckstein, A.M., Elad, M.: Dictionaries for sparse representation modeling. Proc. IEEE **98**(6), 1045–1057 (2010)
4. Tariyal, S., Majumdar, A., Singh, R., Vatsa, M.: Greedy Deep Dictionary Learning, arXiv: 1602.00203v1
5. Engan, K., Aase, S., Hakon-Husoy, J.: Method of optimal directions for frame design. In: IEEE ICASSP (1999)
6. Jain, P., Netrapalli, P., Sanghavi, S.: Low-rank matrix completion using alternating minimization. In: Symposium on Theory of Computing (2013)
7. Agarwal, A., Anandkumar, A., Jain, P., Netrapalli, P.: Learning sparsely used overcomplete dictionaries via alternating minimization. In: International Conference on Learning Theory (2014)
8. Spielman, D.A., Wang, H., Wright, J.: Exact recovery of sparsely-used dictionaries. In: International Conference on Learning Theory (2012)
9. Arora, S., Bhaskara, A., Ge, R., Ma, T.: More Algorithms for Provable Dictionary Learning, arXiv:1401.0579v1
10. Courville, A., Bergstra, J., Bengio, Y.: An empirical evaluation of deep architectures on problems with many factors of variation. In: ICML (2007)

Bi-directional LSTM Recurrent Neural Network for Chinese Word Segmentation

Yushi Yao and Zheng Huang(✉)

School of Electronic Information and Electrical Engineering,
Shanghai Jiaotong University, Shanghai, China
{yys12345,huangzheng}@sjtu.edu.cn

Abstract. Recurrent neural network (RNN) has been broadly applied to natural language process (NLP) problems. This kind of neural network is designed for modeling sequential data and has been testified to be quite efficient in sequential tagging tasks. In this paper, we propose to use bi-directional RNN with long short-term memory (LSTM) units for Chinese word segmentation, which is a crucial task for modeling Chinese sentences and articles. Classical methods focus on designing and combining hand-craft features from context, whereas bi-directional LSTM network (BLSTM) does not need any prior knowledge or pre-designing, and is expert in creating hierarchical feature representation of contextual information from both directions. Experiment result shows that our approach gets state-of-the-art performance in word segmentation on both traditional Chinese datasets and simplified Chinese datasets.

Keywords: Long short-term memory · Chinese word segmentation · Neural network

1 Introduction

Chinese word segmentation (CWS) is directly related to the performance of other tasks such as part-of-speech (POS) tagging, Named entity recognition (NER) and machine translation [1]. In order to indicate the boundaries of Chinese words, CWS can be converted into tagging task of Chinese character, which means every Chinese character is labeled to indicate the segmentation. Traditional methods for CWS basically fall into two categories: statistical models that model the probability distribution of a possible segmentation for a sentence, such as Hidden Markov Model (HMM) [3], conditional random field (CRF) [4] and neural probabilistic models [5], and dictionary-based methods which detect words by matching lexicon. Recently deep neural networks start to show its great capability in integrating inference process for language modeling [2] and recent research reveals that recurrent neural networks (RNN) significantly outperforms popular statistical algorithms.

As a special kind of RNN, LSTM neural network [6] is proved to be efficient in modeling sequential data like speech and text [7]. More over, BLSTM neural

A. Hirose et al. (Eds.): ICONIP 2016, Part IV, LNCS 9950, pp. 345–353, 2016.
DOI: 10.1007/978-3-319-46681-1_42

network [8], which is derived from LSTM network, has advantages in memorizing information for long sequences along both directions, and makes great improvement in linguistic computation and other fields as well. Huang and Xu combined LSTM with CRF and verified the efficiency and robustness of their model in sequential tagging [9]. Ling et al. focus on constructing compact vector representations of words with bi-directional LSTM, which yield state-of-the-art performance in contrast with other word-to-vector models like CBOW and skip-n-gram [10]. The study that is close to ours is Chen et al., which introduced LSTM neural network into Chinese word segmentation [11], while LSTM just memorize contextual information from past contextual. Since Chinese sentences have sophisticated structure and in order to exploit more discriminative features of characters, we argue that both future and past information need to be considered when training segmentation model.

In this study, in order to improve the performance of the Chinese word segmentation, we apply bi-directional LSTM network with character embeddings to word segmentation task. Our contributions can be summarized as follows:

(1) Our work is the first to apply bi-directional LSTM network to Chinese word segmentation benchmark datasets and improve it, and BLSTM is efficient for capturing complex and non-local features of text from both forward and backward direction
(2) The training framework can be regarded as an integration of generating embeddings and tagging characters and it does not need any external datasets.

The paper is organized as follows. Section 2 describes BLSTM network and its variants as well as its application in CWS task. Next, we introduce our training approaches and segmentation framework based on BLSTM network in Sect. 3. Section 4 details our experiments on Chinese dataset and summarizes our experimental results with previous research. Finally, in Sect. 5 we summarize key conclusions.

2 BLSTM Network Architecture

BLSTM neural network is a variant of LSTM network [8]. The special unit of this network is capable of learning long-term dependencies without keeping redundant context information. They work tremendously well on a large variety of sequential problems, and are now widely used in NLP tasks.

2.1 LSTM Unit and BLSTM Network

LSTM. The basic structure of LSTM memory unit is composed of three essential gates and a cell state. As shown in Fig. 1, the memory cell contains the information it memorized at time t, the state of the memory cell is bound up together with three gates, each gate is composed of input part and recurrent part.

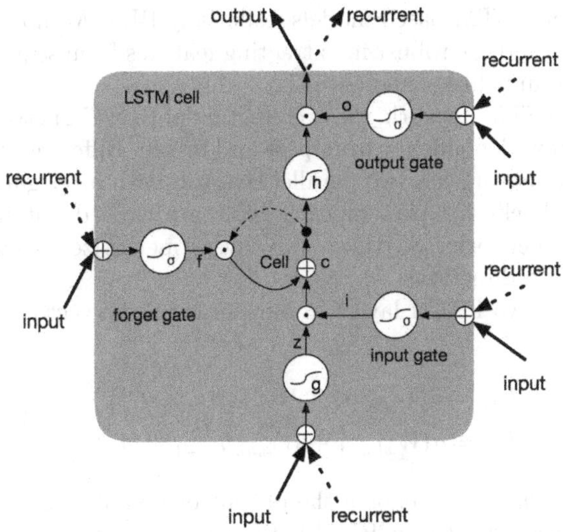

Fig. 1. Schematic of LSTM unit

The formal formulas for updating each gate and cell state are defined as below

$$z^t = g(W_z x^t + R_z y^{t-1} + b_z) \tag{1}$$

$$i^t = \sigma(W_i x^t + R_i y^{t-1} + b_t) \tag{2}$$

$$f^t = \sigma(W_f x^t + R_f y^{t-1} + b_f) \tag{3}$$

$$c^t = i^t \odot z^t + f^t \odot c^{t-1} \tag{4}$$

$$o^t = \sigma(W_o x^t + R_o y^{t-1} + b_o) \tag{5}$$

$$y^t = o^t \odot h(c^t) \tag{6}$$

Here $x^t \in R^d$ denotes a d-dimensional input vector at time t, W are weight matrices for input part, and R are weight matrices for recurrent part. b denotes bias vectors and the functions σ, g and h are point-wise non-linear functions like sigmoid or tanh (we used sigmoid in our experiments), $\odot$ means point-wise multiplication of two vectors. The output vector y^t is also a d-dimension vector. Due to the three controlling gates and the cell state, LSTM network is robust with respect to exploding and vanishing gradient problems [6], thus a LSTM network can be designed to learn long-range dependencies. This superiority allows LSTM to gather more history information and make it possible to model training data without hand-generated features.

BLSTM Network. When modeling Chinese sentences for CWS, statistical methods like CRF model the probability distributions of tags for every characters within a small context window. The size of window restricts the learning ability of model. However, LSTM based models, especially BLSTM models have stable learning dynamics and are robust in extracting features from sequence data with complicated structures like sentences [12].

Similar to the LSTM network, the architecture of BLSTM network is designed to model the context dependency from past and future. Different from the LSTM network, BLSTM network has two parallel layers in both propagation directions, the forward and backward pass of each layer are carried out in similar way of regular neural networks [8], these two layers memorize the information of sentences from both directions.

Since there are two LSTM layers in our network, the vector formula should be also adjusted.

$$h_{f_t} = \sigma(W_{xh_f}x_t + W_{h_f h_f}h_{f_{t-1}} + b_{h_f}) \tag{7}$$

$$h_{bt} = \sigma(W_{xh_b}x_t + W_{h_b h_b}h_{bt-1} + b_{h_b}) \tag{8}$$

The $h_f \in R^d$ and $h_b \in R^d$ denotes the output of LSTM unit of forward layer and backward layer respectively, BLSTM first compute the sub-result of the two hidden layers and the final output y_t is the concatenation of these two parts, which means y_t is a 2d-dimension vector. In this paper, we define the forward and backward layers as a single BLSTM layer.

3 Training Method

In order to convert the segmentation problem into a tagging problem, we assign a label for each character. There are four kinds of labels: *B, M, E, S*, corresponding to the beginning, middle, end of a word, and a single-character word, respectively.

3.1 Loss Function

For CWS, the output layer is designed as a multi-classification task for every single character in our work. Different from the work of Chen et al. [11], we use softmax loss to train the BLSTM model and believe that the softmax function is more suitable for illustrating the probability distribution of the character label, as it introduce discrimination among the labels. Suppose the input character is x_i, and $\hat{y_i}$ means the output vector of the network, where θ means the parameters of the network, we get the probability vector $(\hat{p_i^1}, \hat{p_i^2}, ..., \hat{p_i^k})$ as follows:

$$P(\hat{y_i^j}|x_i;\theta) = \frac{e^{\hat{y_i^j}}}{\sum_{k=1}^{K} e^{-\hat{y_i^k}}}, \quad j \in [1, K] \tag{9}$$

Since the ground truth vector is one-hot, the total loss is defined as follows:

$$J(\theta) = -\frac{1}{m}\sum_{i=1}^{m}\sum_{j=1}^{k}[1\{y_i = j\} logP(\hat{y_i^j}|x_i;\theta)] \tag{10}$$

We use $1\{\cdot\}$ to denote the indicator function that equals one if its argument is true and zero otherwise, y_i denotes the ground truth label of word x_i and m denotes the number of samples, which in our case, is equivalent to the size of training mini-batch.

3.2 Training Framework

The basic procedure of language modeling in our study is as follows (Fig. 2).

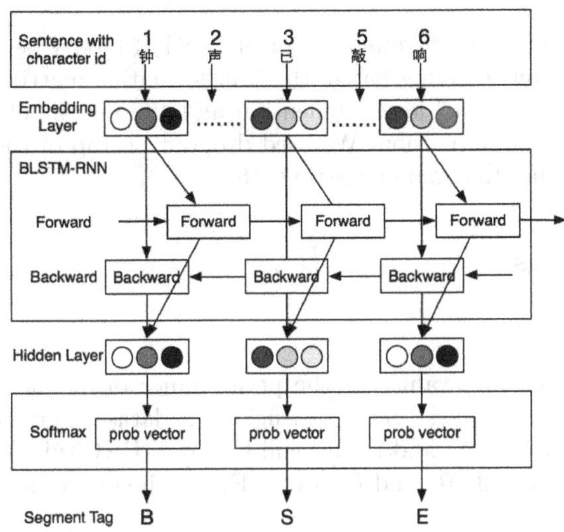

Fig. 2. Illustration of BLSTM language model for CWS

Each character has an id which is defined in a lookup dictionary, the dictionary is constructed by collecting the unique characters in the training set. Instead of one-hot representation, the characters are projected into a d-dimensional space and initialized as dense vectors $v \in R^d$, we regard this initialization step as constructing embeddings for characters. An embedding is stored in a matrix and can be retrieved by its id. As embeddings are efficient in describing word-level features [13], and experiment results show that character-level embeddings can also achieve good performance in CWS.

Secondly, the embeddings are fed into BLSTM network and the final output of BLSTM network is then passed into a hidden layer, followed by a softmax layer, which determines the tag with maximum probability of the character.

3.3 Model Variant

In order to further exploit benefits and improve the structure of BLSTM network, we also propose some variants of BLSTM network based on the method of constructing RNN [14]. Inspired by the structure of convolutional neural network (CNN), deeper network improves performance on extracting features of images (VGG) [15], we stack BLSTM layers and expect to extract contextual features in higher level. However, the output of a BLSTM layer is in double size of the input vector since it is composed of two LSTM layers, and the dimension of the embedding vectors will expand dramatically when the network goes deeper, here we use a transformation matrix to compress the dimension of output vectors, and keep it the same size with input vectors.

$$M_{tran} = W_{tran} \times M_o \tag{11}$$

Assume that the output embedding vector of BLSTM layer $v_o \in R^{2d}$ and there are n characters in one training batch, the transformation matrix W_{tran} project the embedding vectors into lower dimension and the output of each BLSTM layer is kept the same dimension. We used dropout on top of the output layer during in the training to avoid overfitting [16].

4 Experiments

4.1 Setup

The dataset we used for evaluating the performance of our network on CWS is from Backoff 2005, which contains benchmark datasets for both simplified Chinese (PKU and MSRA) and traditional Chinese (HKCityU) CWS task. We used precision (P), recall (R) and F1-score (F1) as the evaluation metric of our model.

We keep the length of BLSTM network to 30 which is the average sentence length of our training data. At the stage of data preprocessing, we split the text into groups of 30-word sequences starting with the first "START" token All the models are trained on NVIDIA GTX Geforce 970, it took about 16 to 17 h to train a model on GPU while more than 4 days to train on CPU, in contrast. We also changed batch size during our training process because of the limit of GPU memory.

4.2 Character Representation

The dimension of embedding vector is a crucial hyper parameter of our network, because it will directly influence the number of parameters and model complexity. The first step of our experiment is to get a suitable size of embedding.

Table 1 shows the performance of BLSTM networks on HKCityU dataset with embedding vectors of different dimensions. When the dimension grows higher, the error rate also gets higher and unstable. From the result we conclude that an efficient representation of Chinese character should not be too long, and we set the embedding dimension to 200 in our following experiments.

Table 1. Performance of BLSTM networks with different embedding dimensions

Embedding size	P	R	F
100	92.2	91.6	91.9
128	92.6	92.1	92.3
200	**94.1**	**93.5**	**93.8**
300	90.6	90.2	90.4

4.3 Stack LSTM Layers

Secondly, we tested the performance of our model with different number of BLSTM layers.

Table 2. Performance of our models on four test sets, add simple RNN and LSTM network for comparison.

Models	PKU			MSRA			AS			HKCityU		
	P	R	F	P	R	F	P	R	F	P	R	F
RNN	92.4	92.1	92.2	95.8	95.3	95.0	95.5	94.5	95.0	96.3	95.3	95.8
Bi-RNN	94.2	92.5	93.3	95.7	94.8	95.2	96.1	96.4	96.2	96.8	95.9	96.3
LSTM*	95.3	94.6	94.9	96.1	95.3	95.7	94.2	93.2	93.7	97.2	96.6	96.9
BLSTM*	96.5	95.3	95.9	96.6	97.1	96.9	97.3	96.9	97.1	97.4	97.2	97.3
BLSTM2*	96.6	95.9	96.2	97.3	97.1	97.2	97.9	97.5	97.7	97.5	**97.4**	97.4
BLSTM3*	**96.8**	**96.3**	**96.5**	**97.4**	**97.3**	**97.3**	**98.0**	**97.6**	**97.8**	**97.7**	97.3	**97.5**

BLSTM2 means a bi-directional LSTM network with two layer, and BLSTM3 means a bi-directional LSTM network with three BLSTM layers, * denotes that the model is trained with dropout.

Table 2 shows that as we stack more BLSTM layers, the performance gets slight improvement, while adding layers becomes not so effective when the number of BLSTM layers exceeds three, which also takes quite long time to train. The result shows that LSTM units comes to be less effective in higher layers so we believe that there is no need to build very deep network for extracting contextual information.

4.4 Comparison with Existing Methods

In this section, we will state the procedure of our experiments and how we get the model with the best performance, and we will also compare the performance of our network with state-of-the-art approaches.

Table 3 lists the performances of our model as well as previous research. [17] designed rich feature templates for CRF, [18] improved supervised word segmentation by exploiting features of unlabeled data. [19] applied semi-supervised approach to extract representations of label distributions from unlabeled and labeled datasets. Nevertheless, all the models or systems above focus on feature engineering, while our approach do not depend on any predesigned features thanks to the strong ability of BLSTM network in automatic feature learning.

Table 3. Comparison of our model with previous research, the results with * symbol are from our reproduction result of their methods

Models	PKU	MSRA	CityU
(Zhao et al. 2006)	-	-	**97.7**
(Sun and Xu 2011)	95.1	97.2	-
(Zhang et al. 2013)	96.1	97.4	-
(Chen et al. 2015)	96.3*	97.3*	-
Ours	**96.5**	**97.6**	97.5

Our model slightly outperforms the work of Chen's [11], which also used character embeddings but applied LSTM network to CWS and use pre-trained character embeddings. In comparison, our training framework do not need any pre-trained information and the results suggest that BLSTM gets better performance than LSTM on segmentation, and indicates that both past and future information should be taken into account for segmentation task.

5 Conclusions

In this paper, we propose to apply bi-directional LSTM neural network to Chinese word segmentation because of its high efficiency and robustness for sequential modeling task. The model learns to extract discriminative character-level features automatically and it does not require any hand-craft features for segmentation or prior knowledge. Experiments conducted on Backoff 2005 datasets show that our model has good performance and generalization on both simplified Chinese and traditional Chinese. Our results suggest that deep neural networks work well on segmentation tasks and BLSTM networks with character embedding is an effective tagging solution and worth further exploration.

References

1. Chang, P.-C., Galley, M., Manning, C.D.: Optimizing Chinese word segmentation for machine translation performance. In: Proceedings of the Third Workshop on Statistical Machine Translation, pp. 224–232. Association for Computational Linguistics (2008)
2. Auli, M., Galley, M., Quirk, C., Zweig, G.: Joint language and translation modeling with recurrent neural networks. In: EMNLP, vol. 3 (2013)
3. Zhang, H.-P., Hong-Kui, Y., Xiong, D.-Y., Liu, Q.: HHMM-based Chinese lexical analyzer ICTCLAS. In: Proceedings of the Second SIGHAN Workshop on Chinese Language Processing, vol. 17, pp. 184–187. Association for Computational Linguistics (2003)
4. Peng, F., Feng, F., McCallum, A.: Chinese segmentation and new word detection using conditional random fields. In: Proceedings of the 20th International Conference on Computational Linguistics, p. 562. Association for Computational Linguistics (2004)

5. Bengio, Y., Ducharme, R., Vincent, P., Janvin, C.: A neural probabilistic language model. J. Mach. Learn. Res. **3**, 1137–1155 (2003)
6. Hochreiter, S., Schmidhuber, J.: Long short-term memory. Neural Comput. **9**(8), 1735–1780 (1997)
7. Sundermeyer, M., Ney, H., Schluter, R.: From feedforward to recurrent LSTM neural networks for language modeling. IEEE/ACM Trans. Audio Speech Lang. Process. **23**(3), 517–529 (2015)
8. Schuster, M., Paliwal, K.K.: Bidirectional recurrent neural networks. IEEE Trans. Signal Process. **45**(11), 2673–2681 (1997)
9. Huang, Z., Wei, X., Kai, Y.: Bidirectional LSTM-CRF models for sequence tagging (2015). arXiv preprint: arXiv:1508.01991
10. Ling, W., Luís, T., Marujo, L., Astudillo, R.F., Amir, S., Dyer, C., Black, A.W., Trancoso, I.: Finding function in form: compositional character models for open vocabulary word representation (2015). arXiv preprint: arXiv:1508.02096
11. Chen, X., Qiu, X., Zhu, C., Liu, P., Huang, X.: Long short-term memory neural networks for Chinese word segmentation. In: Proceedings of the Conference on Empirical Methods in Natural Language Processing (2015)
12. Sundermeyer, M., Schlüter, R., Ney, H.: LSTM neural networks for language modeling. In: INTERSPEECH (2012)
13. Mikolov, T., Sutskever, I., Chen, K., Corrado, G.S., Dean, J.: Distributed representations of words and phrases and their compositionality. In: Advances in Neural Information Processing Systems, pp. 3111–3119 (2013)
14. Pascanu, R., Gulcehre, C., Cho, K., Bengio, Y.: How to construct deep recurrent neural networks (2013). arXiv preprint: arXiv:1312.6026
15. Simonyan, K., Zisserman, A.: Very deep convolutional networks for large-scale image recognition (2014). arXiv preprint: arXiv:1409.1556
16. Srivastava, N., Hinton, G., Krizhevsky, A., Sutskever, I., Salakhutdinov, R.: Dropout: a simple way to prevent neural networks from overfitting. J. Mach. Learn. Res. **15**(1), 1929–1958 (2014)
17. Zhao, H., Huang, C.-N., Li, M.: An improved Chinese word segmentation system with conditional random field. In: Proceedings of the Fifth SIGHAN Workshop on Chinese Language Processing, Sydney, vol. 1082117, July 2006
18. Sun, W.: A stacked sub-word model for joint Chinese word segmentation and part-of-speech tagging. In: Proceedings of the 49th Annual Meeting of the Association for Computational Linguistics: Human Language Technologies, vol. 1, pp. 1385–1394. Association for Computational Linguistics (2011)
19. Zhang, L., Houfeng, W., Sun, X., Mansur, M.: Exploring representations from unlabeled data with co-training for Chinese word segmentation. In: Proceedings of the Conference on Empirical Methods in Natural Language Processing (2013)

Alternating Optimization Method Based on Nonnegative Matrix Factorizations for Deep Neural Networks

Tetsuya Sakurai[1,2(✉)], Akira Imakura[1], Yuto Inoue[1], and Yasunori Futamura[1]

[1] Department of Computer Science, University of Tsukuba, Tsukuba 305-8573, Japan
sakurai@cs.tsukuba.ac.jp
[2] CREST, Japan Science and Technology Agency, Kawaguchi 332-0012, Japan

Abstract. The backpropagation algorithm for calculating gradients has been widely used in computation of weights for deep neural networks (DNNs). This method requires derivatives of objective functions and has some difficulties finding appropriate parameters such as learning rate. In this paper, we propose a novel approach for computing weight matrices of fully-connected DNNs by using two types of semi-nonnegative matrix factorizations (semi-NMFs). In this method, optimization processes are performed by calculating weight matrices alternately, and backpropagation (BP) is not used. We also present a method to calculate stacked autoencoder using a NMF. The output results of the autoencoder are used as pre-training data for DNNs. The experimental results show that our method using three types of NMFs attains similar error rates to the conventional DNNs with BP.

Keywords: Nonnegative matrix factorizations · Alternating optimization method · Deep neural networks

1 Introduction

Deep neural networks (DNNs) attracted a great deal of attention for their high efficiency in various fields, such as speech recognition, image recognition, object detection, materials discovery. By using a backpropagation (BP) technique proposed by Rumelhart et al. [15], computational performance is improved for training multilayer neural networks. However, learning often takes a long time to converge, and it may fall into a local minimum. Bengio et al. [1] proposed a method to improve general performance by pre-training with an autoencoder. Moreover, selection of appropriate learning rates [12] and restriction of weights as dropout [16] have also used to minimize the expected error. Hinton et al. discussed initialization of weights in [7].

Neural networks have variations such as fully-connected networks, convolutional networks and recurrent networks. LeCun et al. [12] showed that convolutional neural networks attain high efficiency for image recognition. In DNNs,

A. Hirose et al. (Eds.): ICONIP 2016, Part IV, LNCS 9950, pp. 354–362, 2016.
DOI: 10.1007/978-3-319-46681-1_43

activation functions are used to attain nonlinear properties. Recently, the rectified linear function (ReLU) [6,13] has often been used.

Feedforward neural networks are computed by multiplying weight matrices and input matrices. Thus, the main computations are matrix–matrix multiplications (GEMM), and accelerators such as GPUs are employed to obtain high performance [2]. However, a large computational cost of neural networks is still a problem.

In this paper, we propose a novel computing method for fully-connected DNNs that uses two types of semi-nonnegative matrix factorizations (semi-NMFs). In this method, optimization processes are performed by calculating weight matrices alternately, and BP is not used. We also present a method to calculate a stacked autoencoder using a NMF [11,14]. The output results of the autoencoder are used as pre-training data for DNNs. Here we note that our proposed method does not constrain the weight matrices to be non-negative. Therefore, our proposed method is different strategy with neural networks using non-negative constraints [3].

In the presented method, computations are represented by matrix–matrix computations, and accelerators such as GPUs and MICs can be employed like in BP computations. In BP computations, mini-batches are used to avoid stagnations of the optimization precess. The use of small mini-batch sizes decreases matrix sizes and gains reductions in computations. The presented method also uses partitioned matrices; however, the matrix size is larger than that of conventional BP, and we expect high performance.

This paper is organized as follows. In Sect. 2, we review the conventional method of computing DNNs. In Sect. 3, we present a method for computing weights in DNNs using two types of semi-NMFs. We also present a method to calculate a stacked autoencoder using NMF. In Sect. 4, we show some experimental results of our proposed approach. Section 5 presents our conclusions.

We use MATLAB colon notations throughout. Moreover, let $A = \{a_{ij}\} \in \mathbb{R}^{m\times n}$, then $A \geq 0$ denotes that all entries are nonnegative: $a_{ij} \geq 0$.

2 Computation of Deep Neural Networks

Let $n_{\text{in}}, n_{\text{out}}, m$ be sizes of input and output units and the training data, respectively. Moreover, let $X \in \mathbb{R}^{n_{\text{in}}\times m}$ and $Y \in \mathbb{R}^{n_{\text{out}}\times m}$ be input and output data. Using a weight matrix $W \in \mathbb{R}^{n_{\text{out}}\times n_{\text{in}}}$ and a bias vector $\boldsymbol{b} \in \mathbb{R}^{n_{\text{out}}}$, the objective function of one layer of neural networks can be written as

$$E(W, \boldsymbol{b}, X, Y) = D\left(Y, f(WX + \boldsymbol{b}\boldsymbol{e}^{\mathrm{T}})\right) + h(W, \boldsymbol{b}),$$

where $D(\cdot,\cdot)$ is a divergence function, $\boldsymbol{e} = [1, 1, \ldots, 1]^{\mathrm{T}} \in \mathbb{R}^m$, $f(U)$ is an activation function and $h(W, \boldsymbol{b})$ is a regularization term. There are several activation functions such as sigmoid functions like the logistic function and the hyperbolic tangent function. Recently, rectified linear unit (ReLU) has been widely used.

The objective function of DNNs with $d-1$ hidden units of size $n_i, i = 1, 2, \ldots, d-1$ is written as

$$\begin{aligned} & E(W_1, \ldots, W_d, \boldsymbol{b}_1, \ldots, \boldsymbol{b}_d, X, Y) \\ & \quad = D\left(Y, W_d f(W_{d-1} \cdots f(W_1 X + \boldsymbol{b}_1 \boldsymbol{e}^{\mathrm{T}}) \cdots + \boldsymbol{b}_{d-1} \boldsymbol{e}^{\mathrm{T}}) + \boldsymbol{b}_d \boldsymbol{e}^{\mathrm{T}}\right) \\ & \qquad + h(W_1, \ldots, W_d, \boldsymbol{b}_1, \ldots, \boldsymbol{b}_d). \end{aligned} \tag{1}$$

Here, $W_i \in \mathbb{R}^{n_i \times n_{i-1}}, \boldsymbol{b}_i \in \mathbb{R}^{n_{i-1}}$, $n_0 = n_{\mathrm{in}}, n_d = n_{\mathrm{out}}$. BP algorithms, which are based on the gradient descent method using derivatives, are one of the most standard algorithms used to minimize the objective function.

3 An Alternating Optimization Method Based on Nonnegative Matrix Factorization

In this paper, we consider solving the following minimization problem

$$\min_{W_1, \ldots, W_d} E(W_1, \ldots, W_d, X, Y), \tag{2}$$

where the objective function simplifies the objective function (1) using the square error of DNNs and is defined by

$$E(W_1, \ldots, W_d, X, Y) := \frac{1}{2} \| Y - W_d f(W_{d-1} \cdots f(W_1 X) \cdots) \|_{\mathrm{F}}^2, \tag{3}$$

where $\| \cdot \|_{\mathrm{F}}$ is the Frobenius norm. Here, the activation function $f(U)$ is set as ReLU.

The basic concept of our algorithm to solve (2) is an alternating optimization that (approximately) optimizes each weight matrix W_i for $i = d, d-1, \ldots, 1$, one by one. Let $W_1^{(0)}, W_2^{(0)}, \ldots, W_d^{(0)}$ be initial guesses of $W_1, \ldots, W_d$, respectively. An autoencoder to set the initial guesses will be discussed in Sect. 4. In each iteration k, we also define objective functions

$$E_i^{(k)}(W_i, X, Y) := E(W_1^{(k-1)}, \ldots, W_{i-1}^{(k-1)}, W_i, W_{i+1}^{(k)}, \ldots, W_d^{(k)}, X, Y),$$

as for the i-th weight matrix W_i. Then, we (approximately) solve the minimization problems

$$W_i^{(k)} = \arg\min_{W_i} E_i^{(k)}(W_i, X, Y)$$

for $i = d, d-1, \ldots, 1$. The basic concept of our proposed method is shown in Algorithm 1.

Let matrices $Z_i^{(k)} \in \mathbb{R}^{n_i \times m}$ be defined as

$$\begin{aligned} Z_0^{(k)} &:= X, \\ Z_i^{(k)} &:= f(W_i^{(k)} f(W_{i-1}^{(k)} \cdots f(W_1^{(k)} X) \cdots)), \quad i = 1, 2, \ldots, d-1. \end{aligned}$$

Then, in what follows, we derive our alternating optimization algorithm using semi-NMF [4] and nonlinear semi-NMF.

Algorithm 1. The basic concept of the proposed method

Set initial guesses $W_1^{(0)}, W_2^{(0)}, \ldots, W_d^{(0)}$
for $k = 1, 2, \ldots$ **do:**
  for $i = d, d-1, \ldots, 1$ **do:**
    Minimize (approx.) $E_i^{(k)}(W_i, X, Y)$ for W_i with an initial guess $W_i^{(k-1)}$, and get $W_i^{(k)}$
  end for
end for

3.1 Optimization for W_d Using Semi-NMF

Using the matrix $Z_{d-1}^{(k-1)}$, the objective function for the weight matrix W_d is rewritten as

$$E_d^{(k)}(W_d, X, Y) = \frac{1}{2}\|Y - W_d Z_{d-1}^{(k-1)}\|_{\mathrm{F}}^2.$$

Here, we note that $Z_{d-1}^{(k-1)} \geq 0$ from the definition of $Z_i^{(k-1)}$. Therefore, we can obtain $W_d^{(k)}$ and $\widehat{Z}_{d-1}^{(k)}$ by (approximately) solving nonnegative constraint minimization problem of the form

$$[W_d^{(k)}, \widehat{Z}_{d-1}^{(k)}] = \arg \min_{W_d, Z_{d-1}} \|Y - W_d Z_{d-1}\|_{\mathrm{F}}, \quad \text{s.t. } Z_{d-1} \geq 0, \tag{4}$$

using initial guesses $W_d^{(k-1)}, Z_{d-1}^{(k-1)}$. This minimization problem is known as semi-NMF.

3.2 Optimization for $W_i, i = d-1, \ldots, 1$ Using Nonlinear Semi-NMF

From the definition of $Z_i^{(k-1)}$, we expect

$$\widehat{Z}_{d-1}^{(k)} \approx f(W_{d-1} Z_{d-2}) \tag{5}$$

to minimize the objective function (3). Then, we consider (approximately) solving the minimization problem

$$[W_i^{(k)}, \widehat{Z}_{i-1}^{(k)}] = \arg \min_{W_i, Z_{i-1}} \|\widehat{Z}_i^{(k)} - f(W_i Z_{i-1})\|_{\mathrm{F}}, \quad \text{s.t. } Z_{i-1} \geq 0 \tag{6}$$

for W_i with $i = d-1, d-2, \ldots, 1$. This minimization problem (6) is a nonnegative constraint minimization problem like (4). However, (6) has a nonlinear activation function. In this paper, we call this problem nonlinear semi-NMF.

In order to solve this nonlinear semi-NMF, we introduce an alternating minimization algorithm that minimizes nonlinear least squares problems

$$\min_{W_i} \|\widehat{Z}_i^{(k)} - f(W_i Z_{i-1}^{(k-1)})\|_{\mathrm{F}} \tag{7}$$

Algorithm 2. An iteration method for solving nonlinear least squares $\min_X \|B - f(XA)\|_{\mathrm{F}}$

```
Set initial guess X_0 and parameter ω
for s = 0, 1, ... do:
    R_s = B − f(X_s A)
    X_{s+1} = X_s + ω R_s A^†
end for
```

Algorithm 3. An iteration method for solving nonnegative constrain nonlinear least squares $\min_{X \geq 0} \|B - f(AX)\|_{\mathrm{F}}$

```
Set initial guess X_0 and parameter ω
for s = 0, 1, ... do:
    R_s = B − f(A X_s)
    X_{s+1} = f(X_s + ω A^† R_s)
end for
```

and

$$\min_{Z_{i-1} \geq 0} \|\widehat{Z}_i^{(k)} - f(W_i^{(k)} Z_{i-1})\|_{\mathrm{F}}, \tag{8}$$

one by one. Here, we note that (8) has a nonnegative constraint on Z_i. We also note that, for $i = 1$, we do not require a solution of (8), because $Z_0 = X$. The nonlinear least squares problems (7) and (8) are solved by stationary iteration-like methods as shown in Algorithms 2 and 3, where $A^\dagger$ is a pseudo-inverse of A. In practice, the pseudo-inverse of A is approximated using a low-rank approximation of A.

3.3 An Alternating Optimization Method

Using semi-NMF (4) and nonlinear semi-NMF (5), the algorithm of the proposed method is summarized in Algorithm 4. In practice, the input data X is approximated using a low-rank approximation based on the singular value decomposition:

$$X = [U_1, U_2] \begin{bmatrix} \Sigma_1 & \\ & \Sigma_2 \end{bmatrix} \begin{bmatrix} V_1^{\mathrm{T}} \\ V_2^{\mathrm{T}} \end{bmatrix} \approx U_1 \Sigma_1 V_1^{\mathrm{T}}.$$

Here, we assume that all hidden units have almost the same size: $n \approx n_i$, then the computational cost of the proposed method is $O(mn^2 + dn^3)$.

The proposed method can also use the mini-batch technique. Let $X_\ell := X(:, \mathcal{J}_\ell)$ be a submatrix of the input data X corresponding to each mini-batch, where $\mathcal{J}_\ell$ is the index set in the mini-batch. Then, in order to use the mini-batch technique for the proposed method, we need to compute the low-rank approximation of $X_\ell \approx U_{\ell,1} \Sigma_{\ell,1} V_{\ell,1}^{\mathrm{T}}$, in each iteration. We can reduce the required computational cost by reusing the results of the low-rank approximation of X as follows:

$$X(:, \mathcal{J}_\ell) \approx U_{\ell,1} \Sigma_{\ell,1} V_{\ell,1}^{\mathrm{T}} \approx U_1 \Sigma_1 V_1(\mathcal{J}_\ell, :)^{\mathrm{T}}.$$

Algorithm 4. A proposed method

```
Set initial guess W_1^(0), W_2^(0), ..., W_d^(0)
for k = 1, 2, ... do:
    Solve (approx.) semi-NMF (4)
    with initial guesses W_d^(k-1), Z_{d-1}^(k-1) and get W_d^(k), \hat{Z}_{d-1}^(k)
    for i = d - 1, ..., 2 do:
        Solve (approx.) nonlinear LSQ (7) by Algorithm 2
        with an initial guess W_i^(k-1), and get W_i^(k)
        Solve (approx.) nonnegative constrain nonlinear LSQ (8) by Algorithm 3
        with an initial guess Z_{i-1}^(k-1), and get \hat{Z}_{i-1}^(k)
    end for
    Solve (approx.) nonlinear LSQ (7) for i = 1 by Algorithm 2
    with an initial guess W_1^(k-1), and get W_1^(k)
    Set Z_i^(k) for i = 1, 2, ..., d - 1
end for
```

Algorithm 5. A proposed stacked autoencoder

```
for i = 1, 2, ..., d - 1 do:
    Set initial guess \tilde{W}_i^(0), Z_i^(0)
    for k = 1, 2, ..., iter_max do:
        Solve (approx.) NMF (9) with initial guesses \tilde{W}_i^(k-1), Z_i^(k-1), and get
        \tilde{W}_i^(k), \hat{Z}_i^(k)
        Solve (approx.) nonlinear LSQ min_{W_i} ||\hat{Z}_i^(k) - f(W_i Z_{i-1})||_F by Algorithm 2
        with initial guess W_i^(k-1), and get W_i^(k)
        Set Z_i^(k) = f(W_i^(k) Z_{i-1})
    end for
end for
```

Other improvement techniques used for BP are also expected to improve the performance of the proposed method.

4 An Alternating Optimization-Based Stacked Autoencoder Using NMF

In this section, we propose an alternating optimization-based stacked autoencoder using NMF for computing initial guesses $W_i^{(0)}$ of the proposed method (Algorithm 4). Let $\widehat{Z}_0 = X$. Then, for the stacked autoencoder, we compute the initial guesses $W_i^{(0)}$ by (approximately) minimizing

$$\min_{W_i, \widetilde{W}_i} \|\widehat{Z}_{i-1} - \widetilde{W}_i f(W_i Z_{i-1})\|_{\mathrm{F}}$$

for $i = 1, 2, \ldots, d-1$ like as for the DNNs. Each minimization problem is solved by NMF as shown below

$$[\widetilde{W}_i, \widehat{Z}_i] = \arg \min_{\widetilde{W}_i, Z_i \geq 0} \|\widehat{Z}_{i-1} - \widetilde{W}_i Z_i\|_{\mathrm{F}} \tag{9}$$

and Algorithm 2 is used to solve the nonlinear least squares problem as

$$W_i = \min_{W_i} \|\widehat{Z}_i - f(W_i Z_{i-1})\|_{\mathrm{F}}. \tag{10}$$

By solving (9) and (10) alternatively, the algorithm for the proposed stacked autoencoder is summarized by Algorithm 5.

5 Performance Evaluations

In this section, we evaluate the performance of the proposed method (Algorithm 4) using the stacked autoencoder (Algorithm 5) for fully-connected DNNs for MNIST [10] and CIFAR10 [9]. There are several techniques for improving the performance of BP such as affine/elastic distortions and denoising autoencoder. These techniques are also expected to improve the performance of our algorithm. Therefore, in this section, we just make a comparison with a simple BP.

For the proposed method, the number of iterations of the autoencoder and the LSQs and the threshold of the low-rank approximation of the input data X were set as $(5, 10, 4.0 \times 10^{-2})$ for MNIST and $(20, 25, 5.0 \times 10^{-3})$ for CIFAR10, respectively. The size of the mini-batches was set as 5000 and the autoencoder was computed using only 5000 random samples. For optimizing parameters of BP, we used ADAM optimizer [8]. For ADAM optimizer, initial learning rates for the stacked autoencoder and for the fine tuning were set as $(10^{-3}, 10^{-3})$ for MNIST and $(5.0 \times 10^{-4}, 10^{-3})$ for CIFAR10, respectively. Other parameters β_1, β_2 and ε of ADAM were set as the default parameters of TensorFlow. We used the normalized initialization [5] for initial guesses of the stacked autoencoder. The size of the mini-batches was set as 100 and the autoencoder was computed using only 5000 random samples.

The performance evaluations were carried out using double precision arithmetic on Intel(R) Xeon(R) CPU E5-2667 v3 (3.20 GHz). The proposed method was implemented in MATLAB and the BP was implemented using TensorFlow [17].

Table 1. The 95 % confidence interval of the error rate and the computation time of 10 epoch of the proposed method (MATLAB) and BP (TensorFlow) with several hidden units for MNIST.

Hidden units	BP			Proposed		
	Error rate [%]		Time [sec.]	Error rate [%]		Time [sec.]
	Train data	Test data		Train data	Test data	
500	0.30 ± 0.014	1.77 ± 0.03	153	3.68 ± 0.043	3.73 ± 0.09	99
1000-500	0.06 ± 0.007	1.39 ± 0.08	330	0.04 ± 0.004	1.50 ± 0.06	310
1500-1000-500	0.32 ± 0.086	1.90 ± 0.15	739	0.01 ± 0.004	1.35 ± 0.03	737
2000-1500-1000-500	0.48 ± 0.132	1.84 ± 0.18	1589	0.00 ± 0.001	1.29 ± 0.04	1581

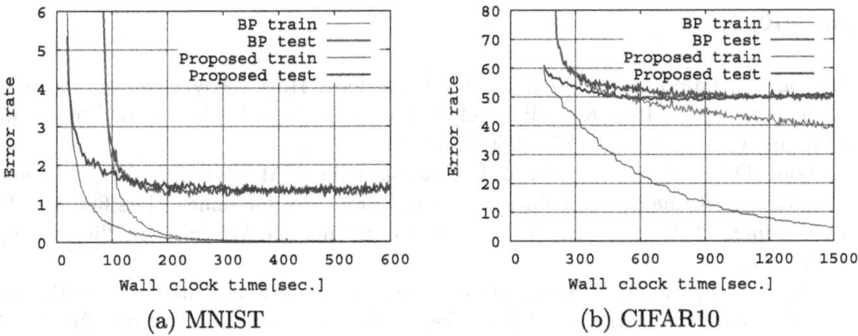

(a) MNIST (b) CIFAR10

Fig. 1. Convergence history of the proposed method and BP with [1000-500] hidden units for MNIST and CIFAR10.

Figure 1 shows the convergence history of the proposed method and BP with [1000-500] hidden units for MNIST and CIFAR10. Table 1 shows the 95 % confidence interval of the error rate and the computation time of 10 epoch of both methods with several hidden units for MNIST.

These experimental results show that our method attains a similar error rate, for several hidden units, as conventional DNNs with BP. Specifically, the proposed method achieves better error rates with deeper hidden units. Moreover, the proposed method needs a smaller computation time for the stacked autoencoder and almost the same computation time for the fine tuning.

6 Conclusions

In this paper, we proposed an alternating optimization algorithm for computing weight matrices of fully-connected DNNs by using the semi-NMF and the nonlinear semi-NMF. We also presented a method to calculate a stacked autoencoder by using NMF. The experimental results showed that our method using NMF attains a similar error rate and a similar computation time to conventional DNNs with BP. Almost the all computations of the proposed method are represented by matrix–matrix computations, and accelerators such as GPUs and MICs are employed like in BP computations. The proposed method also uses mini-batch technique; however, the matrix size is larger than that of conventional BP. Therefore, we expect that the proposed method achieves high performance on recent computational environments.

For future work, we will consider a bias vector, sparse regularizations and other activation functions. Moreover, we will extend our algorithm to convolutional neural networks. We will also consider parallel computation implementation and evaluate the performance in recent parallel environments.

References

1. Bengio, Y., Lamblin, P., Popovici, D., Larochelle, H.: Greedy layer-wise training of deep networks. In: Proceedings of Advances in Neural Information Processing Systems, vol. 19, pp. 153–160 (2006)
2. Ciresan, D.C., Meier, U., Masci, J., Gambardella, L.M., Schmidhuber, J.: Flexible, high performance convolutional neural networks for image classification. In: Proceedings of 22nd International Joint Conference on Artificial Intelligence, pp. 1237–1242 (2011)
3. Chorowski, J., Member, S.: Learning understandable neural networks with non-negative weight constraints. IEEE Trans. Neural Netw. Learn. Syst. **26**, 62–69 (2015)
4. Ding, D., Li, T., Jordan, M.I.: Convex and semi-nonnegative matrix factorizations. IEEE Trans. Pattern Anal. Mach. Intell. **32**, 45–55 (2010)
5. Glorot, X., Bengio, Y.: Understanding the difficulty of training deep feedforward neural networks. In: International Conference on Artificial Intelligence and Statistics, pp. 249–256 (2010)
6. Glorot, X., Bordes, A., Bengio., Y.: Deep sparse rectifier neural networks. In: Proceedings of 14th International Conference on Artificial Intelligence and Statistics, pp. 315–323 (2011)
7. Hinton, G.E., Deng, L., Yu, D., Dahl, G.E., Mohamed, A., Jaitly, N., Senior, A., Vanhoucke, V.: Deep neural networks for acoustic modeling in speech recognition. IEEE Signal Process. Mag. **29**, 82–97 (2012)
8. Kingma, D.P., Ba, J.: ADAM: a method for stochastic optimization. In: The International Conference on Learning Representations (ICLR), San Diego (2015)
9. Krizhevsky, A., Hinton, G.: Learning multiple layers of features from tiny images. Technical report, Computer Science Department, University of Toronto, vol. 1, p. 7 (2009)
10. LeCun, Y.: The MNIST database of handwritten digits. http://yann.lecun.com/exdb/mnist
11. Lee, D.D., Seung, H.S.: Learning the parts of objects by non-negative matrix factorization. Nature **401**, 788–791 (1999)
12. LeCun, Y., Bottou, L., Bengio, Y., Huffier, P.: Gradient-based learning applied to document recognition. Proc. IEEE **86**, 2278–2324 (1998)
13. Nair, V., Hinton, G.E.: Rectified linear units improve restricted Boltzmann machines. In: Proceedings of ICML (2010)
14. Paatero, P., Tapper, U.: Positive matrix factorization: a non-negative factor model with optimal utilization of error estimates of data values. Environmetrics **5**, 111–126 (1994)
15. Rumelhart, D.E., Hinton, G.E., Williams, R.J.: Learning representations by back-propagating errors. Nature **323**, 533–536 (1986)
16. Srivastava, N., Hinton, G.E., Krizhevsky, A., Sutskever, I., Salakhutdinov, R.: Dropout: a simple way to prevent neural networks from overfitting. J. Mach. Learn. Res. **15**, 1929–1958 (2014)
17. TensorFlow. https://www.tensorflow.org/

Fissionable Deep Neural Network

DongXu Tan[3(✉)], JunMin Wu[1,2], HuanXin Zheng[2], Yan Yin[2], and YaXin Liu[3]

[1] Suzhou Institute for Advanced Study, University of Science and Technology of China, Suzhou, China
jmwu@ustc.edu.cn

[2] Department of Computer Science and Technology, University of Science and Technology of China, Suzhou, China
{zhxfl, yyl001}@mail.ustc.edu.cn

[3] School of Software Engineering, University of Science and Technology of China, Suzhou, China
{sa614149, sa614246}@mail.ustc.edu.cn

Abstract. Model combination nearly always improves the performance of machine learning methods. Averaging the predictions of multi-model further decreases the error rate. In order to obtain multi high quality models more quickly, this article proposes a novel deep network architecture called "Fissionable Deep Neural Network", abbreviated as FDNN. Instead of just adjusting the weights in a fixed topology network, FDNN contains multi branches with shared parameters and multi Softmax layers. During training, the model divides until to be multi models. FDNN not only can reduce computational cost, but also overcome the interference of convergence between branches and give an opportunity for the branches falling into a poor local optimal solution to re-learn. It improves the performance of neural network on supervised learning which is demonstrated on MNIST and CIFAR-10 datasets.

Keywords: Model combination · Neural network · Shared parameters · Fission

1 Introduction

Currently, researches about neural network are focusing on static neural network. Mishkin et al. (2016) [4] proposed a method to pre-initialize weights of each convolution or inner-product layer with orthonormal matrices and normalize the variance of the output of each layer to equal to one. [5–8] are the evolution of sparse network structure. Springenberg et al. (2014) [9] re-evaluated the state of the art for object recognition from small images with convolutional networks. [10–12] are activation functions, which improve the neural network's performance.

All these methods constructed the neural networks by confirmed layers. However, Elman (1993) [3] underlined that the successful learning may depend on starting small. It is a striking fact that in humans the greatest learning occurs precisely in childhood, when the most dramatic maturational changes also occur. The children master the knowledge while the neurons of brain are dividing. Instead of just adjusting the weights in a network of fixed topology, Cascade-Correlation [13] begins with a minimal network, then

A. Hirose et al. (Eds.): ICONIP 2016, Part IV, LNCS 9950, pp. 363–371, 2016.
DOI: 10.1007/978-3-319-46681-1_44

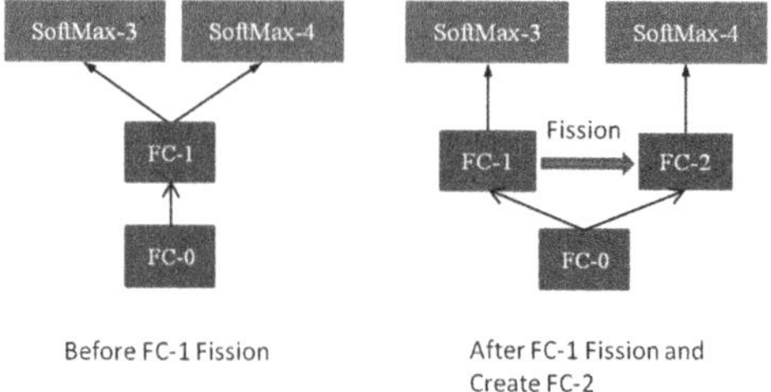

Fig. 1. Fissionable node

automatically trains and adds new hidden units one by one, creating a multi-layer structure. Once a new hidden unit has been added to the network, its input-side weights are frozen. This unit then becomes a permanent feature-detector in the network, available for producing outputs or for creating other, more complex feature detectors. LeCun et al. (1989) proposed [14] "Optimal Brain Damage" by removing unimportant weights from a network. By this way, several improvements can be expected: better generalization, fewer training examples required and faster speed of learning.

FDNN is a technique containing these advantages, which divides during training. Differ from Cascade-Correlation, FDNN contains multi Softmax layers and shares most of the weights, some layers divide when their branches converge into a poor local optimal solution. As Fig. 2 shows, FDNN is a tree data structure with shared parameters. The root node is an input layer of data and all the leaf nodes are Softmax layers. The voting layer is only for test. Every path from root node to a leaf node is a linear structure neural network. The numbers behind the names of layers are just to distinguish layers more clearly.

The nodes with multi branches output in the tree are called fissionable nodes, such as FC-1, as the sub neural network shown in Fig. 1. During training, layer FC-1 splits

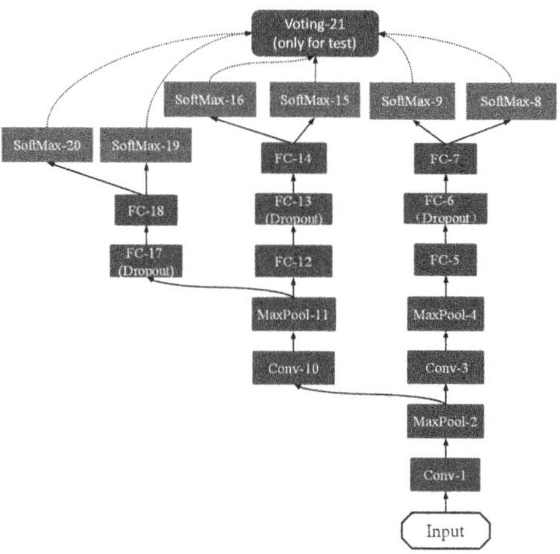

Fig. 2. Fissionable deep neural network

when Softmax-3 or Softmax-4 converges into a poor local optimal solution. FC-2 is a new layer splits from FC-1, its parent and children nodes are inherited from FC-1. The parameters of FC-2 can differ from FC-1 and initialize independently.

2 Motivation

A motivation for FDNN comes from model combination. MCDNN [2] shows how combining several DNN columns into a multi-column DNN further decreases the error rate by 30–40 %. The predictions of all columns are averaged. With unlimited computation, the best way to "regularize" a fixed-sized model is to average the predictions of all possible settings of the parameters, weighting each setting by its posterior probability given the training data. Dropout [1] proposes to do this by approximating an equally weighted geometric mean of the predictions of an exponential number of learned models that share parameters. The idea about shared parameters from dropout is useful to reduce the computation cost. The computation cost of multi models obtaining from MCDNN is almost equal to train each column independently, so FDNN begins with multi branches that share most of the parameters.

Another motivation which is different for FDNN comes from cell division. Cell division is the process by which a parent cell divides into two or more daughter cells. The growth of organisms is mainly due to cell division and cell growth. Similarly, some layers of FDNN divide when their branches converge into a poor local optimal solution. The new layers initialize independently and become new feature-detectors in the network, which makes the differences between each branch more remarkable. New neurons within layers can learn different features, which makes every branch more robust and more powerful.

Certainly, fission is also a way to overcome the interference of convergence. As Fig. 2 shows, the nodes in the path from root to leaves compose multi models and all of these models share parameters. Due to multi Softmax layers, this optimization problem involves multi objective function. However, neural network is an effective solution of a non-convex optimization problem. At the beginning of training, every branch contributes to the faster convergence of shared nodes and all of the shared nodes extract features to feed the unshared nodes. Every model distinguishes from each other by its different unshared nodes, which guarantees that we can get different models composed by diverse feature extractors. After further training, every objective function tries to find the best gradient descent direction and the frequently update of shared parameters interferes each other, so we propose FDNN to avoid the interference in the latter part of training.

3 Model Description

3.1 The Way of Training

The way to train FDNN is DFS (depth-first search). When the iteration reaches one leaf node, the nodes in the path involve in forward propagation, backward propagation, updating the weights operations. After next iteration of another branch, the shared weights are updated again so the shared nodes can converge faster. After six iterations, all of the nodes can be covered in Fig. 2 and we call six iterations as one cover-iteration.

As the Fig. 3 shows, every node in FDNN is traversed difference times within one cover-iteration. Obviously, traversal times of each node are equal to the number of its leaves. So its learning rate is $(\frac{1}{\text{traversal times}})$ times of the original.

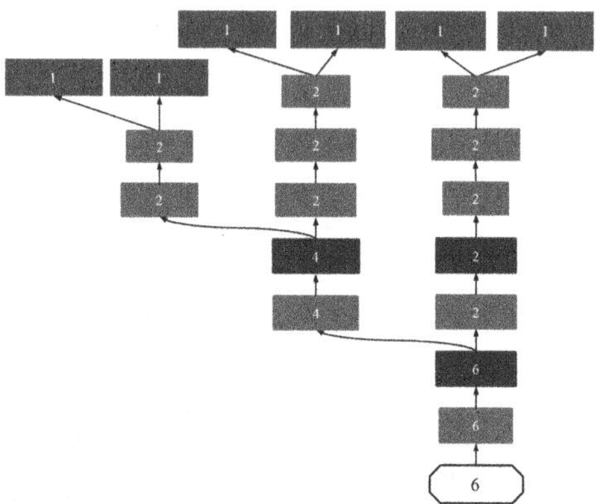

Fig. 3. Traversal times within one cover-iteration during training

3.2 The Mechanism of Fission

Fission happens after the speed of convergence slows down. If the minimum loss value isn't updated in N epochs, the fissionable nodes divide before next epoch. N can be chosen using a validation set or can simply be set at 10, which seems to be close to optimal for a wide range of networks and tasks. This variable controls the degree of interference of convergence with subsequent training. When the N increases, the layers with shared parameters can converge faster, but if N becomes larger, these layers may interfere with each other and the convergence may be slower. The value N is depended on the complexity of the dataset and the architecture of network.

The algorithm to find the fissionable node is called LCA (lowest common ancestor). As Fig. 4 shows, after every N epochs, it chooses the worst convergence leaf node and then backtracks to find the first node with more than one child to be a fissionable node.

3.3 The Way of Prediction

Voting layer is designed for the prediction of neural network. Every sample should be tested several times which are equal to the number of leaves. Voting layer combines multi branches and the predictions of all branches are averaged:

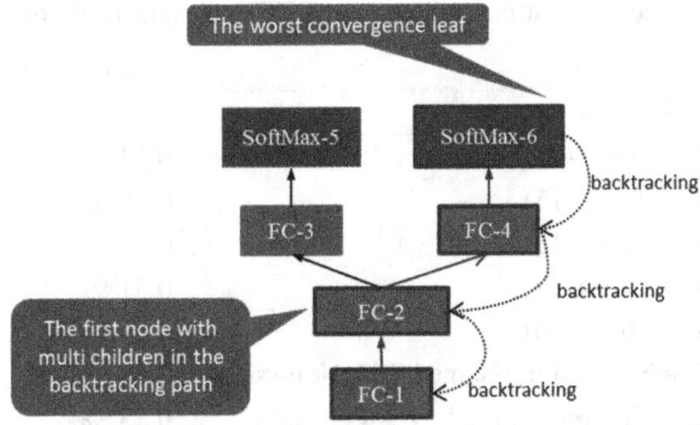

Fig. 4. The fissionable node in backtracking path

$$y^i = \frac{1}{N}\sum_j^N y_j^i \tag{1}$$

As the formula (1) shows, N presents the number of branches and y^i is the probability of sample i.

4 Experiments

4.1 Overview

In order to prove that FDNN can improve the performance of neural network, we evaluate FDNN on two benchmark datasets: MNIST [15] and CIFAR-10 [16]. The networks use for the datasets consist of a tree data structure with shared parameters and multi Softmax layers. As far as possible, each branch is different at the beginning. Dropout is applied in each branch, which is marked in the network structure diagram. We manually set proper initializations for the weights and the learning rates. The experiments use ReLU activation function after convolution layers and fully connected layers. Without data augmentation, all of the experiments use mini-batch SGD with the momentum parameter fixed at 0.9. According to the complexity of the data set and the architecture of network, the corresponding fission mechanism is set up.

4.2 MNIST

For our first experiment on this dataset, the whole network architecture is shown in Fig. 2. The root node takes the image pixels as input. There are six Softmax layers. The path from root node to one leaf is called a model. As shown in Table 1, there are six models, we call them model-1, model-2, model-3, model-4, model-5, model-6 respectively. This article tests our method on this dataset without data augmentation.

The mechanism of fission during training the dataset is shown as follows: The first 40 epochs does not divide, which is to speed up the convergence in the way of sharing parameters to train, then if the minimum loss value of branches isn't updated in 10 epochs, the fissionable nodes begin to divide.

Table 1. Test error rates for MNIST of eight methods

Method	Test error (Voting)
model-1 (Softmax-8)	0.53 %
model-2 (Softmax-9)	0.53 %
model-3 (Softmax-15)	0.65 %
model-4 (Softmax-16)	0.65 %
model-5 (Softmax-19)	0.51 %
model-6 (Softmax-20)	0.51 %
Figure 2 + No Fission (Named "F2_NoFission")	0.48 %
Figure 2 + Fission (Named "F2_Fission")	**0.43 %**

In the manner described above, we obtain a test error of 0.43 % on this dataset, which improves respectively 0.10 %, 0.10 %, 0.22 %, 0.22 %, 0.08 %, 0.08 % compared to the six models of training independently. It proves that the combination of multi models can improve the performance of neural network. A comparison with eight methods is shown in Table 1.

As Table 1 shows, the method of F2_NoFission means that network is trained according to the architecture of Fig. 2, but in the process of training it does not divide. However, F2_Fission is based on the above fission mechanism to train. In Fig. 5, the comparison of these two methods shows: the introduction of fission during training reduced the test error by 0.05 %. The improvement is not particularly obvious, since MNIST has been tuned to a very low error rate.

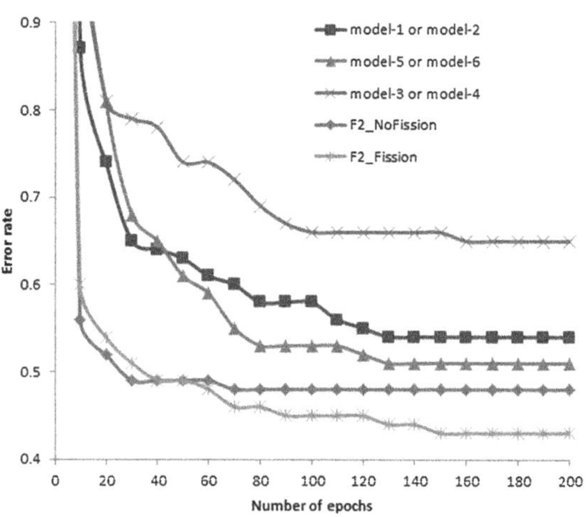

Fig. 5. Testing error of eight methods in the first 200 epochs of training

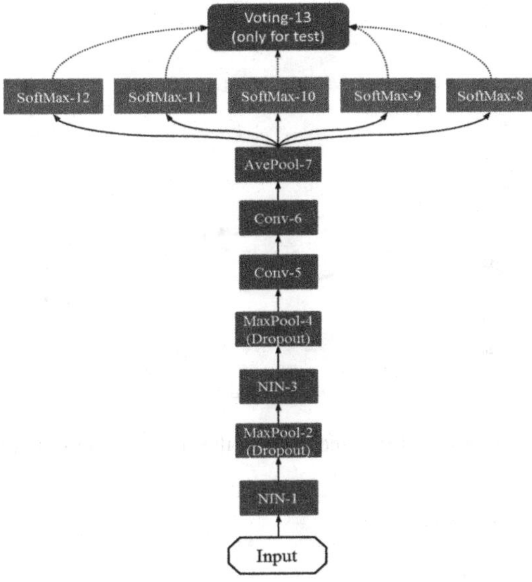

Fig. 6. Fission deep neural network of CIFAR-10

4.3 CIFAR-10

For this dataset, the network architecture is different from the architecture used for MNIST, as shown in Fig. 6. There are five Softmax layers. All of the Softmax classifiers share all of the parameters. NIN [17] is used in the network. Dropout is applied after max-pool layers. This article tests on this dataset without data augmentation.

The mechanism of fission during training the dataset is shown as follows: In the first 15 epochs of training, each epoch divides once, which is to generate five different branches with shared parameters, then every ten epochs FDNN divides once after epochs greater than 60. No fission between 16th epoch and 60th epoch is mainly to speed up the convergence in the way of sharing parameters training.

In this manner, we achieve a test error of 13 % on CIFAR-10 dataset, which improves more than three percent compared to the F6_NoFission. Details of the performance comparison are shown in Table 2. Comparison of two methods in the first 1000 epochs is shown in Fig. 7.

Table 2. Test error rates for CIFAR-10 of two methods

Method	Test error (Voting)
Figure 6 + No Fission (Named "F6_NoFission")	16.18 %
Figure 6 + Fission (named "F6_ Fission")	**13 %**

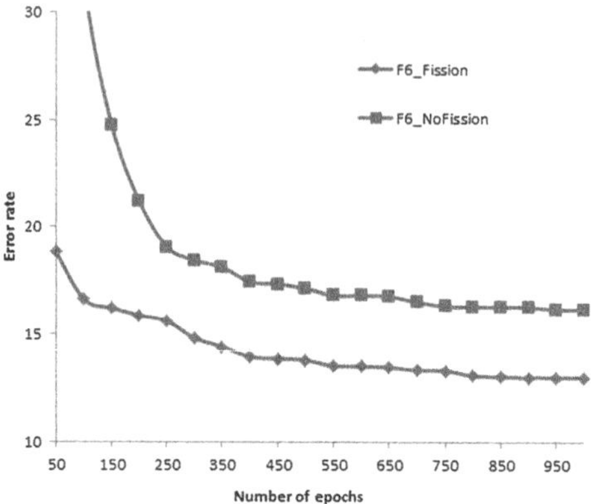

Fig. 7. Testing error of two methods in the first 1000 epochs of training

5 Conclusions

The experiments turn out that FDNN boost the performance of network by improving of the generalization ability of the model. FDNN is a new architecture for neural networks. Multi different branches can reduce computational cost by shared most of parameters. In order to get N network models, the layers of traditional model can be updated only once in an iteration. However, the FDNN layers with sharing parameters can be updated multi-times in the same computational cost, which converges faster. FDNN is useful for incremental learning, in which new information is added to already-trained network. Once fission, a feature detector is created to produce different features and overcome interference of convergence. The most important is that the experiments can get multiple high quality network models finally.

References

1. Srivastava, N., Hinton, G., Krizhevsky, A., et al.: Dropout: a simple way to prevent neural networks from overfitting. J. Mach. Learn. Res. **15**(1), 1929–1958 (2014)
2. Ciresan, D., Meier, U., Schmidhuber, J.: Multi-column deep neural networks for image classification. In: 2012 IEEE Conference on Computer Vision and Pattern Recognition (CVPR), pp. 3642–3649. IEEE (2012)
3. Elman, J.L.: Learning and development in neural networks: the importance of starting small. Cognition **48**(1), 71–99 (1993)
4. Mishkin, D., Matas, J.: All you need is a good init (2015). arXiv preprint arXiv:1511.06422
5. Szegedy, C., Liu, W., Jia, Y., et al.: Going deeper with convolutions. In: Proceedings of the IEEE Conference on Computer Vision and Pattern Recognition, pp. 1–9 (2015)
6. Ioffe, S., Szegedy, C.: Batch normalization: accelerating deep network training by reducing internal covariate shift (2015). arXiv preprint arXiv:1502.03167
7. Szegedy, C., Vanhoucke, V., Ioffe, S., et al.: Rethinking the Inception Architecture for Computer Vision (2015). arXiv preprint arXiv:1512.00567

8. Szegedy, C., Ioffe, S., Vanhoucke, V.: Inception-v4, inception-resnet and the impact of residual connections on learning (2016). arXiv preprint arXiv:1602.07261
9. Springenberg, J.T., Dosovitskiy, A., Brox, T., et al.: Striving for simplicity: The all convolutional net (2014). arXiv preprint arXiv:1412.6806
10. He, K., Zhang, X., Ren, S., et al.: Delving deep into rectifiers: surpassing human-level performance on imagenet classification. In: Proceedings of the IEEE International Conference on Computer Vision, pp. 1026–1034 (2015)
11. Maas, A.L., Hannun, A.Y., Ng, A.Y.: Rectifier nonlinearities improve neural network acoustic models. In: Proceedings of ICML, vol. 30, no. 1 (2013)
12. Goodfellow, I.J., Warde-Farley, D., Mirza, M., et al.: Maxout networks (2013). arXiv preprint arXiv:1302.4389
13. Fahlman, S.E., Lebiere, C.: The cascade-correlation learning architecture (1989)
14. LeCun, Y., Denker, J.S., Solla, S.A., et al.: Optimal brain damage. In: NIPs (1989)
15. LeCun, Y., Bottou, L., Bengio, Y., Haffner, P.: Gradient-based learning applied to document recognition. Proc. IEEE **86**(11), 2278–2324 (1998)
16. Krizhevsky, A., Hinton, G.: Learning multiple layers of features from tiny images. Master's thesis, Department of Computer Science, University of Toronto (2009)
17. Lin, M., Chen, Q., Yan, S.: Network in network (2013). CoRR, abs/1312.4400

A Structural Learning Method of Restricted Boltzmann Machine by Neuron Generation and Annihilation Algorithm

Shin Kamada[1(✉)] and Takumi Ichimura[2]

[1] Graduate School of Information Sciences, Hiroshima City University, Hiroshima, Japan
da65002@e.hiroshima-cu.ac.jp
[2] Faculty of Management and Information Systems, Prefectural University of Hiroshima, Hiroshima, Japan
ichimura@pu-hiroshima.ac.jp

Abstract. Restricted Boltzmann Machine (RBM) is a generative stochastic energy-based model of artificial neural network for unsupervised learning. The adaptive learning method that can discover the optimal number of hidden neurons according to the input space is important method in terms of the stability of energy as well as the computational cost although a traditional RBM model cannot change its network structure during learning phase. Moreover, we should consider the regularities in the sparse of network to extract explicit knowledge from the network because the trained network is often a black box. In this paper, we propose the combination method of adaptive and structural learning method of RBM with Forgetting that can discover the regularities in the trained network. We evaluated our proposed model on MNIST and CIFAR-10 datasets.

Keywords: Restricted Boltzmann Machine · Neuron generation and annihilation · Structural learning · Regularity of RBM Network

1 Introduction

Deep Learning leads a lot of advances in methodology research of artificial intelligence such as machine learning [1]. Especially, the industrial world is deeply impressed by the outcome to increase the capability of image processing. The learning architecture has an advantage of not only multi-layered network structure but also pre-training. The pre-training is that the architecture of Deep Learning accumulates prior knowledge of the features for input patterns. Restricted Boltzmann Machine (RBM) [2] is one of popular methods in Deep Learning for unsupervised learning and it has the capability of representing an probability distribution of input datasets.

Recently, the information technology can collect various kinds of datasets called big data, because the recent tremendous technical advances such as IoT in

A. Hirose et al. (Eds.): ICONIP 2016, Part IV, LNCS 9950, pp. 372–380, 2016.
DOI: 10.1007/978-3-319-46681-1_45

processing power, storage capacity, and network connected to cloud computing. Such sample includes not only image data but also numerical values such as signals, natural language, numerical evaluation, binary data, and so on. The technical methods including Deep Learning to discover knowledge from big data are required as a new method of data mining by industry and business.

The learning of RBM employs in the feature extraction from the input data on some hidden neurons. We may meet the problem how to determine the definition of an optimal initial structure such as the number of hidden neurons according to the features of input pattern since a traditional RBM model cannot change its structure during learning phase. In our previous research, we proposed the adaptive learning method of RBM that can discover an optimal number of hidden neurons according to the training situation by applying the neuron generation and annihilation algorithm [3]. Our proposed RBM model shows the high classification capability, but the trained network is still a black box that cannot extract some explicit knowledge from the trained network. The sparse coding algorithm is known to be effective in terms of regularities of a network structure [4]. Ishikawa proposed 3 stages of structural learning algorithms called Structural Learning Method with Forgetting (SLF) in multi-layerd neural network [5]. The algorithm tackled the difficulty of interpretation of hidden neurons in terms of the extraction of knowledge. We showed that 3 kinds of forgetting algorithms are also effective for the traditional RBM model to the benchmark test [6]. In this paper, we propose the combination method of adaptive and structural learning of RBM. After an optimal size of network structure with stable energy is determined by the neuron generation and annihilation, the network structure with regularities is constructed by forgetting algorithms. To verify our proposed model for the analysis of big data, we show the experimental results for some benchmark datasets.

2 Adaptive Learning Method of RBM

RBM has the network structure with 2 kinds of layers where one is a visible layer for input data and the other is a hidden layer for representing the features of given data space. Each layer consists of some binary neurons and there are the connections between neurons except the neurons in same layer. The RBM learning employs to train the weights and some parameters for visible and hidden neurons till the energy function becomes to a certain small value. The trained RBM can represent a probability for the distribution of input data.

Let $v_i (0 \leq i \leq I)$ and $h_j (0 \leq j \leq J)$ be a binary variable of visible neuron and hidden neuron, respectively. I and J are the number of visible and hidden neurons, respectively. The energy function $E(\boldsymbol{v}, \boldsymbol{h})$ for visible vector $\boldsymbol{v} \in \{0,1\}^I$ and hidden vector $\boldsymbol{h} \in \{0,1\}^J$ is given by Eq. (1). $p(\boldsymbol{v}, \boldsymbol{h})$ is the joint probability distribution of $\boldsymbol{v}$ and $\boldsymbol{h}$ as shown in Eq. (2).

$$E(\boldsymbol{v}, \boldsymbol{h}) = -\sum_i b_i v_i - \sum_j c_j h_j - \sum_i \sum_j v_i W_{ij} h_j, \tag{1}$$

$$p(\boldsymbol{v}, \boldsymbol{h}) = \frac{1}{Z} \exp(-E(\boldsymbol{v}, \boldsymbol{h})), \ Z = \sum_{\boldsymbol{v}} \sum_{\boldsymbol{h}} \exp(-E(\boldsymbol{v}, \boldsymbol{h})), \tag{2}$$

where b_i and c_j are the parameters for v_i and h_j, respectively. W_{ij} is the weight between v_i and h_j. Z is the partition function which is given by summing over all possible pairs of visible and hidden vectors.

The parameters of RBM can be learned by maximum likelihood estimation for $p(\boldsymbol{v}) = \sum_{\boldsymbol{h}} p(\boldsymbol{v}, \boldsymbol{h})$ which is the probability of $\boldsymbol{v}$. However, the computational elements increase exponentially because the optimal configuration for all possible pairs is required to obtain the maximum likelihood estimation. Therefore, the Contrastive Divergence (CD) learning procedure has been proposed as RBM training [7]. CD method is a faster algorithm of Gibbs sampling based on Markov chain Monte Carlo methods. Then CD method is known to make a good performance even in a few sampling steps.

Generally, the solution will be found by using machine learning if and only if the convexity and continuous conditions for an objective function are satisfied. However RBM learning with CD sampling method meets the situation that may cause the slight error and it may not satisfy the continuous condition because of the use of binary neuron. Even if the network has a small error in the initial step, but the total energy after a certain period of iterations will be fluctuated seriously. Therefore, we should consider the convergence situation of RBM under the continuous condition. Carlson et al. discussed the upper bounds on the log partition function for each parameter of RBM by the convexity and Lipschitz continuous [8]. The paper derived that the RBM learning by CD method will be converged if the variance for 3 kinds of parameters $\boldsymbol{\theta} = \{\boldsymbol{b}, \boldsymbol{c}, \boldsymbol{W}\}$ falls into a certain range during training (please see Eq.(18)-(20) in [8] for details). Moreover, we investigated the change of gradients for these parameters during learning in [9]. As a result, we selected 2 parameters $\boldsymbol{c}$ and $\boldsymbol{W}$ which have an influence on the convergence situation of RBM except the parameter $\boldsymbol{b}$ because the gradient for parameter $\boldsymbol{b}$ may be affected by many features of input patterns.

In multi-layerd neural network, the neuron generation and annihilation algorithm during learning phase was proposed [10]. Typically, a hidden neuron tries to learn input features by mapping original input data into feature vector. If a neural network does not have enough neurons to be satisfied to infer, then an input weight vector will tend to fluctuate greatly even after a certain period of the training process, because some hidden neurons may not represent an ambiguous patterns due to lack of the number of hidden neurons. In such a case, we can solve this problem by dividing a neuron which tries to represent the ambiguous patterns into 2 neurons by inheriting the attributes of the parent hidden neuron. The process is called the neuron generation. On the other hand, after an optimal number of neurons are generated, we shall be able to find unnecessary neurons from the network. In such a case, we proposed the neuron annihilation algorithm which can annihilate the redundant neuron if the variance of output signal for a neuron is larger then a certain threshold.

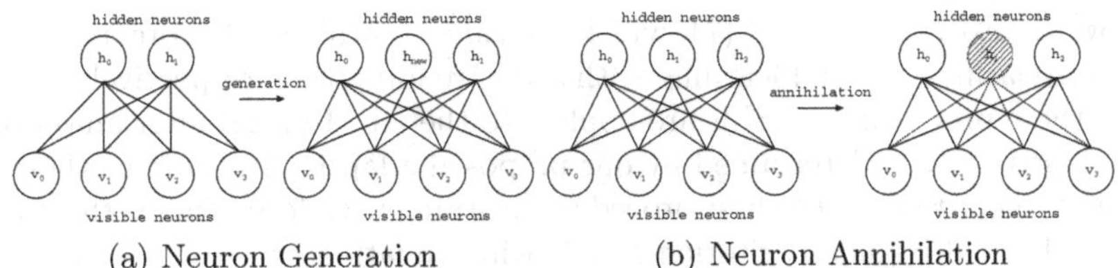

(a) Neuron Generation (b) Neuron Annihilation

Fig. 1. Adaptive RBM

Inspired by these ideas, we proposed the adaptive learning method of RBM that can discover an optimal number of hidden neurons by applying the neuron generation and annihilation algorithm. As mentioned above, we selected 2 parameters $\boldsymbol{c}$ and $\boldsymbol{W}$ which have an influence on the convergence situation of RBM except the parameter $\boldsymbol{b}$ since the variance of $\boldsymbol{b}$ changes large according to the distribution of input data. Then we define the condition of neuron generation with inner product of variance of $\boldsymbol{c}$ and $\boldsymbol{W}$ to consider both 2 kinds of variance simultaneously as shown in Eq. (3).

$$(\alpha_c \cdot dc_j) \cdot (\alpha_W \cdot dW_{ij}) > \theta_G, \tag{3}$$

where dc_j and dW_{ij} are the gradient vectors of the hidden neuron j and the weight vector i, j, respectively. α_c and α_W are the constant values for the adjustment of the range of each parameter. θ_G is an appropriate threshold value. A new hidden neuron will be generated and inserted into the neighborhood of the parent neuron as shown in Fig. 1(a) if Eq. (3) is satisfied. The initial structure of RBM should be set arbitrary neurons according to the dataset before training.

After the neuron generation process, some inactivated hidden neurons that satisfies with Eq. (4) are annihilated.

$$\frac{1}{N}\sum_{n=1}^{N} p(h_j = 1|\boldsymbol{v}_n) < \theta_A, \tag{4}$$

where $\boldsymbol{v}_n = \{\boldsymbol{v}_1, \boldsymbol{v}_2, \cdots, \boldsymbol{v}_N\}$ is a given input data, N is the number of samples of input data. $p(h_j = 1|\boldsymbol{v}_n)$ means a conditional probability of $h_j \in \{0, 1\}$ under a given $\boldsymbol{v}_n$. θ_A is an appropriate threshold value. Figure 1(b) shows the structure of neuron annihilation. The shaded circle in Fig. 1(b) is the redundant neuron removed by the annihilation algorithm. We discussed the effectiveness of proposed method and showed good performance for some benchmark tests [3].

3 Structural Learning Method with Forgetting

Ishikawa proposed the structural learning method with forgetting (SLF) to discover the regularities of the structure in multi-layerd neural network [5]. SLF is implemented by adding the 3 kinds of penalty terms into an objective function as shown in Eqs. (5), (6) and (7), which are called learning with forgetting, hidden units clarification and learning with selective forgetting, respectively. In this paper, these procedures of SLF are applied to the RBM model after an optimal network structure is determined by our proposed adaptive learning method.

The penalty term of the first procedure, learning with forgetting is the weight norm multiplied by small criterion as following equation.

$$J_f = J + \epsilon_1 \| \boldsymbol{W} \|, \tag{5}$$

where J is an original objective function (the CD sampling cost is usually used for RBM learning), J_f is an total objective function with forgetting. ϵ_1 is a penalty criterion for the weight norm $\| \boldsymbol{W} \|$. We can calculate the gradients of the parameters that make unnecessary weights decay and construct the network structure with regularities by partial derivative of Eq. (5).

Although the learning with forgetting makes the weight structure with regularities, the outputs of hidden neurons may not be represented under the binary code. Therefore, we may meet difficulties to extract characteristic patterns from the outputs of network. The second procedure, the hidden units clarification helps the difficulties by forcing each hidden unit to be fully 0 or 1 as the following equation.

$$J_h = J + \epsilon_2 \sum_i \min\{1 - h_i, h_i\}, \tag{6}$$

where h_i is output value of a hidden unit j in $[0, 1]$. ϵ_2 is an appropriate penalty term.

The learning with selective forgetting is used to prevent an objective function to be larger than an original one. The algorithm applies the first algorithm selectively as the following equation. θ is an appropriate threshold value.

$$J_s = J - \epsilon_3 \| \boldsymbol{W}' \|, \; W'_{ij} = \begin{cases} W_{ij}, \, if \; |W_{ij}| < \theta \\ 0, \quad otherwise \end{cases}. \tag{7}$$

Ishikawa proposed the guideline of the learning procedure of SLF with 3 kinds of algorithms. According to [5], both the learning with forgetting and hidden units clarification should be applied simultaneously during learning phase. Alternatively, the learning with selective forgetting is used instead of the learning with forgetting at final learning phase in order to prevent an objective function to be larger.

4 Experimental Results

The 2 kinds of benchmark datasets, "MNIST [11]" and "CIFAR-10 [12]" were used in this experiments. MNIST is a popular dataset of handwritten digits and has 60,000 cases for training set and 10,000 cases for test set. They are categorized into 10 classes. CIFAR-10 is 60,0000 color images dataset included in 50,000 training cases and 10,000 test cases. They are categorized into 10

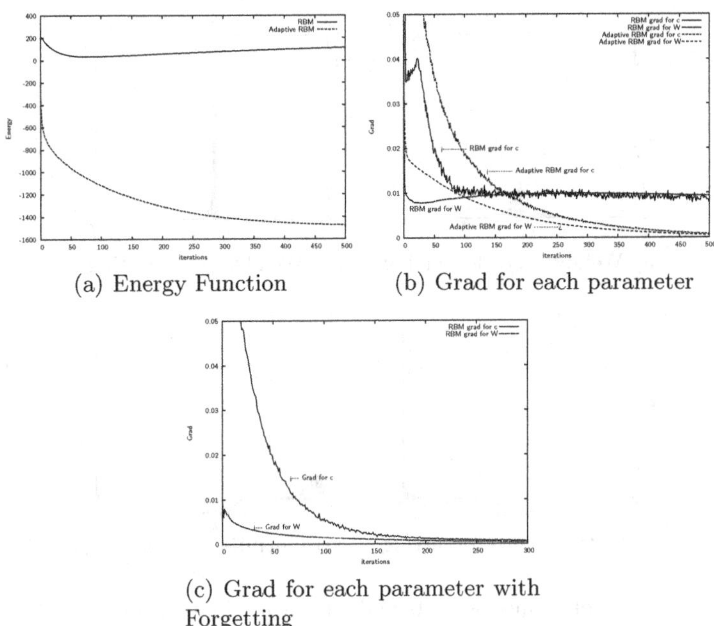

(a) Energy Function (b) Grad for each parameter

(c) Grad for each parameter with Forgetting

Fig. 2. Experimental result in CIFAR-10 (initial hidden neurons = 300)

classes. The original image data in CIFAR-10 only was preprocessed by ZCA whitening as reported in [13].

The following parameters were used for RBM: the training algorithms is Stochastic Gradient Descent (SGD), the batch size is 100, and the learning rate is 0.1. In this simulation, we set parameters as the following values for MNIST: the initial number of hidden neurons is 10, $\theta_G = 0.005$, $\theta_A = 0.3$. We also set parameters as the following values for CIFAR-10: the initial number of hidden neurons is 300, $\theta_G = 0.015$, $\theta_A = 0.01$. In order to evaluate the classification capability with the trained RBM, the output layer for classification is added into the trained hidden layer of RBM, then the network between them is fine-tuned as Hinton introduced [7].

Figure 2(a) and (b) show the energy curve and the gradients for each parameter in the experiment of CIFAR-10, respectively. The proposed RBM generated about 70 additional neurons gradually and then annihilated about 10 redundant neurons. As a result, the energy curve and gradients for each parameter were converged with smaller value than the traditional RBM as shown in Fig. 2(a) and (b). The experimental results for MNIST also showed the same characteristics as the result of CIFAR-10.

After the optimal number of hidden neuron was determined by the adaptive RBM, the structural learning with forgetting was applied. The gradients for parameters $\boldsymbol{W}$ and $\boldsymbol{c}$ were also converged with smaller value as shown in Fig. 2(c). Figure 3 shows the histogram of trained weight value and outputs of

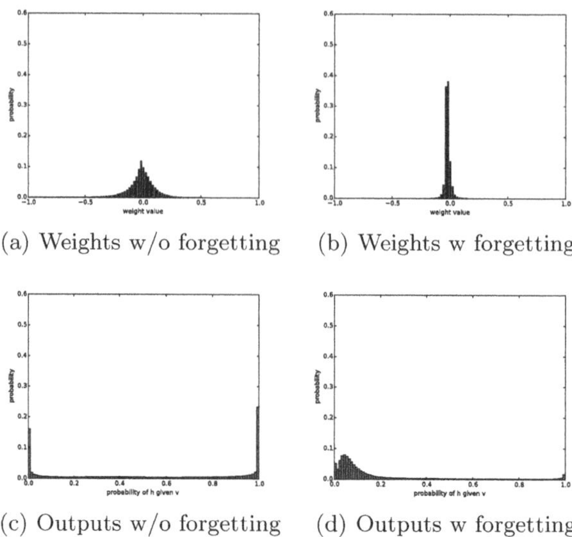

(a) Weights w/o forgetting (b) Weights w forgetting

(c) Outputs w/o forgetting (d) Outputs w forgetting

Fig. 3. Distribution of trained weights and output in hidden neurons on CIFAR-10

Table 1. Classification accuracy

	MNIST		CIFAR-10	
	Training	Test	Training	Test
Traditional RBM [13]	-	-	-	63.0 %
Adaptive RBM	97.9 %	95.4 %	99.9 %	81.2 %
Adaptive RBM with Forgetting	**99.9 %**	**97.1 %**	99.9 %	**85.8 %**

	Adaptive RBM w/o forgetting	Adaptive RBM w forgetting
truck	1.0 0.0 ··· 0.3 ··· 0.6	1.0 0.0 ··· 0.0 ··· 1.0
automobile	0.8 0.7 ··· 0.5 ··· 0.5	1.0 0.0 ··· 0.0 ··· 0.0
ship	0.8 0.9 ··· 0.9 ··· 0.3	0.0 1.0 ··· 1.0 ··· 0.0
airplane	0.0 0.5 ··· 0.9 ··· 0.2	0.0 0.0 ··· 1.0 ··· 0.0
	h_0 h_1 ··· h_j ··· h_J	h_0 h_1 ··· h_j ··· h_J
	v_0 v_1 ··· v_i ··· v_I	v_0 v_1 ··· v_i ··· v_I

Fig. 4. An example of characteristic pattern for some categories in CIFAR-10

hidden neurons in CIFAR-10. As shown in Fig. 3(a) and (b), the adaptive RBM with forgetting made the network structure with more sparse weights and outputs of hidden neurons by Eqs. (5) and (6). Moreover, in comparison to the model without forgetting, the model with forgetting acquired the characteristic binary patterns for each category in hidden neurons as shown in Fig. 4. As a result, the classification accuracy for the test set was increased by applying the forgetting as shown in Table 1.

5 Conclusion

The adaptive RBM that proposed in our previous research is the method to improve the problem of the optimal structure during learning phase. In order to solve the difficulty to extract explicit knowledge from the trained network, we developed the structural and adaptive learning method of RBM with Forgetting. In the simulation, our proposed RBM model with forgetting showed the higher classification capability with sparse network structure in comparison to the other RBM model without forgetting, then the characteristic binary patterns were represented on hidden neurons as knowledge. On the other hand, we found that the learning with forgetting in Eq. (5) and hidden units clarification in Eq. (6) should be also applied simultaneously during learning phase in RBM model. If these equations are applied separately at different learning period, the gradient for one parameter may be bigger seriously due to the change of the other parameter. Moreover, we found that the convergence curve for parameters $\boldsymbol{W}$ and $\boldsymbol{c}$ can be controlled with the decay factors of ϵ_1 and ϵ_2. In future, the deep architecture that can represent more multiple features of input patterns with hierarchical level such as Deep Belief Network [14] will be investigated [15,16].

References

1. Bengio, Y.: Learning deep architectures for AI. Found. Trends Mach. Learn. Arch. **2**(1), 1–127 (2009)
2. Hinton, G.E.: A practical guide to training restricted boltzmann machines. In: Montavon, G., Orr, G.B., Müller, K.-R. (eds.) Neural Networks: Tricks of the Trade, 2nd edn. LNCS, vol. 7700, pp. 599–619. Springer, Heidelberg (2012)
3. Kamada, S., Ichimura, T.: A learning method of adaptive deep belief network by using neuron generation and annihilation algorithm. In: Proceedings of 17th Annual Meeting of Self-Organizing Maps in Japanese, pp. 12.1–12.6 (2016)
4. Ranzato, M., Boureau, Y., LeCun, Y.: Sparse feature learning for deep belief networks. In: Advances in Neural Information Processing Systems 20 (NIPS 2007), pp. 1185–1192 (2007)
5. Ishikawa, M.: Structural learning with forgetting. Neural Netw. **9**(3), 509–521 (1996)
6. Kamada, S., Fujii, Y., Ichimura, T.: Structural learning method of restricted Boltzmann machine with forgetting. In: Proceedings of 17th Annual Meeting of Self-Organizing Maps in Japanese, pp. 13.1–13.6 (2016)
7. Hinton, G.E.: Training products of experts by minimizing contrastive divergence. Neural Comput. **14**, 1771–1800 (2002)
8. Carlson, D., Cevher, V., Carin, L.: Stochastic spectral descent for restricted Boltzmann machines. In: Proceedings of the Eighteenth International Conference on Artificial Intelligence and Statistics, pp. 111–119 (2015)
9. Kamada, S., Ichimura, T., Fujii, Y.: A consideration of convergence of energy function in restricted Boltzmann machine by Lipschitz continuity. In: Proceedings of IEEE SMC Hiroshima Chapter YRW 2015 in Japanese, pp. 53–56 (2015)
10. Ichimura, T., Yoshida, K. (eds.): Knowledge-Based Intelligent Systems for Health Care. Advanced Knowledge International (ISBN 0-9751004-4-0) (2004)

11. LeCun, Y., et al.: THE MNIST DATABASE of handwritten digits (2015). http://yann.lecun.com/exdb/mnist/
12. Krizhevsky, A.: Learning multiple layers of features from tiny images. Master of thesis, University of Toronto (2009)
13. Dieleman, S., Schrauwen, B.: Accelerating sparse restricted Boltzmann machine training using non-Gaussianity measures. In: Deep Learning and Unsupervised Feature Learning (NIPS-2012) (2012)
14. Hinton, G.E., Osindero, S., Teh, Y.: A fast learning algorithm for deep belief nets. Neural Comput. **18**(7), 1527–1554 (2006)
15. Kamada, S., Ichimura, T.: An adaptive learning method with forgetting in deep belief network. In: Proceedings of 9th SICE Symposium on Computational Intelligence, pp. 92–96 (2016)
16. Kamada, S., Ichimura, T.: An adaptive learning method of restricted Boltzmann machine by neuron generation and annihilation algorithm. In: Proceedings of 2016 IEEE SMC (SMC 2016) (accepted)

Semi-supervised Learning for Convolutional Neural Networks Using Mild Supervisory Signals

Takashi Shinozaki[1,2](✉)

[1] CiNet, National Institute of Information and Communications Technology, Koganei, Japan
[2] Graduate School of Information Science and Technology, Osaka University, 1-4 Yamadaoka, Suita, Osaka 565-0871, Japan
tshino@nict.go.jp

Abstract. We propose a novel semi-supervised learning method for convolutional neural networks (CNNs). CNN is one of the most popular models for deep learning and its successes among various types of applications include image and speech recognition, image captioning, and the game of 'go'. However, the requirement for a vast amount of labeled data for supervised learning in CNNs is a serious problem. Unsupervised learning, which uses the information of unlabeled data, might be key to addressing the problem, although it has not been investigated sufficiently in CNN regimes. The proposed method involves both supervised and unsupervised learning in identical feedforward networks, and enables seamless switching among them. We validated the method using an image recognition task. The results showed that learning using non-labeled data dramatically improves the efficiency of supervised learning.

Keywords: Unsupervised learning · Convolutional neural network · Deep learning

1 Introduction

Deep learning is a powerful machine learning method that uses a deep neural network (DNN) and involves many more layers than traditional multilayer neural networks. Many studies have proposed several types of deep learning models. The convolutional neural network (CNN) [1] is one of the most popular models for deep learning, and demonstrates surprising results for higher-level recognition tasks (e.g., visual and speech recognition) [2,3].

To train such a huge-scale neural network, a vast amount of training data is required. However, preparing the vast amount of training data is very difficult, especially for labeled data, and the requirement obstructs the application of deep learning to real-world tasks.

To mitigate the problem, unsupervised learning, which uses the information in unlabeled data, is required. Some previous studies on deep learning have proposed unsupervised learning methods, for example, neocognitron [4], restricted

A. Hirose et al. (Eds.): ICONIP 2016, Part IV, LNCS 9950, pp. 381–388, 2016.
DOI: 10.1007/978-3-319-46681-1_46

Boltzmann machine (RBM) [5], and Google Brain [6]. However, the mechanisms of these methods are fundamentally different from CNNs, and some of them do not have sufficient ability to reduce information, which results in a system size that is too large to apply to real-world tasks. Radford *et al.* proposed an image generative model, which uses unsupervised learning in a CNN [7], but it is difficult to use for discrimination tasks. Furthermore, Goroshin *et al.* proposed a novel unsupervised learning method in CNNs to process movies [8]. However, the method strongly depends on the temporal continuity of movie data, which results in difficulties for the application of general data, for example, static images and medical data. Moreover, the method can only be applied in a very shallow CNN that has only two layers and is only a part of the full network for discrimination. Therefore, transfer learning is required to use the result of unsupervised learning, and cannot apply the learning across the entire final network.

To address these problems, this study proposes a novel unsupervised learning method. The method can apply both supervised and unsupervised learning in an identical network seamlessly to enable semi-supervised learning in CNNs. We verified the effectiveness of the method using Modified National Institute of Standards and Technology (MNIST) handwritten digit recognition.

2 Proposed Learning Method

The proposed method implements unsupervised learning with unlabeled data and supervised learning with labeled data in identical CNNs.

For supervised learning, the supervisory signal is based on the labels of the data, and the information is naturally reliable. The supervisory signal vector has only a single value, one, which corresponds to the location of the correct label, and the remaining values are zero because the correct answer is reliably correct. Conversely, for unsupervised learning, the pseudo-correct label is determined using the unit that has the maximum value in the output layer, and is first set to the supervisory signal vector. The remaining values of the vector are set to a bias value, which corresponds to unreliableness with the range of 0.0 to 1.0. The vector is normalized, with the norm equal to 1.0, which is the same as it is for supervised learning. For example, if the number of units in the output layer is 100 and the bias value is 0.8, this results in 0.0125 for the pseudo-correct label value and 0.0100 for the remaining values.

Regarding the loss function and optimization, because of high reliability, supervised learning uses averaged cross entropy (ACE) and Adam [10] for sharp discrimination characteristics. Because the reliability of unsupervised learning is low, the mean squared error (MSE) and plain stochastic gradient descent (SGD) [11] are used for relatively milder discrimination.

The structure of the CNN for the proposed method should have redundancy regarding the number of units required to process the diverse learning representation, especially in the output layer. Unsupervised learning does not have label information and one learning representation could be diversely represented by several sub-learning representations. Supervised learning reorganizes the diverse

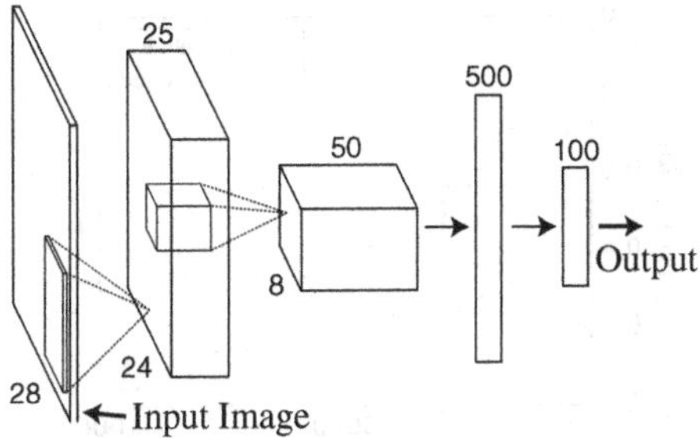

Fig. 1. Structure of the CNN for verification. It consists of two sets of convolutional layers and two fully connected layers.

representation using the information from labeled data, which results in fast and efficient learning of the target discrimination task.

The traditional use of the proposed method has two steps: pre-training and fine tuning. First, unsupervised pre-training is performed with unlabeled data to obtain the basic structure of the target dataset (e.g., continuity in two-dimensional space for image data and harmonics for audio data) as a learning representation of the network. Then, supervised learning is performed to rebind each learning representation and corresponding label, which results in fast and effective fine tuning.

Another style of learning, semi-supervised learning, is much simpler. It has only one mixed process of supervised and unsupervised learning. The learning method is selected according to whether the input data are labeled or unlabeled for every mini-batch. The network performs unsupervised learning when unlabeled data arrive, and performs supervised learning only for labeled data.

3 Verification with MNIST Handwritten Digits

To verify the effectiveness of the proposed method, we performed an image recognition task using the MNIST handwritten image dataset [12].

We used a simple CNN with four layers: the first two layers were convolutional layers and the subsequent two layers were fully connected layers (Fig. 1). The number of units of the convolutional layers were 24×24 and 8×8, the sizes of the corresponding receptive fields were 5×5 and 2×2, respectively, and the numbers of filters were 25 and 50, respectively. The size of max-pooling was 2×2 for both layers. The fully connected layers had 500 and 100 units. These numbers were sufficiently redundant for unsupervised learning, especially for the output layer, which represented only 10 labels by 100 units. All methods were implemented in custom Python code using the Chainer deep learning framework [13].

The weights of the network were initialized uniformly distributed random numbers. The training and test dataset consisted of 60,000 and 10,000 sets of 28×28 pixel images with 10 labels. Unsupervised pre-training was first applied with five time iterations for the entire training dataset, which resulted in 300,000

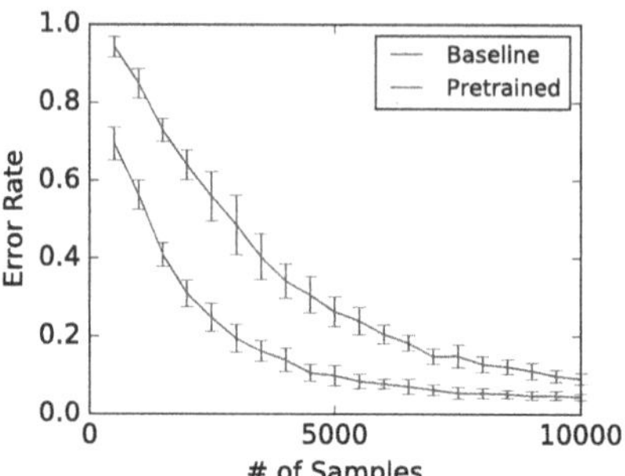

Fig. 2. Temporal sequence of error rates of supervised learning without or with unsupervised learning. The blue line is the baseline result, which had no unsupervised pre-training. The green line is the result with unsupervised pre-training. Error bars show the standard deviations ($n = 10$). (Color figure online)

samples. The error rate of the supervised learning method was calculated every 500 samples. The learning rate of SGD was fixed to 0.01 without momentum, and the parameters of Adam optimization were the following default settings: $\alpha - 0.001, \beta_1 = 0.9, \beta_2 = 0.999$, and $\epsilon = 10^{-8}$. The default value of the bias for unsupervised learning was set to 0.8.

Figure 2 shows the temporal sequences of the error rates of supervised learning without or with unsupervised pre-training. The number of samples that achieved a less than 10 % error rate for the baseline was approximately 9,100, and that with unsupervised pre-training was approximately 4,700, which means that the acceleration of the learning speed was almost double (Table 1). Moreover, the shrinkage of the variances of the error rates in the medium range (error rates $= 0.3 - 0.6$) exhibited a more robust learning process than the baseline.

To clarify the effect of the number of trials of unsupervised pre-training, we examined the error rates for supervised learning with various conditions. Figure 3(a) shows the conditions for a small numbers of trials. The error rates decreased, accompanied by an increase of the number of trials of unsupervised pre-training. Figure 3(b) shows the conditions for a larger numbers of trials. Learning did not improve substantially over five epochs (300,000 trials), which demonstrates the low efficiency of unsupervised pre-training.

The proposed method used a "self-training" method for unsupervised learning. It is sometimes trapped in a type of over-learning condition. We called this the "dominated" state, where most of the outputs are dominated by one label despite that fact that the actual input labels vary. We defined the "dominated" state, in which the fraction of the domination by one output units exceeds 0.5, and quantify the epoch count to achieve the "dominated" state. Figure 4(a) shows the number of epochs required to achieve the "dominated" state for various bias value, loss function, and optimizer combinations. The result suggests that the combination of MSE and SGD is the best and only choice for stable unsupervised pre-training.

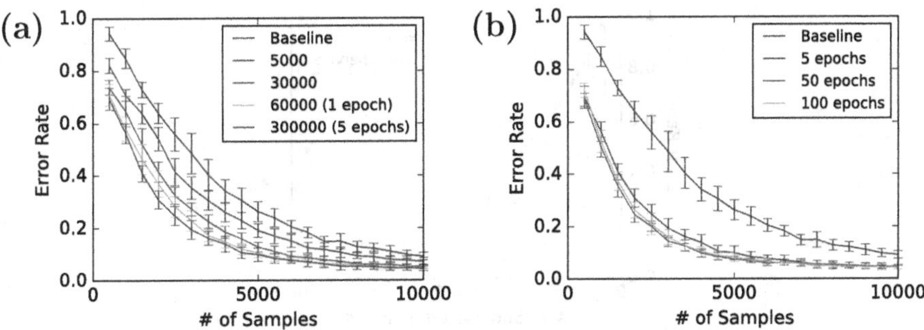

Fig. 3. Effect of the number of trials of unsupervised pre-training: (a) for a small number of trials and (b) for a larger number of trials. One "epoch" means one entire set of the training set with 60,000 samples. Error bars show the standard deviations ($n = 10$).

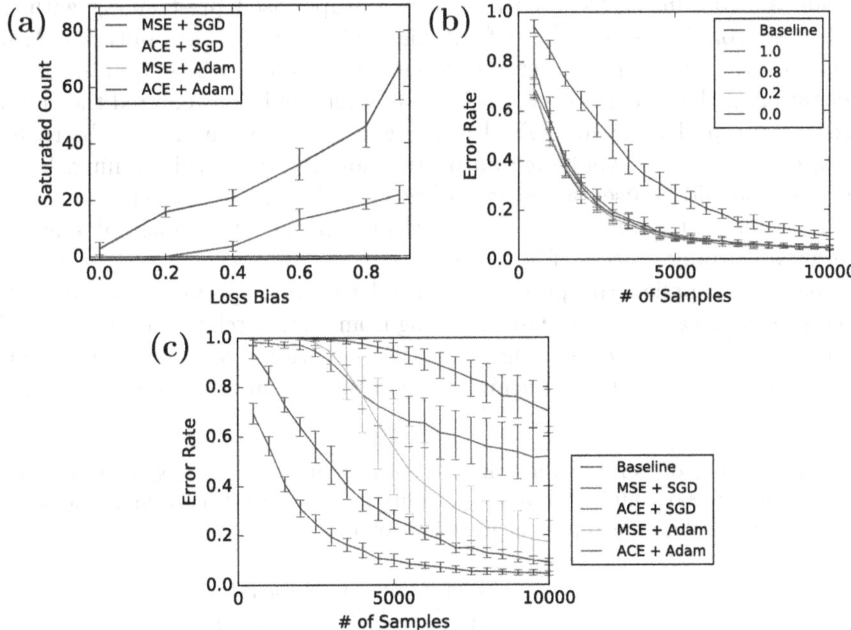

Fig. 4. (a) Comparison of the number of epochs that caused the "dominated" condition among various bias, loss function, and optimizer combinations. (b) Effect of the bias value on learning. Error bars show the standard deviations ($n = 10$). (c) Effect of saturation among the various combinations. The blue line shows the baseline. The green line shows the accelerated condition using the MSE + SGD combination. The cyan line is using MSE + Adam conditions, the red line is using ACE + SGD conditions, and the purple line is using ACE + Adam conditions. (Color figure online)

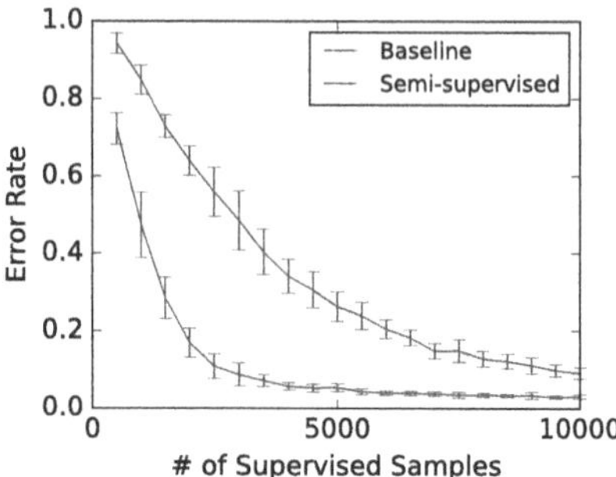

Fig. 5. Effect of the proposed semi-supervised learning method. The error rate of supervised learning for each sample number was drastically improved by the proposed method. The blue line is the baseline and the green line is the condition with semi-supervised learning. Error bars show the standard deviations ($n = 10$).

We also verified the effect of the "dominated" condition on the subsequent supervised learning method. Figure 4(c) shows the error rates for several loss function and optimizer combinations after unsupervised pre-training with 30 epochs. All conditions use the default bias value 0.8. In fact, only the MSE + SGD condition did not enter the "dominated" state, which resulted in an acceleration of the learning speed. All of the other conditions entered the "dominated" state, and their supervised learning processes were clearly obstructed (see also Table 1). To avoid such an obstruction of supervised learning, larger bias values should be used as shown in Fig. 4(a). Thus, we checked the effect of the bias value on learning. There was no clear influence of the bias value on the supervised learning process (Fig. 4(b)).

Finally, we applied the proposed method to semi-supervised learning. We mixed supervised and unsupervised learning from mini-batch to mini-batch, with the fraction of supervised learning as 0.2. We used Adam optimization for both supervised and unsupervised learning using a restriction of the software frame-

Table 1. Number of iterations at which the error rates achieve less than 10 % and error rates for 10,000 samples for various conditions. Some conditions do not achieve a 10 % error rate and do not represent iteration counts.

	# of iterations	Error rate
Control	9150 ± 867	9.2 ± 1.4
MSE + SGD	4700 ± 714	4.5 ± 0.9
ACE + SGD	-	51.9 ± 12.0
MSE + Adam	-	17.3 ± 8.5
ACE + Adam	-	70.1 ± 8.8
Semi-supervised	2400 ± 490	3.1 ± 0.5

work. The effect of semi-supervised learning was verified using the error rate of supervised learning for each number of samples, as in previous sections. Figure 5 shows that the proposed semi-supervised learning method drastically improved the learning speed by using unlabeled data.

4 Discussion

In this study, we proposed a novel unsupervised and semi-supervised learning method that could perform both supervised and unsupervised learning in identical CNNs seamlessly. The proposed method enabled the representation learning of image features using unsupervised learning with a vast amount of unlabeled data, and drastically accelerated the subsequent supervised learning method with a relatively small amount of labeled data.

The method required margins for the number of units in networks, and the margin signified the robustness of the network parameters. One of serious problems of DNNs is the parameter tuning of the network structure. In CNNs, the number of filters in convolutional layers should be dependent on the amount of information from the labeled data, and should fit the number of meaningful learning representations. If filters are also shared with a vast amount of unlabeled data, they might be mainly occupied by relatively meaningless or potentially meaningful learning representations. In such a scenario, the parameter could be managed loosely, which results in a reduction of the effort of parameter tuning.

The information for label discrimination itself does not modulate learning substantially using the proposed method; however, the bias value does it as the mildness of the supervisory signal. If mildness is considered as a reward signal, the proposed method is almost identical to reinforcement learning [14]. We suggest that the method could combine a CNN and reinforcement learning in a very natural way.

The proposed method only uses a basic learning framework of a CNN and could be applicable to many applications, such as image recognition and speech recognition. In particular, the proposed method could be useful for the data for which it is difficult to obtain a vast amount of labeled data (e.g., medical data and personal internet of things data). Further research applications are required.

References

1. LeCun, Y., Boser, B., Denker, J.S., Henderson, D., Howard, R.E., Jackel, L.D.: Backpropagation applied to hand-written zip code recognition. Neural Comput. **1**(4), 541–551 (1989)
2. Krizhevsky, A., Sutskerver, I., Hinton, G.E.: ImageNet classification with deep convolutional neural networks. Adv. Neural Inf. Process. Syst. **25**, 1106–1114 (2012)
3. Dahl, G.E., Yu, D., Deng, L., Acero, A.: Context-dependent pre-trained deep neural networks for large vocabulary speech recognition. IEEE Trans. Audio Speech Lang. Process. **20**(1), 30–42 (2012)

4. Fukushima, K.: Neocognitron: a self organizing neural network model for a mechanism of pattern recognition unaffected by shift in position. Biol. Cybern. **36**(4), 193–202 (1980)
5. Hinton, G.E., Salakhutdinov, R.: Reducing the dimensionality of data with neural networks. Science **313**, 504–507 (2006)
6. Le, Q.V., Ranzato, M.A., Monga, R., Devin, M., Chen, K., Corrado, G.S., Dean, J., Ng, A.Y.: Building high-level features using large scale unsupervised learning. In: Proceedings of the 29th International Conference on Machine Learning (2012)
7. Radford, A., Metz, L.: Unsupervised representation learning with deep convolutional generative adversarial networks. In: IPLR 2016 (2016)
8. Goroshin, R., Bruna, J., Tompson, J., Eigen, D., LeCun, Y.: Unsupervised Learning of Spatiotemporally Coherent Metrics, arXiv:1412.6056 (2015)
9. Bengio, Y., Lambling, P., Popovici, D., Larochelle, H.: Greedy layer-wise training of deep networks. Adv. Neural Inf. Process. Syst. **19**, 153–160 (2007)
10. Kingma, D.P., Ba, J.L.: Adam: a method for stochastic optimization. In: Proceedings of the 4th International Conference on Learning Representations (2015)
11. Bottou, L.: Online algorithms and stochastic approximations. In: Online Learning and Neural Networks. Cambridge University Press (1998)
12. LeCun, Y., Cortes, C., Barges, C.J.C.: The MNIST database of handwritten digits (1998)
13. Tokui, S., Oono, K., Hido, S., Clayton, J.: Chainer: a next-generation open source framework for deep learning. In: NIPS (2015)
14. Sutton, R.S., Barto, A.G.: Reinforcement Learning: An Introduction. The MIT Press, Cambridge (1998)

On the Singularity in Deep Neural Networks

Tohru Nitta(✉)

Human Informatics Research Institute,
National Institute of Advanced Industrial Science and Technology (AIST),
AIST Tsukuba Central 2, 1-1-1 Umezono, Tsukuba, Ibaraki 305-8568, Japan
tohru-nitta@aist.go.jp

Abstract. In this paper, we analyze a deep neural network model from the viewpoint of singularities. First, we show that there exist a large number of critical points introduced by a hierarchical structure in the deep neural network as straight lines. Next, we derive sufficient conditions for the deep neural network having no critical points introduced by a hierarchical structure.

Keywords: Critical point · Deep learning · Neural networks · Singular point

1 Introduction

It has been reported that training deep neural networks is more difficult than training neural networks with one hidden layer [1]. Deep neural networks have been extensively studied since Hinton et al. proposed Deep Belief Networks with a learning algorithm that trains one layer at a time [2]. Much better generalization could be achieved when pre-training each layer with an unsupervised learning algorithm. However, little is known about the reason why the pre-training method works well for training deep neural networks. It is often only said that the pre-training could avoid local minima and plateaus.

A point $\boldsymbol{\omega}_*$ satisfying $\partial E(\boldsymbol{\omega}_*)/\partial \boldsymbol{\omega} = \mathbf{0}$ is referred to as the critical point of the error function E. A critical point can be a local minimum, a local maximum, or a saddle point. Fukumizu et al. have mathematically proved that critical points in a three-layered neural network with $H-1$ hidden neurons behave as critical points in a three-layered neural network with H hidden neurons, and that they can be local minima or saddle points according to conditions [3]. This kind of critical points is the singular point of neural networks, and could cause a standstill of learning.

There are a few literatures on researches of deep neural networks from a theoretical perspective [4,5]. This paper presents deep neural networks having no critical points introduced by a hierarchical structure from a theoretical side. Such deep neural networks do not get stuck in apparent local minima or plateaus, resulting in good generalization comparatively. First, it is shown that there exist a large number of critical points introduced by a hierarchical structure in deep

A. Hirose et al. (Eds.): ICONIP 2016, Part IV, LNCS 9950, pp. 389–396, 2016.
DOI: 10.1007/978-3-319-46681-1_47

neural networks as straight lines. Next, sufficient conditions for deep neural networks having no critical points introduced by a hierarchical structure are derived. All the mathematical proofs are omitted due to limitations of space (see [6] for the proofs, which is an extended version of this paper).

2 Deep Neural Network

This section describes the $(N+2)$-layered deep neural network used in the analysis. It has $L = M_0$ input neurons, one output neuron, and M_k hidden neurons in the hidden layer k $(k = 1, \cdots, N)$. Let $M_{N+1} \equiv 1$ for convenience.

For any $1 \leq k \leq N$, the net input $X_{j_k}^{(k)}$ to the hidden neuron j_k in the hidden layer k is defined as:

$$X_{j_k}^{(k)} = \sum_{j_{k-1}=1}^{M_{k-1}} w_{j_k j_{k-1}}^{(k)} y_{j_{k-1}}^{(k-1)} + \nu_{j_k}^{(k)} = \boldsymbol{w}_{j_k}^{(k)^T} \boldsymbol{y}^{(k-1)} + \nu_{j_k}^{(k)} = \tilde{\boldsymbol{w}}_{j_k}^{(k)^T} \tilde{\boldsymbol{y}}^{(k-1)} \quad (1)$$

where $w_{j_k j_{k-1}}^{(k)}$ is the weight connecting the neuron j_{k-1} in the preceding layer $k-1$ and the neuron j_k in the hidden layer k, $y_{j_{k-1}}^{(k-1)}$ is the output signal from the neuron j_{k-1} in the preceding layer $k-1$, $\nu_{j_k}^{(k)}$ is the threshold value of the neuron j_k in the hidden layer k, $\boldsymbol{w}_{j_k}^{(k)} = (w_{j_k 1}^{(k)} \cdots w_{j_k M_{k-1}}^{(k)})^T$, $\tilde{\boldsymbol{w}}_{j_k}^{(k)} = (\boldsymbol{w}_{j_k}^{(k)} \ \ \nu_{j_k}^{(k)})^T$, $\boldsymbol{y}^{(k)} = (y_1^{(k)} \cdots y_{M_k}^{(k)})^T$, $\tilde{\boldsymbol{y}}^{(k)} = (\boldsymbol{y}^{(k)^T} \ \ 1)^T$, $\tilde{\boldsymbol{y}}^{(0)} = (\boldsymbol{y}^{(0)^T} \ \ 1)^T$, $\boldsymbol{y}^{(0)} = \boldsymbol{x}$, $\tilde{\boldsymbol{x}} = (\boldsymbol{x}^T \ \ 1)^T$, $\boldsymbol{x} \in \boldsymbol{R}^L$ is the input vector to the input layer, $\boldsymbol{R}$ is the set of real numbers, and T denotes transposition. The output signal of the hidden neuron j_k in the hidden layer k is described as $y_{j_k}^{(k)} = \varphi(X_{j_k}^{(k)})$ for $1 \leq j_k \leq M_k$ and $1 \leq k \leq N$ where $\varphi : \boldsymbol{R} \to \boldsymbol{R}$ is a differentiable activation function.

The output value of the deep neural network is defined to be

$$f(\boldsymbol{x}; \boldsymbol{\theta}) = \sum_{j_N=1}^{M_N} v_{j_N}^{(N)} y_{j_N}^{(N)} + \nu^{(N+1)} = \boldsymbol{v}^{(N)^T} \boldsymbol{y}^{(N)} + \nu^{(N+1)} = \tilde{\boldsymbol{v}}^{(N)^T} \tilde{\boldsymbol{y}}^{(N)} \quad (2)$$

where $\boldsymbol{x} \in \boldsymbol{R}^L$ is the input vector to the deep neural network, $v_{j_N}^{(N)} \in \boldsymbol{R}$ is the weight between the hidden neuron j_N in the hidden layer N and the output neuron $(1 \leq j_N \leq M_N)$, $\nu^{(N+1)} \in \boldsymbol{R}$ is the threshold of the output neuron, $\boldsymbol{v}^{(N)} = (v_1^{(N)} \cdots v_{M_N}^{(N)})^T$, $\tilde{\boldsymbol{v}}^{(N)} = (\boldsymbol{v}^{(N)^T} \ \ \nu^{(N+1)})^T$, and $\boldsymbol{\theta}$ is one large vector which summarizes all the parameters of the deep neural network. The activation function $\psi : \boldsymbol{R} \to \boldsymbol{R}$ of the output neuron is linear, that is, $\psi(x) = x$ for any $x \in \boldsymbol{R}$.

Given K training data $\{(\boldsymbol{x}^{(m)}, y^{(m)}) \in \boldsymbol{R}^L \times \boldsymbol{R} \mid m = 1, \cdots, K\}$, we use the deep neural network to realize the relation expressed by the data. The objective of the training is to find the parameters that minimize the error function defined by

$$E(\boldsymbol{\theta}) = \sum_{m=1}^{K} l(y^{(m)}, f(\boldsymbol{x}^{(m)}; \boldsymbol{\theta})) \in \boldsymbol{R} \quad (3)$$

where $l(y, z) : \boldsymbol{R} \times \boldsymbol{R} \longrightarrow \boldsymbol{R}$ is a loss function such that $l(y, z) \geq 0$ and the equality holds if and only if $y = z$.

Here, let $\hat{k}$ denote the number for a specific hidden layer. Then, we call the deep neural network with $M_{\hat{k}}$ hidden neurons in the hidden layer $\hat{k}$ a $M_{\hat{k}}$*-deep network* explicitly indicating that it has $M_{\hat{k}}$ hidden neurons in the hidden layer $\hat{k}$, and use the notations $f^{(M_{\hat{k}})}(\boldsymbol{x}; \boldsymbol{\theta}^{(M_{\hat{k}})})$ and $E_{M_{\hat{k}}}(\boldsymbol{\theta}^{(M_{\hat{k}})})$ for Eqs. (2) and (3), respectively.

3 Critical Points of the Deep Neural Network

This section clarifies the properties of the critical points introduced by the hierarchical structure of the deep neural network described in Sect. 2.

We define the critical point of the $M_{\hat{k}}$-deep network defined in Sect. 2.

Definition 1. *A parameter* $\boldsymbol{\theta}^{(M_{\hat{k}})} = (\theta_1^{(M_{\hat{k}})}, \cdots, \theta_R^{(M_{\hat{k}})}) \in \boldsymbol{\Theta}_{M_{\hat{k}}}$ *is called a critical point of the error function* $E_{M_{\hat{k}}}(\boldsymbol{\theta}^{(M_{\hat{k}})})$ *if the following equation holds:*

$$\frac{\partial E_{M_{\hat{k}}}(\boldsymbol{\theta}^{(M_{\hat{k}})})}{\partial \boldsymbol{\theta}^{(M_{\hat{k}})}} = \left(\frac{\partial E_H(\boldsymbol{\theta}^{(M_{\hat{k}})})}{\partial \theta_1^{(M_{\hat{k}})}}, \cdots, \frac{\partial E_H(\boldsymbol{\theta}^{(M_{\hat{k}})})}{\partial \theta_R^{(M_{\hat{k}})}} \right)^T = \boldsymbol{0} \tag{4}$$

where R is the number of the parameters of the $M_{\hat{k}}$-deep network, and $\boldsymbol{\Theta}_{M_{\hat{k}}}$ is the set of all the parameters (weights and thresholds) of the $M_{\hat{k}}$-deep network with $M_{\hat{k}}$ hidden neurons in the specific hidden layer $\hat{k}$. □

Next, we prepare a $(M_{\hat{k}}-1)$-deep network where the number of hidden neurons in the $\hat{k}$-th hidden layer is one fewer than that of the $M_{\hat{k}}$-deep network for the analysis. Critical points introduced by a hierarchical structure will be constructed by adding one hidden neuron to the $\hat{k}$-th hidden layer of the $(M_{\hat{k}} - 1)$-deep network using the embeddings defined in Definition 3.

Definition 2. *Consider a $(M_{\hat{k}} - 1)$-deep network which has $M_{\hat{k}} - 1$ hidden neurons numbered from 2 to $M_{\hat{k}}$ in the specific hidden layer $\hat{k}$. The output value of the $(M_{\hat{k}} - 1)$-deep neural network is defined to be*

$$f^{(M_{\hat{k}}-1)}(\boldsymbol{x}; \boldsymbol{\theta}^{(M_{\hat{k}}-1)}) = \sum_{j_N=1}^{M_N} v_{j_N}^{(N)} y_{j_N}^{(N)} + \nu^{(N+1)} \tag{5}$$

where

$$y_{j_N}^{(N)} = \varphi(X_{j_N}^{(N)}), \tag{6}$$

$$X_{j_{k+1}}^{(k+1)} = \begin{cases} \sum_{j_k=1}^{M_k} w_{j_{k+1}j_k}^{(k+1)} y_{j_k}^{(k)} + \nu_{j_{k+1}}^{(k+1)} \\ \qquad (1 \leq j_{k+1} \leq M_{k+1}, 1 \leq k \leq N-1, k \neq \hat{k}), \\ \sum_{j_{\hat{k}}=2}^{M_{\hat{k}}} s_{j_{\hat{k}+1}j_{\hat{k}}}^{(\hat{k})} \varphi(\tilde{\boldsymbol{p}}_{j_{\hat{k}}}^{(\hat{k})^T} \tilde{\boldsymbol{y}}^{(\hat{k}-1)}) + \xi_{j_{\hat{k}+1}}^{(\hat{k}+1)} \qquad (k = \hat{k}). \end{cases} \tag{7}$$

The meaning of the variables in Eqs. (5), (6) and (7) are the same as Eqs. (1) and (2). The lower equation in Eq. (7) expresses the net input to the hidden neuron $\hat{k}+1$ *where* $s_{j_{\hat{k}+1}j_{\hat{k}}}^{(\hat{k})}$ *is the weight connecting the hidden neuron* $j_{\hat{k}+1}$ *in the hidden layer* $\hat{k}+1$ *and the hidden neuron* $j_{\hat{k}}$ *in the hidden layer* $\hat{k}$, $\tilde{\boldsymbol{p}}_{j_{\hat{k}}}^{(\hat{k})} = (\boldsymbol{p}_{j_{\hat{k}}}^{(\hat{k})^T} \quad \tau_{j_{\hat{k}}}^{(\hat{k})})^T$, $\boldsymbol{p}_{j_{\hat{k}}}^{(\hat{k})}$ *is the weight vector between the hidden neuron* $j_{\hat{k}}$ *in the hidden layer* $\hat{k}$ *and the hidden layer* $\hat{k}-1$, $\tau_{j_{\hat{k}}}^{(\hat{k})}$ *is the threshold value of the neuron* $j_{\hat{k}}$ *in the hidden layer* $\hat{k}$, *and* $\xi_{j_{\hat{k}+1}}^{(\hat{k}+1)}$ *is the threshold value of the neuron* $j_{\hat{k}+1}$ *in the hidden layer* $\hat{k}+1$. *And also, define* $\boldsymbol{\xi}^{\hat{k}+1} = (\xi_1^{\hat{k}+1} \cdots \xi_{M_{\hat{k}+1}}^{\hat{k}+1})^T$ *and* $\boldsymbol{s}_{j_{\hat{k}}}^{(\hat{k})} = (s_{1j_{\hat{k}}}^{(\hat{k})} \cdots s_{M_{\hat{k}+1}j_{\hat{k}}}^{(\hat{k})})^T (2 \leq j_{\hat{k}} \leq M_{\hat{k}})$. □

Here, we define the following canonical embeddings from $\boldsymbol{\Theta}_{M_{\hat{k}}-1}$ to $\boldsymbol{\Theta}_{M_{\hat{k}}}$ by which one hidden neuron will be added to the $\hat{k}$-th hidden layer, keeping the output value of the neural network unchanged.

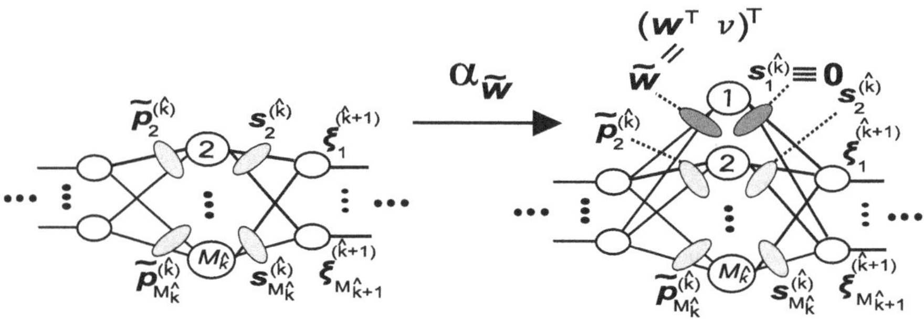

Fig. 1. Embedding $\alpha_{\tilde{w}}$.

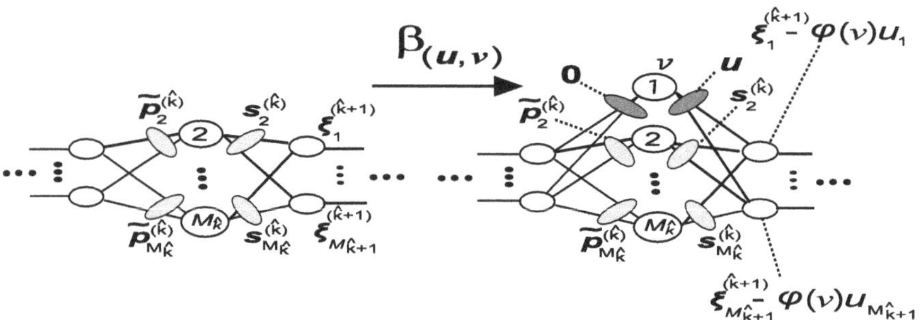

Fig. 2. Embedding $\beta_{(\boldsymbol{u},\nu)}$.

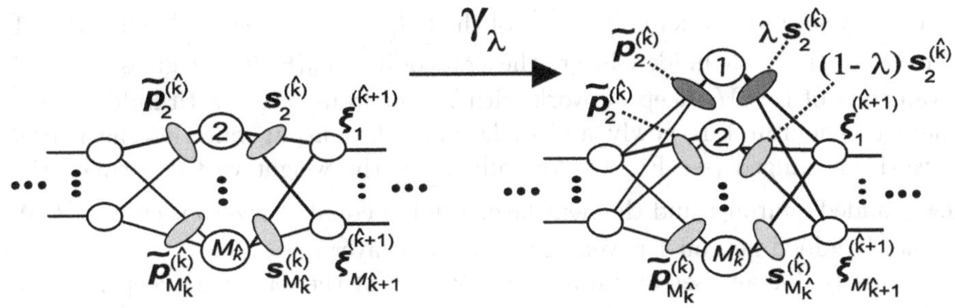

Fig. 3. Embedding γ_λ.

Definition 3. (See Figs. 1, 2 and 3.) (i) *For any* $\tilde{\boldsymbol{w}} \in \boldsymbol{R}^{M_{\hat{k}-1}+1}$, *define*

$$\begin{aligned}\alpha_{\tilde{\boldsymbol{w}}} : \boldsymbol{\Theta}_{M_{\hat{k}}-1} &\longrightarrow \boldsymbol{\Theta}_{M_{\hat{k}}},\\ \boldsymbol{\theta}^{(M_{\hat{k}}-1)} &\longmapsto (\cdots, \boldsymbol{\xi}^{(\hat{k}+1)^T}, \boldsymbol{s}_1^{(\hat{k})^T} \equiv \boldsymbol{0}^T, \boldsymbol{s}_2^{(\hat{k})^T}, \cdots, \boldsymbol{s}_{M_{\hat{k}}}^{(\hat{k})^T},\\ &\qquad \tilde{\boldsymbol{w}}^T, \tilde{\boldsymbol{p}}_2^{(\hat{k})^T}, \cdots, \tilde{\boldsymbol{p}}_{M_{\hat{k}}}^{(\hat{k})^T}, \cdots)^T \end{aligned} \tag{8}$$

where

$$\boldsymbol{\theta}^{(M_{\hat{k}}-1)} = (\cdots, \boldsymbol{\xi}^{(\hat{k}+1)^T}, \boldsymbol{s}_2^{(\hat{k})^T}, \cdots, \boldsymbol{s}_{M_{\hat{k}}}^{(\hat{k})^T}, \tilde{\boldsymbol{p}}_2^{(\hat{k})^T}, \cdots, \tilde{\boldsymbol{p}}_{M_{\hat{k}}}^{(\hat{k})^T}, \cdots)^T. \tag{9}$$

(ii) *For any* $\boldsymbol{u} \in \boldsymbol{R}^{M_{\hat{k}+1}}$, $\nu \in \boldsymbol{R}$, *define*

$$\begin{aligned}\beta_{(\boldsymbol{u},\nu)} : \boldsymbol{\Theta}_{M_{\hat{k}}-1} &\longrightarrow \boldsymbol{\Theta}_{M_{\hat{k}}},\\ \boldsymbol{\theta}^{(M_{\hat{k}}-1)} &\longmapsto (\cdots, \boldsymbol{\xi}^{(\hat{k}+1)^T} - \varphi(\nu)\boldsymbol{u}^T, \boldsymbol{u}^T,\\ \boldsymbol{s}_2^{(\hat{k})^T}, \cdots, \boldsymbol{s}_{M_{\hat{k}}}^{(\hat{k})^T}, &(\boldsymbol{0}^T, \nu), \tilde{\boldsymbol{p}}_2^{(\hat{k})^T}, \cdots, \tilde{\boldsymbol{p}}_{M_{\hat{k}}}^{(\hat{k})^T}, \cdots)^T \end{aligned} \tag{10}$$

where

$$\boldsymbol{\theta}^{(M_{\hat{k}}-1)} = (\cdots, \boldsymbol{\xi}^{(\hat{k}+1)^T}, \boldsymbol{s}_2^{(\hat{k})^T}, \cdots, \boldsymbol{s}_{M_{\hat{k}}}^{(\hat{k})^T}, \tilde{\boldsymbol{p}}_2^{(\hat{k})^T}, \cdots, \tilde{\boldsymbol{p}}_{M_{\hat{k}}}^{(\hat{k})^T}, \cdots)^T. \tag{11}$$

(iii) *For any* $\lambda \in \boldsymbol{R}$, *define*

$$\begin{aligned}\gamma_\lambda : \boldsymbol{\Theta}_{M_{\hat{k}}-1} &\longrightarrow \boldsymbol{\Theta}_{M_{\hat{k}}},\\ \boldsymbol{\theta}^{(M_{\hat{k}}-1)} &\longmapsto (\cdots, \boldsymbol{\xi}^{(\hat{k}+1)^T}, \lambda \boldsymbol{s}_2^{(\hat{k})^T}, (1-\lambda)\boldsymbol{s}_2^{(\hat{k})^T},\\ \boldsymbol{s}_3^{(\hat{k})^T}, \cdots, \boldsymbol{s}_{M_{\hat{k}}}^{(\hat{k})^T}, &\tilde{\boldsymbol{p}}_2^{(\hat{k})^T}, \tilde{\boldsymbol{p}}_2^{(\hat{k})^T}, \tilde{\boldsymbol{p}}_3^{(\hat{k})^T}, \cdots, \tilde{\boldsymbol{p}}_{M_{\hat{k}}}^{(\hat{k})^T}, \cdots)^T \end{aligned} \tag{12}$$

where

$$\boldsymbol{\theta}^{(M_{\hat{k}}-1)} = (\cdots, \boldsymbol{\xi}^{(\hat{k}+1)^T}, \boldsymbol{s}_2^{(\hat{k})^T}, \cdots, \boldsymbol{s}_{M_{\hat{k}}}^{(\hat{k})^T}, \tilde{\boldsymbol{p}}_2^{(\hat{k})^T}, \cdots, \tilde{\boldsymbol{p}}_{M_{\hat{k}}}^{(\hat{k})^T}, \cdots)^T. \tag{13}$$

□

For any set of parameters $\theta^{(M_{\hat{k}}-1)}$ of the $(M_{\hat{k}}-1)$-deep network with $M_{\hat{k}}-1$ neurons in the $\hat{k}$-th hidden layer, the embedding $\alpha_{\tilde{w}}(\theta^{(M_{\hat{k}}-1)})$ gives a set of parameters of the $M_{\hat{k}}$-deep network with $M_{\hat{k}}$ neurons in the $\hat{k}$-th hidden layer where the neuron 1 is newly added, keeping the output value of the neural network unchanged (see Fig. 1). Actually, since the weight vector between the newly added neuron 1 and the next layer $\hat{k}+1$ is equal to a zero vector: $\boldsymbol{s}_1^{(\hat{k})} \equiv \boldsymbol{0}$, the newly added neuron 1 never affect the next layer $\hat{k}+1$.

Similarly, for any set of parameters $\theta^{(M_{\hat{k}}-1)}$ of the $(M_{\hat{k}}-1)$-deep network with $M_{\hat{k}}-1$ neurons in the $\hat{k}$-th hidden layer, the embedding $\beta_{(\boldsymbol{u},\nu)}(\theta^{(M_{\hat{k}}-1)})$ gives a set of parameters of the $M_{\hat{k}}$-deep network with $M_{\hat{k}}$ neurons in the $\hat{k}$-th hidden layer where the neuron 1 is newly added, keeping the output value of the neural network unchanged (see Fig. 2). In this case, since the weight vector between the newly added neuron 1 and the preceding layer $\hat{k}-1$ is equal to a zero vector: $(\boldsymbol{0}^T, \nu)$ where the threshold ν is not always zero, the newly added neuron 1 never affect the next layer $\hat{k}+1$ where the value related on the threshold $\varphi(\nu)u_l$ is subtracted from each threshold $\xi_l^{(\hat{k}+1)}$ of the neurons in the next layer $\hat{k}+1$.

And also, for any set of parameters $\theta^{(M_{\hat{k}}-1)}$ of the $(M_{\hat{k}}-1)$-deep network with $M_{\hat{k}}-1$ neurons in the $\hat{k}$-th hidden layer, the embedding $\gamma_\lambda(\theta^{(M_{\hat{k}}-1)})$ gives a set of parameters of the $M_{\hat{k}}$-deep network with $M_{\hat{k}}$ neurons in the $\hat{k}$-th hidden layer where the neuron 1 is newly added, keeping the output value of the neural network unchanged (see Fig. 3). The embedding γ_λ proportionally distributes the weight vector $\boldsymbol{s}_2^{(\hat{k})}$ between the hidden neuron 2 in the $\hat{k}$-th layer and the next layer $\hat{k}+1$ in the $(M_{\hat{k}}-1)$-deep network into $\lambda \boldsymbol{s}_2^{(\hat{k})}$ and $(1-\lambda)\boldsymbol{s}_2^{(\hat{k})}$ using the constant λ, and assigns them to the weight vector between the newly added neuron 1 and the next layer $\hat{k}+1$, and the weight vector between the neuron 2 and the next layer $\hat{k}+1$ in the $M_{\hat{k}}$-deep network, respectively. In addition, the weight vector between the newly added neuron 1 and the preceding layer $\hat{k}-1$ is equal to the weight vector $\tilde{\boldsymbol{p}}_2^{(\hat{k})}$ between the neuron 2 and the preceding layer $\hat{k}-1$ in the $M_{\hat{k}}$-deep network. That is the reason why newly adding neuron 1 never change the output value of the $(M_{\hat{k}}-1)$-deep network.

Theorem 1. *Consider a $(M_{\hat{k}}-1)$-deep network defined in Definition 2 which has $M_{\hat{k}}-1$ hidden neurons numbered from 2 to $M_{\hat{k}}$ in the specific hidden layer $\hat{k}$. Let $\boldsymbol{\theta}_*^{(M_{\hat{k}}-1)} \in \boldsymbol{\Theta}_{M_{\hat{k}}-1}$ be a critical point of the error function $E_{M_{\hat{k}}-1}$ of the $(M_{\hat{k}}-1)$-deep network.*

(i) Let $\alpha_{\tilde{\boldsymbol{w}}}$ be as in Definition 3 (i). *If $\tilde{\boldsymbol{w}} = (\boldsymbol{0}^T, \nu)^T$, then the point $\boldsymbol{\theta}_*^{(M_{\hat{k}})} = \alpha_{\tilde{\boldsymbol{w}}}(\boldsymbol{\theta}_*^{(M_{\hat{k}}-1)})$ is a critical point of the error function $E_{M_{\hat{k}}}$ of the $M_{\hat{k}}$-deep network for any $\nu \in \boldsymbol{R}$ where ν is the threshold of the hidden neuron 1 newly added in the hidden layer $\hat{k}$ (see Fig. 1).*

(ii) Let $\beta_{(\boldsymbol{u},\nu)}$ be as in Definition 3 (ii). *Then, the point $\boldsymbol{\theta}_*^{(M_{\hat{k}})} = \beta_{(\boldsymbol{0},\nu)}(\boldsymbol{\theta}_*^{(M_{\hat{k}}-1)})$ is a critical point of the error function $E_{M_{\hat{k}}}$ of the $M_{\hat{k}}$-deep network for any*

$\nu \in \boldsymbol{R}$ *where* ν *is the threshold of the hidden neuron 1 newly added in the hidden layer* $\hat{k}$ *(see Fig. 2).*

(iii) Let γ_λ *be as in Definition 3* (iii)*. Then, the point* $\boldsymbol{\theta}_*^{(M_{\hat{k}})} = \gamma_\lambda(\boldsymbol{\theta}_*^{(M_{\hat{k}}-1)})$ *is a critical point of the error function* $E_{M_{\hat{k}}}$ *of the* $M_{\hat{k}}$*-deep network for any* $\lambda \in \boldsymbol{R}$ *(see Fig. 3).*

The critical point in Theorem 1 is called a *critical point introduced by the hierarchical structure* in this paper. Critical points introduced by the hierarchical structure consist of straight lines if we move $\nu \in \boldsymbol{R}$ and $\lambda \in \boldsymbol{R}$, respectively. For any hidden layer k, there are one and M_{k+1} lines formed by moving $\nu \in \boldsymbol{R}$ and $\lambda \in \boldsymbol{R}$, respectively $(1 \le k \le N)$ where $M_{N+1} \equiv 1$. Therefore, there are $\sum_{k=1}^{N} M_{k+1} + N$ lines in the deep neural network defined in Sect. 2. In this connection, there are only 2 lines in a shallow $L - M_1 - 1$ three-layered neural network $(N = 1)$.

4 Sufficient Conditions to Have No Critical Points

This section presents sufficient conditions for deep neural networks having no critical points introduced by the hierarchical structure.

The following Theorem 2 is a sufficient condition for a deep neural network to have no critical points introduced by the hierarchical structure.

Theorem 2. *Assume the following conditions for the hidden layer* $\hat{k}$ *of a* $M_{\hat{k}}$*-deep network defined in Sect. 2* $(1 \le \hat{k} \le N)$*: (a)* $\boldsymbol{w}_l^{(\hat{k})} \neq \boldsymbol{0}, \boldsymbol{v}_l^{(\hat{k})} \neq \boldsymbol{0}$ *for any* $1 \le l \le M_{\hat{k}}$*, and (b)* $\boldsymbol{w}_{l_1}^{(\hat{k})} \neq \pm \boldsymbol{w}_{l_2}^{(\hat{k})}$ *for any* $1 \le l_1, l_2 \le M_{\hat{k}}$ $(l_1 \neq l_2)$ *where* $\boldsymbol{w}_l^{(\hat{k})} = (w_{l1}^{(\hat{k})} \cdots w_{lM_{\hat{k}-1}}^{(\hat{k})})^T$, $\boldsymbol{v}_l^{(\hat{k})} = (w_{1l}^{(\hat{k}+1)} \cdots w_{M_{\hat{k}+1}l}^{(\hat{k}+1)})^T$*(if* $1 \le \hat{k} \le N-1$*) and* $\boldsymbol{v}_l^{(\hat{k})} = (v_l^N)$*(if* $\hat{k} = N$*). Then* $M_{\hat{k}}$*-deep network has no critical points introduced by the hierarchical structure for the hidden layer* $\hat{k}$. □

When the number of the hidden neurons $M_{\hat{k}}$ is not more than that of the hidden neurons in the preceding hidden layer $\hat{k} - 1$ and that of the hidden neurons in the succedent hidden layer $\hat{k} + 1$, the conditions can be simplified as follows.

Corollary 1. *Assume* $M_{\hat{k}} \le M_{\hat{k}-1}$ *and* $M_{\hat{k}} \le M_{\hat{k}+1}$ *for a* $M_{\hat{k}}$*-deep network defined in Sect. 2* $(1 \le \hat{k} \le N)$*. If rank* $\boldsymbol{W}^{(\hat{k})} =$ *rank* $\boldsymbol{V}^{(\hat{k})} = M_{\hat{k}}$*, then the* $M_{\hat{k}}$*-deep network has no critical points introduced by the hierarchical structure for the hidden neuron* $\hat{k}$ *where* $\boldsymbol{W}^{(\hat{k})} = (\boldsymbol{w}_1^{(\hat{k})}, \cdots, \boldsymbol{w}_{M_{\hat{k}}}^{(\hat{k})})^T \in \boldsymbol{R}^{M_{\hat{k}} \times M_{\hat{k}-1}}$ *is the weight matrix between the hidden layers* $\hat{k} - 1$ *and* $\hat{k}$*,* $\boldsymbol{V}^{(\hat{k})} = (\boldsymbol{v}_1^{(\hat{k})}, \cdots, \boldsymbol{v}_{M_{\hat{k}}}^{(\hat{k})}) \in \boldsymbol{R}^{M_{\hat{k}+1} \times M_{\hat{k}}}$ *is the weight matrix between the hidden layers* $\hat{k}$ *and* $\hat{k} + 1$ *where*

$$\boldsymbol{v}_l^{(\hat{k})} = \begin{cases} (w_{1l}^{(\hat{k})} \cdots w_{M_{\hat{k}}l}^{(\hat{k})})^T & (if\ 1 \le \hat{k} \le N-1) \\ (v_l^{(N)}) & (if\ \hat{k} = N) \end{cases} \tag{14}$$

for any $1 \le l \le M_{\hat{k}}$. □

When the number of hidden neurons of all hidden layers is the same, the conditions can still be simplified.

Corollary 2. *Assume* $M_k = L$ $(1 \leq k \leq N)$ *for a deep network defined in Sect. 2. If the weight matrices* $\boldsymbol{W}^{(k)}$ $(1 \leq k \leq N)$ *are all regular and* $v_l^{(N)} \neq 0$ $(1 \leq l \leq L)$*, then the deep neural network has no critical points introduced by the hierarchical structure for all hidden layers from 1 to* N *where* $\boldsymbol{W}^{(k)}$ *is the weight matrix between the hidden layer* $k-1$ *and the hidden layer* k *defined in Corollary 1, and* $v_l^{(N)}$ *is the weight between the hidden neuron* l *in the hidden layer* N *and the output neuron* $(1 \leq l \leq L)$. □

5 Conclusions

This paper presented deep neural networks having no critical points introduced by a hierarchical structure from a theoretical side. First, it was shown that there exist a large number of critical points introduced by a hierarchical structure in deep neural networks as straight lines, depending on the number of hidden layers and the number of hidden neurons. Specifically, there are $\sum_{k=1}^{N} M_{k+1} + N$ critical lines in a deep neural network where N is the number of hidden layers and M_{k+1} is the number of hidden neurons in the hidden layer $k+1$, whereas there are only 2 critical lines in a shallow $L-M_1-1$ three-layered neural network $(N=1)$. Next, sufficient conditions for deep neural networks having no critical points introduced by a hierarchical structure were derived.

Acknowledgments. The author would like to give special thanks to the anonymous reviewers for valuable comments. This work was supported by the Japan Society for the Promotion of Science through the Grants-in-Aid for Scientific Research (KAKENHI) under Grant JP16K00347.

References

1. Bengio, Y.: Learning deep architectures for AI. Found. Trends Mach. Learn. **2**(1), 1–127 (2009)
2. Hinton, G.E., Osindero, S., Teh, Y.: A fast learning algorithm for deep belief nets. Neural Comput. **18**, 1527–1554 (2006)
3. Fukumizu, K., Amari, S.: Local minima and plateaus in hierarchical structures of multilayer perceptrons. Neural Netw. **13**(3), 317–327 (2000)
4. Montúfar, G., Pascanu, R., Cho, K., Bengio, Y.: On the number of linear regions of deep neural networks. In: Advances in Neural Information Processing Systems 27 (NIPS 2014), pp. 2924–2932 (2014)
5. Szymanski, L., McCane, B.: Deep networks are effective encoders of periodicity. IEEE Trans. Neural Netw. Learn. Syst. **25**(10), 1816–1827 (2014)
6. Nitta, T.: Resolution of singularities introduced by hierarchical structure in deep neural networks. IEEE Trans. Neural Netw. Learn. Syst. (accepted). doi:10.1109/TNNLS.2016.2580741

A Deep Neural Network Architecture Using Dimensionality Reduction with Sparse Matrices

Wataru Matsumoto[1(✉)], Manabu Hagiwara[2], Petros T. Boufounos[3], Kunihiko Fukushima[1,4], Toshisada Mariyama[1], and Zhao Xiongxin[1]

[1] Information Technology R&D Center, Mitsubishi Electric Corporation, Kamakura, Kanagawa, Japan
Matsumoto.Wataru@aj.MitsubishiElectric.co.jp
[2] Chiba University, Chiba, Japan
[3] Mitsubishi Electric Research Laboratories, Cambridge, MA, USA
[4] Fuzzy Logic System Institute, Fukuoka, Japan

Abstract. We present a new deep neural network architecture, motivated by sparse random matrix theory that uses a low-complexity embedding through a sparse matrix instead of a conventional stacked autoencoder. We regard autoencoders as an information-preserving dimensionality reduction method, similar to random projections in compressed sensing. Thus, exploiting recent theory on sparse matrices for dimensionality reduction, we demonstrate experimentally that classification performance does not deteriorate if the autoencoder is replaced with a computationally-efficient sparse dimensionality reduction matrix.

Keywords: Deep learning · Deep neural network · Autoencoder · Compressed sensing · Sparsity recovery · Sparse random matrices

1 Introduction

Image and signal classification is one of the most fundamental problems in computer vision and pattern recognition. Recently, deep learning has become a popular solution in this area owing to its superior performance [1, 2]. One of the most commonly used deep learning architectures is the feedforward deep neural network (DNN) [2]. Typical DNNs include a stacked autoencoder as an initial component in their structure. A stacked autoencoder is a multilayer network, trained to reduce data dimensionality while preserving information, thus providing an effective feature extraction and signal representation mechanism for further processing [3].

In a similar vein, compressed sensing (CS) has attracted considerable attention for suggesting that it is possible to overcome the traditional limits of sampling theory. Specifically, Candes et al. [4, 5] and Donoho [6] demonstrated that a signal having a sparse representation can be recovered exactly from a small set of linear, nonadaptive measurements. The CS measurement mechanism is related to Johnson-Lindenstrauss (JL) embeddings [7, 8], and has been extended to other signal sets, such as

A. Hirose et al. (Eds.): ICONIP 2016, Part IV, LNCS 9950, pp. 397–404, 2016.
DOI: 10.1007/978-3-319-46681-1_48

manifolds [9]. The results suggest that it may be possible to sense structured signals by taking far fewer measurements. Furthermore, Berinde and Indyk [10] demonstrated information embedding and algorithms for sparse recovery, based on sparse random matrices. Such matrices are attractive because of their low computational complexity.

In the remainder of this paper, we demonstrate that efficient randomized dimensionality reduction, using a sparse measurement matrix, can be used to replace the autoencoder stage of a DNN. Thanks to its embedding properties, this dimensionality reduction still preserves the information necessary for classification. The sparsity of the matrix significantly reduces the classification complexity. This paper is organized as follows. Section 2 discusses the motivation and develops the proposed sparse construction using recent theoretical work. Section 3 presents an experimental performance comparison of the proposed sparse construction to conventional networks. Section 4 concludes the document.

2 Deep Neural Network with Sparse Construction

To establish notation, we first define a deep neural network structure. We use the N-dimensional vector $\mathbf{x} = (x_1, x_2, \cdots, x_N)^{\mathrm{T}} \in \mathbb{R}^N$ to denote the input signal. We assume a multilayer structure, with l denoting the layer's index as $l = 1, 2, 3, \cdots$. The vector $\mathbf{u}^{(l)} = \left(u_1^{(l)}, u_2^{(l)}, \cdots, u_J^{(l)}\right)^{\mathrm{T}} \in \mathbb{R}^J$ denotes the weighted sum for the l-th layer, computed as $u_j^{(l)} = \sum_{i=1}^{I} w_{ij}^{(l)} x_i^{(l)} + b_i^{(l)}$, where $\mathbf{W}^{(l)} = \begin{bmatrix} w_{11}^{(l)} & \cdots & w_{1I}^{(l)} \\ \vdots & \ddots & \vdots \\ w_{J1}^{(l)} & \cdots & w_{JI}^{(l)} \end{bmatrix}$ is the weight matrix and $\mathbf{b}^{(l)} = \left(b_1^{(l)}, b_2^{(l)}, \cdots, b_J^{(l)}\right)^{\mathrm{T}} \in \mathbb{R}^J$ is the bias vector. Given $u_j^{(l)}$, the activation function f produces the input vector $x_j^{(l+1)}$ for the next (i.e., $l+1$-th) layer, using the element-wise computation $x_j^{(l+1)} = f\left(u_j^{(l)}\right)$. To simplify discussion, in the remainder of this paper we assume $b_j^{(l)} = 0$ and $f(u) = u$ (Fig. 1).

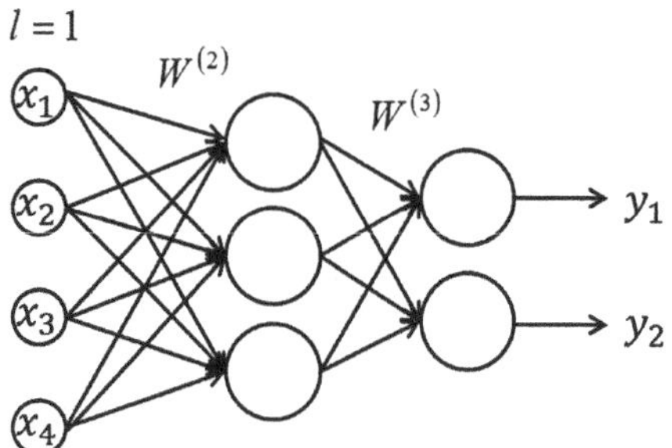

Fig. 1. Multilayer neural network

2.1 Similarity of Autoencoder and Compressed Sensing

A stacked autoencoder is pre-trained using unsupervised learning, separately from the DNN used for classification. The goal is to convert high-dimensional input data to low dimensional features that capture the salient information of the input and can be used for classification. Each layer is typically trained by minimizing the discrepancy between the input data and its reconstruction using the autoencoder. This training is applied layer by layer from bottom to top using gradient descent and backpropagation. However, this works well only if the initial weights are close to a reasonable solution.

In this section, motivated by sparse matrix theory, we describe an effective method of initializing the weights to construct deep neural networks. Instead of a dense autoencoder, we learn and use low-dimensional sparse codes to reduce data dimensionality. The implicit regularization also removes the requirement to use dropout during learning.

Given a network layer, $\mathbf{x}^{(l+1)} = \mathbf{W}^{(l)}\mathbf{x}^{(l)}$, the recovered vector $\hat{\mathbf{x}}^{(l)}$ is generated from $\mathbf{x}^{(l)}$ using a weight matrix $\hat{\mathbf{W}}^{(l)}$ by computing $\mathbf{x}^{(l)} = \hat{\mathbf{W}}^{(l)}\mathbf{x}^{(l+1)}$. Training estimates the weight matrices $\mathbf{W}^{(l)} \approx \hat{\mathbf{W}}^{(l)}$ and $\mathbf{x}^{(l+1)}$ by solving the optimization

$$\min_{\{\hat{\mathbf{x}}^{(l)}\}} \left\| \mathbf{x}^{(l)} - \hat{\mathbf{x}}^{(l)} \right\|_2^2 = \min_{\{\hat{\mathbf{x}}^{(l)}\}\{\mathbf{x}^{(l+1)}\}} \left\| \mathbf{x}^{(l)} - \hat{\mathbf{x}}^{(l)} \right\|_2^2.$$

We use $J^{(l)}$ to denote the length of $\mathbf{x}^{(l)}$. In general, $J^{(l+1)} \leq J^{(l)}$, so that the autoencoder reduces the dimensionality of the data, similar to CS measurements. That is, given a measured signal $\mathbf{x}^{(l+1)}$, the autoencoder problem can be regarded as a problem to recover the original signal $\mathbf{x}^{(l)}$ using a measurement matrix $\mathbf{W}^{(l)}$.

An accurate reconstruction means that the exact signal is reconstructed. The problem is how to design the measurement matrix $\mathbf{W}^{(l)}$. To solve this problem, we first review the CS literature for desirable properties of $\mathbf{W}^{(l)}$ (Fig. 2).

Compressed sensing can be regarded as a generalization of Nyquist-Shannon sampling for structured signals - typically sparse signals that have very few non-zero entries. In the Nyquist-Shannon sampling theorem, signals, images, videos, and other

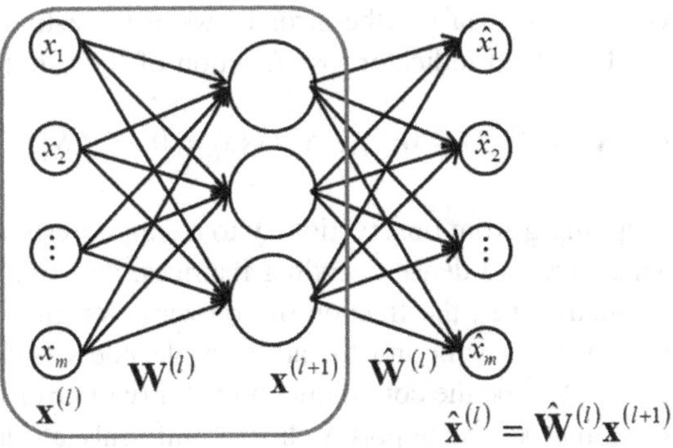

Fig. 2. Structure of autoencoder

data can be exactly represented with a set of uniformly spaced samples taken at the Nyquist rate, i.e., twice the highest frequency present in the signal of interest. Rather than sampling at specific points in times, CS, instead, uses a few generalized linear measurements of the signal, exploiting the fact that most signals of interest exhibit structure such as sparsity that can be used in the reconstruction.

For example, the majority of natural images are characterized by large smooth or textured regions and relatively few sharp edges. In the wavelet transform of a typical natural image, most coefficients are very small [11]. Hence, we can obtain a good approximation of the signal by setting the small coefficients to zero to obtain a K-sparse representation. We say that a vector $\mathbf{x}$ is K-sparse if it contains at most K non-zero entries.

Recognition for a neural network can be considered as a classification problem, under a signal model that exhibits some structure such as sparsity. In this sense, the autoencoder has the same goal as compressive measurements: reduce the dimensionality of the original signal while preserving the appropriate information.

2.2 Sparse Recovery Using Sparse Random Matrices

In the subsequent discussion, we consider a single layer and sparse signal models. We first review sparse recovery guarantees that ensure the preservation of information. To simplify the notation, we replace $\mathbf{x}^{(l+1)} = \mathbf{W}^{(l)}\mathbf{x}^{(l)}$ with $\mathbf{y} = \mathbf{W}\mathbf{x}$. We assume deterministic signals, in which $\mathbf{x} \in \mathbb{R}^N$ is a fixed but unknown vector with exactly K non-zero entries. We use the set $S := \text{supp}(\mathbf{x})$ to denote the support of $\mathbf{x}$, i.e., the location of non-zeros. Note that there are $M = \binom{N}{K}$ possible support sets, corresponding to the M possible K-dimensional subspaces in which $\mathbf{x}$ may lie. We are given a vector of m noisy observations $\mathbf{y} \in \mathbb{R}^m$, of the form

$$\mathbf{y} = \mathbf{W}\mathbf{x} + \mathbf{V},$$

where $\mathbf{W} \in \mathbb{R}^{m \times N}$ is the measurement matrix, and $\mathbf{V} \sim N(0, \sigma^2 \mathbf{I}_{M \times M})$ is additive Gaussian noise. Throughout this paper, we assume, without loss of generality, that $\sigma^2 = 1$, since any scaling of σ can be accounted for in the scaling of $\mathbf{x}$.

We consider exact recovery of the support set S, which corresponds to model selection. More precisely, we measure the error between the estimate $\hat{\mathbf{x}}$ and the true signal $\mathbf{x}$ using the $\{0, 1\}$-valued indicator loss function of the support:

$$\rho(\hat{\mathbf{x}}, \mathbf{x}) := \mathbb{I}\left[\{\hat{x}_i \neq 0, \forall i \in S\} \cap \{\hat{x}_j = 0, \forall j \notin S\}\right].$$

A decoder is a mapping g from observations $\mathbf{y}$ to a support estimate $\hat{S} = \mathrm{g}(\mathbf{y})$. We are interested in both sparse and dense encoding matrices, denoting by γ the sparsification fraction of the matrix, i.e., the fraction of non-zero elements.

A general performance lower bound for arbitrary decoders is developed in [12]. Specifically, let $\mathbb{P}[\mathrm{g}(\mathbf{y}) \neq S|S]$ be the conditional probability of error given that the true support is S. Assuming that $\mathbf{x}$ has support S chosen uniformly at random over the M possible subsets of size K, the average probability of error is given by

$$p_{err} = \frac{1}{\binom{N}{K}} \sum_S \mathbb{P}[g(\mathbf{y}) \neq S|S].$$

We say that sparsity recovery is asymptotically reliable if $p_{err} \to 0$ as $m \to \infty$. Since our goal is *exact* support recovery from noisy measurements, the minimum value of $\mathbf{x}$ on its support is important,

$$x_{\min} := \min_{i \in S} |x_i|.$$

Thus, given $x_{\min}$, we can define a signal class

$$C(x_{\min}) := \{\mathbf{x} \in \mathbb{R}^N | |x_i| \geq x_{\min} \forall i \in S\}.$$

For this class, [12] derives the necessary conditions on parameters $(m, N, K, x_{\min}, \gamma)$, such that a decoder with asymptotically reliable recovery can exist, regardless of its computational complexity. The lower bounds describe the required number of measurements m, in general settings, where both the signal sparsity K, and the measurement sparsity γ are allowed to scale with the signal dimension N. The analysis in [12] applies to random ensembles of measurement matrices $\mathbf{W} \in \mathbb{R}^{m \times N}$, where each entry w_{ij} is drawn i.i.d. from some underlying distribution. The most commonly used ensemble is the standard Gaussian distribution, in which $w_{ij} \sim N(0, 1)$. This choice generates a dense measurement matrix $\mathbf{W}$, with mN non-zero entries. Theorem 1 in [12] applies to more general ensembles satisfying the moment conditions $\mathbb{E}[x_{ij}] = 0$ and $\text{var}(x_{ij}) = 1$, allowing for a variety of non-Gaussian distributions (e.g., uniform, Bernoulli). Further, Theorem 2 in [12] is derived for γ-sparsified matrices $\mathbf{W}$, in which each entry w_{ij} is i.i.d., drawn according to

$$w_{ij} = \begin{cases} N\left(0, \frac{1}{\gamma}\right) & \text{w.p. } \gamma \\ 0 & \text{w.p. } 1 - \gamma \end{cases} \tag{1}$$

Note that when $\gamma = 1$, $\mathbf{W}$ is the standard Gaussian ensemble. We refer to the sparsification parameters $0 \leq \gamma \leq 1$ as the measurement sparsity. The analysis allows this parameter to vary as a function of (m, N, K).

2.3 Bounds on Dense Ensembles and Sparse Ensembles

Theorem 1 in [12] provides a necessary condition for asymptotically reliable recovery on measurement matrices with dense ensembles.

Theorem 1 [12] (General Ensembles). Let the measurement matrix $\mathbf{W} \in \mathbb{R}^{m \times N}$ be drawn with i.i.d. elements from any distribution with zero-mean and variance one. Then a necessary condition for asymptotically reliable recovery over the signal class $C(x_{\min})$ is

$$m > \max\{f_1(N,K,x_{\min}), f_2(N,K,x_{\min}), k-1\},$$

where

$$f_1(N,K,x_{\min}) := \frac{\log\binom{N}{K} - 1}{\frac{1}{2}\log\left(1 + Kx_{\min}^2\left(1 - \frac{K}{N}\right)\right)}$$
$$f_2(N,K,x_{\min}) := \frac{\log(N-K+1) - 1}{\frac{1}{2}\log\left(1 + x_{\min}^2\left(1 - \frac{1}{N-K+1}\right)\right)}.$$

Furthermore, Theorem 2 in [12] provides a necessary condition for asymptotically reliable recovery on measurement matrices with sparse ensembles.

Theorem 2 and Corollary 2 [12] (Sparse Ensembles). Let the measurement matrix $\mathbf{W} \in \mathbb{R}^{m \times N}$ be drawn with i.i.d. elements from the γ-sparsified Gaussian ensemble (1). A necessary condition for asymptotically reliable recovery over the signal class $C(x_{\min})$ is

$$m > \max\{g_1(N,K,x_{\min},\gamma), g_2(N,K,x_{\min,}\gamma), k-1\},$$

where in general

$$g_1(N,K,x_{\min},\gamma) \geq \frac{\log\binom{N}{K} - 1}{\frac{1}{2}\log(1 + Kx_{\min}^2)}$$
$$g_2(N,K,x_{\min},\gamma) \geq \frac{\log(N-K+1) - 1}{\frac{1}{2}\log(1 + x_{\min}^2)}.$$

In general, $N \gg K$, hence $1 \gg \frac{K}{N}, 1 \gg \frac{1}{N-K+1}$. Under this assumption, from Theorems 1 and 2, and Corollary 2, we can derive the following necessary condition:

$$\begin{aligned} m &> \max\{f_1(N,K,x_{\min}), f_2(N,K,x_{\min}), k-1\} \\ &\approx \max\{g_1(N,K,x_{\min},\gamma), g_2(N,K,x_{\min,}\gamma), k-1\}, \end{aligned}$$

where in general

$$g_1(N,K,x_{\min},\gamma) \geq \frac{\log\binom{N}{K} - 1}{\frac{1}{2}\log(1 + Kx_{\min}^2)}$$
$$g_2(N,K,x_{\min},\gamma) \geq \frac{\log(N-K+1) - 1}{\frac{1}{2}\log(1 + x_{\min}^2)}.$$

That is, the number of observations necessary is almost independent of whether we use dense or sparse ensembles. Thus, even if we use sparse matrices as measurement matrices, in principle, we can recover the original signals. We exploit this to provide an effective method of initializing the weights that uses sparse random matrices to construct deep neural networks that use low-dimensional codes as a tool to reduce the dimensionality of the data, instead of using an autoencoder.

3 Performance Evaluation and Comparison

We performed experiments that compared the performance of sparse matrices with those of dense, Gaussian matrices. In our experiments we generated γ-sparsified matrices $\mathbf{W}$ of m rows and N columns, in which each entry w_{ij} was i.i.d. drawn according to (1). Note that when $\gamma = 1$, $\mathbf{W}$ is exactly the standard Gaussian ensemble. d is the average non-zero entries in each column for $\mathbf{W}$, so that $\gamma = \frac{d}{m}$. To evaluate our experiment we used the MNIST handwritten digit dataset containing ten classes (0–9) consisting of 50,000 training, 10,000 validation and 10,000 test images [13]. The digits were size-normalized and centered in 28 × 28 grayscale images.

3.1 Unsupervised Feature Learning Results

To compare the quality of the features learnt by sparse matrices with dense matrices, we first extracted features using the unsupervised learning algorithm and compared the reconstruction mean square error (MSE) $\frac{1}{N}\|\hat{\mathbf{x}} - \mathbf{x}\|_2^2$. The network had 784 input units (28 × 28 grayscale image, normalized to values ranging from [0,1]). These were connected to 500 hidden units, so that $N = 784, m = 500$. For the unsupervised training, we set the momentum to $\tau_t = 0.5$, the learning rate to $\eta_t = 0.1$, and trained for 10,000 epochs.

The MSE results for dense and different sparse matrices are compared in Table 1. We can see that the final MSE of the unsupervised learning algorithm with dense and sparse matrices is almost the same, approximately 5.0e-4.

Table 1. MSE of unsupervised learning algorithm with dense and sparse matrices

Sparsity	$\gamma = 1$	$\gamma = 0.33$	$\gamma = 0.16$	$\gamma = 0.081$	$\gamma = 0.04$
	$d = 784$	$d = 256$	$d = 128$	$d = 64$	$d = 32$
MSE	5.29e-4	5.12e-4	4.95e-4	5.19e-4	5.88e-4

3.2 Supervised Learning Results

In supervised learning, it is a common practice to use the weights generated by the unsupervised learning method to initialize the early layers of a multilayer network. A discriminative algorithm is then used to adjust the weights of the last hidden layer and also to fine tune the weights in the previous layers.

The accuracy results for dense and different sparse matrices are compared in Table 2. As is evident, the performance of supervised learning with dense and sparse matrices is almost the same, with average accuracy approximately 98 % for the ten digit classes (0–9).

Table 2. Average accuracy of supervised learning algorithm with dense and sparse matrices

Sparsity	$\gamma = 1$	$\gamma = 0.33$	$\gamma = 0.16$	$\gamma = 0.081$	$\gamma = 0.04$
	d = 784	d = 256	d = 128	d = 64	d = 32
Average accuracy	98.38 %	98.34 %	98.13 %	98.17 %	97.9 %

4 Conclusion

In this work, we proposed a sparse coding method with sparse measurement matrices for deep neural networks. We described an effective method of initializing the weights that permits sparse random matrices to construct deep neural networks to learn low-dimensional codes as a tool to reduce the dimensionality of data for autoencoder. The resultant neural network was constructed with extremely sparse edge connecting units for each layer. This proposal can provide benefits such as the reduction of training and inference time, reduction of the number of free parameters, and the enhancement of the generalization capability.

References

1. Fukushima, K.: Neocognitron: a self-organizing neural network model for a mechanism of pattern recognition unaffected by shift in position. Biol. Cybern. **36**(4), 193–202 (1980)
2. Hinton, G.E., Osindero, S., Teh, Y.: A fast learning algorithm for deep belief nets. Neural Comput. **18**, 1527–1554 (2006)
3. Hinton, G.E., Salakhutdinov, R.R.: Reducing the dimensionality of data with neural networks. Science **313**, 504–507 (2006)
4. Candes, E., Romberg, J., Tao, T.: Stable signal recovery from incomplete and inaccurate measurements. Commun. Pure Appl. Math. **59**(8), 1207–1223 (2006)
5. Candes, E., Tao, T.: Near optimal signal recovery from random projections: universal encoding strategies? IEEE Trans. Inf. Theory **52**(12), 5406–5425 (2006)
6. Donoho, D.: Compressed sensing. IEEE Trans. Inf. Theory **52**(4), 1289–1306 (2006)
7. Johnson, W.B., Lindenstrauss, J.: Extensions of Lipschitz mappings into a Hilbert space. Contemp. Math. **26**, 189–206 (1984)
8. Baraniuk, R.G., Davenport, M., DeVore, R., Wakin, M.: A simple proof of the restricted isometry property for random matrices. Constr. Approx. **28**(3), 253–263 (2008)
9. Baraniuk, R.G., Michael, B.W.: Random projections of smooth manifolds. Found. Comput. Math. **9**(1), 51–77 (2009)
10. Berinde, R., Indyk, P.: Sparse recovery using sparse random matrices. CSAIL Technical report, MIT-CSAIL-TR-2008-001 (2008)
11. Mallat, S.: A Wavelet Tour of Signal Processing. Academic Press, San Diego (1999)
12. Wang, W., Wainwright, M.J., Ramchandran, K.: Information-theoretic limits on sparse recovery: dense versus sparse measurement matrices. Technical report, Department of Statistics, UC, Berkeley, May 2008
13. LeCun, Y., Bottou, L., Bengio, Y., Haffner, P.: Gradient-based learning applied to document recognition. Proc. IEEE **86**(11), 2278–2324 (1998)

Noisy Softplus: A Biology Inspired Activation Function

Qian Liu[(✉)] and Steve Furber

Advanced Processor Technologies Group, School of Computer Science, University of Manchester, M13 9PL Manchester, UK
{qian.liu-3,steve.furber}@manchester.ac.uk

Abstract. The Spiking Neural Network (SNN) has not achieved the recognition/classification performance of its non-spiking competitor, the Artificial Neural Network(ANN), particularly when used in deep neural networks. The mapping of a well-trained ANN to an SNN is a hot topic in this field, especially using spiking neurons with biological characteristics. This paper proposes a new biologically-inspired activation function, Noisy Softplus, which is well-matched to the response function of LIF (Leaky Integrate-and-Fire) neurons. A convolutional network (ConvNet) was trained on the MNIST database with Noisy Softplus units and converted to an SNN while maintaining a close classification accuracy. This result demonstrates the equivalent recognition capability of the more biologically-realistic SNNs and bring biological features to the activation units in ANNs.

Keywords: Noisy softplus · Biologically-inspired · Spiking neural network · Activation function · LIF neurons

1 Introduction

Deep Neural Networks (DNNs) are the most promising research field in computer vision, even exceeding human-level performance on image classification tasks [8]. To investigate whether brains might work similarly on vision tasks, these powerful DNN models have been converted to spiking neural networks (SNNs). In addition, the spiking DNN offers the prospect of neuromorphic systems that combine remarkable performance with energy-efficient training and operation.

Theoretical studies have shown that biologically-plausible learning, e.g. Spike-Timing-Dependent Plasticity (STDP), could approximate a stochastic version of powerful machine learning algorithms such as Contrastive Divergence [13], Markov Chain Monte Carlo [1] and Gradient Descent [14]. Stochasticity, in contrast with the continuously differentiable functions used by ANNs, is intrinsic to the event-based spiking process, making network training difficult. In practice, ANNs use neuron and synapse models very different from biological neurons, and it remains an unsolved problem to develop SNNs with equivalent performance.

On the other hand, the offline training of an ANN, which is then mapped to an SNN, has shown near loss-less conversion and state-of-the-art classification

A. Hirose et al. (Eds.): ICONIP 2016, Part IV, LNCS 9950, pp. 405–412, 2016.
DOI: 10.1007/978-3-319-46681-1_49

accuracy. This research aims to prove that SNNs are equally capable as their non-spiking rivals of pattern recognition, and at the same time are more biologically realistic and energy-efficient. Jug et al. [10] first proposed the use of the Siegert function to replace the sigmoid activation function in Restricted Boltzmann Machine (RBM) training. The Siegert units map incoming currents driven by Poisson spike trains to the response firing rate of a Leaky Integrate-and-Fire (LIF) neuron. The ratio of the spiking rate to its maximum is equivalent to the output of a sigmoid neuron. A spiking Deep Belief Network (DBN) [15] was implemented on neuromorphic hardware, SpiNNaker [6], to recognise hand written digits in real time. However, cortical neurons seldom saturate their firing rate. Thus Rectified Linear Units (ReLU) were proposed and surpassed the performance of other popular activation units thanks to their advantage of sparsity [7]. Recent work [9] proposed the Soft LIF response function, which is equivalent to Softplus activation.

Even better performance [2,4] has been demonstrated in Spiking Convolutional Networks (ConvNets), but this employed simple integrate and fire neurons. The training used only ReLUs and zero bias to avoid negative outputs, and applied a deep learning technique, dropout, to increase the classification accuracy. Normalising the trained weights for use on an SNN was relatively straightforward and maintained the classification accuracy. This work was extended to a Recursive Neural Network (RNN) [5] and run on the TrueNorth [12] neuromorphic hardware platform.

The Noisy Softplus activation function proposed here is based on LIF neurons with biological characteristics, and is the first attempt to map a spiking neural response accurately to the activation unit of an ANN. The resulting classification accuracy was tested on a spiking ConvNet; the performance was close to that of the original ConvNet, and was better than using Softplus. This study brings a significant biological feature, noise, to the activation units of an ANN, in the hope of promoting research into noise-based computation.

2 Methods

2.1 Neural Science Background

This paper proposes a new activation function, Noisy Softplus, which is inspired by neuroscience observations of LIF neurons. The LIF neuron model follows the following membrane potential dynamics:

$$\tau_m \frac{\mathrm{d}V}{\mathrm{d}t} = V_{rest} - V + R_m I(t). \tag{1}$$

The membrane potential V changes in response to the input current I, starting at the resting membrane potential V_{rest}, where the membrane time constant is $\tau_m = R_m C_m$, R_m is the membrane resistance and C_m is the membrane capacitance. The central idea in converting spiking neurons to activation units lies in the response function of a neuron model. Given a constant current injection I, the response function, i.e. firing rate, of the LIF neuron is:

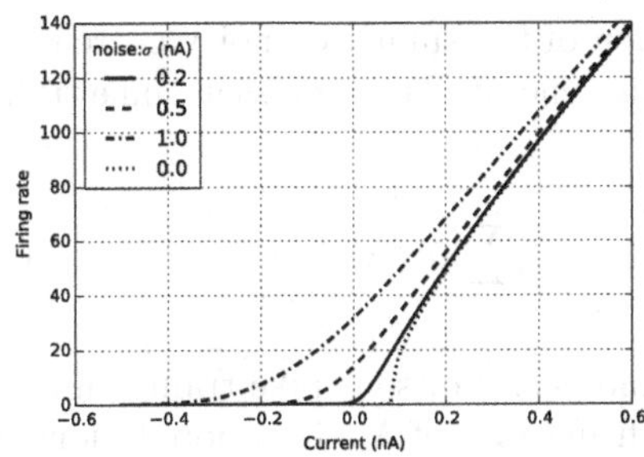

(a) Response function with noisy currents.

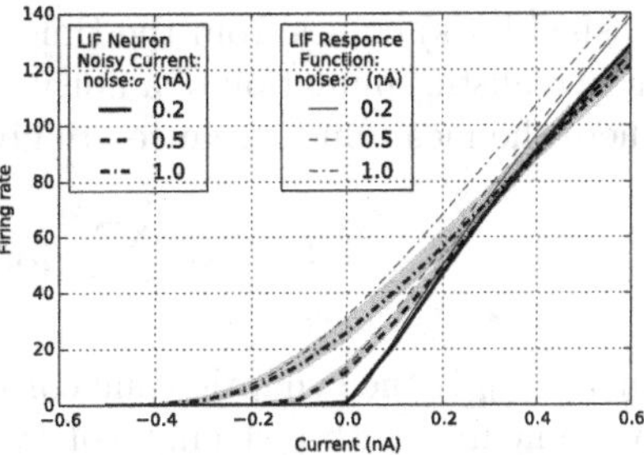

(b) Recorded response firing rate.

Fig. 1. (a) Response function of the LIF neuron with noisy input currents with different standard deviations. (b) Comparing the recorded firing rates of the LIF neuron simulation driven by noisy currents to the response function shown in (a).

Table 1. Parameter setting for the current-based LIF neurons using PyNN.

Parameters	cm	tau_m	tau_refrac	tau_syn_E	tau_syn_I	v_rest	v_thresh	i_offset
Values	0.25	20.0	1.0	5.0	5.0	−65.0	−50.0	0.1
Units	nF	ms	ms	ms	ms	mV	mV	nA

$$\lambda_{out} = \left[t_{ref} - \tau_m \log \left(1 - \frac{V_{th} - V_{rest}}{IR_m} \right) \right]^{-1}, \text{ when } IR_m > V_{th} - V_{rest}, \quad (2)$$

otherwise the membrane potential cannot reach the threshold V_{th} and the output firing rate is zero. The absolute refractory period t_{ref} is included, where all input during this period is ignored. The dotted (zero noise) line in Fig. 1(a) illustrates the response function of an LIF neuron, which inspired the proposal of ReLUs. The parameters of the LIF neuron are all biologically valid (see the listed values in Table 1), and the same parameters are used throughout this paper. In practice, a noisy current generated by the arrival of spike trains, rather than a constant current, flows into the neurons. The response function of the LIF neuron to a noisy current is as follows, where μ and σ are the mean and variance of the current:

$$\lambda_{out} = \left[t_{ref} + \tau_m \int_{\frac{V_{rest} - \mu\tau_m}{\sigma\sqrt{\tau_m}}}^{\frac{V_{th} - \mu\tau_m}{\sigma\sqrt{\tau_m}}} \sqrt{\pi} \exp(u^2)(1 + erf(u)) \mathrm{d}u \right]^{-1}. \quad (3)$$

2.2 LIF Neuron Simulation

To verify the response function, a simulation was carried out using PyNN [3] to compare with the analytical results. A noisy current with a particular μ and σ was injected into an LIF neuron for 10 s. The firing rate was the average among 10 trials, see Fig. 1(b). The slight difference compared to the analytical results (dashed lines) comes from the time resolution of the simulated noisy current. A more realistic simulation of a noisy current is generated by a Poisson spike train, where the mean and variance are given by:

$$\mu = \tau_{syn} \sum_i w_i \lambda_i \ , \ \sigma^2 = \frac{1}{2} \tau_{syn} \sum_i w_i^2 \lambda_i \ , \tag{4}$$

where τ_{syn} is the synaptic time constant, and each Poisson spike train connects to the neuron with a strength of w_i and a firing rate of λ_i. Two populations of Poisson spike sources, for excitatory and inhibitory synapses respectively, were connected to a single LIF neuron to mimic the noisy currents. The firing rates of the Poisson spike generators were determined by the given μ and σ. Figure 2 illustrates the recorded firing rates responding to the spike trains compared to the result driven by noisy currents. The use of noisy currents assumes that the post-synaptic potential (PSP) is a delta function, e.g. τ_{syn} tends to the limits of 0. However, in practice the release of neurotransmitter takes time and the noise added to the mean current is not pure white noise. Thus the experiments show that a longer τ_{syn} increases the level of noise and widens the variance of the output firing rate.

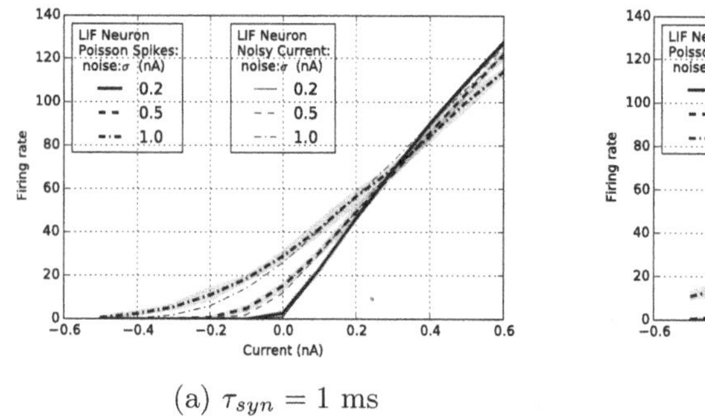

(a) $\tau_{syn} = 1$ ms

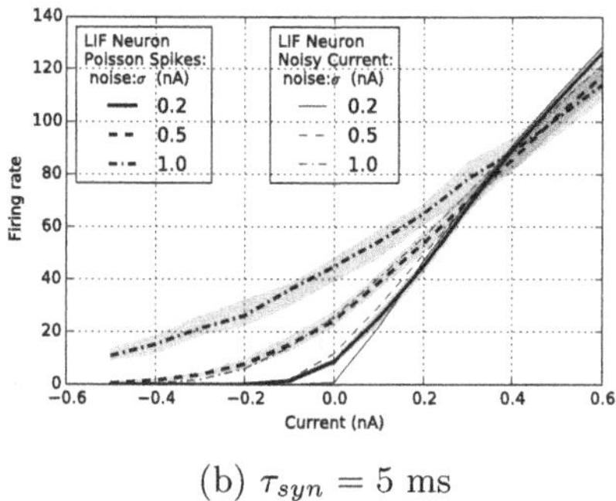

(b) $\tau_{syn} = 5$ ms

Fig. 2. Recorded response firing rate of two LIF neurons with different synaptic constants. The driving noisy current is simulated with Poission spike trains and the results are compared to the noisy current source.

2.3 Noisy Softplus

Inspired by the set of response functions triggered by different levels of noise, we propose the Noisy Softplus activation function:

$$y = f_{ns}(x, \sigma) = k\sigma \log[1 + \exp(\frac{x}{k\sigma})], \tag{5}$$

where x refers to the mean current, y is the normalised output firing rate, σ plays an important role to define the noise level, and k, which is determined by

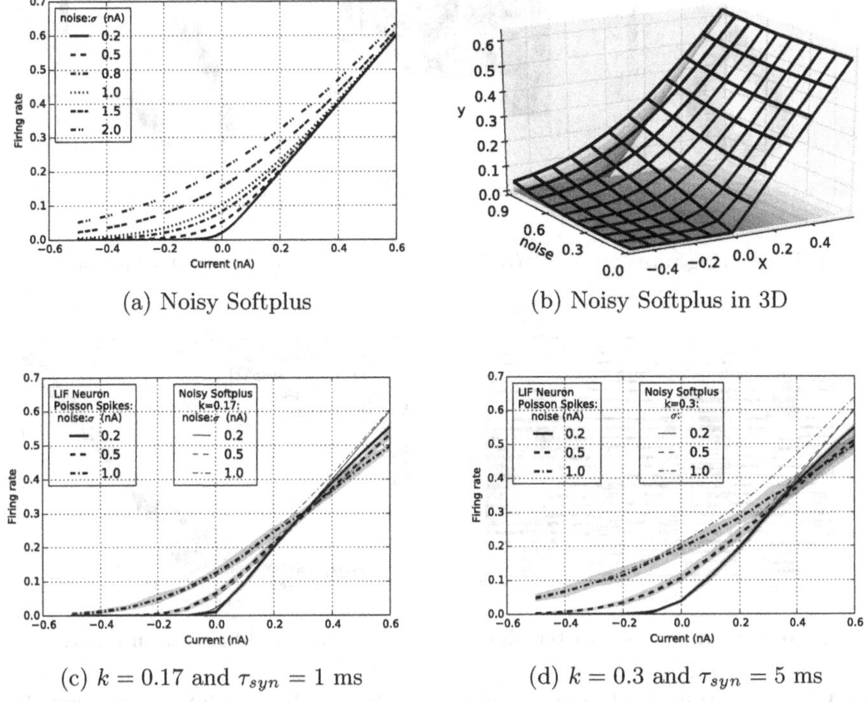

(a) Noisy Softplus

(b) Noisy Softplus in 3D

(c) $k = 0.17$ and $\tau_{syn} = 1$ ms

(d) $k = 0.3$ and $\tau_{syn} = 5$ ms

Fig. 3. Noisy Softplus fits to the response function of the LIF neuron. Noisy Softplus in (a) curve sets and (b) 3D. (c) and (d) show how Noisy Softplus fits to the response firing rates of LIF neurons with different synaptic constants.

the neuron parameters, controls the curve scaling. Note that the novel activation function we propose contains two parameters, the current and its noise; both are naturally obtained in spiking neurons. Figure 3(a) and (b) show the activation function in curve sets and in a 3D plot. The Noisy Softplus fits well to the recorded response firing rate of the LIF neuron with suitable calibration of k, see Fig. 3(c) and (d). The derivative is the logistic function scaled by $k\sigma$:

$$\frac{\partial f_{ns}(x, \sigma)}{\partial x} = \frac{1}{1 + exp(-\frac{x}{k\sigma})}. \tag{6}$$

3 Results

A ConvNet model was trained on MNIST, a popular database in neuromorphic vision, using Noisy Softplus neurons. The architecture contains 28×28 input units, followed by two convolutional layers c5-2s-12c5-2s, and the 10 output neurons represents the classified digit. All the convolution and average sampling neurons use Noisy Softplus units with no bias, while the output neurons are

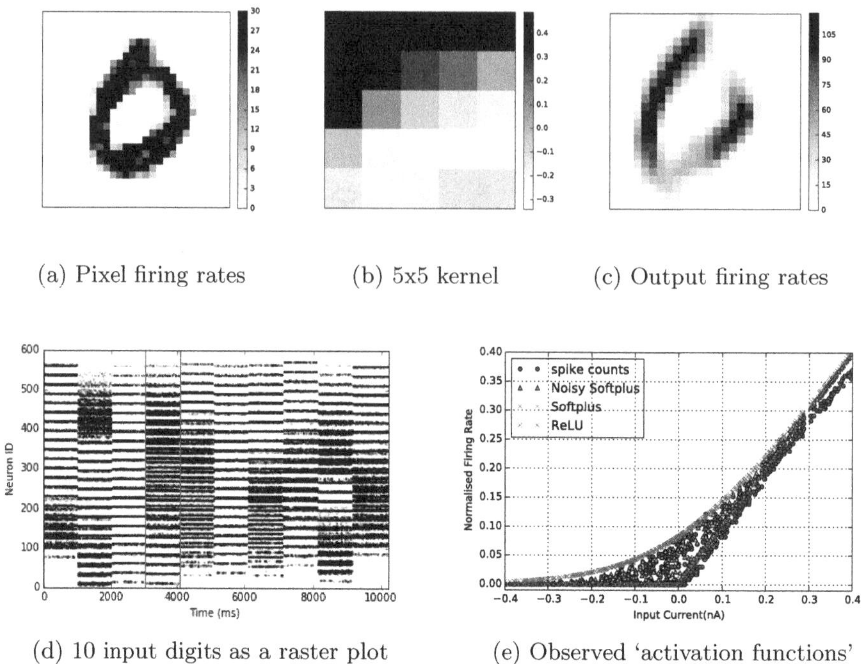

(a) Pixel firing rates (b) 5x5 kernel (c) Output firing rates

(d) 10 input digits as a raster plot (e) Observed 'activation functions'

Fig. 4. Noisy Softplus fits to the neural response firing rate in an SNN simulation. The 28×28 Poisson spike trains in (a) firing rate, and (d) raster plots, are convolved with a 5×5 kernel (b). (c) the convolved map with the firing rates of each neuron. (e) the normalised firing rate compared with Noisy Softplus, Softplus and ReLU activation functions.

softmax units converting a vector of values into the range (0, 1) that add up to 1. The weights were updated using a fixed learning rate, 50 images per batch and 10 epochs. Before testing on spiking LIF neurons, the weights of each layer were scaled to ensure that input synaptic currents stay within a valid range.

To validate how well the Noisy Softplus activation fits to the response firing rate of LIF neurons in a real application, we simulated the model on Nest using the Poisson MNIST dataset [11] and the neurons of a convolutional map were observed. Figure 4 shows the convolution of a 5×5 kernel with an input digit '0' represented by spike trains. The estimated spike counts using Noisy Softplus fit to the real recorded firing rate much more accurately than the Softplus and ReLU activation functions. In Fig. 4(e), we manually selected a suitable scaling factor for Softplus which located on the top slope of the response activity. However, the scale factor remains static for all the neurons, thus resulting in a mismatch at different level of noise. Noisy Softplus adapts to noise automatically.

We compared the training using ReLU, Softplus, and Noisy Softplus by their loss during training averaged over 6 trials, see Fig. 5. The trained networks were scaled to SNNs and compared on recognition rates, 93.34 %, 96.43 % and 97.03 %

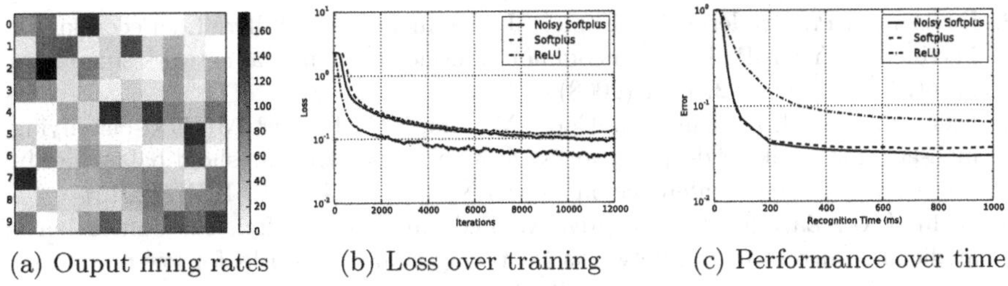

(a) Ouput firing rates (b) Loss over training (c) Performance over time

Fig. 5. Classification performance is calculated by the firing rate of output neurons (a). (b) shows how the loss varies over training. (c) illustrates the accuracy over short response times.

with a conversion loss of 4.76 %, 0.91 % and 0.74 %. As it is a major concern in neuromorphic vision, the recognition performance over short response times is also estimated in Fig. 5(c).

4 Discussion

The biologically-inspired activation function, Noisy Softplus, adapts to the noise level of input currents automatically, and is the first attempt to map activation units accurately to the firing response of LIF neurons. Noisy Softplus not only brings more biological features to the activation function, but also proves capable of performing well in a spiking ConvNet recognition task. The spiking version of Noisy Softplus wins on accuracy over the sigmoid neuron, compared to the result [15] of using Siegert units. As a result of its more accurate mapping, Noisy Softplus outperforms Softplus.

Future work on SNNs will include constraints during training to limit the function within the active range, which is equivalent to constraining the maximum firing rate of an LIF neuron. As a result there should be no need for the scaling process after training. For more accurate mapping, the scale factor k should be (numerically) derived to avoid calibration. In ANNs, it could be useful to study noise as extra information to be gathered by Softplus activation to enhance classification.

Acknowledgments. The research leading to these results has received funding from the European Research Council (FP/2007-2013)/ERC Grant Agreement no. 320689 and from the EU Flagship Human Brain Project (FP7-604102).

References

1. Buesing, L., Bill, J., Nessler, B., Maass, W.: Neural dynamics as sampling: a model for stochastic computation in recurrent networks of spiking neurons. PLoS Comput. Biol. **7**(11), e1002211 (2011)
2. Cao, Y., Chen, Y., Khosla, D.: Spiking deep convolutional neural networks for energy-efficient object recognition. Int. J. Comput. Vis. **113**(1), 54–66 (2015)

3. Davison, A.P., Brüderle, D., Eppler, J., Kremkow, J., Muller, E., Pecevski, D., Perrinet, L., Yger, P.: PyNN: a common interface for neuronal network simulators. Front. Neuroinform. **2**, 1–11 (2008)
4. Diehl, P.U., Neil, D., Binas, J., Cook, M., Liu, S.C., Pfeiffer, M.: Fast-classifying, high-accuracy spiking deep networks through weight and threshold balancing. In: International Joint Conference on Neural Networks (IJCNN). IEEE (2015)
5. Diehl, P.U., Zarrella, G., Cassidy, A., Pedroni, B.U., Neftci, E.: Conversion of artificial recurrent neural networks to spiking neural networks for low-power neuromorphic hardware. arXiv preprint (2016)
6. Furber, S.B., Galluppi, F., Temple, S., Plana, L., et al.: The SpiNNaker project. Proc. IEEE **102**(5), 652–665 (2014)
7. Glorot, X., Bordes, A., Bengio, Y.: Deep sparse rectifier neural networks. In: International Conference on Artificial Intelligence and Statistics, pp. 315–323 (2011)
8. He, K., Zhang, X., Ren, S., Sun, J.: Delving deep into rectifiers: surpassing human-level performance on ImageNet classification. In: Proceedings of the IEEE International Conference on Computer Vision, pp. 1026–1034 (2015)
9. Hunsberger, E., Eliasmith, C.: Spiking deep networks with LIF neurons. arXiv preprint (2015)
10. Jug, F., Lengler, J., Krautz, C., Steger, A.: Spiking networks and their rate-based equivalents: does it make sense to use Siegert neurons? Swiss Soc. Neurosci. (2012). https://www1.ethz.ch/cadmo/as/people/members/fjug/personal_home/preprints/2012_SiegertAbstract.pdf
11. Liu, Q., Garibaldia, P.-G., Stromatias, E., Serrano-Gotarredona, T., Furber, S.: Benchmarking spike-based visual recognition: a dataset and evaluation. Front. Neurosci. (2016, under review)
12. Merolla, P.A., Arthur, J.V., Alvarez-Icaza, R., Cassidy, A.S., Sawada, J., Akopyan, F., Jackson, B.L., Imam, N., Guo, C., Nakamura, Y., et al.: A million spiking-neuron integrated circuit with a scalable communication network and interface. Science **345**(6197), 668–673 (2014)
13. Neftci, E., Das, S., Pedroni, B., Kreutz-Delgado, K., Cauwenberghs, G.: Event-driven contrastive divergence for spiking neuromorphic systems. Front. Neurosci. **7**, 272 (2013)
14. O'Connor, P., Welling, M.: Deep spiking networks. arXiv preprint (2016)
15. Stromatias, E., Neil, D., Galluppi, F., Pfeiffer, M., Liu, S.C., Furber, S.: Scalable energy-efficient, low-latency implementations of trained spiking deep belief networks on SpiNNaker. In: International Joint Conference on Neural Networks (IJCNN). IEEE (2015)

Compressing Word Embeddings

Martin Andrews(✉)

Red Cat Labs, Singapore, Singapore
Martin.Andrews@RedCatLabs.com
http://www.RedCatLabs.com

Abstract. Recent methods for learning vector space representations of words have succeeded in capturing fine-grained semantic and syntactic regularities using large-scale unlabelled text analysis. However, these representations typically consist of dense vectors that require a great deal of storage and cause the internal structure of the vector space to be opaque. A more 'idealized' representation of a vocabulary would be both compact and readily interpretable. With this goal, this paper first shows that Lloyd's algorithm can compress the standard dense vector representation by a factor of 10 without much loss in performance. Then, using that compressed size as a 'storage budget', we describe a new GPU-friendly factorization procedure to obtain a representation which gains interpretability as a side-effect of being sparse and non-negative in each encoding dimension. Word similarity and word-analogy tests are used to demonstrate the effectiveness of the compressed representations obtained.

1 Introduction

Distributed representations of words have been shown to benefit NLP tasks like parsing, named entity recognition, and sentiment analysis, as well as being used as the raw material for other deep learning tasks.

Surprisingly, these word vector embeddings can be derived directly from raw, unannotated corpora. Once created, the vector embedding E can be expressed simply as a list of vocabulary words (of size V), and a matrix of size $V \times d$, where d is the dimensionality of the embedding space.

As argued in [1], while this dense matrix representation may be handled with ease by computers, there are cognitive arguments against such a representation being the basis of language (see [2] for broader discussion on this point). For instance, it seems unlikely that the same small set of features are sufficient and necessary to describe all semantic domains of a full adult vocabulary. It would also be uneconomical for people to store all negative properties of a concept, such as the fact that dogs do not have wheels, or that airplanes are not used for communication. Indeed, in feature-norming exercises (for example [3]) where participants are asked to list the properties of a word, the aggregate descriptions are typically limited to approximately 10–20 characteristics for a given concrete concept.

So, for cognitive plausibility, we claim that a feature set should have three characteristics: it should only store positive facts; it should have a wide range of feature types, to cover all semantic domains in the typical mental lexicon; and only a small number of these should be active to describe each word/concept.

A. Hirose et al. (Eds.): ICONIP 2016, Part IV, LNCS 9950, pp. 413–422, 2016.
DOI: 10.1007/978-3-319-46681-1_50

2 Models

In this work, a 300-dimensional GloVe [4] embedding was used as a concrete baseline for a number of different lossy compression methods[1].

2.1 Simple Compression

A number of different compression methods were explored to establish a 'storage budget' for the subsequent experiments in sparse encodings. These intial methods focussed on discarding data, while resulting in an approximately equivalent level of performance in the 'Google' word analogy task (described later).

The methods that were explored included (i) directly discarding a fixed proportion of the vector dimensions; (ii) thresholding the data; (iii) quantising the data based on curves of the form $\pm|v_i|^\alpha$ for each dimension i of a given vector v; and (iv) adaptive level encoding. This last approach, efficiently implemented via Lloyd's algorithm [5], was shown to be capable of approximating each element v_i of the entire vocabulary with only 8 discrete levels (each dimension i being independently calibrated) while still performing acceptably on the analogy task.

Thus, for the 300-dimensional embedding, a 900-bit 'storage budget' was established, and used as the storage bound for the sparsification experiments.

2.2 Lloyd's Algorithm

For a given number of quantisation levels n and an element $e \sim E$, we want to create a set of quantisation levels e_q that minimises:

$$\sum_{e \in E} \min_q (e - e_q)^2$$

This process can be performed iteratively, starting with the e_q placed uniformly within the values of E sorted numerically.

The algorithm then iteratively updates the e_q such that the centroid of each neighbouring cluster is found, and the e_q are then updated to these positions. The algorithm stops when no more changes are required.

This optimisation is done for each element/column of the embedding independently, and a list of the respective levels of e_q is stored, in addition to the index within each list for each element of the representation.

2.3 Sparse Compression

Non-negative Sparse Embedding ("NNSE"). A very promising approach to sparse embedding was taken in [1], where the following optimisation was solved iteratively to generate a sparse embedding A (which is V vectors each

[1] The techniques described in this paper can be applied to any embedding, since nothing specific to GloVe has been used.

with dimensionality k), and also creates a 'dictionary' D (which maps the sparse elements of A onto the real-valued embedding E):

$$argmin_{A \in \mathbb{R}^{m \times k}, D \in \mathbb{R}^{k \times n}} \sum_{i=1}^{V} ||E_{i,:} - A_{i,:} \times D||_2 + \lambda ||A_{i,:}||_1$$
$$\text{where} : D_{i,:}, D_{i,:}^T \leq 1, 1 \leq \forall i \leq k$$
$$\text{and } A_{i,j} \geq 0, 1 \leq \forall i \leq V, 1 \leq \forall j \leq k$$

Although this produced excellent embedding results, one requirement is to vary the parameters k and λ such that the sparse embedding A had a 'reasonable' number of positive entries. Since one of the targets of this paper is compression, targetting a concrete compression level, this hyperparameter search is not acceptable.

Winner-Take-All Autoencoders. Another approach to sparse encoding, applied chiefly to visual tasks, was described in [6]. There, "Winner-Take-All Autoencoders" obtained strict sparsity targets by interposing a drop-out layer that only allows the top-α fraction of the layer's inputs to be passed through, while the rest are set to zero. An important implementation detail is that this drop-out is done on a *per dimension* basis over each minibatch of training examples. This implicitly causes the training to balance the occurrences of non-zero entries across the different dimensions of the embedding space, while allowing the $A_{i,:}$ vectors to have varying numbers of non-zero entries.

Unfortunately, obtaining the top-α fraction of a set of numbers requires sorting the list, which means that including this layer in an neural network will move data-intensive computations off the GPU. In order to produce a sparse embedding in reasonable time, however, the GPU is required, so an algorithmic short-cut was implemented that can be efficiently executed inside the GPU.

GPU-Friendly Top-α. Clearly, determining the top-α percentile of a distribution is equivalent to determining the hurdle value h^* above which the fraction α of the realised values in the minibatch lie[2].

For a given embedding dimension j, and assuming that $\alpha \ll 50\%$ and that the distribution is not pathelogical, this hurdle can be bounded above by $h_j^+ = max(A_{:,j})$ and below by $h_j^- = mean(A_{:,j})$. A binary section search can then be made for h^*, since the fractional percentile for any given h can be determined by taking the mean of a simple matrix $A_{:,j} > h$ indicator function[3].

Although a thorough implementation of this binary section process would include stopping criteria, that would not make sense within the GPU's cores' execution flow. So a fixed number of bisection iterations was used (specifically 5 for our 16,384 minibatch size). This was found to give robust results, with a value of h being found that identified (approximately) the top-α percentile consistently. The net result of avoiding CPU operations was a 39× speed-up.

[2] A minibatch has 16,384 examples – large enough for distribution approximations.

[3] Also, $\alpha_0^+ = (1/\text{batchsize})$ initially, since it is the maximum value in $A_{:,j}$.

Experiments were also conducted to try and optimise the placement of each h estimate, using distributional properties of $N()$, or linear interpolation. These were not successful, however, since it appears that the neural network learns to exploit the distribution assumptions being made, so as to win artificially low h (and thus a higher accepted α than required).

Sparse Autoencoding for Word Embeddings. To avoid learning *from scratch* the entire sparse matrix A, an autoencoding scheme was set up, so that the target embedding E was mapped onto intermediate layers, then quantised, then mapped back to itself (via a trainable dictionary D).

The layers of the implemented autoencoder are described in Table 1, where concrete sizes are given, assuming a vocabulary size of $2^{17} = 131,072$ words, 300-dimensional underlying embedding, and a sparse embedding of size $k = 1024$.

Table 1. Sparse Auto-encoding model structure

Parameter set	Intra-layer operations	Output shape
Word embedding (all tokens)	$\mathbf{E}$	$(2^{17}, 300) \in \mathbb{R}$
Hidden layer	$\max(\mathbf{W}^H\mathbf{x} + \mathbf{b}^H, 0)$	$(2^{17}, 300 \times 8) \in \mathbb{R}^+$
Pre-binary linear layer	$\mathbf{W}^L\mathbf{x} + \mathbf{b}^L$	$(2^{17}, 1024) \in \mathbb{R}$
Batch normalization layer	batchnormalize($\mathbf{x}$)	$(2^{17}, 1024) \in \mathbb{R}$
Rectification	$\max(\mathbf{x}, 0)$	$(2^{17}, 1024) \in \mathbb{R}^+$
Gaussian noise	$\mathbf{x} + N(0, \beta_t)$	$(2^{17}, 1024) \in \mathbb{R}$
Top-α sparsification	drop-all-but-$\alpha(\mathbf{x}, \alpha_t)$	$(2^{17}, 1024) \in \mathbb{R}^+$
Sparse encoded version	$\mathbf{A} = \mathbf{x}$	$(2^{17}, 1024) \in \mathbb{R}$
Output (=$\mathbf{E}^*$)	$\mathbf{W}^F\mathbf{x} + \mathbf{b}^F$	$(2^{17}, 300) \in \mathbb{R}$

The quantity being optimised via gradient descent was the l_2 error between the output embedding E^*, and the input E, and this scheme was accelerated using ADAM [7].

Note that no regularisation terms were used. There is regularisation being implicitly applied due to the top-α sparsity constraint, coupled with the batch normalization that occurs in the layer before the sparse encoding. These combine to constrain the sparse encoding A within reasonable bounds.

In order to improve convergence, a scheme was used whereby a variable σ (initially zero) was incremented (by 0.01) after each epoch *only if* the l_2 error over the epoch averaged less than 0.01. The variables α_t and β_t were then dynamically adjusted as follows:

$$\alpha_t = 50\,\% \times (e^{10\sigma}) + \alpha \times (1 - e^{10\sigma})$$
$$\beta_t = 0.2 \times e^{0.01\sigma}$$

Having this path-dependent speed regulator enabled learning pressure to be applied incrementally through training in a way that was sensitive to current training progress. In particular, the sparsity factor α_0 was started out at 'easy' values, so that the initial weights could migrate to areas where they were at least tackling the underlying problem - and α_t then moved asymptotically closer to α over the course of a thousand epochs. Similarly, the Guassian noise layer seemed to improve search in later periods, and this was decayed more slowly.

After training, the noise-layer was switched off, and the network-learned A and E^* were used for the performance analyses below.

Representation of Sparse Encodings. Suppose a non-negative sparse encoding over a k-dimensional space is required, and a specific 'bit budget' of n bits is given. Since the majority of encoding values will be zero, the representation need only store the addresses of the non-zero elements, and their values.

Moreover, because the ordering of the list of values is an additional degree of freedom, one can increase the information content by storing (a) the length of the list, (b) the locations of each of the values in declining numerical order, (c) the highest value, and (d) the percentage ratio between each sucessive pair (typically in the range [70 %...100 %]).

Supposing that 3-bits is sufficient for each ratio (d)[4], then the required sparsity ratio α can be determined:

$$\alpha = \frac{n}{k(\log_2(k) + 3)}$$

In the experiments that follow, the α values are chosen to be within this bound.

In addition to the $n \times V$ bits of storage required, there is the additional overhead of storing the dictionary D - which consists of $k \times d$ elements, which can be a significant factor if k is large.

3 Experiments

3.1 Corpus

The text corpus used here was the concatenation of (a) the 'One Billion Word Benchmark' [8], and (b) a cleaned version of English Wikipedia (August 2013 dump, using only pages with greater than 20 pageviews), which was preprocessed by removing non-textual elements, sentence splitting, and tokenization. After preprocessing the corpus contained 1.2 billion tokens (680 million and 524 million from each respective source)[5].

---

[4] While (c) might be stored with higher fidelity, the remaining ratios are less exacting.

[5] Importantly, these resources have been made freely available without restrictive licenses, and in the same spirit, the code for this paper is being released under a permissive license.

Models were derived using windows of 5 tokens to each side of the focus word. Rather than focus on a word occurrence limit (as is common), a vocabulary-size of $2^{17} = 131,072$ words was chosen, since (a) it made batching more convenient, and (b) the limit is arbitrary either way.

3.2 Test Datasets

We evaluated each word representation on seven datasets covering similarity and analogy tasks, using the test framework of Levy et al. [9], which has code and data available at: https://bitbucket.org/omerlevy/hyperwords.

Word Similarity. Five datasets were used to evaluate word similarity: the popular WordSim353 [10] partitioned into two datasets, WordSim Similarity and WordSim Relatedness [11,12]; Bruni et al.'s MEN dataset [13]; Radinsky et al.'s Mechanical Turk dataset [14]; and Luong et al.'s Rare Words dataset [15]. All these datasets contain word pairs together with human-assigned similarity scores. The word vectors are evaluated by ranking the pairs according to their cosine similarities, and measuring the correlation (Spearman's ρ) with the human ratings.

Word Analogy. The two analogy datasets present questions of the form "a is to a^* as b is to b^*", where b^* is hidden, and must be guessed from the entire vocabulary. MSR's analogy dataset [16] contains 8000 morpho-syntactic analogy questions, such as "good is to best as smart is to smartest". Google's analogy dataset [17] contains 19544 questions, about half of the same kind as in MSR (syntactic analogies), and another half of a more semantic nature, such as capital cities ("Paris is to France as Tokyo is to Japan"). After filtering questions involving out-of-vocabulary words, i.e. words that did not appear in the pruned Corpus, we remain with 7118 instances in MSR and 19296 instances in Google.

As in [18], the analogy questions are answered using both 3CosAdd as well as 3CosMul.

3.3 Results

Embeddings with representations compressed by the Lloyd algorithm and the non-negative sparse encoding methods outlined above were run, with results shown in Tables 2 and 3.

In addition to the reconstructed embeddings E^* for each method, the *raw results* of the similarity and analogy tests were run on the intermediate sparse embeddings A themselves.

As a point of comparison, a random vector approach (binarised locality sensitive hashing: "LSH" [19]) was also tested, with the same 900-bit constraint.

Table 2. Similarity results

Method	WordSim Similarity	WordSim Relatedness	Bruni et al.	Radinsky et al.	Luong et al.
			MEN	M. Turk	Rare words
GloVe baseline	66.2 %	52.1 %	69.1 %	63.2 %	22.8 %
Lloyd-8	65.9 %	51.5 %	68.6 %	62.3 %	22.7 %
$E^*(k = 4096, \alpha = 1.50\,\%)$	65.0 %	50.2 %	68.5 %	62.7 %	22.2 %
$E^*(k = 1024, \alpha = 6.75\,\%)$	64.7 %	51.0 %	67.7 %	62.7 %	21.9 %
$A(k = 4096, \alpha = 1.50\,\%)$	63.7 %	45.2 %	62.4 %	49.5 %	18.0 %
$A(k = 1024, \alpha = 6.75\,\%)$	69.2 %	51.3 %	69.5 %	62.1 %	17.4 %
LSH-900	62.0 %	45.8 %	65.9 %	59.2 %	22.0 %

Table 3. Analogy results

Method	Google	MSR
	Add/Mul	Add/Mul
GloVe baseline	67.1 %/68.5 %	53.4 %/56.6 %
Lloyd-8	65.9 %/67.4 %	51.9 %/54.5 %
$E^*(k = 4096, \alpha = 1.50\,\%)$	62.5 %/66.4 %	51.8 %/54.1 %
$E^*(k = 1024, \alpha = 6.75\,\%)$	62.4 %/62.9 %	49.0 %/50.1 %
$A(k = 4096, \alpha = 1.50\,\%)$	37.6 %/40.8 %	27.3 %/29.9 %
$A(k = 1024, \alpha = 6.75\,\%)$	52.5 %/55.1 %	40.5 %/43.8 %
LSH-900	53.0 %/53.1 %	41.2 %/42.2 %

4 Discussion

4.1 Level Quantisation vs. More Sophisticated Methods

The level-quantisation approaches to compression work extremely well, and are relatively simple to implement. Assuming values would otherwise be stored as 32-bit floats, the Lloyd method achieves very similar scores, with only 3-bits per value (the overhead of storing the quantisation levels is only the equivalent of storing 8 of the original word-vectors). However, they do not accomplish the goal of learning about the underlying embedding through an efficient compression algorithm.

4.2 Performance of Sparse Embeddings

As can be seen from Table 2, the reconstruction (E^*) results for both sparse embedding methods are only marginally below those of the original embeddings, and generally better than the LSH re-representation. This is satisfying, because it shows that the GPU-friendly method outlined here actually reconstructs the embedding without a significant loss in performance, within the same 'bit budget' as the quantisation methods.

The fact that the sparse encodings A can also perform as embeddings on their own (without involving the learned dictionary D) is encouraging, since it implies that there is more information about the underlying language that can be obtained from existing word embeddings 'for free'.

Interestingly, their performance on similarity tasks is far higher than on the analogy ones (particularly in the case of $k = 4096$ which has an α of only 1.50 %). This can be understood by considering the algebra of the sparse non-negative vectors being used. For similarity purposes, non-negative vectors can be scored using the same a cosine measure that works for more general dense, real-valued vectors. However, for the analogy tasks, there are implicit subtractions being done, which result in direction vectors that are not part of the same algebra. This deserves further work, to see whether other operators would be able to make use of the sparse vector spaces' geometry more fully (potentially including ideas from [20]).

4.3 Interpretability of the Sparse "A"

Table 4 (which lists the highest weighted words in each dimension that 'motorbike' is also most highly weighted) clearly demonstrates that the A sparse representation has learned something about the structure of the English language 'for free', using only data obtained from an embedding trained on unlabelled data itself.

Table 4. Top 'motorbike' dimensions

Model	Top words in each of first 7 dimensions
GloVe baseline	· lb., four-bladed, propeller, propellers, two-bladed, ...
	· passerine, 1975-79, rennae, fyrstenberg, edw, coots, ...
	· bancboston, oshiomhole, 30-sept, holmer, smithee, recon, ...
	· http://www.nytimes.com, (888), receival, jamiat, shyi, ...
	· subjunctive, purley, 11-july, broaddus, muharram, ebit, ...
	· proximus, pattani, 31-feb, wgc, 30-nov, crossgen, 2,631, ...
	· officership, tvcolumn, integrable, salticidae, o-157, ...
$A(^{k=1024}_{\alpha=6.75\,\%})$	· vehicles, vehicle, cars, scrappage, car, 4x4, armored, ...
	· prix, races, race, laps, vettel, rikknen, sprint, ...
	· ski, coal, gas, taxicab, nuclear, wine, cellphone, ...
	· kool, electrons, pulpit, efta, gallen, gasol, birdman, ...
	· eric, anglo, tornadoes, rt, asteroids, dera, rim, ...
	· wear, trousers, dresses, jeans, wearing, worn, pants, ...
	· stabbed, kercher, 16-year-old, 15-year-old, 18-year-old, ...

5 Conclusion

The joint goals of good compression rates and cognitive plausibility are realistic and achievable.

Using the GPU-friendly sparsity Winner-Take-All Autoencoder scheme described, sparse, non-negative encodings have been demonstrated that combine high compression with interpretability 'for free'.

Further work will include the investigation of operators that respect the geometry of these sparse vectors, so that analogy tests might perform in-line with the word similarity scores (which are purely direction-based).

Acknowledgments. The author thanks DC Frontiers, the creators of the data-centric service 'Handshakes' (http://www.handshakes.com.sg/), for their willingness to support this on-going research. DC Frontiers is the recipient of a Technology Enterprise Commercialisation Scheme grant from SPRING Singapore, under which this work took place.

References

1. Murphy, B., Talukdar, P.P., Mitchell, T.: Learning effective and interpretable semantic models using non-negative sparse embedding. In: International Conference on Computational Linguistics (COLING 2012), Mumbai, India (2012). http://aclweb.org/anthology/C/C12/C12-1118.pdf
2. Griffiths, T.L., Steyvers, M., Tenenbaum, J.B.: Topics in semantic representation. Psychol. Rev. **114**(2), 211 (2007)
3. Vinson, D.P., Vigliocco, G.: Semantic feature production norms for a large set of objects and events. Behav. Res. Methods **40**(1), 183–190 (2008)
4. Pennington, J., Socher, R., Manning, C.D.: Glove: global vectors for word representation. Proceedings of the Empiricial Methods in Natural Language Processing (EMNLP 2014) **12**, 1532–1543 (2014)
5. Lloyd, S.: Least squares quantization in PCM. IEEE Trans. Inf. Theory **28**(2), 129–137 (2006)
6. Makhzani, A., Frey, B.J.: A winner-take-all method for training sparse convolutional autoencoders (2014). CoRR, abs/1409.2752
7. Kingma, D., Ba, J.: Adam: a method for stochastic optimization (2014). arXiv preprint: arXiv:1412.6980
8. Chelba, C., Mikolov, T., Schuster, M., Ge, Q., Brants, T., Koehn, P.: One billion word benchmark for measuring progress in statistical language modeling (2013). CoRR, abs/1312.3005
9. Levy, O., Goldberg, Y., Dagan, I.: Improving distributional similarity with lessons learned from word embeddings. Trans. Assoc. Comput. Linguist. **3**, 211–225 (2015)
10. Finkelstein, L., Gabrilovich, E., Matias, Y., Rivlin, E., Solan, Z., Wolfman, G., Ruppin, E.: Placing search in context: the concept revisited. ACM Trans. Inf. Syst. **20**(1), 116–131 (2002)
11. Zesch, T., Müller, C., Gurevych, I.: Using wiktionary for computing semantic relatedness. In: Proceedings of the 23rd National Conference on Artificial Intelligence, AAAI 2008, vol. 2, pp. 861–866. AAAI Press (2008)

12. Agirre, E., Alfonseca, E., Hall, K., Kravalova, J., Paşca, M., Soroa, A.: A study on similarity and relatedness using distributional andwordnet-based approaches. In: Proceedings of Human Language Technologies: the 2009 Annual Conference of the North American Chapter of the Association for Computational Linguistics, NAACL 2009, Stroudsburg, PA, USA, pp. 19–27. Association for Computational Linguistics (2009)
13. Bruni, E., Boleda, G., Baroni, M., Tran, N.-K.: Distributional semantics in technicolor. In: Proceedings of the 50th Annual Meeting of the Association for Computational Linguistics: Long Papers, ACL 2012, Stroudsburg, PA, USA, vol. 1, pp. 136–145. Association for Computational Linguistics (2012)
14. Radinsky, K., Agichtein, E., Gabrilovich, E., Markovitch, S.: A word at a time: computing word relatedness using temporal semantic analysis. In: Proceedings of the 20th International Conference on World Wide Web, WWW 2011, New York, NY, USA, pp. 337–346. ACM (2011)
15. Luong, M.-T., Socher, R., Manning, C.D.: Better word representations with recursive neural networks for morphology. In: CoNLL, Sofia, Bulgaria (2013)
16. Mikolov, T., tau Yih, W., Zweig, G.: Linguistic regularities in continuous space word representations. In: Proceedings of the 2013 Conference of the North American Chapter of the Association for Computational Linguistics: Human Language Technologies (NAACL-HLT-2013). Association for Computational Linguistics, May 2013
17. Mikolov, T., Chen, K., Corrado, G., Dean, J.: Efficient estimation of word representations in vector space (2013). arXiv preprint: arXiv:1301.3781
18. Levy, O., Goldberg, Y., Ramat-Gan, I.: Linguistic regularities in sparse and explicit word representations. In: CoNLL, pp. 171–180 (2014)
19. Charikar, M.S.: Similarity estimation techniques from rounding algorithms. In: Proceedings of the Thiry-fourth Annual ACM Symposium on Theory of Computing, STOC 2002, New York, NY, USA, pp. 380–388. ACM (2002)
20. Mahadevan, S., Chandar, S.: Reasoning about linguistic regularities in word embeddings using matrix manifolds (2015). CoRR, abs/1507.07636

An Iterative Incremental Learning Algorithm for Complex-Valued Hopfield Associative Memory

Naoki Masuyama(✉) and Chu Kiong Loo

Faculty of Computer Science and Information Technology, University of Malaya, 50603 Kuala Lumpur, Malaysia
naoki.masuyama17@siswa.um.edu.my, ckloo.um@um.edu.my

Abstract. This paper discusses a complex-valued Hopfield associative memory with an iterative incremental learning algorithm. The mathematical proofs derive that the weight matrix is approximated as a weight matrix by the complex-valued pseudo inverse algorithm. Furthermore, the minimum number of iterations for the learning sequence is defined with maintaining the network stability. From the result of simulation experiment in terms of memory capacity and noise tolerance, the proposed model has the superior ability than the model with a complex-valued pseudo inverse learning algorithm.

Keywords: Associative memory · Complex-valued model · Incremental learning

1 Introduction

In the past decades, numerous types of "Artificial Intelligence" models have been proposed in the field of computer science in order to realize the human-like abilities based on the analysis and modeling of essential functions of a biological neuron and its complicated networks in a computer. It has been noted that one of the interesting and challenging subjects is the imitation of memory function of the human brain. In the past decades, several types of artificial associative memory models and its improvements have introduced such as Hopfield Associative Memory (HAM) [7], and Bi-/Multi-directional Associative Memory (BAM, MAM) [6,11]. However, the majority of models are limited as offline and one-shot learning rules. Storkey and Valabregue [14] proposed an incremental learning algorithm for HAM based on the Hebb/Anti-Hebb learning. Chartier and Boukadoum [3] also proposed and analyzed the model with a self-convergent iterative learning rule based on the Hebb/anti-Hebb approach, and a nonlinear output function Diederich and Opper [4] proposed a Widrow-Hoff type learning. However, the convergence of these models is quite slow when the memory patterns are correlated. The local iterative learning, on the other hand, which is proposed by Blatt and Vergini [2] can be dealt with above kind of problems.

A. Hirose et al. (Eds.): ICONIP 2016, Part IV, LNCS 9950, pp. 423–431, 2016.
DOI: 10.1007/978-3-319-46681-1_51

Moreover, this model is able to define the minimum number of learning iterations with maintaining a network stability.

In regard as events in the real world, information representation using binary or bipolar state is insufficient. Jankowski et al. [9] have introduced complex-valued Hopfield Associative Memory (CHAM) with neurons processing a complex-valued discrete activation function. Conventionally, numerous studies that improve complex-valued models are introduced based on the improvements for real-valued models. Lee [13] applied a complex-valued projection matrix to CHAM (PInvCHAM) and analyzed the stability of the model by using energy function. The learning algorithm of these models, however, are characterized by a batch learning. Isokawa et al. [8] analyzed the stability of complex-valued Hopfield model with from a local iterative learning scheme viewpoint though the learning algorithm is Hebb learning base. In this paper, we introduce a local iterative incremental learning with CHAM based on Blatt and Vergini learning algorithm [2]. Here, we shall call the proposed model as BVCHAM. The noteworthy features of BVCHAM are described as follows; (i) the resultant of weight matrix in BVCHAM is approximated to the projection matrix operation. The network is able to store the patterns though the number of stored patterns p is larger than the number of neurons N (It is known that the memory capacity is limited as $p < N$ with a projection matrix learning model.), and (ii) the learning algorithm is guaranteed to calculate a weight matrix within a finite number of iterations with keeping a network stability.

The paper is divided as follows; Sect. 2 describes the dynamics of BVCHAM. Section 3 presents the network stability analysis for BVCHAM. In Sect. 4, it will be presented simulation experiments of BVCHAM in terms of the memory capacity and noise tolerance comparing with PinvCHAM. Concluding remarks are presented in Sect. 5.

2 Dynamics of a Local Iterative Learning for CHAM

This section presents the dynamics of a proposed model, we shall call as BVCHAM.

Let us suppose that the model stores complex-valued fundamental memory vectors X^p, where $X^p = \left[x_1^p, x_2^p, \ldots, x_N^p\right]^T$, N denotes the number of neurons, and p denotes the number of patterns. The components x_i^p are defined as follows;

$$x_i^p \in \exp\left[j2\pi n/q\right]_{n=0}^{q-1}, \quad i = 1, 2, \ldots, N \tag{1}$$

where q denotes a quantization value on the complex valued unit circle.

Here, supposing a new pattern X^l stores to the network by updating the weight matrix $W^{(l-1)}$ that already has $(l-1)$ patterns embedded. $X^{(l)}$ is presented n_l times to the network as follows;

$$W^{(l)} = W^{(l-1)} + \sum_{d=1}^{n_l} \Delta W_d^{(l)} \tag{2}$$

where, ΔW is defined as a following;

$$\Delta W_d^{(l)} = k^{d-1} \left(x_i^l - h_i\right) \left(x_j^l - h_j\right) / N. \tag{3}$$

The local field h_i which changes $(d-1)$ times is performed as a following;

$$h_i = \sum_{j=1}^{N} \left[W_{ij}^{(l-1)} + \sum_{r=1}^{d-1} \left(W_r^{(l)} \right)_{ij} \right] x_j^l. \tag{4}$$

Similar with Eq. (4), the local field h_j is defined. Here, the minimum number of iterations n_l can be defined as a following;

$$n_l \geq \log_k \left[N / \left(\pi/q - T\right)^2 \right] \tag{5}$$

where, a parameter k, called memory coefficient, is a real number that belongs to the interval $0 < k \leq 4$, and a parameter T is set between $0 \leq T < \pi/q$. It allows to perform in order to achieve aligned local fields with values of at least T [2]. q denotes a quantization value on the complex unit circle, which is described in a following part. The iterative weight update is performed until satisfying the criterion as $T >$ error ϵ, namely;

$$T > \epsilon = \sum_{i=1}^{N} \left\| \arg \left(\frac{h_i}{X_i^l} \right) \right\|. \tag{6}$$

In summary, the process of weight update is described as Algorithm 1. For the association process, the self-connections eliminated weight matrix $W^{'}$ is utilized. On the other hand, the weight matrix W which is calculated by Algorithm 1 is applied to the further incremental learning process.

Algorithm 1. An algorithm for a weight matrix $W^{(l)}$

Require: weight matrix $W^{(l-1)}$, fundamental memory vector X^l, error parameter T, memory coefficient k
Ensure: weight matrix $W^{(l)}$
  if $l = 1$ **then**
    Initialize W as a zero matrix
  end if
  Set $d = 1$
  Calculate an error ϵ using Eq. (6)
  while $d \geq n_l$ and $\epsilon > T$ **do**
    Calculate a weight matrix $W^{(l)}$ using Eqs. (2) and (3)
    Calculate an error ϵ using Eq. (6)
    Set $d \leftarrow d + 1$
  end while

The activation function is a complex projection function that operates on each component of the state vector as a following;

$$\phi(Z) = \begin{cases} \exp(j2\pi n/q), & \text{If } \left| \arg \left\{ \frac{Z}{\exp(j2\pi n/q)} \right\} \right| < \pi/q \text{ and } Z \neq 0 \\ \text{previous state}, & \text{If } Z = 0 \end{cases} \tag{7}$$

where, $\arg(\alpha)$ denotes the phase angle of α which is taken to range over $(-\pi, \pi)$. q denotes a quantization value on the complex unit circle, n takes an integer. We utilized a discrete complex unit circle model to determine recalled signals.

Thus, the dynamics of network is summarized as follows;

$$\begin{cases} h'_{(t)} = W' X_{(t)} & (8) \\ X_{(t+1)} = \phi\left(h'_{(t)}\right) & (9) \end{cases}$$

The stationary conditions for the recalled memory vector are described as a following;

$$0 \leq \arg\left(\frac{h'}{X^p}\right) < \frac{\pi}{q}. \tag{10}$$

3 Network Stability Analysis

In this section, based on Blatt and Vergini algorithm [2], the conditions of network stability in BVCHAM will be discussed, and it will be derived a criterion for the number of iterations n_l to guarantee the stability of X^l. Here, Dirac notation in $\mathbb{C}^N$ is utilized for simplifying the representation of proofs. Thus, Eqs. (3) and (4) can be described as follows;

$$\Delta W_d^l = \frac{k^{d-1}}{N} |X^l - h\rangle\langle X^l - h| \tag{11}$$

$$|h\rangle = \left[W^{(l-1)} + \sum_{r=1}^{d-1} \Delta W_r^l\right] |X^l\rangle \tag{12}$$

here, it satisfies $|X^l - h\rangle^\dagger = \langle X^l - h|$.

First of all, the changes of weight connection when a new memory vector is repeated to the network is analyzed. Here, $\Delta W_d^{(l)} (1 \leq d)$ is defined as a following;

$$\sum_{r=1}^{d} \Delta W_r^{(l)} = Q_d^{(l)} \frac{E^{(l-1)}|X^l\rangle\langle X^l|E^{(l-1)}}{N} \tag{13}$$

where,

$$E^{(l)} = I - W^{(l)} \tag{14}$$

here, I denotes an identity matrix which has same dimension with $W^{(l)}$.

It assumes that Eq. (13) holds for d-th iteration in order to prove $(d+1)$-th iteration. The difference of weight connection between the d-th and $(d+1)$-th iteration with a memory vector X^l and its local field h is expressed by Eqs. (12), (13) and (14), as follows;

$$|X^l - h\rangle = (1 - a^{(l)} Q_d^{(l)}) E^{(l-1)} |X^l\rangle \tag{15}$$

where,

$$a^{(l)} = \frac{\langle X^l | E^{(l-1)} | X^l \rangle}{N}. \tag{16}$$

The change in $(d+1)$-th iteration is expressed by Eqs. (11) and (15) as a following;

$$\Delta W_{d+1}^{(l)} = k^d \left(1 - a^{(l)} Q_d^{(l)}\right)^2 \frac{E^{(l-1)} |X^l\rangle\langle X^l| E^{(l-1)}}{N}. \tag{17}$$

Comparing with Eqs. (13) and (17), it can be regarded as a proportional relation with the same operator. Thus, the following recurrence relations are obtained;

$$Q_1^{(l)} = 1 \tag{18}$$

$$Q_d^{(l)} = Q_{d-1}^{(l)} + k^{d-1} \left(1 - a^{(l)} Q_{d-1}^{(l)}\right)^2. \tag{19}$$

From the Eqs. (13) and (14), Eq. (2) is represented as a following;

$$E^{(l)} = E^{(l-1)} - Q_{n_l}^{(l)} \frac{E^{(l-1)} |X^l\rangle\langle X^l| E^{(l-1)}}{N}. \tag{20}$$

Therefore, when a new memory vector is presented to the network, an operator which is proportional to the projector onto $E^{(l-1)}|X^l\rangle$ is added to the weight connection. According to Eq. (11), ΔW increases exponentially with the number of iterations d. However, if the following condition is satisfied, the elements of ΔW remain finite, i.e.;

$$0 \leq \langle \phi | E^{(l)} | \phi \rangle \leq \langle \phi | \phi \rangle \text{ for all } |\phi\rangle \in \mathbb{C}^N. \tag{21}$$

In the first learning step, it maintains $E^{(0)} = I$ and $Q_d^{(1)} = 1$. Furthermore, based on Cauchy–Schwarz inequality, Eq. (21) is verified for $l = 1$. Here, assuming that Eq. (21) is valid for l. Then, the eigenvalues of $E^{(l)}$ take between 0 to 1. Therefore, the following condition is satisfied;

$$\| E^{(l)} |\phi\rangle \|^2 \leq \langle \phi | E^{(l)} | \phi \rangle. \tag{22}$$

Let us suppose that $a^{(l+1)} = 0$, $E^{(l)}|X^{l+1}\rangle = 0$ is derived from Eqs. (16) and (22), then $E^{(l+1)} = E^{(l)}$ is maintained from Eq. (20). Thus, Eq. (21) is also satisfied for $(l+1)$. In contrast, in case of $a^{(l+1)} \neq 0$, it is proved by the inductive hypothesis $0 < a^{(l+1)} \leq 1$. Furthermore, according to [2], following conditions are maintained;

$$0 < a^{(l)} Q_d^{(l)} \leq 1. \tag{23}$$

$$1 - a^{(l)} Q_d^{(l)} \leq \left(\frac{k^{-d}}{a^{(l)}}\right). \tag{24}$$

From the following part, it focuses the memorized pattern $|X^\mu\rangle (1 \leq \mu < l)$ and its local field $|h\rangle$ are able to be closer than other memorized patterns. The distance between $|X^{\mu}\rangle$ and $|h\rangle$ decreases exponentially with the number of iterations. This is derived by the induction process. First of all, the following condition is maintained by the Eqs. (14) and (22);

$$\left\| |X^{\mu}\rangle - W^{(l)}|X^{\mu}\rangle \right\|^2 \leq \langle X^{\mu}|E^{(l)}|X^{\mu}\rangle. \tag{25}$$

Here, Eq. (20) is generalized as a following;

$$E^{(l)} = E^{(\mu)} - \sum_{\nu=\mu+1}^{l} Q_{n_\nu}^{\nu} \frac{E^{(\nu-1)}|X^{\nu}\rangle\langle X^{\nu}|E^{(\nu-1)}}{N}. \tag{26}$$

Then, $\langle X^{\mu}|$ multiplied on the left hand side, and $|X^{\mu}\rangle$ multiplied on the right hand side to Eq. (26), therefore;

$$\langle X^{\mu}|E^{(l)}|X^{\mu}\rangle \leq \langle X^{\mu}|E^{(\mu)}|X^{\mu}\rangle. \tag{27}$$

For the Eq. (20) with a μ-th memory vector, $\langle X^{\mu}|$ is multiplied on the left hand side, and $|X^{\mu}\rangle$ is multiplied on the right hand side, i.e.;

$$\langle X^{\mu}|E^{(\mu)}|X^{\mu}\rangle = \langle X^{\mu}|E^{(\mu-1)}|X^{\mu}\rangle - Q_{n_\mu}^{(\mu)} \frac{(\langle X^{\mu}|E^{(\mu-1)}|X^{\mu}\rangle)^2}{N}. \tag{28}$$

Considering with Eq. (16), Eq. (28) is described as a following;

$$\langle X^{\mu}|E^{(\mu)}|X^{\mu}\rangle = Na^{(\mu)}\left(1 - a^{(\mu)}Q_{n_\mu}^{(\mu)}\right). \tag{29}$$

Furthermore, N and $a^{(\mu)}$ multiplied to Eq. (24) with a μ-th memory vector, i.e.;

$$Na^{(\mu)}\left(1 - a^{(\mu)}Q_{n_\mu}^{(\mu)}\right) \leq Nk^{-n_\mu}. \tag{30}$$

From Eqs. (25), (27), (29) and (30), the following condition is obtained;

$$\begin{aligned} \left\| |X^{\mu}\rangle - W^{(l)}|X^{\mu}\rangle \right\|^2 &\leq \langle X^{\mu}|E^{(l)}|X^{\mu}\rangle \leq \langle X^{\mu}|E^{(\mu)}|X^{\mu}\rangle \\ &= Na^{(\mu)}\left(1 - a^{(\mu)}Q_{n_\mu}^{(\mu)}\right) \leq Nk^{-n_\mu}. \end{aligned} \tag{31}$$

According to the stability condition as Eq. (10), Eq. (31) is described as a following;

$$\left\| |X^{\mu}\rangle - W^{(l)}|X^{\mu}\rangle \right\|^2 \leq Nk^{-n_\mu} \leq \left(\frac{\pi}{q} - T\right)^2 \tag{32}$$

where, q denotes a quantization value on the complex unit circle. The optimal stability can be controlled by the parameter T [5,12]. In addition, the network stability is guaranteed in case of $T = 0$. Furthermore, from Eq. (32), the minimum number of iterations n_l is derived as Eq. (5).

4 Simulation Experiment

This section presents the simulation experiments comparing with a pseudo inverse learning model (PInvCHAM) [1] and a proposed model (BVCHAM), in terms of memory capacity and noise tolerance.

4.1 Condition

Table 1 shows the simulation conditions to evaluate the memory capacity and the noise tolerance. In this experiment, the number of pairs p is increased from 1 to 220 at intervals of 10 ($p = 1, 10, 20, \ldots, 220$). Noise tolerance is a significant property for the associative memory. Typically, "noise" is roughly divided into two types in associative memory. One is the similarity of the stored patterns, another is the stored patterns contain noise itself. Due to the proposed model has the similar properties with a pseudo inverse learning model, it is expected that the proposed model is able to learn the correlated memory vectors without errors. In this paper, therefore, we consider about a latter type of noise which is contained in an initial input with 0 to 50 [%] by the salt & pepper noise. Furthermore, it is known that the ability of a complex-valued associative memory is dependent upon the number of divisions of a complex-valued unit circle. Here, we set 4, 8 and 16 divisions for this simulation.

Table 1. Simulation condition.

Number of pairs p : 1–220
Number of neurons N: 200
Number of divisions q: 4, 8, 16
Data set configuration: Amplitude: 1.0, Phase: Rand

4.2 Result

In Fig. 1, PInvCHAM maintains the high recall rate, especially $NR = 0.0$. However, due to the limitation of the pseudo inverse learning, the recall rate is suddenly dropped under the condition $N \leq p$. As shown in Fig. 2, due to the weight matrix of BVCHAM can be approximated as the weight matrix by a pseudo inverse learning, the recall rate of BVCHAM shows the similar or superior results (especially in Fig. 2(a)) than PInvCHAM. Here, it focuses on the results with $NR = 0.0$ in Fig. 2. Since the weight matrix of BVCHAM is not equal to the identity matrix at $p = N$, BVCHAM is able to maintain the high recall rate than PInvCHAM even if the condition is under $N \leq p$.

From the above results, it is regarded that BVCHAM has the following noteworthy advantages; although a proposed learning algorithm is incremental, the ability can be comparable to a batch learning as a pseudo inverse learning algorithm, and it is able to overcome the limitation of a pseudo inverse learning algorithm that is characterized by $p < N$.

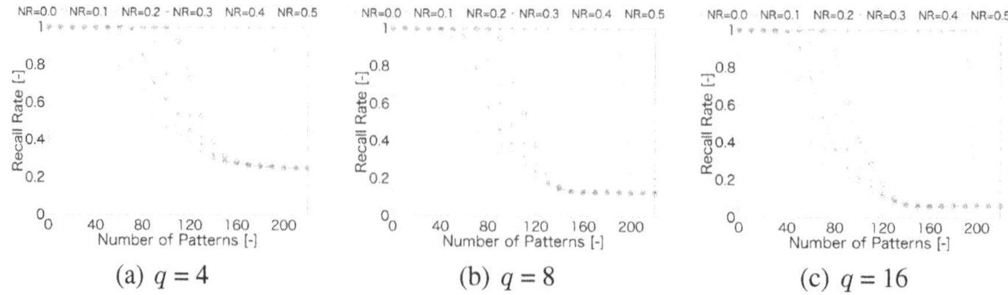

Fig. 1. Results of recall ratio for a pseudo inverse learning model.

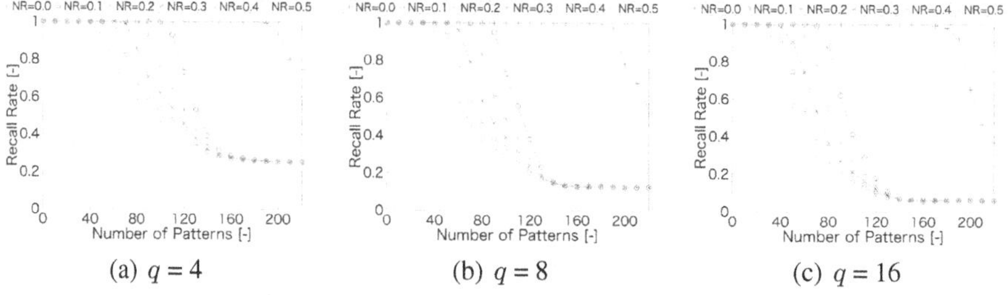

Fig. 2. Results of recall ratio for a BV incremental learning model.

5 Conclusion

This paper introduced a local iterative incremental learning algorithm for a complex-valued Hopfield associative memory. Furthermore, we presented the network stability analysis which derived the weight matrix of BVCHAM is approximated as a weight matrix from a complex-valued pseudo inverse learning algorithm, and the minimum number of iterations with maintaining a network convergence. From the result of simulation experiment in terms of memory capacity and noise tolerance, BVCHAM has the superior ability than the models with the Hebb learning and a complex-valued pseudo inverse learning algorithm. Noteworthy, unlike the model with a pseudo inverse learning algorithm, the proposed model is able to maintain the high recall rate even if the number of stored patterns is larger than the number of neurons.

As a future work, we will extend the proposed incremental learning algorithm to hetero-association models, such as bi/multi-directional associative memory models [10].

Acknowledgments. The authors would like to acknowledge a scholarship provided by the University of Malaya (Fellowship Scheme). This research is supported by High Impact Research UM.C/625/1/HIR/MOHE/FCSIT/10 from University of Malaya.

References

1. Albert, A.: Regression and the Moore-Penrose Inverse. Academic Press, New York (1972)
2. Blatt, M.G., Vergini, E.G.: Neural networks: a local learning prescription for arbitrary correlated patterns. Phys. Rev. Lett. **66**(13), 1793 (1991)
3. Chartier, S., Boukadoum, M.: A bidirectional heteroassociative memory for binary and grey-level patterns. IEEE Trans. Neural Netw. **17**(2), 385–396 (2006)
4. Diederich, S., Opper, M.: Learning of correlated patterns in spin-glass networks by local learning rules. Phys. Rev. Lett. **58**(9), 949 (1987)
5. Gardner, E.: Optimal basins of attraction in randomly sparse neural network models. J. Phys. A: Math. Gen. **22**(12), 1969 (1989)
6. Hagiwara, M.: Multidirectional associative memory. In: International Joint Conference on Neural Networks, vol. 1, pp. 3–6 (1990)
7. Hopfield, J.J.: Neural networks and physical systems with emergent collective computational abilities. Proc. Nat. Acad. Sci. **79**(8), 2554–2558 (1982)
8. Isokawa, T., Nishimura, H., Matsui, N.: An iterative learning scheme for multistate complex-valued and quaternionic Hopfield neural networks (2009)
9. Jankowski, S., Lozowski, A., Zurada, J.: Complex-valued multistate neural associative memory. IEEE Trans. Neural Netw. **7**(6), 1491–1496 (1996)
10. Kobayashi, M., Yamazaki, H.: Complex-valued multidirectional associative memory. Electr. Eng. Jpn. **159**(1), 39–45 (2007)
11. Kosko, B.: Constructing an associative memory. Byte **12**(10), 137–144 (1987)
12. Krauth, W., Mézard, M.: Learning algorithms with optimal stability in neural networks. J. Phys. A: Math. Gen. **20**(11), L745 (1987)
13. Lee, D.L.: Improvements of complex-valued Hopfield associative memory by using generalized projection rules. IEEE Trans. Neural Netw. **17**(5), 1341–1347 (2006)
14. Storkey, A.J., Valabregue, R.: The basins of attraction of a new Hopfield learning rule. Neural Netw. **12**(6), 869–876 (1999)

LDA-Based Word Image Representation for Keyword Spotting on Historical Mongolian Documents

Hongxi Wei(✉), Guanglai Gao, and Xiangdong Su

School of Computer Science, Inner Mongolia University, Hohhot, China
{cswhx, csggl, cssxd}@imu.edu.cn

Abstract. The original Bag-of-Visual-Words approach discards the spatial relations of the visual words. In this paper, a LDA-based topic model is adopted to obtain the semantic relations of visual words for each word image. Because the LDA-based topic model usually hurts retrieval performance when directly employs itself. Therefore, the LDA-based topic model is linearly combined with a visual language model for each word image in this study. After that, the basic query likelihood model is used for realizing the procedure of retrieval. The experimental results on our dataset show that the proposed LDA-based representation approach can efficiently and accurately attain to the aim of keyword spotting on a collection of historical Mongolian documents. Meanwhile, the proposed approach improves the performance significantly than the original BoVW approach.

Keywords: Latent Dirichlet Allocation (LDA) · Topic model · Visual language model · Keyword spotting · Query likelihood model

1 Introduction

Facing a huge number of scanned historical document images, how to retrieve them is still a challenging task. There is a traditional approach for accomplishing the task, which utilizes the *optical character recognition* (OCR) technology to convert document images into texts and the indexing can be created on the OCR'ed texts. But, robust OCR systems especially for handwritten or historical documents are still not available.

When OCR is hard, keyword spotting technology can be taken as an alternative approach. Keyword spotting was originally proposed for speech processing and was firstly introduced into the field of *document image retrieval* (DIR) by Manmatha et al. in [1]. The goal of keyword spotting is to find all word images in the collection that are similar to a given query keyword image by image matching. There are two primary problems in the keyword spotting technology: **(i)** how to represent word images and **(ii)** how to retrieve efficiently and accurately by image matching. In the traditional keyword spotting, profile-based features are widely used to represent word images [2] and compared using *dynamic time warping* (DTW) algorithm [3]. Although the DTW algorithm works well, it is so time-consuming that cannot be competent for real-time

A. Hirose et al. (Eds.): ICONIP 2016, Part IV, LNCS 9950, pp. 432–441, 2016.
DOI: 10.1007/978-3-319-46681-1_52

image matching at the retrieval stage. Hence, there needs a novel approach for representing word images so as to improve efficiency.

Recently, Bag-of-Visual-Words (BoVW) has been attracted more attention and shown superior performance in the field of *content-based image retrieval* (CBIR) [4–6]. The BoVW is derived from *vector space model* (VSM) of information retrieval. In the VSM, a document is represented as a vector by an unordered set of words. The size of each vector is equal to the number of vocabularies. A query should be converted into a corresponding vector in the same way. At the retrieval stage, the cosine similarity between every document and the query can be calculated on their vectors [7].

In the BoVW framework, there are two main steps: *encoding* and *pooling*. In order to realize the encoding and pooling, a codebook composed of a number of visual words should be obtained in advance. The process for creating a codebook on a training set of images is as follows [8]: firstly, local descriptors (e.g. SIFT descriptors) are extracted from each image; secondly, all the local descriptors are collected and divided into different clusters by a clustering algorithm; thirdly, a centroid of one cluster is considered as a visual word, and then a codebook can be formulated by obtaining all visual words. After that, when a new image arrives, its local descriptors would be extracted. In the encoding step, each local descriptor of the new image would be assigned to the nearest visual word (called *hard assignment*). In the following pooling step, the new image would be represented as a histogram (i.e. vector) of the visual words. As well as the VSM, the procedure of image retrieval can be done by computing and ranking the cosine similarities.

However, the BoVW approach has several drawbacks. The first one is that a local descriptor is generally discretized to be the closest visual word (i.e. the closest cluster's centroid) for encoding. The second one is that the spatial order of the neighboring local descriptors is discarded. For handling the first drawback, Wang et al. [9] proposed an effective sparse encoding scheme called *Locality-constrained Linear Coding* (LLC). In the LLC method, k ($k > 1$) nearest visual words were selected for a given local descriptor and then combined in linear fashion. To overcome the second drawback, the *spatial pyramid matching* (SPM) method has been proposed by Lazebnik et al. in [10]. The SPM method partitions an image into increasingly finer spatial sub-regions and computes histograms of local descriptors from each sub-region. Thus, all the histograms can be concatenated together to form the vector representation of an image. Typically, $2^l \times 2^l (l = 0, 1, 2)$ subregions are used in the SPM method for partitioning images such as natural images, face images and so on. Nevertheless, it is difficult to determine the appropriate size of subregions and the number of partition levels for word images. Because the lengths of word images are varying greatly with the number of characters. Consequently, a new kind of spatial order of local descriptors should be proposed for word images.

In this paper, we propose a LDA-based approach to capture semantic relations of visual words, which is integrated into the BoVW framework to realize keyword spotting on historical Mongolian document images. The detailed process is as follows. First of all, a *visual language model* is constructed on visual words of each word image, which is the same as the *language model* used in text information retrieval. And then, a LDA-based topic model is used to extract the semantic relations of visual words. Thus, a *topic model* of one word image is formulated. Finally, the corresponding *visual language model* and *topic model* of a word image are combined together to represent the word image called *LDA-based representation model*. At the retrieval stage, *query likelihood model* is utilized to fulfill the procedure of keyword spotting, in which the probability of generating the query keyword image can be estimated according to the *LDA-based representation model* of each word image. As a result, a ranked list of word images can be returned in descending order of the corresponding query likelihood.

The rest of the paper is organized as follows. The related work is given in Sect. 2. The LDA-based approach for word image representation is described detailedly in Sect. 3. Experimental results of the proposed method are shown in Sect. 4. Section 5 provides the conclusions and future work.

2 Related Work

In the domain of text information retrieval, the *language model* is firstly introduced by Ponte and Croft in [11]. Their approach to retrieval is to infer a language model for each document and to estimate the probability of generating the query according to each of these models. And then, the documents can be ranked by these probabilities. Additionally, Blei et al. [12] introduced *Latent Dirichlet Allocation* (LDA) as a semantically consistent topic model. After that, Wei and Croft [13] combined the original document model (i.e. language model) with the LDA model in a linear fashion and constructed a new LDA-based document model for improving the performance of ad-hoc retrieval.

Inspired by the *language model*, *visual language model* has been presented in [14] for image classification and the co-occurrences of visual words are utilized as the spatial order. However, the *visual language model* does not consider the influence of the codebook size which is usually set to a fixed value. Moreover, the co-occurrence of visual words cannot reflect the real semantic relations of them. Therefore, we concentrate on constructing the semantic relations of visual words in this study. Meanwhile, we also focus on the influence of different sizes of codebook on the performance. And the appropriate size of codebook would be determined. As far as we know, this is the first attempt to apply the LDA-based approach for word image representation to realize the aim of keyword spotting.

3 LDA-Based Word Image Representation

In order to accomplish the aim of keyword spotting, each scanned image of a collection of historical Mongolian documents needs to be segmented into individual word images. The corresponding pre-processing has been proposed in our previous work [15]. Thus, the handling object in this study is the word images. A query-by-example approach is applied in the retrieval procedure and an approach for generating query keyword image for being retrieved has been presented in [16].

The proposed LDA-based word image representation approach is detailedly presented in the following sub-sections.

3.1 Local Descriptor and Codebook

In the literature, a variety of local descriptors have been presented. The most well-known is the *Scale-Invariant Feature Transform* (SIFT), which has been proved effective due to its invariance to scale and rotation as well for the robustness across considerable range of distortion, noise contamination and change in brightness.

In our study, SIFT descriptors are extracted from each word image and considered as the local descriptors. A SIFT descriptor is a 128-dimensional vector. Because the space of SIFT descriptors is continuous, a codebook needs to be obtained by means of vector quantization (VQ). Given a collection of word images, all SIFT descriptors are extracted and the *k-means* algorithm is used to calculate a certain number of clusters. Each centroid of the clusters is regarded as a *visual word*. By this way, a codebook can be formed. After that, every local descriptor will be assigned the label of the closest cluster's centroid as the result of encoding, and then a word image is represented by a set of visual words. Since the performance usually increases when a larger codebook is used, the different codebook sizes are compared with each other. As a result, an appropriate number of visual words for a codebook can be determined (see Sect. 4).

3.2 Visual Language Model of Word Image

After encoding a word image, *n-grams* of the visual words can be determined along with the writing direction of the word. Here, *unigram* and *bigram* are constructed on the visual words of the word image. The *maximum likelihood estimation* (MLE) method is used to estimate the corresponding probabilities by the following equations.

$$P_{\mathrm{unigram}}(\mathrm{t_i}) = \frac{\mathrm{count}(\mathrm{t_i})}{\sum_{\mathrm{t \in d}} \mathrm{count}(\mathrm{t})} \tag{1}$$

$$P_{\mathrm{bigram}}(\mathrm{t_{i+1}} | \mathrm{t_i}) = \frac{\mathrm{count}(\mathrm{t_{i+1}}, \mathrm{t_i})}{\mathrm{count}(\mathrm{t_i})} \tag{2}$$

where $\mathrm{count}(t_i)$ means the occurrence frequency of the $i^{th}(1 \le i \le n)$ visual word in a word image d. And $\sum_{\mathrm{t \in d}} \mathrm{count(t)}$ means the total number of visual words in the word image d. Moreover, $\mathrm{count}(t_{i+1}, t_i)$ means the co-occurrences of the $(i+1)^{th}$ visual word depending on the ith visual word along with the writing direction.

At the retrieval stage, if a visual word of the query is absent from a word image, and then a zero probability will be given. To avoid zero probabilities, a smoothing scheme is usually used to handle the visual language models of word images before being retrieved. In [17], Zhai and Lafferty concluded that the *Jelinek-Mercer smoothing* method is best for long queries. In our query-by-example approach for keyword spotting, query keyword images are equivalent to the long queries. So, the *Jelinek-Mercer smoothing* method is chosen to perform smoothing. For the *unigram*, the formula of the *Jelinek-Mercer smoothing* method is defined as follows:

$$\widehat{P}(t|d) = \lambda \cdot \widehat{P}_{MLE}(t|M_d) + (1-\lambda) \cdot \widehat{P}_{MLE}(t|M_c) \quad (3)$$

where $\lambda (0 < \lambda < 1)$ is a smoothing parameter, M_d is the *unigram* of a word image d and M_c is the *unigram* built from the entire collection of word images. For the *bigram*, the smoothing can be carried out by a linear interpolation between the maximum likelihood estimation of the *bigram* and the smoothed *unigram*.

Because the retrieval performance is generally sensitive to the smoothing parameter, the parameter λ needs to be appropriately set in the *visual language model*.

By using the *visual language model*, each word image can be represented as a distribution of unigrams or bigrams. In Sect. 4, the smoothing parameter λ has been chosen and the performance of the *unigram model* has been compared with the *bigram model*. Furthermore, the suited number of visual vocabularies has also been determined.

3.3 LDA-Based Topic Model of Word Image

In the original LDA-based topic model, the way of modeling the contributions of different topics to a document is to treat each topic as a probability distribution (denoted by φ) over words and a document is viewed as a probabilistic mixture of these topics (denoted by θ). Once the two distributions φ and θ are estimated, the probability of a word w in a document d can be calculated using (4).

$$P\left(w|d, \widehat{\theta}, \widehat{\varphi}\right) = \sum_{z=1}^{T} P(w|z, \widehat{\varphi}) \cdot P(z|\widehat{\theta}, d) \quad (4)$$

where z is a topic from a Dirichlet distribution with parameter β and T is the number of topics. $\widehat{\varphi}$ and $\widehat{\theta}$ are the posterior estimates of φ and θ respectively. φ is the word distribution for each topic and θ is the topic distribution for each document.

Generally, the *Gibbs sampling* approach is used for obtaining $\widehat{\varphi}$ and $\widehat{\theta}$. Their formulas are defined as follows [13]:

$$\widehat{\varphi} = \frac{n_j^{(w)} + \beta}{\sum_{v=1}^{V} n_j^{(v)} + V \cdot \beta}, \quad \widehat{\theta} = \frac{n_j^{(d)} + \alpha}{\sum_{t=1}^{T} n_t^{(d)} + T \cdot \alpha} \quad (5)$$

where $n_j^{(w)}$ is the number of word w assigned to topic j and $n_j^{(d)}$ is the number of words in document d assigned to topic j. V and T are the number of words and topics, separately. And α and β are hyper-parameters. Thus, $\sum_{v=1}^{V} n_j^{(v)}$ is the total number of words assigned to topic j and $\sum_{t=1}^{T} n_t^{(d)}$ is the total number of words in document d.

In order to process easily, in this study, only the *unigram model* is combined with the LDA-based topic model for each word image. Thus, the probability of one visual word (denoted by w) in a word image (denoted by d) can be calculated by the following equation.

$$P(w|d) = \lambda_2 \cdot \widehat{P}(w|d) + (1-\lambda_2) \cdot \left(\sum\nolimits_{j=1}^{T} \frac{n_j^{(w)} + \beta}{\sum\nolimits_{v=1}^{V} n_j^{(v)} + V \cdot \beta} \times \frac{n_j^{(d)} + \alpha}{\sum\nolimits_{t=1}^{T} n_t^{(d)} + T \cdot \alpha} \right) \quad (6)$$

where $\lambda_2 (0 < \lambda_2 < 1)$ is a coefficient and $\widehat{P}(w|d)$ is computed by the Eq. (3).

In this study, for each visual word of one word image, the corresponding probability can be calculated by (6). Eventually, these probabilities of one word image would be used in *query likelihood model* so as to accomplish the aim of keyword spotting. The detailed procedure is shown in the next sub-section.

3.4 Retrieval Scheme

After the above-mentioned processing, one word image is represented as a new formulation, in which the probability of each visual word estimated using (6). In order to realize keyword spotting, *query likelihood model* is utilized. In which, the following formula can rank word images for a given query.

$$P(q|d) = \prod\nolimits_{w \in q} P(w|d) \quad (7)$$

where $P(q|d)$ is the likelihood of a word image d generating the corresponding visual words in the given query q under the bag-of-words assumption, and w is a visual word occurred in the query q. According to (6), thus (7) will be rewritten into (8).

$$P(q|d) = \prod\nolimits_{w \in q} \left(\lambda_2 \cdot \widehat{P}(w|d) + (1-\lambda_2) \cdot \left(\sum\nolimits_{j=1}^{T} \frac{n_j^{(w)} + \beta}{\sum\nolimits_{v=1}^{V} n_j^{(v)} + V \cdot \beta} \times \frac{n_j^{(d)} + \alpha}{\sum\nolimits_{t=1}^{T} n_t^{(d)} + T \cdot \alpha} \right) \right) \quad (8)$$

where $\lambda_2 (0 < \lambda_2 < 1)$ is a coefficient. Thus, a ranking list of word images can be obtained in descending order of the probabilities. In this way, we can spot keyword within a collection of word images in the scenario of query-by-example.

4 Experimental Results

In this section, the original BoVW method is considered as a baseline. Other methods, including *unigram model*, *bigram model* and *LDA-based representation model*, have been tested and compared with the baseline.

4.1 Dataset

To evaluate the performance of the different models, a data set has been collected, which is constitutive of 100 scanned Mongolian Kanjur images and contains 24,827 words. Each page has been transcribed manually to form the ground truth data. Twenty meaningful words are selected and taken as query keywords. The twenty query keywords are the same as in [16].

For obtaining the codebook, all descriptors have been extracted from the 24,827 word images. The number of the SIFT descriptors is 2,283,512. After that, the k-means algorithm has been performed on those descriptors. Therein, the number of the clusters rises from 500 to 10,000 and increases 500 each time. In our experiment, evaluation metric is *mean average precision* (MAP).

4.2 Performance of the Baseline

In the original BoVW framework, each word image was converted into a histogram of visual words and the tf-idf scheme was used for weighting each visual word. So, we tested the performance of the BoVW by varying the number of clusters.

In Fig. 1, the performance of the BoVW is ranged from 9.98 % to 13.43 %. The best performance is **13.43 %** when the number of clusters increases to **5,500**. For the LLC method, the number of a descriptor's neighbors rises from 2 to 10. We can see that the best performance of the LLC is reached at **27.99 %** when the number of neighbors is **2** and the number of clusters is **500**. Consequently, the performance of the LLC method is twice as much as the original BoVW approach.

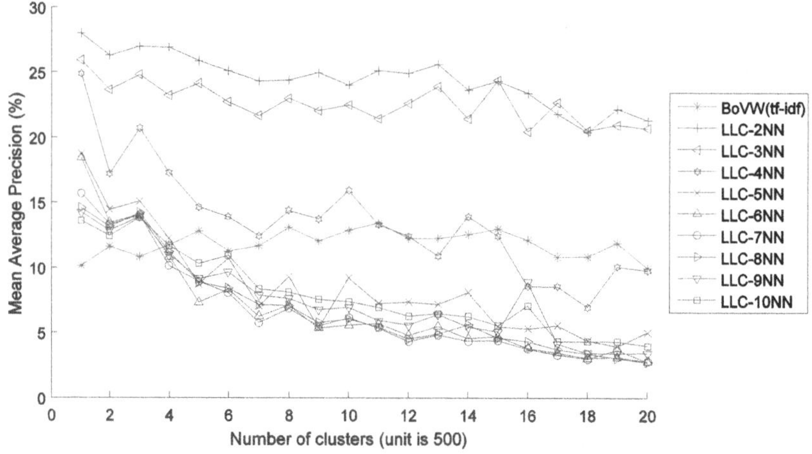

Fig. 1. The performance of BoVW and LLC

4.3 Performance of Unigram and Bigram

Here, we have individually tested the performance of the *unigram model* and the *bigram model*. In the unigram, the number of the clusters rises from 500 to 10,000 and increases 500 each time. Considering scarcity of visual words, the number of the clusters only rises from 500 to 5,000 and increases 500 each time for the *bigram model*. Their MAPs are shown in Figs. 2 and 3, respectively.

From Figs. 2 and 3, we can see that the *unigram model* and the *bigram model* both obtain the best performance when the appropriate number of clusters is **2,000**.

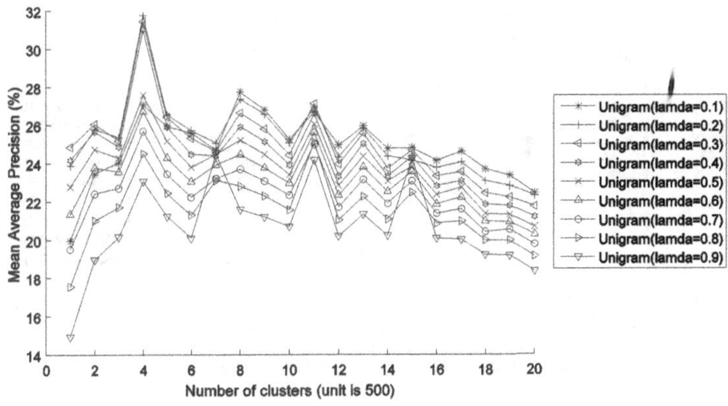

Fig. 2. The performance of unigram

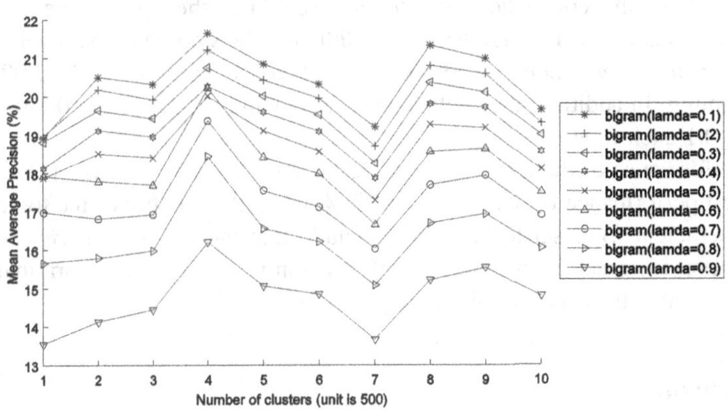

Fig. 3. The performance of bigram

Meanwhile, the smoothing parameter λ is set to **0.2** and **0.1** for the *unigram model* and the *bigram model*, severally.

Although the performance of the original BoVW approach is improved from **13.43 %** to **21.64 %** by the *bigram model*, the LLC method is still superior to the *bigram model*. But, the best performance of the *unigram model* can attain to **31.75 %**. Hence, the *visual language models* of word images can improve the performance significantly.

4.4 The Performance of LDA-Based Representation

For obtaining the two distributions of φ and θ in the LDA-based topic model, the Gibbs sampling method is employed. In the Gibbs sampling, the two hyper-parameters α and β are set to 50/T and 0.01, respectively. T is the number of topics and it rises from 50 to 1,500 with increasing 50 each time. The number of maximum iterations is set to 500.

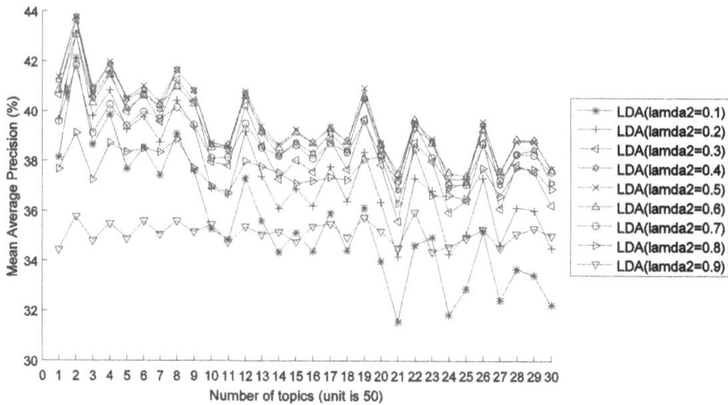

Fig. 4. The performance of LDA-based representation

According to the conclusions of the Sub-sect. 4.3, the appropriate number of clusters (i.e. vocabulary of visual words) is **2,000** and the smoothing parameter λ is **0.2** in the *unigram model*. Therefore, the number of vocabularies is set to **2,000** in the Gibbs sampling. In addition, the coefficient λ_2 in (6) and (8) rises from 0.1 to 0.9 with increasing 0.1 each time.

The performance of the LDA-based representation is shown in Fig. 4. We can see that the best performance is attained to **43.78 %** when the number of the topics is **100** and the coefficient λ_2 is set to **0.4**. We conclude that the *LDA-based representation model* can provide much more semantic information not only than the original BoVW approach, but also than the visual language model.

5 Conclusion

In this paper, the LDA-based topic model has been adopted to extract semantic relations of visual words, which is linearly combined with a visual language model for each word image. In this way, the drawback (i.e. visual words are unordered) of the original BoVW can be overcome.

The performance of our proposed approach increases **226 %** (from 13.43 % to 43.78 %) and **38 %** (from 31.75 % to 43.78 %) against to the BoVW approach and the *unigram model*, respectively. The experimental results prove that the proposed approach can significantly improve the performance of keyword spotting on historical Mongolian document images. In our future work, we will adopt to the word embeddings for obtaining the more deeply semantic relations of visual words.

Acknowledgements. The paper is supported by the National Natural Science Foundation of China under Grant 61463038 and the Research Project of Higher Education School of Inner Mongolia Autonomous Region under Grant NJZY14007.

References

1. Manmatha, R., Han, C., Riseman, E.M., Croft, W.B.: Indexing handwriting using word matching. In: Proceedings of ICDL 1996, pp. 151–159. ACM Press, New York (1996)
2. Rath, T.M., Manmatha, R.: Features for word spotting in historical manuscripts. In: Proceedings of ICDAR 2003, pp. 218–222. IEEE Press, New York (2003)
3. Rath, T.M., Manmatha, R.: Word image matching using dynamic time warping. In: Proceedings of CVPR 2003, pp. 521–527. IEEE Press, New York (2003)
4. Chen, X., Hu, X., Shen, X.: Spatial weighting for bag-of-visual-words and its application in content-based image retrieval. In: Theeramunkong, T., Kijsirikul, B., Cercone, N., Ho, T.-B. (eds.) PAKDD 2009. LNCS, vol. 5476, pp. 867–874. Springer, Heidelberg (2009)
5. Tirilly, P., Claveau, V., Gros, P.: Distance and weighting schemes for bag of visual words image retrieval. In: Proceedings of MIR 2010, pp. 323–332. ACM Press, New York (2010)
6. Zhu, L., Jin, H., Zheng, R., Feng, X.: Weighting scheme for image retrieval based on bag-of-visual-words. IET Image Process **8**(9), 509–518 (2014)
7. Manning, C.D., Raghavan, P., Schutze, H.: Introduction to Information Retrieval. Cambridge University Press, Cambridge (2008). PP. 120–126
8. Lopes-Monroy, A.P., Montes-Y-Gomez, M., Escalante, H.J., Cruz-Roa, A., Gonzalez, F.A.: Improving the BoVW via discriminative visual n-grams and MKL strategies. Neurocomputing **175**, 768–781 (2016)
9. Wang, J., Yang, J., Yu, K., Lv, F., Huang, T., Gong, Y.: Locality-constrained linear coding for image classification. In: Proceedings of CVPR 2010, pp. 3360–3367. IEEE Press, New York (2010)
10. Lazebnik, S., Schmid, C., Ponce, J.: Beyond bags of features: spatial pyramid matching for recognizing natural scene categories. In: Proceedings of CVPR 2006, pp. 2169–2178. IEEE Press, New York (2006)
11. Ponte, J.M., Croft, W.B.: A language modeling approach to information retrieval. In: Proceedings of the 21st Annual International ACM SIGIR Conference on Research and Development in Information Retrieval (SIGIR 1998), pp. 275–281. ACM Press, New York (1998)
12. Blei, D.M., Ng, A.Y., Jordan, M.J.: Latent Dirichlet allocation. J. Mach. Learn. Res. **3**, 993–1022 (2003)
13. Wei, X., Croft, W.B.: LDA-based document models for ad-hoc retrieval. In: Proceedings of the 29th Annual International ACM SIGIR Conference on Research and Development in Information Retrieval (SIGIR 2006), pp. 178–185. ACM Press, New York (2006)
14. Wu, L., Li, M., Li, Z., Ma, W., Yu, N.: Visual language modeling for image classification. In: Proceedings of MIR 2007, pp. 115–124. ACM Press, New York (2007)
15. Wei, H., Gao, G., Bao, Y., Wang, Y.: An efficient binarization method for ancient Mongolian document images. In: Proceedings of the 3rd International Conference on Advanced Computer Theory and Engineering (ICACTE 2010), pp. 43–46. IEEE Press, New York (2010)
16. Wei, H., Gao, G.: A keyword retrieval system for historical Mongolian document images. Int. J. Doc. Anal. Recogn. (IJDAR) **17**(1), 33–45 (2014)
17. Zhai, C., Lafferty, J.: A study of smoothing methods for language models applied to ad hoc information retrieval. In: Proceedings of the 24th Annual International ACM SIGIR Conference on Research and Development in Information Retrieval (SIGIR 2001), pp. 334–342. ACM Press, New York (2001)

Solving the Vanishing Information Problem with Repeated Potential Mutual Information Maximization

Ryotaro Kamimura(✉)

IT Education Center and Graduate School of Science and Technology, Tokai Univerisity, 4-1-1 Kitakaname, Hiratsuka, Kanagawa 259-1292, Japan
ryo@keyaki.cc.u-tokai.ac.jp

Abstract. The present paper shows how to solve the problem of vanishing information in potential mutual information maximization. We have previously developed a new information-theoretic method called "potential learning" which aims to extract the most important features through simplified information maximization. However, one of the major problems is that the potential effect diminishes considerably in the course of learning and it becomes impossible to take into account the potentiality in learning. To solve this problem, we here introduce repeated information maximization. To enhance the processes of information maximization, the method forces the potentiality to be assimilated in learning every time it becomes ineffective. The method was applied to the online article popularity data set to estimate the popularity of articles. To demonstrate the effectiveness of the method, the number of hidden neurons was made excessively large and set to 50. The results show that the potentiality information maximization could increase mutual information even with 50 hidden neurons, and lead to improved generalization performance. In addition, simplified representations could be obtained for better interpretation and generalization.

1 Introduction

Information-theoretic methods have had a strong influence on neural networks [1–7], because one of the fundamental objectives of neural networks is naturally to control the amount of information stored within. However, one of the main problems is that the information-theoretic methods require fairly complicated computational procedures to control information content. Though many methods have been developed to simplify these procedures [3,4,8–11], the problem nevertheless remains. To address this problem, we have previously introduced potential information maximization [12–15] to interpret internal representations and to improve generalization performance. The potentiality is supposed to represent the ability to deal with as many different situations as possible. Hence, potential mutual information maximization aims to extract neurons with higher potentiality. For the first approximation, the potentiality is approximated by the variance of neurons [1].

A. Hirose et al. (Eds.): ICONIP 2016, Part IV, LNCS 9950, pp. 442–451, 2016.
DOI: 10.1007/978-3-319-46681-1_53

In actual learning procedures, the potential information is given to neural networks as initial values. One of the main problems is that the potential information tends to lose its effect considerably in the course of learning. This means that in learning, potentiality becomes weaker and weaker. In addition, when the number of neurons increases, the effect of potentiality naturally diminishes. This problem is called the "vanishing information problem". To solve this problem, we introduce repeated information maximization. In this method, every time the potentiality becomes weaker, a new information maximization procedure begins to enhance the process of information maximization. Though we need to iterate maximization processes many times, the total number of iterations is not excessively large because we use the early stopping method to force learning to stop when over-training occurs.

2 Theory and Computational Methods

2.1 Potential Mutual Information

The potentiality of neurons represents the ability to deal with as many new situations as possible [12–15]. For the first approximation, the potentiality is approximated by the variance of neurons. Linsker [1] already showed that mutual information maximization corresponds to variance maximization with the linear and Gaussian cases.

Now, let us define potentiality for hidden neurons. We can define the potentiality for any neuron in the same way. As shown in Fig. 1, we let w_{jk}^t denote connection weights from the kth input neuron to the jth hidden neuron for the tth data set, then the potentiality in terms of variance is

$$v_j^t = \frac{1}{L-1} \sum_{k=1}^{L} (w_{jk}^t - w_j^t)^2, \tag{1}$$

where L is the number of input neurons and w_j^t is the average weight for the jth hidden neuron

$$w_j^t = \frac{1}{L} \sum_{k=1}^{L} w_{jk}^t. \tag{2}$$

Then, the potentiality is normalized as

$$p(j|t) = \frac{v_j^t}{\sum_{m=1}^{M} v_m^t}, \tag{3}$$

where M is the number of hidden neurons. Then, we have potential mutual information for hidden neurons

$$PI_{hid} = -\sum_{j=1}^{M} p(j) \log p(j) + \sum_{t=1}^{T} p(t) \sum_{j=1}^{M} p(j|t) \log p(j|t), \tag{4}$$

where $p(j)$ denotes the average firing probability for the jth hidden neuron.

This mutual information is increased by the pseudo-information maximization. First, the pseudo-potentiality is defined by

$$\phi_j^{t,r} = \left(\frac{v_j^t}{v_{max}^t}\right)^r, \tag{5}$$

where v_{max} is the maximum potentiality. By normalizing this potentiality, we have the pseudo-firing probability

$$p(j|t;r) = \frac{\phi_j^{t,r}}{\sum_{m=1}^{M} \phi_m^{t,r}} \tag{6}$$

Then, we have pseudo-mutual information

$$PI_{hid} = \log M + \frac{1}{T}\sum_{t=1}^{T}\sum_{j=1}^{M} p(j|t;r)\log p(j|t;r), \tag{7}$$

where for simplicity, the average firing probability $p(j)$ is supposed to be equiprobable. The pseudo-information can be increased just by increasing the parameter r. Thus, when the real mutual information is replaced by the pseudo-potential information, all we have to do to increase mutual information is to increase the parameter r.

2.2 Repeated Information Maximization

As mentioned in the introduction section, the potentiality effect diminishes over the course of learning. Thus, it needs to be repeated several times. As shown in Fig. 1, we can successively assimilate the potentially in neural networks by using that which was computed in the previous learning step. As shown in Fig. 1, connection weights in the gth step are computed by the weights and potentiality of the $g-1$th step. When the learning step increases, the parameter r is gradually increased. Thus, mutual information is forced to be increased by this repeated maximization. In addition, in all steps, the early stopping is used to force learning to stop immediately when the over-training begins.

Now, weights for the gth step are computed by the input and hidden potentiality in the previous $g-1$th step

$${}^{(g)}w_{jk}^t = {}^{(g-1)}\phi_j^{t,r_{g-1}} \; {}^{(g-1)}w_{jk}^t \; {}^{(g-1)}\phi_k^{t,r_{g-1}} \tag{8}$$

where $\phi_k^{t,r_{g-1}}$ denotes the potentiality of the kth input neuron for the tth data set and the potentiality parameter r_{g-1}. In the same way, hidden-output connection weights for the gth step are computed by

$${}^{(g)}w_{ij}^t = {}^{(g-1)}w_{ij}^t \; {}^{(g-1)}\phi_j^{t,r_{g-1}}. \tag{9}$$

This process of potentiality assimilation is repeated until we reach a predetermined number of steps.

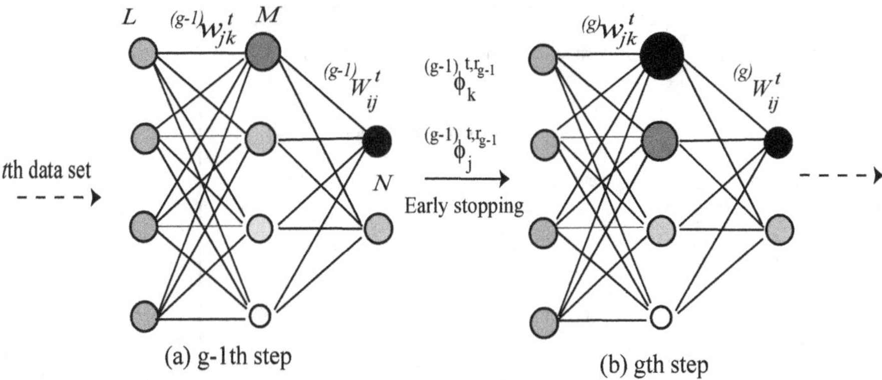

Fig. 1. Repeated mutual information maximization in the gth and $g - 1$th step.

3 Results and Discussion

3.1 Experimental Outline

To conduct the experiment, we used the online news popularity dataset, which contains a set of features about articles [16, 17]. The goal was to predict higher and lower shares in social networks; popularity, in other words. The number of patterns was 16,158, reduced from the original 39,797, to have a data set with an equal amount of high and low popularity. The number of input variables was 58, chosen from the original 59 variables[1] and the higher and lower targets were distinguished based on the average shares. There were 5,000 modeling data, and the remaining data were used exclusively for testing. The number of hidden neurons was increased redundantly to 50. The maximum number of learning steps was set to ten. For evaluating generalization performance, we computed two types of errors, namely, S-average and E-average. The S-average is simply the average of generalization errors over ten different sets of input patterns. On the other hand, E-average is the average obtained by computing the differences between the targets and the average outputs over ten different data sets. We used the BP learning in Matlab with all parameter values set to default, including the tansig activation function and cross entropy function for easy reproduction of the results.

3.2 Mutual Information

The results show that mutual information increased when the number of the learning steps increased. Figure 2 shows that input and hidden information gradually increased. These results show that both types of information could reach sufficiently large points, meaning that the number of activated input and hidden neurons was small, and consequently, the number of strong connection weights

[1] The first variable "timedelta" was deleted from the experiment.

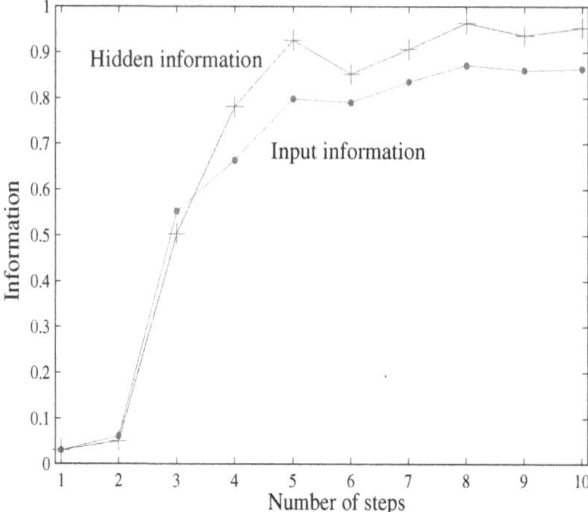

Fig. 2. Input and hidden information with 50 hidden neurons for the on-line data set.

became smaller as well. Usually, input information tends to be smaller than hidden information, because input neurons are directly connected with input patterns, while hidden neurons can be activated more freely.

3.3 Connection Weights

The obtained connection weights were rather simplified and only one feature could be extracted. Figure 3(a) shows input-hidden connection weights for the first step. Though many connection weights were randomly strong, we could see one stronger negative connection weight. This largest weight remained strong when the number of steps increased from two in Fig. 3(b) to ten in Fig. 3(h). Figure 4 shows connection weights from hidden to output neurons. With the first step, almost random weights could be seen in Fig. 4(a). When the step increased from two in Fig. 4(b) to 10 in Fig. 4(h), the number of stronger connection weights became smaller and smaller. Finally, only two connection weights from the seventh hidden neuron were stronger. These results show that an increase in both types of mutual information corresponded to a lower number of strong connection weights.

3.4 Generalization Performance

Generalization performance was improved in direct proportion to the increase in information. The best generalization performance was obtained by the present method. First, we examined how generalization performance could be changed by controlling mutual information. In Fig. 5, generalization errors in terms of S-average decreased sharply in the first three steps and then decreased gradually as the number of steps increased. On the other hand, in terms of E-average, generalization errors decreased more smoothly. The S-average tends to reflect the variation of individual errors more directly, while the E-average can neutralize it.

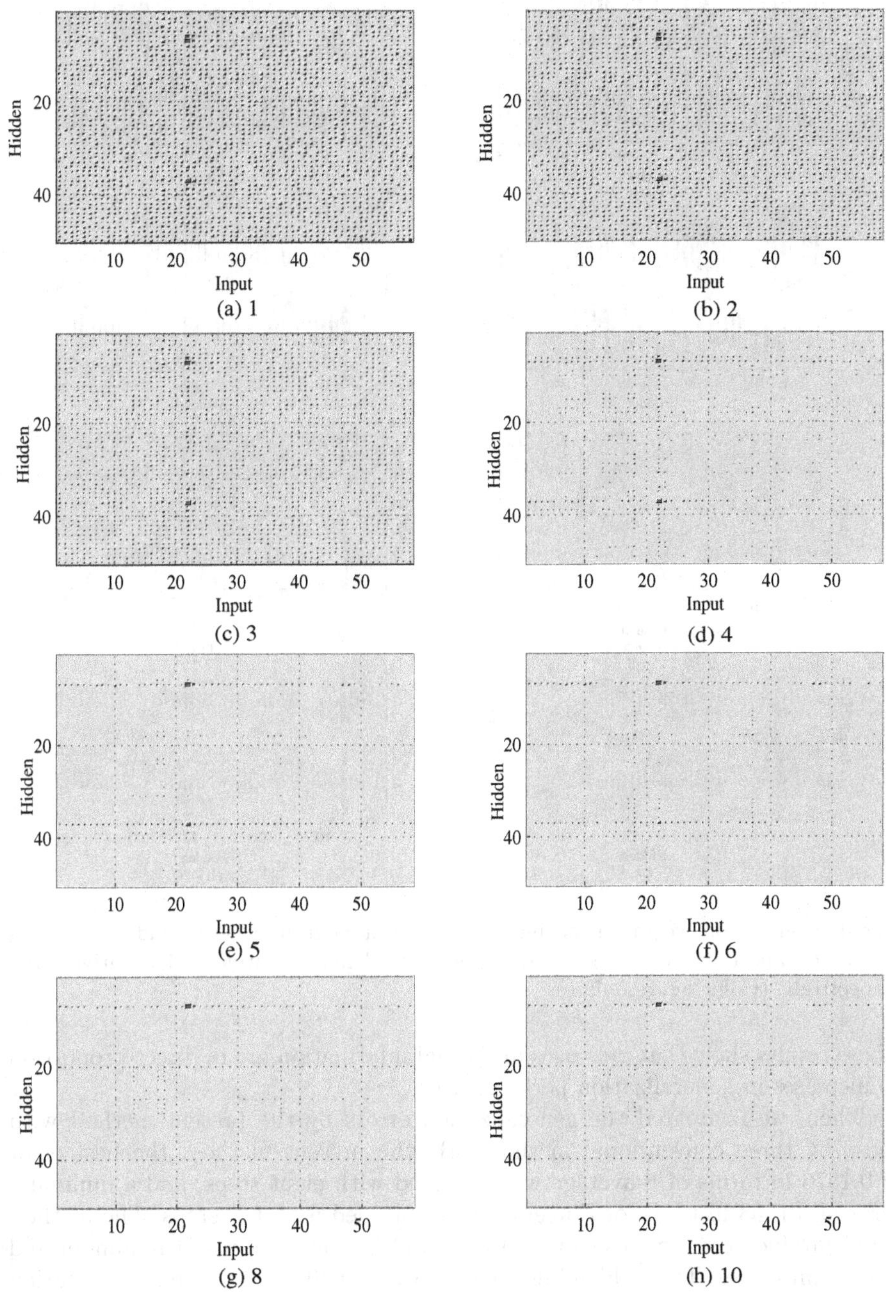

Fig. 3. Connection weights from input to hidden neurons with 50 hidden neurons for the on-line data set. Green and red weights represent positive and negative ones. (Color figure online)

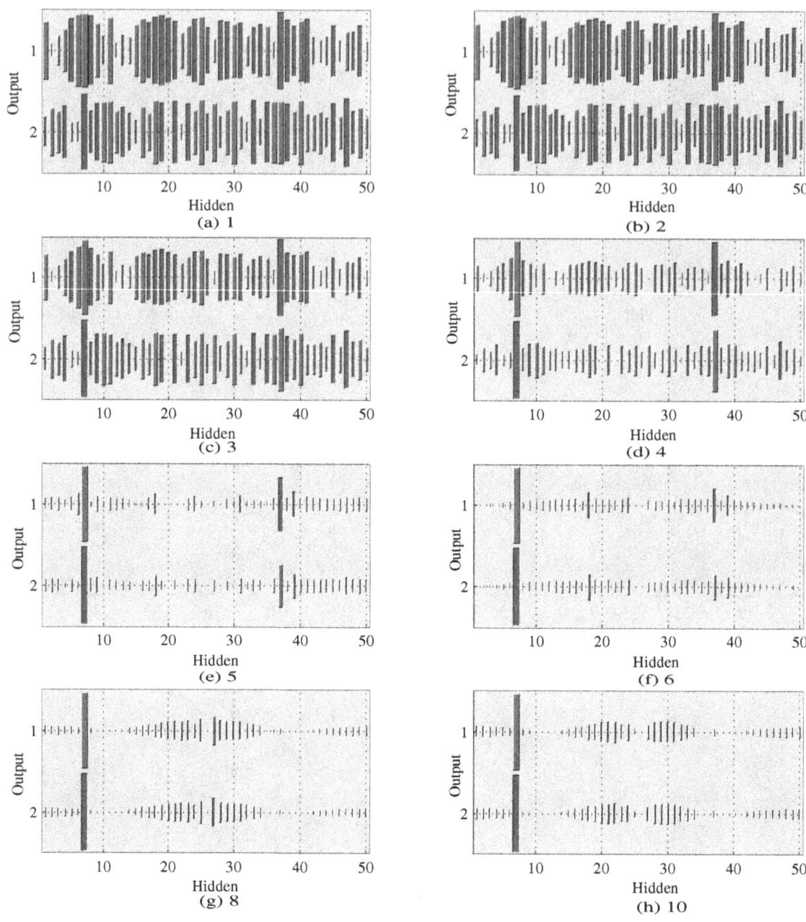

Fig. 4. Connection weights from hidden to output neurons with 50 hidden neurons for the on-line data set. Green and red weights denote positive and negative ones, respectively. (Color figure online)

These results show that increases in mutual information are in direct proportion to increases in generalization performance.

Then, we compared the generalization errors by the present method with those by three conventional methods. By the present method, the best error of 0.1576 in terms of S-average was obtained with eight steps, and a minimum error of 0.1568 in terms of E-average was obtained with ten steps. The method could produce the lowest error of 0.1558 and 0.1598 in terms of minimum and maximum values. The hidden information was 0.9640 and the input information was 0.8723. By the conventional BP with the early stopping, the second best error in terms of E-average was obtained, 0.1617. The third best error of 0.2218 was obtained by the SVM and the worst error of 0.2273 was obtained by the logistic regression analysis (Table 1).

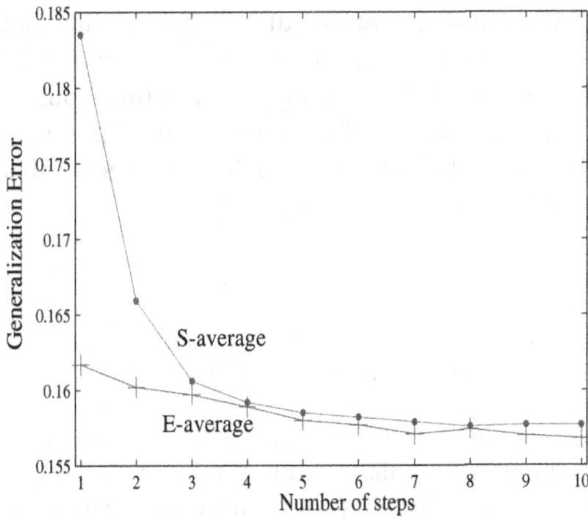

Fig. 5. Generalization errors in terms of S-average and E-average.

Table 1. Summary of experimental results in terms of generalization performance for the on-line news popularity data set.

Method	Step	Hidden	S-average	E-average	Std. dev.	Min	Max	Inp. inf.	Hid. inf.
MI	8	50	0.1576		0.0013	**0.1558**	**0.1598**	0.8723	0.9640
	10	50		**0.1568**	0.0012	0.1562	0.1602	0.8639	0.9539
BP (ES)	1	50	0.1835	0.1617	0.0375	0.1614	0.2745	0.0306	0.0309
Logistic			0.2273		0.0034	0.2205	0.2312		
SVM			0.2218		0.0029	0.2157	0.2252		

The bold face numbers show the best values.

4 Conclusion

The present paper proposed a new type of information-theoretic method to maximize mutual information for input and hidden neurons. Mutual information has played an important role in neural networks since Linsker stated his mutual information maximization principle for perceptual systems. However, his methods have been restricted to simple and linear cases [1]. There have been many attempts to simplify the method of mutual information [3,4,8–11], but we have had much difficulty in computing mutual information. The present method aims to simplify this by using pseudo-mutual information based on the potentiality of neurons. The method has been applied to several problems and is effective in increasing information and improving generalization [18]. However, one of the major problems is the increasing information and number of neurons that come with complex problems. To overcome this problem, we have introduced repeated information maximization, in which mutual information is maximized several times until it reaches sufficiently high levels. The method was applied to the on-line popularity data set from the machine learning database. The experimental results showed that mutual information could be increased by repeating the processes of information maximization and simplified internal representations could be obtained for better interpretation and generalization. One problem with the method is that the average neurons' firing probability was supposed to be uniform in computing mutual information. If it is possible to compute the exact average probability, the method will be more generalizable and can be applied to many practical problems.

References

1. Linsker, R.: Self-organization in a perceptual network. Computer **21**(3), 105–117 (1988)
2. Linsker, R.: How to generate ordered maps by maximizing the mutual information between input and output signals. Neural Comput. **1**(3), 402–411 (1989)
3. Linsker, R.: Local synaptic learning rules suffice to maximize mutual information in a linear network. Neural Comput. **4**(5), 691–702 (1992)
4. Linsker, R.: Improved local learning rule for information maximization and related applications. Neural Netw. **18**(3), 261–265 (2005)
5. Barlow, H.B.: Unsupervised learning. Neural Comput. **1**(3), 295–311 (1989)
6. Barlow, H.B., Kaushal, T.P., Mitchison, G.J.: Finding minimum entropy codes. Neural Comput. **1**(3), 412–423 (1989)
7. Atick, J.J.: Could information theory provide an ecological theory of sensory processing? Netw. Comput. Neural Syst. **3**(2), 213–251 (1992)
8. Principe, J.C., Xu, D., Fisher, J.: Information theoretic learning. Unsuperv. Adapt. Filter. **1**, 265–319 (2000)
9. Principe, J.C.: Information Theoretic Learning: Renyi's Entropy and Kernel Perspectives. Springer Science & Business Media, New York (2010)
10. Nenadic, Z.: Information discriminant analysis: feature extraction with an information-theoretic objective. IEEE Trans. Pattern Anal. Mach. Intell. **29**(8), 1394–1407 (2007)
11. Torkkola, K.: Nonlinear feature transforms using maximum mutual information. In: Proceedings of International Joint Conference on Neural Networks, IJCNN 2001, vol. 4, pp. 2756–2761. IEEE (2001)
12. Kamimura, R.: Self-organizing selective potentiality learning to detect important input neurons. In: 2015 IEEE International Conference on Systems, Man, and Cybernetics (SMC), pp. 1619–1626. IEEE (2015)
13. Kamimura, R., Kitajima, R.: Selective potentiality maximization for input neuron selection in self-organizing maps. In: 2015 International Joint Conference on Neural Networks (IJCNN), pp. 1–8. IEEE (2015)
14. Kamimura, R.: Supervised potentiality actualization learning for improving generalization performance. In: Proceedings on the International Conference on Artificial Intelligence (ICAI), p. 616. The Steering Committee of The World Congress in Computer Science, Computer Engineering and Applied Computing (WorldComp) (2015)
15. Kitajima, R., Kamimura, R.: Simplifying potential learning by supposing maximum and minimum information for improved generalization and interpretation. In: International Conference on Modelling, Identification and Control. IASTED (2015)

16. Fernandes, K., Vinagre, P., Cortez, P.: A proactive intelligent decision support system for predicting the popularity of online news. In: Pereira, F., Machado, P., Costa, E., Cardoso, A. (eds.) EPIA 2015. LNCS, vol. 9273, pp. 535–546. Springer, Heidelberg (2015)
17. Bache, K., Lichman, M.: UCI machine learning repository (2013)
18. Kamimura, R.: Repeated potentiality assimilation: simplifying learning procedures by positive, independent and indirect operation for improving generalization and interpretation. In: Proceedings of IJCNN-2016, Vancouver (2016, in press)

Self-organization on a Sphere with Application to Topological Ordering of Chinese Characters

Andrew P. Papliński(✉)

Monash University, Melbourne, Australia
andrew.paplinski@monash.edu

Abstract. We consider a case of self-organization in which a relatively small number N of data points is mapped on a larger number M of nodes. This is a reverse situation to a typical clustering problem when a node represents a center of the cluster of data points. In our case the objective is to have a Gaussian-like distribution of weights over nodes in the neighbourhood of the winner for a given stimulus. The fact that $M > N$ creates some problem with using learning schemes related to Gaussian Mixture Models. We also show how the objects, Chinese characters in our case, can be topologically ordered on a surface of a 3D sphere. A Chinese character is represented by an angular integral of the Radon Transform (aniRT) which is an RTS-invariant 1-D signature function of an image.

Keywords: Self-organization on a sphere · Probabilistic self-organizing maps · Gaussian Mixture Models · Angular integral of the Radon Transform

1 Introduction

In our ongoing work on multimodal integration of visual and auditory stimuli e.g. [4,15–17] we consider a network of self-organizing modules. Each module performs a self-organizing mapping based, in principle, on Kohonen's algorithm [8,19]. The main difference between our case and a typical clustering algorithm stems from the fact that we have the number of stimuli (data points) N smaller than the number of neurons (nodes) M. In order to maintain the redundancy of the stimuli representation we keep the ratio $M/N \approx 16 \ldots 20$, the number being inferred from the work [12]. In the case on an on-line training [13] additional nodes are generated to maintain the ratio constant. We try to approximate the postsynaptic responses of our module by a Gaussian-like shape over the neurons in the neighbourhood of the winner so that the variants of the stimuli are mapped into the same neighbourhood. The other feature of our model is that the nodes are randomly positioned on a surface of a sphere. In the previous works we used only the northern hemisphere, here we extent the latent space to be the full unity 3-D sphere. The stimuli and the weights are also normalised to be the D-dimensional unity vectors.

A. Hirose et al. (Eds.): ICONIP 2016, Part IV, LNCS 9950, pp. 452–459, 2016.
DOI: 10.1007/978-3-319-46681-1_54

The objectives of this paper are as follows. Firstly, we would like to consider a possibility of applying one of many available versions of probabilistic SOMs stemming from the fundamental Gaussian Mixture Models (GMM). From the significant body of literature on these topics we point to the following recent works [3,7,10,11] and the classics [1]. Secondly, we would like to demonstrate the spherical latent space. As an example of objects to be mapped, we have selected Chinese characters rendered with a specific font and selected from the Unicode CJK table [18].

2 The Stimuli

The Chinese characters from the Unicode table with hex codes from `4E00` to `9FA5` have been rendered using the Microsoft JhengHei UI font of size 32 points. Such rendering produces black-and-white binary images of size 43×43 pixels. For each character image we calculate a signature function, a vector, as an angular integral of the Radon transform (aniRT). Details can be found in [14]. The size of the aniRT vector is equal to the diagonal of the image, 61 in this case. For further processing we select $D = 22$ central components sufficient for a detailed representation. In Fig. 1 we illustrate aniRT vectors for randomly selected 8 Chinese characters. In general, we have pre-calculated aniRT vectors for all Chinese characters from the Unicode table. After that, the aniRT vectors are projected up on the $(D+1)$-dimensional hypersphere. A short note on the selected method of projection is given in the appendix.

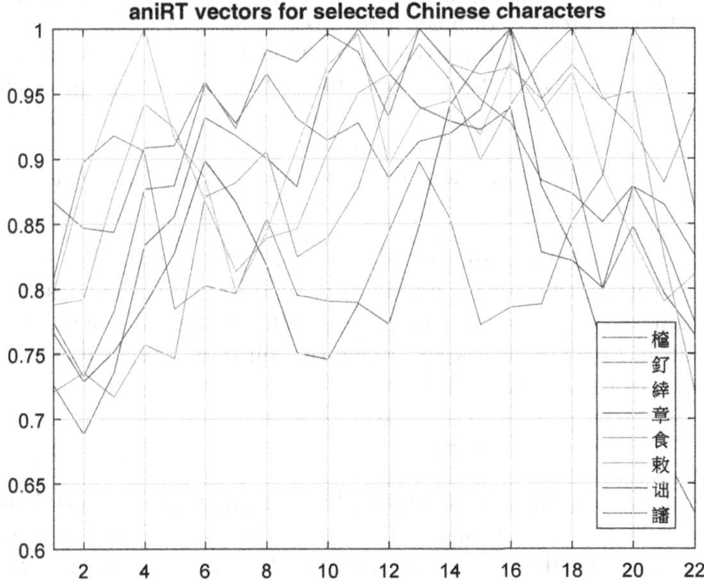

Fig. 1. The aniRT vectors for selected Chinese characters rendered with the Microsoft JhengHei UI font of size 32 points

3 The Self-organizing Module

The self-organizing module that form the networks modelling multimodal integration [15–17] interprets D-dimensional stimuli $\mathbf{x}_n$ in a latent space which is a unity 3-D sphere. There are M neurons/nodes randomly distributed on the surface of the sphere, each node is characterised by the 3-D position vector $\mathbf{v}_m$ and a D-dimensional weight vector $\mathbf{w}_m$. Random, rather than on a regular lattice, distribution of nodes seems to be more biologically plausible. The stimuli are organized into a $D \times N$ matrix, X, one stimulus $\mathbf{x}_n$ per column. The weight vectors, $\mathbf{w}_m$ are organized in a $M \times D$ weight matrix, W, one weight vector per row. Similarly, the position vectors, $\mathbf{v}_m$ are organized in a $M \times 3$ position matrix, V, one location vector per row. All stimuli, weight vectors and position vectors are located on the respective unity hyper-spheres, so that $||\mathbf{x}_n|| = 1$, $||\mathbf{w}_m|| = 1$, $||\mathbf{v}_m|| = 1$. Unlike in many clustering-related applications where number of data points $\mathbf{x}_n$, is greater than the number of nodes, that is $N > M$, in our perceptual modelling case, we maintain a stochastically constant ratio of nodes per stimuli, $\gamma = \frac{M}{N} \approx 16$ to 20 to allow for robustness of the stimuli representation. The specific value has been inferred from the columnar organization of the brain [12].

The fact that the number of stimuli is smaller that the number of weights reverses the classical clustering problem. In our case, for a given stimulus $\mathbf{x}_n$ there should be a central node $\mathbf{v}_m$ with the weight vector at this node $\mathbf{w}_m$ being approximately equal to the stimulus, and the neighbouring nodes having weights "Gaussianly" close to the stimulus.

At this point it would be natural to expect that any Gaussian Mixture Model [1] should solve the problem of placing required weights over the nodes allocated to a specific stimuli. We have started with a model inspired by the Elastic Nets [2,5]. Skipping standard details we ended up with the following log-likelihood to minimize:

$$E(W) = -\sum_{m=1}^{M} \log \sum_{n=1}^{N} \mathcal{N}(\mathbf{x}_n, \mathbf{w}_m, \Sigma_n) + \sum_{m=1}^{M} \frac{1}{2d_m} \sum_{j \in \Lambda_m} \nu_{mj} ||\mathbf{w}_m - \mathbf{w}_j||^2_{C_m} \tag{1}$$

The first term describes a sum of N Gaussians $\mathcal{N}(\mathbf{x}_n, \mathbf{w}_m, \Sigma_n)$, whereas the second term promotes clustering of weights around the winning nodes. The nodes $\mathbf{v}_j$ with weight vector $\mathbf{w}_j$ are located in the neighbourhood Λ_m of the n-th node, see below. From Eq. (1) it is easy to derive the following iterative expression for the next value of the weight matrix, W.

$$W = (I + \beta U)^{-1}(R \cdot X^T + \beta \hat{W}_\Lambda) \tag{2}$$

where R is a $M \times N$ matrix of responsibilities which are formed from ratios of Gaussians, X is a $D \times N$ matrix of stimuli/data points, $\hat{W}_\Lambda$ is a matrix of weighted means of weights in the neighbourhoods Λ, and U is a diagonal matrix formed from the distances between the position vectors in the neighbourhood $\mathbf{v}_j$. Equation (2) converges very quickly to a matrix W, however, instead of spreading winning nodes according to the stimuli $\mathbf{x}_n$ it does the opposite, namely clustering data points to a small number of weights. This behaviour stems from the fact that there are not enough data points for a given number of weights and nodes.

We have adopted the solution originating from the stochastic approximation considerations, which results in solutions close to the original Kohonen learning law, with the additional benefit that the algorithm is obtained by minimization of an energy function. We follow the original paper [6] developed later by [9]. For each stimulus $\mathbf{x}_n$ we calculate the post-synaptic activations, $\mathbf{d}_n = W \cdot \mathbf{x}_n$ and find the node $\mathbf{v}_m^n$ for which the activation attains the maximum. The neighbourhood Λ_n of the winning node is defined by a Gaussian function with the variance $\sigma_{\Lambda_n}^2$ centered at the node $\mathbf{v}_m^n$. The neighbouring nodes are located inside the radius of $r_{\Lambda_n} = 2\sigma_{\Lambda_n}$. Note that the nodes are randomly distributed on a surface of a 3-D sphere. The area of the sphere surface is initially allocated equally to all N stimuli, hence we have $4\pi 1^2/N = \pi r_\Lambda^2$, and $r_\Lambda = 1/\sqrt{N}$, or $\sigma_\Lambda^2 = 1/(4N)$. As a result, we include in the neighbourhood of the winning node $\mathbf{v}_m^n$, all nodes located at $\mathbf{v}_{j_n}$ for which the inner product satisfy the following condition:

$$j_n \in \Lambda_n \text{ if } \mathbf{v}_m^n \mathbf{v}_{j_n}^T > \cos(r_{\Lambda_n}); \quad M_n = |\Lambda_n| \tag{3}$$

where M_n is the number of nodes in the neighbourhood Λ_n. Now, in a way similar to [9], we defined the energy function as:

$$E(W) = \sum_{n=1}^{N} E_n; \;\; E_n = \frac{1}{2} \sum_{j_n \in \Lambda_n} ||\mathbf{w}_{j_n} - \mathbf{x}_n^T||^2 \exp\left(-\frac{||\mathbf{v}_m^n - \mathbf{v}_{j_n}||^2}{2\sigma_{\Lambda_n}^2}\right) \tag{4}$$

To minimize the energy, we calculate the derivative with respect to each weight vector:

$$\frac{\partial E_n}{\partial \mathbf{w}_{j_n}} = (\mathbf{w}_{j_n} - \mathbf{x}_n^T) \exp\left(\frac{\mathbf{v}_m^n \mathbf{v}_{j_n}^T - 1}{\sigma_{\Lambda_n}^2}\right) \tag{5}$$

The above expression results in a Kohonen-like learning law that, for the all weights W_{Λ_n} in the neighbourhood Λ_n specified in Eq. (3), and taking into account the fact that weights need to be kept on the D-dimensional hypersphere, can be written in a way equivalent to the "dot-product" law of the following form

$$W_{\Lambda_n}(t+1) = W_{\Lambda_n}(t) + \eta(t)\Lambda_n(t)(\mathbf{x}_n^T - \mathbf{d}_{\Lambda_n} \cdot W_{\Lambda_n}(t)), \;\; \mathbf{d}_n = W_{\Lambda_n} \cdot \mathbf{x}_n \tag{6}$$

where the neighbourhood function is

$$\Lambda_n(t) = \exp\left(\frac{V_{\Lambda_n} \cdot \mathbf{v}_{j_n}^T - 1}{\sigma_{\Lambda_n}^2(t)}\right) \tag{7}$$

Note that the sizes of the matrices W_{Λ_n} and V_{Λ_n} are $M_n \times D$ and $M_n \times 3$, respectively. Operations between vectors and matrices in Eq. (6) are performed on the row-by-row basis. The learning gain η is reduced according to another Gaussian curve $\eta(t) = \exp(-t^2/(2\sigma_\eta))$, where σ_η is selected so that $\eta(T/2) = 0.5$, T being the total number of epochs. This choice of σ_η ensures a good proportion between the ordering and the convergence phases. The $\sigma_{\Lambda_n}^2(t)$ which describes the narrowing of the neighbourhood is reduced linearly from the initial value $\sigma_{\Lambda_n}^2(1)$ to its final value $\sigma_{\Lambda_n}^2(T)$.

4 Self-organization

In general, the learning law of Eq. (6) can be applied in an on-line/incremental fashion similar to that described in [13]. In this section, we concentrate on demonstrating learning on the sphere, hence, we select randomly $N = 80$ Chinese characters as described in Sect. 2. The results of learning on the sphere are presented in Fig. 2. For $N = 80$ the number of nodes is $M = 20 \times N = 1600$. Since the dimensionality of stimuli is $D = 23$, the sizes of the matrices X, V and W are 23×80, 1600×3 and 1600×23, respectively. In Fig. 2 the neighbourhood for each stimulus $\mathbf{x}_n$ is marked by coloured dots showing the positions of the V_{Λ_n}. The lines indicate the Voronoi tessellation with respect to the positions of the winning nodes. Subject to projection distortion each neighbourhood contains approximately 20 nodes, and the tessellation cells are approximately equal in area. The 3-D view generated by MATLAB uses the orthographic projection, therefore the cells close to the edges appear to be more densely packed. The sphere is semi-transparent, therefore, the characters on the hidden surface are

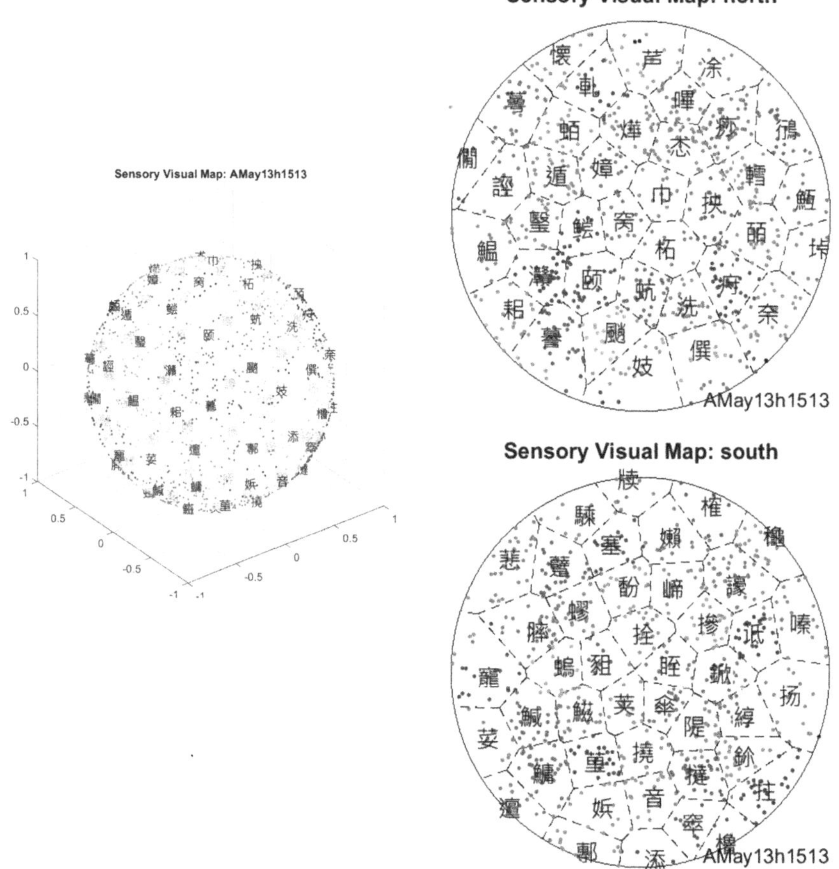

Fig. 2. Learning on a sphere: 3-D view and projection of two hemispheres

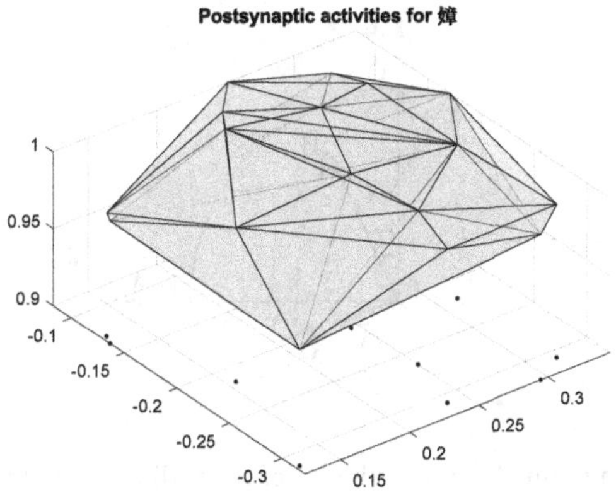

Fig. 3. A map of postsynaptic activities in the neighbourhood of a single stimulus

visible. Our projection method maintains the same proportion along the diameter and the semi-circle, as a consequence looks less distorted.

In Fig. 3 we present a surface of postsynaptic activities for one selected stimulus $\mathbf{x}_n$, namely the character 嫜 (hex '5ADC') that can be found in the upper-central part of the northern hemisphere. The neighbourhood Λ_n consists of 20 nodes V_{Λ_n} with related weights W_{Λ_n}. The components of the vector of postsynaptic activities $\mathbf{d}_{\Lambda_n} = W_{\Lambda_n} \cdot \mathbf{x}_n$ are plotted against the position of the nodes. The positions of the nodes are projected down on a plane $z = 0.9$ and marked with dots in Fig. 3. We use the Delaunay triangulation to create the surface.

As a final comment, please note that the topological ordering of Chinese characters is not done in the order a Chinese speaker might expect, e.g., with respect to radicals. The ordering is strictly based on the aniRT vectors representing the characters as described in Sect. 2.

Conclusion

We have considered problems related to self-organization on a surface of a 3-D sphere in a situation when a number of stimuli is significantly lower than the number of nodes. In this case algorithms originated from the Gaussian Mixture Models might not produce the desired allocation of stimuli to a clusters of nodes. We used an energy function originating from probabilistic considerations which results in a learning law equivalent to the Kohonen-like dot-product law that delivers satisfactory postsynaptic behaviour.

Appendix

Projections between a sphere and a plane related to the problem of creating Earth maps have a long and varied history. The list of available projections is rather long and the Wikipedia[1] is a good point to start.

[1] https://en.wikipedia.org/wiki/List_of_map_projections.

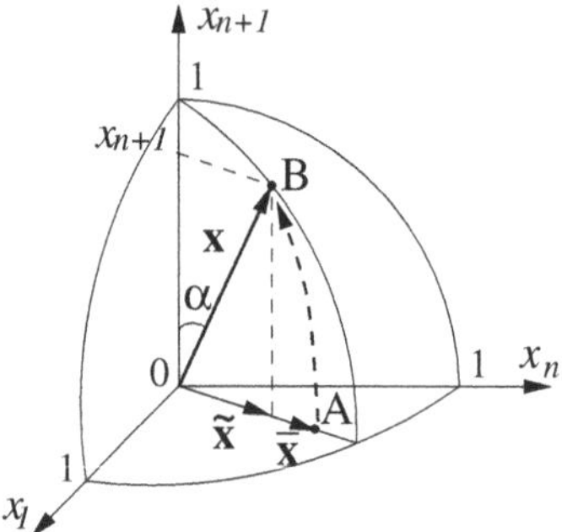

Fig. 4. Projection of a point A from the inside of the n-dimensional hypersphere (shown as a unity circle in 2-D space), onto the $(n+1)$-dimensional unity hypersphere (shown as a point B on the 3-D sphere)

Refer to Fig. 4 and consider an n-dimensional vector $\bar{\mathbf{x}}$ (point A) that needs to be projected onto the $(n+1)$-dimensional unity hypersphere to obtain a unity vector $\mathbf{x}$ (point B). In order to minimise distortions, we select the position of the point B on the grand circle proportional to the position of the point A on the radius. Hence:

$$\alpha = \frac{\pi}{2}\|\bar{\mathbf{x}}\| \tag{8}$$

From this it is easy to obtain the vector $\mathbf{x}$ as:

$$\mathbf{x} = [\tilde{\mathbf{x}}, \cos\alpha] = [k\bar{\mathbf{x}}, \cos\alpha]\,, \quad k = \frac{\sin\alpha}{\|\bar{\mathbf{x}}\|} = \frac{\pi \sin\alpha}{2\alpha} \tag{9}$$

Projecting down the $(n+1)$-dimensional vector $\mathbf{x}$ of Eq. (9) from the surface of the $(n+1)$-dimensional sphere we obtain the related n-dimensional vector $\bar{\mathbf{x}}$ located inside the n-dimensional sphere in the following way:

$$\bar{\mathbf{x}} = \frac{1}{k}\cdot\tilde{\mathbf{x}} = \frac{1}{k}\cdot\mathbf{x}_{1:n} \tag{10}$$

References

1. Bishop, C.M., Svensen, M., Williams, C.K.I.: GTM: the generative topographic mapping. Neural Comput. **10**(1), 215–234 (1998)
2. Carreira-Perpiñán, M.A., Goodhill, G.J.: Generalised elastic nets, pp. 1–52 (2003). arXiv.org: http://arxiv.org/abs/1108.2840
3. Cheng, S.S., Fu, H.C., Wang, H.M.: Model-based clustering by probabilistic self-organizing maps. IEEE Trans. Neural Netw. **20**(5), 805–826 (2009)
4. Chou, S., Papliński, A.P., Gustafsson, L.: Speaker-dependent bimodal integration of Chinese phonemes and letters using multimodal self-organizing networks. In: Proceedings of International Joint Conference on Neural Networks, Orlando, Florida, pp. 248–253, August 2007

5. Cohen, D., Papliński, A.P.: A comparative evaluation of the generative topographic mapping and the elastic net for the formation of ocular dominance stripes. In: Proceedings of WCCI-IJCNN, pp. 3237–3244. IEEE (2012)
6. Heskes, T.: Energy functions for self-organizing maps. In: Kohonen Maps, pp. 303–315. Elsevier (1999)
7. Heskes, T.: Self-organizing maps, vector quantization, and mixture modeling. IEEE Trans. Neural Netw. **12**(6), 1299–1305 (2001)
8. Kohonen, T.: Self-organising Maps, 3rd edn. Springer, Heidelberg (2001)
9. Lau, K., Yin, H., Hubbard, S.: Kernel self-organising maps for classification. Neurocomputing **69**(6), 2033–2040 (2006)
10. Lebbah, M., Jaziri, R., Bennani, Y., Chenot, J.H.: Probabilistic self-organizing map for clusteringand visualizing non-i.i.d data. Int. J. Comput. Intell. Appl. **14**(2), 1–29 (2015)
11. Lopez-Rubio, E.: Probabilistic self-organizing maps for continuous data. IEEE Trans. Neural Netw. **21**(10), 1543–1554 (2010)
12. Mountcastle, V.B.: The columnar organization of the neocortex. Brain **120**, 701–722 (1997)
13. Papliński, A.P.: Incremental Self-Organizing Map (iSOM) in categorization of visual objects. In: Huang, T., Zeng, Z., Li, C., Leung, C.S. (eds.) ICONIP 2012. LNCS, vol. 7664, pp. 125–132. Springer, Heidelberg (2012). doi:10.1007/978-3-642-34481-7_16
14. Papliński, A.P.: The Angular Integral of the Radon Transform (aniRT) as a feature vector in categorization of visual objects. In: Guo, C., Hou, Z.-G., Zeng, Z. (eds.) ISNN 2013. LNCS, vol. 7951, pp. 523–531. Springer, Heidelberg (2013). doi:10.1007/978-3-642-39065-4_63
15. Papliński, A.P., Gustafsson, L., Mount, W.M.: A model of binding concepts to spoken names. Aust. J. Intell. Inf. Process. Syst. **11**(2), 1–5 (2010)
16. Papliński, A.P., Gustafsson, L., Mount, W.M.: A recurrent multimodal network for binding written words and sensory-based semantics into concepts. In: Lu, B.-L., Zhang, L., Kwok, J. (eds.) ICONIP 2011. LNCS, vol. 7062, pp. 413–422. Springer, Heidelberg (2011). doi:10.1007/978-3-642-24955-6_50
17. Papliński, A.P., Mount, W.M.: Bimodal Incremental Self-organizing Network (BiSON) with application to learning Chinese characters. In: Lee, M., Hirose, A., Hou, Z.-G., Kil, R.M. (eds.) ICONIP 2013. LNCS, vol. 8226, pp. 121–128. Springer, Heidelberg (2013). doi:10.1007/978-3-642-42054-2_16
18. Unicode: CJK unified ideographs (2015). http://unicode.org/charts/PDF/U4E00.pdf
19. Yin, H.: The self-organizing maps: background, theories, extensions and applications. Comput. Intell. Compend. **762**, 715–762 (2008)

A Spectrum Allocation Algorithm Based on Optimization and Protection in Cognitive Radio Networks

Jing Gao[1,2(✉)], Jianyu Lv[2], and Xin Song[1,2]

[1] Engineering Optimization and Smart Antenna Institute,
Northeastern University at Qinhuangdao, Qinhuangdao 066004, China
{jinggao166,northeastxinsong}@sina.com

[2] College of Computer Science and Engineering,
Northeastern University, Shenyang 110004, China
jianyulv_1@sina.com

Abstract. Cognitive radio network (CRN) is proposed to solve the problem of the scarce radio spectrum resources. In CRN, primary users (PUs) are allowed to lease out their unused spectrum sharing with cognitive users (CUs). In this paper, we propose a spectrum allocation algorithm based on CUs-demand and PUs-protection in CRN. Our objection is to make the allocated spectrum satisfy CUs demands as much as possible, avoiding the CUs interfering the PUs in the process of spectrum allocation. Simulation results indicate that this algorithm can improve the total spectrum reward, the satisfaction rate of CUs and the protection rate.

Keywords: Cognitive radio network · Spectrum allocation · Cus-demand · Pus-protection

1 Introduction

With the rapid development of the wireless technology, spectrum resources are becoming more and more scarce. Cognitive radio network (CRN) has been a very effective method to solve this problem [1].

Literature [2] proposed a list coloring spectrum allocation algorithm based on graph coloring theory, however, it didn't consider the interference when the different cognitive users (CUs) used the spectrum Simultaneously. Literature [3, 4] proposed a color sensitive graph coloring (CSGC) spectrum allocation algorithm, which not only took the combination of benefit and fairness into account, but also put forward several different label rules based on different target function. But running time was too long. Literature [5, 6] described a parallel spectrum allocation algorithm distributing many spectra in single distribution cycle improving the spectrum efficiency. To enhance the spectrum utilization ratio, aggregation distribution algorithm aggregated small spectrum fragment into a larger one [7–9]. The algorithm allocated the spectrum to users according to their demands and limitation of hardware equipment, comparing with the traditional way of continuous spectrum allocation. Literature [10] took the CUs demand

A. Hirose et al. (Eds.): ICONIP 2016, Part IV, LNCS 9950, pp. 460–469, 2016.
DOI: 10.1007/978-3-319-46681-1_55

and waiting time into account, improving the spectrum reward and the satisfaction rate of CUs. In [11] combining the CUs demand and parallel algorithm reduced the time of spectrum allocation and improved the satisfaction of CUs. However, all of the algorithm above all mentioned only considered the interference among CUs, none of them consider the interference that CUs bring for PUs during the process of spectrum allocation.

In this paper, we put forward a spectrum allocation algorithm based on CUs demand and PUs protection (SAACP). The intention is to improve the total spectrum reward, the satisfaction rate of CUs under the premise of protecting PUs.

2 System Model

We suppose a heterogeneous scenario which consists of I PUs and N CUs. The available spectra are divided into M completely orthogonal channels. The channels are of different bandwidths. There is also a cognitive base station (CBS) in center to handle the spectrum allocation. Each PU has a transmission area with a radius on each channel. So is each CU. A CU can share a channel with a PU only when the CU's transmission area is not overlapped with the PU's. Two CUs cannot use the same channel if there is an overlapped area in their transmission area. An example of the topology is shown in Fig. 1. We set I = 4, N = 5, the four PUs can use channel A–C respectively.

The spectrum allocation problem can be viewed as an equivalent graph coloring problem by mapping each CU into a vertex and each channel into a color. The undirected edge between two vertices represents the interference between the two CUs. When we allocate channel m to user n, it is just similar to assign color m to vertex n. As is shown in Fig. 2.

We consider that the user location and available channels are not changed during the spectrum allocation stage. The spectrum allocation problem can be described by different parameters as below.

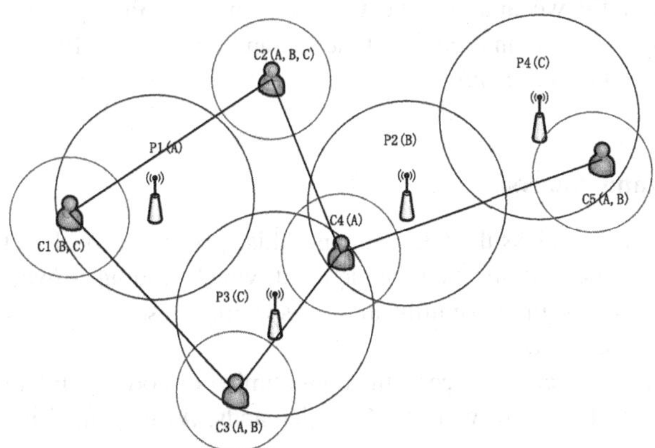

Fig. 1. The topology of CRN

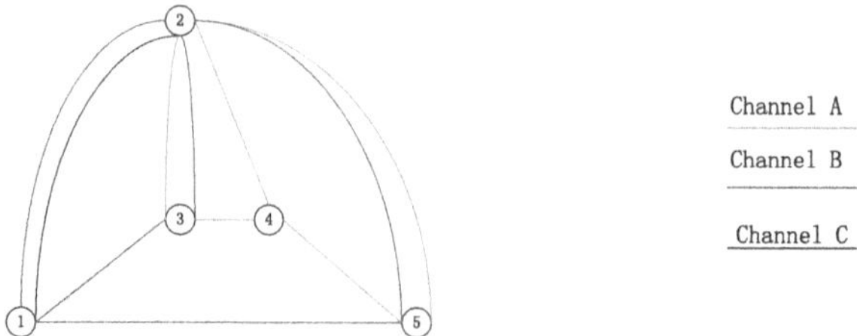

Fig. 2. Graph theory model for CRN (Color figure online)

Channel Availability: $\boldsymbol{L} = \{l_{n,m}|l_{n,m} \in (0,k)\}_{N*M}$, a N by M matrix represents the channel availability. The k stands for minimum number of spectral elements. The channel m is available for CU n if $l_{n,m} = k$. $l_{n,m} = 0$ means channel m is currently used by a PU. Each row total spectral elements of available matrix $\boldsymbol{L}$ is represented by matrix $\boldsymbol{T}$.

CUs demand: $\boldsymbol{D} = \{d_n\}_N$, d_n represents CU n demand. $d_n = [0, a * T_n]$, a represents the CU n demand factor.

Channel Reward: $\boldsymbol{B} = \{b_{n,m}\}_{N*M}$, a N by M matrix represents the channel reward. $b_{n,m}$ describes the maximum transmitting reward, such as the throughput that user n can achieve when using channel m.

Interference Constraint: $\boldsymbol{C} = \{c_{n,t,m}|c_{n,t,m} \in (0,1)\}_{N*N*M}$, a N by M matrix represents the interference constraint among the CUs is determined by the network topology and the channel transmission range. $C_{n,t,m} = 1$ means that CU n and CU t would interfere with each other if they utilize channel m simultaneously, $c_{n,t,m} = 0$ if not. We define $c_{n,n,m} = 1\text{-}l_{n,m}$ if channel m is available to CU n, $l_{n,m} = 1$ and $c_{n,n,m} = 0$.

Channel Allocation: $\boldsymbol{A} = \{a_{n,m}| \in (0,1)\}_{N*M}$, a N by M matrix represent the channel allocation result. If channel m is allocated to user n, we defined $a_{n,m} = 1$.

3 Spectrum Allocation Algorithm

In this section, firstly we analysis the CUs demand and PUs protection. Then our algorithm is proposed with an objective function and a label rule. Finally, we elaborate the spectrum allocation procedure.

3.1 CUs Demand Analysis

If the allocated spectrum less the CUs demand, this spectrum is useless for CUs. If the allocated spectrum much more the CUs demand, which leading to waste of spectrum resources. So in order to improve utilization of spectrum resources, we should take the CUs demand into account.

In the spectrum allocation process, the total number of nodes in the first comparison of the available spectrum unit with the CUs demands. As described in formula (1)

$$\sum_{m=1,m\neq n}^{M} a_{n,m} + l_{n,m} > d_n?. \tag{1}$$

In which $\sum_{m=1,m\neq n}^{M} a_{n,m}$ represents the number of spectral units have been allocated to CU n. The initial state is 0. $l_{n,m}$ Indicates that the sub-graph corresponding to the number of units of the channel. d_n represents CU n demand.

According to formula (1) to decide whether the CU participate in the spectrum allocation. It will be detailed in the spectrum allocation procedure.

3.2 PUs Protection Analysis

In many graph coloring based allocation algorithms, the interference constraint is only defined among the CUs, but the CUs' total interference to the PUs on the same channel is not considered. This is irrational in some network topology, such as Fig. 3 shows.

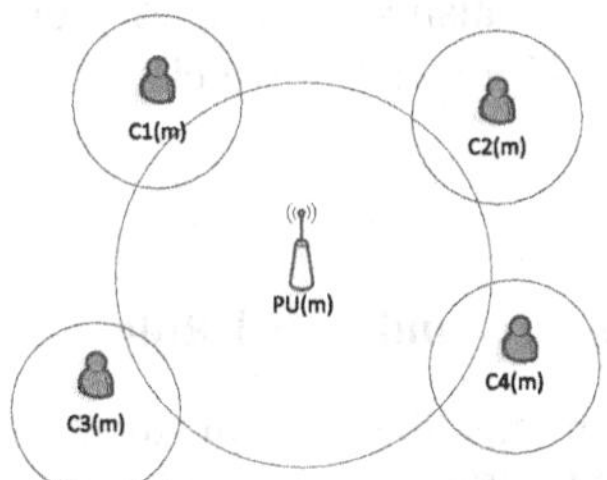

Fig. 3. PU is undetected

In Fig. 4, four CUs all have no overlapped transmission areas with the PU, the PU is detected. The four CUs are all allocated to use the channel m. When the CUs use the channel m simultaneously, although a single CU interference to the PU is faint and ignorant, the total aggregate interference of the four CUs to the PU may be so serious that the PU cannot achieve its SINR threshold. Therefore, we should fully consider the interference avoidance to the PU.

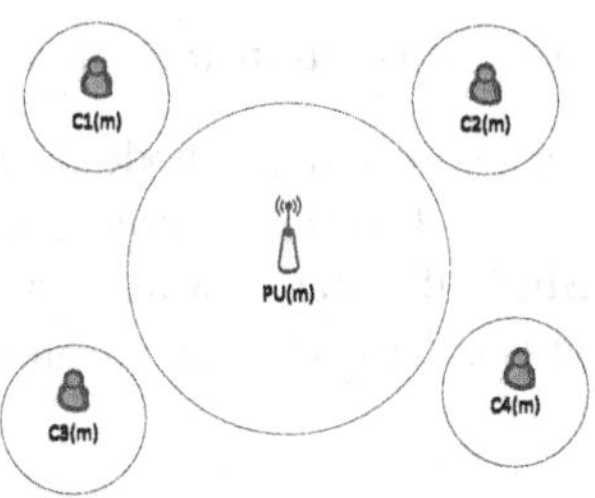

Fig. 4. PU is detected

We assume that each PU has the same SINR threshold γ. Here we denoted the ith PU's transmitting power P_{pi}^{m} on channel m. The lowest transmitting power for CU n to transmit data reliably on channel m is declared as P_{sn}^{m}. The path gain from CU n to PU i on channel m is declared as G_{ni}^{m}. The path gain from PU j to PU i on channel m is declared as G_{ji}^{m}. For the PU i who uses channel m, to achieve its SINR threshold, the following restriction in (2) should be satisfied.

$$\frac{P_{pi}^{m}}{N_0 + \sum_{j=1, j\neq i}^{k} P_{pj}^{m} G_{ji}^{m} + \sum_{n=1}^{N} a_{n,m} P_{sn}^{m} G_{ni}^{m}} \geq \gamma. \tag{2}$$

Here N_0 is the noise power spectral density $\sum_{j=1, j\neq i}^{k} P_{pj}^{m} G_{ji}^{m}$ describes the total interference of the PUs who utilize the channel m (not including PU i) to the PU i, $\sum_{n=1}^{N} a_{n,m} P_{pj}^{m} G_{ji}^{m}$ describes the total interference of the CUs who utilize the channel m to the PU i. In the allocation process, before we assign channel m to CU n, we judge whether the constraint in (2) is satisfied. If not, to protect the PU i, we regard that channel m is not available to CU n and remove channel m from the available channel list of CU n.

3.3 SAACP Objective Function and Label Rule

Our objective function maximums the system spectrum utilization. According to CSGC-CMSB(Collaborative-Max-Sum-Band-width), the SAACP objective function can be expressed as following.

$$F = \max \sum_{n=1}^{N} \sum_{m=1}^{M} a_{n,m} \times \lambda_n \times b_{n,m}. \tag{3}$$

In the equation above, $\lambda_n = \frac{D_n}{\sum_{n=1}^{N} T_n} \leq 1$, and $\sum_{n=1}^{N} T_n$ indicates the total bandwidth of available spectrum for CU n. In order to make sure of the $\lambda_n \leq 1$, the node whose demand is larger than $\sum_{n=1}^{N} T_n$ is deleted from the topology. After deleted, the node would not take part in allocation. So deleting a node does not affect the value of $a_{n,m}$ and objective function F. When the function is used during allocation spectrum, the labeling method CSGC-CMSB(Collaborative-Max-Sum-Bandwidth) will be introduced. As a result, the labeling standards change as follows:

$$Label_n = \max \lambda_n \times \frac{b_{n,m}}{R_{n,m}+1}. \tag{4}$$

$$Color_n = \arg\max \lambda_n \times \frac{b_{n,m}}{R_{n,m}+1}. \tag{5}$$

In which $R_{n,m}$ represents the number of node which can't be shared channel m with CU n.

3.4 Spectrum Allocation Procedure

The SAACP algorithm is shown in Fig. 5.

Step1: Initialize the system parameters, matrices and shield protected the primary user.

Step2: Judge whether the demand of CU n d_n is larger than spectrum channel units T_n. If it is true, it indicates that the spectrum make the CU n needs be never satisfied. So, remove the user from the graph and update the node information. Otherwise, construct spectrum m corresponding sub-graph.

Step3: Calculate the value of CU label, and allocate spectrum to the CU whose label is the largest.

Step4: Judge whether the number of spectrum $\sum_{m=1, m\neq n}^{N} a_{n,m}$ which currently assigned for CU n add the number of spectrum m $l_{n,m}$ which will assign for CU n is

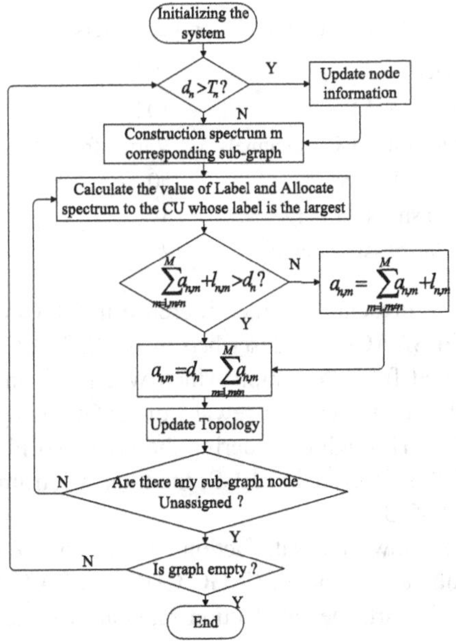

Fig. 5. Flow chart of the allocation algorithm.

larger than CU n spectrum demand d_n. If it is fault, which indicates that CU demand doesn't get met. Then allocate all the spectrum units to CU n, and update node need $a_{n,m} = \sum_{m=1,m\neq n}^{M} a_{n,m} + l_{n,m}$. It will participate in the other sub-graphs spectrum allocation process. If it is true, which indicates CU n spectrum demand d_n gets met. Then allocate the met CU demand spectrum to it. The value is $a_{n,m} = d_n - \sum_{m=1,m\neq n}^{M} a_{n,m}$. The node will exit the process of the sub-graph spectrum allocation and other available spectrum allocation.

Step5: Judge whether there are any sub-graph nod unassigned. If it is true, turn back step 3. If not, continue step6.

Step6: Judge whether the graph is empty. If it is true, it ends here. If not, go back to step2.

4 Simulation Results and Analysis

Throng the following simulation, we compare the three performances among the CSGC, the CSGC based on CU demand (CSGC-CUD), and the SAACP algorithm based on collaboration.

Assume that PUs and SUs are randomly distributed in a 100*100 square area.The parameters need in simulation is in Table 1. The CU demand matrix: $\mathbf{D} = \{d_n | n = 1 : \mathrm{N}\}$, d_n is set randomly within $[0, a * T_n]$. The CU demand factor $\alpha = 0.2 : 0.2 : 2$. The interference constraint is determined by the distribution of CUs. We take the average result of 1000 simulation experiments.

Table 1. Simulation parameters

Parameter	Value
Number of PU I	10
Number of available channels M	Varies from 10 to 30
Number of CU N	50
PU's transmission range	4
CU's transmission range	2

First of all, The Fig. 6 presents that the difference in total spectrum reward among CSGC, CSGC-CUD and SAACP. As described in III-A, The CU whose demand can not be met will be deleted from the graph which will result in reducing the total of spectrum reward. So, the total spectrum reward of CSGC is significantly lower than CSGC-CUD and SAACP. Through considering the protected PUs whose channel will be not available for CUs, as described in III-B, the total spectrum reward of SAACP is sightly lower than CSGC-CUD.

And then the Fig. 7 shows the satisfaction rate when use the three algorithms. Apparently, The CUs satisfaction rate of CSGC-CUD and SAACP is more higher than the CUs of CSGC. In this part, the satisfaction rate can use the expression (6) (7) as follows:

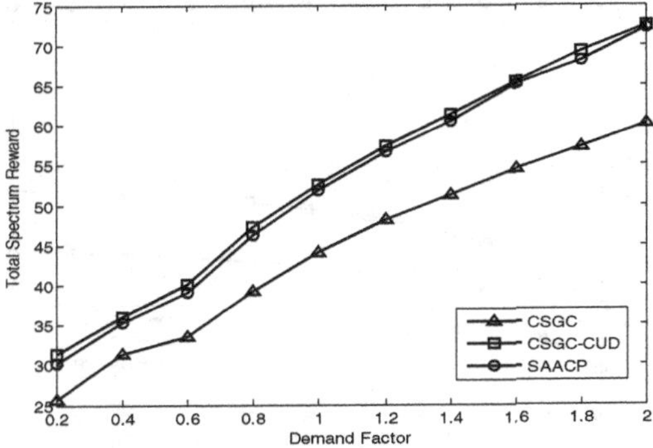

Fig. 6. The comparison of CUs' the overall network utilization.

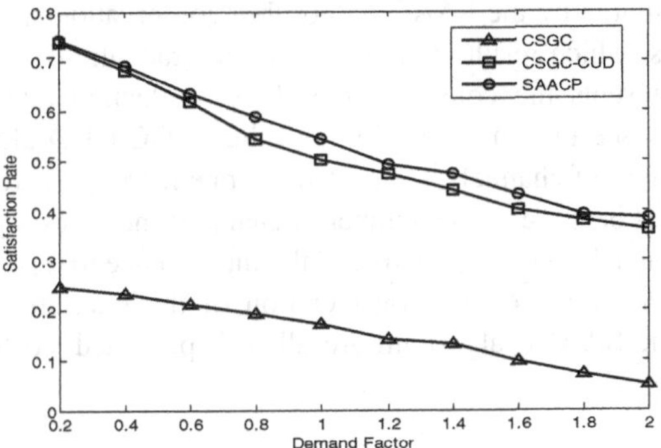

Fig. 7. The comparison of CUs' satisfaction rate.

$$S = \left\{ n | \sum_{m=1}^{M} a_{n,m} \geq D_n \right\}. \tag{6}$$

$$\eta = \frac{\sum\limits_{n \in S} D_n}{\sum\limits_{n=1}^{N} D_n}. \tag{7}$$

In which S represents the number of CUs whose demand have been met. η represents the satisfaction rate of the system. Because the SAACP and CSGC-CUD take the same algorithm, so their satisfaction rates are very close. However, with the demand factor increasing, the advantage of the SAACP and CSGC-CUD algorithm gradually disappear. Because when the demand factor is small, which means CU demand is small, Through the Improve algorithm SAACP can save the portion of the spectrum resource which can be used for other CUs. When the demand factor increases, every CU demand corresponding increase. The competition between CUs enhances, the part

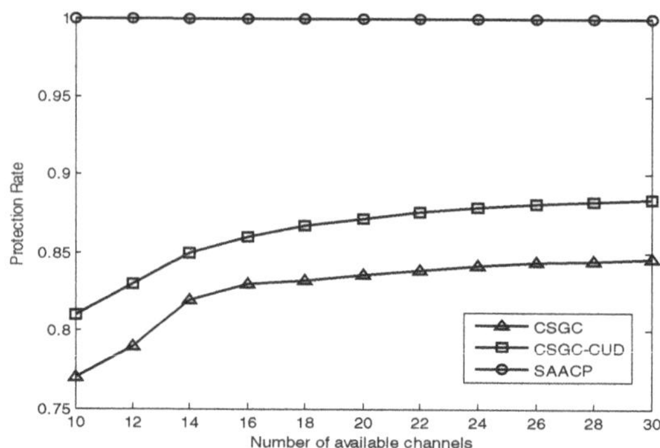

Fig. 8. The comparison of PUs' protection rate.

of the spectrum resource by the SAACP algorithm conservation still can not meet the needs of other CUs, which result in satisfaction rate gradually decline.

Next, Fig. 8 presents the comparison of PUs' protection rate among the three algorithms. We can see that in the CSGC and the CSGC-CUD algorithms, with the increase in the number of channels, the protection rate increases and tends to be stable at about 0.85. This is because as the number of channels increases, the average number of CUs for each channel decreases relatively, the interference to the PU on that channel also becomes lower. However, by consideration of the interference avoidance, the primary users in the SAACP algorithm are all well protected, so the protection rate keeps being 1.

5 Conclusions

In this paper, we proposed an improved algorithm based on CUs' demand and PUs' protection. Through considering CUs' demand can improve the total spectrum reward and satisfaction rate comparing to CSGC, and considering Pus' protection can improve the protection rate comparing to CSGC and CSGC-CUD, it also have some drawbacks, such as it will take longer time, and it also doesn't consider the fairness for CUs and so on.

Acknowledgement. This work has been supported by the National Natural Science Foundation of China under Grant No. 61403069 and No. 61473066, Natural Science Foundation of Hebei Province under Grant No. F2014501055, the Program of Science and Technology Research of Hebei University No. ZD20132003.

References

1. Haykin, S.: Cognitive radio: brain-empowered wireless communications. IEEE J. Sel. Areas Commun. **23**(2), 201–220 (2005). Canada
2. Wang, W., Liu, X.: List-coloring based channel allocation for open-spectrum wireless networks. In: Proceedings of the 62nd IEEE Vehicular Technology Conference (VTC), United States, pp. 690–694 (2006)

3. Peng, C., Zhao, B.Y.: Utilization and fairness in spectrum assignment for opportunistic spectrum access. Mobile Netw. Appl. **11**(4), 555–576 (2006). Beijing, China
4. Zheng, H., Peng, C.: Collaboration and fairness in opportunistic spectrum access. In: IEEE International Conference on Communications ICC, BeiJing, China, pp. 3132–3136 (2005)
5. Liang, Q.L.: Radar sensor wireless channel modeling in foliage environment: UWB versus narrowband. IEEE Sens. J. **11**(6), 1448–1457 (2011). United States
6. Liang, J., Liang, Q.L.: Design and analysis of distributed radar sensor networks. IEEE Trans. Parallel Distrib. Syst. **22**(11), 1926–1933 (2011). United States
7. Poston, J.D., Horne, W.D.: Discontinuous OFDM considerations for dynamic spectrum access networks. In: IEEE Conference Publications, Baltimore, pp. 607–610 (2005)
8. Lee, H., Vahid, S., Moessner, K.: A survey of radio resource management for spectrum aggregation in LTE-advanced. IEEE Commun. Surv. Tutorials **16**(2), 745–760 (2014). United Kingdom
9. Sheng, F., Ma, L., Tan, X., Yin, C., Yu, Y.: Spectrum allocation algorithm aware spectrum aggregation in cognitive radio networks. In: 2013 The Third International Conference on Instrumentation, Measurement, Computer, Communication and Control(IMCCC), Harbin, China, pp. 75–79 (2013)
10. Ping, L., Jinyu, X.: Improvement of CSGC algorithm based on users waiting time and bandwidth demand. In: International Conference on Multimedia Technology(ICMT), Ningbo, China, p. 4 (2010)
11. Yan, B., Wang, S.: Research on an improved parallel algorithm based on user's requirement in cognitive radar network. EURASIP J. Wirel. Commun. Network, UNSP 47. Springer, China (2015)

A Conjugate Gradient-Based Efficient Algorithm for Training Single-Hidden-Layer Neural Networks

Xiaoling Gong[1,2,3,4(✉)], Jian Wang[1,2,3,4], Yanjiang Wang[1,2,3,4], and Jacek M. Zurada[1,2,3,4]

[1] College of Information & Control Engineering, China University of Petroleum, Qingdao 266580, China
s14050610@s.upc.edu.cn

[2] College of Science, China University of Petroleum, Qingdao 266580, China

[3] Department of Electrical and Computer Engineering, University of Louisville, Louisville, KY 40292, USA
{wangjiannl,yjwang}@upc.edu.cn

[4] Information Technology Institute, University of Social Sciences, 90-113 Łódź, Poland
jacek.zurada@louisville.edu

Abstract. A single hidden layer neural networks (SHLNNs) learning algorithm has been proposed which is called Extreme Learning Machine (ELM). It shows extremely faster than typical back propagation (BP) neural networks based on gradient descent method. However, it requires many more hidden neurons than BP neural networks to achieve assortive classification accuracy. This then leads more test time which plays an important role in practice. A novel learning algorithm (USA) for SHLNNs has been presented which updates the weights by using gradient method in the ELM framework. In this paper, we employ the conjugate gradient method to train the SHLNNs on the MNIST digit recognition problem. The simulated experiment demonstrates the better generalization and less required hidden neurons than the common ELM and USA.

Keywords: Neural network · Extreme learning machine · Conjugate gradient · MNIST

1 Introduction

Back propagation (BP) neural networks have been widely used in pattern recognition, computational intelligence. Gradient descent method is one of the most

J. Wang—This work was supported in part by the National Natural Science Foundation of China (No. 61305075), the China Postdoctoral Science Foundation (No. 2012M520624), the Natural Science Foundation of Shandong Province (No. ZR2013FQ004, ZR2013DM015), the Specialized Research Fund for the Doctoral Program of Higher Education of China (No. 20130133120014) and the Fundamental Research Funds for the Central Universities (No. 13CX05016A, 14CX05042A, 15CX05053A, 15CX08011A).

A. Hirose et al. (Eds.): ICONIP 2016, Part IV, LNCS 9950, pp. 470–478, 2016.
DOI: 10.1007/978-3-319-46681-1_56

popular optimal techniques to train BP neural networks. Two main drawbacks such as slow convergence speed and easy trapped in local minimum constrain the BP neural networks efficiently used in some real applications.

An extremely fast learning algorithm for SHLNNs has been proposed in [1] which is named extreme learning machine (ELM). For training process, it first randomly choose the weights between input and hidden layers, then uses Moore-Penrose generalized inverse formula [2,3] to compute the output weights. The singular value decomposition (SVD) is the popular method to compute the Moore-Penrose generalized inverse. Instead of using SVD method, a conjugate gradient method was employed to compute the generalized inverse in [4]. The simulations demonstrated the faster training speed than the common ELM under the condition of the same generalization accuracy.

It is clear to see that there is no iteration step in the whole training process which is significantly different from the common BP neural networks and the support vector machine (SVM). For more details of the comparison between ELM and SVM, we refer to the survey work [5].

As a fact, ELM displays a very faster training speed and better generalization. Unfortunately, one of the main shortcomings of ELM is that the trained network model requires more hidden neurons to achieve the matched performance. An improved algorithm based on ELM, upper-layer-solution-aware (USA) algorithm, has been proposed in [9] which effectively reduces the redundant hidden neurons. The main idea of this algorithm is that it uses the gradient descent method to iteratively update the weights between in put and hidden layers. This is similar to the standard training procedure of BP neural network. The significant difference exists in the expression of the output weights. They are evaluated by using Moore-Penrose generalized inverse, which is analogous to the ELM training. It sounds like a combination of BP algorithm and ELM. Compared with BP algorithm, it enjoys a considerable faster training speed. It requires much less hidden neurons than the typical ELM. This then results in less testing time which plays an important role in real applications.

Inspired by the novelty in Yu's [9], an efficient learning algorithm based on conjugate gradient method has been presented in this paper. Gradient descent method is a special case of conjugate gradient method which engages a faster rate of convergence. Although there is a heavier computational burden of conjugate gradient method, it is restricted in a acceptable range. We compare our algorithm with ELM and USA in the MNIST digit recognition database. The simulation demonstrates that the proposed algorithm, in this paper observes significantly better generalization.

The rest of the paper is organized as follows. In Sect. 2, we give a briefly review of ELM and USA. In Sect. 3, we describe our algorithm which stems from the idea in training the SHLNNs with conjugate gradient method. The experimental results on MNIST database have been demonstrated in Sect. 4. Finally, we conclude the paper in Sect. 5.

2 Related Works

In the community of artificial neural networks, single-hidden-layer neural networks (SHLNNs) have been widely studied which include the popular networks such as radial basis function (RBF) neural networks and multi-layer perceptron (MLP) neural networks.

Given N arbitrary distinct training samples $(\mathbf{x}_j, \mathbf{t}_j)$, where $\mathbf{x}_j = [x_{j1}, x_{j2}, \cdots, x_{jn}]^T \in \mathbf{R}^n$ are input vectors, and $\mathbf{t}_j = [t_{j1}, t_{j2}, \cdots, t_{jm}]^T \in \mathbf{R}^m$ are the desired vectors. Typically, the sigmoid function $g(x)$ is used to be the activation function of the hidden layer, while linear function $f(x) = x$ is the activation function of output layer for SHLNNs. Assume the number of hidden nodes is set to $\widetilde{N}$, then the output of a standard SHLNNs can be evaluated as follows:

$$\mathbf{y}_j = \sum_{i=1}^{\widetilde{N}} \mathbf{u}_i g_i(\mathbf{w}_i \cdot \mathbf{x}_j + b_i), \quad j = 1, \cdots, N, \tag{1}$$

where $\mathbf{w}_i = [w_{i1}, w_{i2}, ..., w_{in}]^T$ is the weight vector from the input nodes to the ith hidden node, $\mathbf{u}_i = [u_{i1}, u_{i2}, ..., u_{in}]^T$ is the weight vector from the ith hidden node to the output nodes, and b_i is the bias of the ith hidden node.

It has beenha proved that SHLNNs with a nonpolynomial activation function can approximate (in measure) any continuous functions [10,11], which can be mathematically modeled as

$$\mathbf{y}_j = \sum_{i=1}^{\widetilde{N}} \mathbf{u}_i g_i(\mathbf{w}_i \cdot \mathbf{x}_j + b_i) = \mathbf{t}_j, \quad j = 1, \cdots, N, \tag{2}$$

In real applications, the neural networks are trained to get the specific $\widehat{\mathbf{w}}_i$, $\widehat{b}_i$ and $\widehat{\mathbf{u}}_i$ such that

$$\min_{\mathbf{w}_i, b_i, \mathbf{u}_i} \sum_{j=1}^{N} \|\mathbf{y}_j - \mathbf{t}_j\|^2. \tag{3}$$

There are numerous methods to implement network models which minimize the above objective function. We first introduce two efficient learning algorithm, and then present a novel one in the next section.

2.1 Extreme Learning Machine

There are increasing interests in SHLNNs with the above least square error function. A popular learning technique, extreme learning machine (ELM), was proposed in [1] which has the advantages both on simple structure and fast training speed.

Consider the total training samples together, (2) can integrated as the following matrix equation

$$\mathbf{HU} = \mathbf{T}, \tag{4}$$

where $\mathbf{H}(\mathbf{w}_i, \mathbf{b}_i)$ is the hidden layer output matrix of the neural network [12,13], $\mathbf{U} = [\mathbf{u}_1, \mathbf{u}_2, \cdots, \mathbf{u}_{\widetilde{N}}]^T$ is the hidden-output layer weights matrix combined $\widetilde{N}$ vectors $\mathbf{u}_i$, and $\mathbf{T} = [\mathbf{t}_1, \mathbf{t}_2, \cdots, \mathbf{t}_N]^T$ is output matrix combined N vectors $\mathbf{t}_j$.

The hidden layer output matrix $\mathbf{H}$ can be determined by randomly chosen input-hidden layer weights $\mathbf{w}_i$ and the biases b_i. Thus, (4) is simply equivalent to a linear system. If the number of hidden layer nodes $\widetilde{N}$ is not greater than the number of training samples N, matrix $\mathbf{H}$ is invertible with probability 1. (4) has unique solution

$$\mathbf{U} = \mathbf{H}^{-1}\mathbf{T}, \tag{5}$$

which means the SHLNNs can approximate these training samples with zero error [12,13].

However, in practice, to find the minimum solution, ELM model employs the least-square method on this linear system

$$\mathbf{U} = \mathbf{H}^{\dagger}\mathbf{T}, \tag{6}$$

where $\mathbf{H}^{\dagger}$ is the Moore-Penrose generalized inverse of matrix $\mathbf{H}$ [2,3],

$$\mathbf{H}^{\dagger} = (\mathbf{H}^T\mathbf{H})^{-1}\mathbf{H}^T. \tag{7}$$

For ELM training, the initial weights between input and hidden layers are randomly chosen instead of iteratively training, the output weights are directly computed by (6). This is significantly different from many other gradient-based learning algorithms. As a result, ELM performs much faster speed than the conventional BP training algorithms.

As a cost, ELM often requires more hidden neurons than those conventional neural networks when it achieves the matched accuracy. Sequentially, this leads to a more time-consuming situation in test procedure. It is hard to be widely used in practice.

2.2 Upper-Layer-Solution-Aware Algorithm

Upper-layer-solution-aware algorithm has been first proposed in [9] which can effectively reach good classification accuracy with a small sized architecture. The essential idea of this algorithm is that the output weights are considered as a function of input weights. The objective function can then be minimized by using gradient descent method.

For convenient, the total error function can be rewritten as follows

$$E = \|\mathbf{Y} - \mathbf{T}\|^2 = \mathrm{Tr}[(\mathbf{Y} - \mathbf{T})(\mathbf{Y} - \mathbf{T})^T]. \tag{8}$$

Note that the bias terms are implicitly represented as weight terms in which the input dimensions are augmented with 1, $\mathbf{H}$ can be simplified as

$$\mathbf{H} = g(\mathbf{W}^T\mathbf{X}), \tag{9}$$

where $\mathbf{W} = [\mathbf{w}_1, \mathbf{w}_2, \cdots, \mathbf{w}_{\widetilde{N}}]$ is the input-hidden weights matrix with a combination of $\widetilde{N}$ vectors $\mathbf{w}_i$.

Similar to ELM model, the weights $\mathbf{U}$ between hidden and output layers can be expressed as follows,

$$\mathbf{U} = \mathbf{H}^{\dagger}\mathbf{T} = (\mathbf{H}^T\mathbf{H})^{-1}\mathbf{H}^T\mathbf{T}. \tag{10}$$

According to (9) and (10), it is easy to see that $\mathbf{U}$ is a function of the weights $\mathbf{W}$, that is, it can be determined by the inner weights $\mathbf{W}$.

For USA algorithm, the typical gradient method was employed to find the optimal weights $\mathbf{U}$. The gradient of error function E with respect $\mathbf{W}$ is calculated as

$$\begin{aligned}\frac{\partial E}{\partial \mathbf{W}} &= \frac{\partial \mathrm{Tr}[(\mathbf{U}^T\mathbf{H} - \mathbf{T})(\mathbf{U}^T\mathbf{H} - \mathbf{T})^T]}{\partial \mathbf{W}} \\ &= 2\mathbf{X}[\mathbf{H}^T \circ (1 - \mathbf{H})^T \circ [\mathbf{H}^{\dagger}(\mathbf{H}\mathbf{T}^T)(\mathbf{T}\mathbf{H}^{\dagger} - \mathbf{T}^T(\mathbf{T}\mathbf{H}^{\dagger}))]],\end{aligned} \tag{11}$$

where $\circ$ represents the Hadamard product.

The weight updating sequence of $\mathbf{W}$ is given by

$$\mathbf{W}_{k+1} = \mathbf{W}_k - \eta\frac{\partial E}{\partial \mathbf{W}}, \quad k = 0, 1, 2, \cdots, \tag{12}$$

where η is the learning rate.

3 Conjugate Gradient-Based Efficient Algorithm

Motivated by the USA algorithm, we use the conjugate gradient method to updata the weights. In comparison with gradient descent method, conjugate gradient method has a faster convergence speed through choosing conjugate decrease direction instead of gradient direction. More importantly, it does not require computing the second order Hessian matrix which is necessary for Newton method.

For any given input weights $\mathbf{W}$, the output of the network is

$$\mathbf{Y} = \mathbf{U}^T\mathbf{H}. \tag{13}$$

The error function is defined as below

$$E = \|\mathbf{Y} - \mathbf{T}\|^2 = \mathrm{Tr}[(\mathbf{Y} - \mathbf{T})(\mathbf{Y} - \mathbf{T})^T]. \tag{14}$$

Similar to USA algorithm, the gradient of E with respect to W can be written

$$\begin{aligned}\frac{\partial E}{\partial \mathbf{W}} &= \frac{\partial \mathrm{Tr}[(\mathbf{U}^T\mathbf{H} - \mathbf{T})(\mathbf{U}^T\mathbf{H} - \mathbf{T})^T]}{\partial \mathbf{W}} \\ &= 2\mathbf{X}[\mathbf{H}^T \circ (1 - \mathbf{H})^T \circ [\mathbf{H}^{\dagger}(\mathbf{H}\mathbf{T}^T)(\mathbf{T}\mathbf{H}^{\dagger} - \mathbf{T}^T(\mathbf{T}\mathbf{H}^{\dagger}))]].\end{aligned} \tag{15}$$

In this paper, a specific conjugate gradient method, Fletcher-Reeves [14], is used to determine the decreasing direction.

$$\mathbf{d}^k = \begin{cases} -\dfrac{\partial E(\mathbf{W}_k)}{\partial \mathbf{W}}, & \text{if} \quad k = 0, \\ -\dfrac{\partial E(\mathbf{W}_k)}{\partial \mathbf{W}} + \beta_k \mathbf{d}^{k-1}, & \text{if} \quad k \geq 1, \end{cases} \tag{16}$$

where $\frac{\partial E(\mathbf{W}_k)}{\partial \mathbf{W}}$ is the gradient of the k-th iteration, β_k is the conjugate coefficient,

$$\beta_k = \frac{\left(\frac{\partial E(\mathbf{W}_k)}{\partial \mathbf{W}}\right)^T \frac{\partial E(\mathbf{W}_k)}{\partial \mathbf{W}}}{\left(\frac{\partial E(\mathbf{W}_{k-1})}{\partial \mathbf{W}}\right)^T \frac{\partial E(\mathbf{W}_{k-1})}{\partial \mathbf{W}}} \tag{17}$$

Then the updating weight sequence of $\mathbf{W}$ is with

$$\mathbf{W}_{k+1} = \mathbf{W}_k + \eta \mathbf{d}^k \tag{18}$$

where η is a constant learning rate.

This training procedure is summarized as follows

Procedure of CGE Algorithm

Require: $\mathbf{X} \in \mathbb{R}^{n\times N}$ and $\mathbf{T} \in \mathbb{R}^{m\times N}$, $g(x)$, $\widetilde{N}$, K

$\mathbf{W}_0 \leftarrow rand(n, \widetilde{N})$

for $k = 0; k < K; k++$

$\mathbf{H}_k = g(\mathbf{W}_k^T \mathbf{X})$

$\frac{\partial E_k}{\partial \mathbf{W}_k} = 2\mathbf{X}[\mathbf{H}_k^T \circ (1-\mathbf{H}_k)^T \circ [\mathbf{H}_k^\dagger(\mathbf{H}_k\mathbf{T}^T)(\mathbf{T}\mathbf{H}_k^\dagger - \mathbf{T}^T(\mathbf{T}\mathbf{H}_k^\dagger))]]$

$$\mathbf{d}^k = \begin{cases} -\dfrac{\partial E_k}{\partial \mathbf{W}_k}, & \text{if} \quad k = 0 \\ -\dfrac{\partial E_k}{\partial \mathbf{W}_k} + \beta_k^{FR}\mathbf{d}^{k-1}, & \text{if} \quad k \geq 1 \end{cases}$$

$\mathbf{W}_{k+1} = \mathbf{W}_k + \eta \mathbf{d}^k$

end for

$\mathbf{U}_K = (\mathbf{H}_{K-1}^T)^\dagger \mathbf{T}^T$

4 Simulation

To verify the good performance of the proposed CGE algorithm, simulations have been done on the MNIST handwritten digital database. The results have been compared with those of ELM and USA algorithm.

4.1 Database Description

The MNIST handwritten digital database is collected from the NIST database of the National Institute of Standards and Technology, and its each digital image has been normalized to an image for 28×28. The data set is stored in the form of matrix, in which the each sample, i.e., each handwritten number, is a 1×784 vector. Each element in the vector is a number between 0∼255, representing the gray levels of each pixel. The MNIST database has a total number of 60000 for training samples and 10000 for testing samples.

Table 1. Accuracy (%) comparison for different algorithms

Algorithms	Hidden neurons	Training accuracy	Testing accuracy
ELM	64	67.13	67.88
ELM	128	78.32	78.99
ELM	256	85.08	85.55
ELM	512	89.50	89.65
ELM	1024	92.74	92.65
ELM	2048	95.32	94.68
USA	64	85.80	84.27
USA	128	89.71	88.06
USA	256	92.97	90.82
USA	512	95.18	93.79
USA	1024	97.62	95.82
USA	2048	98.93	97.86
CGE	64	86.70	85.89
CGE	128	90.40	89.62
CGE	256	93.35	92.68
CGE	512	96.03	95.58
CGE	1024	97.87	97.63
CGE	2048	98.99	98.95

4.2 Experimental Results

We compared the ELM, USA and the proposed CGE algorithms with same number of hidden neurons: 64, 128, 256, 512, 1024 and 2048. Each network configuration was run 10 times to calculate the mean values of classification accuracies both on training and test sets. Starting with the same initial weights **W** for these three different algorithms, ELM evaluated the output weights **U** at once and with non-iteration. For USA and CGE, the identical stop criteria are configured to guarantee fairly comparison.

The results summarized in Table 1 are compared from two aspects, training and testing accuracies, based on same hidden neurons. It is clear to see that the accuracies are monotonically grown with the increasing number of hidden neurons. We observe that CGE and USA perform much better than the typical ELM mode. And CGE is slightly better than USA algorithm.

To make a simple observation, the testing accuracies have been graphed in Fig. 1 which intuitively describe the generalization abilities of these three algorithms. We can obtain different interesting conclusions from the following two perspectives. (1) If we draw vertical lines in Fig. 1, we can observe th different accuracies with same number of hidden neurons. Obviously, the proposed CGE algorithm performs much better than USA and ELM. (2) If we draw horizontal

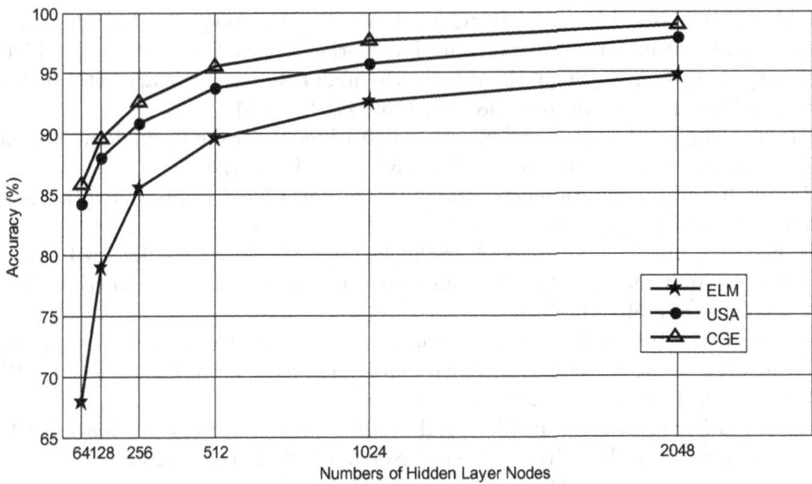

Fig. 1. The testing accuracy of different learning algorithms on the MNIST database.

lines in Fig. 1, we can get that different number of hidden neurons are required to reach the matched accuracy for these three algorithm. It apparently indicates that CGE algorithm may attain the simplest network architecture among these algorithms.

5 Conclusions

In this paper, a novel efficient learning algorithm, CGE, has been proposed for SHLNNs which is motivated by the USA algorithm. Instead of using gradient descent method, a specific conjugate gradient method, F-R, has been employed to train the networks. The popular digital dataset, MNIST, has been used to verify the advantages of CGE. The simulations demonstrate that CGE performs much better than its counterparts, ELM and USA, and results in simplest network.

References

1. Huang, G.B., Zhu, Q.Y., Siew, C.K.: Extreme learning machine: theory and applications. Neurocomputing **70**(123), 489–501 (2006)
2. Serre, D.: Matrices: Theory and Applications. Springer, New York (2002)
3. Rao, C.R., Mitra, S.K.: Generalized Inverse of Matrices and Its Applications. Wiley, New York (1971)
4. Zhang, P., Wang, X., Gu, D., Zhao, S.: Extreme learning machine based on conjugate gradient. J. Comput. Appl. **35**(10), 2757–2760 (2015)
5. Chorowski, J., Wang, J., Zurada, J.M.: Review and performance comparison of SVM- and ELM-based classifiers. Neurocomputing **128**(5), 507–516 (2014)
6. Bartlett, P.L.: The sample complexity of pattern classification with neural networks: the size of the weights is more important than the size of the network. IEEE Trans. Inf. Theor. **44**(2), 525–536 (1998)

7. Widrow, B., Greenblatt, A., Kim, Y., Park, D.: The no-prop algorithm: a new learning algorithm for multilayer neural networks. Neural Netw. **37**, 182–188 (2013)
8. Zhu, Q.Y., Qin, A.K., Suganthan, P.N., Huang, G.B.: Evolutionary extreme learning machine. Pattern Recogn. **38**(10), 1759–1763 (2005)
9. Yu, D., Deng, L.: Efficient and effective algorithms for training single-hidden-layer neural networks. Pattern Recogn. Lett. **33**(5), 554–558 (2012)
10. Hornik, K.: Approximation capabilities of multilayer feedforward networks. Neural Netw. **4**(2), 251–257 (1991)
11. Leshno, M., Lin, V.Y., Pinkus, A., Schocken, S.: Multilayer feedforward networks with a nonpolynomial activation function can approximate any function. Neural Netw. **6**(6), 861–867 (1993)
12. Huang, G.B., Babri, H.A.: Upper bounds on the number of hidden neurons in feedforward networks with arbitrary bounded nonlinear activation functions. IEEE Trans. Neural Netw. **9**(1), 224–229 (1998)
13. Huang, G.B.: Learning capability and storage capacity of two hidden-layer feedforward networks. IEEE Trans. Neural Netw. **14**(2), 274–281 (2003)
14. Fletcher, R., Reeves, C.M.: Function minimization by conjugate gradients. Comput. J. **7**, 149–154 (1964)

The Ability of Learning Algorithms for Fuzzy Inference Systems Using Vector Quantization

Hirofumi Miyajima[1], Noritaka Shigei[2], and Hiromi Miyajima[2](✉)

[1] Graduate School of Biomedical Sciences, Nagasaki University, Sakamoto, Nagasaki, Japan
k3768085@kagoshima-u.ac.jp

[2] Graduate School of Science and Engineering, Kagoshima University, Korimoto, Kagoshima, Japan
{shigei,miya}@eee.kagoshima-u.ac.jp

Abstract. Many studies on learning of fuzzy inference systems have been made. Specifically, it is known that learning methods using VQ (Vector Quantization) and SDM (Steepest Descend Method) are superior to other methods. We already proposed new learning methods iterating VQ and SDM. In their learning methods, VQ is used only in determination of parameters for the antecedent part of fuzzy rules. In order to improve them, we added the method determining of parameters for the consequent part of fuzzy rules to processing of VQ and SDM. That is, we proposed a learning method composed of three stages as VQ, GIM(Generalized Inverse Matrix) and SDM in the previous paper. In this paper, the ability of the proposed method is compared with other ones using VQ. As a result, it is shown that the proposed method outperforms conventional ones using VQ in terms of accuracy and the number of rules.

Keywords: Fuzzy inference systems · Vector quantization · Neural gas · Generalized inverse method

1 Introduction

There have been many studies on learning of fuzzy systems [1,2]. Their aim is to construct self-turning systems from learning data based on SDM. Some novel methods on them have been developed which (1) create fuzzy rules one by one starting from any number of rules, or delete fuzzy rules one by one starting from a sufficiently large number of rules, (2) use GA (Genetic Algorithm) and PSO (Particle Swarm Optimization) to determine the structure of fuzzy systems [2], (3) use fuzzy inference systems composed of small number of input rule modules, such as SIRMs (Single Input Rule Modules) and DIRMs (Double Input Rule Modules) methods [3], (4) use a self-organization or a vector quantization technique to determine the initial assignment [4–8,11]. Specifically, it is known that learning methods using VQ (Vector Quantization) and SDM (Steepest Descend Method) are superior to other methods [9,11]. With their studies,

A. Hirose et al. (Eds.): ICONIP 2016, Part IV, LNCS 9950, pp. 479–488, 2016.
DOI: 10.1007/978-3-319-46681-1_57

the first learning methods using VQ are ones using VQ only in determining of the initial assignment of parameters [1,4–7]. As the second learning methods using VQ, we proposed learning methods iterating the processing of VQ and SDM, where VQ is used in determination of the assignment of parameters in iterating steps [8,11]. In both the first and second learning methods, VQ is used only in determination of parameters, i.e., center and width ones, for the antecedent part of fuzzy rules. In order to improve them, we added the method determining of parameters, i.e., weight parameters, for the consequent part of fuzzy rules to the second learning methods. It is called the interpolation method using GIM [1]. That is, the third learning method is one composed of three stages. In the first stage, the center parameters are determined by VQ and width parameters are computed from center parameters. In the second stage, weight parameters are determined by solving the interpolation problem using GIM. In the third stage, all parameters are updated using SDM. In iterating processes, parameters of the result of SDM are set to initial ones of the next process. In the previous paper, we proposed a learning method with VQ, GIM and SDM as the third method and showed the effectiveness [9].

In this paper, we will compare the ability of the proposed method with one of conventional methods using VQ and show that the proposed method outperforms conventional methods in terms of accuracy, the number of rules and learning time in the simulation of function approximation problems.

2 Preliminaries

2.1 The Conventional Fuzzy Inference Model

The conventional fuzzy inference model using SDM is described [1]. Let $Z_j = \{1, \cdots, j\}$ and $Z_j^* = \{0, 1, \cdots, j\}$ for the positive integer j. Let R be the set of real numbers. Let $\boldsymbol{x} = (x_1, \cdots, x_m)$ and y be input and output variables, respectively, where $x_i \in R$ for $i \in Z_m$ and $y \in R$. Then the rule of simplified fuzzy inference model is expressed as

$$R_j \; : \text{ if } x_1 \text{ is } M_{1j} \text{ and } \cdots \text{ and } x_m \text{ is } M_{mj} \text{ then } y \text{ is } w_j, \tag{1}$$

where $j \in Z_n$ is a rule number, $i \in Z_m$ is a variable number, M_{ij} is a membership function of the antecedent part, and w_j is the weight of the consequent part.

A membership value of the antecedent part μ_j for input $\boldsymbol{x}$ is expressed as

$$\mu_i = \prod_{j=1}^{m} M_{ij}(x_j). \tag{2}$$

Let c_{ij} and b_{ij} denote the center and the width values of M_{ij}, respectively. If Gaussian membership function is used, then M_{ij} is expressed as

$$M_{ij}(x_j) = \exp\left(-\frac{1}{2}\left(\frac{x_j - c_{ij}}{b_{ij}}\right)^2\right). \tag{3}$$

The output y^* of fuzzy inference is calculated as

$$y^* = \frac{\sum_{i=1}^{n} \mu_i \cdot w_i}{\sum_{i=1}^{n} \mu_i}. \tag{4}$$

In order to construct the effective model, the conventional learning is introduced [1]. The objective function E is determined to evaluate the inference error between the desirable output y^r and the inference output y^*.

Let $D = \{(x_1^p, \cdots, x_m^p, y_p^r) | p \in Z_P\}$ and $D^* = \{(x_1^p, \cdots, x_m^p) | p \in Z_P\}$be the set of learning data and the set of input data of D, respectively. The objective of learning is to minimize the following mean square error(MSE) as

$$E = \frac{1}{P} \sum_{p=1}^{P} (y_p^* - y_p^r)^2. \tag{5}$$

where y_p^* means output for the p-th input $\boldsymbol{x}^p$.

In order to minimize the objective function E, each parameter $\alpha \in \{\boldsymbol{c}, b, w\}$ is updated based on SDM as

$$\alpha(t+1) = \alpha(t) - K_\alpha \frac{\partial E}{\partial \alpha} \tag{6}$$

where t is iteration time and K_α is a learning constant [1].

See Refs. [1,9] about the detailed learning algorithm for the conventional fuzzy inference model.

2.2 Neural Gas and K-Means Methods

Vector quantization techniques encode a data space $V \subseteq R^m$, utilizing only a finite set $C = \{\boldsymbol{c}_i | i \in Z_r\}$ of reference vectors, where m and r are positive integers [10].

Let the winner vector $\boldsymbol{c}_{i(\boldsymbol{v})}$ be defined for any vector $\boldsymbol{v} \in V$ as

$$i(\boldsymbol{v}) = \arg\min_{i \in Z_r} ||\boldsymbol{v} - \boldsymbol{c}_i|| \tag{7}$$

From the finite set C, V is partitioned as

$$V_i = \{\boldsymbol{v} \in V | ||\boldsymbol{v} - \boldsymbol{c}_i|| \leq ||\boldsymbol{v} - \boldsymbol{c}_j|| \ for \ j \in Z_r\} \tag{8}$$

The evaluation function for the partition is defined as

$$E = \sum_{i=1}^{r} \sum_{\boldsymbol{v} \in V_i} \frac{1}{n_i} ||\boldsymbol{v} - \boldsymbol{c}_i||^2, \tag{9}$$

where n_i means the cardinality of dataset V_i.

For neural gas method [10], the following method is used:

Given an input data vector $\boldsymbol{v}$, we determine the neighborhood-ranking $\boldsymbol{c}_{i_k}$ for $k \in Z_{r-1}^*$, being the reference vector for which there are k vectors $\boldsymbol{c}_j$ with

$$||\boldsymbol{v} - \boldsymbol{c}_j|| < ||\boldsymbol{v} - \boldsymbol{c}_{i_k}|| \tag{10}$$

If we denote the number k associated with each vector $\boldsymbol{c}_i$ by $k_i(\boldsymbol{v}, \boldsymbol{c}_i)$, then the adaption step for adjusting the $\boldsymbol{c}_i$'s is given by

$$\triangle \boldsymbol{c}_i = \varepsilon \cdot h_\lambda(k_i(\boldsymbol{v}, \boldsymbol{c}_i)) \cdot (\boldsymbol{v} - \boldsymbol{c}_i) \tag{11}$$

$$h_\lambda(k_i(\boldsymbol{v}, \boldsymbol{c}_i)) = \exp(-k_i(\boldsymbol{v}, \boldsymbol{c}_i)/\lambda) \tag{12}$$

where $\varepsilon \in [0, 1]$ and $\lambda > 0$.

If $\lambda \to 0$, Eq. (11) becomes equivalent to the K-means method [10]. Otherwise, not only the winner $\boldsymbol{c}_{i_0}$ but the second, third nearest reference vector $\boldsymbol{c}_{i_1}$, $\boldsymbol{c}_{i_2}$, etc., are also updated.

Let the probability of $\boldsymbol{v}$ selected from V be denoted by $p(\boldsymbol{v})$.

See Refs. [9,10] about the detailed algorithm for NG and K-means.

2.3 The Relation Between the Proposed Algorithm and Related Works

The fundamental learning algorithm is shown in Fig. 1(a). Initial parameters of $\boldsymbol{c}$, $\boldsymbol{b}$ and $\boldsymbol{w}$ are set randomly and all parameters are updated using SDM until the inference error become sufficiently small [1]. The first learning methods using VQ are ones that the initial assignment of center parameters is determined by NG using learning data D^*, width parameters $\boldsymbol{b}$ are computed from Eq. (18) using center parameters and weight parameters are set randomly. Further, all parameters are updated by SDM [4,7]. Further, the initial assignment of parameters $\boldsymbol{c}$ and $\boldsymbol{b}$ using D instead of D^* in above methods was also proposed [6]. In the former, the initial parameters of $\boldsymbol{c}$ are determined by the result of VQ using only input part D^* for learning data. Therefore, in the later, the initial parameters of $\boldsymbol{c}$ are determined by the probability $p(\boldsymbol{x})$ obtained by considering the rate of output change for learning data D. We will explain about the function $p(\boldsymbol{x})$ later. The second methods using VQ are ones that not only the initial assignment of parameters but also the assignment of parameters in iterating step (output loop of Fig. 1(b), (c), (d)) is also determined by NG using D^*. That is, it is learning method composed of two stages. The center parameters $\boldsymbol{c}$ are determined using D^* by VQ, $\boldsymbol{b}$ is computed by Eq. (18) using the result of center parameters and weight parameters $\boldsymbol{w}$ is set to the results of SDM, where the initial values of $\boldsymbol{w}$ are set randomly. Further, all parameters are updated using SDM for the definite number of learning time. In iterating processes, parameters of the result obtained by SDM are set to initial ones of the next process. Outer iterating process is repeated until the inference error become sufficiently small. Further, the same methods as the above were also proposed using D instead of D^* [8,11]. See Fig. 1(c). The third learning methods using VQ are ones that parameters $\boldsymbol{w}$ are determined using GIM after parameters $\boldsymbol{c}$ and $\boldsymbol{b}$ are determined by VQ using D(or D^*) and parameters are updated based on SDM. See Fig. 1(d). That is, it is learning method composed of three stages. In the first stage, the center parameters $\boldsymbol{c}$ are determined using D by VQ and $\boldsymbol{b}$ is computed from the result of center parameters. In the second stage, weight parameters $\boldsymbol{w}$ are determined by solving the interpolation problem using GIM [1,9]. In the third

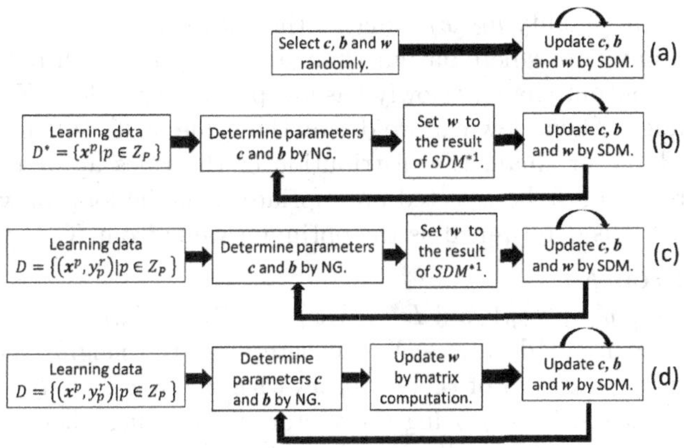

Fig. 1. Concept of conventional and proposed algorithms: the mark $*1$ means that initial values of $\boldsymbol{w}$ are selected randomly and parameters $\boldsymbol{w}$ are set to the result of SDM after the second step.

stage, all parameters are updated using SDM for the definite number of learning time. In iterating processes, parameters of the result of SDM are set to initial ones of the next process. Outer process is repeated until the inference error become sufficiently small.

3 The Proposed Learning Method Using VQ

Let us explain the detailed algorithm of Fig. 1(d). It uses three techniques as follows:

(1) Determine the assignment of $\boldsymbol{c}$ using the probability $p_K(\boldsymbol{x})$.
(2) Determine the assignment of weight parameters $\boldsymbol{w}$ by solving the interpolation problem using GIM.
(3) The processes (1), (2) and learning of SDM are iterated.

The general scheme of the proposed method is shown as follows:

It consists of five phases:

In the step 1, all values for algorithm are initialized. In the step 4, the probability $p_K(\boldsymbol{x})$ is obtained for the size of range K. In the step 5, parameters $\boldsymbol{c}$ are determined by NG using $p_K(\boldsymbol{x})$ and parameters $\boldsymbol{b}$ are computed from parameters $\boldsymbol{c}$. In the step 6, parameters $\boldsymbol{w}$ are determined by GIM. In the step 7 to 10, all parameters are updated by SDM.

Let us explain about Algorithms Prob(K), Weight($\boldsymbol{c}$, $\boldsymbol{b}$) and Center(p) in Table 1.

First, Algorithm Prob in the step 4 is considered as follows:

It is known that many rules are needed to the places where output data change quickly in learning of fuzzy inference systems. Then how can we find the variation of output? The probability $p_K(\boldsymbol{x})$ is one method to perform it. As shown in Eqs. (13) and (14), any data where output changes quickly has high probability and any data where output changes slowly has low probability, where K is the size of considering range. Then, how can we decide the value of K. When K for $p_K(\boldsymbol{x})$ is sufficiently large or small, the distribution of the rules approaches uniform one. Therefore, K must be selected appropriately. In the loop of while2 of the general scheme, it is determined as the optimum value from K_{min} to K_{max}.

Algorithm Prob(K)
$D = \{(x_1^p, \cdots, x_m^p, y_p^r) | p \in Z_p\}$ and $D^* = \{(x_1^p, \cdots, x_m^p) | p \in Z_p\}$:
Step 1: Give an input data $\boldsymbol{x}_i \in D^*$, we determine the neighborhood-ranking $(\boldsymbol{x}^{i_0}, \boldsymbol{x}^{i_1}, \cdots, \boldsymbol{x}^{i_k}, \cdots, \boldsymbol{x}^{i_{P-1}})$ of the vector $\boldsymbol{x}^i$ with $\boldsymbol{x}^{i_0} = \boldsymbol{x}^i$, $\boldsymbol{x}^{i_1}$ being closest to $\boldsymbol{x}^i$ and $\boldsymbol{x}^{i_k} (k = 0, \cdots, P-1)$ being the vector $\boldsymbol{x}^i$ for which there are k vectors $\boldsymbol{x}^j$ with $||\boldsymbol{x}^i - \boldsymbol{x}^j|| < ||\boldsymbol{x}^i - \boldsymbol{x}^{i_k}||$.

Step 2: Determine $H(\boldsymbol{x}^i)$ which shows the rate of change of inclination of the output around output data y^i to input data $\boldsymbol{x}^i$, by the following equation:

$$H(\boldsymbol{x}^i) = \sum_{l=1}^{K} \frac{|y^i - y^{i_l}|}{||\boldsymbol{x}^i - \boldsymbol{x}^{i_l}||}, \tag{13}$$

where $\boldsymbol{x}^{i_l}$ for $l \in Z_M$ means the l-th neighborhood-ranking of $\boldsymbol{x}^i$, $i \in Z_P$ and y^i and y^{i_l} are output for input $\boldsymbol{x}^i$ and $\boldsymbol{x}^{i_l}$, respectively. The number K means the size of the range considering $H(\boldsymbol{x})$.
Step 3: Determine the probability $p_K(\boldsymbol{x}^i)$ for $\boldsymbol{x}^i$ by normalizing $H(\boldsymbol{x}^i)$.

$$p_K(\boldsymbol{x}^i) = \frac{H(\boldsymbol{x}^i)}{\sum_{j=1}^{P} H(\boldsymbol{x}^j)} \tag{14}$$

Second, Algorithm Weight($\boldsymbol{c}$, $\boldsymbol{b}$) in step 6 of Table 1 is considered as follows:

The optimum values of parameters $\boldsymbol{c}$ and $\boldsymbol{b}$ are determined by using $p_K(\boldsymbol{x})$. Then how can we decide weight parameters $\boldsymbol{w}$? We can determine them as the interpolation problem for parameters $\boldsymbol{c}$, $\boldsymbol{b}$, and $\boldsymbol{w}$. That is, it is the method that membership values for antecedent part of rules are computed from $\boldsymbol{c}$ and $\boldsymbol{b}$, and weight parameters $\boldsymbol{w}$ are determined by solving the interpolation problem. So far, the method was used as a determination problem of weight parameters for RBF neural networks [1].

Algorithm Weight($\boldsymbol{c}, b$)
$\boldsymbol{c}$, $\boldsymbol{b}$, $\boldsymbol{w}$: parameters of fuzzy inference system
$D = \{(x_1^p, \cdots, x_m^p, y_p^r) | i \in Z_P\}$
$\boldsymbol{y}^r = (y_1^r, \cdots, y_p^r)^T$
Step 1: Calculate μ_i based on Eq. (2)
Step 2: Calculate the matrix Φ and Φ^+ as follows:

$$\Phi = \begin{bmatrix} \phi_{11} & \phi_{12} & \cdots & \phi_{1n} \\ \phi_{21} & \phi_{22} & \cdots & \vdots \\ \vdots & \vdots & \ddots & \vdots \\ \phi_{P1} & \phi_{P2} & \cdots & \phi_{Pn} \end{bmatrix} \tag{15}$$

Table 1. Algorithm NEWLEARNING($T_{max1}, T_{max2}, \theta_1, \theta_2, K_0, K_{max}$)

```
   c_min, b_min, w_min : the optimal parameters for c, b, w.
   T_max1 and T_max2 : the maximum numbers of learning time for NG and SDM.
   θ1 : the threshold for the rate of change for MSE.
   θ2 : threshold for MSE.
   K0, K_max : the size of initial and final of ranges
   ΔK : the rate of change of the range
   D and D* : learning data D = {(x^i, y_i^r)|i∈Z_p} and D* = {x^i|i∈Z_p}
   R_f : the number of rules
   E(t) : MSE of inference error at step t
   E_min : the minimum number of E
   ΔE(t) : ΔE(t) = |E(t) − E(t − 1)|
1.  Initialize( )
2.  while1 E_min < θ2 do {Determination of the number of rules}
3.   while2 K≠K_max do {Determination of the optimal number of K}
4.    Determine the probability p_K(x) for x∈D using Algorithm Prob(K)
5.    Determine the center and width parameters of membership function
      using Algorithm Center(p) and p_K(D) for x∈D
6.    Determine the weight parameters using Algorithm Weight(c, b)
7.    while3 ΔE(t)≤θ1 or t = T_max2 do {Learning loop by SDM}
8.     Learning of the parameters c, b, w using Algorithm SDM in Sec. 2.1.
9.     Calculate E(t) and ΔE(t)
10.    t←t + 1
11.   end while3
12.   If E(t)≤E_min
13.    E_min←E(t), c_min←c, b_min←b, w_min←w
14.    K←K + 1
15.  end while2
16.  R_f←R_f + 1
17. end while1
```

$$\phi_{pi}(||\boldsymbol{x}^p - \boldsymbol{c}_i||) = \frac{\mu_i^p}{\sum_{j=1}^n \mu_j^p},\ \mu_i^p = \Pi_{j=1}^N \exp\left(-\frac{1}{2}\left(\frac{x_j^p - c_{ij}}{b_{ij}}\right)\right)$$

and

$$\Phi^+ \triangleq [\Phi^T \Phi]^{-1} \Phi^T \tag{16}$$

Step 3: Determine the weight vectors $\boldsymbol{w}$ as follows:

$$\boldsymbol{w} = \Phi^+ \boldsymbol{y}^r \tag{17}$$

Based on Refs. [4,5,9], Algorithm Center in the step 5 of Table 1 is introduced as follows:

In the algorithm, center parameters are determined from NG using $p(\boldsymbol{x})$ and width parameters are computed from the result of center parameters.

Algorithm Center(p)
$D^* = \{(x_1^p, \cdots, x_m^p) | p \in Z_p\}$
$p(\boldsymbol{x})$: the probability $\boldsymbol{x}$ for $\boldsymbol{x} \in D^*$.
Step 1 : By using $p(\boldsymbol{x})$ for $\boldsymbol{x} \in D^*$, NG method [9,10] is performed. As a result, the set C of reference vectors for D^* is made, where $|C| = n$.
Step 2: Each value for center parameters $\boldsymbol{c}$'s is set to a reference vector.
Let

$$b_{ij} = \frac{1}{n_i} \sum_{\boldsymbol{x}_k \in C_i} (c_{ij} - x_{kj})^2, \tag{18}$$

where C_i and n_i are set and the number of learning data belonging to the i-th cluster C_i, and $C = \bigcup_{i=1}^{r} C_i$ and $n = \sum_{i=1}^{r} n_i$.

As a result, center parameters $\boldsymbol{c}$'s are determined from $p(\boldsymbol{x})$ and width parameters $\boldsymbol{b}$'s are determined from center parameters.

4 Numerical Simulations

In order to compare the ability of Learning Algorithm (d) with Learning Algorithms (a), (b) and (c) in Fig. 1, numerical simulations of function approximation are performed. In the following, (a), (b), (c) and (d) mean Learning Algorithms (a), (b), (c) and (d) in Fig. 1, respectively. See Refs. [9,11] about detailed algorithms of (a), (b) and (c). The systems are identified by fuzzy inference systems. This simulation uses four systems specified by the following functions with two-dimensional input space $[0,1]^2$(Eqs. (19), (20), (21) and (22)), and one output with the range $[0,1]$;

$$y = \sin(\pi x_1^3) x_2 \tag{19}$$

$$y = \frac{\sin(2\pi x_1^3)\cos(\pi x_2) + 1}{2} \tag{20}$$

$$y = \frac{1.9((1.35 + \exp(x_1))\sin(13(x_1 - 0.6)^2)\exp(-x_2)\sin(7x_2))}{7} \tag{21}$$

$$y = \frac{\sin(10(x_1 - 0.5)^2 + 10(x_2 - 0.5)^2) + 1}{2} \tag{22}$$

In this simulation, $T_{max1} = 100000$ and $T_{max2} = 50000$ for (a) and $T_{max1} = 10000$ and $T_{max2} = 5000$ for (b), (c) and (d), and $\theta_1 = 1.0 \times 10^{-4}$, $K_0 = 100$, $K_{max} = 190$, $\triangle K = 10$, $K_c = 0.01$, $K_b = 0.01$, $K_c = 0.1$, the number of learning data is 200 and the number of test data = 2500.

Table 2 shows the results for simulations. In Table 2, the number of rules, MSE's for learning and test, and learning time(second) are shown, where the number of rules means one when the threshold θ_1 of inference error is achieved

Table 2. The results of simulations for function approximation

		Eq. (19)	Eq. (20)	Eq. (21)	Eq. (22)
(a)	The number of rules	8.3	22.5	52.4	6.1
	MSE for learning($\times 10^{-4}$)	0.47	0.35	0.65	0.41
	MSE of test($\times 10^{-4}$)	2.29	21.12	2.83	7.37
	Learning time(s)	208.2	1505.6	6438.7	120.3
(b)	The number of rules	4.7	6.8	9.6	4.0
	MSE of learning($\times 10^{-4}$)	0.44	0.38	0.84	0.35
	MSE of test($\times 10^{-4}$)	0.70	2.96	2.34	0.48
	Learning time(s)	559.4	823.9	1725.9	370.6
(c)	The number of rules	5.4	7.4	11.1	3.5
	MSE of learning($\times 10^{-4}$)	0.24	0.54	0.65	0.33
	MSE of test($\times 10^{-4}$)	0.65	1.36	4.48	0.44
	Learning time(s)	157.3	274.5	586.6	68.7
(d)	The number of rules	4.3	6.1	9.7	3.5
	MSE of learning($\times 10^{-4}$)	0.28	0.39	0.6	0.29
	MSE of test($\times 10^{-4}$)	0.57	1.93	1.78	0.36
	Learning time(s)	358.8	657.6	826.8	251.6

in learning. The result of simulation is the average value from twenty trials. As a result, the ability of the proposed method (d) is superior in the number of rules, accuracy and learning time to conventional methods as (a), (b) and (c).

Otherwise, we also performed numerical simulation of other learning methods such as ones using GIM and SDM, and using VQ and SDM. It is shown that the proposed method is superior to these methods. From the result, it is valid to use VQ, GIM and SDM in order to improve the ability of learning methods using VQ.

Further, we performed another simulation for some problems on pattern classification in UCI [12] to compare with other methods as BP [13]. As a result, it was shown that fuzzy inference systems with VQ is superior in terms of the number of rules (parameters) to fuzzy inference system without VQ, back propagation method (BP) and hybrid BP.

5 Conclusion

In this paper, we investigated the ability of the proposed learning algorithm composed of three stages, VQ, GIM and SDM compared to other learning methods using VQ. As a result, it was shown that the proposed method outperformed conventional ones using VQ in terms of accuracy and the number of rules in numerical simulation of function approximation. In the future works, we will apply the idea using VQ to other systems using SDM such as BP(including deep learning).

References

1. Gupta, M.M., et al.: Static and Dynamic Neural Networks. IEEE Press, Hoboken (2003)
2. Cordon, O.: A historical review of evolutionary learning methods for Mamdani-type fuzzy rule-based systems, designing interpretable genetic fuzzy systems. J. Approx. Reason. **52**, 894–913 (2011)
3. Miyajima, H., et al.: SIRMs fuzzy inference model with linear transformation of input variables and universal approximation, advances in computational intelligence. In: Proceedings of the 13th International Work Conference on Artificial Neural Networks, Part I, Spain, pp. 561–575 (2015)
4. Kishida, K., et al.: A self-tuning method of fuzzy modeling using vector quantization. In: Proceedings of the FUZZ-IEEE 1997, pp. 397–402 (1997)
5. Kishida, K., et al.: Destructive fuzzy modeling using neural gas network. IEICE Trans. Fundam. **E80–A**(9), 1578–1584 (1997)
6. Kishida, K., et. al.: A learning method of fuzzy inference rules using vector quantization. In: Proceedings of the International Conference on Artificial Neural Networks, vol. 2, pp. 827–832 (1998)
7. Fukumoto, S., et al.: A decision procedure of the initial values of fuzzy inference system using counterpropagation networks. J. Sig. Process. **9**(4), 335–342 (2005)
8. Pedrycz, W., et al.: Cluster-centric fuzzy modeling. IEEE Trans. Fuzzy Syst. **22**(6), 1585–1597 (2014)
9. Miyajima, H., et. al.: Fast learning algorithm for fuzzy inference systems using vector quantization. In: International MultiConference of Engineers and Computer Scientists 2016, Hong Kong, vol. I, pp. 1–6, March 2016
10. Martinetz, T.M., et al.: Neural gas network for vector quantization and its application to time-series prediction. IEEE Trans. Neural Netw. **4**(4), 558–569 (1993)
11. Miyajima, H., et al.: An improved learning algorithm of fuzzy inference systems using vector quantization. Adv. Fuzzy Sets Syst. **21**(1), 59–77 (2016)
12. UCI Repository of Machine Learning Databases and Domain Theories. ftp://ftp.ics.uci.edu/pub/machinelearning-Databases
13. Miyajima, H., et al.: Performance comparison of hybrid electromagnetism-like mechanism algorithms with descent method. J. Artif. Intell. Soft Comput. Res. **5**(4), 271–282 (2015)

An Improved Multi-strategy Ensemble Artificial Bee Colony Algorithm with Neighborhood Search

Xinyu Zhou(✉), Mingwen Wang, Jianyi Wan, and Jiali Zuo

School of Computer and Information Engineering, Jiangxi Normal University, Nanchang 330022, China
xyzhou@whu.edu.cn

Abstract. Artificial bee colony (ABC) algorithm has been shown its good performance over many optimization problems. Recently, a multi-strategy ensemble ABC (MEABC) algorithm was proposed which employed three distinct solution search strategies. Although its such mechanism works well, it may run the risk of causing the problem of premature convergence when solving complex optimization problems. Hence, we present an improved version by integrating the neighborhood search operator of which object is to perturb the global best food source for better balancing the exploration and exploitation. Experiments are conducted on a set of 22 well-known benchmark functions, and the results show that both of the quality of final results and convergence speed can be improved.

Keywords: Artificial bee colony · Solution search strategy · Neighborhood search · Exploration and exploitation

1 Introduction

Many difficult problems can be expressed as optimization problems in real world. Among these problems, however, most of them are often characterized as non-convex, discontinuous or non-differentiable. It is difficult to solve such problems with traditional optimization methods. As one of the most popular evolutionary algorithms (EAs), artificial bee colony (ABC) algorithm has been shown its superior performance in dealing with some optimization problems [1], such as the flowshop scheduling problem [2], filter design problem [3], and vehicle routing problem [4].

Although ABC has been shown good performance, it also suffers from some knotty shortcomings when solving complex optimization problems, such as slow convergence speed and easily being trapped by local optimum. These shortcomings are mainly caused by its solution search strategy which does well in exploration but badly in exploitation. To fix this insufficiency, during the last few years, some new improved search strategies are designed. For instance, Zhu *et al.* [5] proposed a gbest-guided ABC (GABC) based on the inspiration of

A. Hirose et al. (Eds.): ICONIP 2016, Part IV, LNCS 9950, pp. 489–496, 2016.
DOI: 10.1007/978-3-319-46681-1_58

particle swarm optimization (PSO) algorithm. In GABC, the global best individual (gbest) is incorporated into the solution search strategy for enhancing the exploitation. Gao *et al.* [6] designed an ABC/best/1 search strategy in their modified ABC (MABC), which also utilizes the search information of gbest. Unlike these two representative ABCs, recently, Wang *et al.* [7] proposed a multi-strategy ensemble ABC (MEABC) algorithm. In MEABC, a strategy pool is constructed to contain the original search strategy, the GABC's search strategy and the MABC's search strategy. These three search strategies compete to produce offspring during different stages of the search process. The experimental results have shown the superiority of MEABC.

Although MEABC employs three different search strategies with distinct characteristics, it is not difficult to observe that both of GABC and MABC use the gbest for better exploitation, and this may result in that MEABC is too greedy. Because if the gbest gets stuck in the area of a local optimum in the search space, the entire population would quickly converge to its location and get stagnation. As a potential solution to fix this, the gbest should be perturbed to jump out of the local optimum, so that it can lead the population for further search. Based on this idea, we propose an improved MEABC (iMEABC) by integrating a neighborhood search operator which has been shown its effectiveness in other EAs [8,9]. The structure of our approach is very simple and easy to implement. Twenty-two well-known benchmark functions are used to verify our approach, and the experimental results show that iMEABC has better performance in terms of the quality of solutions and convergence speed.

2 Basic ABC Algorithm

The basic ABC simulates the intelligent foraging behavior of a honeybee swarm which consists of three different kinds of bees: employed bees, onlooker bees and scout bees. Accordingly, the search process of ABC can be divided into three phases. Similar to other EAs, at first, ABC starts with an initial population of *SN* randomly generated food sources. Each food source $X_i = (x_{i,1}, x_{i,2}, \cdots, x_{i,D})$ represents a candidate solution, and D denotes the dimension size. After initialization, these three phases can be described as follows [1].

(1) Employed bee phase
In this phase, each employed bee generates a new food source $V_i = (v_{i,1}, v_{i,2}, \cdots, v_{i,D})$ in the neighborhood of its parent position $X_i = (x_{i,1},_{i,2}, \cdots, x_{i,D})$ by using the following solution search strategy.

$$v_{i,j} = x_{i,j} + \phi_{i,j} \cdot (x_{i,j} - x_{k,j}) \tag{1}$$

where $k \in \{1, 2, \cdots, SN\}$ and $j \in \{1, 2, \cdots, D\}$ are randomly chosen indexes, k has to be different from i. $\phi_{i,j}$ is a random number in the range $[-1, 1]$. If the new food source V_i is better than its parent X_i, then X_i is replaced with V_i.

(2) Onlooker bee phase
After the employed bees finish their search work, the onlooker bees would continue to select part of the food sources to exploit by using the same solution

search equation listed in Eq. (1). The select probability p_i depends on the nectar amounts of a food source, the following Eq. (2) is usually used to calculate the probability.

$$p_i = \frac{f(X_i)}{\sum_{j=1}^{SN} f(X_j)} \tag{2}$$

where $f(X_i)$ is the fitness value of the ith food source. As in the case of the employed bees, the greedy selection method is also employed to retain a better one from the old food source and the new food source.

(3) Scout bee phase
If a food source cannot be further improved for at least *limit* times, it is considered to be exhausted. In ABC, *limit* is the only one single specific control parameter needed to be tuned. For the scout bee, the Eq. (3) is used to generate a new food source to replace the abandoned one.

$$x_{i,j} = a_j + rand_j \cdot (b_j - a_j) \tag{3}$$

where $[a_j, b_j]$ is the boundary constraint for the jth variable, and $rand_j \in [0, 1]$ is a random number.

3 Improved Multi-strategy Ensemble ABC (iMEABC)

3.1 MEABC

In the search process of ABC, different evolution stages usually require different search strategies for maximal performance. These strategies should have diverse characteristics, so they can exhibit distinct search behaviors for different fitness landscapes. In MEABC, a strategy pool is constructed in which different strategies coexist throughout the search process and compete to produce better offspring [7]. The involved strategies are the original one (Eq. (1)), the GABC, and the MABC, respectively, and they are listed as follows.

$$V_i = \begin{cases} X_i + \phi_i \cdot (X_i - X_k) \\ X_i + \phi_i \cdot (X_i - X_k) + \Psi_i \cdot (X_{best} - X_i) \\ X_{best} + \phi_i \cdot (X_{best} - X_k) \end{cases} \tag{4}$$

where X_{best} is the gbest which has the best fitness value among the entire population. In the Eq. (4), the first row is the original search strategy, the second row denotes GABC, and MABC is listed in the last row. Initially, each food source (solution) is randomly assigned a strategy from the strategy pool. When searching a food source in the evolution, each bee generates offspring according to its assigned strategy. Then if the new generated food source is worse than the old one, this indicates the current strategy does not work well, so one out of another two strategies is randomly selected in the next time. The procedure of MEABC is illustrated in Algorithm 1, where FEs is the number of fitness evaluations, $MaxFEs$ denotes the maximum number of function evaluations, and S_i represents the assigned strategy.

Algorithm 1. The procedure of MEABC

```
Randomly generate SN candidate solutions {X_i | i = 1, 2, ..., SN} as food sources;
Calculate their fitness values and set FEs = SN;
Randomly select a strategy S_i from the strategy pool for each solution;
while FEs ≤ MaxFEs do
  for i = 1 to SN do
    switch(S_i)
      According to Eq. (4), generate a new food source V_i by the assigned S_i;
    end switch
    Calculate the fitness value of V_i and set FEs = FEs + 1;
    if f(V_i) < f(X_i) then
      Replace X_i with V_i;
    else
      Randomly select one of the other two search strategies for S_i;
    end if
  end for
  Update the gbest;
end while
```

3.2 The Neighborhood Search Operator

When solving complex optimization problems, ABC often suffers from the problem of premature convergence that its individuals easily be trapped into local optima. Because in the fitness landscape of these problems, there exist a number of local optima which disperse in different regions of the fitness landscape. However it is worth to note that some of the local optima are very close to the global optimum, if an individual is unfortunately trapped by one of them, searching the neighborhoods of this individual would be helpful to find better solutions or even the global optimum. After observing this, different neighborhood search operators have been designed, and they also have been shown superior performance in some EAs [8–11]. Recently, Wang *et al.* [8] presented an efficient global neighborhood search operator for improving PSO algorithm. In this operator, when searching the individual X_i, a trial individual TX_i would be generated as follows.

$$TX_i = r_1 \cdot X_i + r_2 \cdot gbest + r_3 \cdot (X_a - X_b) \tag{5}$$

where r_1, r_2, and r_3 are three mutually exclusive numbers between 0 and 1, they have to meet another condition: $r_1+r_2+r_3 = 1$. What's more, they are randomly generated anew in each generation, and kept the same for all dimensions in each generation. X_a and X_b are two randomly selected food sources and they have to be different from X_i.

3.3 Our Approach

From the strategy pool of MEABC (Eq. (4)), we can observe that although it consists of three different strategies, both of the last two strategies employ the gbest to guide the search of new food sources. Meanwhile, in the search process, each strategy has the equal probability 1/3 of being selected by each bee to search for new food sources. It implies that each bee has 1/3 chance to select the first original search strategy which is good at exploration, while 2/3 chance to select the other two ones which are good at exploitation. To some extent, this mechanism may run the risk of making MEABC too greedy and causing premature convergence. Because if the gbest gets stuck in the area of a local optimum, the entire population would quickly converge to its location and get stagnation. To fix this drawback, we draw inspiration from the neighborhood search operator to disturb the gbest for jumping out of the local optimum, so that the gbest may lead the population toward the direction of the global optimum.

To be specific, after the procedure of MEABC, the neighborhood search operator is carried out. It is worth to point out that we randomly select a food source X_i as the first component of the right hand of Eq. (4), and X_i is different from the gbest, X_a and X_b. This is helpful to extend the available search region around the gbest and improve the exploration. What's more, we also make a small modification for the first search strategy of MEABC. A randomly selected food source is used to replace the first component of the right hand of the first search strategy for further strengthening the exploration. The procedure of our approach is illustrated in Algorithm 2.

Algorithm 2. The procedure of our approach

Randomly generate SN candidate solutions $\{X_i \mid i = 1, 2, \cdots, SN\}$ as food sources;
Calculate their fitness values and set $FEs = SN$;
Randomly select a strategy S_i from the strategy pool for each solution;
while $FEs \leq MaxFEs$ **do**
  Run the procedure of MEABC according to Algorithm 1 to search for new food sources;
  Randomly select a food source X_i, and generate its trial solution TX_i for searching the neighborhood of the gbest according to Eq. (5);
  Calculate the fitness value of TX_i and set $FEs = FEs + 1$;
  if $f(TX_i) < f(X_{best})$ **then**
    Replace the gbest with TX_i;
  end if
end while

4 Experiments and Discussions

4.1 Benchmark Functions and Parameter Settings

To verify the performance of our approach, a set of 22 well-known benchmark functions is used in the experiments. The first 11 functions are unimodal type, while the remaining ones are multimodal, and all of these functions have the same global optimum zero. The name of the involved functions are listed in Table 1, and their definitions can be referred to the literatures [12,13]. For the parameter settings, we follow the original settings of MEABC [7], i.e. $SN = 50$. What's more, the basic ABC is also included for providing a comparison baseline, and it shares the same setting for SN, but its parameter *limit* is set to 100 which is not contained in both of MEABC and iMEABC. In the following experiments, all the tested functions are set to $D = 30$, and the corresponding $MaxFEs$ is set to 100 000. Each algorithm is run 30 times on each test function, and the mean value and standard deviation (mean value ± standard deviation) are recorded.

Table 1. The comparative results of ABC, MEABC, and iMEABC.

Functions	ABC	MEABC	iMEABC
Sphere	4.60E−10±3.72E−10†	3.21E−24 ± 2.65E−24†	1.08E−31 ± 9.98E−32
Schwefel 2.22	1.23E−06 ± 3.48E−07†	2.43E−13 ± 9.86E−14†	7.15E−17 ± 3.41E−17
Schwefel 1.2	7.70E+03 ± 1.41E+03†	1.08E+04 ± 2.32E+03†	1.14E−09 ± 1.60E−09
Schwefel 2.21	4.28E+01 ± 6.74E+00†	1.06E+01 ± 1.07E+00†	5.13E−06 ± 2.55E−06
Rosenbrock	9.20E+00 ± 1.20E+01‡	4.13E+00 ± 5.90E+00‡	2.81E+01 ± 5.65E−01
Step	8.57E+00 ± 9.01E+00†	0.00E+00 ± 0.00E+00≈	0.00E+00 ± 0.00E+00
Quartic	3.95E−01 ± 1.17E−01†	4.30E−02 ± 8.84E−03†	1.50E−03 ± 6.77E−04
Elliptic	1.06E−03 ± 1.93E−03†	3.91E−21 ± 3.38E−21†	3.38E−28 ± 2.54E−28
SumSquare	1.66E−11 ± 1.56E−11†	4.41E−25 ± 3.42E−25†	1.03E−32 ± 6.48E−33
SumPower	1.28E−10 ± 2.79E−10†	8.13E−31 ± 2.62E−30†	1.55E−37 ± 4.29E−37
Exponential	2.27E−05 ± 1.53E−05†	1.53E−06 ± 2.67E−06†	2.45E−14 ± 5.15E−14
Schwefel 2.26	2.91E+01 ± 4.54E+01†	3.82E−04 ± 3.27E−13≈	3.82E−04 ± 3.27E−13
Rastrigin	1.19E−04 ± 6.40E−04†	0.00E+00 ± 0.00E+00≈	0.00E+00 ± 0.00E+00
Ackley	4.38E−06 ± 2.02E−06†	2.16E−12 ± 8.01E−13†	4.94E−15 ± 1.57E−15
Griewank	8.15E−08 ± 2.80E−07†	1.77E−12 ± 9.50E−12†	0.00E+00 ± 0.00E+00
Penalized_1	9.79E−12 ± 9.21E−12†	2.58E−26 ± 1.63E−26†	2.16E−28 ± 2.53E−28
Penalized_2	6.91E−10 ± 6.11E−10†	7.64E−25 ± 6.38E−25†	3.36E−27 ± 3.08E−27
NCRastrigin	2.51E−04 ± 1.24E−03†	0.00E+00 ± 0.00E+00≈	0.00E+00 ± 0.00E+00
Alpine	1.08E−04 ± 6.80E−05†	3.68E−13 ± 2.22E−13†	1.07E−13 ± 8.51E−14
Levy	6.17E−08 ± 5.87E−08†	1.13E−23 ± 1.03E−23†	2.57E−26 ± 2.02E−26
Bohachevsky_2	1.21E−07 ± 1.93E−07†	0.00E+00 ± 0.00E+00≈	0.00E+00 ± 0.00E+00
Weierstrass	4.39E−04 ± 1.12E−04†	1.63E−14 ± 1.02E−14†	0.00E+00 ± 0.00E+00
†/‡/≈	21/1/0	16/1/5	– –

4.2 Comparison Results and Discussions

Table 1 presents the final results among ABC, MEABC, and iMEABC, in which the paired Wilcoxon signed-rank test is used to compare the significance between two algorithms. The signs "†", "‡", and "≈" indicate our approach is better than,

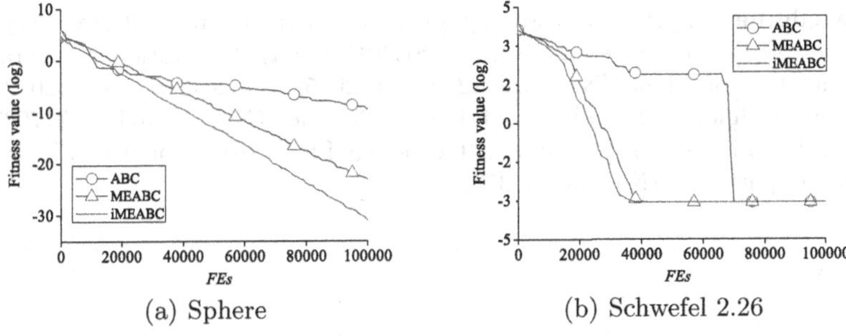

(a) Sphere (b) Schwefel 2.26

Fig. 1. The convergence curves of ABC, MEABC, and iMEABC on two test functions.

worse than, and similar to its competitor according to the Wilcoxon signed-ranked test at $\alpha = 0.05$, respectively. The last row summarizes the comparison results. It can be seen that compared with the basic ABC, our approach wins on all the used test functions except for the Rosenbrock problem. For the MEABC, our approach also obtains promising results. Specifically, our approach wins on 16 test functions, ties on 5 ones while loses on only one test function. The comparison results demonstrate that our approach has good overall performance, and also verify that the idea of utilizing the neighborhood search operator to improve the performance of MEABC can indeed work well.

In addition to the comparison of quality of final results, we further investigate the convergence speed of our approach. Due to the limit of paper space, we only present the convergence curves of the above three algorithms on two representative test functions in Fig. 1. From Fig. 1, it can be seen that our approach can achieve the fastest convergence speed among the involved three algorithms, this also illustrates the neighborhood search operator can speed up the convergence except for improving the quality of final results.

5 Conclusions

In this paper, we presented an improved MEABC (iMEABC) algorithm by integrating the neighborhood search operator. In the MEABC algorithm, it employs a pool of three distinct search strategies: the basic ABC, GABC, and MABC. Although these strategies can exhibit different search behaviors at different evolution stages, it may run a risk of causing the problem of premature convergence. Because both of the strategies of GABC and MABC contain the gbest. If the fitness landscape of the optimization problem at hand is very rugged, the gbest can be easily be trapped into local optima, and this would result in the entire population getting stagnation. However, by integrating the neighborhood search operator, the gbest can be perturbed to some extent, and the chance of jumping out of the local optima can be highly increased. In experiments, the comparison results on a set of 22 well-known benchmark functions show that the idea of integrating the neighborhood search operator works well, and both of the quality of final results and convergence speed of MEABC are improved simultaneously.

Acknowledgments. This work was supported by the Foundation of State Key Laboratory of Software Engineering (No. SKLSE2014-10-04), the National Natural Science Foundation of China (Nos. 61272212, 61462045, 61462043 and 61562042), the Science and Technology Foundation of Jiangxi Province (Nos. 20151BAB217007 and 20151BAB217014), and the Science and Technology Plan Projects of Jiangxi Provincial Education Department (No. GJJ150318).

References

1. Karaboga, D., Gorkemli, B., Ozturk, C., Karaboga, N.: A comprehensive survey: artificial bee colony (ABC) algorithm and applications. Artif. Intell. Rev. **42**, 21–57 (2014)
2. Pan, Q.K., Wang, L., Li, J.Q., Duan, J.H.: A novel discrete artificial bee colony algorithm for the hybrid flowshop scheduling problem with makespan minimisation. Omega **45**, 42–56 (2014)
3. Bose, D., Biswas, S., Vasilakos, A.V., Laha, S.: Optimal filter design using an improved artificial bee colony algorithm. Inf. Sci. **281**, 443–461 (2014)
4. Szeto, W., Wu, Y., Ho, S.C.: An artificial bee colony algorithm for the capacitated vehicle routing problem. Eur. J. Oper. Res. **215**, 126–135 (2011)
5. Zhu, G., Kwong, S.: Gbest-guided artificial bee colony algorithm for numerical function optimization. Appl. Math. Comput. **217**, 3166–3173 (2010)
6. Gao, W., Liu, S.: A modified artificial bee colony algorithm. Comput. Oper. Res. **39**, 687–697 (2012)
7. Wang, H., Wu, Z., Rahnamayan, S., Sun, H., Liu, Y., Pan, J.S.: Multi-strategy ensemble artificial bee colony algorithm. Inf. Sci. **279**, 587–603 (2014)
8. Wang, H., Sun, H., Li, C., Rahnamayan, S., Pan, J.S.: Diversity enhanced particle swarm optimization with neighborhood search. Inf. Sci. **223**, 119–135 (2013)
9. Zhou, X., Wang, H., Wang, M., Wan, J.: Enhancing the modified artificial bee colony algorithm with neighborhood search. Soft Comput. 1–11 (2015). doi:10.1007/s00500-015-1977-x
10. Das, S., Abraham, A., Chakraborty, U.K., Konar, A.: Differential evolution using a neighborhood-based mutation operator. IEEE Trans. Evol. Comput. **13**, 526–553 (2009)
11. Gao, W., Chan, F.T., Huang, L., Liu, S.: Bare bones artificial bee colony algorithm with parameter adaptation and fitness-based neighborhood. Inf. Sci. **316**, 180–200 (2015)
12. Yao, X., Liu, Y., Lin, G.: Evolutionary programming made faster. IEEE Trans. Evol. Comput. **3**, 82–102 (1999)
13. Xiong, G., Shi, D., Duan, X.: Enhancing the performance of biogeography-based optimization using polyphyletic migration operator and orthogonal learning. Comput. Oper. Res. **41**, 125–139 (2014)

Gender-Specific Classifiers in Phoneme Recognition and Academic Emotion Detection

Arnulfo Azcarraga, Arces Talavera, and Judith Azcarraga(✉)

College of Computer Studies, De La Salle University, Manila, Philippines
{arnulfo.azcarraga, arces_talavera, judith.azcarraga}@dlsu.edu.ph

Abstract. Gender-specific classifiers are shown to outperform general classifiers. In calibrated experiments designed to demonstrate this, two sets of data were used to build male-specific and female-specific classifiers. The first dataset is used to predict vowel phonemes based on speech signals, and the second dataset is used to predict negative emotions based on brainwave (EEG) signals. A Multi-Layered-Perceptron (MLP) is first trained as a general classifier, where all data from both male and female users are combined. This general classifier recognizes vowel phonemes with a baseline accuracy of 91.09 %, while that for EEG signals has an average baseline accuracy of 58.70 %. The experiments show that the performance significantly improves when the classifiers are trained to be gender-specific – that is, there is a separate classifier for male users, and a separate classifier for female users. For the vowel phoneme recognition dataset, the average accuracy increases to 94.20 % and 95.60 %, for male only users and female-only users, respectively. As for the EEG dataset, the accuracy increases to 65.33 % for male-only users and to 70.50 % for female-only users. Performance rates using recall and precision show the same trend. A further probe is done using SOM to visualize the distribution of the sub-clusters among male and female users.

Keywords: Multi-layered perceptron · Gender-specific classifier · EEG · Academic emotions · Phoneme recognition · Self-organizing map

1 Introduction

In certain application domains that seek to cluster and classify physiological signals from male and female users, the signals must be processed separately in order to increase the classification performance. Being physically and biologically different, males and females would naturally produce quite distinct physiological signals. For example, it is relatively often easy to distinguish whether a given voice signal comes from a male or female speaker. As such, phoneme recognition would benefit from an additional information as to the gender of the speaker, so that a specialized, gender-specific classifier can be called in to process the voice signals.

Gender-specific classifiers, built for male-only or female-only users, are well expected to outperform general classifiers, where all data from both male and female users are combined. In calibrated experiments designed to demonstrate such a claim,

A. Hirose et al. (Eds.): ICONIP 2016, Part IV, LNCS 9950, pp. 497–504, 2016.
DOI: 10.1007/978-3-319-46681-1_59

two sets of data are used in this study. The first dataset is used to recognize the vowel phonemes “a, e, i, o, u” based on speech signals, while the second dataset is used to predict the negative emotions “confused”, “bored” and “frustrated” based on brainwave (EEG) signals. We show that for both datasets, if we want the prediction rates to be higher, one solution is to build separate classifiers, one for male users and another for female users.

2 Experimental Set-up

The phonemes dataset from [1] contains articulatory and acoustic signals of nineteen (19) male and female speakers reading different English phrases. The dataset used the recordings in the MOCHA-TIMIT database [2] which contains the phoneme recordings, where 9 of those speakers are female and 10 are male. The sounds of the vowel phonemes “a, e, i, o, and u” are extracted from the recordings at different time intervals and were fed to the Mel Frequency Cepstral Coefficient (MFCC) feature extractor in MATLAB in order to get the MFCCs that constitute the speech sounds of the phoneme recordings. The feature extraction yields a total 15 features. Refer to [3] for a thorough review of machine learning techniques for phoneme recognition.

It is well-known that different sounds for vowel phonemes are produced as unobstructed air flow from the mouth, and more specifically according to the different positions and configurations of the tongue and the shape of the lips. The positions of the tongue go from high, to middle, to low, and from front to back, while the shape of the lips also influences the final sound produced when pronouncing specific vowel phonemes. Lax vowels are short vowels such as the phoneme a, while tense vowels are the long vowels which include the phonemes e, i, o, and u. Although both the phonemes e and i are formed with the tongue in the front position, the phoneme e is formed in the middle position of the tongue while the phoneme i is formed at the high position of the tongue. Phonemes o and u are formed with the tongue at the back while also forming round lips, with the phoneme u requiring that the lips are more tightly protruding out of the mouth. Similar to the phoneme i, the phoneme u is formed with the tongue in the high position, while the phoneme o is similar to the phoneme e where both are formed with the tongue in the middle position.

As to the brainwaves dataset, EEG signals were collected from forty-nine (49) students of ages 12 to 16. Thirty one (31) were male, while eighteen (18) were female. The assumption is that students experience negative emotions such as frustration, boredom and confusion while engaged in some learning activities, and these emotions are expressed in terms of different brainwaves or electroencephalogram (EEG) signals [4–6]. The 49 academically-gifted gifted learners were subjected to a calibrated, video-taped learning session using Aplusix, a learning system for algebra [7]. Lasting for about 10 min, the Aplusix problem-solving session involves 6 algebra problems ordered from easy, to moderate, to difficult. The participants were made to wear an Emotiv EPOC headset [8] equipped with 14 channels that capture signals on frontal, temporal, occipital and parietal regions of the brain. Software modules for collecting simultaneous data signals from the EEG sensor, the user screen, and for self-reporting of emotions were deployed. Each student was asked to report the level of their confusion, boredom, frustration as well as the difficulty of the task by clicking on a sliding bar.

Each session begins with each participant being asked to rest for about 3 min. The EEG signals collected during this 'resting-state' are used as baseline data of the said participant. The data signals collected from the experiments are then segmented into 2-second window samples and all the pre-processed EEG data and self-reported emotion tags are synchronized, merged and uniformly segmented into 2-second windows with a 1-second overlap. Artifacts are then removed by EEG data transformation, i.e. frequency transform followed by the application of low-pass and high-pass filters. Each segment is treated as a single instance in the dataset.

For the supervised learning techniques that are used to predict the academic emotion associated to each test instance, the self-reported emotions are used as tags (or label) during training. A slide-bar provided for self-reporting allowed the learners to rate their emotion (confused, frustrated, bored, interest) from 0 to 100. The emotion rate is then discretized to either low or high, which is what is then used to label each instance. The label is low (0) if the emotion value is less than 50, otherwise, the label is set to high (1).

A total of 126 features are extracted from five brainwave frequencies, namely, alpha, theta, beta, delta and gamma waves. The feature values are computed as deviations from the baseline EEG of the 'resting-state' session. The processed data are then normalized and standardized into z-scores within the range of [−3,3]. Extreme z-scores are treated as aberrations and were clipped to −3 or +3. The details are provided in [7].

3 Comparison of Classification Accuracies

To establish that building separate classifiers for each gender would indeed result improved classification accuracy, we prepare three distinct datasets for the vowel phonemes, and likewise three datasets for each of the three academic emotions. The first dataset has all the data samples, the second dataset has only the data samples from the male users, and the third dataset has samples from just the female users. As such, there are a total of 12 datasets (3 for the vowel phonemes and 9 for the three academic emotions), and we compare the performance of male-only and female-only classifiers to that of the classifiers were all male and female users were mixed together.

To evaluate the prediction performance of the classifiers for each of the 3 academic emotions and for the vowel phonemes, RapidMiner is used to train and test the classifiers and to then measure their precision, recall, f-measure and accuracy. For the vowel phoneme dataset, one hidden layer with 13 nodes was used as the architecture of a Multi-Layered Perceptron (MLP). A five-fold cross validation on each of the three datasets was conducted and the resulting baseline confusion matrix, for the combined dataset, is shown in Table 1. The average performance rates for the baseline classifier of the combined dataset are then compared with those of the gender-specific classifiers, and the comparison is summarized in Table 2. Note the significant increase in average accuracies, recall, precision, and f-measures when using gender-specific classifiers. The female-only classifier also performs slightly better than the male-only classifier. In particular, the general classifier recognizes vowel phonemes with a baseline accuracy

Table 1. Baseline confusion matrix of the MLP classifier for vowel phoneme recognition. Note the large number (underlined frequency counts) of erroneous predictions between the vowel phonemes i and e, plus some e sounds being mistaken as the phoneme a. Note also the confusion between o and u.

		Predicted						
		a	**e**	**i**	**o**	**u**	**total**	**Recall**
	a	2070	30	15	4	14	2133	97.0%
Actual	**e**	<u>72</u>	1849	<u>71</u>	2	14	2008	92.1%
	i	14	<u>90</u>	1876	1	14	1995	94.0%
	o	19	5	11	1585	<u>292</u>	1912	82.9%
	u	32	22	19	<u>135</u>	1581	1789	88.4%
total		2207	1996	1992	1727	1915	9837	
Precision		93.8%	92.6%	94.2%	91.8%	82.6%		

Table 2. Comparison of average performance rates of MLP built using the combined and gender-specific datsets for vowel phoneme recognition

	Precision	Recall	F-Measure	Accuracy
All	90.99%	90.89%	90.94%	91.09%
Male-only	94.07%	94.10%	94.09%	94.20%
Female-only	95.68%	95.55%	95.62%	95.60%

of 91.09 %. The performance significantly improves to 94.20 %, for male-only speakers, and to 95.60 % for female only speakers.

In building MLP classifiers for the EEG data, the student-level cross validation technique was employed. In this technique, all the data samples from a given student is set aside for a given run, and used for testing, while all the rest of the samples are used to train the classifier. This is a far-stricter method than a simple cross-validation method were randomly selected samples of a given student are placed in the train set, while the rest are in the test set. Indeed, in the student-level cross validation technique, the classifier will not have seen a sample signal from the same student during training.

Prior to subjecting to student-level cross validation, the train set is balanced by repeating randomly-selected instances until the number of instances in both low and high classes is the same. And this is done for every emotion. Balancing the train set is a critical data preparation step for the proper use of Multi-Layered Perceptron in the case of the EEG dataset since the "low" (L) tag occurs much more frequently among the examples for both the male and female learners. In other words, there were not as many occasions where the students felt bored, confused, or frustrated during the experiments conducted, which is why the number of samples denoting those times when they were bored, confused, and frustrated had to be increased in number.

For each of the negative emotions, the average performance rates for the baseline classifier of the combined dataset are also compared with those of the corresponding

Table 3. Comparison of average performance of MLP classifiers based on the combined and the gender-specific models in predicting the level of negative emotions using EEG signals

		Precision	Recall	F-measure	Accuracy
Frustrated	All	59.50%	48.80%	53.60%	52.10%
	Male	67.20%	52.70%	59.10%	61.30%
	Female	80.10%	80.20%	80.20%	74.90%
Confused	All	73.80%	58.50%	65.30%	64.50%
	Male	64.50%	63.50%	64.00%	66.10%
	Female	81.00%	70.70%	75.50%	65.90%
Bored	All	52.40%	57.90%	55.00%	59.50%
	Male	56.40%	67.80%	61.60%	68.60%
	Female	71.70%	73.90%	72.80%	70.70%
AVERAGE	All	**61.90%**	**55.07%**	**57.97%**	**58.70%**
	Male	**62.70%**	**61.33%**	**61.57%**	**65.33%**
	Female	**77.60%**	**74.93%**	**76.17%**	**70.50%**

gender-specific classifiers. The comparison is summarized in Table 3. Just as with the vowel phonemes, note the significant increase in average accuracies, recall, precision, and f-measures when using gender-specific classifiers. The average accuracy, for example, increases from the baseline 58.70 % to 65.33 % for male-only users, and to 70.50 % for female-only users. Clearly, it is important to remember that EEG signals may vary significantly between male and female learners, hence it is better to build separate classifiers for each gender in order to enhance prediction performance.

4 Data Visualization Using SOM

Clustering and visualization of the data samples of the combined phonemes dataset (for both genders) are also performed using the well-known Self-Organizing Maps (SOM) [9, 10]. The SOM has been trained for 100,000 training iterations during the global ordering phase, and another 100,000 cycles for the fine adjustment phase, with a learning rate that starts at 1.0 and drops to 0.1 by the end of the global ordering phase. The learning rate stays at 0.1 throughout the fine-adjustment phase. Once trained, each SOM is then labeled with the vowel phoneme (A,E,I,O,U) and gender (m,f) of the data sample that is nearest to the node. To do this, we use the Euclidean distance between the data samples and the node weight vectors.

The labeled SOM for the phoneme dataset is shown in Fig. 1. Visual inspection of the labeled SOM affirms the fact that the vowel sounds o and u are somewhat similar, and that the vowels i and e are likewise similar to each other. Indeed, we can see that the left region of the map have nodes that are labeled as i and e, with the upper-left quadrant being labeled mostly as i, while the bottom-left quadrant is labeled as mostly e. The bottom-middle to right regions, on the other hand, are mostly labeled as vowel a. Reflecting the relative classification accuracies shown in Table 1, where the lowest accuracies are for the vowels o and u, we see from the map of Fig. 1 that the nodes labeled as o and u are the ones that have regions of nodes in the labeled SOM that are somewhat mixed together in the central and upper-middle regions of the map.

More importantly, and even more related to the discussion in Sect. 3 about gender specific classifiers for vowel phoneme recognition, Fig. 1 shows that regardless of the gender, most of the vowel phonemes are clustered based on what phonemes they are. That is, there is a contiguous region for the phoneme a, another for the phoneme i, and another for e. Within a phoneme region, however, there are sub-regions, with a sub-region for male speakers (of the same phoneme) and a sub-region for female speakers. This is most notable for the phonemes a, i, and e, and also still somewhat noticeable for phonemes o and u. In other words, the SOM is able to visually render the insight that both male and female speakers are producing highly similar phoneme sounds - in a manner that allows the listener to distinguish the phoneme that is being produced, regardless of whether the speaker is male or female. However, it is also true that for a given vowel phoneme, it is often also possible to discern whether the speaker is male or female. This leads to a quite intuitive expectation that gender-specific classifiers would perform better than general classifiers for phoneme recognition.

SOMs were similarly trained and then labeled for each of the three negative academic emotions. Although not as evident as the distribution of nodes that are sensitive to the five vowel phonemes in Fig. 1, there are also clusters of male-specific as well as female-specific, within node regions of high (H) or low emotion (L). As it can be seen from Fig. 2 for the case of the "bored" emotion, there are sub-regions of nodes that are male, and just as there are sub-regions that are female, among all the nodes that are sensitive to EEG signals pertaining to the "low" bored emotion. The same is true for the sub-regions of male and sub-regions of females among the nodes that are sensitive to the EEG signals for the "high" bored emotion.

If	If	If	If	If	If	If	Om	Uf	Of	Of	Of	Of	Of	Om	Om	Of	Of	Of	Of
If	If	If	If	If	If	If	Uf	Of	Of	Of	Of	Om	Of	Uf	Om	Of	Of	Of	Of
If	If	If	If	If	If	If	Um	Um	Of	Of	Of	Um	Um	Of	Of	Of	Of	Of	Of
If	If	If	If	If	If	If	Om	Om	Om	Um	Um	Of	Uf	Uf	Of	Of	Of	Of	Of
Im	Im	If	If	If	Ef	Ef	Om	Om	Om	Om	Um	Um	Of	Uf	Uf	Uf	Uf	Uf	Uf
Im	Im	If	If	Im	Ef	Ef	Uf	Om	Um	Um	Um	Of	Om	Om	Om	Uf	Uf	Uf	Uf
Im	Im	Im	Im	Im	Im	Im	Im	Om	Om	Um	Um	Um	Om	Om	Om	Uf	Uf	Uf	Uf
Im	Im	Im	Im	Im	Im	Im	Im	Om	Uf	Um	Um	Om	Um	Um	Uf	Uf	Uf	Uf	Uf
Im	Im	Im	Im	Im	Im	Im	Em	Uf	Uf	Uf	Om	Om	Om	Uf	Uf	Um	Um	Um	Om
Im	Im	Im	Im	Em	Em	Em	Em	Uf	Om	Om	Um	Um	Uf	Uf	Um	Um	Um	Om	Om
Im	Im	Im	Em	Em	Em	Em	Em	If	If	Am	Uf	Uf	Uf	Uf	Um	Um	Um	Om	Om
Em	Em	Em	Em	Em	Em	Em	Em	If	Om	Om	Uf	Uf	Um	Uf	Um	Um	Um	Om	Om
Em	Ef	Ef	Ef	Ef	Ef	Em	Em	Im	Im	Om	Um	Um	Um	Uf	Uf	Om	Om	Um	Um
If	If	Ef	Ef	Ef	Ef	Ef	Em	If	If	Um	Um	Am	Af	Am	Am	Am	Am	Um	Um
If	If	Ef	Ef	Ef	Ef	Ef	Ef	Af	Af	Af	Am	Am	Am	Am	Am	Am	Am	Am	Am
Ef	If	Ef	Ef	Ef	Ef	Ef	Ef	Af	Af	Af	Am	Am	Am	Am	Am	Am	Am	Af	Af
Em	Em	Ef	Ef	Ef	Ef	Ef	Af	Af	Af	Af	Af	Am	Am	Af	Am	Am	Am	Af	Af
Em	Em	Ef	Em	Em	Af	Af	Af	Af	Af	Af	Af	Af	Am	Af	Af	Am	Am	Af	Af
Em	Em	Ef	Em	Af	Af	Af	Af	Af	Af	Af	Af	Af	Af	Am	Am	Af	Am	Am	Am
Em	Em	Em	Em	Af	Af	Af	Af	Af	Af	Af	Af	Af	Af	Am	Am	Am	Am	Am	Am

Fig. 1. Self-Organizing Map for the combined dataset of both genders. Each node is labeled with the vowel phoneme (A, E, I, O,U) and gender (m, f) of the data sample that is nearest to the node, based on the Euclidean distance between the data samples and the node weights.

Lm	Lm	Lm	Lm	Lm	Lm	Lm	Lm	Hm	Lm	Lm	Lm	Lm	Lm	Lm
Lm	Lf	Lm	Hm	Hm	Hm	Lm	Lm	Lm	Lm	Lm	Lm	Lm	Lm	Lm
Lf	Lm	Lm	Hm	Hm	Lm	Lm	Lm	Lf	Lf	Lf	Lf	Hm	Lf	Lf
Lm	Lm	Lm	Lm	Hm	Hm	Lm	Lm	Lf	Lf	Lf	Lm	Lm	Lm	Lf
Lm	Lm	Lm	Lm	Hf	Lf	Lm	Lm	Lm	Lf	Lf	Hm	Lm	Lm	Lf
Lf	Lm	Lm	Lm	Hm	Hm	Hm	Lm	Lm	Lm	Lm	Lm	Lm	Lm	Lm
Lf	Lm	Lm	Lm	Hm	Hm	Hm	Lf	Lm	Lm	Lm	Lm	Lm	Hm	Lm
Hf	Lm	Lm	Lm	Lm	Lm	Hm	Lf	Lm	Lf	Lm	Lf	Lm	Lf	Lm
Hf	Lm	Lm	Lm	Hm	Lm	Lm	Lm	Lf	Lf	Lf	Lf	Hf	Hf	Hm
Lf	Lm	Lm	Lm	Lm	Lm	Hm	Lm	Hm	Lf	Lf	Lf	Hf	Hf	Lm
Lm	Lf	Lf	Lm	Lm	Lm	Lm	Lm	Hm	Hm	Hf	Hf	Lf	Lm	Lm
Lf	Lf	Lm	Lf	Lf	Lf	Lf	Lf	Lm	Lm	Hf	Lf	Lf	Lm	Lm
Lf	Lf	Lm	Lm	Lm	Lf	Lf	Lf	Lm	Lm	Lf	Lf	Lf	Lm	Lm
Lf	Lf	Lm	Lm	Lm	Lm	Lf	Lm	Lf	Lf	Lm	Lm	Lm	Lm	Lm
Hf	Hf	Lf	Lf	Lm	Lm	Lm	Lm	Lf	Lf	Lm	Lm	Lm	Lm	Lm

Fig. 2. Self-organizing map for "bored" EEG dataset including both genders. Each node is labeled with the level of boredom (L for low, H for high) and gender (m, f) of the data sample that is nearest to the node.

5 Summary and Conclusion

Gender-specific classifiers, built for male-only or female-only users, are well expected to outperform general classifiers, where all data from both male and female users are combined. In calibrated experiments designed to demonstrate this, two sets of data were used to build male-specific and female-specific classifiers. The first dataset is used to predict the vowel phonemes "a, e, i, o, u" based on speech signals, while the second dataset is used to predict the negative emotions "confused", "bored" and "frustrated" based on brainwave (EEG) signals.

Using RapidMiner, multi-layered perceptrons (MLP) are trained as gender-specific classifiers for each of the datasets, and their classification performance rates are compared to the classifiers built using the combined data from both male and female users. For the vowel phonemes dataset, nineteen (19) different speakers, 9 of which are female and 10 are male, were asked to read English phrases. Articulatory and acoustic features were extracted from the sounds of the vowel phonemes. For the EEG data, thirty one (31) male and eighteen (18) female learners, of ages 12 to 16, answered algebra problems and reported their academic emotion while their brainwaves were being captured using brainwave sensors.

The general MLP classifier that predicts the 5 vowel phonemes shows an accuracy rate of 91.09 %, while the general MLP classifier that is trained for the combined male and female learners can predict the level of confusion, frustration and boredom with an average accuracy of 58.7 %. For both datasets, the performance significantly improves when training and testing are restricted to datasets where the samples come from learners of the same gender. For the vowel phoneme dataset, the average accuracies increase to 94.20 % and 95.60 % for male only users and female-only users, respectively. As for the EEG dataset, the average accuracy increases to 65.33 % for male-only learners, and to 70.5 % for female-only learners. Performance rates using recall and precision also manifest the same trend.

Indeed, in both the case of vowel phoneme recognition and emotion detection based on EEG signals, the male users produce quite distinct signals as compared to female users. The male voice is usually distinguishable from a female voice, even without prior acquaintance or familiarity with the specific male or female speaker. Apparently, brain signals also differ according to gender. Clearly, therefore, if we want the prediction rates to be higher in these kinds of application domains, we have to build separate classifiers for male users and female users.

A further probe is also done using Self-Organizing Maps to visualize the distribution of the sub-clusters among male and female users within a single cluster of a specific vowel phoneme. The trained SOMs clearly illustrate why gender-specific classifiers would perform better than general classifiers for such application domains where males and females, being biologically different, may have to be treated differently since the physiological signals they produce may be quite distinct.

References

1. Agustin, N.: Using self-organizing maps and regression to solve the acoustic-to-articulatory inversion as input to a visual articulatory feedback system, DLSU (2014)
2. Wrench, A.: MOCHA-TIMIT. www.cstr.ed.ac.uk/research/projects/artic/mocha.html
3. Deng, L., Li, X.: Machine learning paradigms for speech recognition: an overview. IEEE Trans. Audio Speech Lang. Process. **21**(5), 1060–1089 (2013)
4. Pekrun, R., Goetx, T., Titz, W., Perry, R.P.: Academic emotions in students' self-regulated learning and achievement: a program of qualitative and quantitative research. Educ. Psychol. **37**(2), 91–105 (2002)
5. Azcarraga, J., Suarez, M.: Recognizing student emotions using brainwaves and mouse behavior data. Int. J. Distance Educ. Technol. **11**(2), 1–15 (2013)
6. Azcarraga, J., Marcos, N., Suarez, M.: Modelling EEG signals for the prediction of academic emotions. In: Workshop on Utilizing EEG Input in Intelligent Tutoring Systems of the 12th International Conference on Intelligent Tutoring Systems (2014)
7. Azcarraga, J.: Analysis and visualization of EEG data towards academic emotion recognition, Ph.D. dissertation, DLSU-Manila (2014)
8. Emotiv EPOC Headset. http://www.emotiv.com
9. Kangas, J.A., Kohonen, T.K., Laaksonen, J.T.: Variants of self-organizing maps. IEEE Trans. Neural Netw. **1**(1), 93–99 (1990)
10. Kohonen, T., Somervuo, P.: Self-organizing maps of symbol strings. Neurocomputing **21**(1), 19–30 (1998)

Local Invariance Representation Learning Algorithm with Multi-layer Extreme Learning Machine

Xibin Jia[1(✉)], Xiaobo Li[1], Hua Du[1], and Bir Bhanu[2]

[1] Beijing Key Laboratory on Integration and Analysis of Large-scale Stream Data, Beijing University of Technology, Beijing 100124, China
jiaxibin@bjut.edu.cn, {lixiaobo,dhmj2012}@emails.bjut.edu.cn

[2] Center for Research in Intelligent Systems, University of California at Riverside, Riverside, CA 92521, USA
bhanu@ee.ucr.edu

Abstract. Multi-layer extreme learning machine (ML-ELM) is a stacked extreme learning machine based auto-encoding (ELM-AE). It provides an effective solution for deep feature extraction with higher training efficiency. To enhance the local-input invariance of feature extraction, we propose a contractive multi-layer extreme learning machine (C-ML-ELM) by adding a penalty term in the optimization function to minimize derivative of output to input at each hidden layer. In this way, the extracted feature is supposed to keep consecutiveness attribution of an image. The experiments have been done on MNIST handwriting dataset and face expression dataset CAFÉ. The results show that it outperforms several state-of-art classification algorithms with less error and higher training efficiency.

Keywords: Local invariant representation learning · Multi-layer extreme learning · Contractive auto-encoder

1 Introduction

Discriminative features play important role in improving the performance of image understanding [1]. However, image feature is selected empirically. For example, local binary pattern is widely used in face recognition field [2]. Histogram of gradient feature is usually selected for pedestrian detection [3]. Additionally, the performance of image understanding is impacted by the selected classifiers. In our earlier work, we made some analysis of the combination performance of the feature and classifier based on several typical features: Gabor and Geometric feature and classifiers: 1-NN (1-nearest neighboring), SMO (Sequential Minimal Optimization), MLP (Multilayer Perceptron) and NB (Naive Bayes). We find that a different combination displays different recognition performance in different expression recognition datasets [4]. This indicates for some extent that proper selection of the feature and classifier combination for a

A. Hirose et al. (Eds.): ICONIP 2016, Part IV, LNCS 9950, pp. 505–513, 2016.
DOI: 10.1007/978-3-319-46681-1_60

certain dataset will enhance the image perception ability. But it is not practical to get the optimal ones based on comprehensive experiments. Fusion of multiple features is one of solution [5]. In some earlier works, the iterative approaches for the feature automatical selection is proposed to improve the closeness of the features to the model. Bir Bhanu et al. proposes a reinforcement learning for adaptive feature in image segmentation problem [6]. Shang liu et al. employs the iterative matching approach to update the feature weights to select the discriminative features in image classification and retrieval problem [7]. Currently, the deep learning algorithms are employed widely in the image understanding [8]. It has obvious advantage in extracting meaningful feature by data-driven learning rather than handmade construction. It is much effective to the complex objects such as images, where the discriminative capability of feature depends largely on the recognition target and disturbed by tangled factors. Recently, Huang et al. proposed a multi-layer extreme learning (ML-ELM) [9]. Same as most deep networks, ML-ELM displays good performance in extracting the meaningful representation with the gradual abstraction characteristic for complex objects. On the other hand, like most extreme learning machine (ELM), ML-ELM has the property of universal approximation and high speed learning advantage, comparing to deep networks, such as DBNs (Deep Boltzmann machine) [10], SAEs (Stacked auto-encoder) [11]. In the deep feature learning methods, definition of the optimization function is important. For ML-ELM, it sets to minimum error between estimated input and the input at each layer under unsupervising learning. However, it This method is not good at processing the variance of input, such as image deformation, noise disturbance. In order to improve input variance robustness, this paper proposes adding a penalty term at the optimal function, which aims to minimize the first-order derivative of the output to the input. This strategy is enlightened by the principle of Contractive Auto-Encoder (CAE) [12]. Rifai et al. proves empirically it results in a localized space contraction and yields robust features at the activation layer by adding the penalty term with the derivation of output to the input by computing the Frobenius norm of the Jacobian matrix of the encoder activations at each hidden layer. Actually, this is reasonable that realistic objects usually keep the smooth structure. In view of this strategy, this paper proposes adding this additional penalty in the optimal function at each layer to improve the local invariance of learning representation. Inheriting the naming of contractive auto encoding algorithm, we call our method contractive multi-layer extreme learning machine, abbreviated as C-ML-ELM.

In the rest of the paper, we organize the paper as follows. Section 2 introduces our proposed method, addresses the relative principle and elaborates parameter learning. Section 3 gives the experimental results and analysis.

2 Local Invariance Representation Learning Algorithm with ML-ELM

Our approach is based on the modification the optimization function of each ELM-AE at ML-ELM. By minimizing the derivation of output to the input, the local contractive feature is obtained. The detail of the proposed algorithm is illustrated as follows.

2.1 Structure of the Proposed C-ML-ELM

As shown at the bottom part of Fig. 1, C-ML-ELM is multi-layer network. Each hidden layer is a full connected network which is constituted of the encoding network of a contractive ELM-AE. The weight β is reflection of the decoding model to the encoding model. The illustration of a complete contractive ELM-AE is showed in the upper part of Fig. 1. It contains two stages: encoding and decoding. At the encoding stage, the input vector X is processed and the feature is obtained. At the coding stage, the estimated signal of input is calculated. Following one of widespread assumption of auto-encoding algorithm, we assume that the encoding and decoding network are symmetric. Therefore, we mapping the trained parameters at decoding network to the encoding network. This is illustrated with the circle mark of parameter β in Fig. 1. The principle and training of parameters is described in the following two sections.

2.2 Local Variance Robust Performance of C-ML-ELM

As illustrated in Fig. 1, the C-ML-ELM is a stacked contractive ELM-AE. The parameters of each hidden layer are initialized independently in unsupervised way. In contrast to other deep networks such as DBM, C-ML-ELM doesn't

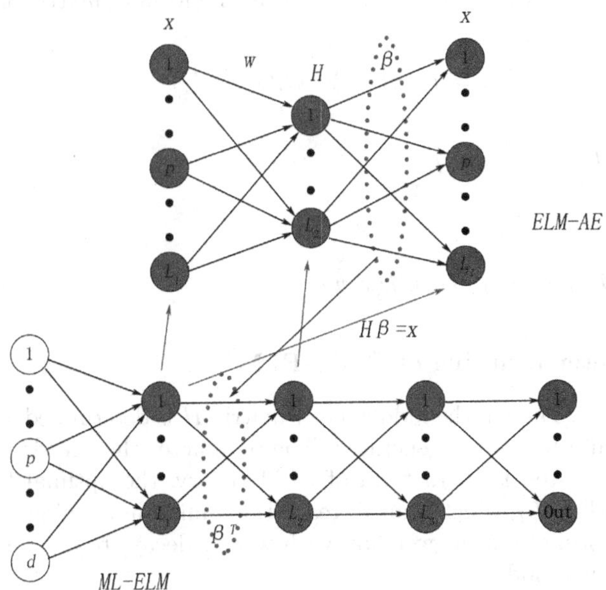

Fig. 1. Illustration of C-ML-ELM's structure and parameter training

require fine-tuning using back propagating (BP). Each contractive ELM-AE layer includes two parts: encoding and decoding. The input vector of this single layer x is either output of the former hidden layer or the input of C-ML-ELM, with dimension of d or L_{i-l}, viz. the node number of input or hidden layer. Extracted representation at the current hidden layer L_i is denoted as h. It is the function of input x of this hidden layer. As shown in Eq. 1.

$$h(x) = G(wx + b_h) \tag{1}$$

where G uses the typical logistic sigmoid $G(z) = \frac{1}{1+e^{-z}}$. The weights w and random biases b_h of the hidden nodes are required to be orthogonal subjecting to $w^T w = I; b_h^T b_h = I$. The orthogonalization of randomly generated weights and biases helps to improve the generalization performance of ELM-AE [9]. To perform unsupervised feature learning, that labeled output vector T is asked to equal to the input vector according to the principle of auto-encoding, viz. $T = x$. This principle is reasonable because the estimated x is supposed to equal to the original input x. For the estimated x at the decoding stage is calculated from the calculated feature, which is transformation from the input x at the encoding stage. The output is calculated as in Eq. 2.

$$\hat{x} = f_L(h) = s_g(w^T h + b_y) \tag{2}$$

where s_g typically is either the identity or a logistic sigmoid function. In the paper we use the identity function to have the linear mapping relationship at the output layer. Therefore, the output to the input for each contractive ML-ELM is as in Eq. 3, which is parameterized in the form of matrix as in Eq. 4.

$$\hat{x} = f_L(h) = \sum_{i=1}^{L} G_i(x, w_i, b_i) \cdot \beta_i \tag{3}$$

where $w_i \in R^d, b_i \in R, \beta_i \in R^o$.

$$\hat{x} = H \cdot \beta \tag{4}$$

where $H = [G_1(x, w_1.b_1), \cdots, G_L(x, w_L, b_L)]$.

2.3 Parameter Learning of C-ML-ELM

As shown in Eqs. 3 and 4, the hidden layer matrix H is determined, once the random weights and biases are generated. The parameter that needs to be learned in the following step is β. In term of ELM theory, the optimal function J is designed as in Eq. (5), which aims to reach the minimum training error and also the smallest norm of output weights (with weight decay) to ensure the generalization of learned model.

$$J_{ELM}(\beta) = ||H\beta - T||_2^2 + \lambda||\beta||_2^2 \tag{5}$$

where λ is the hyper-parameter controls to balance the weight decay, whilst the output T equals to the input x, viz. $T = x$ for this unsupervising leaning case.

In the paper, we aim to improve the robustness of output with respect to the variety of input by adding an extra penalty term at the optimal function. This is achieved by adding a first-order derivative of output to the input. Referring to [12], we use the Frobenius norm of the Jacobian matrix of the encoder activations, as shown in Eq. (6).

$$J_{C-M-ELM}(\beta) = ||H\beta - T||_2^2 + \lambda||J_f(x)||_F^2 \tag{6}$$

where $||J_f(x)||_F^2 = \sum_{ij}(\frac{\partial \sum_{k=1}^{L} G_k(x,w_k,b_k)\cdot\beta_{kj}}{\partial x_i})^2$

Then β is calculated with the least square method by minimizing $F2$ norm of this optimal function as shown in Eq. (7).

$$2H^T(H\beta - x) + 2\lambda||\frac{\partial H}{\partial x}||_f^2\beta = 0 \tag{7}$$

Accordingly, the output weight matrix β is derived as in Eq. (8).

$$\hat{\beta} = H^T H + \lambda||\frac{\partial H}{\partial x}||_F^2 H^T x \tag{8}$$

Computing $||\frac{\partial H}{\partial x}||$ contains two steps. Firstly, we compute the norm of partial derivative of each columns of H with respect to x to get the Frobenius norm of the Jacobian matrix of each sample. Secondly, combine the each sample's result of the first step to a diagonal matrix for matching the matrix shape.

In this way, each hidden layer of C-ML-ELM is trained with this unsupervised approach by assigning the output weights to that of the corresponding layer. Under this constraint of our optimal function, the learned C-ML-ELM not only meet the targets of having the minimum prediction error but also ensures local direction variance robust of the output of each hidden layer to the input at this layer. This result ensures the local consecutiveness of image at each layer. With less propagation of influence of variety impact, the further classification or approximation solution will have less error.

3 Experiment Results and Analysis

To test the effectiveness of our proposed C-ML-ELM approach, we perform pilot experiments for the classification problem on the public MNIST handwriting dataset [13], face expression dataset: CAFÉ [14]. The codes are running on the workstation with a core Intel(R) Xeon(R) CPU E5-2687W 3.10 GHz processor and 32.0 GB RAM.

3.1 Datasets

In the paper, we did the pilot experiments on the public dataset: MNIST handwriting dataset. The MNIST is a handwritten digits dataset, which has 60000 training samples and 10000 testing samples with 28 * 28 size of each grayscale image. We test the proposed algorithm C-ML-ELM on the original MNIST without any preprocessing. Moreover, we test the local invariance representation performance on a more complex problem: expression recognition using a public expression datasets: CAFÉ. CAFÉ (the CAlifornia Facial Expression database) contains Seventy-seven images of 11 out of 60 posers (5 female, 6 male). The 11 posers included 7 Caucasians, 2 east Asians (one of Korean descent and one of Chinese descent), and 2 of unknown descent (most likely from the Pacific region).

3.2 Configuration of Deep Networks for Comparison

The parameters of hidden layers have impact on the performance of deep algorithms. Therefore, we make some adjustment and choose the best configuration for each deep algorithm and do the comparison. For example, for the MNIST dataset, the configuration is set as follows. The input node number is 784. Output node number is 10. To algorithms C-ML-ELM and ML-ELM, there are 3 hidden layers and the node number for each hidden layer are set as 700, 700, 6500, respectively. The balance weight parameters of Jacobian matrix λ at each hidden layer are set empirically at 10e−1, 1e3, 1e8, respectively. DBN and SDA use 3 hidden layers with hidden layer node numbers 1000, 1000, 1000, respectively. The iteration number for initializing training is 1000 and learning rate is 0.01, while the iteration number for fine-tuned training is 15 and learning rate is 0.1.

3.3 Classification and Computing Efficiency Comparison

To test the classification performance of our proposed C-ML-ELM, we make the comparison with ML-ELM, ELM, SAE (sparse auto-encoder) [10] and DBN [11]. Note: for the top layer of C-ML-ELM, we employ the supervised ELM. The classification accuracy rate and training time for the MNIST handwriting dataset are listed at the Table 1.

Table 1. Experiments on the MNIST dataset

	C-ML-ELM	ML-ELM	ELM	DBN	SAE
Accuracy (%)	99.775 ± 0.05	99.037	98.595	98.870	98.623
Training times (s)	705.79	823.294	783.864	20580	-

From the above results, we can see that our C-ML-ELM outperforms the ML-ELM and basic ELM in classification accuracy with comparable training speed. Comparing to the typical deep learning algorithm DBN and the stacked auto-encoding algorithm, our proposed C-ML-ELM achieves higher classification accuracy and significant fast learning speed.

To test our C-ML-ELM's performance with more complex objects, we do the comparison test on the expression dataset: CAFÉ. Here we mainly make comparison between the ML-ELM and basic ELM algorithms, which our method extends derived from. From the results, we can see that our method achieves better recognition rate than that of ML-ELM and basic ELM as shown in Table 2.

For the problem of facial expression recognition, the interfering factors such as facial appearance, race and pose have great influence of the expression recognition, therefore learning the discriminative representation of expression and removing the interference plays an important role in improving the recognition results. Actually, the training samples for the expression recognition problem contain the expression information, which should be stable and consecutive whilst the other factors such as facial appearance are varied among samples. Therefore, by adding this first-order derivation as constraint in optimizational process, it ensures that the extracted representation revealing emotion attributes rather than other face factors. As shown in Table 2, our extension C-ML-ELM displays better recognition results for expression recognition than that of ML-ELM and basic ELM.

Table 2. Experiments on the CAFE dataset

	C-ML-ELM	ML-ELM	ELM
CAFE	82.304 %	81.634 %	80.543 %

3.4 Representation Performance Comparison

As described earlier, our C-ML-ELM is proposed based on the M-ELM by adding the constraint term with minimizing the Frobenius norm of the Jacobian matrix for the output with respect to the input at each layer. By adding this penalty term, it is beneficial to alleviate the local invariance impact and enhance the consecutiveness in image representation. To verify the effectiveness, we make the comparison empirically by visualizing the weight network of the hidden layer between C-ML-ELM and ML-ELM. One example of results is shown in Fig. 2.

From this result, we can see that our proposed C-ML-ELM has better performance in mitigating the impact of invariance, which emphasizes on the salient features and removes the random noise.

3.5 Analyzing the Impact of Number of Hidden Layer Node

To have some idea of the impact of the node number of the hidden layer, we did a test by analyzing the accuracy rate changing with the hidden layer node number. As shown in Fig. 3, the accuracy rate rise up with increasing of node number. Then the values of the accuracy rate are gradually flat, when the node number arrives at a certain number. Here it is about 5000.

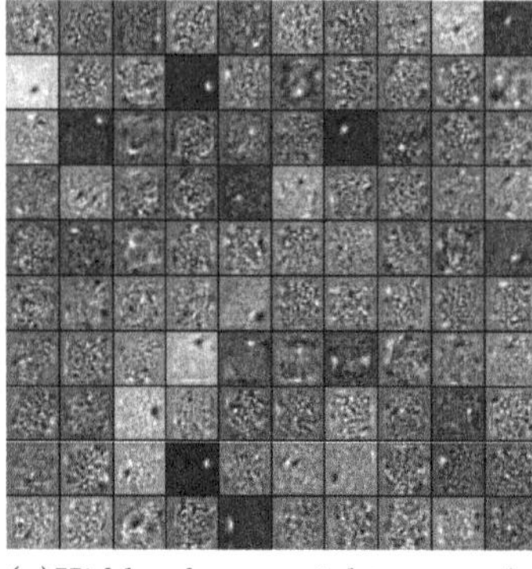

(a)Hidden layer weight network for ML-ELM

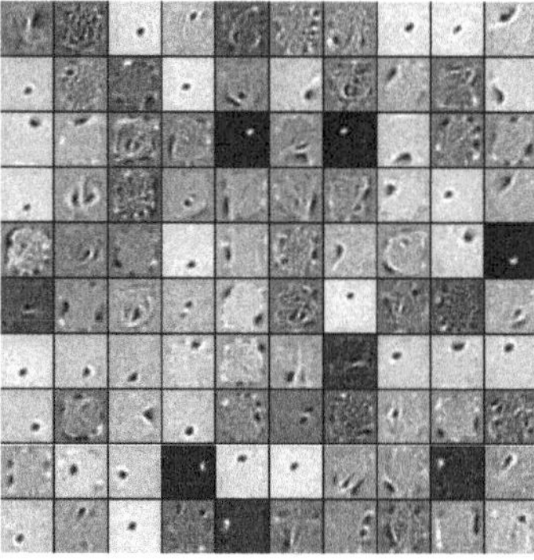

(b)Hidden layer weight network for C-ML-ELM

Fig. 2. The visualization of hidden layer weight network

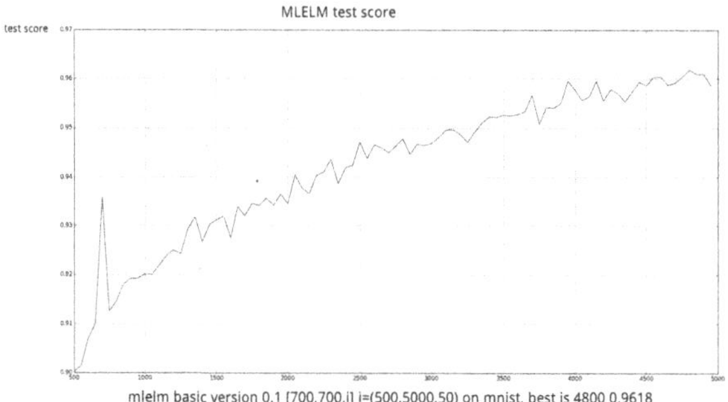

Fig. 3. Impact of hidden layer node number

4 Conclusions

This paper proposes an algorithm called contractive ML-ELM which aims to extract discriminative feature with enhanced optimizing function. By adding this the consecutive constraint, it is beneficial to extract local invariant robust features that enhance the discriminative representation of complex objects and reduce the influence of local changes. This results in the improvement of classification performance. Experiments show that our method achieves state of the art classification error on MNIST dataset, comparing with the ML-ELM and deep learning methods (DBN). As all the other ML-ELM, the C-ML-ELM has significant fast learning speed without parameter fine-tuning, comparing to most typical BP based deep networks. Future work will be done in improving the optimization strategy to extract more meaningful feature for problems such as face recognition, expression recognition to alleviate the inference factor tangled with emotion factors in each recognition objective.

Acknowledgments. This research is partially sponsored by the National Nature Science Foundation of China (Nos. 61672070, 61370113, 91546111), Beijing Municipal Natural Science Foundation (No. 4152005), Key projects of Beijing Municipal Education Commission (No. KZ201610005009).

References

1. Kumar, G., Bhatia, P.K.: A detailed review of feature extraction in image processing systems. In: IEEE Fourth International Conference on Advanced Computing and Communication Technologies, pp. 5–12 (2014)
2. Ahonen, T., Hadid, A., Pietikainen, M.: Face description with local binary patterns: application to face recognition. IEEE Trans. Pattern Anal. Mach. Intell. **28**(12), 2037–2041 (2006)
3. Dalal, N., Triggs, B.: Histograms of oriented gradients for human detection. In: IEEE Conference on Computer Vision and Pattern Recognition, pp. 886–893 (2013)
4. Jia, X., Zhang, Y., Powers, D., Ali, H.B.: Multi-classifier fusion based facial expression recognition approach. KSII Trans. Internet Inf. Syst. **8**(1), 196–212 (2014)
5. Zavaschi, T.H.H., Britto Jr., A.S., Oliveira, L.E.S., Koerich, A.L.: Fusion of feature sets and classifiers for facial expression recognition. Expert Syst. Appl. **40**(2), 646–655 (2013)
6. Jing, P., Bhanu, B.: Delayed reinforcement learning for adaptive image segmentation and feature extraction. IEEE Trans. Syst. Man Cybern. Part C Appl. Rev. **28**(3), 482–488 (1998)
7. Liu, S., Bai, X.: Discriminative features for image classification and retrieval. In: IEEE International Conference on Image and Graphics, pp. 744–751 (2011)
8. Chen, M., Zhang, L., Allebach, J.P.: Learning deep features for image emotion classification. In: IEEE International Conference on Image Processing. IEEE (2015)
9. Kasun, L.L.C., Zhou, H., Huang, G.-B., Vong, C.M.: Representational learning with extreme learning machine for big data. IEEE Intell. Syst. **28**(6), 31–34 (2013)
10. Vincent, P., et al.: Stacked denoising autoencoders: learning useful representations in a deep network with a local denoising criterion. J. Mach. Learn. Res. **11**(6), 3371–3408 (2010)
11. Hinton, G.E., Salakhutdinov, R.R.: Reducing the dimensionality of data with neural networks. Science **313**(5786), 504–507 (2006)
12. Rifai, S., Vincent, P., Muller, X., et al.: Contractive auto-encoders: explicit invariance during feature extraction. In: Proceedings of the 28th International Conference on Machine Learning (ICML-2011), Bellevue, WA, USA, pp. 833–840 (2011)
13. LeCun, Y., Bottou, L., Bengio, Y., et al.: Gradient-based learning applied to document recognition. Proc. IEEE **86**(11), 2278–2324 (1998)
14. Dailey, M.N., Cottrell, G.W., Reilly, J.: CAlifornia Facial Expressions (CAFE) (2001). Computer Science and Engineering Department, UCSD, La Jolla (2011). http://www.cs.ucsd.edu/users/gary/CAFE/

Two-Dimensional Soft Linear Discriminant Projection for Robust Image Feature Extraction and Recognition

Yu Tang[1,2], Zhao Zhang[1,2(✉)], and Weiming Jiang[1,2]

[1] School of Computer Science and Technology and Joint International Research Laboratory of Machine Learning and Neuromorphic Computing, Soochow University, Suzhou 215006, China
cszzhang@gmail.com

[2] Collaborative Innovation Center of Novel Software Technology and Industrialization, Nanjing 210023, China

Abstract. In this study, we propose a Robust Soft Linear Discriminant Projection (RS-LDP) algorithm for extracting two-dimensional (2D) image features for recognition. RS-LDP is based on the soft label linear discriminant analysis (SL-LDA) that is shown to be effective for semi-supervised feature learning, but SLDA works in the vector space and thus extract one-dimensional (1D) features directly, so it has to convert the two-dimensional (2D) image matrices into the 1D vectorized representations in a high-dimensional space when dealing with images. But such transformation usually destroys the intrinsic topology structures of the images pixels and thus loses certain important information, which may result in degraded performance. Compared with SL-LDA for representation, our RS-LDP can effectively preserve the topology structures among image pixels, and more importantly it would be more efficient due to the matrix representations. Extensive simulations on real-world image datasets show that our proposed RS-LDP can deliver enhanced performance over other state-of-the-arts for recognition.

Keywords: Robust two-dimensional projection · Linear discriminant analysis · Feature extraction · Soft label · Label propagation

1 Introduction

In practical applications, more and more informative and valuable features are minded from the high-dimensional real-world data of huge volume, e.g. image. To improve the representation performance of high-dimensional data effectively and get rid of redundant information or noise, dimensionality reduction (DR) or feature learning are used. Based on whether supervised class information of samples is available, existing DR methods can be divided into supervised, unsupervised and semi-supervised [17, 18] models roughly. Both *Principal Component Analysis* (PCA) [1] and *Locality Preserving Projection* (LPP) [2] are popular unsupervised models for local or global geometry structure preservation. In contrast, Representative supervised models include *Linear Discriminant Analysis* (LDA) [3] and its extensions [4, 5], etc. that aim at inter-class discrimination.

A. Hirose et al. (Eds.): ICONIP 2016, Part IV, LNCS 9950, pp. 514–521, 2016.
DOI: 10.1007/978-3-319-46681-1_61

In most real applications, obtaining enough labeled number of samples for leanring is not a workable solution because of high cost. So, many semi-supervised learning methods have been proposed in the past few years. They have been shown to be effective for image representation, including *A soft label based Linear Discriminant Analysis* (SL-LDA) [6], *L1-Norm Driven Semi-Supervised Local Discriminant Projection* (SSLDP) [7], *Semi-Supervised Maximum Margin Criterion* (SSMMC) and *Semi-Supervised Linear Discriminant Analysis* (SSLDA) [8], etc.

Compared with the LDA, SL-LDA has delivered enhanced performance, since it firstly uses the predicted class labels of unlabeled samples by label propagation and define the scatter matrices over soft labels. But it should be noticed it performs in 1D vector space. Because of the fact that images are essentially two-dimensional (2D) matrices or second-order tensor [11, 19, 20], the vectorized representation of images may lose important topology information among pixels in given images.

We thus propose a new method called Robust Soft Linear Discriminant Projection (RS-LDP) to solve this issue effectively. Different form SL-LDA, our proposed RS-LDP works in the 2D matrix space, so it can effectively maintain the spatial topology structures and intrinsic correlation between image pixels. In addition, our RS-LDP uses the recent *Nonnegative Sparse Neighborhood Propagation* (SparseNP) [12] for estimating the labels of unlabeled samples for more accurate prediction so that the resulted scatter matrices over the soft lables would be more accurate for image featue extraction. Simulations on real image datasets verified the effectiveness of RS-LDP.

We outline the paper as follows. Section 2 briefly reviews the formulation of SL-LDA. In Sect. 3, we proposes our RS-LDP model. Section 4 shows the experiment settings and results. Finally, in Sect. 5, we draw the conclusions.

2 SL-LDA Revisited

Firstly, this section reviews a semi-supervised method SL-LDA [6] based on a popular supervised method LDA. In addition, it enhances the performance via incorporating the soft labels into the construction of scatter matrixes to find a transformation for dimensionality reduction.

Let $X = [X_L, X_U] \in R^{m\times(l+u)}$ be a set of samples and $Y = [y_1, y_2, \ldots, y_c]$ be associated class labels of $X_L = [x_1, x_2, \ldots, x_l]$ that denotes the labeled samples, where $l + u = N$ is the number of samples, each sample $x \in R^{m\times 1}$ denotes an m-dimensional column vector, $X_U = [x_{l+1}, x_{l+2}, \ldots, x_{l+u}]$ is the unlabeled set and c is the number of classes.

We call fixed label of each sample hard label, i.e., the probability f_{ij}^{l} of x_j belonging to the class i is either 0 or 1, while the predicted labels of unlabeled samples $f_{ij}^{u}\left(0 \le f_{ij}^{u} \le 1\right)$ are called soft labels that they have key probability information for discriminative learning. Via label propagation process that propagates the label information from labeled set to unlabeled set, we can get soft labels.

First of all, the total-class, within-class and between-class scatter matrixes, i.e. $\widetilde{S_t}$, $\widetilde{S_w}$ and $\widetilde{S_b}$, need to be defined as:

$$\begin{aligned}\widetilde{S_t} &= \sum_{i=1}^{c}\sum_{j=1}^{l+u} f_{ij}\left(x_j - \widetilde{\mu}\right)\left(x_j - \widetilde{\mu}\right)^T \\ \widetilde{S_w} &= \sum_{i=1}^{c}\sum_{j=1}^{l+u} f_{ij}\left(x_j - \widetilde{\mu}_i\right)\left(x_j - \widetilde{\mu}_i\right)^T \\ \widetilde{S_b} &= \sum_{i=1}^{c}\sum_{j=1}^{l+u} f_{ij}(\widetilde{\mu}_i - \widetilde{\mu})(\widetilde{\mu}_i - \widetilde{\mu})^T\end{aligned} \tag{1}$$

where $\widetilde{\mu}_i$ and $\widetilde{\mu}$ are the soft means of dataset in the ith class and in all classes as:

$$\widetilde{\mu}_i = \sum\nolimits_{j=1}^{l+u} f_{ij}x_j \Big/ \sum\nolimits_{j=1}^{l+u} f_{ij} \quad and \quad \widetilde{\mu} = \sum_{i=1}^{c}\sum\nolimits_{j=1}^{l+u} f_{ij}x_j \Big/ \sum_{i=1}^{c}\sum\nolimits_{j=1}^{l+u} f_{ij} \tag{2}$$

SL-LDA aims to try to find a linear transformation matrix $P_{SL-LDA} \in R^{m \times d}$, where d is the number of reduced dimensions, that minimizes the within-class scatter and maximizes the between-class scatter at the same time in the reduced subspace as:

$$P_{SL-LDA} = \underset{P \in R^{m \times d}}{\arg\max}\, tr\left(\left(P^T \widetilde{S_w} P\right)^{-1} P \widetilde{S_b} P\right) \tag{3}$$

More details about solving this obective function cn be referred to [6].

3 Proposed Formulation

Let $A=[A_L, A_U] \in R^{(m \times n) \times (l+u)}$ be a set of training images, where $A_i \in R^{m \times n}$ denotes an image, m and n are image width and length respectively, $A_L=[A_1, A_2, \ldots, A_l] \in R^{(m \times n) \times l}$ is a labeled set and $A_U=[A_{l+1}, A_{l+2}, \ldots, A_{l+u}] \in R^{(m \times n) \times u}$ is an unlabeled set. Let $Y = [y_i](i = 1, 2, \ldots, l+u) \in R^{c \times (l+u)}$ be the initial labels of all samples, for the labeled sample.

3.1 The Objective Function

We first perform SparseNP [12] to estimate the labels of unlabeled samples by obtaining the soft label matrix F_c [12]. Then, similar to SL-LDA, we first define the total-class, within-class and between-class scatter matrixes over the soft labels:

$$\begin{aligned}
\widetilde{S_t} &= \sum_{i=1}^{c}\sum_{j=1}^{l+u} f_{ij}(A_j - \widetilde{\mu})(A_j - \widetilde{\mu})^T = A(E - Eee^TE/eEe^T)A^T \\
\widetilde{S_w} &= \sum_{i=1}^{c}\sum_{j=1}^{l+u} f_{ij}(A_j - \widetilde{\mu_i})(A_j - \widetilde{\mu_i})^T = A(E - F_c^TG^{-1}F_c)A^T \\
\widetilde{S_b} &= \sum_{i=1}^{c}\sum_{j=1}^{l+u} f_{ij}(\widetilde{\mu_i} - \widetilde{\mu})(\widetilde{\mu_i} - \widetilde{\mu})^T = A(F_c^TG^{-1}F_c - Eee^TE/eEe^T)A^T
\end{aligned} \tag{4}$$

where $E_{jj} \in \sum_{i=1}^{c} f_{ij}$, $G_{ii} \in \sum_{j=1}^{l+u} f_{ij}$, e is a unit matrix and $\widetilde{\mu_i}$ and $\widetilde{\mu}$ are the soft means of dataset in the *ith* class and in all classes as:

$$\widetilde{\mu_i} = \sum\nolimits_{j=1}^{l+u} f_{ij}A_j \Big/ \sum\nolimits_{j=1}^{l+u} f_{ij} \quad \textit{and} \quad \widetilde{\mu} = \sum_{i=1}^{c}\sum\nolimits_{j=1}^{l+u} f_{ij}A_j \Big/ \sum_{i=1}^{c}\sum\nolimits_{j=1}^{l+u} f_{ij} \tag{5}$$

By maximizing the between-class scatter matrix while minimizing the within-class scatter matrix in the reduced subspace, we obtain the object function as:

$$J(P) = \max_P tr\left(\left(P^T\widetilde{S_w}P\right)^{-1} P^T\widetilde{S_b}P\right) \tag{6}$$

Then, the optimal solution of RS-LDP can be obtained by the generalized eigenvalue decomposition as:

$$\left(\widetilde{S_w} + aI\right)^{-1}\widetilde{S_b}P^* = P^*\Lambda \quad or \quad \left(\widetilde{S_t} + aI\right)^{-1}\widetilde{S_b}P^* = P^*\Lambda \tag{7}$$

where Λ denotes diagonal matrix of the eigenvalue, aI is a multiply of identity matrix. Next, we discuss the optimization procedure for our RS-LDP.

3.2 An Efficient Approach of RS-LDP

Resembling SL-LDA, RS-LDP and W-LS also have the similar relationship [6]. Let us consider a weighted and regularized least square problem:

$$J(P,b) = \min_P \sum\nolimits_{i=1}^{c}\sum\nolimits_{j=1}^{l+u} f_{ij}\left\|P^TA_j + b^T - t_j\right\|_F^2 + a\|P\|_F^2 \tag{8}$$

In matrix form, Eq. (8) is wrote as

$$\begin{aligned}
J(P,b) =& tr\left(\left(P^TA + b^Te\right)E\left(P^TA + b^Te\right)^T\right) \\
& - 2tr\left(TE\left(P^TA + b^Te\right)^T\right) + a\left(tr\left(P^TP\right)\right)
\end{aligned} \tag{9}$$

By setting the derivative w.r.t. P and b to zero, we have:

$$\begin{cases} AEA^T P + AE(e^T b - T^T) + \alpha P = 0 \\ eEe^T b + eE(A^T P - T^T) = 0 \end{cases} \tag{10}$$

Then, the optimal solution of W-LS as:

$$P^*_{W-LS} = (AL_s A^T + \alpha I)^{-1} AL_s T^T \tag{11}$$

where $L_s = E - Eee^T E / eEe^T$ and following Eq. (9), we have $AL_s A^T = \widetilde{S}_t$. To connect to RS-LDP, a class indicator is chosen as $T = G^{-1/2} FE^{-1}$, and each t_j meets:

$$t_{ij} = w_{ij} / \left(E_{jj}\sqrt{F_{ii}}\right) \tag{12}$$

The verification of formulation $AL_s T^T TL_s A^T = \widetilde{S}_b$ is easy. Therefore, if we let $\widetilde{H_b} = AL_s T^T \in R^{m \times c}$, i.e. $\widetilde{H_b}\widetilde{H_b}^T = \widetilde{S}_b$. According to some related theorems [6], we can know that P^*_{RS-LDP}, the optimal solution of RS-LDP, can be obtained by computing the d eigenvectors of $\left(\widetilde{S}_t + aI\right)^{-1} \widetilde{S}_b = \left(\widetilde{S}_t + aI\right)^{-1} \widetilde{H_b}\widetilde{H_b}^T$. It has the same nonzero eigenvalues to the auxiliary matrix $M = \widetilde{H_b}^T \left(\widetilde{S}_t + aI\right)^{-1} \widetilde{H_b}$. According to theorem, if U are the eigenvectors of M, then $\left(\widetilde{S}_t + aI\right)^{-1} \widetilde{H_b} U$ are the eigenvectors of $\left(\widetilde{S}_t + aI\right)^{-1} \widetilde{H_b}\widetilde{H_b}^T$. Hence, we can obtain that $P^*_{RS-LDP} = P^*_{W-LS} U$. If we embed new sample A onto the projection matrix P^*_{RS-LDP} obtained, the feature matrix Z of sample A will be computed by $Z = P^*_{RS-LDP^T} A$. We describe the basic steps of our proposed algorithm as follows in Table 1.

Table 1. The algorithm of the proposed method

Input: Data $A \in R^{(m \times n) \times (l+u)}$, reduced matrix d and other related parameters.
Output: The projection matrix $P^*_{RS-LDP} \in R^{m \times d}$.
Algorithm:

1. Perform the label propagation and obtain the soft labels of unlabeled samples F^u via SparseNP.
2. Calculate the class indicator using soft labels as in Eq. (12). Solve the optimal solution P^*_{W-LS} as in Eq. (11).
3. Output $P^*_{RS-LDP} \leftarrow P^*_{W-LS}$

4 Experimental Results and Analysis

In this section, we conduct simulations to evaluate the performance of our RS-LDP for discriminative representation and classification. Furthermore, the performance of our method is compared with the other related algorithms, i.e., LDA, SSLDA, SL-LDA,

Table 2. List of used datasets and dataset information

Dataset name	#Samples	#Dim	#Class	#Samples per class	#Labeled set
Mixed Georgia and Yale	915	10	65	15/11	2, 4, 6, 8
Mixed ORL and Yale	565	10	55	10/11	2, 4, 6, 8

2DPCA [9] and 2DLPP [10]. In this study, two real image databases are involved, and the detailed descriptions of used datasets are described in Table 2. As is common practice the images in all databases are resized to 32 × 32 pixels due to computational consideration. Because of the difference in data quality, we average the results over 10 random splits of training/test sets. We fix the number d to 10 in all simulations. A one-nearest-neighbor classifier with the Euclidean distance [13] is used to evaluate the test accuracy. Besides, we implement all algorithms in MATLAB R2014a and perform experiments on a PC with Intel (R) Core i5 4 GB RAM.

4.1 Object Recognition on COIL-20 Database

We first test each approach for recognizing object images using the *COIL-20* image database [14]. In this simulation, we randomly choose certain number of training images from each class as labeled to form the training set and test on the rest. Four settings based on the number of labeled data in each class Lab = 6, 12, 18 and 24. Some original image examples are showed in Fig. 1. The results are illustrated in Table 3, from which we can see clearly that our method RS-LDA performs better than others by delivering higher accuracies in most cases.

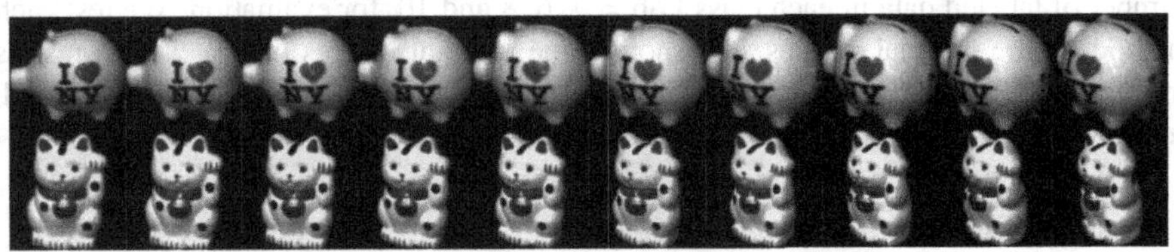

Fig. 1. Original sample images of COIL-20 database.

Table 3. Performance comparisons on COIL-20 dataset (d = 10)

Methods	Dataset							
	N = 1440, Lab = 6		N = 1440, Lab = 12		N = 1440, Lab = 18		N = 1440, Lab = 24	
	Mean	Best	Mean	Best	Mean	Best	Mean	Best
LDA	0.6326	0.6909	0.7152	0.8899	0.7502	0.9081	0.7652	0.9252
SSLDA	0.6756	0.7053	0.7525	0.9026	0.7639	0.9160	0.7906	0.9350
SL-LDA	0.7941	0.8348	0.7292	0.8953	0.7578	0.9109	0.7690	0.9318
2DPCA	0.8753	0.8902	0.8017	0.9102	0.7907	0.9212	0.7850	0.9435
2DLPP	0.8775	0.8867	0.8767	0.9155	0.8835	0.9463	0.9000	0.9554
RS-LDP	0.8958	0.9071	0.8975	0.9387	0.9000	0.9554	0.9083	0.9656

Fig. 2. Original sample images of Mixed Georgia and Yale face database.

Table 4. Performance comparisons on Mixed Georgia and Yale Dataset

Methods	Dataset							
	N = 915, Lab = 2		N = 915, Lab = 4		N = 915, Lab = 6		N = 915, Lab = 8	
	Mean	Best	Mean	Best	Mean	Best	Mean	Best
LDA	0.5098	0.5316	0.5310	0.6486	0.5131	0.6601	0.5013	0.7035
SSLDA	0.5197	0.5592	0.5664	0.6641	0.5238	0.7038	0.5366	0.7229
SL-LDA	0.5423	0.5366	0.6131	0.6588	0.6617	0.7001	0.6911	0.7231
2DPCA	0.5856	0.5610	0.6473	0.6717	0.6781	0.7250	0.7014	0.7372
2DLPP	0.5623	0.5812	0.6299	0.6658	0.6620	0.7215	0.7015	0.7458
RS-LDP	0.6002	0.6002	0.6501	0.6968	0.6815	0.7395	0.7131	0.7668

4.2 Face Recognition on Mixed Georgia and Yale Dataset

In this subsection, we evaluate each method for representing and recognizing face images of different persons. The used dataset is called *Mixed Georgia and Yale dataset* that is a mixed dataset that contains two datasets, i.e., *Georgia Tech face dataset* [15] and *Yale face dataset* [16]. We also prepare four configurations that based on the number of labeled data in each class Lab = 4, 6, 8 and 10, for evaluation. We test each method on the original images. We show some original images in Fig. 2. The results are reported in Table 4. From the simulation results, similar findings can be obtained, i.e., our RS-LDP outperforms other methods in most cases.

5 Conclusion

A semi-supervised feature extraction problem is discussed in two-dimensional matrix space. Our RS-LDP can effectively preserve the topology structures among image pixels, and more importantly it would be more efficient due to the matrix representations. As a result, our algorithm has delivered enhanced performance compared with some related methods. In future, we will investigate more efficient approach to solve the problem. Besides, extending the 2D feature extraction to the semi-supervised scenario is also needed.

Acknowledgment. This work is partially supported by the National Natural Science Foundation of China (61402310, 61373093), Major Program of Natural Science Foundation of Jiangsu Higher Education Institutions of China (15KJA520002), Special Funding of China Postdoctoral Science Foundation (2016T90494), Postdoctoral Science Foundation of China (2015M580462), Postdoctoral Science Foundation of Jiangsu Province of China (1501091B), and the Natural Science Foundation of Jiangsu Province of China (BK20140008 and BK20141195).

References

1. Jolliffe, I.: Principal Component Analysis. Wiley, Hoboken (2002)
2. Niyogi, X.: Locality preserving projections. Neural Inf. Process. Syst. **16**, 153 (2014)
3. Yu, H., Yang, J.: A direct LDA algorithm for high-dimensional data-with application to face recognition. Pattern Recogn. **34**(10), 2067–2070 (2001)
4. Sugiyama, M.: Local fisher discriminant analysis for supervised dimensionality reduction. In: Proceedings of the 23rd International Conference on Machine Learning (ICML), pp. 905–912 (2006)
5. Zhang, Z., Chow, T.W.: Robust linearly optimized discriminant analysis. Neurocomputing **79**, 140–157 (2012)
6. Zhao, M., Zhang, Z., Zhang, H.: A soft label based linear discriminant analysis for semi-supervised dimensionality reduction. In: Proceedings of the 2013 International Joint Conference on Neural Networks (IJCNN), pp. 1–8. IEEE (2013)
7. Wang, X., Zhang, Z., Tang, Y., et al.: L1-Norm driven semi-supervised local discriminant projection for robust image representation. In: Proceedings of the 27th IEEE International Conference on Tools with Artificial Intelligence (ICTAI), pp. 391–397. IEEE (2015)
8. Song, Y., Nie, F., Zhang, C., et al.: A unified framework for semi-supervised dimensionality reduction. Pattern Recogn. **41**(9), 2789–2799 (2008)
9. Yang, J., Zhang, D., Frangi, A.F., Yang, J.Y.: Two-dimensional PCA: a new approach to appearance-based face representation and recognition. IEEE Trans. Pattern Anal. Mach. Intell. **26**(1), 131–137 (2004)
10. Chen, S., Zhao, H., Kong, M., et al.: 2D-LPP: a two-dimensional extension of locality preserving projections. Zeurocomputing **70**(4), 912–921 (2007)
11. Zhang, Z., Chow, T.W.: Maximum margin multisurface support tensor machines with application to image classification and segmentation. Expert Syst. Appl. **39**(1), 850–861 (2012)
12. Zhang, Z., Zhang, L., Zhao, M., et al.: Semi-supervised image classification by nonnegative sparse neighborhood propagation. In: Proceedings of the 5th ACM on International Conference on Multimedia Retrieval, pp. 139–146 (2015)
13. Hastie, T., Tibshirani, R.: Discriminant adaptive nearest neighbor classification. IEEE Trans. Pattern Anal. Mach. Intell. **18**(6), 607–616 (1996)
14. Nene, S.A., Nayar, S.K, et al.: Columbia object image library (COIL-20). Technical report CUCS-005-96 (1996)
15. Nefian, A.V.: Georgia tech face database (2013)
16. Sim, T., Kanade, T.: Combining models and exemplars for face recognition: an illuminating example. In: Proceedings of the CVPR 2001 Workshop on Models versus Exemplars in Computer Vision, vol. 1 (2001)
17. Zhu, X., Ghahramani, Z, et al.: Semi-supervised learning using Gaussian fields and harmonic functions. In: Proceedings of the International Conference on Machine Learning (ICML) (2003)
18. Cai, D., He, X., Han, J.: Semi-supervised discriminant analysis. In: Proceedings of ICCV (2007)
19. Zhang, Z., Chow, W.S.: Tensor locally linear discriminative analysis. IEEE Sig. Process. Lett. **18**(11), 843–846 (2011)
20. Wang, Z., Chen, S., Liu, J., et al.: Pattern representation in feature extraction and classifier design: matrix versus vector. IEEE Trans. Neural Netw. **19**(5), 758–769 (2008)

Asymmetric Synaptic Connections in Z(2) Gauge Neural Network

Atsutomo Murai and Tetsuo Matsui(✉)

Department of Physics, Kindai University, Higashi-Osaka 577-8502, Japan
mizunohigasi@gmail.com, matsui@phys.kindai.ac.jp

Abstract. We consider Z(2) gauge neural network which involves neuron variables $S_i(=\pm 1)$ and synaptic connection (gauge) variables $J_{ij}(=\pm 1)$. Its energy consists of the Hopfield term $-c_1 S_i J_{ij} S_j$ and the reverberation term $-c_2 J_{ij} J_{jk} J_{ki}$ for signal propagation on a closed loop. The model of symmetric couplings $J_{ij} = J_{ji}$ has been studied; its phase diagram in the $c_2 - c_1$ plane exhibits Higgs, Coulomb and confinement phases, each of which is characterized by the ability of learning and/or recalling patterns. In this paper, we consider the model of asymmetric coupling (J_{ij} and J_{ji} are independent), and examine the effect of asymmetry on a partially connected random network.

Keywords: Hopfield model · Gauge neural network · Asymmetric synaptic connection

1 Introduction

The Hopfield model [1] of neural network has served for long time as a canonical model of associative memory. In this model, the state of the i-th neuron is described by the time-dependent Z(2) variable, $S_i(t) = \pm 1$, while the synaptic weights J_{ij} connecting from the j-th neuron to the i-th neuron are kept constants.

Then, various models have been proposed as models of learning patterns of neurons by treating the plasticity(time-dependence) of synaptic connections [2]. Among others, the gauge neural network [3] regards the synaptic weight $J_{ij}(t)$ as a gauge-field variable and postulates the system energy to satisfy the local Z(2) gauge symmetry.

To understand the role of J_{ij} as a gauge variable in the context of the Hopfield model itself, let us consider an electric signal which starts from the neuron j with the electric potential S_j (in some unit) and arrives at the neuron i, where the potential at i is converted from S_j to $J_{ij}S_j$. Namely, J_{ij} measures the relation, relative difference, of two local frames of potential at j and i. Such quantity is to be called a gauge field. The gauge symmetry implies that observable quantities such as energy should be independent of change of local frame as $S_i \to -S_i$ [See Eq. (7) below].

A. Hirose et al. (Eds.): ICONIP 2016, Part IV, LNCS 9950, pp. 522–530, 2016.
DOI: 10.1007/978-3-319-46681-1_62

This Z(2) gauge symmetry is also viewed as a remnant of the full U(1) gauge symmetry of Maxwell's electromagnetism [4], and therefore quite natural to postulate for neural networks whose underlying processes are based on electromagnetic signals.

By treating these gauge models as models in statistical mechanics, the phase structure and the efficiency of learning and recalling are calculated explicitly in the Z(2) gauge neural network on the 3D lattice [5], and a sparsely connected network [6]. Reflecting that it is a gauge theory, there appear three phases known as typical ones of lattice gauge theory [7–9]; (1) Higgs (2) Coulomb (3) confinement phases. The averages $\langle J_{ij}\rangle$ and $\langle S_i\rangle$ (with some gauge-fixing [9] as in the mean-field theory) and the abilities of learning and recalling patterns are collected in Table 1.

Table 1. Each phase of gauge neural network and its ability.

Phase	$\langle J_{ij}\rangle$	$\langle S_i\rangle$	Ability
Higgs	$\neq 0$	$\neq 0$	Learning and recalling
Coulomb	$\neq 0$	0	Learning
Confinement	0	0	N.A.

The ability of learning patterns requires stable $\langle J_{ij}\rangle$ with small fluctuations as the Hebb's rule [10] shows; $J_{ij} = \sum_\alpha \xi_i^\alpha \xi_j^\alpha$ for N_L learned patterns $S_i = \xi_i^\alpha$ ($\alpha = 1, \cdots, N_L$). Large fluctuations of J_{ij} imply $\langle J_{ij}\rangle = 0$ and loss of this ability. The ability of recalling patterns requires stable $\langle S_i\rangle$ with small fluctuations as the Hopfield model shows; $S_i = \xi_i^\alpha$.

These studies on the Z(2) gauge neural networks have been confined to assuming that the synaptic connections are symmetric, i.e.,

$$J_{ij} = J_{ji}. \tag{1}$$

[See Fig. 1 (left).] The restriction to the symmetric connection (1) is traced back to the Hopfield model. In fact, by writing J_{ij} as

$$J_{ij} = J_{ij}^{\mathrm{S}} + J_{ij}^{\mathrm{A}},\ J_{ij}^{\mathrm{S}} \equiv \frac{1}{2}(J_{ij} + J_{ji}),\ J_{ij}^{\mathrm{A}} \equiv \frac{1}{2}(J_{ij} - J_{ji}), \tag{2}$$

one confirms that the Hopfield energy $\propto \sum_{i,j} S_i J_{ij} S_j$ picks up only the symmetric part J_{ij}^{S} and the asymmetric part J_{ij}^{A} has no effects upon system dynamics.

From neurophysiological point of view, J_{ij} and J_{ji} describe primarily two independent synaptic connections that reside on a different locations and function with different neurotransmitters [See Fig. 1 (right)]. So to study the case of effect of asymmetric synaptic connections is quite welcome. Although implementation of such asymmetry to the Hopfield energy is impossible as we discussed above, the additional interaction allowed in the gauge-invariant energy reflects

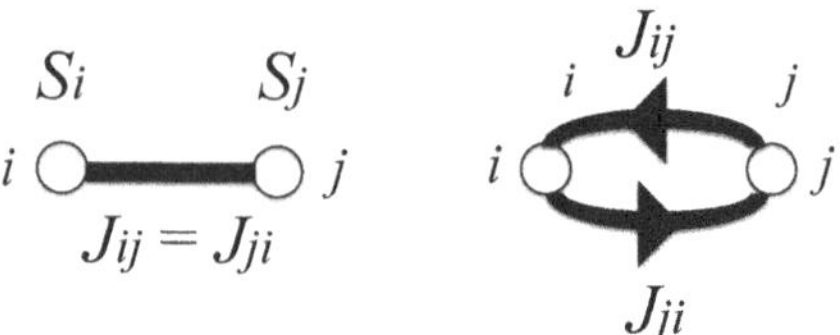

Fig. 1. S_i and J_{ij} for a pair of two sites i and j. Left: symmetric synaptic connection; $J_{ij} = J_{ji}$. Right: asymmetric synaptic connection; J_{ij} and J_{ji} are independent.

the antisymmetric part $J^{\rm A}_{ij}$ [See, e.g., the discussion below Eq. (6)] and open the way to study the asymmetry effect.

In this paper, we consider a neural network described by $S_i(=\pm 1)$ and $J_{ij}(=\pm 1)$. Its energy is Z(2) gauge-invariant and do have a dependence on $J^{\rm A}_{ij}$. Explicitly, we consider a network of random connections with the connectivity p.

2 Symmetric and Asymmetric Z(2) Gauge Neural Network

In this section we introduce two models on two kinds of networks.

2.1 Undirected and Directed Networks with Random Connection

Let us introduce n nodes and connect them randomly with the probability (connectivity) p. If the i-th node x_i and j-th node x_j are connected, we first distinguish the two cases; connection starting from x_i to x_j and connection starting from x_j to x_i. That is the link between two nodes is directed. One may introduce the connection index

$$c(i,j) = \begin{cases} 1 & \text{if link from } x_j \text{ to } x_i \text{ exists} \\ 0 & \text{if } x_j \text{ and } x_i \text{ are disconnected.} \end{cases} \tag{3}$$

Below we consider two cases of network. For $i \neq j$,

$$\begin{aligned} \textit{undirected network}: &\ c(i,j) = c(j,i), \\ \textit{directed network}: &\ c(i,j) \text{ and } c(j,i) \text{ are independent} \end{aligned} \tag{4}$$

In both case, $c(i,j)$ is a random number taking 1 with the probability p, and we avoid self coupling by setting $c(i,i) = 0$.

2.2 Models on the Random Network

To build the Z(2) gauge neural network, we put the neuron variable $S_i = \pm 1$ on the node x_i $(i = 1, \cdots, N)$ and the synaptic connection variable $J_{ij} = \pm 1 (i,j = 1, \cdots, N)$ on the link $c(i,j) = 1$. We define $J_{ij} = 0$ for $c(i,j) = 0$. Let us define the model by giving its energy $E(S,J)$.

The symmetric model (SM) with symmetric synaptic connections is build upon the undirected network, and its energy and the partition function are given by

$$
\begin{aligned}
E_{\mathrm{SM}} &= -c_1 \sum_{1\le i<j\le N} S_i J_{ij} S_j - \frac{c_2}{N} \sum_{1\le i<j<k\le N} J_{ij} J_{jk} J_{ki}, \\
J_{ij} &= J_{ji},\ J_{ii} = 0, \\
Z_{\mathrm{SM}} &= \sum_S \sum_J \exp(-E_{\mathrm{SM}}), \\
\sum_S &\equiv \prod_{i=1}^{N} \frac{1}{2} \sum_{S_i=\pm 1},\ \sum_J \equiv \prod_{i<j,c(i,j)=1} \frac{1}{2} \sum_{J_{ij}=\pm 1},
\end{aligned} \tag{5}
$$

where we consider the canonical ensemble in statistical mechanics. c_1, c_2 are real parameters that characterize the network (each brain). They are inversely proportional to the effective temperature T which controls the size of fluctuations of S_i and J_{ij} generated by noises in signals and their propagations. The first term is the Hopfield energy and the second term is the reverberation energy representing the current running the circuit $x_i \to x_k \to x_j \to x_i$ [10].

The asymmetric model (AM) with asymmetric synaptic connections is build upon the directed network, and its energy and the partition function are given by

$$
\begin{aligned}
E_{\mathrm{AM}} &= -\frac{c_1}{2} \sum_{i,j=1}^{N} S_i J_{ij} S_j - \frac{c_2}{6N} \sum_{i,j,k=1}^{N} J_{ij} J_{jk} J_{ki}, \\
&J_{ij} \text{ and } J_{ji} (i \ne j) \text{ independent},\ J_{ii} = 0, \\
Z_{\mathrm{AM}} &= \sum_S \sum_J{}' \exp(-E_{\mathrm{AM}}),\ \sum_J{}' \equiv \prod_{i,j,c(i,j)=1} \frac{1}{2} \sum_{J_{ij}=\pm 1}.
\end{aligned} \tag{6}
$$

The factors $1/2, 1/6$ are introduced so that the number of terms in each summation becomes the same as E_{SM}. As explained in Sect. 1, the second c_2-term has contribution from the antisymmetric component J^{A}_{ij} through the combination such as $J^{\mathrm{A}}_{ij} J^{\mathrm{A}}_{jk} J^{\mathrm{S}}_{ki}$.

Both energy E_{SM} and E_{AM} are invariant under the following Z(2) local(i-dependent) gauge transformation,

$$
S_i \to S'_i = V_i S_i,\ J_{ij} \to J'_{ij} = V_i J_{ij} V_j,\ V_i = \pm 1. \tag{7}
$$

The thermodynamic quantities such as the internal energy U and the specific heat C are define by the averages as

$$
\langle O(S,J) \rangle \equiv \frac{1}{Z} \sum_{S,J} O(S,J) e^{-E},\quad U = \frac{1}{N} \langle E \rangle,\quad C = \frac{1}{N} \left[\langle E^2 \rangle - \langle E \rangle^2 \right]. \tag{8}
$$

3 Phase Diagram

3.1 Phase Diagram of SM and AM

To study the phase structure of the AM, let us first recall the result of SM of Eq. (5). Figure 2 is the phase diagram in the c_2-c_1 plane of the SM with $p = 1$ [6]. There are three phases as announced, and the magnitude of fluctuation of J_{ij} becomes larger in the order of Higgs → Coulomb → confinement.

In Fig. 3, we show the phase diagram for $p = 1.0$ and $p = 0.8$ in (a) SM and (b) AM. The points on the curves are locations of the peak of the specific heat. The effect of partial connectivity $p < 1$ makes the fluctuation of J_{ij} larger and the area of confinement phase (J_{ij}-disordered phase) becomes larger.

(We note that Coulomb phase is also a J_{ij}-disordered phase as shown in Table 1, but its area (hight) diminishes as $O(1/N)$ as $N \to \infty$; See the figure caption of Fig. 2(b)).

For AM in Fig. 3(b), the p dependece has the same tendency as Fig. 3(a), but the overall area of the confinement phase is larger than the SM. In Fig. 4, we superpose the two phase diagrams of SM and AM for (a) $p = 1.0$ and (b) $p = 0.8$. Certainly, the AM has the wider confinement phase. In short, this is because the number of independent variables of J_{ij} is doubled in the AM from the SM, and the AM has larger amount of the total fluctuations. More explicitly, let us note that the phase space of the AM model includes the special set of configurations $J_{ij} = J_{ji}$, which is just those of the AM; $E_{\rm AM}|_{J_{ij}=J_{ji}} = E_{\rm SM}$. If only this set is allowed in the AM model, it gives rise to the same phase diagram as the SM. The remaining configurations of the AM express additional fluctuations of J_{ij} for the system and certainly extend the region of the disordered phase such as the confinement phase.

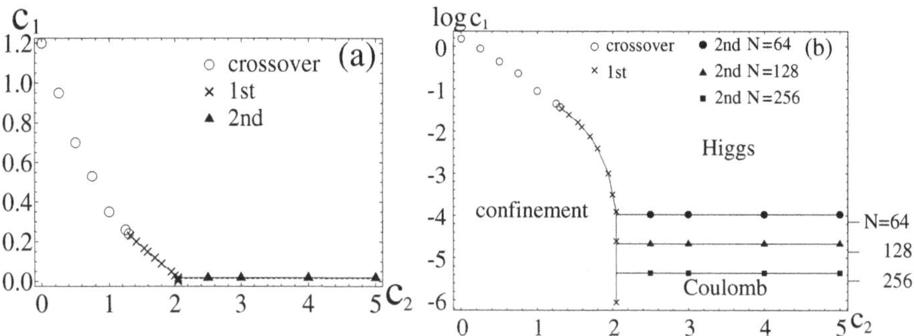

Fig. 2. Phase diagram for SM with $p = 1.0$ [6]. The transition line emanating from the triple point and separating the confinement and Higgs phases is a first-order line, which terminates at the critical point near $c_2 = 1.4$ followed by the cross overs. The confinement-Coulomb transition is of first order. The Higgs-Coulomb transition is of second-order, and the critical value of c_1 approaches $1/N$ as $c_2 \to \infty$ as shown in Fig. 2(b).

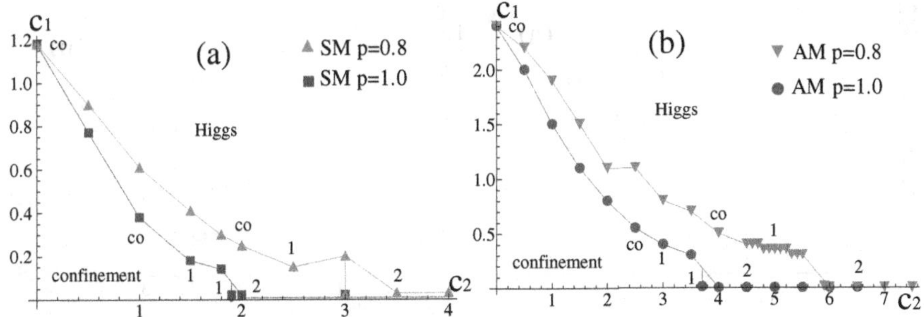

Fig. 3. Phase diagram in the c_2-c_1 plane with $p = 1.0, 0.8$ ($N = 64$). (a) SM and (b) AM. The lines are determined by the peak of the specific heat. Each point on the line indicates a crossover, 1st, or 2nd transition as marked by co, 1 and 2. The $p = 0.8$ case has a wider confinement phase due to the partial connectivity.

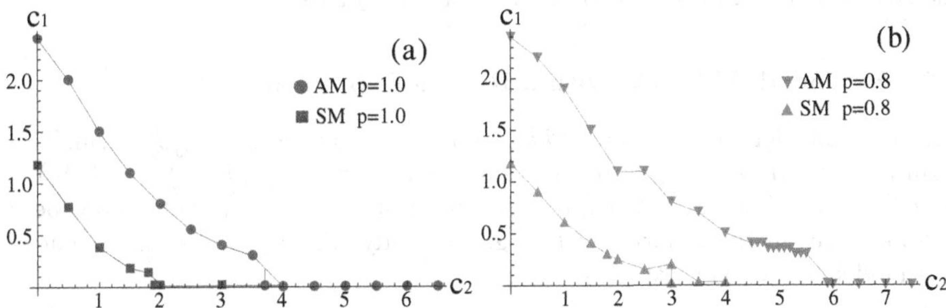

Fig. 4. Phase diagram for SM and AM with (a) $p = 1.0$ and (b) $p = 0.8$. The AM model has the wider confinement phase both in the c_2 and c_1 directions.

As an explicit example, let us consider the line along $c_2 = 0$. There no couplings among J_{ij} exist, and therefore each J_{ij} fluctuates independently according to the decoupled Boltzmann factor, $\exp(-c_1 J_{ij}^2/2)$ in Eq. (6). In fact, one may obtains the exact solution at $c_2 = 0$, $Z_{\rm SM} = (2\tanh(c_1/2))^{pN(N-1)/2}$, $Z_{\rm AM} = (2\tanh(c_1))^{pN(N-1)}$. Therefore, it has the peak of the specific heat at the doubled value of the SM with the Boltzmann factor $\exp(-c_1 J_{ij}^2)$ in Eq. (5). This consideration is confirmed by the critical value $c_1 \simeq 2.4$ at $c_1 = 0$ in Fig. 3(b), which is just twice with $c_1 \simeq 1.2$ in Fig. 3(a).

To make the comparison in a quantitative manner, we reduce the AM phase diagram by the factor 1/2 in both directions, i.e., we replace the coordinate of the transition point $(c_2, c_1) \to \frac{1}{2}(c_2, c_1)$, and then make a superposition with the SM phase diagram. Figure 5 is the result; the two lines almost coincide. The agreement in the c_1 direction is somewhat expected from the example for $c_2 = 0$ given above, as long as $c_2 \sim 0$. Concerning to the agreement in the c_2 direction, it should certainly reflect that the number of variables J_{ij} is doubled in the AM, but we have found no convincing and precise explanation for it. The coincidence

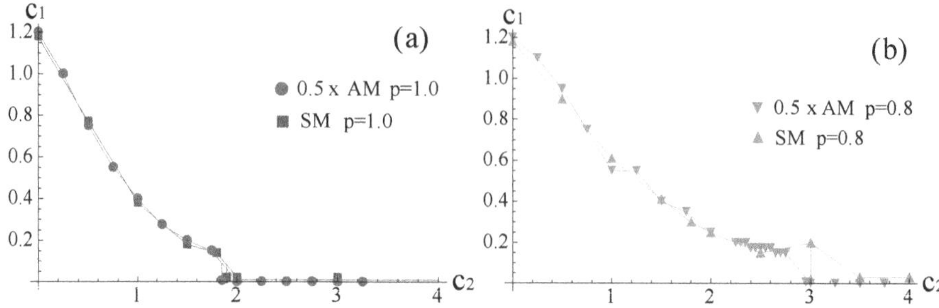

Fig. 5. Phase diagram for SM and the half size of AM with (a) $p = 1.0$ and (b) $p = 0.8$. They almost agree each other.

may be broken for the location of the strong first-order transition point between the confinement and the Coulomb phase on the c_2 axis.

3.2 Extended AM: AM with a Round-Trip Term

One may consider an extension of the AM of Eq. (6) by adding $c' J_{ij} J_{ji}$ term. This term in induced as a double processes of the c_1 term as $(S_i J_{ij} S_j) \cdot (S_j J_{ji} S_i) = J_{ij} J_{ji}$ due to $S_i^2 = 1$, and expresses a round-trip of signals. This may be a relevant term in the renormalization-group study. The energy of the extended AM model (EAM) is written as

$$E_{\mathrm{EAM}} = -\frac{c_1}{2} \sum_{i,j} S_i J_{ij} S_j - \frac{c'}{2} \sum_{i,j} J_{ij} J_{ji} - \frac{c_2}{6N} \sum_{i,j,k} J_{ij} J_{jk} J_{ki}. \tag{9}$$

We note that the similar term for the symmetric coupling becomes $J_{ij} J_{ji} = (J_{ij})^2 = 1$, and so irrelevant in the SM.

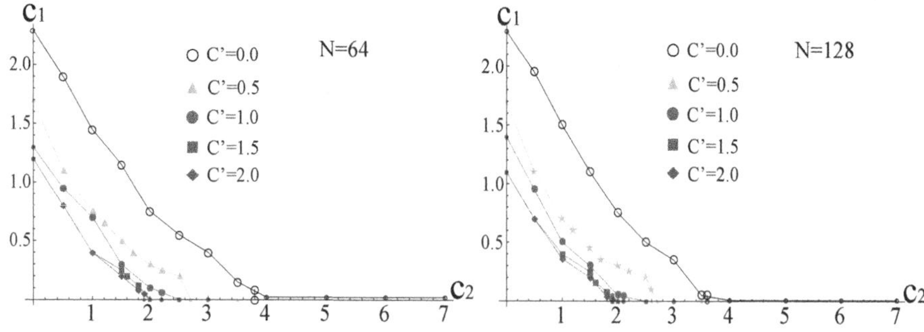

Fig. 6. Phase diagram in the $c_2 - c_1$ plane of the EAM of Eq. (9) for various values of c'; $N = 64$ (left), $N = 128$ (right).

The additional energy $-c' J_{ij} J_{ji}$ controls the relevance of the antisymmetric part $J^{\rm A}_{ij}$. In fact, as $c'(> 0)$ is increased, the product $J_{ij} J_{ji}$ favors +1 energetically, i.e., $J_{ij} = J_{ji}$, that is the case of the SM. So we have $\lim_{c' \to \infty} E'_{\rm AM} = E_{\rm SM}$. The modified model AM′ interpolates AM and SM; its phase transition curve in the c_2-c_1 plane starts from Fig. 4(a) at $c' = 0$ and approaches to the curve of SM in Fig. 4(b) as $c' \to \infty$. In Fig. 6 we show the phase diagram of EAM for various values of c' with $p = 1$, which confirms this interpretation. It shows that the region of Higgs phase increases as the coefficient c' increases. That is the effect of the round-trip term enhances the ability of learning and recalling patterns.

4 Conclusion and Future Problems

We considered the Z(2) gauge neural-network model of learning and recalling patterns, which contains the Z(2) neuron variable S_i and the synaptic connection variable J_{ij}, a Z(2) gauge-field variable. The energy consists of the Hopfield term $c_1 SJS$ and the reverberating term $c_2 JJJ$, both of which are gauge invariant. The c_2 term is capable to reflect the antisymmetricity of the synaptic connections. The system is treated as a canonical ensemble with noises controlled by c_1, c_2 and the phase diagram is obtained by Monte Carlo simulation. By comparing the phase diagram of symmetric model and asymmetric model, we confirmed that the antisymmetric part $J^{\rm A}_{ij}$ certainly extends the region of the confinement(disordered) phase. This implies that the antisymmetric part reduces both abilities of learning patterns of S_i and recalling them. The phase diagram of the AM model is almost the doubled one of the SM in both c_1-c_2 directions. Its precise understanding is a future problem. Calculation of the efficiency of learning patterns of S_i and recalling them as done in Ref. [5] may extract some other effects of antisymmetric part.

Concerning to the network structure, we considered oriented and unoriented networks with random connections. Neurons in the human brain is known to have complicated network structures, such as left and right hemispheres, multilayer-structure, column-structure, small-world network, etc. It is interesting to set up gauge neural net-work models on these networks and compare their phase diagrams. Although we expect any phase may be classified into the basic three phases in Table I, the details of phase structure should be structure-dependent and shed some light on the study of brain architecture.

Finally, we comment on the role of parameters in the energy, c_1, c_2, c'. They distinguish system by system (brain by brain), but even within one system, they may change spontaneously or compulsory. For example, they should take different values while one is awake or asleep, with or without anesthesia, etc. Such change may give rise to a move into another phase. It is certainly interesting to think about the rule and/or mechanism that control their changes.

Acknowledgment. The authors would like to thank Dr. Y. Nakano for discussion.

References

1. Hopfield, J.J.: Proc. Nat. Acd. Sci. USA **79**, 2554 (1982)
2. Haykin, S.: Neural Networks; A Comprehensive Foundation. Macmillan, London (1994)
3. Matsui, T.: In: Janke, W. et al. (eds.) Fluctuating Paths and Fields, p. 271. World Scientific (2001) (cond-mat/0112463)
4. Fujita, Y., Matsui, T.: Wang, L., et al. (eds.) Proceedings of 9th International Conference on Neural Information Processing, pp. 1360–1367 (2002). cond-mat/0207023; Fujita, Y., Hiramatsu, T., Matsui, T.: Proceedings of International Joint Conference on Neural Networks, Montreal, Canada, p. 1108 (2005)
5. Kemuriyama, M., Matsui, T., Sakakibara, K.: Phys. A **356**, 525 (2005) (cond-mat/0203136)
6. Takafuji, Y., Nakano, Y., Matsui, T.: Physica A **391**, 5258–5304 (2012). arXiv:1206.1110
7. Wilson, K.: Phys. Rev. D **10**, 2445 (1974); Kogut, J.B.: Rev. Mod. Phys. **51**, 659 (1979)
8. Rothe, H.J.: Lattice Gauge Theories: An Introduction. World Scientific, Singapore (2005)
9. Ichinose, I., Matsui, T.: Mod. Phys. Lett. B **28**, 1430012 (2014)
10. Hebb, D.O.: The Organization of Behavior: A Neuropsychological Theory. Wiley, New York (1949)

SOMphony: Visualizing Symphonies Using Self Organizing Maps

Arnulfo Azcarraga(✉) and Fritz Kevin Flores

Computer Technology, De La Salle University,
2401 Taft Avenue, Manila, Philippines
arnie.azcarraga@delasalle.ph, fritz.flores@dlsu.edu.ph
http://www.dlsu.edu.ph/

Abstract. Symphonies are musical compositions played by a full orchestra which have evolved in style since the 16th Century. Self-Organizing Maps (SOM) are shown to be useful in visualizing symphonies as a musical trajectory across the nodes in a trained map. This allows for some insights about the relationships and influences between and among composers in terms of their composition styles, and how the symphonic compositions have evolved over the years from one major music period to the next. The research focuses on Self Organizing Maps that are trained using 1-second music segments extracted from 45 different symphonies, from 15 different composers, with 3 composers from each of the 5 major musical periods. The trained SOM is further processed by doing a k-means clustering of the node vectors, which then allows for the quantitative comparison of music trajectories between symphonies of the same composer, between symphonies of different composers of the same music period, and between composers from different music periods.

Keywords: Symphony · Self Organizing Maps · k-means clustering · Music trajectory

1 Introduction

Dating back from the 16th Century, a symphony is defined as an elaborate musical composition in about 3 to 5 movements, which is played by a full orchestra. We try to visualize the relationships of the different music periods, from the 16th century Baroque, to the Classical period, into the 19th century, and then the Romantic period, and the early 20th Century, in terms of how the symphonies have evolved. For each period, we choose 3 famous composers and processed 3 of their symphonic compositions, giving a total of 45 symphonies. In particular, we explore the possibility of using Self-Organizing Maps (SOM) [1–5] to encode the musical trajectory of each of the 45 symphonies as a basis for comparing the similarity between symphonic compositions of the same composer, as well as comparisons between compositions by different composers of the same period, and also across music periods. This way, we can, for example, confirm whether great composers, such as Wolfgang Amadeus Mozart, have had lasting influence on music compositions even centuries after his first compositions

A. Hirose et al. (Eds.): ICONIP 2016, Part IV, LNCS 9950, pp. 531–537, 2016.
DOI: 10.1007/978-3-319-46681-1_63

have been played. This we do by confirming whether indeed the musical trajectories of symphonic compositions from composers after Mozart would still resemble the original Mozart pieces.

In this study, SOM and k-means clustering are used in tandem in order to quantify the trajectory of a symphony from type of music/sound to another, as it progresses through the 3 to 5 movements. Once a trained SOM has been subjected to k-means clustering, we can visualize on a per-second basis the trajectory of each symphonic composition – and this trajectory is the basis for judging whether two symphonies are similar or not. Although we could also have used the U-matrix, we opted to use k-means clustering, as this would allow us to easily tag the nodes and group them [8].

2 Experiment Set-Up

We train Self Organizing Maps using 1-second music segments extracted from 45 different symphonies. These symphonies are from 15 different composers, with 3 composers from each of the 5 major musical periods, and 3 symphonies per composer. The trained SOM is further processed by doing a k-means clustering of the node vectors, which then allows for the quantitative comparison of music trajectories between symphonies of the same composer, between symphonies of different composers of the same music period, and between composers from different music periods.

Each symphony is prepared by splitting the audio into 1-second segments. After which, jAudio 1.0.1 is used to extract the features from each segment, amounting to 78 features per segment, following the pre-processing steps from [1]. Once completed, the entire dataset is then normalized and clustered through Self-Organizing Maps.

The SOM is a 16 by 16-rectangular map, trained with a learning rate of 0.9 and a neighborhood distance of 16 during the initial learning phase. A global ordering phase of 10,000 cycles and a fine- tuning phase of another 10,000 cycles are used. The learning rate and neighborhood use a linear decay function decreasing to a value of 0.1 and 1 respectively by the completion of the global ordering phase. For the fine-tuning phase, the values of the learning rate and neighborhood are kept constant at 0.1 and 1, respectively.

Once a trained SOM is completed, the weights of each of the 256 nodes are clustered using k-means clustering. For the experiments conducted, we used k = 21, which resulted in clusters with at least 6 nodes in each of them. Figure 1 is the clustered trained SOM, which we playfully refer to as a SOMphony.

Each cluster in the SOMphony is composed of nodes that are sensitive to specific types of music segments, with slight variations among the node weight vectors in each cluster. In other words, each 1-second music segment of a symphony would have a best-matching-unit (BMU) in one of the 21 clusters. Hence, if we play a given symphony that has been spliced into 1-second music segments, we can trace the sequence of BMU for the entire length of the musical piece and refer to this sequence as the "musical trajectory" of the symphony. Two symphonies that are similar would thus produce similar trajectories in the SOMphony.

In the next section, we do a subjective, qualitative comparison of the 45 different trajectories by visually inspecting them and comparing them pairwise to see whether

Fig. 1. SOMphony map using K = 21

certain symphonies "appear" to be similar to each other. For the visual inspection of the trajectories, the result of the k-means clustering is not yet used, as we pinpoint the exact BMU among the 256 nodes for each 1-second music segment. We then complement the qualitative assessment with a more objective quantitative comparison of the trajectories. It is this quantitative approach that directly involves the 21 clusters.

3 Comparison of Music Trajectories

3.1 Visual Rendering of the Music Trajectory on the Map

After obtaining the SOMphony Map, the symphony compositions are played in a way that for each 1-second segment of the composition, the BMU in the map is determined. The BMU is the one node among the 256 nodes in the map that is closest, based on the Euclidean distance, between the node weights and the features of the music segment. As explained earlier, this sequence of BMU nodes in the map is the music trajectory associated with the symphony. Figure 2 shows the individual trajectories for each of the 45 compositions, grouped by musical period.

In the visual inspection of the music trajectories, we denote using a color-encoding scheme the time sequence of music segments (blue are the earlier segments and the color moves towards red as the music progresses). This way, we can see where the music trajectory started and how the music traverses the SOMphony as the music unfolds. The notion of time is not used for the quantitative comparison of symphonies using cluster frequency counts. We simply use the normalized frequency counts once an entire symphony has been played out, and these cluster frequency counts are what we use to compare the similarity between symphonies. These are explained in greater detail in the next section.

For the first symphonies during the baroque period, the compositions were typically complex and being religious in nature, typically have repetitive sections and patterns, however they do not show any similarity in visualization and style between composers and their compositions as opposed to that of other periods. In contrast, the classical period hosts one of the most influential names in music, Mozart, where Fig. 2 reveals that his Symphony 41, "Jupiter", has a diagonal movement which can also be seen

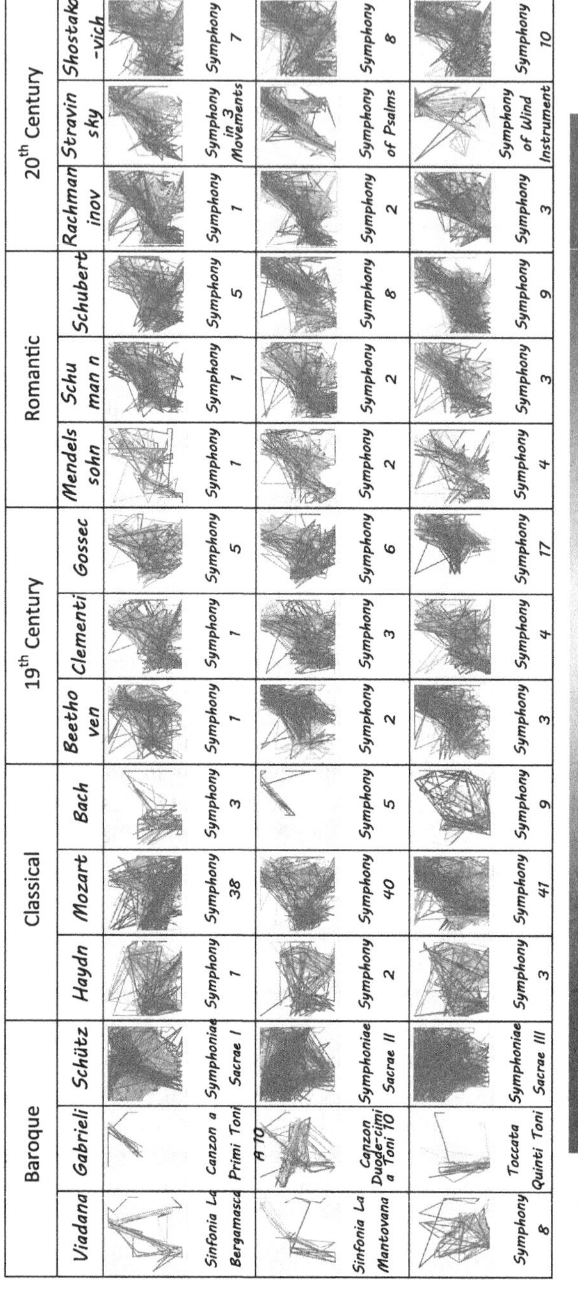

Fig. 2. SOMphony trajectories and color spectrum to designate time in the SOMphony map (Color figure online)

from the compositions of almost all of the composers of the succeeding periods such as Mendelssohn, Schumann, Schubert, and Rachmaninov. Although these patterns could be seen from multiple composers after Mozart, the visualization does not allow any quantitative evidence that they are indeed related to each other.

3.2 Quantitative Comparison Among Music Trajectories

For each symphony, we keep track of the frequency count of each of the 21 clusters of nodes found using k-means clustering. Each time a 1-second music segment has a BMU within a certain cluster c(i), we increase the frequency count for cluster c(i). This way, if the symphony mainly uses music segments that have BMUs in among 3 clusters only, for example, then only these 3 clusters will have high frequency counts. We then normalize these frequency counts by dividing the counts of a given composition by its total number of 1-second music segment. A normalized count of s% for cluster c(i), for example, means that s% of the music segments had BMUs among the nodes contained in cluster c(i). Once these normalized frequencies have been summarized, it is now possible to use these percentages as basis for doing a pair-wise quantitative comparison among the 45 compositions to see which compositions can be considered to be similar with each other. We expect that some symphonies of the same composer would tend to be similar, if the composer has some coherent style. Also, there can be similarities among compositions of a given composer if another composer had heavily influenced him. Or a pair of compositions by different composers can turn out to be similar if these compositions have both been influenced heavily by the same (earlier) composer.

The pairwise comparisons between music trajectories are shown in Table 1. The comparisons are based on a simple Euclidean distance between the normalized frequency counts of composition 1 with those of composition 2. The smaller the entry, the more similar are the two compositions. In Table 1, we tag those distances of up to 0.2, in order to quickly see which are the pairs of symphonies with a relatively low distance between them.

We have opted to only include the quantitative comparisons for three musical periods, as these are where most of the pairs of similar symphonies can be seen. The compositions from the baroque period do not share similar trajectories with compositions of the same composer, nor with compositions of other composers of the same period. Neither do they resemble the compositions that appeared in the later centuries.

From Table 1, we discern two composers who seemed to have been very influential. During the Classical period, Mozart became one of the most famous composers of his time and is known to have been a huge influence in music. Table 1 indeed reveals similarities between Mozart's own compositions, implying some coherent style that Mozart may have developed. Furthermore, his compositions are also similar to those of a good number of composers during the 19th century following the classical period, namely Beethhoven, Clementi and Gossec, as well as the symphonic compositions of the Romantic and even the 20th century impressionist and post-impressionist era, including Mendelssohn, Schumann, Schubert, Rachmaninov and Shostakovich.

For the 19th Century musical period after Mozart, Beethoven became the most famous composer of his time and is also known to have a huge influence in music.

Table 1. Pairwise Euclidean distance comparison of symphony clusters

Distance		Classical									19th Century									Romantic									20th Century								
		H1	H2	H3	M1	M2	M3	B1	B2	B3	B1	B2	B3	C1	C2	C2	G1	G2	G3	M1	M1	M3	SM1	SM2	SM3	SB1	SB2	SB3	R1	R2	R3	ST1	ST2	ST3	SH1	SH2	SH3
Classical	Haydn 1		14	7	25	25	30	30	76	31	26	27	31	25	27	30	35	29	27	26	28	28	24	22	22	29	33	31	32	34	32	46	48	65	32	34	33
	Haydn 2			15	29	29	32	37	76	36	29	31	34	30	34	38	36	34	31	27	32	32	30	30	29	34	35	36	33	39	38	47	50	65	34	32	33
	Haydn 3				25	24	28	28	75	30	28	30	31	23	24	28	36	29	29	25	27	27	27	22	24	28	32	30	31	33	29	45	47	64	30	32	32
	Mozart 1					15	27	30	79	41	32	36	35	18	31	30	41	23	25	31	27	25	32	18	28	17	33	24	31	27	23	48	51	66	31	34	34
	Mozart 2						16	32	70	42	22	28	23	14	27	29	33	16	15	23	19	18	23	12	20	10	23	11	20	15	21	36	39	55	21	26	25
	Mozart 3							38	58	43	21	27	19	20	29	31	24	20	16	15	12	11	26	24	27	20	14	15	15	19	22	26	28	45	10	16	15
	Bach 1								84	25	44	47	45	30	32	33	53	34	39	42	38	35	40	29	37	31	43	34	42	36	34	55	56	72	40	44	44
	Bach 2									82	58	57	52	74	77	78	57	74	67	57	63	63	67	74	69	75	50	65	53	63	74	38	41	39	52	50	47
	Bach 3										45	47	48	42	41	43	53	43	44	45	44	41	46	42	45	44	47	46	46	48	46	56	57	70	46	48	46
19th Century	Beethoven 1											10	8	29	35	38	25	29	21	19	25	26	16	25	20	29	19	23	16	24	36	25	29	46	21	23	20
	Beethoven 2												12	34	39	42	27	35	27	22	30	31	16	28	18	35	23	29	20	29	41	28	33	49	26	27	24
	Beethoven 3													29	36	38	25	30	21	18	24	25	18	27	21	30	14	22	11	22	36	19	24	41	18	20	16
	Clementi 1														16	16	35	18	19	23	17	16	27	16	25	13	25	17	25	22	13	41	42	60	23	29	28
	Clementi 2															9	41	30	30	28	26	25	33	26	31	28	32	29	33	33	22	47	44	63	30	36	35
	Clementi 3																42	29	30	31	27	25	36	28	35	27	33	30	35	34	18	48	46	66	32	39	37
	Gossec 1																	30	24	21	21	23	29	36	32	36	24	33	28	35	35	33	36	55	26	28	26
	Gossec 2																		11	27	16	16	26	18	27	12	28	18	28	22	20	41	43	61	27	33	32
	Gossec 3																			21	15	15	21	20	24	17	21	17	20	21	24	33	37	55	22	27	25
Romantic	Mendelssohn 1																				15	17	23	27	25	28	13	24	16	27	28	28	29	48	15	15	14
	Mendelssohn 2																					7	25	23	26	19	18	19	21	23	17	34	35	55	17	22	21
	Mendelssohn 3																						27	22	28	18	17	18	21	22	15	33	35	54	16	23	21
	Schumann 1																							19	10	27	26	23	22	24	35	35	39	57	28	31	29
	Schumann 2																								14	13	29	16	26	17	24	42	45	63	27	32	31
	Schumann 3																									25	28	23	24	23	33	39	42	60	28	31	30
	Schubert 1																										28	13	26	16	17	42	44	62	25	31	31
	Schubert 2																											20	9	22	28	19	22	41	11	14	9
	Schubert 3																												18	6	22	31	34	51	17	24	23

Indeed, Table 1 reveals similarities between Beethoven's symphonies with those of the three Romantics (Mendelssohn, Schumann and Schubert), as well as Rachmaninov and even Stravinsky. And similar to Mozart, Beethoven also has a distinct style, and indeed, Table 1 also shows that Beethoven's compositions are quite similar to each other.

As to the Romantic period, it is interesting to note the similarity between compositions of the same composers, as well as the compositions among the three famous composers of this period. From Table 1, the pairwise comparisons among the compositions reveal that the music trajectories during this period were all of similar style. Worthwhile also to note is the fact that Gossec's symphonies of the 19th century resemble closely the works of the Romantics. And all the compositions from the Romantic period seemed to have carried over to the 20th century impressionist composers, especially Rachmaninov.

Table 1 also shows that some composers stick to their style of music, such as those of Beethoven, Clementi, Mendelssohn, and Schumann. However since the data is limited to the 3 compositions for each composer, not all possibilities are shown.

4 Conclusion

Self-Organizing Maps are shown to be indeed useful in visualizing symphonies as a musical trajectory across the nodes in a trained map. This allows for some insights about the relationships and influences between and among composers in terms of their composition styles, and how the symphonic compositions have evolved over the years from one major music period to the next. Even though the results presented a qualitative/subjective comparison of symphonies as well as an objective/quantitative mechanism for comparing symphonies, a larger dataset would be needed to confirm whether the approach is indeed valid. If proven to be so, it is clear that the approach described here may also be applied to other forms of music.

References

1. Azcarraga, A., Caronongan, A., Setiono, R., Manalili, S.: Validating the Stable Clustering of Songs in a Structured 3D SOM (2016)
2. Corrêa, D.C., Rodrigues, F.A.: A survey on symbolic data-based music genre classification. Expert Syst. Appl. **60**, 190–210 (2016)
3. Hartigan, J.A., Wong, M.A.: Algorithm AS 136: a k-means clustering algorithm. J. Roy. Stat. Soc.: Ser. C (Appl. Stat.) **28**(1), 100–108 (1979)
4. Kangas, J.A., Kohonen, T.K., Laaksonen, J.T.: Variants of self-organizing maps. IEEE Trans. Neural Netw. **1**(1), 93–99 (1990)
5. Kohonen, T., Somervuo, P.: Self-organizing maps of symbol strings. Neurocomputing **21**(1), 19–30 (1998)
6. Scaringella, N., Zoia, G., Mlynek, D.: Automatic genre classification of music content: a survey. IEEE Signal Process. Mag. **23**(2), 133–141 (2006)
7. Toussaint, G.T.: Algorithmic, geometric, and combinatorial problems in computational music theory. In: Proceedings of X Encuentros de Geometria Computacional, pp. 101–107 (2003)
8. Ultsch, A.: Self-organizing neural networks for visualisation and classification. In: Opitz, O., Lausen, B., Klar, R. (eds.) Information and Classification, pp. 307–313. Springer, Heidelberg (1993)

Online EM for the Normalized Gaussian Network with Weight-Time-Dependent Updates

Jana Backhus[1(✉)], Ichigaku Takigawa[1,2], Hideyuki Imai[1], Mineichi Kudo[1], and Masanori Sugimoto[1]

[1] Department of Computer Science and Information Technology, Graduate School of Information Science and Technology, Hokkaido University, Kita 14 Nishi 9, Kita-ku, Sapporo 060-0814, Japan
jana@main.ist.hokudai.ac.jp

[2] JST PRESTO, 4-1-8 Honcho, Kawaguchi, Saitama 332-0012, Japan

Abstract. In this paper, we propose a weight-time-dependent (WTD) update approach for an online EM algorithm applied to the Normalized Gaussian network (NGnet). WTD aims to improve a recently proposed weight-dependent (WD) update approach by Celaya and Agostini. First, we discuss the derivation of WD from an older time-dependent (TD) update approach. Then, we consider additional aspects to improve WD, and by including them we derive the new WTD approach from TD. The difference between WD and WTD is discussed, and some experiments are conducted to demonstrate the effectiveness of the proposed approach. WTD succeeds in improving the learning performance for a function approximation task with balanced and dynamic data distributions.

Keywords: Normalized Gaussian networks · Online EM · Local model · Weight-dependent forgetting

1 Introduction

In sequential learning schemes, only one data sample is observed at any time, and applied neural networks need to be trained incrementally. In many real applications, training data is however not independent and identically distributed (i.i.d.), which can give rise to the problem of negative interference. Negative interference refers to unwanted forgetting of already learned information. Neural networks generally change their model in favor of newly arriving training samples what makes them prone to negative interference. Therefore, training methods need to be considered in sequential learning schemes that achieve robust learning performance.

In this paper, we consider an incremental learning approach for the Normalized Gaussian network (NGnet), a network for stochastic function approximation in continuous domains. The NGnet is an instance of Mixture-of-Experts, and Xu et al. have proposed to estimate model parameters with an offline Expectation-Maximization (EM) algorithm [6]. For incremental learning, an online EM algorithm has been later proposed by Sato and Ishii [5], which applies time-dependent

A. Hirose et al. (Eds.): ICONIP 2016, Part IV, LNCS 9950, pp. 538–546, 2016.
DOI: 10.1007/978-3-319-46681-1_64

(TD) updates including a discount factor. The discount factor controls the amount of forgetting of old learning results for each NGnet component based exclusively on time. Time-dependent forgetting speeds up convergence [4] and performs well when sample data are i.i.d., but learning performance is unstable otherwise. Therefore, Celaya and Agostini have proposed a modification of TD, which we will call the weight-dependent (WD) update approach [1]. WD aims to forget based on the amount of newly received information instead of time, and it is more stable in environments prone to negative interference.

A comparison of TD and WD shows that the latter improves the approximation accuracy and exhibits much greater stability in its learning performance [1]. WD has however been derived from an assumption based exclusively on the update weights, and we can further improve WD by including time dependency in the derivation. In this paper, we shortly discuss the derivation of Celaya and Agostini's WD approach and then derive a new weight-time-dependent (WTD) approach from TD. The newly proposed WTD approach is compared to WD: numerical differences are discussed, and some experiments are conducted for a function approximation task with balanced and dynamic data distributions. The experimental results show that our proposed WTD approach has an overall better learning performance and is therefore preferable to WD.

2 Normalized Gaussian Network and Online EM

The Normalized Gaussian network (NGnet) [3] is a universal function approximator that transforms an N-dimensional input vector x to a D-dimensional output vector y with

$$y = \sum_{i=1}^{M} N_i(x)\tilde{W}_i\tilde{x}. \tag{1}$$

Here, M is the number of units, $\tilde{x}' \equiv (x', 1)$ where prime ($'$) denotes a transpose, and $\tilde{W}_i \equiv (W_i, b_i)$ is a $D \times (N+1)$-dimensional linear regression matrix. Normalized Gaussian functions N_i are used as activation functions with

$$N_i(x) \equiv G_i(x) / \sum_{j=1}^{M} G_j(x) \tag{2}$$

$$G_i(x) \equiv (2\pi)^{-N/2}|\Sigma_i|^{-1/2} \exp\left[-\frac{1}{2}(x-\mu_i)'\Sigma_i^{-1}(x-\mu_i)\right]. \tag{3}$$

The i-th multivariate Gaussian function $G_i(x)$ has an N-dimensional center μ_i and an $N \times N$-dimensional covariance matrix Σ_i. The model then softly partitions the input space into M local units.

2.1 Online EM Algorithm

A stochastic interpretation of the NGnet model has been first proposed by Xu et al. [6]. The unknown model parameters are then estimated by maximum likelihood estimation based on the log-likelihood of the observed in- and output data

(x, y). Xu et al. applied an offline Expectation-Maximization (EM) algorithm [6] that has been later adopted to an online EM algorithm by Sato and Ishii [5]. The stochastic model is defined for a complete event (x, y, i) with the probability distribution

$$P(x, y, i|\theta) = (2\pi)^{-\frac{D+N}{2}} \sigma_i^{-D} |\Sigma_i|^{-\frac{1}{2}} M^{-1} \times \exp\left[-\frac{1}{2}(x-\mu_i)'\Sigma_i^{-1}(x-\mu_i) - \frac{1}{2\sigma_i^2}|y - \tilde{W}_i\tilde{x}|^2\right], \quad (4)$$

where $\theta \equiv \{\mu_i, \Sigma_i, \sigma_i^2, \tilde{W}_i | i = 1, ..., M\}$ is the set of model parameters that have to be estimated. For the online EM algorithm, the parameters are updated with the following E- and M-step at every time step t.

E (Estimation) Step: Given the current estimator $\theta(t-1)$, the posterior probability $P_i(t) \equiv P(i|x(t), y(t), \theta(t-1))$ is calculated for the i-th unit to evaluate how likely this unit generates the current observation $(x(t), y(t))$:

$$\begin{aligned} P_i(t) &\equiv P(i|x(t), y(t), \theta(t-1)) \\ &= P(x(t), y(t), i|\theta(t-1)) / \sum_{j=1}^{M} P(x(t), y(t), j|\theta(t-1)). \end{aligned} \quad (5)$$

M (Maximization) Step: Then, the model parameters θ are updated with

$$\mu_i(t) = \langle\langle x\rangle\rangle_i(t) / \langle\langle 1\rangle\rangle_i(t) \quad (6)$$

$$\Sigma_i^{-1}(t) = [\langle\langle xx'\rangle\rangle_i(t) / \langle\langle 1\rangle\rangle_i(t) - \mu_i(t)\mu_i'(t)]^{-1} \quad (7)$$

$$\tilde{W}_i(t) = \langle\langle y\tilde{x}'\rangle\rangle_i(t)[\langle\langle \tilde{x}\tilde{x}'\rangle\rangle_i(t)]^{-1} \quad (8)$$

$$\sigma_i^2(t) = \frac{[\langle\langle |y|^2\rangle\rangle_i(t) - Tr(\tilde{W}_i(t)\langle\langle \tilde{x}y'\rangle\rangle_i(t))]}{D\langle\langle 1\rangle\rangle_i(t)} \quad (9)$$

2.2 Time-Dependent Update (TD)

The updates (6)–(9) include weighted accumulators of the form $\langle\langle\cdot\rangle\rangle_i(t)$. Originally, these were weighted *means* in the time-dependent (TD) approach [5], but they have been changed to weighted *sums* [1] which we will use. First, we consider a TD update [5] of the weighted sum for unit i over the whole time T

$$\langle\langle f\rangle\rangle_i(T) \equiv \sum_{t=1}^{T}\left(\prod_{s=t+1}^{T}\lambda(s)\right) P_i(t) f_t, \quad (10)$$

where $f \equiv f(x, y)$ and $f_t \equiv f(x(t), y(t))$ are used as abbreviations. The discount factor $\lambda(t)$ is time-dependent and plays an important role in discarding the effect of old learning results that were employed to an earlier inaccurate estimator. The factor has to be chosen so that $\lambda \to 1$ when $t \to \infty$ for fulfilling the Robbins-Monro condition for convergence of stochastic approximations [2]. For online updates, a stepwise update is obtained from (10) for each time step t

$$\langle\langle f \rangle\rangle_i(t) = \lambda(t)\langle\langle f \rangle\rangle_i(t-1) + P_i(t) f_t. \tag{11}$$

2.3 Weight-Dependent Update (WD)

The weight-dependent (WD) update approach is a modification of TD, proposed by Celaya and Agostini [1] to improve learning robustness against negative interference. TD can be problematic in applications where data is not i.i.d., because the network model updates only units in the same region as the received data sample while it forgets old learning results over the whole input space. Therefore, WD [1] has been introduced, and we shortly review its derivation from TD in the following.

For WD, learned information should be forgotten only as much as new updates are received. Therefore, model updates were based exclusively on weights, and the notation of (11) has been changed to use weight-dependent indexing with $\omega_k = P_i(t)$ as the k-th received weight:

$$\langle\langle f \rangle\rangle(\omega_{k-1} + \omega_k) = \Lambda(\omega_k)\langle\langle f \rangle\rangle(\omega_{k-1}) + \Omega(\omega_k) f_t. \tag{12}$$

Here, $\Lambda(\omega_k)$ and $\Omega(\omega_k)$ are a weight-dependent forgetting and update factor respectively. The two factors are unknown functions of ω_k that were determined based on the following conditions. For a full update, $\Lambda(\omega_k)$ and $\Omega(\omega_k)$ should be $\Lambda(1) = \lambda(t)$ and $\Omega(1) = 1$, reducing to the same values as for a full update of TD. On the other hand, the weighted sum $\langle\langle f \rangle\rangle$ should remain unaltered when $\omega_k = 0$. Additionally, a consistency condition need to be fulfilled, imposing that a weighted sum $\langle\langle f \rangle\rangle$ updated once with value f_t and weight $(\omega_k + \omega_{k+1})$ must be the same as $\langle\langle f \rangle\rangle$ updated twice with f_t and weights ω_k and ω_{k+1}. In the following, the two cases are expressed based on (12):

$$\langle\langle f \rangle\rangle(\omega_{k-1} + (\omega_k + \omega_{k+1})) = \Lambda\left(\omega_k + \omega_{k+1}\right)\langle\langle f \rangle\rangle(\omega_{k-1}) + \Omega\left(\omega_k + \omega_{k+1}\right) f_t, \tag{13}$$

$$\begin{aligned} \langle\langle f \rangle\rangle((\omega_{k-1} + \omega_k) + \omega_{k+1}) &= \Lambda(\omega_{k+1})\langle\langle f \rangle\rangle(\omega_{k-1} + \omega_k) + \Omega(\omega_{k+1}) f_t \\ &= \Lambda(\omega_{k+1})\Lambda(\omega_k)\langle\langle f \rangle\rangle(\omega_{k-1}) + \left(\Lambda(\omega_{k+1})\Omega(\omega_k) + \Omega(\omega_{k+1})\right) f_t. \end{aligned} \tag{14}$$

Based on (13) and (14), the following functional equations were obtained

$$\Lambda(\omega_k + \omega_{k+1}) = \Lambda(\omega_{k+1})\Lambda(\omega_k), \tag{15}$$

$$\Omega(\omega_k + \omega_{k+1}) = \Lambda(\omega_{k+1})\Omega(\omega_k) + \Omega(\omega_{k+1}). \tag{16}$$

Including the full update condition, these functional equations were used to obtain a weight-dependent forgetting and update factor, where $\Lambda(\omega_k) = \lambda(t)^{\omega_k}$ and $\Omega(\omega_k) = \frac{1-\lambda(t)^{\omega_k}}{1-\lambda(t)}$. Finally, the notation was returned to a time based index, and ω_k was replaced with $P_i(t)$. The weight-dependent stepwise update becomes

$$\langle\langle f \rangle\rangle_i(t) = \lambda(t)^{P_i(t)}\langle\langle f \rangle\rangle_i(t-1) + \left(\frac{1-\lambda(t)^{P_i(t)}}{1-\lambda(t)}\right) f_t. \tag{17}$$

3 Weight-Time-Dependent Update (WTD)

In the following, we propose a new modification of TD, which we will call the weight-time-dependent (WTD) update approach. The modification is motivated by the dependency of WD's update factor $\Omega(P_i(t)) = \frac{1-\lambda(t)^{P_i(t)}}{1-\lambda(t)}$ on the discount factor $\lambda(t)$. We think this to be bad practice, because $\lambda(t)$ controls forgetting and originally has been introduced to speed up convergence [4]. If we consider a case without discount, where $\lambda(t) = 1$ for all t, then the WD update factor becomes $\frac{1-1^{P_i(t)}}{1-1} = \frac{0}{0}$ and learning is impossible. So, we derive a new update approach from TD where not only dependency on weight but also time and unit division is considered.

Our derivation is similar to the one of WD, but we keep the time index notation of TD over the whole derivation to emphasize the new time dependency. The stepwise update (11) is then rewritten to

$$\langle\langle f\rangle\rangle_i(t) = \Lambda(P_i(t))\langle\langle f\rangle\rangle_i(t-1) + \Omega(P_i(t))f_t, \tag{18}$$

where a new forgetting factor $\Lambda(P_i(t))$ and update factor $\Omega(P_i(t))$ have to be determined by considering the following conditions.

First, we consider a full update condition, where $\Lambda(1) = \lambda(t)$ and $\Omega(1) = 1$, with additional dependencies on time t and unit division. The local units divide the input space between each other according to their representation ability of the current sample $(x(t), y(t))$. A full update can therefore only happen at a same time step t for a weighted sum $\langle\langle f\rangle\rangle_M$ that accumulates information over all M units, or when only one unit represents the data sample to 100 % which is rarely happening. If we rewrite the batch update (10) to an update over all M units, then we get

$$\langle\langle f\rangle\rangle_M(T) = \sum_{t=1}^{T}\left(\prod_{s=t+1}^{T}\lambda(s)\right)\sum_{i=1}^{M}P_i(t)f_t = \sum_{t=1}^{T}\left(\prod_{s=t+1}^{T}\lambda(s)\right)f_t \tag{19}$$

as the full update for TD. Considering the same for WTD, we would get

$$\langle\langle f\rangle\rangle_M(T) = \sum_{t=1}^{T}\left(\prod_{s=t+1}^{T}\left(\prod_{j=1}^{M}\Lambda(P_j(s))\right)\right)\sum_{i=1}^{M}\Omega(P_i(t))f_t = \sum_{t=1}^{T}\left(\prod_{s=t+1}^{T}\lambda(s)\right)f_t. \tag{20}$$

Then, TD and WTD reduce to the same full update.

Additionally, we consider a consistency condition for a shared weighted sum $\langle\langle f\rangle\rangle_{i+j}(t-1)$ by unit i and j. Here, an update with value f_t should be the same when updated once with weight $(P_i(t) + P_j(t))$ or twice with weights $P_i(t)$ and $P_j(t)$ at a time step t. For a single update with $(P_i(t) + P_j(t))$, we get

$$\langle\langle f\rangle\rangle_{i+j}(t) = \Lambda(P_i(t) + P_j(t))\langle\langle f\rangle\rangle_{i+j}(t-1) + \Omega(P_i(t) + P_j(t))f_t. \tag{21}$$

If $\langle\langle f\rangle\rangle_{i+j}(t-1)$ is updated twice with $P_i(t)$ and $P_j(t)$, then

$$\langle\langle f\rangle\rangle_{i+j}(t) = \Lambda(P_i(t))\langle\langle f\rangle\rangle_{i+j}(t-1) + \Omega(P_i(t))f_t \tag{22}$$

$$\langle\langle f\rangle\rangle_{i+j}(t+1) = \Lambda(P_j(t))\langle\langle f\rangle\rangle_{i+j}(t) + \Omega(P_j(t))f_t. \tag{23}$$

We can insert (22) into (23) and get

$$\langle\langle f\rangle\rangle_{i+j}(t+1) = \Lambda(P_j(t))\Lambda(P_i(t))\langle\langle f\rangle\rangle_{i+j}(t-1) + (\Lambda(P_j(t))\Omega(P_i(t)) + \Omega(P_j(t)))f_t, \tag{24}$$

which is similar to (14). (24) results however in an update that is one step ahead of time and ignores the unit division of the current data sample $(x(t), y(t))$ at time step t. Furthermore, the update $\Omega(P_i(t))f_t$ is already partially forgotten by $\Lambda(P_j(t))$. Because we want to update twice on the same time step t, we have to prevent forgetting of partial updates which reduces (24) to

$$\langle\langle f\rangle\rangle_{i+j}(t) = \Lambda(P_j(t))\Lambda(P_i(t))\langle\langle f\rangle\rangle_{i+j}(t-1) + (\Omega(P_i(t)) + \Omega(P_j(t)))f_t. \tag{25}$$

The same equation can also be obtained when (20) is rewritten to a stepwise update equation and considered as a weighted sum shared by two units i and j.

Finally, we get the following two functional equations from (21) and (25)

$$\Lambda(P_i(t) + P_j(t)) = \Lambda(P_j(t))\Lambda(P_i(t)), \tag{26}$$

$$\Omega(P_i(t) + P_j(t)) = \Omega(P_i(t)) + \Omega(P_j(t)). \tag{27}$$

For (26), the solution is well known to be of the form $\Lambda(P_i(t)) = c^{P_i(t)}$, where c can be determined by using $\Lambda(1) = \lambda(t)$. The same solution is derived as for WD with $\Lambda(P_i(t)) = \lambda(t)^{P_i(t)}$. But for the update factor, we get the new solution $\Omega(P_i(t)) = P_i(t)$ from (27) which is the same as for TD. The newly derived WTD update approach complies with all considered dependencies and updates units independently from $\lambda(t)$. The new stepwise WTD update equation is

$$\langle\langle f\rangle\rangle_i(t) = \lambda(t)^{P_i(t)}\langle\langle f\rangle\rangle_i(t-1) + P_i(t)f_t. \tag{28}$$

4 Comparison

In the following, we discuss differences between the former weight-dependent (WD) and the newly proposed weight-time-dependent (WTD) approach.

4.1 Numerical Difference

The numerical difference between WD and WTD is mainly affected by the update factors. The WD update factor $\Omega(P_i(t)) = \frac{1-\lambda(t)^{P_i(t)}}{1-\lambda(t)}$ depends on the current discount factor $\lambda(t)$ and unit weight $P_i(t)$, while the WTD update factor is equal to $P_i(t)$. The numerical difference between them is then highly dependent on the value of $\lambda(t)$. In Table 1, we present some sample values for the WD update factor that show how it is affected by different values for $P_i(t)$ and $\lambda(t)$. The difference becomes bigger as further $\lambda(t)$ is from one and $P_i(t)$ is near to 0.5. Again, it is worth noting that the WD update factor is not defined when $\lambda(t) = 1$ over all $P_i(t)$, while WTD is defined over all values of $\lambda(t)$.

Table 1. Sample Values for WD Update Factor

Discount $\lambda(t)$	Posterior Probability $P_i(t)$				
	0.01	0.2	0.4	0.6	0.8
0.9	0.01053	0.20852	0.41268	0.61260	0.80834
0.99	0.01005	0.20080	0.40121	0.60121	0.80080
0.999	0.01000	0.20008	0.40012	0.60012	0.80008

4.2 Experiments

In order to evaluate the effectiveness of the proposed method WTD, we compare it with WD for the following function approximation task ([1,5])

$$g(x_1, x_2) = \max\{e^{-10x_1^2}, e^{-50x_2^2}, 1.25e^{-5(x_1^2+x_2^2)}\} \quad (-1 \leq x_1, x_2 \leq 1). \tag{29}$$

Additionally, a normally distributed random noise $\epsilon(t) \sim N(0, 0.01)$ is added to the data output.

Two experiments are conducted for that we obtain training data with different data distributions. For the first experiment *Balanced*, 100,000 data samples are i.i.d. over the whole input space. In the second experiment *Dynamic*, the data distribution is slowly changing over time, starting at input interval $[-1, -0.2]$ and ending at interval $[0.2, 1]$ after 250,000 samples. In addition, test data samples are obtained from a regular grid in the input domain formed by 21×21 points to evaluate the learning performance. The change in learning performance is then evaluated with the normalized mean square error (NMSE) every 500 time steps. Also, we use an identical model initialization over all test runs, where 25 units are equally distributed in the input space, and the presented results are the average of 50 test runs.

The discount factor $\lambda(t)$ has to approach one with $t \rightarrow \infty$ so that learning converges, which is accomplished here with $\lambda(t) = 1 - \frac{1-a}{at+b}$ ([1,5]). Here, a and b are parameters that influence how fast $\lambda(t) \rightarrow 1$ and the initial value of $\lambda(t)$ respectively. We use $b = 10$, which is the same as $\lambda(0) = 0.9$, a rather big discount that makes the performance difference between WD and WTD more evident. Also, we apply three different convergence schedules with $a = \{0, 0.001, 0.01\}$. When $a = 0$, $\lambda(t)$ is constant over all t otherwise $\lambda(t) \rightarrow 1$ with faster convergence for bigger a.

In Fig. 1(a) to (c), the changing learning performance over time is presented for the *Balanced* testbed. Here, WTD performs better for all three discount schedules. Yet, the difference is less visible for bigger a, because the numerical difference between WTD and WD becomes smaller. A similar observation can be made for the *Dynamic* testbed in Fig. 2(a) to (c). Without convergence, learning is very unstable, and both approaches perform badly for $a = 0$. But for the other two discount schedules where $\lambda(t) \rightarrow 1$, both approaches are able to converge after seeing the bigger part of the input space. Here, WTD again performs better than WD. In Table 2, final performance results are presented as an averaged

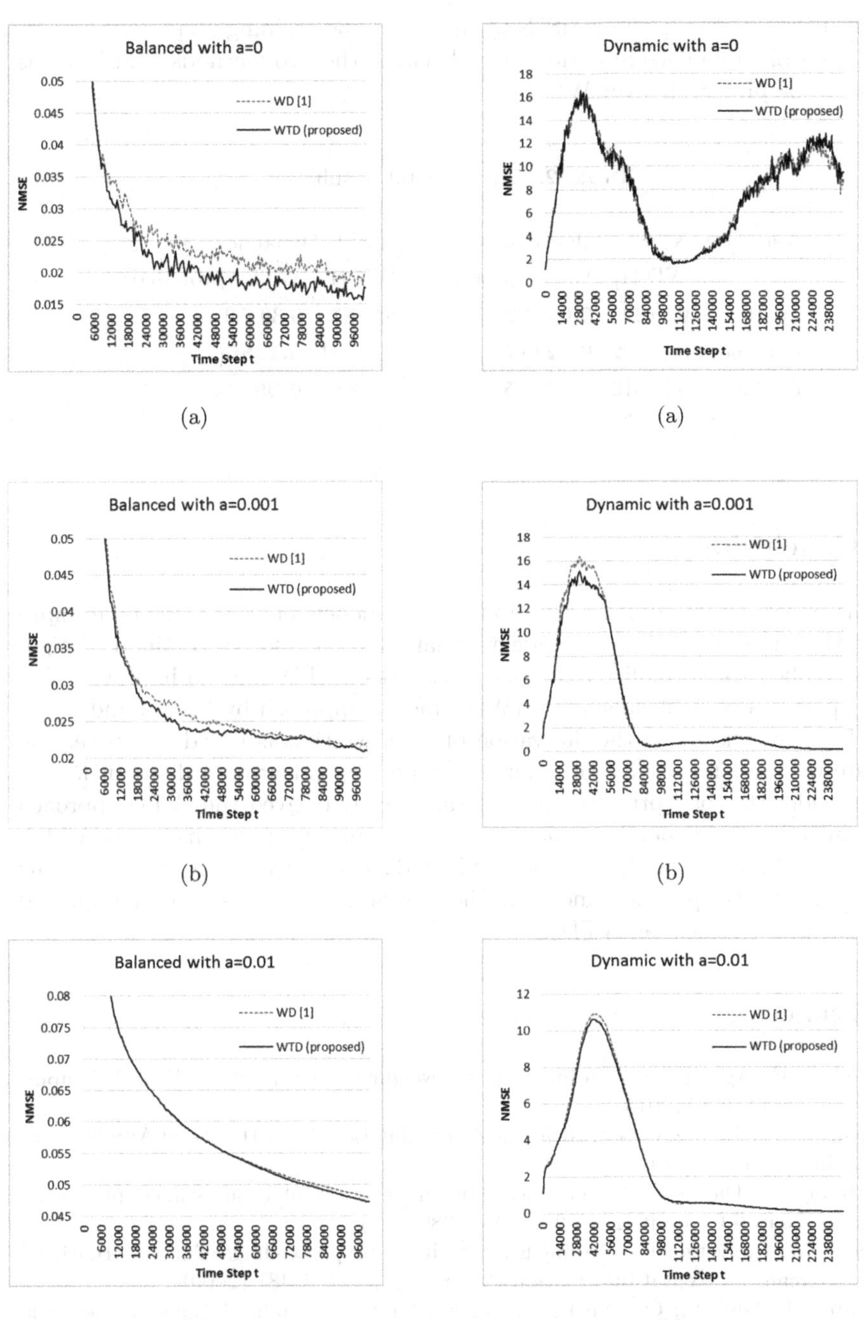

Fig. 1. Balanced testbed: comparison of WD [1] with WTD (proposed) for (a) $a = 0$, (b) $a = 0.001$ and (c) $a = 0.01$

Fig. 2. Dynamic testbed: comparison of WD [1] with WTD (proposed) for (a) $a = 0$, (b) $a = 0.001$ and (c) $a = 0.01$

NMSE calculated over the NMSEs of the last 5,000 training steps. The results show a visible difference in performance between the two methods, and WTD is shown to be preferable over WD.

Table 2. Experimental Results

Test Case	NMSE Balanced		NMSE Dynamic	
	WD [1]	WTD (proposed)	WD [1]	WTD (proposed)
a = 0	0.01896	**0.01632**	8.78918	9.39159
a = 0.001	0.02192	**0.02117**	0.15041	**0.12866**
a = 0.01	0.04819	**0.04745**	0.09843	**0.09724**

5 Conclusion

In this paper, we discussed a new weight-time-dependent (WTD) update approach that modifies a time-dependent update approach for the online EM algorithm applied to Normalized Gaussian networks. WTD aims to improve a formerly proposed weight-dependent (WD) update approach by Celaya and Agostini. First, we discussed the derivation of WD, which was based exclusively on update weights and results in updates that are dependent on a discount factor. We consider this bad practice and therefore newly derived the WTD approach with an additional dependency on time. In our comparison, we discussed WTD's numerical differences with WD and conducted some experiments to evaluate the effectiveness of the proposed method. The experimental results show an improved learning performance for WTD.

References

1. Celaya, E., Agostini, A.: On-line EM with weight-based forgetting. Neural Comput. **27**(5), 1142–1157 (2015)
2. Kushner, H.J., Yin, G.G.: Stochastic Approximation Algorithms and Applications. Springer, New York (1997)
3. Moody, J., Darken, C.J.: Fast learning in networks of locally-tuned processing units. Neural Comput. **1**(2), 281–294 (1989)
4. Sato, M.: Convergence of on-line EM algorithm. In: proceedings of the International Conference on Neural Information Processing, pp. 476–481 (2000)
5. Sato, M., Ishii, S.: On-line EM algorithm for the normalized Gaussian network. Neural Comput. **12**(2), 407–432 (2000)
6. Xu, L., Jordan, M., Hinton, G.: An alternative model for mixtures of experts. In: Cowan, J.D., Tesauro, G., Alspector, J. (eds.) Advances in Neural Information Processing Systems, vol. 7, pp. 633–640. MIT Press, Cambridge (1995)

Learning Phrase Representations Based on Word and Character Embeddings

Jiangping Huang[1], Donghong Ji[1(✉)], Shuxin Yao[2], Wenzhi Huang[1], and Bo Chen[1]

[1] Computer School, Wuhan University, Wuhan 430072, China
{hjp,dhji,hwz208,chenbo}@whu.edu.cn

[2] Language Technologies Institute, Carnegie Mellon University, Pittsburgh, PA 15213, USA
shuxiny@cs.cmu.edu

Abstract. Most phrase embedding methods consider a phrase as a basic term and learn embeddings according to phrases' external contexts, ignoring the internal structures of words and characters. There are some languages such as Chinese, a phrase is usually composed of several words or characters and contains rich internal information. The semantic meaning of a phrase is also related to the meanings of its composing words or characters. Therefore, we take Chinese for example, and propose a joint words and characters embedding model for learning phrase representation. In order to disambiguate the word and character and address the issue of non-compositional phrases, we present multiple-prototype word and character embeddings and an effective phrase selection method. We evaluate the effectiveness of the proposed model on phrase similarities computation and analogical reasoning. The empirical result shows that our model outperforms other baseline methods which ignore internal word and character information.

Keywords: Embedding · Phrase representation · Semantic composition · Analogical reasoning

1 Introduction

Representing the semantics of phrases and words are fundamental tasks in knowledge representation. Numerous methods for learning distributed word representations have been proposed in the Natural Language Processing (NLP) community [4,7,10]. Distributed word representations have shown to improve performance in a wide-range of tasks such as, machine translation [2], semantic similarity measurement [5,7], semantic composition [11], syntactic parsing [9] and word sense disambiguation [3]. Beyond word representations, it is also essential to find appropriate representations for phrases or longer utterances. Hence, distributional semantic representation models [4,8] have been proposed to constructed the representations of phrases or sentences based on the representations of the words they contain.

A. Hirose et al. (Eds.): ICONIP 2016, Part IV, LNCS 9950, pp. 547–554, 2016.
DOI: 10.1007/978-3-319-46681-1_65

Most existing phrase embedding methods can be divided into the following two typical types. (1) Semantic composition. These models use element-wise composition operations on word vectors for phrase vectors. For example, the additive model ($\mathbf{z} = \mathbf{x} + \mathbf{y}$) and multiplicative model ($\mathbf{z} = \mathbf{x} \odot \mathbf{y}$) [6]. However, both of the operations are naive, which maybe unreasonable for semantic composition since the order of word sequences in a phrase may influence its meaning. For instance, *machine learning* and *learning machine* have different meanings, while the commutative functions will return the same representation for them. (2) External contexts. These methods typically learn phrase embeddings according to the external contexts of phrases in large-scale corpora. However, in some languages such as Chinese, a phrase, usually composed of several words or characters, contains rich internal information. Take a Chinese phrase "机器学习" (machine learning) for example. The semantic meaning of the phrase can be learned form its context in text corpora. Meanwhile, we emphasize that its semantic meaning can also be inferred from the meanings of its word "机器" (machine) and "学习" (learning). What's more, the semantic meanings of the words "机器" and "学习" can also be inferred from the meanings of its characters "机" (machine) "器" (implement) "学" (learning) "习" (practice) further.

Due to the linguistic nature of semantic composition, the semantic meanings of internal words and characters may also play an important role in modeling semantic meanings of phrases. Hence an intuitive idea is to take internal words and characters into account for learning phrase embeddings. In this paper, we consider Chinese as a typical language. We take advantages of both internal words and characters and external contexts, and propose a new model which based on word and character embeddings for phrase representations.

2 Proposed Approach

In this section we describe three strategies for learning phrase representation. Different strategies will learn different phrase vector and generate dissimilar representations for the same phrase.

2.1 Character-Based Phrase Representation

We considered character embeddings in an effort to improve phrase embeddings. We denote the Chinese character set as C and use Z as the Chinese phrase set. Each character $c_i \in C$ is represented by vector $\mathbf{c}_i$, and each phrase $z_i \in Z$ is represented by vector $\mathbf{z}_i$. As we learn to maximize the average log probability in Eq. (1) with a phrase sequence $S = \{z_1,...,z_M\}$, we represent context phrases with both character embeddings and phrase embeddings to predict target phrases. Formally, a context phrase z_j is represented as

$$z_j = \mathbf{z}_j \oplus \frac{1}{N_j} \sum_{k=1}^{N_j} \mathbf{c}_k, \qquad (1)$$

where $\mathbf{z}_j$ is the phrase embedding of z_j, N_j is the number of characters in z_j, $\mathbf{c}_k$ is the embedding of the k-th character c_k in z_j, and $\oplus$ is the composition operation.

There are two options for the operation $\oplus$, addition and concatenation. For the addition operation, we require the dimensions of phrase embeddings and character embeddings to be equal (i.e., $|\mathbf{z}_j| = |\mathbf{c}_k|$). We simply add the phrase embedding with the average of character embeddings to obtain $\mathbf{z}_j$. On the other hand, we can also concatenate the phrase embedding and the average of character embeddings into the embedding $\mathbf{z}_j$ with a dimension of $|\mathbf{z}_j| + |\mathbf{c}_k|$. In this case the dimension of phrase embeddings is not necessarily equal to that of character embeddings. Technically, we also use

$$z_j = \alpha(\mathbf{z}_j + \frac{1}{N_j}\sum_{k=1}^{N_j} \mathbf{c}_k). \tag{2}$$

The coefficient α is crucial because it maintains similar length between embeddings of compositional and non-compositional phrases, it will be determined in experiments. Moreover, we ignore the character embeddings on the side of target phrases in negative sampling and hierarchical softmax for simplicity. The pivotal idea is to replace the stored vectors $\mathbf{z}$ in continuous bag-of-words model (CBOW) [5] with real-time compositions of $\mathbf{z}$ and $\mathbf{c}$, but shares the same objective in Eq. (1). As a result, the represent of phrase z_i will change due to the change of character embeddings $\mathbf{c}$ even when the phrase is not inside the context window.

2.2 Word-Based Phrase Representation

Although phrase representation can be learnt with character embeddings, most of the phrases also can be represented with word embeddings. Following the mentioned Chinese word set V and the Chinese phrase set Z, each word $w_i \in V$ is represented by vector $\mathbf{w}_i$, and each phrase $z_i \in Z$ is represented by vector $\mathbf{z}_i$. As we learn to maximize the average log probability in Eq. (1) with a phrase sequence $S = \{z_1,...,z_M\}$, we represent context phrases with both word embeddings and phrase embeddings to predict target phrases. Formally, a context phrase z_j is represented as

$$z_j = \mathbf{z}_j \oplus \frac{1}{N_j}\sum_{k=1}^{N_j} \mathbf{w}_k, \tag{3}$$

where $\mathbf{z}_j$ is the phrase embedding of z_j, N_j is the number of words in z_j, $\mathbf{w}_k$ is the embedding of the k-th word w_k in z_j, and $\oplus$ is the composition operation.

We have also two options for the operation $\oplus$, addition and concatenation. For the addition operation, we require the dimensions of phrase embeddings and word embeddings to be equal (i.e., $|\mathbf{z}_j| = |\mathbf{w}_k|$). We simply add the phrase embedding with the average of word embeddings to obtain $\mathbf{z}_j$. On the other hand, we can also concatenate the phrase embedding and the average of word embeddings into the embedding $\mathbf{z}_j$ with a dimension of $|\mathbf{z}_j| + |\mathbf{w}_k|$. In this case the dimension of phrase embeddings is not necessarily equal to that of word embeddings. Technically, we use

$$z_j = \beta(\mathbf{z}_j + \frac{1}{N_j}\sum_{k=1}^{N_j}\mathbf{w}_k). \tag{4}$$

The coefficient β is crucial because it maintains similar length between embeddings of compositional and non-compositional phrases. Moreover, we ignore the phrase embeddings on the side of target phrases in negative sampling and hierarchical softmax for simplicity. The pivotal idea is to replace the stored vectors $\mathbf{z}$ in CBOW with real-time compositions of $\mathbf{z}$ and $\mathbf{w}$, but shares the same objective in Eq. (3). Therefore, the represent of word z_i will be adjusted following the change of word embeddings $\mathbf{w}$ even though the phrase is not inside the context window.

2.3 Joint Learning for Phrase Representation

As a joint word and character embedding for phrase representation model, it take word and character into account for the phrase embedding. We still denote the Chinese character set as C, the Chinese word vocabulary as V and the Chinese phrase set as Z. Each character $c_i \in C$ is represented by vector $\mathbf{c}_i$, and each word $w_l \in V$ is represented by vector $\mathbf{w}_l$, and each phrase $z_i \in V$ is represented by vector $\mathbf{z}_i$. As we learn to maximize the average log probability in Eq. (1) with a phrase sequence $S = \{z_1,...,z_M\}$, we represent context phrase with character embeddings, word embeddings and phrase embeddings to predict target phrases. Formally, a context phrase z_j is represented as

$$z_j = \mathbf{z}_j \oplus \frac{1}{N_j}\sum_{k=1}^{N_j}\mathbf{c}_k \oplus \frac{1}{N_t}\sum_{l=1}^{N_t}\mathbf{w}_l, \tag{5}$$

where $\mathbf{w}_l$ is the word embedding of w_l, N_j is the number of characters in z_j, N_t is the number of words in z_j, $\mathbf{c}_k$ is the embedding of the k-th character c_k in z_j, and $\oplus$ is the composition operation.

We have also two options for the operation $\oplus$, addition and concatenation. For the addition operation, we require the dimensions of phrase embeddings, word embeddings and character embeddings to be equal (i.e., $|\mathbf{z}_j| = |\mathbf{w}_l| = |\mathbf{c}_k|$). We simply add the phrase embedding with the average of character embeddings and word embeddings to obtain $\mathbf{z}_j$. On the other hand, we can also concatenate the phrase embedding and the average of character embeddings word embeddings into the embedding $\mathbf{z}_j$ with a dimension of $|\mathbf{w}_l| + |\mathbf{c}_k| + |\mathbf{z}_j|$. In this case the dimension of phrase embeddings is not necessarily equal to that of word embeddings and character embeddings. In the experiments, we find the concatenation operation, although being more time consuming, does not outperform the addition operation significantly, hence we only consider the addition operation for simplicity in this work. Technically, we use

$$z_j = \phi(\mathbf{z}_j + \frac{1}{N_j}\sum_{k=1}^{N_j} \mathbf{c}_k) + \varphi(\mathbf{z}_j + \frac{1}{N_t}\sum_{l=1}^{N_t} \mathbf{w}_l). \quad (6)$$

The ϕ and φ are two important coefficients because they maintain similar length between embeddings of compositional and non-compositional phrases, which will be determined in experiments.

3 Experiments and Analysis

In this paper, we evaluate our methodology by judging similarities between phrases and analogical reasoning.

3.1 Dataset and Evaluation

We select a human-annotated corpus with encyclopedia articles from the Chinese Wikipedia[1] for embedding learning, and the corpus has 247604 articles after pruning articles shorter than 50 words. We use OpenCC[2] package to convert the Traditional Chinese to Simplified Chinese. The corpus has 91 million Chinese words and 244 million Chinese characters after filtered the non-Chinese symbols. We set vector dimension as 200 and context window size as 5. For optimization, we use both hierarchical softmax and 10-word negative sampling. We perform word and phrase selection for the proposed model and use pre-trained character and word embeddings as well. We introduce CBOW, Skip-gram and GloVe as baseline methods, suing the same vector dimension and default parameters. We evaluate the effectiveness of our model on phrase similarity computation and analogical reasoning.

3.2 Similarity Computation

In this task, each strategies is required to compute semantic similarity of given phrase pairs. The correlations between results of models and human judgements are reported as the model performance. In this work, we select two datasets, phrasesim-138 and phrasesim-182 for evaluation referring to wordsim work [1]. In phrasesim-138, there are 138 pairs of Chinese phrases and human-labeled related scores. Of the 138 phrase pairs, the phrases in 132 phrase pairs have appeared in the learning corpus and there are new phrase in the left 6 phrase pairs. In phrasesim-182, the phrases in 174 phrase pairs have appeared in the learning corpus and the left 8 pairs have new phrases.

We compute the Spearman correlation ρ between relatedness scores from a model and the human judgements for comparison. For our model and other

[1] https://dumps.wikimedia.org/zhwiki/latest/zhwiki-latest-pages-articles.xml.bz2.

[2] https://github.com/BYVoid/OpenCC.

baseline embedding methods, the relatedness score of two phrases are computed via cosine similarity of phrase embeddings. In our study, the proposed model is implemented based on CBOW and obtains phrase embeddings via Eqs. (2), (4) and (6). In every equation, the coefficient α, β, ϕ and φ are set to $\frac{1}{2}$ in our experiments. For a phrase pair with new phrases, we assume its similarity is 0 in baseline methods since we can do nothing more, while the proposed model can generate embeddings for these new phrases from their word embeddings and character embeddings for relatedness computation. The evaluation results of our model and baseline methods on phrasesim-138 and phrasesim-182 are shown in Table 1.

Table 1. Evaluating the similarity on phrasesim-138 and phrasesim-182 ($\rho \times 100$)

Dataset	phrasesim-138		phrasesim-182	
Method	132 Pairs	138 Pairs	174 Pairs	182 Pairs
CBOW	47.23	47.45	49.76	47.12
Skip-gram	48.61	48.15	47.44	45.32
GloVe	41.68	43.35	46.49	45.92
Model(c)	47.81	47.95	50.29	50.82
Model(w)	48.68	48.75	50.23	50.77
Model(cw)	**50.42**	**49.41**	**50.98**	**51.05**

From the evaluation results on phrasesim-138, we observe the following several results. Firstly, our model which based on character embeddings (Model(c)) and word embeddings (Model(w)) all significantly outperform baseline methods on both 132 phrase pairs and 138 pairs. Secondly, joint word embeddings and character embeddings (Model(cw)) obtained better performance than (Model(c)) and (Model(w)). The results indicate that the joint character and word embeddings is very helpful for phrase representation.

3.3 Analogical Reasoning

The analogical reasoning task consists of analogies such as "巴黎 (Paris)" + "法国 (Paris)" - "柏林 (Berlin)" = ?. Embedding methods are expected to find a word w such that its vector $\mathbf{w}$ is closest to vec(巴黎) + vec(法国) - vec(柏林) according to the cosine similarity. If the word "德国 (Germany)" is found, the model is considered having answered the problem correctly. In Chinese, there is no existing phrase analogical reasoning dataset, we manually build a Chinese phrase dataset consisting of 683 analogies. It contains 3 analogy types: (1) political parties of countries (247 groups); (2)states/provinces of cites (255 groups); and (3) beauty spot (181 groups). The learning corpus covers more than 95 % of all the testing phrases. Since many existing word embedding models can be easily integrated into our model, we implement our model based on CBOW,

Skip-gram and GloVe models. We show their evaluation results on analogical reasoning in Table 2. In the experimental results, joint word embeddings and character embeddings phrase representation models are reported for their stability of performance.

Table 2. Evaluation accuracies (%) on analogical reasoning

Method	Party	State	Spot	Average
Model_CBOW_c	54.66	57.25	55.25	55.78
Model_CBOW_w	58.3	58.04	59.67	58.57
Model_CBOW_wc	59.51	59.61	58.01	59.15
Model_Skip_c	62.75	65.88	72.93	66.62
Model_Skip_w	**67.61**	67.06	75.14	69.4
Model_Skip_wc	66.8	**70.2**	**77.9**	**71.01**
Model_GloVe_c	58.3	59.61	65.19	60.61
Model_GloVe_w	61.13	61.96	68.51	63.4
Model_GloVe_wc	62.75	63.14	71.27	65.15

From Table 2, we observe that for each of CBOW, Skip-gram and GloVe, the models combined character and word embedding for phrase representation ourperform the models based character or word embeddings.

3.4 Analysis

In this work, we take the phrase similarity computation and analogical reasoning tasks for example to investigate the influence of different phrase representation learning strategies. As shown in the Tables 1 and 2, we list the experimental results of joint word embeddings and character embeddings for phrase representation. We find that the joint method is better than character-based embeddings phrase representation and word-based embeddings phrase representation. This indicates the necessity of considering the internal information of phrase by joint word and character embeddings for phrase representation.

4 Conclusions

In this paper we introduce internal word and character information into phrase embedding methods to alleviate excessive reliance on external information. We present the framework of joint word and character embedding model, which can integrate both word and character embedding into phrase representation. In experiments of phrase similarity computation and analogical reasoning, we have shown that the employing of character embeddings and word embeddings can consistently and significantly improve the quality of phrase embeddings. This work indicates the necessity of taking internal information into account phrase representations. In our future work, we may explore more sophisticated composition models to build phrase or sentence embeddings from both word embeddings and character embeddings, which is motivated by semantic composition models based on matrices or tensors, this will enhance our model with more powerful capacity of encoding internal structure information.

Acknowledgements. We thank the reviewers for their valuable comments and suggestions. This work is supported by grants: State Key Program of National Natural Science Foundation of China (61133012), National Natural Science Foundation of China (61373108 and 61170148), Humanities and Social Science Foundation of Ministry of Education of China (16YJCZH004), China Postdoctoral Science Foundation (2013M540593, 2014T70722).

References

1. Chen, X., Xu, L., Liu, Z., Sun, M., Luan, H.: Joint learning of character and word embeddings. In: Proceedings of the Twenty-Fourth IJCAI, pp. 1236–1242 (2015)
2. Cho, K., van Merrienboer, B., Gulcehre, C., Bahdanau, D., Bougares, F., Schwenk, H., Bengio, Y.: Learning phrase representations using rnn encoder-decoder for statistical machine translation. In: Proceedings of the 2014 Conference on EMNLP, pp. 1724–1734. Association for Computational Linguistics (2014)
3. Huang, E., Socher, R., Manning, C., Ng, A.: Improving word representations via global context and multiple word prototypes. In: Proceedings of the 50th Annual Meeting of the ACL, pp. 873–882. Association for Computational Linguistics (2012)
4. Mikolov, T., Sutskever, I., Chen, K., Corrado, G.S., Dean, J.: Distributed representations of words and phrases and their compositionality. In: Advances in Neural Information Processing Systems 26, pp. 3111–3119. Curran Associates, Inc. (2013)
5. Mikolov, T., Yih, W.t., Zweig, G.: Linguistic regularities in continuous space word representations. In: Proceedings of the 2013 Conference of the NAACL: HLT, pp. 746–751. Association for Computational Linguistics (2013)
6. Mitchell, J., Lapata, M.: Composition in distributional models of semantics. Cogn. Sci. **34**(8), 1388–1429 (2010)
7. Pennington, J., Socher, R., Manning, C.: GloVe: global vectors for word representation. In: Proceedings of the 2014 Conference on EMNLP, pp. 1532–1543. Association for Computational Linguistics, Doha (2014)
8. Quoc, L., Tomas, M.: Distributed representations of sentences and documents. In: Proceedings of the 31st International Conference on Machine Learning, pp. 1188–1196. JMLR.org, Beijing (2014)
9. Socher, R., Bauer, J., Manning, C.D., Andrew, Y., N.: Parsing with compositional vector grammars. In: Proceedings of the 51st Annual Meeting of the ACL, pp. 455–465. Association for Computational Linguistics (2013)
10. Turian, J., Ratinov, L.A., Bengio, Y.: Word representations: a simple and general method for semi-supervised learning. In: Proceedings of the 48th Annual Meeting of the ACL, pp. 384–394. Association for Computational Linguistics, July 2010
11. Yu, Z., Zhiyuan, L., Maosong, S.: Phrase type sensitive tensor indexing model for semantic composition. In: Proceedings of the Twenty-Ninth AAAI Conference on Artificial Intelligence, pp. 2195–2201 (2015)

A Mobile-Based Obstacle Detection Method: Application to the Assistance of Visually Impaired People

Manal Abdulaziz Alshehri[1(✉)], Salma Kammoun Jarraya[1,2], and Hanene Ben-Abdallah[1,2]

[1] Faculty of Computing and Information Technology, King Abdulaziz University, Jeddah, Saudi Arabia
malshehri0247@stu.kau.edu.sa, {smohamad1,hbenabdallah}@kau.edu.sa
[2] MIRACL-Laboratory, Sfax, Tunisia

Abstract. Visual impairments suffer many difficulties when they navigate from one place to another in their daily life. The biggest problem is obstacle detection. In this work, we propose a new smartphone-based method for obstacle detection. We aim to detect static and dynamic obstacles in unknown environments while offering maximum flexibility to the user and using the least expensive equipment possible. Detecting obstacles is based on the analysis of different regions of video frames and using a new decision algorithm. The analysis uses prediction model for each region that generated by a supervised learning process. The user is notified about the existing of an obstacle by alert message. The efficiency of the work is measured by many experiments studies on different complex scenes. It records low false alarm rate in the range of [0.2 % to 11 %], and high accuracy in the range of [86 % to 94 %].

Keywords: Visually impaired people · Obstacle detection · Prediction model · Supervised learning

1 Introduction

Visual impairment is the decrease or lack of visual information, which hinders the performance of daily activities such as moving around, reading and socializing. According to statistics from WHO[1] (The World Health Organization) more than 285 million people are estimated to be visually impaired worldwide: 39 million are blind and 246 have low vision; 23.50 million of them are from Middle East.

Among the main challenges facing visually impaired people are obstacle detection and path identification. So far, the white cane is the most evolved tool used for obstacle detection, but one must memorize all locations to become familiar with the place. Hence, in unfamiliar settings, a visually impaired person is completely dependent on a sighted person to reach the desired destination. Assistive technologies such as laser feedback on nearby obstacles, reading

[1] http://www.who.int/mediacentre/factsheets/fs282/en/.

A. Hirose et al. (Eds.): ICONIP 2016, Part IV, LNCS 9950, pp. 555–564, 2016.
DOI: 10.1007/978-3-319-46681-1_66

of RFIDs have been developed to aid in navigation either in indoor or outdoor environments(e.g. [1]). In addition, there has been an increasing interest towards computer vision and camera-based solutions as an assistive technology that overcomes the limits of the white cane.

In fact, obstacle detection remains a challenge against the performance and success of several computer vision systems. Its complexity is mainly attributed, on the one hand, to dynamic changes in the natural scene and, on the other hand, to the visual features of obstacles (color, shape, size). The development of an efficient method for detecting obstacles should therefore consider performance under unknown scene conditions and independently of obstacle features. Due to the difficulty of these challenges, several techniques have been proposed in the literature. According to our survey, most so-far proposed computer vision based systems and methods for obstacle detection need special equipment such as Microsoft Kinect sensors (e.g. [2]), ultrasonic sensors (e.g. [3,4]) and wearable cameras (e.g. [5]). In addition, all the existing systems impose that the user carries a camera (usually of a special type) that is mounted in a fixed position; such requirements make the system inflexible and possibly costly. Therefore, there is a need for a new low-cost obstacle detection method that is more flexible and imposes minimum constrains on the user movement and the position of the camera. We take these issues in our consideration to propose mobile-based obstacle detection method that performs with high precision under minimum restrictions on the mobile (i.e. camera) position and movement.

The remainder of this paper is structured as follows. In Sect. 2, we review state of art on obstacle detection. In Sect. 3, we present our proposed method for obstacle detection. Followed that, Sect. 4, we explain how prediction models are generated. The efficiency and accuracy of our work are illustrated by an exhaustive experimental evaluation and comparison results in Sect. 5. Finally, a summary and ongoing works are presented in Sect. 6.

2 State of Art on Obstacle Detection

Obstacle detection is a fundamental processing step in almost all navigation assistance systems which are based on computer vision. A great number of obstacle detection methods have been proposed for various computer vision applications [2,5–7]. It can be classify contributions reported in the literature in two classes with a categorization based on the number of used cameras: computer stereo vision based approach (CSV), and single camera based approach. The methods that adopt CVS approach [2,6,8,9] are costly because they are use more than one camera to capture the scene and have a high computational complexity. There are many obstacle detection methods that are based on a single camera, e.g. [5,7–13]. A comparison between them is illustrated in Table 1.

From Table 1, we can conclude that, except [5], all obstacle detection methods require special or expensive equipment. In addition, all these methods impose that the camera be mounted on the chest of the user; this may make the user feel uncomfortable and seem unnatural. The methods in [5,7,13] rely on prior

Table 1. A comparison between the methods based on a single camera

Method	Required equipment	Required prior information	Environment	Used technique	Distance estimation
[5]	Smart-phone	No	Indoor and outdoor	HOG descriptors and SVM classifier	No
[7]	A body-mounted camera	Saliency maps of pre-learned objects. Ex: chair and table	Indoor only	Saliency map	No
[10]	A body-mounted camera	No	Indoor and outdoor	Depth estimation	Yes
[11]	A body-mounted camera and gyroscopic sensor	No	Indoor and outdoor	Collision risk estimation	Yes
[12]	A body-mounted camera	Prior images of the background	Indoor and outdoor	The basic video processing tools	Yes
[13]	Glasses-type vision camera and WIFI network	No	Indoor and outdoor	Deformable grid	Yes

knowledge, consequently, their robustness drops dramatically if the environment changes. Although distance estimation is a significant information of obstacle detection methods to avoid collisions, the methods of [5,7] do not take distance in their consideration. In this research, we propose a new mobile-based obstacle detection method that imposes minimum restrictions on the mobile (i.e. camera) position and movement to detect static and dynamic obstacles in both indoor and outdoor environments, with no prior knowledge about the background and objects.

3 Proposed Smartphone-Based Method to Detect Static and Dynamic Obstacles

Our obstacle detection method is based on dividing each frame into regions and classify each one of them into Obstacle and non-Obstacle (it indicate that the region involves obstacle or not) depending on a prediction model that built specially for that region. Thus, determining whether the frame contains obstacle or not is based on the obstacle detection of their regions. We build region-based prediction models instead of one frame-based prediction model to decrease the faults and propose a new region-based decision algorithm to increase the accuracy of the method. In the proposed method, the user has just to direct the phone camera about 45-degree angle on the floor. Then, each frame is analyzed independently to determine whether it involves an obstacle or not. Thus, we

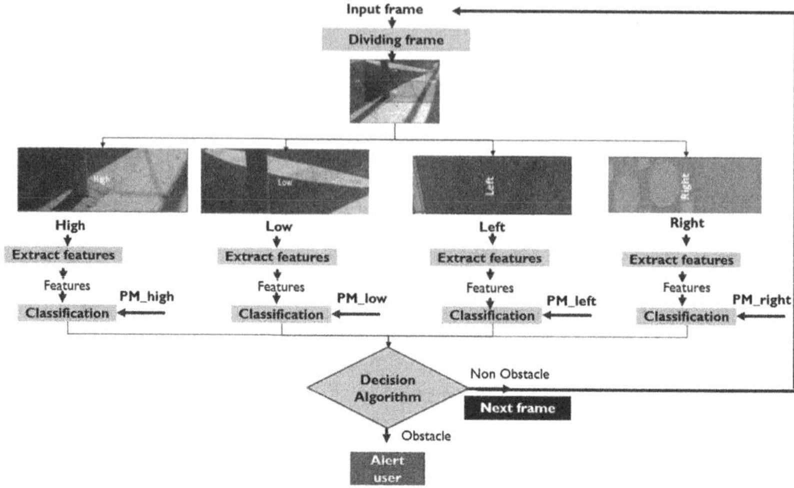

Fig. 1. Flowchart of the proposed method

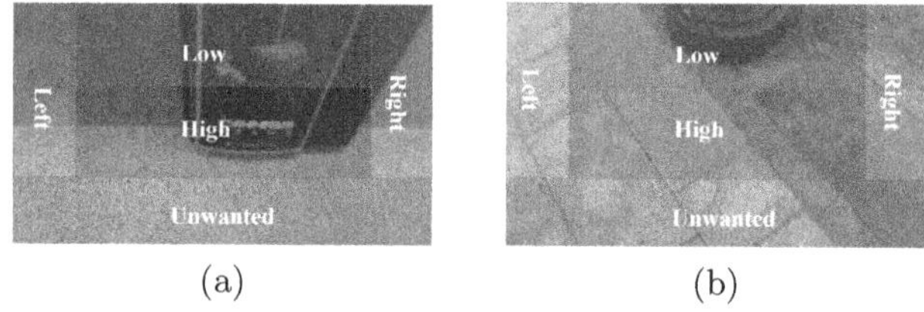

(a) (b)

Fig. 2. Frame Division; (a) Obstacle exist in *High*, *Low* and *Right* regions (b) Obstacle exist only in *Low* region. (Color figure online)

attempt to increase the usability as much as possible, by decreasing the number and the complexity of steps that users have to follow.

As presented in Fig. 1, the first step of the method is frame division. The input frame is divided into five regions: *High*, *Low*, *Left*, *Right* and *Unwanted*. The most significant region is *High* because it exists immediately in front of the user. *Low* is less significant, while *Left* and *Right* are minor regions that used to support the final decision. The method ignores *Unwanted* region since it may contains the user feet. Fig. 2 shows the frame regions.

The second step is extracting the feature vectors *(Gray, HSV and RGB Color Space Histograms)* from the four regions to use them in the classification process. Feature selection step is well descibed in Sect. 4.2. After that, the regions are classified into *Obstacle* or into *non-Obstacle* based on their associated prediction models (PM_{High}, PM_{Low}, PM_{Left} and PM_{Right}). The four prediction models are generated by a supervised learning process described in Sect. 4. The last step is deciding whether the frame involves an obstacle or not. As Algorithm 1 illustrates, it is based on the four regions; thus, if all of them have obstacles or at least *High* and *Low*, that indicates the frame contains obstacle and the user have alerted. We got this condition based on our observations, since all argent obstacles exist in both *High* and *Low* and sometimes in all regions. These steps are repeated for each input frame.

Algorithm 1. Obstacle detection

Input : $\text{Class}_{\text{High}}$, $\text{Class}_{\text{Low}}$, $\text{Class}_{\text{Left}}$, $\text{Class}_{\text{Right}}$
Output: Alert message

if $Class_{High} == obst$ *and* $Class_{Low} == obst$ *and* $Class_{Left} == obst$ *and* $Class_{Right} == obst$ **then**
    return decision = Obstacle
else if $Class_{High} == obst$ *and* $Class_{Low} == obst$ **then**
    return decision = Obstacle
else
    return decision = No obstacle

---

4 Prediction Model Generation

Prediction model generation constitutes a considerable work to determine whether the images contain obstacle or not based on the visual features that extracted from these images. Prediction models are generated based on supervised machine learning techniques to analyze all the pertinent features and extract useful knowledge. Our choice of these techniques is due to two of their multiple advantages which are applicable to our context: Firstly, the number of classes is predetermined (*Obstacle/non-Obstacle*), and the targets (image pixels) can be labeled. Secondly, classification of images into *Obstacle* or *non-Obstacle* is a critical problem; unsupervised learning techniques split the data into different clusters without accounting for their classes. Furthermore, based on the comparison of several supervised learning techniques regarding comprehensibility of their learned models [14], we will adopt the well-established technique "induction of decision trees" [15].

In fact, we build four distinct prediction models, one for each of the four mentioned regions (*High, Low, Left and Right*) of video frames. For these, firstly, we build four learning set (huge number of *Obstacle* and *non-Obstacle* images for each region) from the frames of videos, which recorded the path in natural conditions in both indoor and outdoor environments. Secondly, we identify the effectiveness features for describing obstacle to building an n-dimensional table from our training data. Thirdly, we select the appropriate learning algorithm to generate each prediction model.

4.1 Learning Data Preparing

To include all the circumstances that may expose our method. We prepare a large and representative corpus (fourteen long videos of paths recorded in nature seen).

The videos cover all conditions related to environment (indoor and outdoor), ground (reflection, shadow, color and texture contrast), and obstacle (static and dynamic). Then we use them to construct the learning datasets and test datasets for the prediction models.

In order to prepare the learning data for the four prediction models, we extract their associated regions (*High, Low, Left* and *Right*) from all videos frames. After that, the following steps are implemented for each region. (i) Dividing the images into *Obstacle* group and *non-Obstacle* group. The first one indicates there are obstacles while the second group is free from them. We make the number of images of the two groups are equal to increase the accuray of the prediction models. Thus, the number of images in each region are: *High* region: *Obstacle*: (17,419) images and *non-Obstacle* (17,419) images, *Low* region: *Obstacle*: (17,451) images and *non-Obstacle* (17,451) images, *Left* region: *Obstacle*: (14,964) images and *non-Obstacle* (14,964) images, *Right* region: *Obstacle*: (15,259) images and *non-Obstacle* (15,259) images. (ii) Extracting the feature vectors of the images of both groups. Thus, *Obstacle* and *non-Obstacle* datasets, which used for prediction model are obtained. (iii) Splitting the datasets into (70 %) as training set and the other (30 %) as a test set.

4.2 Extracting Features

We collect from the literature the features that are most commonly used in obstacle and object detection methods which achieve these two pre-conditions: (i) they do not describe moving. (ii) They do not define the shape. Because detecting the shape or moving of obstacles is not useful in our method since we based on detecting obstacles from frames regions. Throw exploring features we found that *Histogram of Oriented Gradients (HOG), Local Binary Patterns (LBP), Scale-Invariant Feature Transform (SIFT), Speeded Up Robust Features (SURF), Gray, HSV and RGB Color Space Histograms* are most appropriate to our settings. Since the method have to work in real time manner, the used features should be effeciant. We asssest the mentiond features by extracting them from the regions of a certain image frame and compute the execution time. According to the results in Table 2, we select *Gray, HSV and RGB Color Space Histograms.* Since these histograms produced very long vectors, we normalize them into vectors of 8 elements. Thus, each image in the datasets is described by a feature vector of 56 elements.

Table 2. The execution time of extracting some features from the four regions in (ms)

	Gray histogram	HSV histogram	RGB histogram	HOG	LBP	SIFT	SURF
Laptop	15	50	18	141	80		184
Smartphone (Android)	5	13	9	269		956	

4.3 Generating Prediction Model

To select the appropriate learning algorithm to generate the prediction models, we test and evaluate eleven decision trees based algorithms according to the Precision and Recall rates for both *Obstacle* and *non-Obstacle* datasets. These algorithms are *Limited Search Tree Algorithm, ID3-IV, GID3, ASSISTANT 86, ChAID, C4.5, Improved C4.5, Improved ChAID, Cost-sensitive C4.5, one-VS-All Decision tree and Multithreaded ChAID*. We found that *C4.5* is proper to our work because of two reasons. Firstly, it gives the best Precision and Recall values for both *non-Obstacles (A)* and *Obstacles (B)* datasets in all regions. For *High*, class A: Precession (98 %) and Recall (98 %), class B: Precession (98 %) and Recall (98 %). For *Low*, class A: Precession (99 %) and Recall (99 %), class B: Precession (99 %) and Recall (99 %). For *Left*, class A: Precession (98 %) and Recall (98 %), class B: Precession (98 %) and Recall (98 %). For *Right*, class A: Precession (98 %) and Recall (98 %), class B: Precession (98 %) and Recall (98 %). Secondly, it generates simple decision trees that can quickly convert its rule into conditional statements.

5 Experimental Results

In order to evaluate our proposed method, we carried out a series of experiments. Firstly, we present the results of the four prediction models in both learning set and inactive set. Secondly, we test them throw several recorded videos, each video includes some of the challenges. Thirdly, we assess the performance of the predection models by comparing them with a well-known vision-based method of obstacle detection [16]. For the performance evaluation of the prediction models, we have used Confusion Matrix to compute the Precision and Recall of both *non-Obstacle* (A) and *Obstacle* (B) datasets. Fourthly, we test the effectiveness and accuracy of the proposed method by applying it into the five test videos in addition to the dataset of [16], then we compute True Positive Rate (TPR), True Negative Rate (TNR), False negative rate (FNR), false alarm rate (FAR) and Classification Accuracy (AC) for each video.

The results of the prediction models in both learning set and inactive set are as the following. In learning set for *High* region, class (A): Precession (89 %) and Recall (87 %), class (B): Precession (87 %) and Recall (89 %), for *Low* region, class (A): Precession (89 %) and Recall (90 %), class (B): Precession (90 %) and Recall (88 %), for *Left* region, class (A): Precession (89 %) and Recall (88 %), class (B): Precession (88 %) and Recall (89 %), for *Right* region, class (A): Precession (89 %) and Recall (89 %), class (B): Precession (89 %) and Recall (89 %). In inactive set, for *High* region, class (A): Precession (87 %) and Recall (86 %), class (B): Precession (86 %) and Recall (87 %), for *Low* region, class (A): Precession (87 %) and Recall (87 %), class (B): Precession (87 %) and Recall (87 %), for *Left* region, class (A): Precession (87 %) and Recall (86 %), class (B): Precession (86 %) and Recall (87 %), for *Right* region, class (A): Precession (87 %) and Recall (88 %), class (B): Precession (88 %) and

Table 3. Test videos

		Location	No. of frames	Challenges					
				Ground color contrast	Ground texture contrast	Shadow	Reflection	Static obstacle	Dynamic obstacle
Indoor	Video.1	Supermarket	359			✓	✓	✓	✓
	Video.2	Mall 1	276	✓	✓	✓	✓	✓	✓
	Video.3	Mall 2	236	✓		✓	✓	✓	✓
Outdoor	Video.4	Beach	763	✓	✓	✓		✓	✓
	Video.5	Wallkway	295	✓	✓	✓		✓	

Recall (87 %). It indicates that they are efficient and robust throw the challenges (since the training data involves all expected challenges).

In order to test the prediction models on new data, we applied them on five test videos. The locations, number of frames and type of challenges in each video is presented in Table 3. The performance of the prediction models on the five videos is illustrated in Fig. 3.

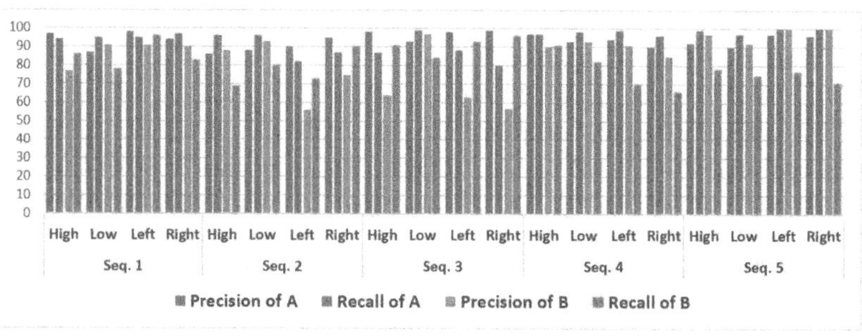

Fig. 3. The performance of the prediction models on the five test videos

The prediction models achieve high Precision and Recall on *non-Obstacle* dataset (A) for all videos (avg. Precision: 90 % – 97 %, avg. Recall: 89 % – 99 %). Since the aim of our method is obstacle detection, we should focus on *Obstacle* dataset (B). The prediction models archive a significant Precision and Recall percentages (avg. precision: 70 % – 97 %, avg. recall: 77 % – 91 %) on *Obstacle* dataset (B). Which indicates that despite all the challenges that are exist in the videos, the prediction models can detect almost obstacle in the seen.

In order to make the assessment more realistic, it should be comparing our prediction models with other obstacle detection methods. There are a lot of them that publish their results. However, lack of the access to their data or codes makes the comparison with them is not possible. Throw the searching, we found just one vision-based method of obstacle detection [16] that publish its test dataset. So we apply the prediction models on these data, Fig. 4 gives the results of this process. Although the experimental results of [16] is not well discussed in their paper, it can say that our method exceeds the highest performance of it, since the average Precision (84 %) and Recall (80 %) of the four prediction models of *Obstacle* dataset exceeds the highest value of Precision (84 %)and Recall (74 %) of [16].

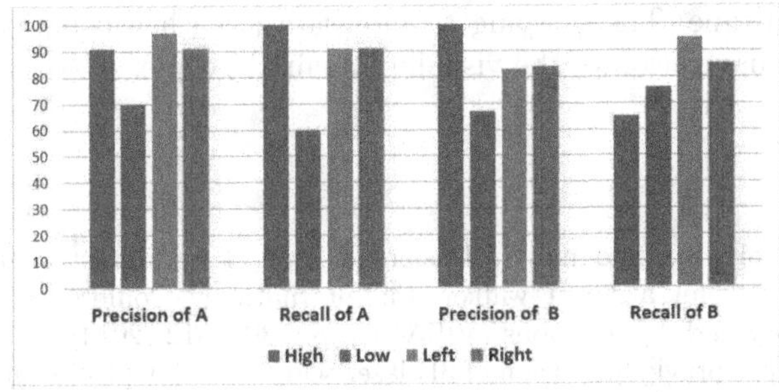

Fig. 4. The performance of the prediction models on the dataset of [16]

Finally, we present the results in term of True Positive Rate (TPR), True Negative Rate (TNR), False Negative Rate (FNR), False Alarm Rate (FAR) and Classification Accuracy (AC) for each video for the proposed method based on the decision algorithm. We apply the proposed method on the five test videos in addition to dataset of [16]. The results of this experimental study are presented in Table 4.

Table 4. The evaluation of implementing the proposed method on test data

	TPR	TNR	FNR	FAR	Accuracy
Video.1	83	95	17	5	92
Video.2	65	96	35	4	86
Video.3	90	90	10	11	89
Video.4	87	97	13	3	94
Video.5	65	99	35	0.4	91
Dataset of [16]	70	99	30	0.2	94

As we can see from Table 4, the accuracy of our method is between (86 % – 94 %) while the false alarm rate is in the range between (0.2 % – 11 %). That indicate the effectiveness of our obstacle detection method and its robustness throw different kind of environment challenges.

6 Conclusion

In this paper, we presented a smartphone-based method for obstacle detection based on a supervised learning process. Prediction models that generated for each frame region in addition to the proposed method was evaluated by a series of experiments with various videos against different conditions. We obtain significant results from each experiment study, that indicate the efficient of our proposed method. Future works will focus on applying and evaluating the proposed method on real-time. Then, trying to integrate it with a complete assistance system for safe navigation of the visually impaired people.

References

1. Weiss, V., Cloix, S., Bologna, G., Hasler, D., Pun, T.: A robust, real-time ground change detector for a smart walker. In: International Conference on Computer Vision Theory and Applications (VISAPP), pp. 305–312 (2014)
2. Budzan, S., Kasprzyk, J.: Fusion of 3D laser scanner and depth images for obstacle recognition in mobile applications. Optics Lasers Eng. **77**, 230–240 (2016)
3. Cao, Z., Cheng, L., Zhou, C., Gu, N., Wang, X., Tan, M.: Spiking neural network-based target tracking control for autonomous mobile robots. Neural Comput. Appl. **26**(8), 1839–1847 (2015)
4. Wang, X., Hou, Z., Zou, A., Tan, M., Cheng, L.: A behavior controller based on spiking neural networks for mobile robots. Neurocomputing **71**(4–6), 655–666 (2008)
5. Tapu, R., Mocanu, B., Bursuc, A., Zaharia, T.: A smartphone-based obstacle detection and classification system for assisting visually impaired people. In: IEEE International Conference on Computer Vision Workshops (ICCVW), pp. 444–451 (2013)
6. Sez, J.M., Escolano, F., Lozano, M.A.: Aerial obstacle detection with 3-D mobile devices. Biomed. Health Inform. IEEE J. **19**(1), 74–80 (2015)
7. Muthulakshmi, L., Ganesh, A.B.: Bimodal based Environmental Awareness System for visually impaired people. Procedia Eng. **38**, 1132–1137 (2012)
8. Petrovai, A., Costea, A., Oniga, F., Nedevschi, S.: Obstacle detection using stereovision for Android-based mobile devices. In: IEEE International Conference on Intelligent Computer Communication and Processing (ICCP), pp. 141–147 (2014)
9. Bourbakis, N., Makrogiannis, S.K., Dakopoulos, D.: A system-prototype representing 3D space via alternative-sensing for visually impaired navigation. IEEE Sensors J. **13**(7), 2535–2547 (2013)
10. Praveen, R.G., Paily, R.P.: Blind navigation assistance for visually impaired based on local depth hypothesis from a single image. Procedia Eng. **64**, 351–360 (2013)
11. Pundlik, S., Tomasi, M., Luo, G.: Collision detection for visually impaired from a body-mounted camera. In: IEEE Conference on Computer Vision and Pattern Recognition Workshops (CVPRW), pp. 41–47 (2013)
12. Bangar, S., Narkhede, P., Paranjape, R.: Vocal vision for visually impaired. Int. J. Eng. Sci. (IJES) **2**, 1–7 (2013)
13. Kang, M., Chae, S., Sun, J., Yoo, J.: A novel obstacle detection method based on deformable grid for the visually impaired. IEEE Trans. Consum. Electron. **61**, 376–383 (2015)
14. Hammami, M., Chahir, Y., Chen, L.: WebGuard: a web filtering engine combining textual, structural, and visual content-based analysis. IEEE Trans. Knowl. Data Eng. **18**(2), 272–284 (2006)
15. Quinlan, J.R.: Induction of decision trees. Mach. Learn. **1**(1), 81–106 (1986)
16. Ess, A., Leibe, B., Schindler, K., Gool, L.: Moving obstacle detection in highly dynamic scenes. In: IEEE International Conference on Robotics and Automation, ICRA 2009, pp. 56–63 (2009)

t-SNE Based Visualisation and Clustering of Geological Domain

Mehala Balamurali[(✉)] and Arman Melkumyan

Australian Centre for Field Robotic, University of Sydney, Sydney, Australia
{m.balamurali, a.melkumyan}@acfr.usyd.edu.au

Abstract. Identification of geological domains and their boundaries plays a vital role in the estimation of mineral resources. Geologists are often interested in exploratory data analysis and visualization of geological data in two or three dimensions in order to detect quality issues or to generate new hypotheses. We compare PCA and some other linear and non-linear methods with a newer method, t-Distributed Stochastic Neighbor Embedding (t-SNE) for the visualization of large geochemical assay datasets. The t-SNE based reduced dimensions can then be used with clustering algorithm to extract well clustered geological regions using exploration and production datasets. Significant differences between the nonlinear method t-SNE and the state of the art methods were observed in two dimensional target spaces.

Keywords: Visual analytics · Dimensionality reduction · Clustering · Geological domain · Geochemical data

1 Introduction

Geological interpretations are carried out based on information obtained by drilling and supported by other field data such as surface mapping. At the scale of the drill hole, the geological strands are interpreted based on natural gamma traces. Mineralised intervals in the drill holes are identified based on assays and logging data and these are domained into different geozone codes. The geozone field is a numeric field that is based on geological unit and whether the interval is a non-hydrated mineralisation, hydrated mineralisation or un-mineralised. The geozone field is used to ensure that all the samples that are spatially located inside a modelling domain are grouped together and used in the grade estimation of that domain (Sommerville et al. 2014). A major challenge in geological modelling is to reliably assign domains to all the samples including production data. Our previous studies (Balamurali and Melkumyan 2015) deal with application of supervised learning to the identification of different combinations of geological features which are better suited for classification of the samples spatially located between adjacent domains.

This study instead, focuses on clustering geochemical production and exploration data and its two dimensional visualization in unsupervised manner. Two and three dimensional visualizations are often a good way to get a first impression of properties or the quality of a dataset or of special patterns within the data by showing clusters such as mineral regions and waste regions, revealing outliers, a high level of noise or to generate

A. Hirose et al. (Eds.): ICONIP 2016, Part IV, LNCS 9950, pp. 565–572, 2016.
DOI: 10.1007/978-3-319-46681-1_67

hypotheses for further experimentation. It can also be used as an input for further automated classification. On the other hand, applications like the visualization of high-dimensional data may benefit from extracting information from all features. In this field several other methods have been developed since PCA, such as Sammon mapping (Sammon 1969), Isomap (Tenenbaum et al. 2000), Locally Linear Embedding (Roweis and Saul 2000), Classical multidimensional scaling (Torgerson 1952), Laplacian Eigenmap (Belkin and Niyogi 2002), m-SNE (Xie et al. 2010), t-SNE (Maaten and Hinton 2008), and others.

In this article we will focus on t-SNE and compare it with linear methods such as PCA and LLE and thenon-linear method k-PCA on mine geology production and exploration datasets. It has been suggested (Maaten and Hinton 2008) that most of the nonlinear dimensionality reduction techniques (except t-SNE) perform strongly on artificial data sets but their visualization of real, high-dimensional data sets is poor. t-SNE often outperforms the other techniques and captures most of the local structure while revealing global structure like presence of clusters at various scales. It uses Gaussian distribution for calculating the probability of data points in higher dimension and uses Student-t distribution for representing relationships between points in the lower dimension used for visualisation. It maps the points in higher dimension into lower dimension by minimizing the Kullback-Leibler divergence and employs the gradient descent technique for numerical optimisation.

In order to evaluate the quality of the visualization and the insight it can provide we (Von Luxburg 2007) and used the known geozone labels of exploration data for checking the validity of the clusters. We also provide a comparison between the t-SNE and other methods on geo chemical datasets.

2 T_SNE

Given a set of N high-dimensional objects $X_1, \ldots, X_N$, t-SNE first computes probabilities $p_{i,j}$ that are proportional to the similarity of objects X_i and X_j, as follows:

$$p_{j|i} = \frac{\exp(-\|X_i - X_j\|^2/2\sigma_i^2}{\sum_{k \neq i} \exp(-\|X_i - X_k\|^2/2\sigma_i^2} \tag{1}$$

$$p_{i,j} = \frac{p_{j|i} + p_{i|j}}{2N} \tag{2}$$

The bandwidth of the Gaussian kernels σ_i, is set in such a way that the perplexity of the conditional distribution equals a predefined perplexity using a binary search. As a result, the bandwidth is adapted to the density of the data: smaller values of σ_i are used in denser parts of the data space.

t-SNE aims to learn a d-dimensional map $Y_1, \ldots, Y_N$ (with $Y_i \in \mathbb{R}^d$) that reflects the similarities $p_{i,j}$ as well as possible. To this end, it measures similarities $q_{i,j}$ between two points in the map y_i and y_j, using a very similar approach. Specifically, $q_{i,j}$ is defined as:

$$q_{i,j} = \frac{\left(1 + \|y_{i} - y_{j}\|^{2}\right)^{-1}}{\sum_{k \neq i}\left(1 + \|y_{k} - y_{j}\|^{2}\right)^{-1}} \tag{3}$$

Herein a heavy-tailed Student-t distribution is used to measure similarities between low-dimensional points in order to allow dissimilar objects to be modeled far apart in the map.

The locations of the points y_i in the map are determined by minimizing the (non-symmetric) Kullback–Leibler divergence of the distribution Q from the distribution P, that is:

$$KL(P||Q) = \sum_{i \neq j} p_{ij} log \frac{p_{ij}}{q_{ij}} \tag{4}$$

The minimization of the Kullback–Leibler divergence with respect to the points y_i is performed using gradient descent. The result of this optimization is a map that reflects the similarities between the high-dimensional inputs well.

For t-SNE the matlab reference implementation is used (Maaten and Hinton 2014).

3 Experimental Setup

3.1 Datasets

The data used in this study area is collected from the drill holes from an iron ore deposit located in the Brockman Iron Formation of the Hamersley Province, Western Australia. The methods were tested on both exploration and the production assay data sets. The exploration assay data consists of 2639 measurements of mineralised, waste and hydrated data of nearly 150 drill holes that have undergone drilling and chemical assays analysis to determine its composition. The holes are generally 25-100 m apart and tens to hundreds of meters deep. Within each hole, data is collected at an interval of 2 m. The measurements include the position (east, north, elevation) data along with the ten variables of interest: iron (Fe), silica (SiO_2), alumina (Al_2O_3), phosphorus (P), manganese (Mn), loss on ignition (LOI), sulphur (S), titanium oxide (TiO_2), calcium oxide (CaO) and magnesium oxide (MgO) (Sommerville et al. 2014), the corresponding geological domains and the drill-hole names. The data includes mineralised samples, waste samples and hydrated samples. The production data consists of 14906 measurements of blast holes which are generally 5 m apart. The measurements include the position (east, north, elevation) data along with the grades of ten chemical elements. Production data is sparse in the vertical direction while the exploration data is dense in that direction. Production data do not have pre assigned labels.

3.2 Benchmark

Our benchmark is divided into three parts. First, the dimensionality reduction methods were applied to exploration and production assay dataset. The low-dimensional datasets were then assessed by cluster validation and by its visual interpretation. The compactness of clusters within the low-dimensional representations was tested with the exploration data included within each clusters.

3.3 Dimensionality Reduction

Four unsupervised dimensionality reduction techniques were compared within this study: Principal Component Analysis (PCA), Kernel PCA (KPCA), Locally Linear Embedding (LLE) and t-distributed Stochastic Embedding (t-SNE). These method were applied on both exploration and production geochemical assay which initially has 10 chemical species. The final results has been projected on their two dimensional reduced features.

3.4 Parameters

There are two parameters for the implementation of t-SNE: initial dimensions and perplexity. Initial dimensions are a preprocessing reduction with PCA to eliminate the most likely noise with skipping components with virtually no variance; it makes the computation faster. Perplexity is used as defined in information theory, for example in (Brown et al. 1992). Perplexity can be interpreted in this method as a smooth measure of the effective number of neighbors. The values for initial dimensions is set to 10 as there are 10 grades used in this study and the perplexity of the Gaussian kernel that is employed is set to the default value 30.

LLE uses 10 nearest neighbors.

3.5 Cluster Validity

We used normalized spectral clustering algorithm defined on the k-nearest neighbor similarity graph, where k is 20. Since there are already known three domains (mineralized, waste and hydrated) we set the cluster number to four. As if the samples don't belong to any of the three domains we expect them to go to the fourth cluster. As result from clustering the labels would be assigned through the clustering method, whereas in our case the labels of exploration data are the true (external) classes and the values of the variables are 'generated' (=transformed original values).

4 Results and Discussion

Figure 1 compares their two dimensional embedding of the mineral and waste exploration datasets to the first two principal components of PCA, kPCA, LLE and t-SNE. The results (Fig. 1) show that the structure of geochemical assay data is often too complex to be captured well in very low dimensional target spaces using a linear method such as PCA and LLE. Nonlinear methods t-SNE and kPCA preserve more information in the data than the first few principle components of a PCA and LLE are able to cover.

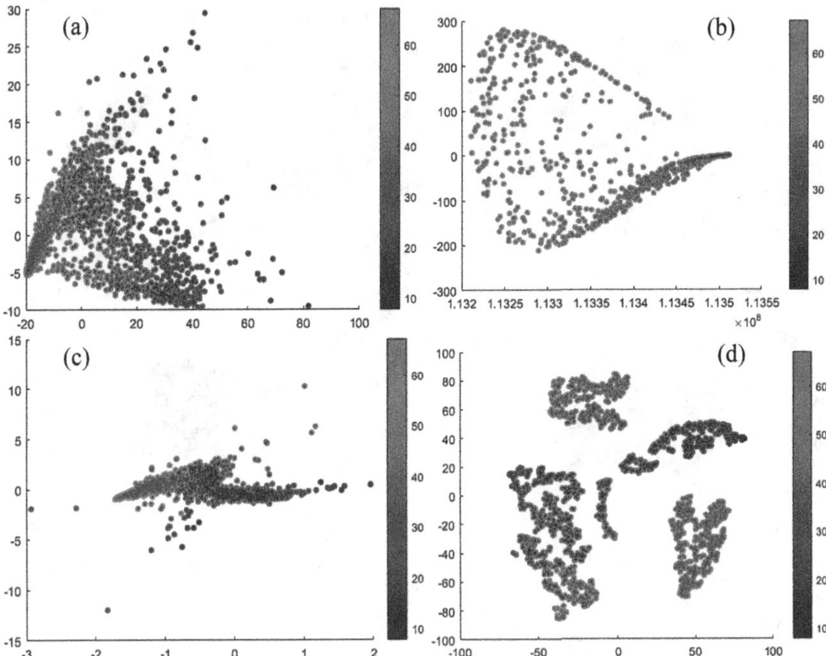

Fig. 1. Two dimensional visualization on Mineral and Waste exploration dataset. The (d) 2D t-SNE embedding shows more distinct clusters than the (a) PCA, (b) kPCA and (c) LLE. Colourr bars show the Fe distribution. Blue and red dots show the waste and mineral samples respectively. (Color figure online)

In addition, the t-SNE-based visualization allows users to easily see patterns of shape/trend that are not apparent in traditional visualizations. This was further tested on the combination of production and exploration chemical assay data as shown in the Fig. 2a. It has been clearly seen that the production assay data which need to be domained, cluster together as they share the same domain assay values of pre zoned assay values of exploration data with the same trend (Fig. 2a). The waste boundaries of production data which is low in Fe grade exhibit a high association with the waste regions of exploration data. Also, in the Fig. 2d and e a distinct zone roughly follows the shape of the boundary. However, identifying the boundaries of mineral and hydrated regions is ambiguous in Figs. 2a and b.

Spectral clustering algorithm is further used to examine the relationships among the assay distributions in clusters. The clustering results can be seen in the Fig. 2b and each cluster is colour coded as Cluster1-pink, Cluster2-black, Cluster3-blue and Cluster4-green. Figure 2c, d, e and f show how the clusters are arranged in their transformed spatial coordinates. It can be seen from the Fig. 2d and f that the assay data in Cluster 3 and 4 are spatially closed within their domain. Box plots show the grades distribution within each cluster of both production and exploration data.

Figure 3 presents the statistical evaluation of data assigned to each cluster. As we already know the geozone labels for the exploration data we used them as a ground truth to validate the clustering results. Figure 3a shows the proportion of each cluster

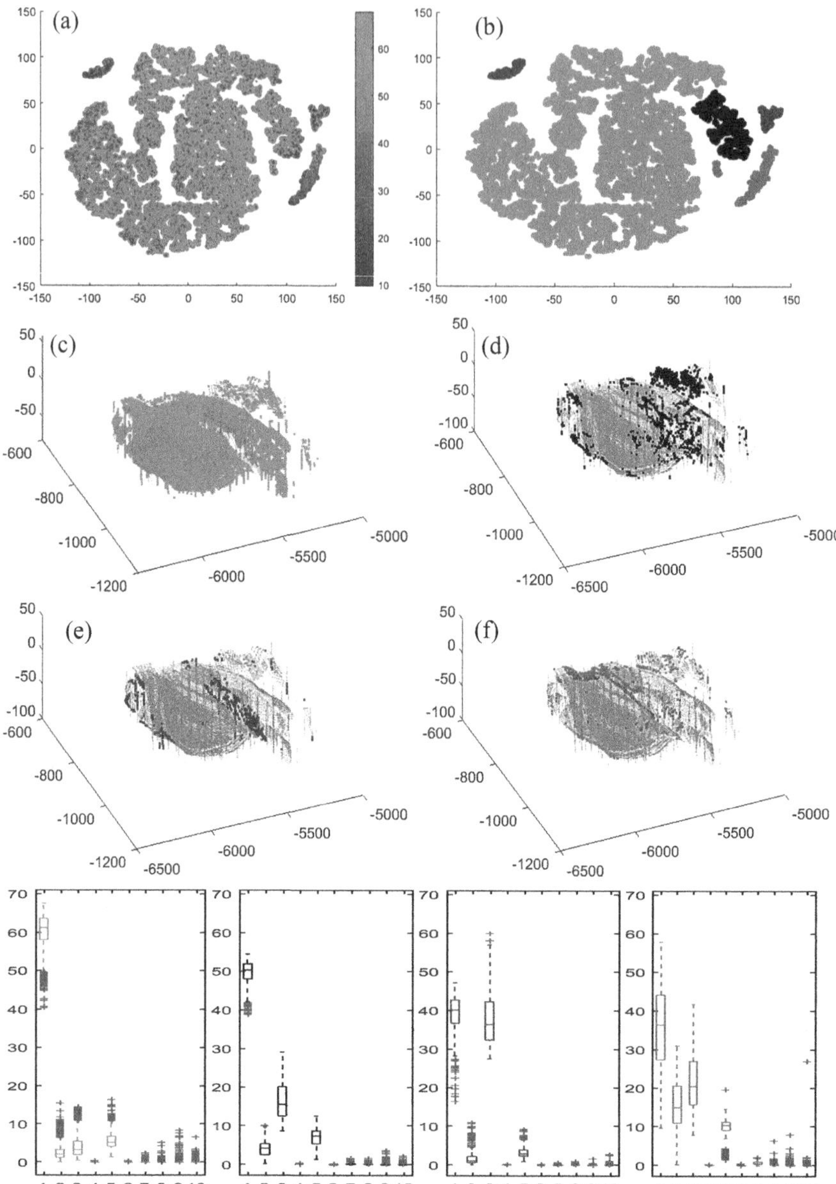

Fig. 2. (a) Visualisation of assay data on t-SNE coordinates. Colour bar shows the Fe distribution. Blue, red and green dots show the waste, mineral and hydrated samples respectively. (b) Spectral clustering results on the reduced 2D using t-SNE. (c), (d), (e) and (f) Corresponding clustering on their transformed spatial coordinates of production and exploration data. Box plots show the chemical distribution corresponding to each cluster in the order 1–10: Fe, Al_2O_3, SiO_2, P, LOI, S, TiO_2, MgO, CaO and Mn. (Color figure online)

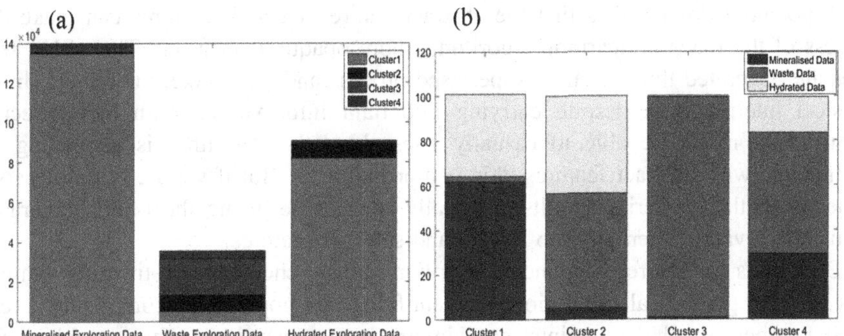

Fig. 3. The exploration assays (known geozone labels) of each geological region grouped into diffrent clusters.

that is present in the mineralized, waste and hydrated exploration samples. Figure 3b shows the percentage of mineralised, waste and hydrated samples out of total cluster samples. The plots indicate that the 50.8 % of waste exploration samples were included in Cluster 3 which represents the 98.4 % of the size of the Cluster 3. As we see in the Fig. 3a, 97.7 % of mineral samples and 90.6 % of hydrated samples were included in Cluster1. However the mineral samples occupy the 61.9 % of the total size of Cluster1. This is nearly double the size of the hydrated sample available in the cluster. Similarly, in Cluster 2 the waste and the hydrated samples nearly equally (51.2 %, 43.5 %) occupy the total cluster size. Our results suggest that the approach proposed in this paper works very well in most places when the samples are located between the mineralised and waste regions which have significantly different distribution for chemical elements. However, for the samples from the geological domains: mineralised and hydrated mineralised which have limited differences in the variations of the chemical elements doesn't show significant differences. Because of the inconsistency in the measurements and the high variation in the hydrated sample it was highly distributed between the clusters.

5 Conclusion

In this work, we evaluate the domain interpretation quality and visualising the spatially correlated geological regions with the t-SNE based reduced dimension applied to geological datasets. We compared the application of t-SNE non- linear dimensionality reduction method to LLE, PCA and kPCA dimensionality reduction methods and showed that t-SNE outperforms the other techniques and captures most of the local structure while revealing global structure like presence of clusters at various scales. T-SNE also helps in displaying the data in a visual format which is more understandable and would help in improving inferences, comprehension and decision making.

Further, application of spectral clustering with the reduced dimension using t-SNE showed that the most significant information within geo-chemical assays data can be well captured.

A potential downside is that the abstract nature of t-SNE features can make the process of the machine learning algorithms more opaque than usual. The t-SNE features are generated through an unsupervised process and are unlikely to have a direct physical interpretation despite carrying important information. With hand-selected features, we might be able to casually guess how the algorithm is generating its predictions; with abstract features, this is more difficult. But if we are primarily concerned with the clustering results as usually is the case, using the t-SNE generated features is a viable alternative to using hand-selected features.

Directions for future work include applying other schemes for estimating similarities between geological sub regions or identifying rare populations comprising a very small number of all data points, combining clusters based on other combinatorial algorithms or geological criteria.

Acknowledgement. This work has been supported by the Australian Centre for Field Robotics and the Rio Tinto Centre for Mine Automation.

References

Roweis, S.T., Saul, L.K.: Nonlinear dimensionality reduction by locally linear embedding. Science **290**, 2323–2326 (2000)

Tenenbaum, J.B., De Silva, V., Langford, J.C.: A global geometric framework for nonlinear dimensionality reduction. Science **290**(5500), 2319–2323 (2000)

Sammon, J.W.: A nonlinear mapping for data structure analysis. IEEE Trans. Comput. **18**(5), 401–409 (1969)

Torgerson, W.S.: Multidimensional scaling: I. Theory and method. Psychometrika **17**(4), 401–419 (1952)

Belkin, M., Niyogi, P.: Laplacian eigenmaps and spectral techniques for embedding and clustering. In: Advances in Neural Information Processing Systems 14, vols. 1, 2 and 14, pp. 585–591 (2002)

Xie, B., Mu, Y., Tao, D.: m-SNE: multiview stochastic neighbor embedding. In: Mendis, B.U., Bouzerdoum, A., Wong, K.W. (eds.) ICONIP 2010, Part I. LNCS, vol. 6443, pp. 338–346. Springer, Heidelberg (2010)

Maaten, L.V.D., Hinton, G.: Visualizing data using t-SNE. J. Mach. Learn. Res. **9**, 2579–2605 (2008)

Sommerville, B., Boyle, C., Brajkovich, N., Savory, P., Latscha, A.: Mineral resource estimation of the Brockman 4 iron ore deposit in the Pilbara region. Appl. Earth Sci. **123**(2), 135–145 (2014)

Balamurali, M., Melkumyan, A.: Geological domain classification using multiple chemical elements and spatial information. In: Application of Computers and Operations Research in the Mineral Industry (APCOM), Fairbanks, AK, pp. 195–203 (2015)

Maaten, L.V.D., Hinton, G., Geoffrey, E.: User's Guide for t-SNE Software (2014)

Brown, P.L., Della Pietra, V., Lai, J.C., Mercer, R.L.: An estimate of an upper bound for the entropy of English. Comput. Linguist. **18**, 31–40 (1992)

Von Luxburg, U.: A tutorial on spectral clustering. Stat. Comput. **17**(4), 395–416 (2007)

Data-Based Optimal Tracking Control of Nonaffine Nonlinear Discrete-Time Systems

Biao Luo[1(✉)], Derong Liu[2], Tingwen Huang[3], and Chao Li[1]

[1] The State Key Laboratory of Management and Control for Complex Systems, Institute of Automation, Chinese Academy of Sciences, Beijing 100190, China
{biao.luo,lichao2012}@ia.ac.cn
[2] School of Automation and Electrical Engineering, University of Science and Technology Beijing, Beijing 100083, China
derong@ustb.edu.cn
[3] Texas A&M University at Qatar, PO Box 23874, Doha, Qatar
tingwen.huang@qatar.tamu.edu

Abstract. The optimal tracking control problem of nonaffine nonlinear discrete-time systems is considered in this paper. The problem relies on the solution of the so-called tracking Hamilton-Jacobi-Bellman equation, which is extremely difficult to be solved even for simple systems. To overcome this difficulty, the data-based Q-learning algorithm is proposed by learning the optimal tracking control policy from data of the practical system. For its implementation purpose, the critic-only neural network structure is developed, where only critic neural network is required to estimate the Q-function and the least-square scheme is employed to update the weight of neural network.

Keywords: Optimal tracking control · Data-based · Q-learning · Critic-only

1 Introduction

Reinforcement learning is a machine learning technique that has been widely studied from the computational intelligence and machine learning scope in the artificial intelligence community [1,2]. It refers to an actor or agent that interacts with its environment and aims to learn the optimal actions or control policies, by observing their responses from the environment. Over the past few years, the idea of reinforcement learning techniques has been introduced to control community to solve optimal control problems and many works have been reported [3–12]. Reinforcement learning methods were proposed to solve control problems of continuous-time systems [4,11], discrete-time systems [10], distributed parameter systems [5], constrained input [11], H_∞ control [5,7], event-triggered control [13], and so on.

It is found that most of works are for regulation control problems, while the tracking control problem still not receives much attention and only few results

A. Hirose et al. (Eds.): ICONIP 2016, Part IV, LNCS 9950, pp. 573–581, 2016.
DOI: 10.1007/978-3-319-46681-1_68

have been reported. Based on the desired reference trajectories and the system model, the expected control is derived firstly, which is then used to define the the error performance index. The optimal tracking control problem can be reformulated as the optimal regulation problem of the error system with the error performance index, and then some reinforcement learning methods were proposed [14–18]. These methods are all model-based as the analytic expression of the expected control requires systems' mathematical model. To alleviate the dependence of the system model, partially model-based and data-based reinforcement learning methods [19–23] were also developed to solve the optimal tracking control problems. Most of these were presented for affine nonlinear systems [14,16,17,20,21] or linear systems [19,22,24]. To solve the data-based optimal tracking control problem of general nonaffine nonlinear discrete-time systems, the critic-only Q-learning (CoQL) method is developed in this paper.

2 Problem Description

Let us consider the following nonaffine nonlinear discrete-time system:

$$x(k+1) = f(x(k), u(k)), \tag{1}$$

where $x(k) \in \mathbb{R}^n$ is the state and $u(k) \in \mathbb{R}^m$ is the control input. It is assumed that the system (1) is stabilizable on the set $\mathcal{X}$ and $f(0,0) = 0$.

Let $r(k) \in \mathbb{R}^n$ be the desired reference trajectory. For the optimal tracking control problem, the objective is to design the control input $u(k)$ for the system (1), such that the state $x(k)$ tracks $r(k)$ and minimize the performance index. Assume that $r(k)$ be is bounded and generated by the command system

$$r(k+1) = h(r(k)), \tag{2}$$

where $h(r)$ is a Lipschitz continuous vector function with $h(0) = 0$. Denoting the tracking error as $e(k) \triangleq x(k) - r(k)$, it follows from (1) and (2) that

$$e(k+1) = f(e(k) + r(k), u(k)) - h(r(k)). \tag{3}$$

Define the state of the augmented system as $y(k) \triangleq [e^{\mathsf{T}}(k) \quad r^{\mathsf{T}}(k)]^{\mathsf{T}}$. Then, combining (2) and (3) yields the following augmented system:

$$y(k+1) = F(y(k), u(k)), \tag{4}$$

where $y(0) = [e^{\mathsf{T}}(0) \quad r^{\mathsf{T}}(0)]^{\mathsf{T}}$ and

$$F(y(k), u(k)) \triangleq \begin{bmatrix} f(e(k) + r(k), u(k)) - h(r(k)) \\ h(r(k)) \end{bmatrix}. \tag{5}$$

Consider the model-free optimal tracking control problem of the system (1) with the following discounted performance index:

$$J(y(0), u) \triangleq \sum_{l=0}^{\infty} \gamma^l \mathcal{R}(y(l), u(l)), \tag{6}$$

where $0 < \gamma \leqslant 1$ is the discount factor and $\mathcal{R}(y,u) \triangleq W(e) + R(u)$ with $W(e)$ and $R(u)$ are positive definite functions, i.e., $W(e) > 0$, $R(u) > 0$ for $\forall e \neq 0$, $u \neq 0$, and $W(e) = 0$, $R(u) = 0$ only when $e = 0$, $u = 0$. Then, the optimal tracking control problem of the system (1) is converted to an optimal regulation problem, i.e., finding the following optimal control

$$u^*(y) \triangleq \arg\min_u J(y(0), u) \tag{7}$$

with respected to the augmented system (4) and the performance index (6). In this paper, we aim to solve the model-free optimal tracking control problem, i.e., the explicit mathematical expressions of $F(y,u)$ and $\mathcal{R}(y,u)$ are unknown.

3 Q-learning Algorithm

Let $y \in \mathcal{Y}$, $u \in \mathcal{U}$, where $\mathcal{Y}$ and $\mathcal{U}$ be two compact sets, and denote $\mathcal{D} \triangleq \{(y,u)|y \in \mathcal{Y}, u \in \mathcal{U}\}$. For an admissible control policy $u(y)$, define its value function as:

$$V_u(y(k)) \triangleq \sum_{l=k}^{\infty} \gamma^{l-k} \mathcal{R}(y(l), u(l)), \tag{8}$$

which can be rewritten as the following equation:

$$\begin{aligned} V_u(y(k)) &= \mathcal{R}(y(k), u(k)) + \sum_{l=k+1}^{\infty} \gamma^{l-k} \mathcal{R}(y(l), u(l)) \\ &= \mathcal{R}(y(k), u(k)) + \gamma V_u(y(k+1)). \end{aligned} \tag{9}$$

The optimal control law (7) can be rewritten as

$$u^*(y(k)) \triangleq \arg\min_u V_u(y(k)). \tag{10}$$

Denoting the optimal value function as $V^*(y) \triangleq V_{u^*}(y)$, the tracking Hamilton-Jacobi-Bellman equation (HJBE) is given as follows:

$$\begin{aligned} V^*(y(k)) &= \min_u \{\mathcal{R}(y(k), u(k)) + \gamma V^*(y(k+1))\} \\ &= \mathcal{R}(y(k), u^*(k)) + \gamma V^*(y(k+1)), \end{aligned} \tag{11}$$

which is a nonlinear difference equation. Note that the optimal control policy $u^*(y)$ depends on the solution of the tracking HJBE (11), which is difficult to solve for nonlinear systems. Even worse, the unavailability of the explicit expressions $F(y,u)$ and $\mathcal{R}(y,u)$ prevents using model-based methods to solve the tracking HJBE for the optimal value function V^*. To overcome those difficulties, we propose a Q-learning algorithm for direct optimal tracking control design with real system data.

For an admissible control policy $u(y)$, define its Q-function as

$$Q_u(y(k), a) = \mathcal{R}(y(k), a) + \gamma V_u(y(k+1)) \tag{12}$$

where $Q_u(0,0) = 0$. For the optimal control policy $u^*(y)$, it follows from (12) that the associated optimal Q-function $Q^*(y,a) \triangleq Q_{u^*}(y,a)$ is given by

$$Q^*(y(k),a) = \mathcal{R}(y(k),a) + \gamma V^*(y(k+1)). \tag{13}$$

Thus, the optimal control policy $u^*(y)$ can also be represented as

$$u^*(y) = \arg\min_u V_u(y) = \arg\min_a Q^*(y,a). \tag{14}$$

To learn the optimal Q-function $Q^*(y,a)$ and optimal tracking control $u^*(y)$, the following Q-learning algorithm is proposed:

Algorithm 1 . Q-learning

▶ *Step 1:* Let $u^{(0)}(y)$ be an initial admissible control policy, and $i = 0$;
▶ *Step 2:* (**Policy evaluation**) Solve the equation

$$Q^{(i)}(y(k),a) = \mathcal{R}(y(k),a) + \gamma Q^{(i)}(y(k+1),u^{(i)}) \tag{15}$$

for the unknown Q-function $Q^{(i)} \triangleq Q_{u^{(i)}}$;
▶ *Step 3:* (**Policy improvement**) Update control policy with

$$u^{(i+1)}(y) = \arg\min_a Q^{(i)}(y,a); \tag{16}$$

▶ *Step 4:* Let $i = i+1$, go back to Step 2 and continue. □

---

4 Critic-Only Q-learning for Adaptive Tracking Control

Based on Algorithm 1, the CoQL method is developed for adaptive tracking control design. It is known that NN is an universal approximator [25,26] for estimating continuous function. To solve the equation (15), a critic NN is employed for estimating the unknown Q-function $Q^{(i)}(y,a)$ on $\mathcal{D}$. Then, the output of the critic NN is given by:

$$\hat{Q}^{(i)}(y,a) = \sum_{j=1}^{L} \hat{\theta}_j^{(i)} \psi_j(y,a) = \Psi_L^{\mathsf{T}}(y,a)\hat{\theta}^{(i)} \tag{17}$$

where $\hat{\theta}^{(i)} \triangleq [\hat{\theta}_1^{(i)},...,\hat{\theta}_L^{(i)}]^{\mathsf{T}}$ and $\Psi_L(x,a) \triangleq [\psi_1(y,a),...,\psi_L(y,a)]^{\mathsf{T}}$ is the critic NN activation function vector. With $\hat{Q}^{(i)}(y,a)$, it follows from (16) that

$$\hat{u}^{(i+1)}(y) = \arg\min_a \hat{Q}^{(i)}(y,a). \tag{18}$$

For $\forall y \in \mathcal{Y}$, based on the gradient descent method, we have

$$\begin{aligned}\hat{u}^{(i+1)}(y) &= \hat{u}^{(i)}(y) - \alpha \frac{\partial \hat{Q}^{(i)}(y,a)}{\partial a}\Big|_{a=\hat{u}^{(i)}(y)} \\ &= \hat{u}^{(i)}(y) - \alpha \frac{\partial \Psi_L^{\mathsf{T}}(y,a)}{\partial a}\Big|_{a=\hat{u}^{(i)}(y)} \hat{\theta}^{(i)},\end{aligned} \tag{19}$$

where $\alpha > 0$.

To compute $\hat{\theta}^{(i)}$ for $\hat{Q}^{(i)}(y,a)$, a least-square scheme is developed using real system data. For notation simplicity, denote $(y, a, y', \mathcal{R}(y,a))$ be a data measured from the real system (4), where y' represents the next state under the control action a at state y. The system data set is denoted as $\mathcal{S}_M \triangleq \{(y_{[l]}, a_{[l]}, y'_{[l]}, \mathcal{R}_{[l]}) | (y_{[l]}, a_{[l]}) \in \mathcal{D}, l = 1, 2, ..., M\}$ with its size be M. For each data $(y_{[l]}, a_{[l]}, y'_{[l]}, \mathcal{R}_{[l]})$ in $\mathcal{S}_M$, it follows from (15), (17) and (18) that its residual error is given by

$$\epsilon_{[l]}^{(i)} \triangleq [\Psi_L(y_{[l]}, a_{[l]}) - \gamma \Psi_L(y'_{[l]}, \hat{u}^{(i)}(y'_{[l]}))]^{\mathsf{T}} \hat{\theta}^{(i)} - \mathcal{R}_{[l]}, \tag{20}$$

where $\mathcal{R}_{[l]} \triangleq \mathcal{R}(y_{[l]}, a_{[l]})$. The critic NN weight vector $\hat{\theta}^{(i)}$ can be computed with a least-square scheme by minimizing the sum of residual errors, i.e.,

$$\min \sum_{l=1}^{M} (\epsilon_{[l]}^{(i)})^2. \tag{21}$$

Then, the least-square scheme is given by

$$\hat{\theta}^{(i)} = [(Z^{(i)})^{\mathsf{T}} Z^{(i)}]^{-1} [Z^{(i)}]^{\mathsf{T}} \eta, \tag{22}$$

where $\eta \triangleq [\mathcal{R}_{[1]} \cdots \mathcal{R}_{[M]}]^{\mathsf{T}}$ and $Z^{(i)} \triangleq [z_{[1]}^{(i)} \cdots z_{[M]}^{(i)}]^{\mathsf{T}}$, with $z_{[l]}^{(i)} \triangleq \Psi_L(y_{[l]}, a_{[l]}) - \gamma \Psi_L(y'_{[l]}, \hat{u}^{(i)}(y'_{[l]}))$. By using the least-square scheme (22), the following CoQL algorithm is developed to learn the optimal Q-function.

Algorithm 2 . Critic-only Q-learning

- ▶ *Step 1:* Let $\hat{u}^{(0)} = u^{(0)}(y)$ be an initial admissible control policy, and $i = 0$;
- ▶ *Step 2:* Compute critic NN weight vector $\hat{\theta}^{(i)}$ with (22).
- ▶ *Step 3:* If $i \geqslant 1$ and $\|\hat{\theta}^{(i)} - \hat{\theta}^{(i-1)}\| \leqslant \varepsilon$ ($\varepsilon > 0$ is a small parameter), stop iteration; else, $i = i + 1$, go back to Step 2 and continue. □

After the CoQL Algorithm 2 is terminated, the convergent Q-function is used for adaptive tracking control design. Denote the convergent critic NN weight vector as θ_c and the convergent Q-function as $Q_c(y,a) = \Psi_L^{\mathsf{T}}(y,a)\theta_c$. Then, according to (14), the tracking control law is given by $u(k) = \arg\min_a Q_c(y(k), a)$.

To solve this optimization problem at each time instant k, the adaptive tracking control is designed via the gradient descent method as follows:

$$u(k) = u(k-1) - \alpha \frac{\partial \Psi_L^{\mathsf{T}}(y(k), a)}{\partial a}\bigg|_{a=u(k-1)} \theta_c. \tag{23}$$

5 Simulation Studies

To verify the effectiveness of the developed CoQL method, consider the following system

$$\begin{cases} x_1(k+1) = 0.9926x_1(k) + 0.0486x_2(k) \\ x_2(k+1) = -0.2919x_1(k) + 0.9440x_2(k) + \sin(u) \end{cases} \tag{24}$$

with $x(0) = [3, -2]^{\mathsf{T}}$. Let the desired trajectory $r(k)$ is generated by the following command system

$$\begin{cases} r_1(k+1) = 0.9950r_1(k) + 0.0499r_2(k) \\ r_2(k+1) = -0.1997r_1(k) + 0.9950r_2(k) \end{cases}. \tag{25}$$

The desired trajectory $r(k)$ are sinusoidal signals, which are shown in Figs. 1 and 2 with black doted lines. For the discounted performance index (6), let $\mathcal{R}(y(l), u(l)) = e_1^2(l) + 2e_2^2(l) + u^2$ and $\gamma = 0.95$.

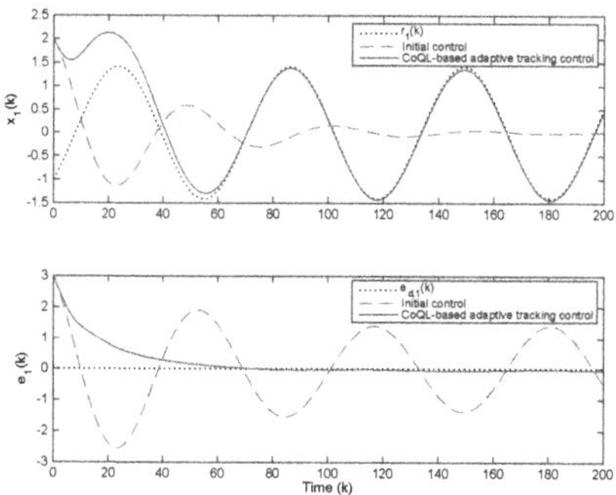

Fig. 1. The trajectories of $r_1(k)$, state $x_1(k)$ and tracking error $e_1(k)$.

To learn the optimal Q-function with the developed CoQL method (i.e., Algorithm 2), let the initial control $u^{(0)}(y) = 0$, the terminate condition $\varepsilon = 10^{-5}$ and the critic NN activation function $\Psi_L(x, a) = [e_1^2, e_1e_2, e_1r_1, e_1r_2, e_1a, e_2^2, e_2r_1, e_2r_2, e_2a, r_1^2, r_1r_2, r_1a, r_2^2, r_2a, a^2, \sin(a), e_1\sin(a), e_2\sin(a), r_1\sin(a), r_2\sin(a)]^{\mathsf{T}}$. After the convergence of Algorithm 2, the convergent critic NN weight vector θ_c is employed to design the CoQL-based adaptive control law with (23). To show how the CoQL method improves the control performance, comparative studies are also conducted with the initial control $u^{(0)}(y)$. The comparative results are given in Figs. 1 and 2. It is indicated from figures that with the initial

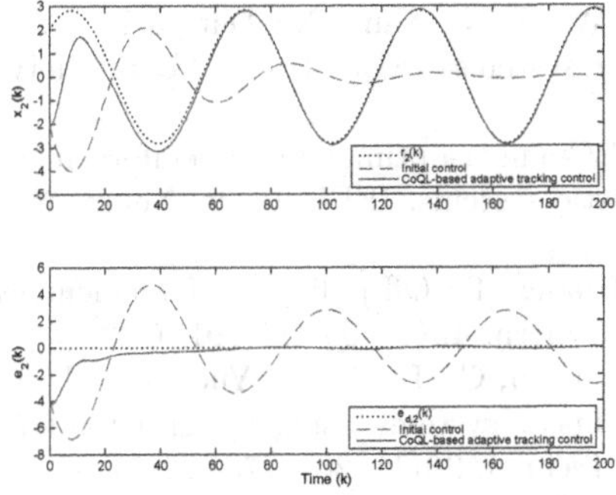

Fig. 2. The trajectories of $r_2(k)$, state $x_2(k)$ and tracking error $e_2(k)$.

control, the system state $x(k)$ can not track the desired trajectory $r(k)$. Note that the CoQL-based adaptive tracking control achieves a much better tracking performance.

6 Conclusions

In this paper, the data-based optimal tracking control problem of general non-affine nonlinear discrete-time systems is solved with the CoQL method. By introducing the Q-function, the Q-learning algorithm was proposed. With the use of only one critic neural network to approximate the Q-function, the CoQL based adaptive tracking control method was developed. To verify the developed method, the simulation study was conducted on a numerical example. The results demonstrate that good tracking performance was achieved.

Acknowledgements. This work was supported in part by the National Natural Science Foundation of China under Grants 61233001, 61273140, 61304086, 61374105, 61503377, 61533017, and U1501251, in part by the Early Career Development Award of SKLMCCS and in part by the NPRP grant #NPRP 7-1482-1-278 from the Qatar National Research Fund (a member of Qatar Foundation).

References

1. Sutton, R.S., Barto, A.G.: Reinforcement Learning: An Introduction. The MIT Press, Cambridge (1998)
2. Hafner, R., Riedmiller, M.: Reinforcement learning in feedback control. Mach. Learn. **84**(1–2), 137–169 (2011)
3. Lewis, F.L., Liu, D.: Reinforcement Learning and Approximate Dynamic Programming for Feedback Control, vol. 17. Wiley, Hoboken (2013)
4. Luo, B., Wu, H.N., Huang, T., Liu, D.: Data-based approximate policy iteration for affine nonlinear continuous-time optimal control design. Automatica **50**(12), 3281–3290 (2014)

5. Luo, B., Huang, T., Wu, H.N., Yang, X.: Data-driven H_∞ control for nonlinear distributed parameter systems. IEEE Trans. Neural Netw. Learn. Syst. **26**(11), 2949–2961 (2015)
6. Zhao, D., Zhu, Y.: MEC-a near-optimal online reinforcement learning algorithm for continuous deterministic systems. IEEE Trans. Neural Netw. Learn. Syst. **26**(2), 346–356 (2015)
7. Luo, B., Wu, H.N., Huang, T.: Off-policy reinforcement learning for H_∞ control design. IEEE Trans. Cybern. **45**(1), 65–76 (2015)
8. Zhu, L., Modares, H., Peen, G., Lewis, F., Yue, B.: Adaptive suboptimal output-feedback control for linear systems using integral reinforcement learning. IEEE Trans. Control Syst. Technol. **23**(1), 264–273 (2015)
9. Luo, B., Wu, H.N., Li, H.X.: Adaptive optimal control of highly dissipative nonlinear spatially distributed processes with neuro-dynamic programming. IEEE Trans. Neural Netw. Learn. Syst. **26**(4), 684–696 (2015)
10. Liu, Y.J., Tang, L., Tong, S., Chen, C., Li, D.J.: Reinforcement learning design-based adaptive tracking control with less learning parameters for nonlinear discrete-time mimo systems. IEEE Trans. Neural Netw. Learn. Syst. **26**(1), 165–176 (2015)
11. Luo, B., Wu, H.N., Huang, T., Liu, D.: Reinforcement learning solution for HJB equation arising in constrained optimal control problem. Neural Netw. **71**, 150–158 (2015)
12. Kamalapurkar, R., Andrews, L., Walters, P., Dixon, W.E.: Model-based reinforcement learning for infinite-horizon approximate optimal tracking. IEEE Trans. Neural Netw. Learn. Syst. **PP**(99), 1–6 (2016)
13. Zhong, X., He, H.: An event-triggered ADP control approach for continuous-time system with unknown internal states. IEEE Trans. Cybern. **PP**(99), 1–12 (2016)
14. Zhang, H., Wei, Q., Luo, Y.: A novel infinite-time optimal tracking control scheme for a class of discrete-time nonlinear systems via the greedy HDP iteration algorithm. IEEE Trans. Syst. Man Cybern. Part B: Cybern. **38**(4), 937–942 (2008)
15. Zhang, H., Song, R., Wei, Q., Zhang, T.: Optimal tracking control for a class of nonlinear discrete-time systems with time delays based on heuristic dynamic programming. IEEE Trans. Neural Netw. **22**(12), 1851–1862 (2011)
16. Wei, Q., Liu, D.: Neural-network-based adaptive optimal tracking control scheme for discrete-time nonlinear systems with approximation errors. Neurocomputing **149, Part A**, 106–115 (2015)
17. Kamalapurkar, R., Dinh, H., Bhasin, S., Dixon, W.E.: Approximate optimal trajectory tracking for continuous-time nonlinear systems. Automatica **51**(1), 40–48 (2015)
18. Zhang, H., Cui, L., Zhang, X., Luo, Y.: Data-driven robust approximate optimal tracking control for unknown general nonlinear systems using adaptive dynamic programming method. IEEE Trans. Neural Netw. **22**(12), 2226–2236 (2011)
19. Modares, H., Lewis, F.L.: Linear quadratic tracking control of partially-unknown continuous-time systems using reinforcement learning. IEEE Trans. Autom. Control **59**(11), 3051–3056 (2014)
20. Liu, D., Yang, X., Li, H.: Adaptive optimal control for a class of continuous-time affine nonlinear systems with unknown internal dynamics. Neural Comput. Appl. **23**(7–8), 1843–1850 (2013)
21. Kiumarsi, B., Lewis, F.: Actor-critic-based optimal tracking for partially unknown nonlinear discrete-time systems. IEEE Trans. Neural Netw. Learn. Syst. **26**(1), 140–151 (2015)

22. Kiumarsi, B., Lewis, F.L., Modares, H., Karimpour, A., Naghibi-Sistani, M.B.: Reinforcement Q-learning for optimal tracking control of linear discrete-time systems with unknown dynamics. Automatica **50**(4), 1167–1175 (2014)
23. Qin, C., Zhang, H., Luo, Y.: Online optimal tracking control of continuous-time linear systems with unknown dynamics by using adaptive dynamic programming. Int. J. Control **87**(5), 1000–1009 (2014)
24. Kiumarsi, B., Lewis, F., Naghibi-Sistani, M.B., Karimpour, A.: Optimal tracking control of unknown discrete-time linear systems using input-output measured data. IEEE Trans. Cybern. **45**(12), 2770–2779 (2015)
25. Spooner, J.T., Maggiore, M., Ordonez, R., Passino, K.M.: Stable Adaptive Control and Estimation for Nonlinear Systems: Neural and Fuzzy Approximator Techniques, vol. 43. Wiley, New York (2004)
26. Hornik, K., Stinchcombe, M., White, H.: Universal approximation of an unknown mapping and its derivatives using multilayer feedforward networks. Neural Netw. **3**(5), 551–560 (1990)

Time Series Classification Based on Multi-codebook Important Time Subsequence Approximation Algorithm

Zhiwei Tao, Li Zhang(✉), Bangjun Wang, and Fanzhang Li

School of Computer Science and Technology & Joint International Research Laboratory of Machine Learning and Neuromorphic Computing, Soochow University, Suzhou 215006, Jiangsu, China
20144227051@stu.suda.edu.cn, {zhangliml, wangbangjun, lfzh}@suda.edu.cn

Abstract. This paper proposes a multi-codebook important time subsequence approximation (MCITSA) algorithm for time series classification. MCITSA generates a codebook using important time subsequences for each class based on the difference of categories. In this way, each codebook contains the class information itself. To predict the class label of an unseen time series, MCITSA needs to compare the similarities between important time subsequences extracted from the unseen time series and codewords of each class. Experimental results on time series datasets demonstrate that MCITSA is more powerful than PVQA in classifying time series.

Keywords: Dimensionality reduction · Piecewise vector quantized approximation · Codebook · Time series classification · Label information

1 Introduction

Time series classification has been attracted considerable attention in data mining for its wide application, such as economics, medicine, and meteorology [1, 2, 19]. Many traditional techniques for classification tasks may be not suitable for time series classification when we deal with the time series directly. Therefore, dimensionality reduction becomes one of core issues for time series classification.

Many methods have been proposed for reducing the dimensionality of time series, such as discrete wavelet transformation (DWT) [3, 4], discrete Fourier transformation (DFT) [5], symbolic aggregate approximation (SAX) [6, 7], piecewise aggregate approximation (PAA) [6, 8, 14], singular value decomposition (SVD) [9], and piecewise vector quantized approximation (PVQA) [1, 10, 12, 13]. Nearest neighbor (NN) can be used as a simple classifier to classify the reduced time series. The most commonly used similarity measurements in the NN classifier for time series include the Euclidean distance (ED), the cosine similarity (CS), and the dynamic time warping (DTW) [2, 11].

Since PVQA is one of the most efficient methods for dimensionality reduction which can reduce both the time complexity and the space complexity, we pay our

A. Hirose et al. (Eds.): ICONIP 2016, Part IV, LNCS 9950, pp. 582–589, 2016.
DOI: 10.1007/978-3-319-46681-1_69

attention to it. First, PVQA generates a codebook using all training subsequences (piecewise time series). Then all training time series are reconstructed by some codewords in the codebook. Finally, a given time sequence is also reconstructed by a part of codewords in the codebook and classified by using the NN classifier. However, PVQA has some limitations, which have seriously affected its performance. In the procedure of generating the codebook, PVAQ ignores the label information of the given time series which would result in a bad classification performance. In addition, PVQA must reconstruct all training and unseen time series in order to make a prediction, which would speed more time.

To remedy it, this paper develops a multi-codebook important time subsequence approximation (MCITSA) method based on PVQA. MCITSA takes into account the important time subsequences (ITSs). In order to dispose the influence generated by different time subsequence belonging to different class, MCITSA constructs a codebook for each of class. Besides, there is no need for MCITSA to reconstruct whole training time series as PVQA does. Our experimental results on the benchmark time series datasets [18] show that MCITSA is significantly superior to PVQA.

2 Multi-codebook Important Time Subsequence Approximation

In order to improve the classification performance of PVQA, we propose a multi-codebook important time subsequence approximation (MCITSA) whose framework is shown in Fig. 1. The proposed algorithm involves four major steps: (a) Segment the time series into important time subsequences (ITSs) according to perceptually important points (PIPs), and record the location information of ITSs, (b) Perform PVQA on time subsequences based on the location information for each class and obtaining multiple codebooks, (c) Reconstruct the important time subsequences of

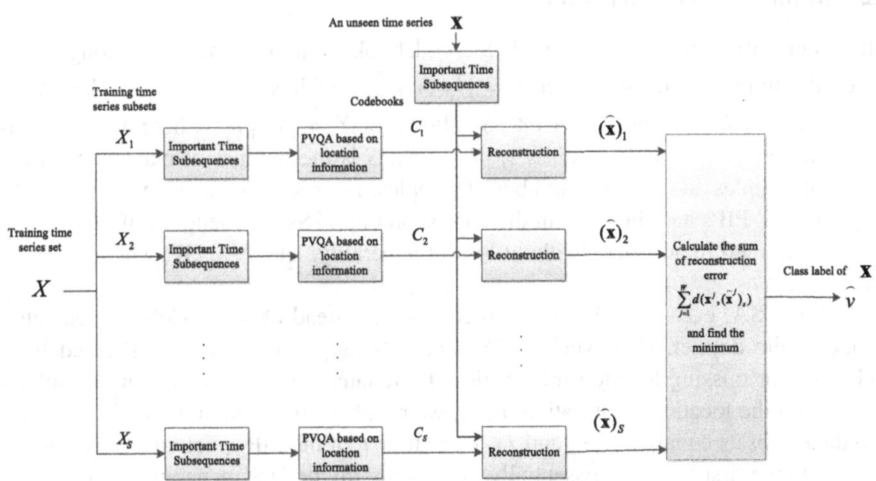

Fig. 1. The framework of MCITSA

unseen time series in terms of the generated codebooks, and (d) Assign the class label for the unseen time series according to the reconstruction error.

2.1 Important Time Subsequences

Time series is composed by a sequence of data points whose amplitudes have different influence on the shape of time series [15, 16]. A data point may represent the shape of time series well while another one may be not. As the most simple mathematical theorem states: Two points can determine a straight line. These two points are perceptually important in the line visual recognition process, and the other points can be generated by them. So we call them perceptually important points (PIPs) [15, 16].

Given a time series **P**, we can calculate its PIPs according to the algorithm described in [15]. Let W be the appointed number of PIPs. An important time subsequence (ITS) is the time subsequence containing one PIP at least. As PIPs contain the most characters of the original time series, each ITS can be taken as the subsequence which represents a sub-shape of original time series. The traditional linear segmentation for time series may ruin the shape feature and affect the subsequent classification. ITSs can reserve the shape feature to a large extent compared to subsequence generated by applying linear segmentation. To generate a codebook, it requires a lot of time subsequences. Thus, we generate three time subsequences for the PIP y_i, $i = 3, \cdots, W$. Let the time of y_i be t_i. Then we have $[x_{t_i}, x_{t_i+1}, \cdots, x_{t_i+l-1}]^T$, $[x_{t_i - \lfloor \frac{l-1}{2} \rfloor}, \cdots, x_{t_i}, \cdots, x_{t_i+l-1-\lfloor \frac{l-1}{2} \rfloor}]^T$, and $[x_{t_i-l+1}, x_{t_i-l+2}, \cdots, x_{t_i}]^T$, where l is the length of time subsequence, and $\lfloor \cdot \rfloor$ denotes the function rounding $\cdot$ to the nearest integers towards minus infinity. For the first two PIPs y_1 and y_2, only one ITS can be obtained, or $[x_{t_1}, x_{t_1+1}, \cdots, x_{t_1+l-1}]^T$ and $[x_{t_2-l+1}, x_{t_2-l+2}, \cdots, x_{t_2}]^T$, respectively.

2.2 Multi-codebook Generation

After segmenting time series into ITSs, codebooks can be generated through VQ. Given the training time series set $X = \{(\mathbf{x}_i, v_i)\}_{i=1}^{N}$, with $\mathbf{x}_i \in R^M$ and its class label $v_i \in \{1, 2, \cdots, S\}$, we separate it into S subsets, or $X = \bigcup_{s=1}^{S} X_s$, where X_s only contains time series belonging to class s, $|X_s| = N_s$, S is the number of classes, M is the length of samples and N is the number of samples. For each time series $\mathbf{x}_i$ in the set X, we choose W PIPs and then obtain the corresponding ITSs. The length l of ITSs needs to be determined in advance. Without loss of generality, let $l = M/W$, where $W \ll M$ and $l > 1$.

In MCITSA, each class has its own codebook instead of one codebook generated by the whole dataset. However an ITS cannot be approximately reconstructed by a codeword for missing location information. Fortunately, this problem could be solved by marking the location information. Here, we roughly divide the time set $\{t_1, \cdots, t_M\}$ into three subsets equally, T_1,T_2 and T_3. If the time t_i of the PIP y_i belongs to T_1, then y_i belongs to the first location. Essentially, we mainly divide ITSs generated by PIPs into three locations. For class s, there are $(3W - 4) * N_s$ ITSs totally. We use the K-means clustering algorithm in different locations and proportionally generate the corresponding codewords. The codebook C_s of class s consists of codewords from three locations. In addition, $|C_s| = K$. Let $\mathbf{c}_q^{s,u}$ be the qth codeword in the uth location for class s, where $\mathbf{c}_q^{s,u} \in R^l$.

2.3 Reconstruction

Once S codebooks are generated, we can approximate important time subsequences of any time series. As the multiple codebooks contain the label information, we do not need to reconstruct training time series. Moreover, by reconstruction here does not mean reconstruct the whole time series. Due to PIPs, all we need to do is to reconstruct important time subsequences.

Given an unseen time series $\mathbf{x} \in R^M$, we need to reconstruct its ITSs with a minimal fitting error. We also find W PIPs for $\mathbf{x}$ and then construct W ITSs to represent it, or $(\mathbf{x}^1, u_1), (\mathbf{x}^2, u_2), \cdots, (\mathbf{x}^W, u_W)$ where $u_j \in \{1, 2, 3\}$ denotes the location information of the jth ITS $\mathbf{x}^j$ of $\mathbf{x}$. We use the S codebooks to represent these ITSs, respectively. Certainly, we only compare codewords which have the same location information with $\mathbf{x}^j$. We find the most similar codeword in the codebook C_s to substitute the ITS $\mathbf{x}^j$. Namely, if

$$q = \arg\min_k \; d(\mathbf{x}^j, \mathbf{c}_k^{s,u_j}) \tag{1}$$

then $\widehat{\mathbf{x}}^{j,s} = \mathbf{c}_q^{s,u_j}$. Therefore, $\mathbf{x}^1, \mathbf{x}^2, \cdots, \mathbf{x}^W$ can be represented as $\widehat{\mathbf{x}}^{1,s}, \widehat{\mathbf{x}}^{2,s}, \cdots, \widehat{\mathbf{x}}^{W,s}$ in class s, respectively.

2.4 Classification

According to the classification criterion of sparse representation classifier [14], we calculate the total distance between the original ITSs and their approximations, and assign the class label with the minimal reconstructed error. If

$$\hat{v} = \arg\min_{s=1,2\cdots,S} \sum_{j=1}^{W} d(\mathbf{x}^j, \widehat{\mathbf{x}}^{j,s}) \tag{2}$$

then $\hat{v}$ is the estimated label for $\mathbf{x}$.

3 Simulation Experiments

This section verifies the classification performance of MCITSA on time series. Since PVQA is better than the Euclidean distance and the piecewise approximation based on PCA and SAX on many real and synthetic datasets [1], we only compare the performance of time series classification using PVQA and MCITSA. All the time series datasets used in this paper can be found in [18]. PVQA needs to reconstruct all the training time series and the unseen time series completely. MCITSA only needs to reconstruct the ITSs of unseen time series using the codebooks. For PVQA, we use the nearest neighbor (NN) algorithm to classify the unseen time series. For MCITSA, the minimal reconstructed error rule (2) is employed to perform classification tasks.

3.1 Trace Dataset

The Trace dataset was generated on the basis of data obtained from EDF (Electricité de France) [17]. In the Trace dataset, there are 100 training time series and 100 test time series. The length of each time series is 275. There are four categories.

Let $W = 11$ and $K = 32$. The total number of codewords for both PVQA and MCITSA is 32. PVQA directly uses the whole training time subsequence to generate a codebook as shown in Fig. 2(a). There is only 8 codewords for each class in MCITSA for a fair comparison. Each row in Fig. 2(b) represents 8 codewords for each class obtained by MCITSA. It is easy to observe that some codewords generated by PVQA are similar to ones of MCITSA since PVQA extracts features from the whole time subsequences. More importantly, some codewords generated by MCITSA have distinct characteristic compared with PVQA, such as the 8th codeword in Fig. 2(b).

The total number of codewords K varies in the set $\{32, 48, 64, 80\}$. Since MCITSA has four codebooks, the number of codewords in each class is 8, 12, 16, and 20, respectively. This process can make the comparison of methods fairer. W only takes values in the set $\{5, 11\}$ on account of the limitation of PVQA. The classification performance is measured by the classification error rate (CER) defined as

$$CER = \frac{\textit{The number of wrong classified time series}}{\textit{The number of time series in the test dataset}} \times 100\,\% \tag{3}$$

The comparison of these two methods is shown in Fig. 3. The classification error decreases as the increase of both K and W, which is consistent with the conclusion summarized in [10]. MCITSA outperforms PVQA in most cases. The best CER on the Trace dataset is 6 % achieved by MCITSA when $K = 20$ for each class and $W = 11$. For PVQA, the best one is 19 % achieved when $K = 64$ and $W = 5$.

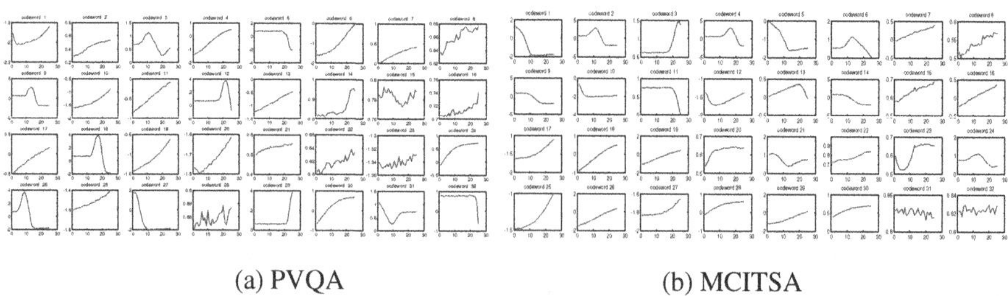

(a) PVQA (b) MCITSA

Fig. 2. Codebook of the trace dataset obtained by PVQA (a), and MCITSA (b), when total number $K = 32$ and $W = 11$

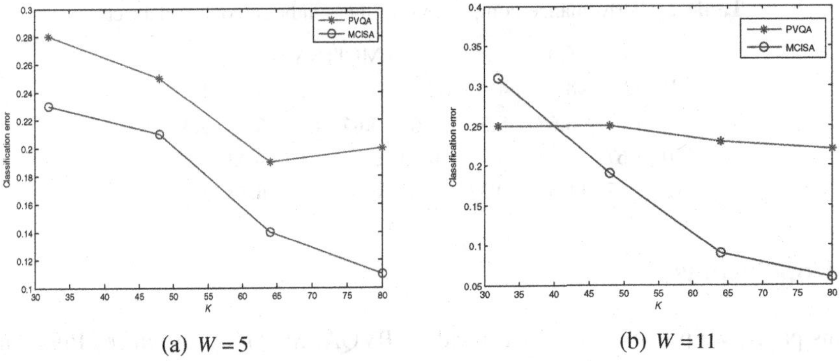

(a) $W = 5$ (b) $W = 11$

Fig. 3. The comparison of classification error vs. K on the trace dataset under different W.

3.2 More Datasets

In order to demonstrate the robust of MCITSA, we also perform experiments on other two time series datasets, CBF and Synthetic Control. In the following, K of PVQA also varies in the set $\{32, 48, 64, 80\}$. For MCITSA, the number of codewords in each class is $\{8, 12, 16, 20\}$, respectively.

In the CBF dataset, there are 30 training time series and 900 test time series. The length of each time series is 128. The number of categories for CBF is 3. And the total number of codewords for MCITSA is $\{24, 36, 48, 60\}$, respectively. Now, we segment the time series into $W \in \{4, 8, 16\}$ subsequences. The comparison of these two methods is shown in Table 1. Obviously, MCITSA also outperforms PVQA as a whole. For example, MCITSA has the better performance over $11/12$ comparison while PVQA achieves the best CER only when $W = 8$ and $K = 32$.

Table 1. Performance comparison on the CBF dataset

	PVQA(%)				MCITSA(%)			
W	32	48	64	80	24	36	48	60
4	12.1	11.5	11.33	14.11	**11.56**	**6.56**	**6.33**	**5.22**
8	**10.5**	8.67	10.33	10.78	13.33	**6.44**	**3.67**	**3.33**
16	7.78	8.44	7.67	10.11	**7.67**	**5.89**	**6**	**2.78**

The Synthetic Control dataset contains 300 training time series and 300 test time series. The length of each time series is 60. The number of categories in this dataset is 6. Here, the total number of codewords in MCITSA are $\{48, 72, 96, 120\}$, respectively. Now, we segment the time series into $W \in \{6, 10, 15\}$ subsequences. The comparison of these two methods is shown in Table 2. In this dataset, MCITSA outperforms PVQA in all case.

Table 2. Performance comparison on the synthetic control dataset

	PVQA(%)				MCITSA(%)			
W	32	48	64	80	48	72	96	120
6	7	6.67	8.33	7.67	**5.67**	**4.67**	**5**	**6.33**
10	8.67	7	9	9.33	**5**	**5.67**	**4.33**	**3.67**
15	9.67	11.3	10.33	10.67	**6**	**6.33**	**6.33**	**5**

4 Conclusions

In this paper, we propose MCITSA based on PVQA. MCITSA introduces PIPs when segmenting time series in order to extract shape features. The location information of ITSs is considered in this stage. Furthermore, MCITSA considers the difference between time subsequences in different classes which may affect the generation of the codebook. Thus, MCITSA generates a codebook using ITSs in the same location and belonging to the same class. Meanwhile, MCITSA does not need to reconstruct the whole time series as PVQA does. The reconstruction error minimization rule is taken as the classification rule. Experimental results on three time series datasets indicate that MCITSA is much better than PVQA on the classification performance of time series.

Since the location information is user-defined, we will try to confirm the benefit of the local information the following research. In addition, since MCITSA uses the K-means clustering to generate codebook, the performance can be affected because of the instability of K-means. In the future, we plan to develop a method to generate codebook stably and effectively.

Acknowledgments. This work was supported in part by the National Natural Science Foundation of China under Grant Nos. 61373093 and 61402310, by the Natural Science Foundation of Jiangsu Province of China under Grant No. BK20140008, by the Natural Science Foundation of the Jiangsu Higher Education Institutions of China under Grant No. 13KJA520001, and by the Soochow Scholar Project.

References

1. Wang, Q., Megalooikonomou, V.: A dimensionality reduction technique for efficient time series similarity analysis. Inf. Syst. **33**, 115–132 (2008)
2. Serrà, J., Arcos, J.L.: An empirical evaluation of similarity measures for time series classification. Knowl.-Based Syst. **67**, 305–314 (2014)
3. Chan, K.P., Fu, W.C.: Efficient time series matching by wavelets. In: 15th IEEE International Conference on Data Engineering, pp. 126–133, Sydney (1999)
4. Chan, K.P., Fu, W.C., Yu, C.: Haar wavelets for efficient similarity search of time series: with and without time warping. IEEE Trans. Knowl. Data Eng. **15**, 686–705 (2003)
5. Faloutsos, C., Ranganathan, M., Manolopoulos, Y.: Fast subsequence matching in time series databases, In: ACM Sigmode Record, pp. 419–429. ACM Press (1994)
6. Li, H., Yang, L.: Time series visualization based on shape features. Knowl.-Based Syst. **41**, 43–53 (2013)

7. Keogh, E., Lin, J., Fu, A.: Hot SAX: efficiently finding the most unusual time series subsequence. In: 5th IEEE International Conference on Data Mining, pp. 226–233 (2005)
8. Lin, J., Keogh, E., Li, W., Lonardi, S.: Experiencing SAX: a novel symbolic representation of time series. Data Min. Knowl. Discov. **15**, 107–144 (2007)
9. Lathauwer, L.D., Moor, B.D., Vandewalle, J.: A multilinear singular value decomposition. SIAM J. Matrix Anal. Appl. **21**, 1253–1278 (2000)
10. Li, H., Yang, L., Guo, C.: Improved piecewise vector quantized approximation based on normalized time subsequences. Measurement **46**, 3429–3439 (2013)
11. Górecki, T., Łuczak, M.: Multivariate time series classification with parametric derivative dynamic time warping. Expert Syst. Appl. **42**, 2305–2312 (2014)
12. Sasikala, I., Banu, I.N.: Privacy preserving data mining using piecewise vector quantization (PVQ). Int. J. Adv. Res. Comput. Sci. Technol. **2**, 302–306 (2014)
13. Lu, Z.M., Wang, J.X., Liu, B.B.: An improved lossless data hiding scheme based on image VQ-index residual value coding. J. Syst. Softw. **82**, 1016–1024 (2009)
14. Zhang, L., Zhou, W.D., Chang, P.C.: Kernel sparse representation-based classifier. IEEE Trans. Sig. Process. **60**, 1684–1695 (2012)
15. Fu, T.C., Chung, F.L., Kwok, K.Y.: Stock time series visualization based on data point importance. Eng. Appl. Artif. Intell. **21**, 1217–1232 (2008)
16. Tsinaslanidis, P.E., Kugiumtzis, D.: A prediction scheme using perceptually important points and dynamic time warping. Expert Syst. Appl. Intell. J. **41**, 6848–6860 (2014)
17. Roverso, D.: Multivariate temporal classification by windowed wavelet decomposition and recurrent neural networks. In: 3rd ANS International Topical Meeting on Nuclear Plant Instrumentation, Control and Human-Machine Interface, vol. 20 (2000)
18. Welcome to the UCR time series classification/clustering page (2003). http:// www.cs.ucr.edu/~eamonn/time_series_data/
19. Wang, Q., Megalooikonomou, V., Faloutsos, C.: Time series analysis with multiple resolutions. Inf. Syst. **35**, 56–74 (2010)

Performance Improvement via Bagging in Ensemble Prediction of Chaotic Time Series Using Similarity of Attractors and LOOCV Predictable Horizon

Mitsuki Toidani, Kazuya Matsuo, and Shuichi Kurogi(✉)

Kyushu Institute of Technology, Tobata, Kitakyushu, Fukuoka 804-8550, Japan
toidani@kurolab.cntl.kyutech.ac.jp, {matsuo,kuro}@cntl.kyutech.ac.jp
http://kurolab.cntl.kyutech.ac.jp

Abstract. Recently, we have presented a method of ensemble prediction of chaotic time series. The method employs strong learners capable of making predictions with small error and usual ensemble mean does not work for long term prediction owing to the long term unpredictability of chaotic time series. Thus, the method uses similarity of attractors to select plausible predictions from original predictions generated by strong leaners, and then calculates LOOCV (leave-one-out cross-validation) measure to estimate predictable horizons. Finally, it provides representative prediction and an estimation of the predictable horizon. We have used CAN2s (competitive associative nets) for learning piecewise linear approximation of nonlinear function as strong learners in the previous study, and this paper employs bagging of them to improve the performance, and shows the validity and the effectiveness of the method.

Keywords: Ensemble prediction of chaotic time series · Long-term unpredictability · Attractors of chaotic time series · Leave-one-out cross-validation · Estimation of predictable horizon

1 Introduction

Recently, we have presented a method of ensemble prediction of chaotic time series [1]. Here, from [2], the probabilistic prediction has come to dominate the science of weather and climate forecasting, mainly because the theory of chaos at the heart of meteorology shows that for a simple set of nonlinear equations (or Lorenz's equations shown below) with initial conditions changed by minute perturbations, there is no longer a single deterministic solution and hence all forecasts must be treated as probabilistic. Although most of the methods shown in [2] use ensemble mean for representative forecast, our method in [1] selects representative individual prediction from a set of plausible predictions because our method employs strong learners capable of making predictions with small error and we have empirically found individual predictions showing better performance than ensemble mean. This is because ensemble mean does not work for long term prediction due to the long term unpredictability of chaotic time series.

A. Hirose et al. (Eds.): ICONIP 2016, Part IV, LNCS 9950, pp. 590–598, 2016.
DOI: 10.1007/978-3-319-46681-1_70

Our method in [1] employs similarity of attractors to select plausible predictions, and an LOOCV (leave-one-out cross-validation) method to estimate predictable horizon. Comparing with our previous methods embedding model selection techniques using MSE (mean square prediction error) [3,4], the method in [1] has an advantage that it provides representative prediction which is estimated to have the longest predictable horizon among plausible predictions for each start time of prediction. Furthermore it has provided long predictable horizons on average, while there are several cases providing short predictable horizons. It seems hard to have longer predictable horizons for such cases mainly because the analysis of LOOCV predictable horizon has not been done sufficiently, so far.

This paper describes performance improvement to have longer predictable horizons by means of using bagging learning machines, and show the analysis of LOOCV predictable horizon. Here, the bagging is known to use ensemble mean with an ability to reduce the variance of predictions by single learning machines, and then we can expect that the performance in time series prediction becomes more stable. Note that the bagging ensemble is employed for iterated one-step ahead (IOS) prediction of time series, and the present ensemble is developed for longer term prediction of time series. Furthermore, we use CAN2 (competitive associative net 2) as a learning machine, where CAN2 has been introduced for learning piecewise linear approximation of nonlinear function and the high performance in regression problems has been shown in Evaluating Predictive Uncertainty Challenge [5].

We show the present method of ensemble prediction of chaotic time series in Sect. 2, experimental results and analysis in Sect. 3, and the conclusion in Sect. 4.

2 Ensemble Prediction of Chaotic Time Series

2.1 IOS Prediction of Chaotic Time Series

Let y_t ($\in \mathbb{R}$) denote a chaotic time series for a discrete time $t = 0, 1, 2, \cdots$ satisfying

$$y_t = r(\boldsymbol{x}_t) + e(\boldsymbol{x}_t), \tag{1}$$

where $r(\boldsymbol{x}_t)$ is a nonlinear target function of a vector $\boldsymbol{x}_t = (y_{t-1}, y_{t-2}, \cdots, y_{t-k})^T$ generated by k dimensional delay embedding from a chaotic differential dynamical system (see [6] for the theory of chaotic time series). Here, y_t is obtained not analytically but numerically, and then y_t involves an error $e(\boldsymbol{x}_t)$ owing to an executable finite calculation precision. This indicates that there are a number of plausible target functions $r(\boldsymbol{x}_t)$ with allowable error $e(\boldsymbol{x}_t)$. Furthermore, in general, a time series generated with higher precision has small prediction error for longer duration of time from the initial time of prediction. Thus, let a time series generated with a high precision (or 128 bit precision; see Sect. 3 for details), be ground truth time series $y_t^{[\mathrm{gt}]}$, while we examine the predictions generated with standard 64 bit precision.

Let $y_{t:h} = y_t y_{t+1} \cdots y_{t+h-1}$ denote a time series with the initial time t and the horizon h. For a given training time series $y_{t_\mathrm{g}:h_\mathrm{g}} (= y_{t_\mathrm{g}:h_\mathrm{g}}^{[\mathrm{train}]})$, we are supposed to predict succeeding time series $y_{t_\mathrm{p}:h_\mathrm{p}}$ for $t_\mathrm{p} \geq t_\mathrm{g} + h_\mathrm{g}$. Then, we make the training dataset $D^{[\mathrm{train}]} = \{(\boldsymbol{x}_t, y_t) \mid t \in I^{[\mathrm{train}]}\}$ for $I^{[\mathrm{train}]} = \{t \mid t_\mathrm{g} + k \leq t < t_\mathrm{g} + h_\mathrm{g}\}$ to train a learning machine. After the learning, the machine executes IOS prediction by

$$\hat{y}_t = f(\boldsymbol{x}_t) \tag{2}$$

for $t = t_\mathrm{p}, t_{\mathrm{p}+1}, \cdots$, recursively, where $f(\boldsymbol{x}_t)$ denotes prediction function of $\boldsymbol{x}_t = (x_{t1}, x_{t2}, \cdots, x_{tk})$ whose elements are given by $x_{tj} = y_{t-j}$ for $t - j < t_\mathrm{p}$ and $x_{tj} = \hat{y}_{t-j}$ for $t - j \geq t_\mathrm{p}$. Here, we suppose that y_t for $t < t_\mathrm{p}$ is known as the initial state for making the prediction $\hat{y}_{t_p:h_p}$. As explained above, we execute the prediction with standard 64 bit precision, and we may say that there are a number of plausible prediction functions $f(\boldsymbol{x}_t)$ with small error for a duration of time from the initial time of prediction by means of using strong learning machines.

2.2 Single CAN2 and the Bagging for IOS Prediction

We use CAN2 as a learning machine. A single CAN2 has N units. The jth unit has a weight vector $\boldsymbol{w}_j \triangleq (w_{j1}, \cdots, w_{jk})^T \in \mathbb{R}^{k\times 1}$ and an associative matrix (or a row vector) $\boldsymbol{M}_j \triangleq (M_{j0}, M_{j1}, \cdots, M_{jk}) \in \mathbb{R}^{1\times(k+1)}$ for $j \in I^N \triangleq \{1, 2, \cdots, N\}$. The CAN2 after learning the training dataset $D^{[\mathrm{train}]} = \{(\boldsymbol{x}_t, y_t) \mid t \in I^{[\mathrm{train}]}\}$ approximates the target function $r(\boldsymbol{x}_t)$ by

$$\widehat{y}_t = \widetilde{y}_{c(t)} = \boldsymbol{M}_{c(t)} \widetilde{\boldsymbol{x}}_t, \tag{3}$$

where $\widetilde{\boldsymbol{x}}_t \triangleq (1, \boldsymbol{x}_t^T)^T \in \mathbb{R}^{(k+1)\times 1}$ denotes the (extended) input vector to the CAN2, and $\widetilde{y}_{c(t)} = \boldsymbol{M}_{c(t)} \widetilde{\boldsymbol{x}}_t$ is the output value of the $c(t)$th unit of the CAN2. The index $c(t)$ indicates the unit who has the weight vector $\boldsymbol{w}_{c(t)}$ closest to the input vector $\boldsymbol{x}_t$, or $c(t) \triangleq \underset{j\in I^N}{\mathrm{argmin}} \, \|\boldsymbol{x}_t - \boldsymbol{w}_j\|$. Note that the above prediction performs piecewise linear approximation of $y = r(\boldsymbol{x})$ and N indicates the number of piecewise linear regions. We use the learning algorithm shown in [7] whose high performance in regression problems has been shown in Evaluating Predictive Uncertainty Challenge [5].

We obtain bagging prediction by means of using a number of single CAN2s as follows (see [8,9] for details); let $D^{[n\alpha^\sharp, j]} = \{(\boldsymbol{x}_t, y_t) \mid t \in I^{[n\alpha^\sharp, j]})\}$ be the jth bag (multiset, or bootstrap sample set) involving $n\alpha$ elements, where the elements in $D^{[n\alpha^\sharp, j]}$ are resampled randomly with replacement from the training dataset $D^{[\mathrm{train}]}$ involving $n = |D^{[\mathrm{train}]}|$ elements. Here, α (> 0) indicates the bag size ratio to the given dataset, and $j \in J^{[\mathrm{bag}]} \triangleq \{1, 2, \cdots, b\}$. Here, note that $\alpha = 1$ is used in many applications (see [9,10]), which we use in the experiments shown below after the tuning of α (see [9] for validity and effectiveness of using variable α). Using multiple CAN2s employing N units after leaning $D^{[n\alpha^\sharp,j]}$, which we denote $\theta_N^{[j]}$ ($\in \Theta_N \triangleq \{\theta_N^{[j]} \mid j \in J^{[\mathrm{bag}]}\}$), the bagging for predicting the target value $r_t = r(\boldsymbol{x}_t)$ is done by

$$\hat{y}_t^{[\theta_N]} \triangleq \frac{1}{b} \sum_{j\in J^{[\mathrm{bag}]}} \hat{y}_t^{[j]} \equiv \left\langle \hat{y}_t^{[j]} \right\rangle_{j\in J^{[\mathrm{bag}]}} \tag{4}$$

where $\hat{y}_t^{[j]} \triangleq \hat{y}^{[j]}(\boldsymbol{x}_t)$ denotes the prediction by the jth machine $\theta_N^{[j]}$. The angle brackets $\langle\cdot\rangle$ indicate the mean, and the subscript $j \in J^{\mathrm{bag}}$ indicates the range of the mean. For simple expression, we sometimes use $\langle\cdot\rangle_j$ instead of $\langle\cdot\rangle_{j\in J^{[\mathrm{bag}]}}$ in the following.

2.3 Ensemble Prediction and Estimation of Predictable Horizon

Similarity of Attractors to Select Plausible Predictions. First, we make a number of IOS predictions $\hat{y}_{t_p:h_p}$ by means of using learning machines with different parameter values, i.e., we use $\hat{y}_{t_p:h_p} = y_{t_p:h_p}^{[\theta_N]}$ generated by CAN2s with different number of units $\theta_N \in \Theta$, where Θ indicates the set of all learning machines. We suppose that there are a number of plausible prediction functions $f(\cdot) = f^{[\theta_N]}(\cdot)$, and we have to remove implausible ones. To have this done, we select the following set of plausible predictions:

$$Y_{t_p:h_p}^{[S_{\mathrm{th}}]} = \left\{ y_{t_p,h_p}^{[\theta_N]} \;\middle|\; S\left(y_{t_p,h_p}^{[\theta_N]}, y_{t_g:h_g}^{[\mathrm{train}]}\right) \geq S_{\mathrm{th}}, \theta_N \in \Theta \right\} \tag{5}$$

where

$$S\left(y_{t_p,h_p}^{[\theta_N]}, y_{t_g:h_g}^{[\mathrm{train}]}\right) \triangleq \frac{\sum_i \sum_j a_{ij}^{[\theta_N]} a_{ij}^{[\mathrm{train}]}}{\sqrt{\sum_i \sum_j \left(a_{ij}^{[\theta_N]}\right)^2} \sqrt{\sum_i \sum_j \left(a_{ij}^{[\mathrm{train}]}\right)^2}} \tag{6}$$

denotes the similarity of two-dimensional attractor (trajectory) distributions $a_{ij}^{[\theta_N]}$ and $a_{ij}^{[\mathrm{train}]}$ of time series $y_{t_p,h_p}^{[\theta_N]}$ and $y_{t_g:h_g}^{[\mathrm{train}]}$, respectively, and S_{th} is a threshold. Here, the two-dimensional attractor distribution, a_{ij}, of a time-series $y_{t:h}$ is given by

$$a_{ij} = \sum_{s=t}^{t+h-1} \mathbf{1}\left\{ \left\lfloor \frac{y_s - v_0}{\Delta_a} \right\rfloor = i \wedge \left\lfloor \frac{y_{s+1} - v_0}{\Delta_a} \right\rfloor = j \right\}, \tag{7}$$

where v_0 is a constant less than the minimum value of y_t for all time series and Δ_a indicates a resolution of the distribution. Furthermore, $\mathbf{1}\{z\}$ is an indicator function equal to 1 if z is true, and 0 if z is false, and $\lfloor\cdot\rfloor$ indicates the floor function.

LOOCV Measure to Estimate Predictable Horizons. Let us define predictable horizon between two predictions $y^{[\theta_N]}_{t_p:h_p}$ and $y^{[\theta_{N'}]}_{t_p:h_p}$ in $Y^{[S_{\rm th}]}_{t_p:h_p}$ as

$$h\left(y^{[\theta_N]}_{t_p:h_p}, y^{[\theta_{N'}]}_{t_p:h_p}\right) = \max\left\{h \mid \forall s < h \leq h_p; |y^{[\theta_N]}_{t_p+s} - y^{[\theta_{N'}]}_{t_p+s}| \leq e_y\right\}, \tag{8}$$

where e_y indicates the threshold of prediction error to determine the horizon. Then, we employ LOOCV method to estimate predictable horizon of $y^{[\theta_N]}_{t_p:h_p}$ in $Y^{[S_{\rm th}]}_{t_p:h_p}$. Namely, we use

$$\begin{aligned}\tilde{h}^{[\theta_N]}_{t_p} &= h\left(y^{[\theta_N]}_{t_p:h_p}, Y^{[S_{\rm th}]}_{t_p:h_p} \setminus \{y^{[\theta_N]}_{t_p:h_p}\}\right) \\ &= \left\langle h\left(y^{[\theta_N]}_{t_p:h_p}, y^{[\theta_{N'}]}_{t_p:h_p}\right)\right\rangle_{y^{[\theta_{N'}]}_{t_p:h_p} \in Y^{[S_{\rm th}]}_{t_p:h_p} \setminus \{y^{[\theta_N]}_{t_p:h_p}\}},\end{aligned} \tag{9}$$

which we call LOOCV measure of predictable horizon or LOOCV predictable horizon. In this context, we expect that $h\left(y^{[\theta_N]}_{t_p:h_p}, Y^{[S_{\rm th}]}_{t_p:h_p} \setminus \{y^{[\theta_N]}_{t_p:h_p}\right)$ and $h\left(y^{[\theta_N]}_{t_p:h_p}, y^{[\rm gt]}_t\right)$ have positive correlation because predictions in $Y^{[S_{\rm th}]}_{t_p:h_p}$ are all plausible, although this expectation is ad hoc.

Representative Prediction and Estimation of Predictable Horizon. We sort LOOCV predictable horizons by their lengths as $\tilde{h}^{[\theta_{\sigma(i)}]}_{t_p} \geq \tilde{h}^{[\theta_{\sigma(i+1)}]}_{t_p}$, where $\sigma(i)$ denotes the order for $i = 1, 2, \cdots, |Y^{[S_{\rm th}]}_{t_p:h_p}|$. Then, we select $y^{[\theta_{\sigma(1)}]}_{t_p:h_p}$ as representative prediction, and provide an estimation of $h\left(y^{[\theta_{\sigma(1)}]}_{t_p:h_p}, y^{[\rm gt]}_t\right)$ by

$$\hat{h}^{[\theta_{\sigma(1)}]}_{t_p} = \min\left\{h(y^{[\theta_{\sigma(1)}]}_{t_p:h_p}, y^{[\theta]}_{t_p:h_p}) \mid \forall y^{\theta}_{t_p:h_p} \in Y^{[H_{th}, S_{th}]}_{t_p:h_p} \setminus y^{[\theta_{\sigma(1)}]}_{t_p:h_p}\right\}, \tag{10}$$

where

$$Y^{[H_{\rm th}, S_{\rm th}]}_{t_p:h_p} = \left\{ y^{[\theta_{\sigma(i)}]}_{t_p:h_p} \;\middle|\; \frac{i}{|Y^{[S_{\rm th}]}_{t_p:h_p}|} \leq H_{\rm th} \right\}. \tag{11}$$

Here, the threshold $H_{\rm th}$ $(0 < H_{\rm th} \leq 1)$ indicates the ratio of the number of elements in $Y^{[H_{\rm th}, S\ ]}_{t_p:h_p}$ and $Y^{[S_{\rm th}]}_{t_p:h_p}$, or $H_{\rm th} = \left|Y^{[H_{\rm th}, S\ ]}_{t_p:h_p}\right| / \left|Y^{[S_{\rm th}]}_{t_p:h_p}\right|$. We can tune $H_{\rm th}$ depending on the required safeness of the estimation of predictable horizon. Here, the safe estimation of $\hat{h}^{[\theta_{\sigma(1)}]}_{t_p}$ indicates that $\hat{h}^{[\theta_{\sigma(1)}]}_{t_p}$ is smaller than or equal to the actual predictable horizon $h\left(y^{[\theta_{\sigma(1)}]}_{t_p:h_p}, y^{[\rm gt]}_t\right)$, and we can see that $\hat{h}^{[\theta_{\sigma(1)}]}_{t_p}$ become safer with the increase of $H_{\rm th}$.

3 Numerical Experiments and Analysis

3.1 Experimental Settings

We use the Lorenz time series, as shown in Fig. 1 and [4], obtained from the original differential dynamical system given by

$$\frac{dx_c}{dt_c} = -\sigma x_c + \sigma y_c, \quad \frac{dy_c}{dt_c} = -x_c z_c + r x_c - y_c, \quad \frac{dz_c}{dt_c} = x_c y_c - b z_c, \tag{12}$$

for $\sigma = 10$, $b = 8/3$, $r = 28$. Here, we use t_c for continuous time and t ($= 0, 1, 2, \cdots$) for discrete time related by $t_c = tT$ with sampling time T. We have generated the time series $y(t) = x_c(tT)$ for $t = 1, 2, \cdots, 5000$ from the initial state $(x_c(0), y_c(0), z_c(0)) = (-8, 8, 27)$ with $T = 25\,\text{ms}$ via Runge-Kutta method with 128 bit precision of GMP (GNU multi-precision library). As a result of preliminary experiments as shown in [4], $y(t)$ for each duration of time less than 1,200 steps ($= 30\,\text{s}/25\,\text{ms}$) in Fig. 1, or $y_{t_0:1200}$ for each initial time $t_0 = 0, 1, 2, \cdots$ with initial state $(x(t_0), y(t_0), z(t_0))$, is supposed to be correct, while cumulative computational error may increase exponentially after the duration.

We use $y_{0:2000}$ for training learning machines (single and bagging CAN2s), and execute multistep prediction of $y_{t_p:h_p}$ with the initial input vector $\boldsymbol{x}_{t_p} = (y(t_p - 1), \cdots, y(t_p - k))$ for prediction start time $t_p = 2000 + 100i$ ($i = 0, 1, 2, \cdots, 19$) and prediction horizon $h_p = 500$. We show the results for the embedding dimension $k = 10$, while the results for $k = 8$ are shown in [1] and both are not significantly but slightly different.

3.2 Results and Analysis

For generating original predictions, we use single and bagging CAN2s with the number of units being $N = 5 + 20i$ ($i = 0, 1, 2, \cdots, 14$). For detailed understanding, we show an example of predictions $\hat{y}^{[\theta_N]}_{t_p:h_p}$ for $t_p = 2300$ in Fig. 2(a). Note that, as shown below (see Fig. 3(a)), $t_p = 2300$ is the start time of representative prediction $\hat{h}^{[\theta_{\sigma(1)}]}_{t_p}$ with predictable horizon being smaller than 100 by single CAN2 (actually $\hat{h}^{[\theta_{\sigma(1)}]}_{t_p:h_p} = 72$) and improved by bagging CAN2 as $\hat{h}^{[\theta_{\sigma(1)}]}_{t_p} = 183$.

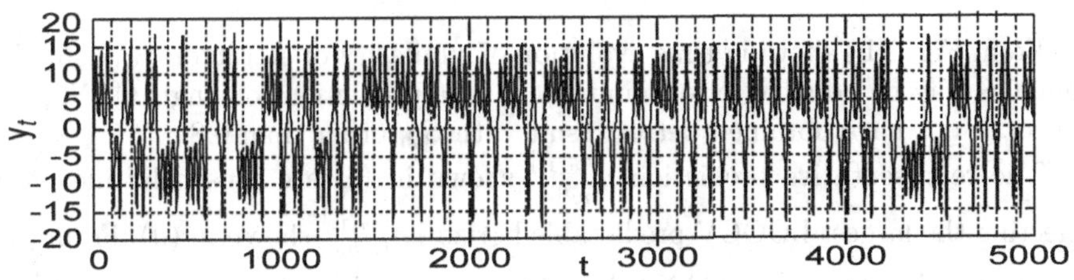

Fig. 1. Lorenz time series $y(t)$ for $t = 0, 1, 2, \cdots, 4999$, or ground truth time series $y^{[\text{gt}]}_{0:5000}$.

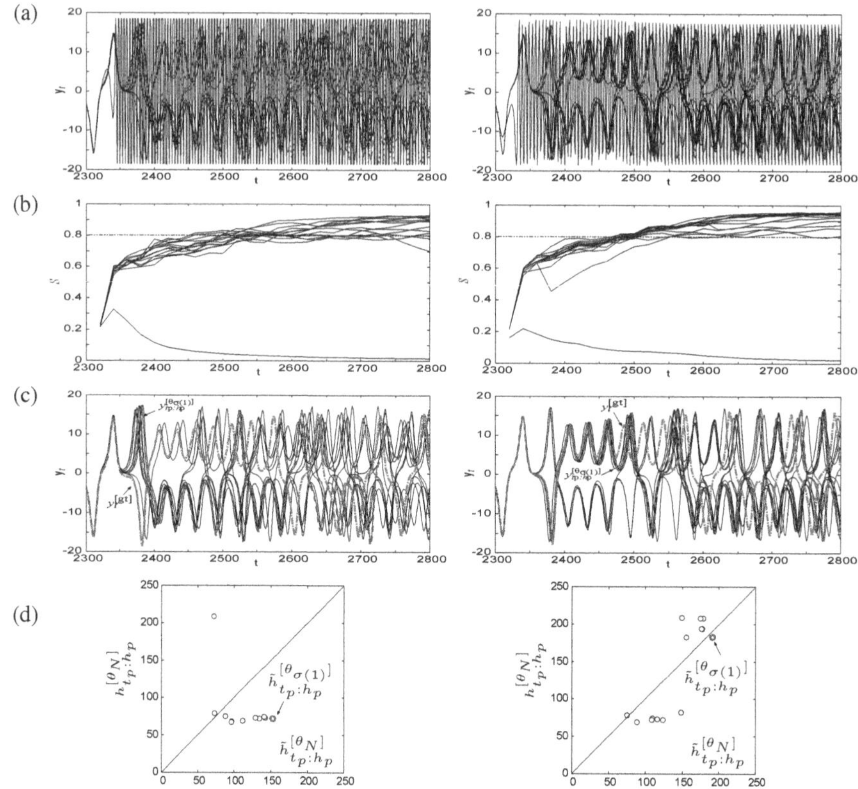

Fig. 2. Experimental results obtained by single CAN2s (left) and bagging CAN2s (right) for the prediction start time $t_p = 2300$ and the horizon $h_p = 500$. The top row, (a), shows superimposed original predictions $y_{t_p:h_p}^{[\theta_N]}$. (b) shows time evolution of similarity S of attractors. The predictions with $S \geq S_{\rm th} = 0.8$ at $t = t_p + h_p - 1 = 2799$ are selected as plausible predictions. (c) shows selected plausible predictions $y_{t_p:h_p}^{[\theta_N]}$ as well as ground truth time series $y_t^{[\rm gt]}$ (red) and representative prediction $y_{t_p:h_p}^{[\theta_{\sigma(1)}]}$ (green). (d) shows the relationship between actual predictable horizons $h_{t_p:h_p}^{[\theta_N]}$ and LOOCV predictable horizons $\tilde{h}_{t_p:h_p}^{[\theta_N]}$ of plausible predictions. (Color figure online)

From Fig. 2(b) showing time evolution of similarity of attractors, $S_{\rm th}$ smaller than 0.8 may be possible for obtaining plausible predictions of single CAN2, but we could not have obtained better result with $S_{\rm th}$ other than 0.8.

The representative prediction $y_{t_p:h_p}^{[\theta_{\sigma(1)}]}$ shown in (c) are chosen by means of selecting the largest LOOCV predictable horizon $\tilde{h}_{t_p:h_p}^{[\theta_{\sigma(1)}]}$ shown in (d). From (d), we can see that the single CAN2 (left) has the actual predictable horizon $h_{t_p:h_p}^{[\theta_N]}$ larger than 200 and LOOCV predictable horizon $\tilde{h}_{t_p:h_p}^{[\theta_N]}$ smaller than 100, actually

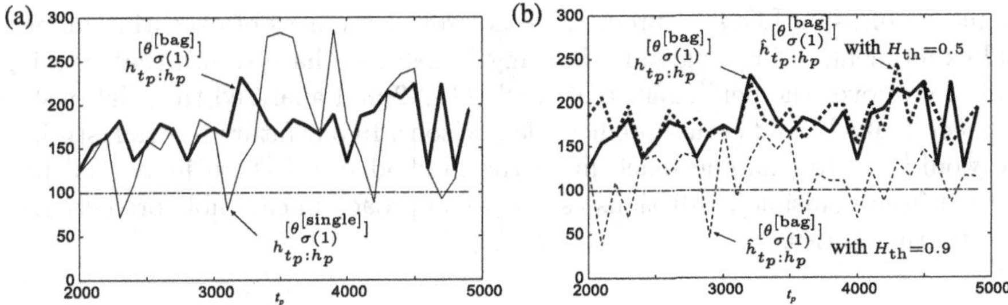

Fig. 3. Experimental result of (a) actual predictable horizons $h_{t_p:h_p}^{[\theta_{\sigma(1)}^{[\text{single}]}]}$ and $h_{t_p:h_p}^{[\theta_{\sigma(1)}^{[\text{bag}]}]}$, and (b) estimated predictable horizon $\hat{h}_{t_p:h_p}^{[\theta_{\sigma(1)}^{[\text{bag}]}]}$ with $H_{\text{th}} = 0.9$ and 0.5 for $t_p = 2300$ and $e_y = 10$.

$(h_{t_p:h_p}^{[\theta_N]}, \tilde{h}_{t_p:h_p}^{[\theta_N]}) = (209, 72.1)$. Since the present method selects the prediction with the largest $\tilde{h}_{t_p:h_p}^{[\theta_N]}$, the prediction with $h_{t_p:h_p}^{[\theta_N]} = 209$ could not have selected. On the other hand, we can see that bagging CAN2 (right in (d)) successfully selects the prediction with $h_{t_p:h_p}^{[\theta_N]}$ larger than 100, actually $(h_{t_p:h_p}^{[\theta_N]}, \tilde{h}_{t_p:h_p}^{[\theta_N]}) = (183, 191)$. Precisely, bagging CAN2s have successfully provided large $\tilde{h}_{t_p:h_p}^{[\theta_N]} = 191$ because there are a number of predictions with long predictable horizons around $h_{t_p:h_p}^{[\theta_N]} = 200$ as shown as the group of points neighboring $h_{t_p:h_p}^{[\theta_N]} = 200$ in (d) on the right hand side. Incidentally, from (c), we can see that ensemble mean does not seem appropriate for the representative prediction in long term prediction of chaotic time series.

In Fig. 3, we show the results of actual and estimated predictable horizons. From (a), we can see that the performance of single CAN2 is improved by bagging CAN2 from the point of view that the former has four actual predictable horizons $h_{t_p:h_p}^{[\theta_{\sigma(1)}^{[\text{single}]}]}$ smaller than 100 among all predictions for $t_{\text{p}} = 2000 + 100i$ ($i = 0, 1, 2, \cdots, 19$) and bagging CAN2 has achieved all $h_{t_p:h_p}^{[\theta_{\sigma(1)}^{[\text{bag}]}]}$ larger than 100. From (b), we can see that the estimated predictable horizon $\hat{h}_{t_p:h_p}^{[\theta_{\sigma(1)}]}$ with $H_{\text{th}} = 0.5$ is almost the same as actual predictable horizon $h_{t_p:h_p}^{[\theta_{\sigma(1)}]}$, while $H_{\text{th}} = 0.9$ has achieved safe estimation, or $\hat{h}_{t_p:h_p}^{[\theta_{\sigma(1)}]} \leq h_{t_p:h_p}^{[\theta_{\sigma(1)}]}$.

4 Conclusion

We have presented a performance improvement of the method for ensemble prediction of chaotic time series by means of using bagging learning machines. The method obtains a set of plausible predictions by means of using similarity of attractors between training and predicted time series. And then, it provides representative prediction which is estimated to have the longest predictable horizon by means of using LOOCV predictable horizon. By means of executing numerical experiments using single and bagging CAN2s, we have shown that bagging CAN2 improves the performance of single CAN2 and analyzed the relationship between LOOCV and actual predictable horizons. In our future research studies, we would like to examine much more the method to select an individual prediction from plausible predictions as a novel approach in ensemble prediction of chaotic time series.

References

1. Kurogi, S., Toidani, M., Shigematsu, R., Matsuo, K.: Probabilistic prediction of chaotic time series using similarity of attractors and LOOCV predictable horizons for obtaining plausible predictions. In: Arik, S., Huang, T., Lai, W.K., Liu, Q. (eds.) ICONIP 2015. LNCS, vol. 9491, pp. 72–81. Springer, Heidelberg (2015). doi:10.1007/978-3-319-26555-1_9
2. Slingo, J., Palmer, T.: Uncertainty in weather and climate prediction. Phil. Trans. R. Soc. A **369**, 4751–4767 (2011)
3. Kurogi, S., Ono, K., Nishida, T.: Experimental analysis of moments of predictive deviations as ensemble diversity measures for model selection in time series prediction. In: Lee, M., Hirose, A., Hou, Z.-G., Kil, R.M. (eds.) ICONIP 2013. LNCS, vol. 8228, pp. 557–565. Springer, Heidelberg (2013). doi:10.1007/978-3-642-42051-1_69
4. Kurogi, S., Shigematsu, R., Ono, K.: Properties of direct multi-step ahead prediction of chaotic time series and out-of-bag estimate for model selection. In: Loo, C.K., Yap, K.S., Wong, K.W., Teoh, A., Huang, K. (eds.) ICONIP 2014. LNCS, vol. 8835, pp. 421–428. Springer, Heidelberg (2014). doi:10.1007/978-3-319-12640-1_51
5. Quiñonero-Candela, J., Rasmussen, C.E., Sinz, F., Bousquet, O., Schölkopf, B.: Evaluating predictive uncertainty challenge. In: Quiñonero-Candela, J., Dagan, I., Magnini, B., d'Alché-Buc, F. (eds.) MLCW 2005. LNCS (LNAI), vol. 3944, pp. 1–27. Springer, Heidelberg (2006). doi:10.1007/11736790_1
6. Aihara, K.: Theories and Applications of Chaotic Time Series Analysis. Sangyo Tosho, Tokyo (2000)
7. Kurogi, S., Sawa, M., Tanaka, S.: Competitive associative nets and cross-validation for estimating predictive uncertainty on regression problems. In: Quiñonero-Candela, J., Dagan, I., Magnini, B., d'Alché-Buc, F. (eds.) MLCW 2005. LNCS (LNAI), vol. 3944, pp. 78–94. Springer, Heidelberg (2006). doi:10.1007/11736790_6
8. Breiman, L.: Bagging predictors. Mach. Learn. **26**, 123–140 (1996)
9. Kurogi, S.: Improving generalization performance via out-of-bag estimate using variable size of bags. J. Jpn Neural Netw. Soc. **16**(2), 81–92 (2009)
10. Efron, B., Tbshirani, R.: Improvements on cross-validation: the.632+ bootstrap method. J. Am. Stat. Assoc. **92**, 548–560 (1997)

A Review of EEG Signal Simulation Methods

Muhammad Izhan Noorzi and Ibrahima Faye(✉)

Centre for Intelligent Signal and Imaging Research (CISIR) and
Department of Fundamental and Applied Sciences, Universiti Teknologi PETRONAS,
32610 Bandar Seri Iskandar, Perak, Malaysia
ibrahima_faye@petronas.com.my

Abstract. This paper describes EEG signal simulation methods. Three main methods have been included in this study: Markov Process Amplitude (MPA), Artificial Neural Network (ANN), and Autoregressive (AR) models. Each method is described procedurally, along with mathematical expressions. By the end of the description of each method, the limitations and benefits are described in comparison with other methods. MPA comprises of three variations; first-order MPA, nonlinear MPA, and adaptive MPA. ANN consists of two variations; feed forward back-propagation NN and multilayer feed forward with error back-propagation NN with embedded driving signal. AR model based filtering has been considered with its variation, genetic algorithm based on autoregressive moving average (ARMA) filtering.

Keywords: ANN · AR · ARMA · EEG simulation · MPA

1 Introduction

The behavior of neurophysical activities could be viewed through a comprehensive signal simulation. Usually, it is represented as a function of time, space, or any other independent variables mathematically [1]. The simulation of electrical activity on the scalp [2] or electroencephalography (EEG) has been proved by the researchers [3–5] in getting the broad and comprehensive sense of the nature of the cerebral activities in term of signal before applying analysis in particular signal processing for future research development. It can also be used for forecasting the future neurological outcome and for data compression [6].

EEG consists of nonstationary [4,7], nonlinearity [8] and stochastic properties. Early researchers conceptualize EEG signal and some of them exhibit simple EEG signal [4]. As time progressed, the computational power become advance and thus, realistically achieving some complex problem which was not technologically possible in the past. It also creates new field of knowledge i.e. parallel computing and high performance computing.

Studies has shown that EEG has its own rhythms classified by certain activities which could be simulated with certain technique. To generate a signal according to certain EEG rhythm, that will resemble the real, measured, and recorded

A. Hirose et al. (Eds.): ICONIP 2016, Part IV, LNCS 9950, pp. 599–608, 2016.
DOI: 10.1007/978-3-319-46681-1_71

signal in real time is very challenging. Nevertheless, there exists method to simulate EEG signals with the mentioned constraints.

The first method is Markov Process Amplitude (MPA). Markov process is a type of stochastic process whereby the conditional probability distribution of future states $x(n+1)$ of a particular system is only depend on the present state $x(n)$, without considering the past state of the system [9,10]. Any past history is omitted in the computation. In particular, the significant of EEG simulation using Markov process could be viewed as interpreting the serial dependencies of the EEG signal between the adjacent periods. By analysing the probability of interdependence between the adjacent periods, the next state could thus be predicted (future state) while exploiting the event that precedes it.

The adaptation of human neurological system in the field of artificial intelligence has brought neural network methods. The nonlinearity, complexity, and parallel computing capabilities of brain inspires scientists and researchers forming a groundwork for Artificial Neural Network based approach in Artificial Intelligence. According to Haykin [11], neural network has the capability to manifest the actual brain learning activity, nonlinearity, input-output mapping, adaptivity, and contextual information [11]. The modeling of EEG signal based on ANN has been extensively researched by Adeli et al. [12,13].

The last method considered in this work is based on Autoregressive (AR) model. This model is one where the current value of a variable depends only upon the values that the variable took in previous periods plus an error term [14]). AR model is a part of linear prediction models with the main objective of finding a set of model parameters that best describe the signal generation system [4]. The signal in AR modelling is described to be linearly related with respect to a number of its previous samples.

The structure of this paper is started with the introduction, followed by each methods with its variations. And lastly, the conclusion.

2 Markov Process Amplitude

In this section, we will describe MPA with its variations; First-Order MPA [7], followed by Nonlinear MPA model 7, and lastly the Adaptive MPA [6].

2.1 First-Order MPA

Nishida et al. pioneered this method in 1986 by exploiting the sinusoidal waves with the Markov process amplitude is utilized to simulate the EEG signal [7]. The estimated MPA EEG output $x(n\Delta t)$ is composed by K different oscillations $(k = 1, 2, \ldots, K)$ as

$$x(n\Delta t) = \sum_{k=1}^{K} a_k(n\Delta t) sin(2\pi m_k n\Delta t + \phi_k) \tag{1}$$

where $a_k(n\Delta t)$ is the model amplitude of the first-order Markov process is, m_k is the kth average frequency, ϕ_k is the initial phase which assumed to be zero. The definition of the following model amplitude estimation

$$a_k((n+1)\Delta t) = \gamma_k a_k(n\Delta t) + \xi_k(n\Delta t) \quad (2)$$
$$0 < \gamma_k < 1, k = 1, 2, \ldots, K$$

where $\xi_k(n\Delta t)$ is the independent increments of Gaussian distribution with zero mean and unity variance. γ_k is the coefficient of the first order Markov process. For stability, it is proved that γ_k satisfies the condition $0 < \gamma_k < 1$.

The electrical amplitude of $(n+1)\Delta t$ this model depends solely on the previous value $n\Delta t$ with the coefficient γ_k and independent for any time before $n\Delta t$. Thus, first-order Markov process is the process with the independent increments. This method is best in simulating stationary EEG signals since the first-order exhibit the linear properties of the signal, and the parameters are determined based on the power spectrum for both spontaneous and mutually coupled components of the EEG signals [7].

2.2 Nonlinear MPA

The existing MPA is further improvised to overcome the limitation of distinguishing within the same frequency band of nonlinearly coupled frequencies from spontaneously excited signals by utilizing the delta and alpha rhythm of recorded EEG signals to determine nonlinear coupling phenomena [7]. The linear spontaneous oscillations of linear MPA EEG model is viewed in [7] while according to Wiener [15] the most general nonlinearity apart from many other complicated nonlinear features; is inspired by quadratic coupling features [16] of the spontaneously activated oscillations in EEG [7]. The nonlinear coupling is composed of two oscillatory waves that pass through a nonlinear square system and thus generates self-coupling frequencies and cross-coupling frequencies [7]

$$\begin{aligned} x(n\Delta t) = & \sum_{k=1}^{K} a_k(n\Delta t) sin(2\pi m_k n\Delta t + \phi_k) \\ & + \sum_{k=1}^{K} \epsilon_k^s a_k(n\Delta t) sin(2\pi 2 m_k n\Delta t + 2\phi_k') \\ & + \sum_{i,j\in K, i\neq j} \{\epsilon_{ij}^{c1} a_i(n\Delta t) a_j(n\Delta t) cos[2\pi(m_i - m_j)n\Delta t + (\phi_i - \phi_j)] \\ & \epsilon_{ij}^{c2} a_i(n\Delta t) a_j(n\Delta t) cos[2\pi(m_i + m_j)n\Delta t + (\phi_i + \phi_j)]\} \end{aligned} \quad (3)$$

The power spectrum is obtained by transforming the digital EEG into Fourier components with Kaiser-Bessel [17,18] by Fast Fourier Transformation with the same length of the window and the FFT as the identically divided EEG data. The Welch method with Kaiser-Bessel window is used to obtain low variance of the EEG power spectrum. Thus, by knowing frequency resolution of the power spectrum, it is then used to determine the parameters and coefficients of the model. Next, the minimization of the square sum of the difference between the nonlinear EEG model and continuous EEG, and power spectrum is done by algorithm in [19], which exhibit rapid convergence for minimization property. The model is evaluated by a criterion of percentage error [7].

This method is used under the consideration of the nonlinearly coupled frequency components of EEG [7]. It is proved that for self-coupling part, this experiment exhibit the same properties of quadratic nonlinear as mentioned by Nunez [20] interactions of delta and alpha rhythm of EEG signal; as the delta and alpha oscillations are separated by the nonlinear MPA EEG model. The regarded nonlinear cross-coupling components could be view as a good match to the power spectrum of continuous EEG signal [7].

2.3 Adaptive MPA

The adaptive MPA model utilizes the least mean square algorithm to determine the parameters adaptively dated in 2004. This model should free from the stationary limitations and the need to repeatedly and manually compute the MPA model parameters. In this method, some parameter in the equation of first-order MPA is simplified [7]. if $s(n)$ is the EEG signal to be modeled, then the instantaneous error of the adaptive system is viewed as

$$e(n) = s(n) - y(n) \tag{4}$$

The core of this method, least-mean-square algorithm uses the mean square error (MSE). It could be viewed as follow

$$\begin{aligned} J &= \frac{1}{2}E(e(n)^2) = \frac{1}{2}E((s(n)-y(n))^2) \\ &= \frac{1}{2}Rs - \sum_{j=1}^{K} a_j(n)Rsx_j + \frac{1}{2}\sum_{i=1}^{K}\sum_{j=1}^{K} a_i(n)a_j(n)Rxx_{i,j} \\ &= \frac{1}{2} - \sum_{j}^{K}\Big(\gamma_j(n-1)a_j(n-1) + \mu_j(n-1)\xi_j(n-1)\Big)Rsx_j \\ &\quad + \frac{1}{2}\sum_{i=1}^{K}\sum_{j=1}^{K}\Big(\gamma_i(n-1)a_j(n-1) + \mu_i(n-1)\xi_i(n-1)\Big) \\ &\quad \times\big(\gamma_j(n-1)a_j(n-1) + \mu_j(n-1)\xi_j(n-1)\big)Rxx_{i,j} \end{aligned} \tag{5}$$

LMS [21] is used to adjust γ and μ adaptively. Due to high degree of nonstationarity in each EEG segment, this model could not track the distinction of the domain. However, it could track transient EEG activities. Also, the calculated NMSE for AMPA is proved to be better than MPA model. Another variation of MPA method is proposed in [22] synergizing the neural network knowledge to determine nonlinearities of the EEG signal variations.

3 Artificial Neural Network

This method comprises of two variations started with multi-layered back-propagation Neural Network [22], and followed by Neural Network based approach with embedded Driving Signal (DS) concept [23] in the existing time-delayed neural network estimation model [24].

3.1 Multi-layered Back Propagation Neural Network

In this method, the same procedure for recording EEG signals as in [6]. It described the approximated EEG output $y(n)$ is a function of K different oscillations. NN architecture is designed with the standard back-propagation learning algorithm with a hidden layer of 80 nodes. The result is analyzed by comparing the EEG segments' power spectral density (PSD) and normalized mean squared error (NMSE) value. NMSE is used to determine the deviation between the predicted values and the actual values [25]. It is reported that this method could liberate the generated signal from stationary constraints and to determine the model nonlinearity parameter [22].

3.2 Multi-layered Feedforward with Back Propagation Neural Network + Driving Signal (DS)

The multilayer NN is chosen due its capabilities to performed tasks in time series modelling, prediction and estimation [24,26]. The recorded EEG signal, $x(t)$ is averaged to set its baseline variations, $u(t)$ [23]. EEG background activity without variations, $r(t)$ is used throughout this research, can be obtained from deducting the baseline variation signal from the recorded signal [23]. Then, the autocorrelation function (ACF) [27] is used to determine the sample length between EEG values $r(t)$. The correlated number of EEG samples, L will be used as the length of ANN input vector is obtained by counting the correlated EEG samples until the ACF dropped for the first time below the correlation threshold.

Next, the commencement of training and validation procedure is done to adjust the weight and bias of the network. The measured EEG data is set as training data set. In the training procedure [24], EEG values are normalized into the range of $[-1,+1]$. For the weight coefficients and biases, the algorithm in [28] is used. The stopping criteria, MSE is calculated between generated and target output EEG values. Then, the incremental training procedure is performed as the weight coefficients is adjusted after every randomly selected input/output pair. Next, the input vector is initialized by randomly segmenting EEG value $r(t)$ with L length. It is then propagated into the EEG simulator with generated DS value as the last element of the input vector. Only one value is generated at a time. The output value is then reiterated into the input vector and is propagated again along with the newly generated DS value as the last element into the EEG simulator to produce a new value. This is called time-delayed neural network [24] or shifting process, in particular the new output will replace the last element while omitting the first value of the input vector.

DS is used to implement the randomness into the algorithm due to ANN output entered stationary state after several iterations in the closed loop prediction. Since DS is also used to navigate the prediction signal, the value generated from DS is used in determining the direction of the predicted signal. DS element consists of two values; -1 and $+1$ with negative and positive value represent the descending and ascending side of slope respectively [23]. DS is generated from DS extractor and DS generator. Extractor composed of a couple of units; difference and transfer. In [23], the function of the first unit is to determine the difference between the two successive EEG values while the second unit transfers the output value from the first unit into one of the conditional values using sign function[1] [29].

Next is DS generator. This part essentially to generate output in the value of either -1 or $+1$ based on preceding values. The sequence of values with determined length is used to predict the output. The conditional probabilities is exploited for this reason [30]. For every combination of the sequence, the probabilities of the next values were computed and stored in the lookup table [23]. The algorithm worked when the input sequence with N length is compared with values in the lookup table to produce certain output. To avoid biasness, the classical roulette wheel method [31] is used with random values in the range of [0, 1]. The next output were calculated based on the first output in the closed loop system. The first output will be included into the sequence of input as the last element, and the first element is then omitted thus the shifting process occurred. Then, the process of determining output based on the stored probability in the lookup table is occurred and determining output in the selection test based on classical roulette wheel method.

Neural Network method approach is better than ARMA filtering method in this research with seven determined criteria. Those are the probability density function (PDF) and cumulative distribution function (CDF) [32], level-crossing rates (LCS) average duration of fades (ADF) [33], correlation properties [34], power margin quality measures [35], weighted mean-square autocorrelation error (WMSAE) [36], representation in time domain, and representation in frequency domain [37].

4 Autoregressive (AR) Model

This section comprises white Gaussian noise filtering based on AR model [38] with its variation, Genetic approach with ARMA filtering [39]. The signal in

---

[1] *signum* function or $sgn\ (x)$

$$sgn(x) = \begin{cases} 1 & \text{if } x \geq 0 \\ -1 & \text{if } x < 0 \end{cases}.$$

AR modeling is described to be linearly related with respect to a number of its previous samples

$$y(n) = -\sum_{k=1}^{p} a_k y(n-k) + x(n) \tag{6}$$

where a_k,$k = 1, 2, \ldots, p$, are the linear parameters, n is the discrete sample time normalized to unity, and $x(n)$ is the noise input.

4.1 Autoregressive (AR) Model

Doležal et al. [40] could not manage to model continuous non-movement related EEG based on Markov model. So, it is done via AR analysis [38]. It started with EEG pre-processing; the recorded EEG data were set to be clean from any artefact, analysis of AR and resting EEG modelling. Next is followed by assessing the quality of the EEG model. For modelling EEG signal, the frequency limitation is done with by filtering for getting new frequency and subtracting the DC component. These is done upon recorded EEG signal. It is done to avoid new poles to the processed spectrum. Then, the autocorrelation function as well as the power of the signal is computed and averaged. The modelling filter coefficients is computed by using Yule-Walker equation [41]. Thus, the model is gained by filtering white noise.

4.2 Genetic Approach to Autoregressive Moving Average (ARMA) Filter Synthesis

This model is inspired by Zetterbergs method for simulating EEG signals. The method is simulated as the result of filtering a white noise source of specific characteristics (statistical characteristics and flat spectrum) with an ARMA filter. The linear differential equation [42] represents the filter at order p (where $p \geq q$). ARMA filter is composed by Autoregressive and the Moving Average filters, could be described as the following transfer function

$$H_{ARMA} = \frac{B(z-1, q)}{A(z-1, p)} \tag{7}$$

where the numerator is related to the MA while the denominator to the AR models. AR filter is suitable for the simulation of signals which spectrum displays sharp peaks whereby MA filter is more suitable to signals with deep valleys in the spectrum [39]. For estimating the order of ARMA filter, the approach proposed by Vaz et al. [43] is used.

This method started with generating Gaussian distribution for the white noise source, then the real EEG signal is recorded from a patient by the expert in order to compare with the simulated signal. The following steps is to transformed fitness function from time domain to frequency domain utilizing Discrete Fourier Transform (DFT). Since GA is used in this method, the variables represented in this method were represented as genes or filter coefficients. The fitness function is used to compare between the spectrum of real EEG and simulated EEG signals. Next is selecting the genetic algorithm parameters via tournament selection for slower convergence. This selection is used using 10 % of the population. The process is continued with the simulation and analysis. The order selection criterion is utilized to analyze and validate for real and simulated EEG signals. In the result section, the real signal has smaller average amplitude than simulated signal represented in time domain. It is also shown that ARMA filters are best in both biological signal processing and control systems [39].

5 Conclusion

The original MPA model which can only model stationary and linear EEG model has evolved to suit the nonlinear element with self-coupling and cross coupling into the EEG artificial model. MPA based approach were using instantaneous error of simulated and actual EEG signal. The improvement lead to more comprehensive model of the actual EEG signal. The closest method that resembles the real EEG signal is Neural Network method due to the nature of NN algorithm which utilizes the actual data set for training purpose. The issue with NN training data set is the data availability and data size. One might need a very large data set. Also, since the training data is big, the NN architecture also need to be optimized to avoid undertraining and overtraining issue. On the other hand, AR model which is based on white Gaussian noise filtering has its own uniqueness. Different approaches could be implemented but this method could only model certain type of EEG signals. This method prioritizes the selection of the model order and filter coefficient.

Acknowledgments. This research is supported by the HiCoE grant for CISIR (0153CA-002) and FRGS/1/2014/SG04/UTP/02/1 from the Ministry of Education (MoE) of Malaysia.

References

1. Proakis, J.G., Manolakis, D.G.: Digital Signal Processing: Principles, Algorithms, and Applications. Prentice-Hall Int., Inc., New Jersey (1996)
2. Sörnmo, L., Laguna, P.: Bioelectrical Signal Processing in Cardiac and Neurological Applications. Elsevier Academic Press, Massachusetts (2005)
3. Barlow, J.S.: The Electroencephalogram: Its Patterns and Origins. MIT Press, Cambridge (1993)
4. Sanei, S., Chambers, J.A.: EEG Signal Processing. Wiley, England (2007)
5. Fabri, S.G., Camilleri, K.P., Cassar, T.: Parametric modelling of EEG data for the identification of mental tasks. In: Biomedical Engineering, Trends in Electronics, Communications and Software, pp. 367–386. InTech (2011)
6. Al-Nashash, H., Al-Assaf, Y., Paul, J., Thakor, N.: EEG signal modeling using adaptive Markov process amplitude. IEEE Trans. Biomed. Eng. **51**(5), 744–751 (2004)

7. Bai, O., Nakamura, M., Ikeda, A., Shibasaki, H.: Nonlinear Markov process amplitude EEG model for nonlinear coupling interaction of spontaneous EEG. IEEE Trans. Biomed. Eng. **47**(9), 1141–1146 (2000)
8. Fell, J., Kaplan, A., Darkhovsky, B., Roschke, J.: EEG analysis with nonlinear deterministic and stochastic methods: a combined strategy. Acta Neurolobiol. **60**, 87–108 (2000)
9. Dynkin, E.B., Brown, D.E., Kövåry, T. (eds.): Theory of Markov Processes. Pergamon Press Ltd., London (1960)
10. Hernandez-Lerma, O.: Adaptive Markov Control Processes. Springer, New York (1989)
11. Haykin, S.: Neural Networks and Learning Machines. Pearson Prentice Hall, New Jersey (2009)
12. Adeli, H., Hung, S.L.: Machine Learning - Neural Networks, Genetic Algorithms, and Fuzzy Systems. Wiley, New York (1995)
13. Ghosh-Dastidar, S., Adeli, H.: Spiking neural networks: spiking neurons and learning algorithms. In: Automated EEG-Based Diagnosis of Neurological Disorders - Inventing the Future of Neurology, pp. 240–270. CRC Press, Boca Raton (2010)
14. Tsay, R.S.: Analysis of Financial Time Series. Wiley, New Jersey (2005)
15. Wiener, N.: Nonlinear Problems in Random Theory. MIT Press, Cambridge (1958)
16. Ning, T., Bronzino, J.D.: Nonlinear analysis of the hippocampal subfiels of CA1 and the dentate gyrus. IEEE Trans. Biomed. Eng. **40**(9), 870–876 (1993)
17. Welch, P.D.: The use of fast fourier transform for the estimation of power spectra: a method based on time average over short, modified periodograms. IEEE Trans. Audio Electroacoust. **AU–15**, 70–73 (1967)
18. Rabiner, L.R., Gold, B.: Theory and Application of Digital Signal Processing. Prentice Hall, New Jersey (1975)
19. Fletcher, R., Powell, M.J.D.: A rapidly convergent descent method for minimization. The Comput. J. **6**, 163–168 (1963)
20. Nunez, P.L.: Neocortical Dynamics and Human EEG Rhythms, pp. 434–436. Oxford University Press, Oxford (1995)
21. Widrow, B., Sterns, S.D.: Adaptive Signal Processing. Prentice-Hall, New Jersey (1985)
22. Al-Nashash, H.A., Zalzala, A.M.S., Thakor, N.V.: A neural networks approach to EEG signals modeling. In: 25th Annual International Conference of the IEEE EMBS, Cancun, Mexico (2003)
23. Tomasevic, N.M., Neskovic, A.M., Neskovic, N.J.: Artifical neural network based approach to EEG signal simulation. Int. J. Neural Syst. **22**(3), 1–16 (2012)
24. Hassoun, M.H.: Fundamentals of Artificial Neural Networks. MIT Press, Detroit (1995)
25. Liu, J.N.K., Yu, Y.X.: Support vector regression with kernal mahalanobis measure for financial forecast. In: Pedrycz, W., Chen, S.M. (eds.) Time Series Analysis, Modeling and Applications 2013. ISRL, vol. 47, pp. 2014–2027. Springer, Heidelberg (2013)
26. Széliga, M.I., Verdes, P.F., Granitto, P.M., Ceccatto, H.A.: Artificial neural network learning of nonstationary behavior in time series. Int. J. Neural Syst. **13**(2), 103–109 (2003)
27. Chatfield, C.: The Analysis of Time Series. CRC Press, Florida (1996)
28. Nguyen, D., Widrow, B.: Improving the learning speed of 2-layer neural networks by choosing initial values of the adaptive weights. In: IEEE First International Joint Conference on Neural Networks, vol. 3 (1990)

29. Bracewell, R.: The sign function, *sgn x*. In: The Fourier Transform and Its Application, pp. 65–66. McGraw-Hill, New York (1999)
30. Frank, H., Althoen, S.C.: Statistics: Concepts and Applications. Cambridge University Press, Cambridge (1994)
31. Čížek, P.: Non linear regression modeling. In: Gentle, J.E., Härdle, W., Mori, Y. (eds.) Handbook of Computational Statistics Concepts and Methods. Springer Handbooks of Computational Statistics, pp. 621–655. Springer, Heidelberg (2004)
32. Leon-Garcia, A.: Probability, Statistics, and Random Processes for Electrical Engineering. Pearson Education Inc., Upper Saddle River (2008)
33. Goldsmith, A.: Wireless Communications. Cambridge University Press, Cambridge (2005)
34. Bendat, J.S., Piersol, A.G.: Random Data: Analysis and Measurement Procedures. Wiley, Hoboken (2010)
35. Young, D.J., Beaulieu, N.C.: Power margin quality measures for correlated random variates derived from the normal distribution. IEEE Trans. Inf. Theory **49**(1), 241–252 (2003)
36. Crespo, P.M., Jimenez, J.: Computer simulation of radio channels using a harmonic decomposition technique. IEEE Trans. Veh. Tech. **44**(3), 414–419 (1995)
37. Salivahanan, S., Vallavaraj, A., Gnanapriya, C.: Digital Signal Processing. Tata Mcgraw-Hill Publishing Company Limited, New Delhi (2000)
38. Doležal, J., Štastný, J., Sovka, P.: Modeling and recognition of movement related EEG signal. In: International Conference on Applied Electronics (2006)
39. Janeczko, C., Lopes, H.S.: A genetic approach to ARMA filter synthesis for EEG signal simulation. In: Proceedings of the 2000 Congress on Evolutionary Computation, La Jolla, California (2000)
40. Doležal, J., Štastný, J., Sovka, P.: Recognition of direction of finger movement from EEG signal using Markov models. In: The 3rd European Medical and Biological Engineering Conference, EMBEC, Prague, Czech Republic (2005)
41. Marple, S.L.: Digital Spectral Analysis with Applications. Prentice-Hall, Englewood Cliffs (1987)
42. Vaz, F., Oliveira, P.G., Principe, J.C.: A study on the best order for autoregressive EEG modeling. Int. J. Bio-Med. Comput. **20**, 41–50 (1987)
43. Manolakis, D.G., Ingle, V.K., Kogon, S.M.: Statistical and Adaptive Signal Processing: Spectral Estimation, Signal Modeling, Adaptive Filtering and Array Processing. Artech House, Norwood (2005)

A New Blind Image Quality Assessment Based on Pairwise

Jianbin Jiang, Yue Zhou(✉), and Liming He

Institude of Image Processing and Pattern Recognition,
Shanghai Jiao Tong University, Shanghai, China
{rz_jiang,zhouyue,heliming}@sjtu.edu.cn

Abstract. Recently, the algorithms of general purpose blind image quality assessment (BIQA) have been an important research area in the field of image processing, but the previous approaches usually depend on human scores image for training and using the regression methods to predict the image quality. In this paper, we first apply the full-reference image quality measure to obtain the image quality scores for training to let our algorithm independent of the judgment of human. Then, we abstract features using an NSS model of the image DCT coefficient which is indicative of perceptual quality, and subsequently, we import Pairwise approach of Learning to rank (machine-learned ranking) to predict the perceptual scores of image quality. Our algorithm is tested on LIVE II and CSIQ database and it is proved to perform highly correlate with human judgment of image quality, and better than the popular SSIM index and competitive with the state-of-the-art BIQA algorithms.

Keywords: Blind image quality assessment · Learning to rank · Natural scene statistics · Full-reference image quality assessment · No reference image quality assessment

1 Introduction

Blind image quality assessment(BIQA) algorithms are classified into two categories: distortion-specific approaches and distortion independent methods. The former category uses the law of distortion to estimate the image quality that applied in the specific application. The later category generally uses training methods based on features extracted from diverse distortion images that can be used to evaluate generally images. Our work focus on the problem of the later category. Recent five years, a large number of successful applications use the later category have been proposed. In [2], the BIQA metric applies local discrete cosine transform coefficients for training, in [7], uses code book created by spatial-domain features obtained by clustering algorithms, and the latest algorithm in [10], adopts deep learning to abstract features from images directly. The mentioned algorithms perform statistically better than SSIM index. However, all the algorithms must rely on the human scores. In this paper, to solve

A. Hirose et al. (Eds.): ICONIP 2016, Part IV, LNCS 9950, pp. 609–617, 2016.
DOI: 10.1007/978-3-319-46681-1_72

the above problems, the key contribution are as follows: (1) We incorporate the FR-IQA into learning dataset generation to obtain the subject scores of the sample images which used in training model. The proposed strategy will let our algorithm independent of image databases and addresses the problem of inaccurate evaluation by human. (2) We use DCT domain features to comprehensively reflect the quality of image to enhance the accuracy of measures. (3) We import the Ranking-SVM algorithm to train the features and predict the score of image for the first time.

2 Overview of the Method

Figure 1 shows our algorithm is consisted of three parts: Learning dataset generation, features abstraction and prediction model. The learning dataset generation obtain the image quality scores S_i by using full-reference image quality assessment and features abstraction obtain images' features f_i, In the prediction model we using the $\{f_i, s_i\}$ to form the model by applying the ranking SVM of Pairwise approach of learning to rank.

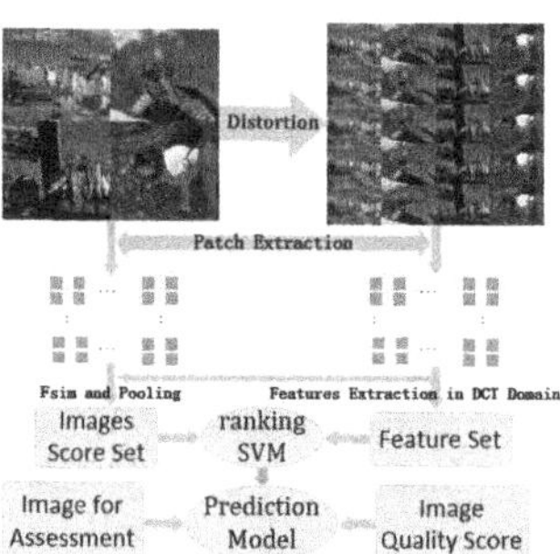

Fig. 1. Flowchart of proposed algorithm.

3 Proposed Algorithm

3.1 Learning Dataset Generation

To obtain the image quality score for training, we use the FR-IQA such as SSIM [1], FSIM [5]. In this paper, we use FSIM. Firstly, the reference and distorted image is partitioned into overlapped patches. The patch of reference image is denoted by R_i and distorted version by D_i. Then the perceptual quality of D_i can be calculated by using FSIM:

$$s_i = \frac{2PC(x_i)PC(d_i) + t_1}{PC(x_i)^2 + PC(d_i)^2 + t_1} \times \frac{2G(x_i)G(d_i) + t_2}{G(x_i)^2 + G(d_i)^2 + t_2} \tag{1}$$

Where $PC(x_i)$ is the phase congruency and $G(x_i)$ is the gradient magnitude at the center of x_i. For numerical stability, the positive constants t_1 and t_2 are imported. S_i is the similarity score of d_i. Now, We can take all d_i to obtain the perceptual quality score of the distorted images. In [6,7], which show the human visual system is more sensitive to the worst patches of image and the worst 10% percent pooling has much better linearity. Based on this knowledge, we normalize S_i as follows [6]:

$$C = \frac{10 S_i \sum_{i \in \Omega_p S_i}}{\sum_{i \in \Omega S_i}} \tag{2}$$

Ω is the similarity score set of all patches of an image and the Ω_p is the set of 10% percent lowest quality patches. Then, the training-scores of distorted images are obtained.

3.2 Abstract NSS Features in DCT Domain

In this paper, image features are extracted in the discrete cosine transform (DCT) domain. The main reason is the observation that the statistics in DCT domain changes when the image quality changes [2]. First, partition the image into $n \times n$ patches which are transformed to DCT domain by using variable-change transform from computationally efficient fast Fourier transform algorithms [3]. Then, the image quality perceptual features are calculated by using univariate generalized Gaussian density. The function of univariate generalized Gaussian density as (3):

$$f(x|\alpha, \beta, \gamma) = \alpha e^{-(\beta|x-\mu|)^{\gamma}} \tag{3}$$

Where the γ is the shape parameter and μ is the mean, α is the normalizing parameter and β is the scale parameter as:

$$\alpha = \frac{\beta\gamma}{2\tau(1/\gamma)} \qquad \beta = \frac{1}{\delta}\sqrt{\frac{\tau(3/\gamma)}{\tau(1/\gamma)}} \qquad \tau(z) = \int_0^{\infty} t^{z-1} e^{-t} dt. \tag{4}$$

where δ is the standard deviation, t is the gamma function, when $\beta = 2$, the distribution donates Gaussian distribution, and we deploy the parameter estimate method given in [4]. We use the features opposed in [2], Shape parameter, Coefficient of Frequency Variation, and Energy Subband Ratio Measure. In the end of this section, we will show the all features changes with the image quality changes.

Generalized Gaussian Model Shape Parameter. In (Eq. 3), the generalized Gaussian density parameterized by μ, β, and γ. The shape features are calculated over all patches. For the pooling strategy, we apply two ways. First, we apply the worst 10 % strategy which has been observed to improve the correlations with image quality. Then, the 100% average strategy.

Coefficient of Frequency Variation. Coefficient of frequency variation feature is the next feature used:

$$\xi = \frac{\delta|x|}{\mu|x|} = \sqrt{\frac{\tau(1/\gamma)\tau(3/\gamma)}{\tau^2(2/\gamma)} - 1} \tag{5}$$

Where the x is the histogrammed DCT coefficients extracted from patches, $\delta|x|$ is standard deviation of x and μ is the mean of x. For the pooling strategy, we use the highest 10th percentile pooling strategy and 100% average strategy.

Energy Subband Ratio Measure. In DCT domain, the lower frequency information of a image is concentrated upon the top-left corner of the image matrix, and the higher is concentrated upon the bottom-right. We define three frequency bands in Fig. 1, Low band (band 1){ C_{21} C_{12} C_{13} C_{31} C_{22} }, middle band (band 2) { C_{41} C_{51} C_{32} C_{42} C_{52} C_{23} C_{33} C_{43} C_{14} C_{24} C_{34} C_{15} C_{25} }, high band (band 3){ C_{53} C_{44} C_{54} C_{35} C_{45} C_{55} }. Where C_{ij} denotes the value of the location(i_{th} cols and j_{th} rows) in 5×5 matrix, we define the average energy denoted by variance δ^2:

Abstracted Features Changes with Image Quality. Figure 2 shows the features of Generalized Gaussian Model Shape Parameter and Energy Subband Ratio Measure increase with the DMOS of image increase, feature of Coefficient of Frequency Variation reduce with DMOS of image increase, so our abstracted features can be used to predict image quality.

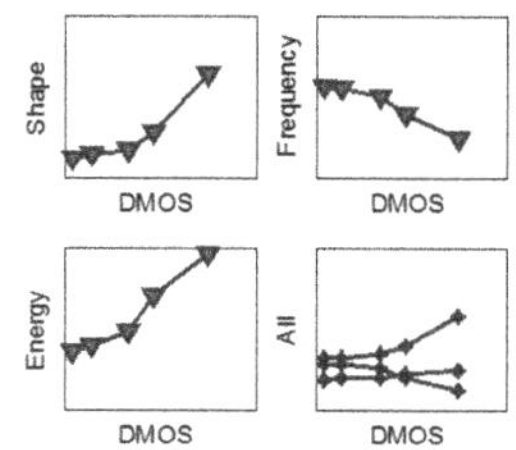

Fig. 2. The value of the abstracted features changes with the image quality, left-top: Generalized Gaussian Model Shape Parameter. right-top: Coefficient of Frequency Variation. left-bottom: Energy Subband Ratio Measure. right-bottom: the all features

3.3 Prediction Model

Learning to rank (LTR) is the key problem in many applications like Information Retrieval [8]. LTR is consisted of PointWise, PairWise, and ListWise. We choose the Ranking SVM of Pairwise. First, we construct the training data of correlational pairs as Fig. 3, x_i is the feature vector, f_i is the score of the i_{th} image. L_{ij} is the label denotes which feature vector of image should be ranked ahead, for examples, $f_1 > f_2$, then the $L_{12} = 1$, $L_{21} = 0$, and the correlational pairs $(x_1 - x_2, 1), (x_2 - x_1, 0)$ are formed. Ranking SVM is formalized as follows:

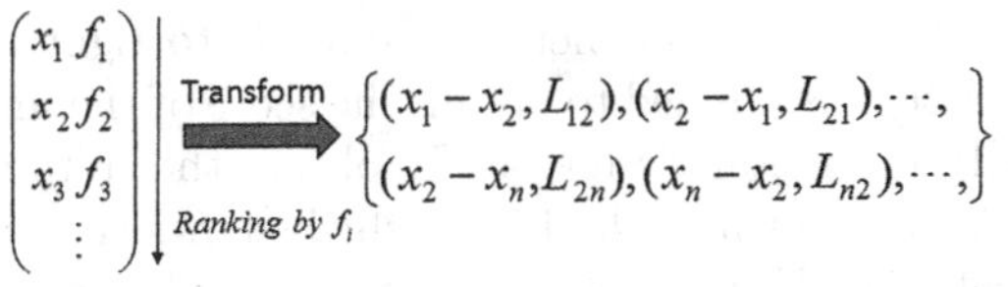

$$\begin{pmatrix} x_1 f_1 \\ x_2 f_2 \\ x_3 f_3 \\ \vdots \end{pmatrix} \xrightarrow{\text{Transform}} \left\{ \begin{matrix} (x_1 - x_2, L_{12}), (x_2 - x_1, L_{21}), \cdots, \\ (x_2 - x_n, L_{2n}), (x_n - x_2, L_{n2}), \cdots, \end{matrix} \right\}$$

Ranking by f_i

Fig. 3. Transfor for Ranking SVM.

$$
\begin{aligned}
& min_{\omega,\xi} \frac{1}{2}||x||^2 + C\sum_{i=1}^{m} \xi_i \\
s.t. \quad & y_i(\omega, x_i - x_j) \geq 1 - \xi_i \\
& \xi \geq 0 \\
& i = 1, ..., m
\end{aligned}
$$

Where x_i denotes the feature vector i, ξ is the slack variable, y_i is the label, ω is the parameter vector.

In this paper, we combine s_i with f_i to form a unit, then we create a sequence including all unit ranged by f_i from highest to lowest. At the end, we use the ranking SVM to train our prediction model by using the sequence. Compared with all the image from sequence when assess a new image's quality. For example, the image quality is lower than first image in the sequence, the 1_{th} of our result sequence will be 0, the image quality is higher than second image in the sequence, the 2_{th} of our result sequence will be 1. Then, the image quality is between 1_{th} image and 2_{th} image of the arranged sequence. However, for the classification exists misclassified, the reality result-sequence as Fig. 4.

The red rectangle is our interested region. In the left of the region are 0, which denotes the assessment image quality is lower than the left set. Similarly, the quality is higher than the right set. Obviously, the quality is belong to the red region. We apply the n vs 10 strategy to obtain the score of the image quality, 10 denotes the continuous 10 images in the sequence, n denotes n image quality in the 10 image set are lower than the assessment image. And the 10 image set

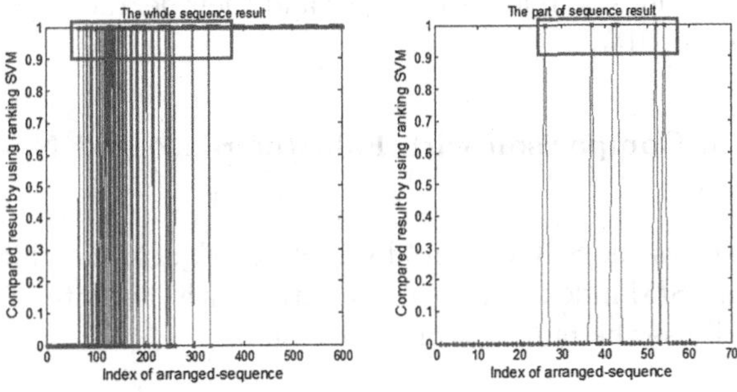

Fig. 4. 0-1 sequence of the assessment image (Color figure online)

is chosen by using sliding window method from left to right in the red region. Average pooling strategy is applied to obtain the score of the image quality when find the first n vs 10 red rectangle region. To obtain the suitable n, we tested it on the Live database as shown in the Fig. 5 which shows the SROCC coefficient and LCC coefficient both highest when $n = 8$, so we choose n=8 to obtain the image quality.

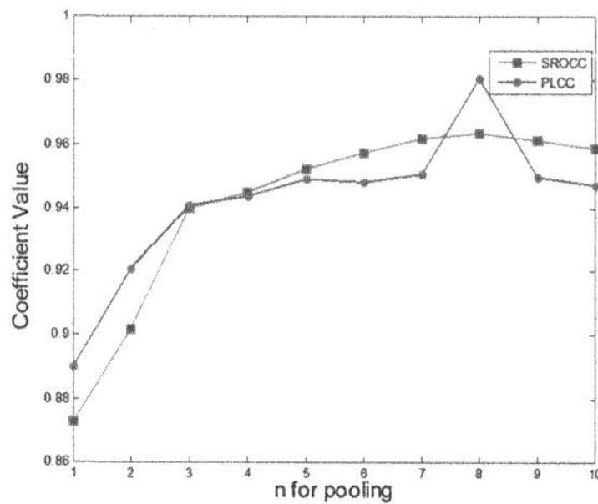

Fig. 5. Transfor for Ranking SVM.

4 Experiments and Results

We train our model by using images include 29 reference images and 80 % of their five different types of distortions, i.e., JPEG2k compression (JP2K), JPEG compression (JPEG), additive white Gaussian noise (WN), Gaussian blurring (BLUR) and fast fading (FF) to train our model in Live II. Then we tested all the images on Live II. In the Sect. 4.1, we use the commonly used performance metrics, Pearson Linear Correlation Coefficient (PLCC), Spearman Rank-Order Correlation Coefficient (SROCC) to compare with other classical and new algorithms. In Sect. 4.2, we plot the scatter plots of DMOS vs. the predicted scores of proposed approach on Live II database. In Sect. 4.3, we test our approach on CSIQ database includes 30 reference images and their degraded versions (JPEG, JP2K, WN, and BLUR).

4.1 Statistical Comparison with Full-Reference and No-Reference Approaches

Tables 1 and 2 shows the SROCC and LCC of our algorithm is high, it is better than the popular SSIM index and can competitive with the state-of-the-art BIQA algorithms which use the human scores for training.

Table 1. LCC of Different Methods on Live II Database

Method	JP2K	JPEG	WN	GBLUR	FF	ALL
PSNR	0.896	0.860	0.986	0.783	0.890	0.824
SSIM	0.937	0.928	0.970	0.874	0.943	0.863
CB100	0.915	0.846	0.953	0.946	0.878	0.839
BLINDS-II	0.963	0.979	0.985	0.948	0.944	0.923
Ours	0.953	0.955	0.977	0.970	0.913	0.950

Table 2. Srocc of Different Methods on Live II Database

Method	JP2K	JPEG	WN	GBLUR	FF	ALL
PSNR	0.890	0.841	0.985	0.782	0.890	0.820
SSIM	0.932	0.903	0.963	0.894	0.941	0.851
CB100	0.918	0.843	0.970	0.947	0.878	0.821
BLINDS-II	0.951	0.942	0.978	0.944	0.927	0.920
Ours	0.950	0.943	0.967	0.955	0.907	0.943

4.2 Robustness Against Distortion Types

The Tables 1 and 2 shows the LCC and SROCC of all different distortion types are all big than 0.9 when using our algorithms, scatter plots of DMOS vs. the predicted scores of proposed approach on the whole Live II database and five different distortion types shown in Fig. 6, which shows our predict image quality score is correlate highly with human judgments of quality and robusstness against distortion types.

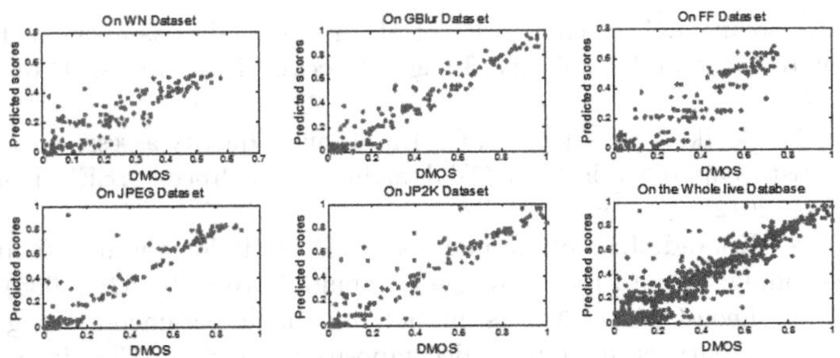

Fig. 6. Scatter plots of DMOS vs. the predicted scores of proposed approach on the whole CSIQ database and five different distortion types of data sets.

4.3 Database Independence

We test our approach on the CSIQ database, we report SROCC and LCC in Table 1, which shows the correlation are consistently high which shows our algorithm is independent of database (Table 3).

Table 3. PLCCS and SROCCon the CSIQ database

	Awgn	Blur	Jpeg	Jepg2000	All
SROCC	0.889	0.951	0.935	0.938	0.920
LCC	0.889	0.963	0.944	0.940	0.930

5 Conclusion

This paper, we innovatively import the Learning to rank method to form the prediction model of a new blind image quality assessment based on pairwise. Experiments prove our approach is better than he most two popular full-reference PSNR and SSIM methods, and it is competitive with the no-reference image quality methods which use the human scores for training such as BLIINDS-II.

Acknowledgment. The work is supported by National High-Tech R&D Program (863 Program) under Grant 2015AA016402 and Shanghai Natural Science Foundation under Grant 14Z111050022.

References

1. Wang, Z., Bovik, A.C., Sheikh, H.R., et al.: Image quality assessment: from error visibility to structural similarity. Image Process. IEEE Trans. **13**(4), 600–612 (2004)
2. Saad, M.A., Bovik, A.C., Charrier, C.: Blind image quality assessment: A natural scene statistics approach in the DCT domain. Image Process. IEEE Trans. **21**(8), 3339–3352 (2012)
3. Boinovi, N., Konrad, J.: Motion analysis in 3D DCT domain and its application to video coding. Signal Process. Image Commun. **20**(6), 510–528 (2005)
4. Sharifi, K., Leon-Garcia, A.: Estimation of shape parameter for generalized Gaussian distributions in subband decompositions of video. Circuits Syst. Video Technol. IEEE Trans. **5**(1), 52–56 (1995)
5. Zhang, L., Zhang, L., Mou, X., et al.: FSIM: a feature similarity index for image quality assessment. Image Process. IEEE Trans. **20**(8), 2378–2386 (2011)
6. Xue, W., Zhang, L., Mou, X.: Learning without human scores for blind image quality assessment. In: Proceedings of the IEEE Conference on Computer Vision and Pattern Recognition, pp. 995–1002 (2013)

7. Ye, P., Kumar, J., Kang, L., et al.: Unsupervised feature learning framework for no-reference image quality assessment. In: 2012 IEEE Conference on Computer Vision and Pattern Recognition (CVPR), pp. 1098–1105. IEEE (2012)
8. Hang, L.I.: A short introduction to learning to rank. IEICE Trans. Inf. Syst. **94**(10), 1854–1862 (2011)
9. Cohen, E., Yitzhaky, Y.: No-reference assessment of blur and noise impacts on image quality. Signal Image Video Process. **4**(3), 289–302 (2010)
10. Hou, W., Gao, X., Tao, D., et al.: Blind image quality assessment via deep learning. Neural Netw. Learn. Syst. IEEE Trans. **26**(6), 1275–1286 (2015)

Self-organizing Maps as Feature Detectors for Supervised Neural Network Pattern Recognition

Macario O. Cordel II(✉), Arren Matthew C. Antioquia, and Arnulfo P. Azcarraga

College of Computer Studies, De La Salle University, 2401 Taft Avenue, 1004 Manila, Philippines
{macario.cordel,arren_antioquia,arnulfo.azcarraga}@dlsu.edu.ph

Abstract. Convolutional neural network (CNN)-based works show that learned features, rather than handpicked features, produce more desirable performance in pattern recognition. This learning approach is based on higher organisms visual system which are developed based on the input environment. However, the feature detectors of CNN are trained using an error-correcting teacher as opposed to the natural competition to build node connections. As such, a neural network model using self-organizing map (SOM) as feature detector is proposed in this work. As proof of concept, the handwritten digits dataset is used to test the performance of the proposed architecture. The size of the feature detector as well as the different arrangement of receptive fields are considered to benchmark the performance of the proposed network. The performance for the proposed architecture achieved comparable performance to vanilla MLP, being 96.93 % using 4×4 SOM and six receptive field regions.

Keywords: Feature detectors · Self-organizing maps · Multilayer perceptron · Pattern recognition

1 Introduction

With the architecture of neural networks that resembles the arrangements of neurons from the receptors, e.g. auditory, optic and olfactory, to central nervous system; it is not surprising that most of the advanced solutions [10,11,14,15] to stimuli processing and identification problems that are typically performed by the human brain are based on neural networks. Some of these problems include speech recognition, performed by the auditory system, odor identification, carried out by olfactory system, and pattern recognition, normally performed by the human visual system.

In an attempt to explain the pattern recognition capability of higher organisms visual system, Fukushima in 1980 presented a neural network architecture based on this hierarchy of the visual cortex, named as neocognitron [3], which is capable of recognizing patterns and invariant to small translation. A work

A. Hirose et al. (Eds.): ICONIP 2016, Part IV, LNCS 9950, pp. 618–625, 2016.
DOI: 10.1007/978-3-319-46681-1_73

by LeCun et al. [6] in 1989 proposed a system similar to the idea of neocognitron that requires position invariant features for pattern recognition, through weight sharing. Unlike neocognitron, the proposed system, later called the convolutional neural network (CNN) [7], uses gradient-based learning rather than the unsupervised nature of the neocognitron. LeCun et al. argue that features for pattern recognition tasks must be learned rather than hand-crafted to eliminate the limitation of the designed system to a specific task and the dependence to the ability of the designer to come up with an appropriate set of features. This later has become successful in handwritten digit recognition [7] achieving 0.35 % training error rate and 0.95 % test error rate.

In the experiment by Hubel and Wiesel [4], the cat visual system exposed to the same patterns, i.e. vertical lines, developed detectors, through self organization, which are sensitive to vertical lines. This competitive nature of the neurons detectors, in Machine Learning (ML) rather than learning via error-correcting teacher, is similar to the unsupervised learning characteristic of an ML model.

Thus, this paper proposes a direct application of SOM as the feature detector for the MLP classifier, to mimic the competitive nature of the visual system during the formative stage of the visual system. Several unsupervised method of training the feature extraction portion of the CNN-based architecture are cited in Sect. 2. Section 3 discusses the details of this proposed architecture and the relative parameters that could affect its performance. The results and analysis are presented in Sect. 4 where the benchmarking tests for this proposed work is presented. Finally, Sect. 5 concludes the paper together with future works as this proposed architecture is a work in progress, thus, has not been tested extensively.

2 Related Works

Mohebi and Bagirov [9] proposes a solution to the handwritten digits recognition problem using Convolutional Recursive Modified SOM. The proposed work performs splitting and merging algorithm to determine locations of high density data. It represents each high density data with cluster centroid which are used as the CNN nodes in the convolution layer. To accommodate this clustering of SOM nodes in high density area data, to the CNN architecture, the SOM topology is modified from neighborhood of nodes to connectivity of nodes, thus the modified SOM. One weakness of their work though is that each category requires its own set of feature detectors.

Dong, Wu Pei and Jia in 2014 presented a vehicle type classification system [2] which takes in high resolution video frames from traffic camera. This system has two levels of CNN. The first level extracts the high-level features for the identification of the global features which provide holistic description of vehicles. The second level aims to extract low-level features which characterize the vehicle parts precisely. This approach, however, requires some labeled data to allow incorporation of prior knowledge to the dataset. Comparing the performance of their system with other supervised system, the CNN-based system achieves 96.1 % accuracy as compared with other vehicle classification scheme which reaches up to 93.7 % only.

Another unsupervised CNN work, proposed by Arevalo et al. [1], is used in detecting basal cell carcinoma in medical images. The system investigated three main unsupervised feature learning (UFL) approaches but only the topographic independent component analysis (TICA) applies to the CNN architecture. TICA is topographic model inspired by the biological visual system, like the CNN, i.e. learned feature detectors such that similar activations are closer together while different activations are farther apart. For 746 training images and 671 test images, the [1] unsupervised CNN was able to achieve 0.975 area under the ROC curve.

One of the challenges in machine vision that is addressed by the work in [12] is the detection of dynamic object in different scenarios. The system presented performs dynamic background modeling to consider illumination issues, camouflage, stopped dynamic objects bootstrapping and camera jitter. The system uses Retinotopic SOM [13] (RESOM) for background modeling. This architecture achieves best Precision, Recall, F1 and Similarity equal to 1, for a moving object while worst Precision, Recall, F1 and Similarity equal to 0.5352, 0.1278, 0.2063 and 0.1150, respectively.

3 Feature Extraction Using Self-organizing Maps

The proposed architecture for this work directly uses the self-organizing map (SOM) [5] as feature detector, instead of the convolution-pooling layer of CNN, as illustrated in Fig. 1. Each SOM in the feature detector layer or the SOM bank layer has corresponding receptive field an area in the input image whose pixel values are used as the input vector to the SOM. There are two training phases for this proposed biologically-inspired architecture: the unsupervised training of the weight vectors of the SOM bank layer and the supervised training of the multilayer perceptron using the developed SOM bank layer as feature detector.

SOM represents each receptive field using nodes in a map as points in the two-dimensional vector space. In the first training phase, each node's weight vector is updated every iteration. Generally, the weight vectors are updated using the equation below

$$\mathbf{w}_i(t+1) = \mathbf{w}_i(t) + N(t)\alpha_i||\mathbf{x}(t) - \mathbf{w}_i(t)|| \quad (1)$$

where t represents the iteration number, $\mathbf{w}_i$ represents the weight vector of the ith node, $\mathbf{x}$ is the input vector chosen randomly from the training set, α_i is the learning rate of the adaptation process, $||\mathbf{x}(t) - \mathbf{w}_i(t)||$ is the Euclidean distance between $\mathbf{x}(t)$ and $\mathbf{w}_i(t)$, and $N(t)$ is the neighborhood function which is typically a rectangular function or a Gaussian function.

The input vector during the SOM training period is composed of the pixel values in the receptive field of a SOM feature detector. An example of these receptive fields is shown in Fig. 1. For a SOM feature detector, k, with receptive field from pixel row 1 to row 10 and pixel column 1 to column 10, for example, denoted for this work as, $\mathbf{P}_{i,j}^{(k)}$, where i and j are the row and column subscripts

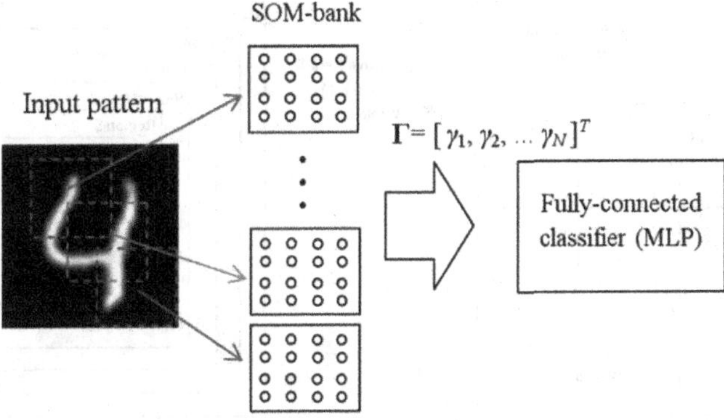

Fig. 1. The proposed architecture with several SOM as feature detectors. The regions defined by the broken lines in the input pattern are the receptive fields assigned to the maps in the SOM-bank layer. Each node (4×4 in the illustration) has weight vector of length equal to the number of pixels in their receptive fields. After the SOM-bank is the classifier of the proposed architecture, whose inputs are the similarity measures of the SOM nodes and the input data.

for the pixel location, then the receptive field $\mathbf{P}^{(k)}_{i,j}$ is such that i and j are from 1 to 10. Thus, the length of the weight vector for this specific SOM feature detector is 100.

There are infinitely many ways to arrange the $\mathbf{P}^{(k)}$ receptive fields of the SOM-bank layer, with different sizes and different number of feature detectors. We started at evaluating our proposed architecture based on the three receptive field arrangements in Fig. 2. The receptive fields on the top left ($k = 1$ to 6) show equal priorities on detecting patterns in four corners and on the center of the input pattern as same number of receptive fields are assigned to each part. For the receptive field arrangement on the top right ($k = 1$ to 6), patterns which are at the center of the input patterns have more priorities in feature detections as all SOMs receptive fields include the center of the input pattern. The third pattern ($k = 1$ to 11) at the bottom is somewhat a combination of the first two arrangements as receptive fields are assigned to the four corners as well as more receptive fields at the center of the input pattern.

After training the SOM-bank layer, the similarities of (1) the receptive field regions in the input pattern with respect to (2) the developed SOM nodes as feature detectors, are then used as input elements to the classifier for the second training phase. The similarity measure, $\gamma_{a,b}$, between the ath input receptive field $\mathbf{P}^{(k)}$ and the weight vector $\mathbf{w}$ of the bth node of the corresponding SOM feature detector for the receptive field $\mathbf{P}^{(k)}$, is computed as,

$$\gamma_{a,b} = \frac{\mathbf{P}^{(k)}_a \cdot \mathbf{w}_b}{||\mathbf{P}^{(k)}_a||||\mathbf{w}_b||} \tag{2}$$

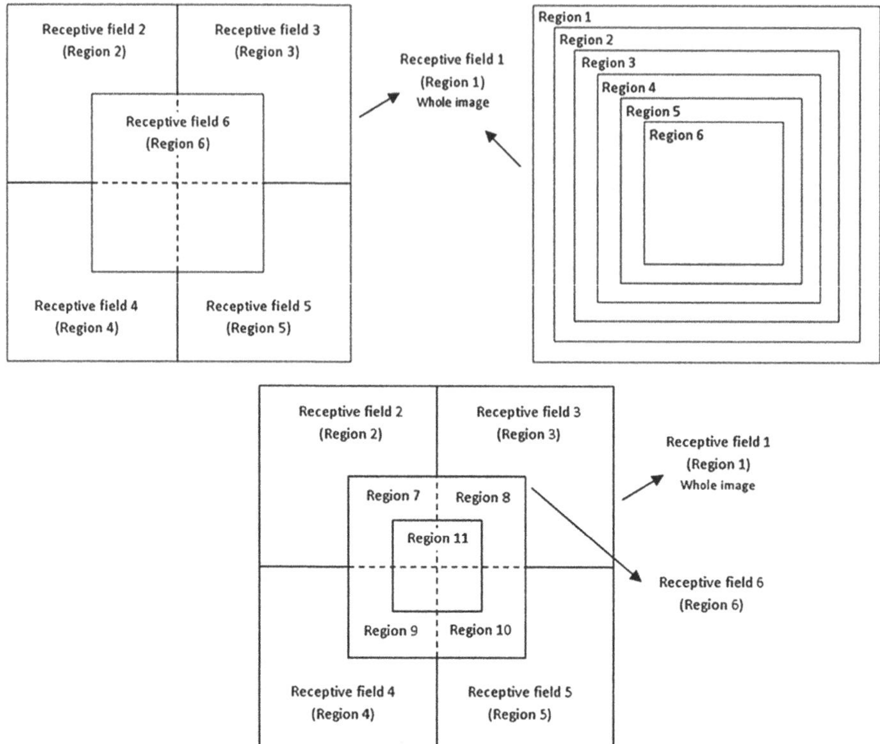

Fig. 2. The three arrangements of receptive field regions $\mathbf{P}^{(k)}$ considered for this work, each region corresponding to a feature detector. The arrangement of receptive fields on the top left (ARR1), with $k = 1$ to 6, is such that the four receptive fields are focused on the four corners of the image. For the arrangement of receptive fields on the top right (ARR2), with $k = 1$ to 6, the receptive fields prioritize the patterns at the center of the input image. The bottom arrangement (ARR3) of receptive fields, with $k = 1$ to 11, is somewhat a combination of the two top arrangements, with additional five regions at the center of the input pattern.

such that the input values to the classifier are from 0 to 1. $\mathbf{P}_a^{(k)} \cdot \mathbf{w}_b$ is the dot product between the ath input pixel vector and the bth node in the SOM feature detector and the $||\cdot||$ is the Frobenius norm operator. That is, using MNIST [8] dataset in this work, a is up to the number of training data being equal to 60,000 samples. Using the number of regions k in Fig. 2 and nodes evaluated from 2×2 to 15×15 SOM, i.e. the maximum value of b for a 2×2 SOM is 4 and 225 for a 15×15, then the combined MLP input is the concatenation of the similarity measure $\gamma_{a,b}$ of all nodes from all the k SOM regions. To be exact, the MLP input is of size $1 \times kB^2$, where B is the SOM size.

4 Results and Analysis

The novelty of this work is on the direct use of SOM as feature detectors for pattern recognition. As opposed to the feature detectors of CNN learned through carefully labeled data, our proposed approach is based on competitive learning of SOM, similar to the development of visual system detectors. Consequently, the developed feature detectors would provide insight as to the kind of input environment in which the feature detectors are exposed to.

To benchmark this proposed architecture, the classification performance for map size from 2 × 2 to 15 × 15, on three arrangements of receptive fields (see Fig. 2) were performed. MNIST dataset with 60,000 training samples and 10,000 test samples were used. The primary classifier that was considered is MLP.

Figure 3 summarizes the results of the benchmarking process for this test for three arrangements of the receptive fields shown in Fig. 2. Note that the receptive field arrangements define which part of the variation in the input image will have representations in the classifier in the form of input features. To reiterate, the first arrangement (ARR1) in Fig. 2 top left provides equal weights to the four corners and the center of the input image. The second arrangement (ARR2) in Fig. 2 top right is more focused on the variations in the center of the input image. Finally, the third arrangement (ARR3) in Fig. 2 bottom which is similar to the first arrangement, with additional receptive fields at the center of the input image.

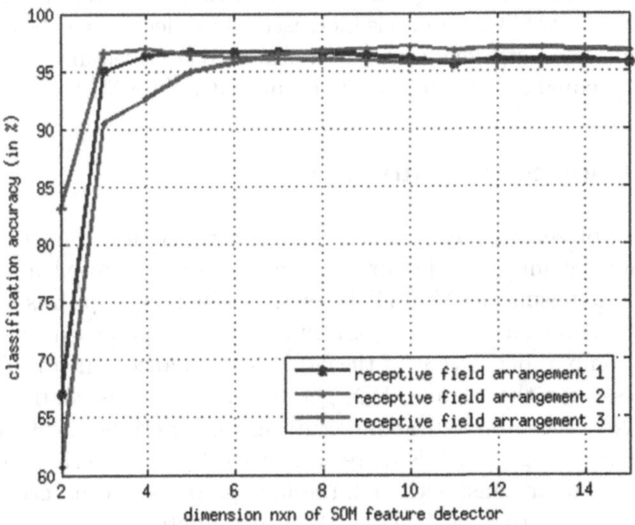

Fig. 3. Classification accuracy of the proposed network in different receptive field arrangements (refer to Fig. 2). For a 3 × 3 SOM, or 99 input features to MLP for ARR3, the performance of the proposed network, 96.54 %, is comparable to that the vanilla MLP (one hidden layer with hidden nodes being 256) with 97.23 %.

For a pure MLP classifier with input parameters being the raw image, i.e. 784 features corresponding to the 28×28 image, one hidden layer and 256 hidden nodes, the performance is 97.23 %. Comparing this to the values in Fig. 3, for ARR3, SOM size 3×3 up will result in a comparable performance with the vanilla MLP, i.e. 96.53 % vs. 97.23 % for the proposed network and vanilla MLP, respectively. This is despite the significant decrease in the number of input features, i.e. 99 features (9 nodes for 11 regions) for a 3×3 for ARR3. The highest performance for this arrangement is 96.93 % that is for SOM feature detectors of size 4×4. In terms of the number of input parameters without sacrificing the performance, our proposed architecture is a desirable alternative. For ARR2, the highest performance for our proposed architecture is achieved at 97.14 % and SOM size 10×10. This is despite of the decrease in the input parameter which is 600 features (100 nodes for 6 regions).

Comparing arrangements of 1 and 2, at lower number of nodes (SOM size $< 9 \times 9$), receptive field ARR1 has higher performance. For higher number of nodes (SOM size $> 7 \times 7$), ARR2 has higher performance. There are two parameters that could explain this, i.e. the number of SOM nodes and the positioning of the SOM receptive fields. SOM receptive fields for ARR1 is said to be more distributed as compared to the receptive fields of ARR2. Furthermore, SOM represents the input data using these nodes, i.e. higher number of nodes implies more vector quantization levels thus less vector quantization error. At lower number of nodes, the high vector quantization error in ARR1 performance is compensated by the distribution of its receptive fields, which is not present in ARR2. Nevertheless, ARR2 performance increases for SOM size greater than 7×7, as the number of SOM nodes is increased, thus, decreasing the quantization error. The high-est performance for our proposed architecture is from ARR2, with SOM size equal to 10×10, or 600 input features to MLP.

5 Conclusion and Future Works

This work has presented a new biologically-inspired architecture using directly the self-organizing maps for feature extraction. The proposed architecture is initially evaluated using the MNIST dataset, on different map sizes and different receptive field arrangements. The performance of the proposed network which is 96.93 % is comparable to that of the MLP performance which is 97.23 %, and almost 1 % less than the 2-stage CNN performance which is 98.01 %.

Unlike CNN, the proposed architecture has competitive mechanism at the feature extraction stage which is more similar to the characteristic of the visual system of higher organisms. Although the idea is similar to the neocognitron in terms of this competitive mechanism, the feature detectors of our work is randomly initialized rather than handpicked. The development of feature detectors is guided by the kind of input patterns in which the feature detectors are exposed to.

However, our proposed architecture is yet to be evaluated on its invariance to small distortion and translation, which are the main characteristics of CNN.

Thus, future works for this architecture include the evaluation on these characteristics as well as applications to more complex patterns e.g. vehicle, face and scene analysis. Further works will also include the behavior of this proposed visual system model to a constrained input environment to test the competitive mechanism in the learning process.

References

1. Arevalo, J., Cruz-Roa, A., Arias, V., Romero, E., Gonzalez, F.: An unsupervised feature learning framework for basal cell carcinoma image analysis. Artif. Intell. Med. **64**(2), 131–145 (2015)
2. Dong, Z., Wu, Y., Pei, M., Jia, Y.: Vehicle type classification using semisupervised convolutional neural network. IEEE Trans. Intell. Transp. Syst. **16**(4), 2247–2256 (2015)
3. Fukushima, K.: Neocognitron: a self-organizing neural network model for a mechanism of pattern recognition unaffected by shift in position. Biol. Cybern. **36**, 193–202 (1980)
4. Hubel, D.H., Wiesel, T.N.: Receptive fields, binocular interaction, and functional architecture in the cat's visual cortex. J. Physiol. **106**, 106–154 (1962)
5. Kohonen, T.: Self-organizing maps. Springer, London (2001)
6. LeCun, Y., Boser, B., Denker, J.S., Henderson, D., Howard, R.E., Hubbard, W., Jackel, L.D.: Backpropagation applied to handwritten zip code recognition. Neural Comput. **1**(4), 541–551 (1989)
7. LeCun, Y., Bottou, L., Bengio, Y., Haffner, P.: Gradient-based learning applied to document recognition. Proc. IEEE **86**, 11 (1998)
8. LeCun, Y., Cortes, C., Burges, C.J.: The MNIST database of handwritten digits (1998). http://yann.lecun.com/exdb/mnist/
9. Mohebi, E., Bagirov, A.: A convolutional recursive modified Self Organizing Map for handwritten digits recognition. Neural Netw. **60**, 104–115 (2014)
10. Omatu, S., Yano, M.: E-nose system by using neural networks. Neurocomputing **172**, 394–398 (2015)
11. Pomerleau, D.A.: ALVINN: An autonomous land vehicle in a neural network. In: Advances in Neural Information Processing Systems 1. Denver, Colorado, USA (1988)
12. Ramirez-Quintana, J.A., Chacon-Murguia, M.I.: Self-adaptive SOM-CNN neural sys-tem for dynamic object detection in normal and complex scenarios. Pattern Recogn. **48**(4), 1137–1149 (2015)
13. Ramirez-Quintana, J.A., Chacon-Murguia, M.I.: Self-organizing retinotopic maps applied to background modeling for dynamic object segmentation in video sequences. In: Proceedings International Joint Conference on Neural Networks, Dallas (2013)
14. Shahamiri, S.R., Salim, S.S.B.: Real-time frequency-based noise-robust automatic speech recognition using multi-nets artificial neural network. Neurocomputing **129**, 199–207 (2014)
15. Siniscalchi, S.M., Svendsen, T., Lee, C.H.: An artificial neural network approach to automatic speech processing. Neurocomputing **140**, 326–338 (2014)

A Review of Electroencephalogram-Based Analysis and Classification Frameworks for Dyslexia

Harshani Perera(✉), Mohd Fairuz Shiratuddin, and Kok Wai Wong

School of Engineering and Information Technology,
Murdoch University, Murdoch, Australia
H.Perera@murdoch.edu.au

Abstract. Dyslexia is a hidden learning disability that causes difficulties in reading and writing despite average intelligence. Electroencephalogram (EEG) is one of the upcoming methods being researched for identifying unique brain activation patterns in dyslexics. This paper examines pros and cons of existing EEG-based analysis and classification frameworks for dyslexia and recommends optimizations through the findings to assist future research.

Keywords: Dyslexia · Electroencephalogram · Feature extraction · Artifact removal · Artifact subspace reconstruction · Support vector machine · Classification

1 Introduction

Dyslexia is a disability with a neurological origin that involves deficiencies in reading and writing capabilities but does not affect intellect [1]. The conventional dyslexia detection techniques are often based on 'behavioral' symptoms which are assessed using standardized tests such as WIAT (Wechsler Individual Achievement Test), CTOPP (Comprehensive Test of Phonological Processing), OWLS (Oral and Written Language Scales), WJ (Woodcock Johnson) are used to assess reading, writing, IQ, memory and phonological processing abilities. Detecting dyslexia based on behavioral aspects can become tedious and time-consuming. Because of the changing nature of symptoms from person to person it can be difficult and challenging to pin point whether a person is dyslexic or not [2].

Studies show [3] that dyslexics have unique brain structures and behaviors. EEG, a 'record of the oscillations of brain electric potential recorded from electrodes on the human scalp' [4:3] is one of the methods that can be used to reveal indicators that are present inside the brains of dyslexics.

This paper evaluates the existing EEG-based analysis and classification frameworks for dyslexia. Each framework is assessed using a pre-defined format to arrange the data in a meaningful manner and to recognize its strengths and weaknesses. These discoveries are then used to propose an improved EEG-based classification framework for dyslexia.

A. Hirose et al. (Eds.): ICONIP 2016, Part IV, LNCS 9950, pp. 626–635, 2016.
DOI: 10.1007/978-3-319-46681-1_74

2 What Are the Existing Frameworks and Their Shortcomings?

Given below is an overview of the review process, which consists of 5 main steps. Each framework will be analyzed taking into account the following criteria (Fig. 1).

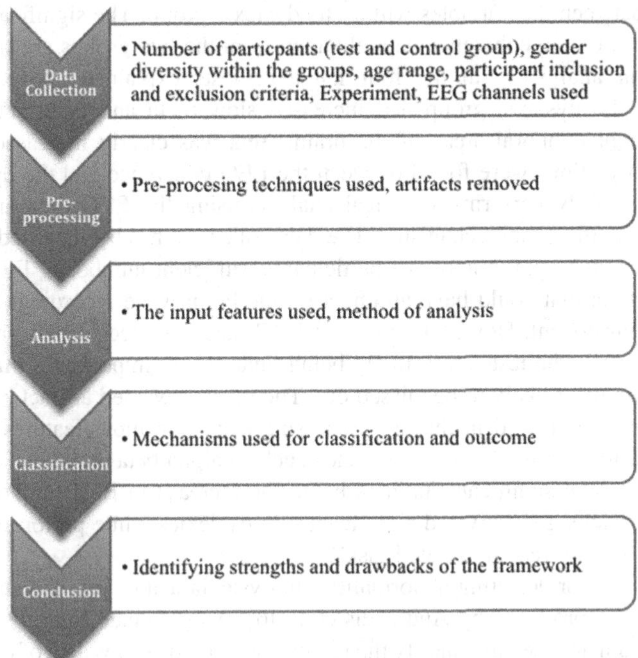

Fig. 1. Overview of the review process

A study carried out by Arns et al. [5] was able to uncover unique brain activation patterns in dyslexic children. A total of 38 participants: 19 Dyslexic (11 males and 8 females) and 19 Control (11 males and 8 females) between the ages of 8 to 16 years took part in this study. The exclusion criteria included mental illness or genetic disorders in person or family history, neurological disorder, brain injury, addiction to drug or alcohol and serious medical conditions. The EEG data was acquired at a sampling rate of 500 Hz using the 10–20 electrode positioning system having 28 channels. The experiment was performed in a sound, light and temperature controlled room. EEG was recorded for 2 min while being seated with eyes open, focusing the attention on a red dot displayed on a computer screen. The group of participants with dyslexia was also given few language tests. These tests consist of articulation, rapid naming of letters, phoneme deletion and spelling. These reading related tasks were collected to find the correlation between EEG and the neurological findings of dyslexia. However, EEGs were not recorded while these tasks were performed, instead the above explained tasks with eyes open was used since the EEG of resting state highly correlated with the tests. The data is EOG corrected prior to the analysis. This data is then examined using the power spectral analysis. The approach followed is that the data is first partitioned to adjacent 4-second sections, next the data is transformed to the frequency domain from the time domain using Fast Fourier Transform (FFT) and finally the average power spectra is calculated for specified frequency bands ranging within the delta, theta and alpha bands. The EEG is then analyzed statistically using one-way ANOVA to find the significant differences between the groups. Further a correlation matrix is acquired for correlations between the variables within the dyslexic group. The significant measures of the EEG power and coherence data obtained from the two groups are submitted for the correlation analysis with the four language tests explained above. The study revealed that the dyslexic group had increased slow theta and delta activity in the frontal and right temporal areas of the brain. Beta was clearly increased at F7 and significant correlations were found between the EEG coherence and the dyslexia tests [5]. This study only performs statistical analysis using the EEG data and does not present any classification mechanisms. The data collection has been carried out wisely, taking into account equal number of participants, sufficient number of EEG channels, excluding criteria that could have an effect on the brain wave recordings and using a controlled environment. However, since the EEG data is collected only in the resting state and not while the tests are actually being undertaken, important artifacts specific to each task are most likely to be missed out. The main unwanted artifact being the eye blinks have been removed in the pre-processing step. The input features include the power spectra for specified frequency bands such as alpha beta and theta at each EEG channel. One of the significant findings being the increase in Beta frequency verifies that the brain waves get activated significantly in dyslexics while performing tasks, in this case specifically reading related tasks.

A framework for detecting abnormalities in dyslexia using approximate entropy of EEG signals was proposed by Andreadis et al. [6]. Approximate Entropy (ApEn) is a 'statistical parameter used to quantify the regularity of a time series data of physiological signals' [6]. This study consisted of a total of 57 participants: 38 Dyslexic (26 males and 12 females) and 19 Control (7 males and 12 females) between the ages of 2 to 13 years. The exclusion criterion comprises of difficulties in hearing, history of head injury, neurological diseases or attention deficit disorders. The EEG was recorded using the international 10–20 system, containing 15 channels. The experiment for this study is that a single sound tone was presented to the participant via earphones, which was of a high frequency of 3000 Hz or low frequency of 500 Hz, followed by numbers that had to be memorized. The brain wave data was collected as EEG signal for 500 ms before the stimulus and as Event Related Potential (ERP) after the stimulus for 1000 ms. The pre-processing mechanisms include two main steps. The first step was recording the EOG and rejecting values higher than 75 μV and the second step was normalizing the waveforms by subtracting the mean value and dividing by the standard deviation of each signal. This data is then analyzed using ApEn and Cross-ApEn (comparing EEG signals from two electrodes). A Support Vector Machine (SVM) classifier was then implemented using the statistical significant electrodes for all subjects obtained using ApEn as input features. This classifier offered promising results achieving a sensitivity of 89.47 % and specificity of 57.89 %. The study was then taken a next step forward to enhance the classifier using the input features from Cross-ApEn. This method looks at significant pairs of electrodes instead of evaluating electrodes on its own. Although is technique delivered better discrimination abilities, no clear pattern has yet been found because there was a very high number of statistically significant pairs of electrodes. Looking at the study as a whole, a successful classifier was developed to differentiate between the groups. However, the experiment used looks into only the working memory abilities and does not involve any reading or writing related elements. The same research team performed another analysis using the same experiment and data by using Wavelet Entropy [7]. The findings revealed that Wavelet Entropy could be used as a quantified measure to observe and analyze EEG and ERP signals to detect brain patterns specific to dyslexia.

A Malaysian research team conducted a frequency analysis of EEG signals generated between dyslexic and normal children during writing [8, 9]. The EEGs were recorded from total of 6 right-handed children: 3 Dyslexic and 3 Control subjects between the ages of 8–12 years using the international 10–20 system. This study uses only 4 EEG channels. EEG is recorded in the relaxed state and while writing. During the pre-processing phase, Electrocardiograms (ECG) and EOG were filtered out. Next the signals containing the writing related data was extracted using a band pass FIR filter ranging from 8 Hz to 30 Hz. For the frequency analysis, the signals are transformed to the frequency domain from the time domain using Fast Fourier Transform (FFT). The study revealed that the dyslexic children consume more energy and resulting in high frequency beta wave relaxed states and well as during writing related activities compared to normal children. The frequency range identified for dyslexic children is between 22–28 Hz whereas for non-dyslexic children it is between 14–22 Hz. Overall this study does only frequency analysis and does not provide any classification mechanism. The number of channels and participants are too small to arrive at a conclusion for using these results for a framework to discriminate between the dyslexic and the non-dyslexic. The study has explicitly used subject that are right-handed, which is important since the handedness has an effect on the EEG [10]. However, excluding factors and that could have an effect on the EEG recordings have not been taken into consideration. Additionally is it not indicated whether a silent and temperature controlled room was used to carry out the experiment. The pre-processing techniques used in this study is similar to previous similar studies, however since this study involves hand movements, it is not specified how the artifacts generated from the hand movements were filtered out. Further the experiment focuses only on the writing related tasks.

Frid, Breznitz [11] proposed a SVM based algorithm for differentiating between dyslexic readers and regular readers using ERPs. The study was carried out with 50 participants: 20 Dyslexic and 30 Control of the ages between 24–40 years. The signals were recorded at a sampling rate of 2048 Hz using the standard 10–20 system with 64 channels. The experiment used in the study is that the subject is required to press a button in response to a target stimulus, which is a tone. The conditions consist of 50 stimuli of target tones at frequencies of 1000 Hz and 50 non-target tones of 2000 Hz. The data collected is first pre-processed using a band pass filter at 0.1–100 Hz, and then a notch filter at 50 Hz is used to remove electric noise and finally filtering out eye and muscle movements. The five features selected are Positive Area (Ap), Maximal Peak

Amplitude/Time ratio (Mp), Spectral Flatness Measure (SFM), Standard Deviation and Skewness, Power Spectral Density (PSD). Although the classification was first attempted using a single classifier for all features, it was not successful. Therefore, the approach followed was to use ensemble SVMs. The classification results were compared for the combinations: the best single feature, an ensemble of three SVMs and only the left or right hemispheres. To recapitulate, the study uses a simple experiment task, which relates to working memory and reasoning abilities but does not engage any stimulus with regard to reading or writing. This may have bypassed on activating vital areas of the brain specific to dyslexia. The study does not indicate whether any inclusion and exclusion criteria were taken into account of participants.

A classification model to distinguish dyslexic children from the normal children during rest state was suggested by [12]. A total of 6 participants: 3 Dyslexic and 3 Control within the ages of 4 to 7 years took part in this study. The EEG is collected using the international 10–20 system using 8 channels with a sampling rate of 250 Hz. Experiment was done in a room with controlled temperature and lighting while the participants are in the resting state with both eyes closed and eyes open. During the pre-processing phase noise and irrelevant artifacts have been removed. Since the data collection is done in the resting state, the frequency band relating to this state is alpha, and this has been extracted using band-pass filtering. The next phase being the Feature Extraction is performed using Kernel Density Estimation (KDE). Finally the classifier is trained using Multilayer Perceptron (MLP) and was able to obtain an accuracy rate of 90 % in the classification. To wrap-up, the study uses EEG data from only the resting state disregarding the essential reading and writing related brain wave data. The number of participants and the number of channels used is quite low. No inclusion or exclusion criteria for participants used is indicated. Further, although the study gave a 90 % accuracy rate since the dataset used is very small it is not very encouraging.

A Wavelet Packet analysis of EEG signals between dyslexic and non-dyslexic children during writing was proposed by [13]. A total of 8 subjects: 4 dyslexics and 4 controls between the ages of 7 to 12 years took part in this study. The EEG was recorded in the temperature-controlled room at 24 °C using the international 10–20 system with 4 channels having a sample rate of 256 Hz. The signals were captured in the relaxed state, writing state and during letter recognition and each task was repeated 6 times. This is then examined using Wavelet packet Analysis for alpha and beta frequency bands. The outcome of the study discovered that there was no significant difference in the alpha band frequencies during the relaxed state and writing state in dyslexics, however for non-dyslexics the alpha band frequency was higher during relaxed state compared to writing state. During writing beta frequency was higher in dyslexics compared to non-dyslexics. This study looks into the brain behaviors during the resting and writing states, but does not look into the reading state. No information is provided about pre-processing the signal to remove unwanted artifacts such as eye blinks. The number of subjects and the number of channels used in the study is low compared to similar research [6, 11]. Finally the study performs only as analysis and does not perform any classifications.

3 Is There a Need for an Improved Framework?

3.1 Data Collection

Number of Participants. In research that deals with a unique dataset, the number of participants are often determined using the sample size of a similar study [14]. Accuracy could be made higher by getting the mean sample size of multiple similar studies (Table 1).

Table 1. Determination of number of subjects

Research	Test group size	Control group size	Total
[5]	19	19	38
[7]	38	19	57
[6]	38	19	57
[9]	3	3	6
[11]	20	30	50
[12]	3	3	6
[13]	4	4	8
Mean sample size (rounded)	18	15	32

Age Range and Gender. Similar studies have carried out on children as well as adults, which means either group can be used. However, it is important to make sure that the subjects of age range selected have parallel reading and writing abilities. Previous studies have not compared brain waves specific to gender. Therefore, for future work, the comparison between the female and male dyslexic brain wave patters is a gap to be filled.

EEG Channels. The popular choice of EEG channels was determined using channels specifically mentioned as prominent and channels that overlaps at least between 2 researches (Table 2).

Table 2. Popular choice of EEG Channels

Research	No of channels	Channels
[5]	28	Fp1, Fp2, F7, F3, Fz, F4, F8, FC3, FCz, FC4, T3, C3, Cz, C4, T4, CP3, CPz, CP4, T5, P3, Pz, P4, T6, O1, Oz, O2
[7]	15	Fp1, F3, C5, C3, Fp2, F4, C6, C4, O1, O2, P4, P3, Pz, Cz, Fz.
[6]	15	Fp1, F3, C5, C3, Fp2, F4, C6, C4, O1, O2, P4, P3, Pz, Cz, Fz.
[9]	4	C3, C4, P3, P4
[11]	64	F3, F4, P6, Pz, F8, CP4, AF7, F3, F5, T7, PO3, FC6, TP7, P7 (Not all)
[12]	8	F3, F4, C2, C3, C4, P3, P4, T3, T4
[13]	4	C3, C4, P3, P4
Popular	Fp1, F3, Fz, F4, F7, F8, T3, C3, Cz, C4, T4, Pz, AF3, TP7, P7	

Inclusion and Exclusion Criteria of the Subjects. The inclusion and exclusion criteria summarized from the reviews are given below. Exclusions: Mental illness, Genetic and Neurological disorders, Brain injuries, Drug or alcohol addiction, serious medical condition, Difficulties in hearing/vision – this would not apply if the subject has corrected vision/hearing, Attention deficit disorders Inclusions: Handedness – The participants' recruited need to be either left handed or right handed and not have a mix of the both. This is because there is a difference in EEG activities between the right-handed and left-handed subjects [10, 15].

Experiment. It can be presumed that dyslexia specific brain wave activation patterns are more prominent during performing reading and writing activities instead of having tasks that are only related to the working memory and reasoning. Reading related tasks and can be drilled down further to find out brain signal patterns while reading regular words against nonsense-words. Research [16] show that dyslexics perform worse in reading irregular and nonsense-words compared to regular words. Therefore including a task to read nonsense-words may show noticeable results. Today writing is often replaced by typing in day-to-day activities; therefore this too could be included in the tasks. Further a task with a combination of reading and writing can be incorporated.

3.2 Pre-processing

When recording EEG signals one of the most commonly seen irrelevant artifacts are the eye-movements and eye blinks. Typically body movements are kept to a minimum during EEG-based experiments. However, new methods have now been introduced making it possible to collect data during real-life activities instead of only collecting data during resting state or simple activities such as button clicks. Artifact Subspace Reconstruction (ASR) is one such method which can be used to filter out body movement and muscle burst artifacts from the EEG signals [17]. ASR 'relies on a sliding-window Principal Component Analysis, which statistically interpolates any high-variance signal components exceeding a threshold relative to the covariance of the calibration dataset. Each affected time point of EEG is then linearly reconstructed from the retained signal subspace based on the correlation structure observed in the calibration data' [17]. Another important aspect to be filtered prior to the analysis is the noise caused by electric power lines. This is often seen at 60 Hz or 50 Hz and this can be filtered out using a notch filter.

3.3 Analysis

One of the common analyses used in EEG-based pattern classification frameworks for dyslexia is the frequency analysis. The raw EEG signal recorded is in the time domain. This waveform is a combination of a number of sinusoidal waves although is it not directly visible. Fast Fourier Transform, commonly known as FFT can be used for the decomposition of the waveform into a sum of sinusoids of different frequencies. Therefore, by performing the FFT it helps detect spikes in the frequency domain which could have not been visible before. Once all the channels have been transformed from the time domain to the frequency domain, this could be decomposed into sub bands namely delta, theta, alpha and beta. This method allows analyzing the frequencies at specific frequency bands instead of analyzing each frequency in isolation. The most important step in the analysis phase is the extraction of features. This helps to analyze the data in terms of a reduced set of features instead of the large original input data set. The input features identified through the review are power spectral density, entropy, positive area, maximal peak amplitude/time ratio, spectral flatness measure, standard deviation and skewness. Energy, average valley amplitude, peak variation, root mean square and power are few of the features used in recent EEG related studies that could be incorporated in EEG-based pattern classification for dyslexia frameworks as well.

3.4 Classification

Linear Discriminant Analysis (LDA), Neural Networks (NN) or SVM. LDA classifies data by first creating 'models of the probability density functions for data generated from each class. Then, a new data point is classified by determining the probability density function whose value is larger than the others' [18]. The algorithm 'assumes that each of the class probability density functions can be modeled as a normal density, and that the normal density functions for all classes have the same covariance' [18]. LDA is known to be a simple classifier that requires very small computations. However, this algorithm is not suitable for complex non-linear EEG classifications since it does not produce good results for such scenarios [19].

NN is 'an assembly of several artificial neurons which enables to produce nonlinear decision boundaries' [19]. NN perform better for EEG classifications compared to LDA since it can be used to implement boundaries for non-linear classifications. Nevertheless, to acquire the desired level of accuracy, it is important to choose suitable number of hidden units, which can become problematic. Having a larger number of hidden units than required results in memorizing the training set which causes poor generalization [20].

SVM is a supervised learning method, which can handle both linear and non-linear classifications. It produces a hyper-plane having the maximal margin to the support vectors. SVM can classify even overlapping and non-separable data sets by mapping into higher dimensional spaces using the kernel functions [20, 21].

Popular Classification Technique. Through the comparison it can be concluded that SVM is a better choice. Further research [19, 20] has recommended SVM as a more appropriate choice for EEG signal classifications.

4 Conclusion

Research shows distinctions in the brain wave patterns and brain structures of dyslexics compared to non-dyslexics and many research have attempted to introduce and improve EEG-based classification frameworks. According to the review it was revealed that frameworks require a minimum of 15 subjects per each group, the studies could be conducted on children or adults and comparison between the female and male dyslexic brain wave patterns need to be conducted. It is also important to identify the inclusion and exclusion criteria prior to the data collection to minimize the amount of outliers. It was discovered that experiments used were simple tasks, which measures working memory and reasoning abilities instead of reading and writing abilities. This could be because to reduce the unwanted artifacts caused by body movements in the EEG signals during reading and writing activities. We have proposed using ASR; a successful method that has been used in recent studies to filter out body movement and muscle burst artifacts from the EEG signals [17]. Finally we have proposed more input features and recommended SVM as the classifier to be used in EEG-based classification frameworks for dyslexia.

References

1. Fletcher, J.M., Lyon, G.R., Fuchs, L.S., Barnes, M.A.: Learning Disabilities: From Identification to Intervention. Guilford Press, New York (2006)
2. Ekhsan, H.M., Ahmad, S.Z., Halim, S.A., Hamid, J.N., Mansor, N.H.: The implementation of interactive multimedia in early screening of dyslexia. In: 2012 International Conference on Innovation Management and Technology Research (ICIMTR), 21–22 May 2012, pp. 566–569 (2012). doi:10.1109/ICIMTR.2012.6236459
3. Mohamad, S., Mansor, W., Lee, K.Y.: Review of neurological techniques of diagnosing dyslexia in children. In: 2013 IEEE 3rd International Conference on System Engineering and Technology (ICSET), 19–20 August 2013, pp. 389–393 (2013). doi:10.1109/ICSEngT.2013.6650206
4. Nunez, P.L., Srinivasan, R.: Electric Fields of the Brain: The Neurophysics of EEG. Oxford University Press, Oxford (2006)
5. Arns, M., Peters, S., Breteler, R., Verhoeven, L.: Different brain activation patterns in dyslexic children: evidence from EEG power and coherence patterns for the double-deficit theory of dyslexia. J. Integr. Neurosci. **6**(1), 175–190 (2007). doi:10.1142/S0219635207001404
6. Andreadis, I.I., Giannakakis, G.A., Papageorgiou, C., Nikita, K.S.: Detecting complexity abnormalities in dyslexia measuring approximate entropy of electroencephalographic signals. In: Annual International Conference of the IEEE Engineering in Medicine and Biology Society, EMBC 2009, 3–6 September 2009, pp. 6292–6295 (2009). doi:10.1109/IEMBS.2009.5332798
7. Giannakakis, G.A., Tsiaparas, N.N., Xenikou, M.F.S., Papageorgiou, C., Nikita, K.S: Wavelet entropy differentiations of event related potentials in dyslexia. In: 8th IEEE International Conference on BioInformatics and BioEngineering, BIBE 2008, 8–10 October 2008, pp. 1–6 (2008). doi:10.1109/BIBE.2008.4696836
8. Che Wan Fadzal, C.W.N.F., Mansor, W., Lee, K.Y., Mohamad, S., Amirin, S.: Frequency analysis of EEG signal generated from dyslexic children. In: 2012 IEEE Symposium on Computer Applications and Industrial Electronics (ISCAIE), 3–4 December 2012, pp. 202–204 (2012). doi:10.1109/ISCAIE.2012.6482096
9. Che Wan Fadzal, C.W.N.F., Mansor, W., Lee, K.Y., Mohamad, S., Mohamad, N., Amirin, S.: Comparison between characteristics of EEG signal generated from dyslexic and normal children. In: 2012 IEEE EMBS Conference on Biomedical Engineering and Sciences (IECBES), 17–19 December 2012, pp. 943–946 (2012). doi:10.1109/IECBES.2012.6498210

10. Andrew Ng, C.R., Leong, W.Y.: An EEG-based approach for left-handedness detection. Biomed. Sig. Process. Control **10**, 92–101 (2014). doi:10.1016/j.bspc.2014.01.005
11. Frid, A., Breznitz, Z.: An SVM based algorithm for analysis and discrimination of dyslexic readers from regular readers using ERPs. In: 2012 IEEE 27th Convention of Electrical & Electronics Engineers in Israel (IEEEI), 14–17 November 2012, pp. 1–4 (2012). doi:10.1109/EEEI.2012.6377068
12. Karim, I., Abdul, W., Kamaruddin, N.: Classification of dyslexic and normal children during resting condition using KDE and MLP. In: 2013 5th International Conference on Information and Communication Technology for the Muslim World (ICT4 M), 26–27 March 2013, pp. 1–5 (2013). doi:10.1109/ICT4M.2013.6518886
13. Fuad. N., Mansor, W., Lee, K.Y.: Wavelet packet analysis of EEG signals from children during writing. In: 2013 IEEE Symposium on Computers & Informatics (ISCI), 7–9 April 2013, pp, 228–230 (2013). doi:10.1109/ISCI.2013.6612408
14. Israel, G.D.: Determining sample size. University of Florida Cooperative Extension Service, Institute of Food and Agriculture Sciences, EDIS (1992)
15. Provins, K.A., Cunliffe, P.: The relationship between E.E.G. activity and handedness. Cortex **8**(2), 136–146 (1972). doi:10.1016/S0010-9452(72)80014-5
16. Ziegler, J.C., Castel, C., Pech-Georgel, C., George, F., Alario, F.X., Perry, C.: Developmental dyslexia and the dual route model of reading: simulating individual differences and subtypes. Cognition **107**(1), 151–178 (2008). doi:10.1016/j.cognition.2007.09.004
17. Mullen, T., Kothe, C., Chi, Y.M., Ojeda, A., Kerth, T., Makeig, S., Cauwenberghs, G., Jung, T.-P.: Real-time modeling and 3D visualization of source dynamics and connectivity using wearable EEG. In: 35th Annual International Conference of the IEEE Engineering in Medicine and Biology Society (EMBC), United States, pp. 2184–2187. IEEE (2013). doi:10.1109/EMBC.2013.6609968
18. Eslahi, S.V., Dabanloo, N.J.: Fuzzy support vector machine analysis in EEG classification (2013)
19. Lotte, F., Congedo, M., Lécuyer, A., Lamarche, F., Arnaldi, B.: A review of classification algorithms for EEG-based brain–computer interfaces. J. Neural Eng. **4**, R1 (2007)
20. Garrett, D., Peterson, D.A., Anderson, C.W., Thaut, M.H.: Comparison of linear, nonlinear, and feature selection methods for EEG signal classification. IEEE Trans. Neural Syst. Rehabil. Eng. **11**(2), 141–144 (2003). doi:10.1109/TNSRE.2003.814441
21. Liu, S., Song, Q., Hu, W., Cao, A.: Diseases classification using support vector machine (SVM). In: Proceedings of the 9th International Conference on Neural Information Processing, ICONIP 2002, 18–22 November 2002, vol. 762, pp. 760–763 (2002). doi:10.1109/ICONIP.2002.1198160

Rule-Based Grass Biomass Classification for Roadside Fire Risk Assessment

Ligang Zhang(✉) and Brijesh Verma

Centre for Intelligent Systems,
Central Queensland University, Rockhampton, Australia
{l.zhang, b.verma}@cqu.edu.au

Abstract. Roadside grass fire is a major hazard to the security of drivers and vehicles. However, automatic assessment of roadside grass fire risk has not been fully investigated. This paper presents an approach, for the first time to our best knowledge, that automatically estimates and classifies grass biomass for determining the fire risk level of roadside grasses from video frames. A major novelty is automatic measurement of grass coverage and height for predicting the biomass. For a sampling grass region, the approach performs two-level grass segmentation using class-specific neural networks. The brown grass coverage is then calculated and an algorithm is proposed that uses continuously connected vertical grass pixels to estimate the grass height. Based on brown grass coverage and grass height, a set of threshold based rules are designed to classify grasses into low, medium or high risk. Experiments on a challenging real-world dataset demonstrate promising results of our approach.

Keywords: Fire risk assessment · Roadside image segmentation · Neural networks

1 Introduction

Roadside fire risk assessment plays an important role for effectiveness management of roadsides, identification of fire-prone road regions, and minimization of hazards imposed to drivers and vehicles. One of the main resources that cause the risk is roadside grasses, which can grow into high, brown and dense stems characterized by high biomass outputs that potentially lead to a high change of fire burning. The biomass is typically defined as the over-dry mass of the above ground portion of a group of vegetation in forestry [1] and has been found as having a positive correlation with the fire risk [2]. Accurately estimating the biomass of roadside grasses plays an important role in assisting the determination of fire risk levels and implementing appropriate treatments to prevent or suppress the occurrence of fire. However, current practice of relevant authorities is heavily dependent on manual measurements by humans, which suffers from big labour, time and cost investments. Thus, it is very important to develop automatic and efficient methods for the estimation of the biomass of roadside grasses and the assessment of fire risk.

In this paper, a rule based grass biomass classification approach is proposed for roadside fire risk assessment on video data collected using vehicle-mounted cameras. Class-specific Artificial Neural Networks (ANNs) are firstly trained based on color and

A. Hirose et al. (Eds.): ICONIP 2016, Part IV, LNCS 9950, pp. 636–644, 2016.
DOI: 10.1007/978-3-319-46681-1_75

texture features to discriminate between grass and non-grass, as well as between green and brown grasses. The grass height and brown grass coverage are then estimated and used to determine fire risk levels, namely low, medium and high, based on a set of threshold-based rules. Our approach is inspired by measurement methods in grass curing, which estimate grass biomass based on a 2D chart comprising of grass coverage and grass height. The main contributions of this paper are as follows: (1) To the best of our knowledge, this is the first study that attempts to automatically assess the fire risk of roadside grasses from ground based images. (2) We propose a Vertical Orientation Connectivity of Grass Pixels (VOCGP) algorithm to estimate the grass height, which is still a challenging issue and has not been investigated previously. (3) We create a natural image evaluation dataset from video data collected by the Department of Transport and Main Roads (DTMR), Queensland, Australia, and demonstrate promising results of the proposed approach.

The rest of the paper is organized as follows. Section 2 discusses related work. Section 3 introduces the proposed approach. The experiments are presented in Sect. 4 and finally Sect. 5 draws the conclusions.

2 Related Work

Existing solutions into estimating grass biomass roughly fall into three groups:

(1) Traditional field tests, which involve destructive samplings of plants at different growth stages, counting of the number of plants contained in the sample and calculating the weighting after over-drying them [3]. This method is generally of high accuracy but requires time- and labour-intensive human efforts and big costs. It may also suffer from the difficulty accessing some areas due to geographical conditions and the requirement of permission from private landowners.

(2) Human visual observations, which visually check growing conditions of grasses (e.g. height and brownness) by human eyes and assign a corresponding value of biomass based on pre-defined relationships between them. Similar to field tests, this method often requires considerable time, labour and cost investments, and expertise in the field, and is infeasible for a large-scale field survey.

(3) Remote sensing methods, which automatically measure the vegetation biomass and identify fire risk primarily using chlorophyll-related vegetation index (VI), including Normalized Difference Vegetation Index (NDVI) [4], Accumulated Relative Normalised Difference (ARND) [5], relative greenness [6], and combination of VIs with factors, such as temperature, vegetation cover, and humidity [7]. These methods often focus on large-scale fields and have difficulty supporting site-specific analysis. They also suffer from high expenses and are prone to atmospheric conditions.

3 Proposed Approach

Figure 1 shows an overview of the framework for the proposed approach. For a sampling grass region in an input roadside image, both color and texture features are extracted to represent visual characteristics of grasses, and they are further used to

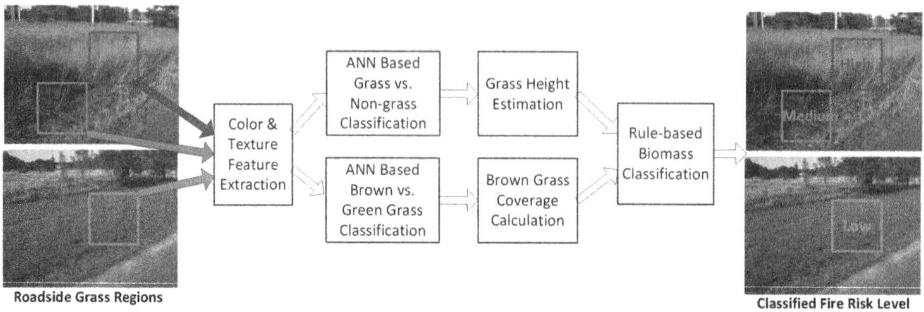

Fig. 1. Framework of the proposed approach for roadside grass fire risk assessment. (Color figure online)

classify grass vs. non-grass, and green vs. brown grasses using class-specific binary ANNs. From segmented grass regions, the brown grass coverage is calculated and a VOCGP algorithm is proposed to estimate the grass height. Based on predicted grass coverage and height, we further propose a set of threshold-based rules to classify the grass region into one of three categories of fire risk, namely low, medium and high.

3.1 Color and Texture Feature Extraction

Effective feature representation of grasses is crucial in grass segmentation. However, there is no commonly accepted set of features that can work for any types of natural grasses. This paper considers two commonly used features - color and texture.

Color: this paper adopts the CIELab, which has high perceptually consistency with human vision and good performance on roadside image analysis [8]. We also include RGB as it may contain complementary information for roadside objects. For a pixel at the coordinate (x, y) in an image, its color features are composed of:

$$\mathrm{V}^{\mathrm{c}}_{\mathrm{x,y}} = [\mathrm{R}, \mathrm{G}, \mathrm{B}, \mathrm{L}, a, b] \tag{1}$$

Texture: this paper uses 17-D filter banks [9], due to their high accuracy for object classification. The filter banks include Gaussians with 3 different scales (1, 2, 4) applied to L, a, and b channels, Laplacians of Gaussians with 4 different scales (1, 2, 4, 8) and the derivatives of Gaussians with two different scales (2, 4) for each axis (x and y) on the L channel. For a pixel at (x,y), its texture features are composed of:

$$\mathrm{V}^{\mathrm{t}}_{\mathrm{x,y}} = [\mathrm{G}^{\mathrm{L}}_{1,2,4}, \mathrm{G}^{\mathrm{a}}_{1,2,4}, \mathrm{G}^{\mathrm{b}}_{1,2,4}, \mathrm{LOG}^{\mathrm{L}}_{1,2,4,8}, \mathrm{DOG}^{\mathrm{L}}_{2,4,\mathrm{x}} \mathrm{DOG}^{\mathrm{L}}_{2,4,\mathrm{y}}] \tag{2}$$

3.2 Grass Region Segmentation

To get site-specific parameters of grasses for the estimation of their biomass, we need to segment grass from non-grass regions and also distinguish brown from green grasses. To complete this task, this section employs two separate class-specific binary neural networks based on the extracted color and texture features.

The ANN takes an input feature vector $V_{x,y} = [\mathrm{V}^{\mathrm{c}}_{\mathrm{x,y}}, \mathrm{V}^{\mathrm{t}}_{\mathrm{x,y}}]$ and outputs a probability for both classes using Eq. 3.

$$p^{i}_{x,y} = tran(w_i V_{x,y} + b_i) \quad (3)$$

where, *tran* stands for the prediction function of a three-layer ANN with a tan-sigmoid activation function, and w_i and b_i are the trained weights and constant parameters for the *i*-th object class. The class with the highest probability across the two classes wins the classification label using Eq. 4.

$$C^{i}_{x,y} = \max_{i \in C} p^{i}_{x,y} \quad (4)$$

where, C stands for two classes (i.e. grass vs. non-grass, or green vs. brown grasses), and $C^{i}_{x,y}$ represents the *i*-th object label for a pixel at (*x*,*y*).

3.3 Grass Coverage and Height Calculation

This part calculates brown grass coverage and grass height to estimate the grass biomass. They have been used as two critical factors in human visual measurement of grass biomass in grass curing (e.g. Country Fire Authority, Victoria, Australia).

(1) Brown grass coverage. The biomass of grasses may vary significantly dependent on the coverage of brown grasses in different seasons. We calculate the percentage of brown grass pixels as an indication of the coverage: N_{bg}/N, where N_{bg} and N are the numbers of brown grass pixels and all pixels in a sampling region, respectively.
(2) Grass height. The grass height has been found having a statistically close relationship with biomass yield [10, 11]. Automatic measurement of the grass height in roadside images is still a difficult task. We propose a VOCGP algorithm that uses vertical orientation connectivity of grass pixels to predict the grass height. The algorithm includes two main steps:
    (a) Detecting the strongest texture orientation at each pixel. For the sampling region W, it is first converted to a grey scale by averaging R, G and B values. The responses of Gabor filters $F^{\theta,\phi}$ at an orientation θ and a sale ϕ can be obtained by convolving the filter with all pixels in W: $G^{\theta,\phi} = W \oplus F^{\theta,\phi}$. The real and imaginary components of the output $G^{\theta,\phi}$ are combined in a square norm to indicate the absolute response strength of Gabor filters. The responses are then averaged over all scales to produce a single response for each orientation. The dominant orientation $O_{x,y}$ of the pixel at (*x*, *y*) can be obtained by performing a vote on responses along all orientations and taking the maximum one using Eq.Â 5.

$$O_{x,y} = j \quad if\, \bar{G}^{j}_{x,y} = \max_{k=1,\ldots,N_\theta} \bar{G}^{k}_{x,y} \quad (5)$$

where, N_θ is the number of orientations θ, $\bar{G}^{k}_{x,y}$ is the average response along the *k*-th orientation for the pixel at (*x*, *y*), $N_\theta \geq k \geq 1$, and $N_\theta \geq j \geq 1$.

(b) Obtaining the length of grass pixels. For the each column in W, we calculate all lengths of continuously connected grass pixels and take the longest length. The longest lengths are averaged over all columns to derive an estimation of the grass height in the sampling region.

3.4 Rule-Based Fire Risk Classification

Once brown grass coverage and grass height are calculated for estimating the biomass, we can proceed to determine the level of fire risk using a set of threshold-based rules. This rule-based method is motivated by human visual measurement in grass curing, which uses a 2D table (height and coverage) to determine the biomass.

Table 1 shows the rules used in the proposed approach, which classifies the fire risk into three levels of low, medium, or high based on brown grass coverage and grass height. Unlike the measurement approach in grass curing that quantizes the values of both grass height and coverage into ten equal ranges, our approach considers only three ranges by setting two thresholds for each of them. Note that the grass coverage is represented in a form of percentages and varies between [0 1], while the grass height can be any value dependent on the size of the sampling region. The values for thresholds T_c^1, T_c^2, T_h^1 and T_h^2 are set experimentally.

Table 1. Threshold-based rules for classifying fire risk based on brown grass coverage and grass height.

		Grass height		
		$< T_h^1$	T_h^1 - T_h^2	$> T_h^2$
Brown grass coverage (%)	0.0 - T_c^1	Low	Low	Medium
	T_c^1 - T_c^2	Low	Medium	High
	T_c^2 - 1.0	Medium	High	High

4 Experiments

4.1 Evaluation Dataset and System Parameters

There is no public image dataset that contains ground truths of fire risk levels of roadside grasses. Thus, we create an image dataset from video data collected by the DTMR, Queensland, Australia. The video was captured using a left-view vehicle-mounted camera and has a resolution of 1632 × 1248 pixels. 100 frames are selected from 22 videos for a state road No. 16A in the Fitzroy region. In each frame, 15 overlapped sampling regions as shown in Fig. 3 are selected and manually annotated into a category of low, medium, or high risk by an expert on grass fire assessment.

The size of Gaussian filters is set to 7 × 7 pixels based on our previous results [12]. The Gabor filters have four orientations θ = (0°, 45°, 90°, 135°), five scales ($\phi_m = f_{max}/(\sqrt{2})^m, m = 0, 1, \ldots, 4,\ f_{max} = 0.25$) and 11 × 11 Gabor kernels.

The ANN has three layers of 23-16-2, and is trained using a resilient backpropagation algorithm with a goal error of 0.001 and a maximum epoch of 200. The training data includes 650 cropped regions extracted from the DTMR data, covering both grass (green and brown grasses) and non-grass (road, tree, sky and soil) regions.

4.2 Global Accuracy

One key task of the proposed approach is to find the best values of the thresholds for brown grass coverage and grass height as listed in Table 1. Figure 2 shows average class accuracy vs. the values of thresholds for brown grass coverage and grass height respectively. For both grass coverage and height, there are substantial accuracy variations when different values of thresholds are used, and this confirms the big impact of the thresholds on the classification accuracy. For grass coverage, the accuracy increases gradually when the values of T_c^1 and T_c^2 are increasing until reaching 0.5 and 0.9 respectively, where the highest accuracy of 51.25 % is obtained. A similar result is also observed for grass height and the highest accuracy of 55.03 % is achieved when T_h^1 and T_h^2 equal to 29 and 36 respectively. Thus, these four values of thresholds are used for determining the fire risk level using the rules in Table 1, which produces global accuracy of 64.0 %. The relatively low global accuracy reflects that automatic grass fire risk assessment from unconstrained ground-based images is still a very challenging task. The grass segmentation and grass height estimation have to overcome problems such as substantial variations in visual appearance of objects and environmental effects such as lighting conditions and shadows of objects.

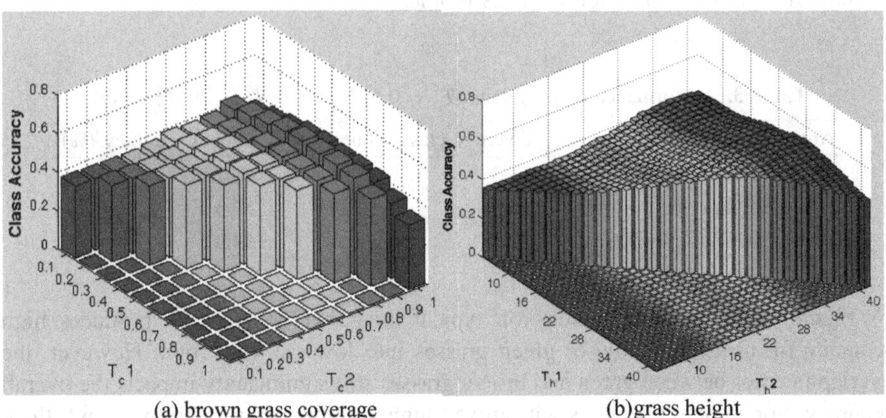

(a) brown grass coverage (b) grass height

Fig. 2. Average class accuracy obtained using different values of thresholds for (a) brown grass coverage and (b) grass height. The values for grass coverage range between [0.1, 1] with an increasing step of 0.1, while those for grass height range between [10, 40] with an increasing step of 1. Accuracy is shown in a half matrix due to conditions of $T_c^1 \leq T_c^2$ and $T_h^1 \leq T_h^2$.

4.3 Class Accuracy

Table 2 displays the confusion matrix for three levels of fire risk. It can be seen that low risk is the easiest one for classification with 74.7 % accuracy, which is followed by high risk with 63.1 % accuracy. By contrast, medium is the most difficult one with only 35.7 % accuracy. More than 36 % of medium regions are misclassified to high risk, and more than 27 % are misclassified to low risk. This is within our expectation as there are significant confusions between medium and high risk grasses, as well as between medium and low risk grasses in real-world roadside images, making it difficult for accurate classification even for humans. This is also observed during the manual annotation of ground truths, where many regions are hard to be classified into a medium risk level with high confidence.

Table 2. Confusion matrix for three levels of fire risk (global accuracy = 64.0 %).

	Low	Medium	High
Low	74.7	18.8	6.5
Medium	27.5	35.7	36.8
High	18.8	18.1	63.1

4.4 Performance Comparisons

Table 3 compares the proposed fusion of Brown Grass Coverage (BGC) and Grass Height (GH) with BGC or GH alone. The proposed approach achieves global accuracy of 64.0 % and average class accuracy of 57.8 %, which are higher than those obtained using BGC or GH alone. The results indicate that both the height and density of grasses are important in the estimation of grass biomass.

Table 3. Performance comparisons ($T_c^1 = 0.5, T_c^2 = 0.9, T_h^1 = 29, T_h^2 = 36$).

Approach	Average class accuracy (%)	Global accuracy (%)
Brown Grass Coverage (BGC)	51.3	48.7
Grass Height (GH)	55.0	59.5
BGC + GH (proposed)	**57.8**	**64.0**

Figure 3 illustrates the results on typical images. Our approach produces high accuracy for classifying low or green grasses into low risk (top row). However, the overlap in color between green and brown grasses still significantly impacts the overall accuracy. For sampling regions with mixed high and low grasses (bottom row), there exist significant confusion with respect to their risk levels which leads to misclassification. The shadows also lead to error in grass segmentation, and as a result, some high risk regions are misclassified into low risk (bottom row). The results indicate typical challenges for fire risk assessment on natural roadside images.

Fig. 3. Outputs of sample images using the proposed approach. Grass regions are represented by white pixels in grass segmention results (L – Low, M – Medium, H – High). (Color figure online)

5 Conclusions

This paper proposes a biomass based approach for fire risk assessment on natural roadside images. A key novelty is that the approach estimates the grass biomass based on automatically measured brown grass coverage and grass height, and then designs a set of threshold-based rules to classify fire risk into low, medium or high. We achieved global accuracy of 64.0 % on a challenging natural dataset with 100 images comprising 1,500 sampling grass regions. It was found that the medium risk is difficult to be classified, due to its significant confusion with low and high risks. For robust fire risk assessment in natural conditions, our results indicate that the overlap in color between green and brown grasses, as well as varied environmental conditions led to substantial misclassification. Our future work will improve grass segmentation results by incorporating a class label smoothing step and conduct larger field tests.

Acknowledgments. We acknowledge the support from ARC and DTMR. This research was supported under Australian Research Council's Linkage Projects funding scheme (project number LP140100939).

References

1. Vazirabad, Y.F., Karslioglu, M.O.: LIDAR for biomass estimation. In: Biomass - Detection, Production and Usage. INTECH Open Access Publisher (2011)
2. Bond, W.J., Van Wilgen, B.W.: Fire and Plants. Springer Science & Business Media, Berlin (2012)
3. Royo, C., Villegas, D.: Field measurements of canopy spectra for biomass assessment of small-grain cereals. In: Biomass - Detection, Production and Usage. INTECH Open Access Publisher (2011)
4. Sannier, C., Taylor, J., Plessis, W.D.: Real-time monitoring of vegetation biomass with NOAA-AVHRR in Etosha National Park, Namibia, for fire risk assessment. Int. J. Remote Sens. **23**, 71–89 (2002)
5. Verbesselt, J., Somers, B., Van Aardt, J., Jonckheere, I., Coppin, P.: Monitoring herbaceous biomass and water content with SPOT VEGETATION time-series to improve fire risk assessment in savanna ecosystems. Remote Sens. Environ. **101**, 399–414 (2006)

6. Schneider, P., Roberts, D.A., Kyriakidis, P.C.: A VARI-based relative greenness from MODIS data for computing the fire potential index. Remote Sens. Environ. **112**, 1151–1167 (2008)
7. Hernandez-Leal, P.A., Arbelo, M., Gonzalez-Calvo, A.: Fire risk assessment using satellite data. Adv. Space Res. **37**, 741–746 (2006)
8. Zhang, L., Verma, B., Stockwell, D.: Spatial contextual superpixel model for natural roadside vegetation classification. Pattern Recognit. **60**, 444–457 (2016)
9. Winn, J., Criminisi, A., Minka, T.: Object categorization by learned universal visual dictionary. In: IEEE International Conference on Computer Vision, pp. 1800–1807 (2005)
10. Tilly, N., Hoffmeister, D., Cao, Q., Lenz-Wiedemann, V., Miao, Y., Bareth, G.: Transferability of models for estimating paddy rice biomass from spatial plant height data. Agriculture **5**, 538–560 (2015)
11. Tilly, N., Aasen, H., Bareth, G.: Fusion of plant height and vegetation indices for the estimation of barley biomass. Remote Sens. **7**, 11449–11480 (2015)
12. Zhang, L., Verma, B., Stockwell, D.: Class-semantic color-texture textons for vegetation classification. In: Arik, S., et al. (eds.) ICONIP 2015. LNCS, vol. 9489, pp. 354–362. Springer, Heidelberg (2015). doi:10.1007/978-3-319-26532-2_39

Efficient Recognition of Attentional Bias Using EEG Data and the NeuCube Evolving Spatio-Temporal Data Machine

Zohreh Gholami Doborjeh(✉), Maryam Gholami Doborjeh, and Nikola Kasabov

Knowledge Engineering and Discovery Research Institute, Auckland University of Technology, Auckland 1010, New Zealand
zgholami@aut.ac.nz

Abstract. Modelling of dynamic brain activity for better understanding of human decision making processes becomes an important task in many areas of study. Inspired by importance of the attentional bias principle in human choice behaviour, we proposed a Spiking Neural Network (SNN) model for efficient recognition of attentional bias. The model is based on the evolving spatio-temporal data machine NeuCube. The proposed model is tested on a case study experimental EEG data collected from a group of subjects exemplified here on a group of moderate drinkers when they were presented by different product features (in this case different features of drinks). The results showed a very high accuracy of discriminating attentional bias to non-target objects and their features when compared with a poor performance of traditional machine learning methods. Potential applications in neuromarketing and cognitive studies are also discussed.

Keywords: Spiking neural networks · NeuCube · Spatiotemporal EEG data · Attentional bias

1 Introduction

Attentional bias is defined as a tendency of human perception to be affected by recurring thoughts. In decision making study, human choice behavior is influenced by several factors including cultural, social, personal, etc. which may arouse a brain attentional bias that would have an effect on decision making. In neuromarketing field [1,2] consumers may be directed by their brain attentional bias towards irrational decisions [3,4]. Although branding is considered as a dominant feature of a product, there are also other external stimuli along with the brand name that may attract our attention. The human brain acts as a complex machine learning system that processes input information through the activity of billions of neurons. So far massive amount of Spatio-Temporal Brain Data (STBD) is measured for the purpose of study the human choice behavior,

A. Hirose et al. (Eds.): ICONIP 2016, Part IV, LNCS 9950, pp. 645–653, 2016.
DOI: 10.1007/978-3-319-46681-1_76

which data has not been efficiently analyzed with the use of traditional machine learning methods such as SVM, MLP, and regression techniques.

Biological neurons communicate with each other through electrical pulses, spikes, that form temporal sequences transferred between spatially distributed neurons. Based on brain-information principles, the third generation of artificial neural networks, called spiking neural networks (SNN) has been developed [5–7], and their neuromorphic highly parallel implementations are advancing very fast [8,9]. In this paper we present a model to study the effect of the human attentional bias using a SNN system called NeuCube [10]. The aim of this study is to test the suitability of the NeuCube model to discriminate subtle spatio-temporal brain activity patterns generated by different non-target stimuli. As a case study we have illustrated our approach on EEG data classification and visualisation while a group of subjects were dealing with a brand name as a text versus when the brand name comes along with a context such as design, color, alcoholic or non-alcoholic content. We have demonstrated that attentional bias towards concerned-related stimuli affects the human attention and this can be detected using the NeuCube models. The case study example reveals a strong attentional bias towards non-target objects that have features of alcoholic drinks.

2 The NeuCube Evolving Spatio-Temporal Data Machine (eSTDM) for Modelling, Learning, Visualization and Classification of Spatio-Temporal Brain Data (STBD)

NeuCube is a generic eSTDM based on SNN for learning, classification/regression, visualization and interpretation of any spatio-temporal data. It was first introduced for STBD [10], and has been successfully applied on different STBD case studies [11–14]. A NeuCube based model for data modelling includes the following five procedures [15]:

STBD Encoding and Mapping: A time series of an input variable is encoded into a sequence of spikes using a Threshold-Based Representation method (TBR) or other methods [16]. The input spike sequences are spatially mapped into spatially allocated spiking neurons in SNNc according to the spatial (x,y,z) locations of the input variables using a brain template such as Talairach [17] or MNI [18].

Deep, Unsupervised Learning in the Evolving 3D SNN Cube [18]**:** The initial neuronal connections in the SNNc are established based on the small-world connectivity [10,19]. The SNNc is trained in an unsupervised mode using Spike-Timing-Dependent Plasticity (STDP) learning rule [18] on the spike sequences that represent the input spatio-temporal data. During the STDP learning, the SNNc connectivity is modified by the transferring spikes across synaptic connections. The SNNcube is a scalable and evolving structure. It can be incrementally trained on new samples, getting more spiking neurons involved and connected.

Supervised Learning and Classification in Evolving SNN Classifier: An evolving SNN output classifier is trained to learn the SNNc spatio-temporal activities that represent STBD patterns and their predefined classes. A dynamic evolving SNN (deSNN) is used in this paper as an output classifier [20], but other classifiers can also be employed [21]. An evolving SNN allows for new output neurons to be added (to evolve) if new samples for new classes are used in an incremental learning, possibly on-line and real time.

Parameter Optimization: In the illustrative case study experiments we have used a grid search algorithm (exhaustive search) to optimize some of the parameters of the model. This optimization uses as an objective function maximum classification accuracy between predefined classes of data that correspond to EEG data measuring attentional bias under different stimuli. Some of the most important parameters of the model are: threshold for STBD encoding; distance threshold for the small world connectivity; STDP learning rate; deSNN classifier parameters mod and drift. The details of the NeuCube parameters is explained in [6,10,20] and also in the NeuCube Manual (www.kedri.aut.ac.nz/neucube/).

Model Visualization and Interpretation: For the interpretation of brain activities and the discovery of new spatio-temporal relationships between STBD variables, a trained NeuCube model of STBD can be visualized in a 3D virtual reality space. The proposed NeuCube-based methodology for mapping, learning, and classification of EEG data is shown graphically in Fig. 1.

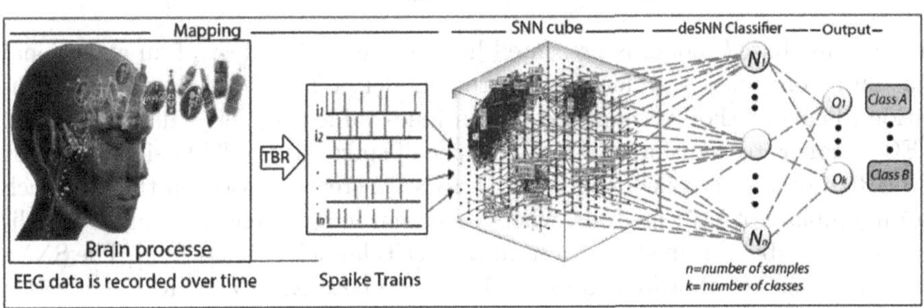

Fig. 1. The Neucube architecture with its main modules for EEG data, encoding, mapping, learning, visualization and classification

3 Understanding Attentional Bias as an Effective Factor in Human Choice Behavior: A Case Study

So far researchers have studied addict people behavior towards addiction related-stimuli in comparison to non-addiction related stimuli [22]. It has been shown that people addicted to alcohol and other drugs pay more attention to addiction-related stimuli [22]. In this paper, we analyzed consumers' brain activity and the effect of their attentional bias toward product features. As a case study, we used EEG data collected from moderate drinkers while they were performing a cognitive task. Nine male volunteers were included in this study with mean age of 36.40. The EEG measurement was performed in the knowledge Engineering and Discovery Research Institute KEDRI of Auckland University of Technology, Auckland, New Zealand under ethical approval. Scalp potentials were recorded from 32 electrodes mounted in an EEG wireless headset (Cognionics) in configuration with the standard 10_20 location system. In this cognitive task, different images of drink features, such as brand name alone and brand name along with design, colour, alcoholic or non-alcoholic contexts were used as stimuli set. As an initial instruction, participants were required to observe stimuli on the screen and make a manual response to target stimuli which in this study was a bottle of water. The duration of each stimulus presentation was 500 ms, and the interval between the stimuli was randomly varied between 1300 and 1500 ms. The target stimulus was appeared 66 times and each of the non-target stimuli (different product features) were presented 26 or 27 times, with random order of presentation. A total number of 264 stimuli for non-target stimuli were presented. During this cognitive task, subjects were asked to concentrate on the target stimuli, therefore they were unconscious to the other stimuli (brand name and brand along with the other context). Therefore, regardless of how the subjects like or dislike the stimuli, we could observe their brain subconscious biases towards different product features.

4 A NeuCube-Based SNN Model

A NeuCube-based model is presented here to study the effect of an attentional bias on a consumer' s preferences. In this model, a Threshold-Based Representation (TBR) method was applied on continues value time series data of each 32 EEG channels to transfer it into a sequence of spikes. If the EEG signal exceeds a TBR threshold, a spike occurs. A 3D SNNc is created based on the Talairach brain template of 1471 neurons, where each neuron represents the center coordinate of one cubic centimeter area from the 3D Talairach Atlas [17,23]. The SNNc input neurons are allocated to the 32 EEG channels. After a NeuCube model is defined, the SNNc is trained with the encoded EEG data using a STDP learning method [18]. After the unsupervised learning process is completed, a supervised learning stage is performed for classification purposes to evaluate how the NeuCube-based model can separate the brain activity patterns against different product features. EEG data was divided into 5 classes according to the attentional bias stimuli used in this case study (class 1: alcoholic content of the drink; class 2: non-alcoholic content; class 3: design; class 4: drink color; and class 5: brand name; to train a NeuCube model. We prepared 9 EEG samples per class. In total we obtained 45 samples. After a grid search parameter optimization, the following parameter values are used in the NeuCube model: TBR: 0.28, distance of small world connectivity: 0.15, STDP rate: 0.01, Threshold of firing: 0.5, mode: 0.4 and drift: 0.25. In Table 1, we reported the confusion table showing the number of correctly classified samples versus the number of miss classified samples across the 5 classes stimuli when 25 samples were used for training and 20 samples for testing the model. Table 2 summarizes the classification accuracy

Table 1. A NeuCube confusion table obtained by classifying EEG data from 20 samples as a test subset into 5 classes: Alc (class1), Non-Alc (class2), design (class 3), drink color (class 4), and brand name (class 5). The number of correctly classified samples in each class are located in the diagonal of the table

EEG data Classes	Alc: C1	Non-Alc: C2	Design: C3	Color: C4	BrandName: C5
Alc: C1	4	0	0	0	0
Non-Alc: C2	0	4	0	0	0
Design: C3	0	0	4	0	0
Drink color: C4	0	0	1	3	0
Brand name: C5	0	0	1	0	3

Table 2. Classification accuracy is obtained via traditional machine learning methods using NeuCom (www.theneucom.com). The parameters of the MLP model are: number of hidden neurons 3; number of training cycles 300; output value precision 0.0001; output function precision 0.0001; output activation function - linear. The SVM model uses polynomial kernel

Methods	MLP	MLR	SVM
Accuracy in percent	17.78	15.56	13.33
Class performance variance	0.95	0.81	1.49

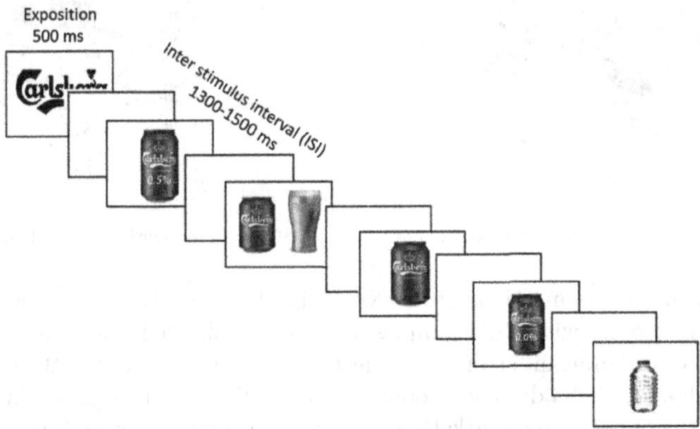

Fig. 2. A diagram of the case study cognitive task scenario. The trial session begins with the presentation of the target stimuli (a bottle of water), which remains on screen for 500 ms to remind that a manual response is required. Then other non-targeted stimuli are presented as combination of different marketing features and the related EEG data of a subject, that measure attentional bias to these stimuli, is recorded

obtained using traditional machine learning methods. Using the NeuCube-based model for learning and classification of EEG data related to all subjects and all stimuli, we obtained higher classification accuracy compared with the traditional

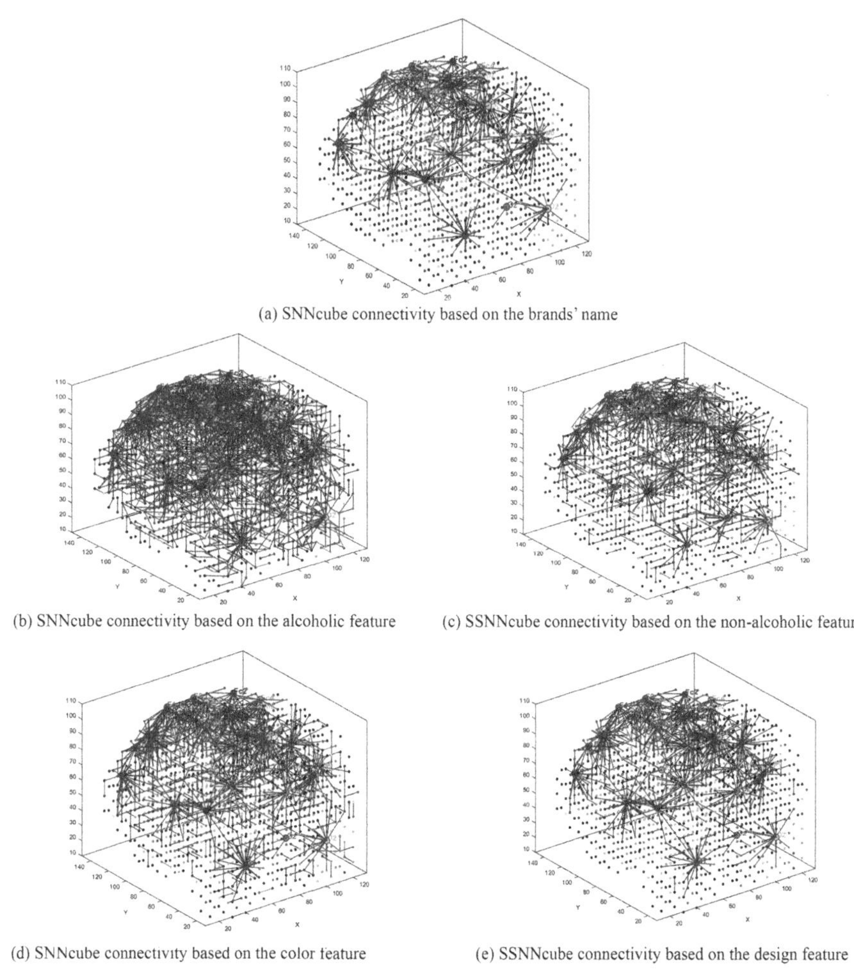

(a) SNNcube connectivity based on the brands' name

(b) SNNcube connectivity based on the alcoholic feature

(c) SSNNcube connectivity based on the non-alcoholic feature

(d) SNNcube connectivity based on the color feature

(e) SSNNcube connectivity based on the design feature

Fig. 3. The neuronal connections of the NeuCube- based SNN models generated after STDP learning process with EEG samples of 5 classes collected from 9 subjects. Across the cortical areas, significant stronger connectivity can be observed in the created SNN model trained on the EEG data of alcoholic feature reflecting stronger brain attentional bias towards alcoholic-related marketing stimuli when compared with the other product features. The EEG channels are mapped into the SNNcube using the Talairach brain template [17]

machine learning methods. Note that the cognitive task used here was complex as each presented image contained all the product features with one dominant feature as shown in Fig. 2. Due to overlapping of the product features during the task presentation, the brain activities were only slightly varied. These slight variations could not be captured and classified properly by using MLP, MLR and SVM methods, which are designed to mainly process static vector data and cannot model both interaction and interrelationship between time and space components of the spatio-temporal data as it is the case here and in many other applications of brain data [10]. Figure 3 illustrates that the SNNc connectivity evolved differently during the unsupervised training with EEG data related to different drink features reflecting on different evoked cognitive functions corresponding to the subject preference. It shows that when a SNN cube was trained on the EEG data related to a brand name, less neuronal connections were evoked in comparison with the other drink features in (b), (c), (d), and (e). The SNNc trained on EEG data of alcoholic feature resulted in stronger neuronal connections which were mostly evolved around the EEG channels located in the Frontal and Parietal areas of the brain. The consumers attention was less affected by the color and the design of the product.

5 Conclusion and Discussions

In this paper we proposed a methodology for classification and analysis of EEG data measuring attentional bias. With the use of the NeuCube SNN evolving spatio-temporal data machine, model was created for a case study cognitive test scenario. The model includes procedures for: spatial mapping of the EEG data into a 3D SNNc; unsupervised learning in the SNNc [18]; visualization of the trained SNNc connectivity; supervised learning in a SNN classifier [20]; parameter optimization; and model validation. The method proposed here is suitable to detect brain attentional bias towards brain-related concerns. As a case study, we have applied this method on EEG data related to marketing product features, but it can be applied on other case studies as well.

Our case study findings suggest that a product brand name may not significantly impress consumers by itself. However, when the name of a brand comes along with a context, such as design, color, alcoholic or non-alcoholic features, etc. it may direct the consumers attention to certain features and lead the consumers to choose a product. In this particular case study, we found that attentional bias towards alcoholic-related stimuli had stronger effects on consumer brain activity as stronger cortical activities were generated in Fig. 3 (b). In the field of neuromarketing, consumer' s attentional bias may increase the consumer' s attention to certain products.

In our experiment, the NeuCube-based model was superior in two aspects when compared with traditional machine learning methods: (1) the model connectivity is interpreted in terms of understanding dynamic interactions between functional areas of the brain during a stimuli presentation (none of the traditional methods reported in Table 2 can be interpreted in such a way); (2) A much higher classification accuracy is achieved.

In future research more experimental data may be needed for a more conclusive findings depending on the type of the attentional bias studied. The proposed approach will be further developed towards: (1) Predicting human choice behavior influenced by different product features: (2) Cognitive studies on memory tasks.

Acknowledgments. The research is supported by the Knowledge Engineering and Discovery Research Institute of the Auckland University of Technology (www.kedri.aut.ac.nz). Z. Gholami was supported by AUT summer research scholarship. A NeuCube software version is available free from: http://www.kedri.aut.ac.nz/neucube/.

References

1. Touhami, Z.O., Benlafkih, L., Jiddane, M., Cherrah, Y., El Malki, H.O., Benomar, A.: Neuromarketing: where marketing and neuroscience meet. Afr. J. Bus. Manag. **5**, 1528–1532 (2011). Academic Journals
2. Dooley, R.: Brainfluence: 100 ways to persuade and convince consumers with neuromarketing (2011)
3. Nazari, M., Doborjeh, Z.G., Oghaz, T.A., Fadardi, J.S., Yazdi, S.A.A.: Evaluation of consumers preference to the brands of beverage by means of ERP precomprehension component. In: Proceedings of the International Conference on Global Economy, Commerce and Service Science (GECSS), Thailand (2014)
4. Harrison, N.R., McCann, A.: The effect of colour and size on attentional bias to alcohol-related pictures. Int. J. Methodol. Exp. Psychol. **35**, 39–48 (2014)
5. Maass, W., Natschlager, T., Markram, H.: Real-time computing without stable states: a new framework for neural computation based on perturbations. Neural Comput. **14**(11), 2531–2560 (2002)
6. Thorpe, S., Gautrais, J.: Rank order coding. In: Bower, J.M. (ed.) Computational Neuroscience, pp. 113–118. Springer, Berlin (1998)
7. Masquelier, T., Guyonneau, R., Thorpe, S.J.: Competitive STDP-based spike pattern learning. Neural Comput. **21**(5), 1259–1276 (2009)
8. Furber, S.B., Galluppi, F., Temple, S., Plana, L.A.: The SpiNNaker project. Proc. IEEE **102**(5), 652–665 (2014)
9. Modha. D.S.: Introducing a brain inspired computer (2016). http://www.research.ibm.com/articles/brain-chip.shtml
10. Kasabov, N.: NeuCube: a spiking neural network architecture for mapping, learning and understanding of spatio-temporal brain data. Neural Netw. **52**, 62–76 (2014). Elsevier
11. Kasabov, N., Capecci, E.: Spiking neural network methodology for modelling, classification and understanding of EEG spatio-temporal data measuring cognitive processes. Inf. Sci. **294**, 565–575 (2015)
12. Doborjeh, M.G, Capecci, E., Kasabov, N.: Classification and segmentation of fMRI spatio-temporal brain data with a NeuCube evolving spiking neural network model. In: IEEE SSCI, Orlando, U.S.A., pp. 73–80 (2014)
13. Doborjeh, M.G., Wang, G.Y., Kasabov, N.K., Kydd, R., Russell, B.: A spiking neural network methodology and system for learning and comparative analysis of EEG data from healthy vs addiction treated vs addiction not treated subjects. IEEE Trans. Biomed. Eng. **63**(9), 1830–1841 (2015)
14. Doborjeh, M.G., Kasabov, N.: Dynamic 3D clustering of spatio-temporal brain data in the NeuCube spiking neural network architecture on a case study of fMRI data. In: Neural Information Processing, Istanbul (2015)
15. Kasabov, N., Scott, N., Tu, E., Marks, S., Sengupta, N., Capecci, E.: Evolving spatio-temporal data machines based on the NeuCube neuromorphic framework: design methodology and selected applications. Neural Netw. **78**, 1–14 (2016)

16. van Schaik, A., Liu, S.C.: AER EAR: a matched silicon cochlea pair with address event representation interface. In: IEEE International Symposium on Circuits and Systems, vol. 5, pp. 4213–4216 (2005)
17. Talairach, T., Tournoux, P.: Co-planar Stereotaxic Atlas of the Human Brain. 3-Dimensional Proportional System: An Approach to Cerebral Imaging. Thieme Medical Publishers, New York (1988)
18. Song, S., Miller, K.D., Abbott, L.F.: Competitive Hebbian learning through spike-timing-dependent synaptic plasticity. Nature Neurosci. **3**(9), 919–926 (2000)
19. Tu, E., Kasabov, N., Yang, J.: Mapping temporal variables into the NeuCube for improved pattern recognition, predictive modeling, and understanding of stream data. IEEE Trans. Neural Netw. Learn. Syst. **99**, 1–13 (2016)
20. Kasabov, N., Dhoble, K., Nuntalid, N., Indiveri, G.: Dynamic evolving spiking neural networks for on-line spatio-and spectro-temporal pattern recognition. Neural Netw. **41**, 188–201 (2013)
21. Schliebs, S., Kasabov, N.: Evolving spiking neural networks: a survey, evolving systems. Evolving Syst. **4**(2), 87–98 (2013)
22. Field, M., Cox, W.M.: Attentional bias in addictive behaviors: a review of its development, causes, and consequences. Drug Alcohol Depend. **1**, 1–20 (2008). Elsevier
23. Koessler, L., Maillard, L., Benhadid, A., Vignal, J.P., Felblinger, J., Vespignani, H., Braun, M.: Automated cortical projection of EEG sensors: anatomical correlation via the international 10–10 system. Neuroimage **46**(1), 64–72 (2009)

Author Index

Printed in the United States
By Bookmasters